Springer Science+Business Media, LLC

Dear Reader

We would very much appreciate receiving your suggestions and criticisms on the *Plant Tissue Culture Manual*. They will be most helpful during our preparations for future supplements.

Would you please answer the questions listed below, and send your comments with any further suggestions you may have, to *Ir. A. Plaizier* at the above-mentioned address.

Thank you for your assistance!

A. Plaizier
Publisher

— —

PLANT TISSUE CULTURE MANUAL

1. What errors have you found? (list page numbers and describe mistakes)
2. What protocols do you find to be confusing or lacking in detail? (list chapter numbers and page numbers and describe problems)
3. What protocols do you feel should be replaced in future supplements with newer (better) methods?
4. What new topics or other material would you like to see included in future supplements?

Please print or type your answers in the space below, and continue overleaf.

Name: Date:

Address:

PLANT TISSUE CULTURE MANUAL

Supplement 7

Edited by:

K. LINDSEY
Department of Biological Sciences, University of Durham, U.K.

Springer Science+Business Media, LLC

Library of Congress Cataloging-in-Publication Data

Plant tissue culture manual: fundamentals and applications / edited by K. Lindsey
    p.   cm.
Includes bibliographical references and index.

   1. Plant tissue culture—Laboratory manuals.   I. Lindsey, K.
QK725.P587   1991
581′.0724—dc20

90–26765

Manual
ISBN 978-94-011-7658-3

ISBN 978-94-011-7658-3   ISBN 978-94-009-0103-2 (eBook)
DOI 10.1007/978-94-009-0103-2

Printed on acid-free paper

Contents

Preface

SECTION A: BASIC TECHNIQUES – CELL & TISSUE CULTURE OF MODEL SPECIES

1. Media preparation
   O.L. Gamborg, Fort Collins, USA

2. The initiation and maintenance of callus cultures of carrot & tobacco
   R.D. Hall, Wageningen, The Netherlands

3. The initiation and maintenance of plant cell suspension cultures
   R.D. Hall, Wageningen, The Netherlands

4. Shoot cultures and root cultures of tobacco
   J.F. Topping and K. Lindsey, Leicester, UK

5. Somatic embryogenesis in orchardgrass
   M.E. Horn, Davis, USA

6. *Arabidopsis* regeneration and transformation (leaf & cotyledon explant system)
   R. Schmidt and L. Willmitzer, Berlin, Germany

7. *Arabidopsis* protoplast transformation and regeneration
   B. Damm and L. Willmitzer, Berlin, Germany

8. *Arabidopsis* regeneration and transformation (root explant system)
   D. Valvekens, M. van Lijsebettens and M. van Montagu, Gent, Belgium

9. Somatic embryogenesis in carrot
   A.D. Krikorian and D.L. Smith, New York, USA

10. Low density cultures: microdroplets and single-cell nurse cultures
    G. Spangenberg, Ladenburg, Germany, and HU Koop, Munich, Germany

11. Tobacco protoplast isolation, culture and regeneration
    I. Negrutiu, Gembloux, Belgium

SECTION B: TISSUE CULTURE & TRANSFORMATION OF CROP SPECIES

1. Embryogenic callus, cell suspension and protoplast cultures of cereals
   I.K. Vasil and V. Vasil, Gainesville, USA

2. Transformation and regeneration of rice protoplasts
   J. Kyozuka and K. Shimamoto, Yokohama, Japan

3. Transformation and regeneration of orchardgrass protoplasts
   M.E. Horn, Davis, USA

4. Transformation and regeneration of oilseed rape protoplasts
   D. Rouan and P. Guerche, Versailles, France

5. Regeneration and transformation of potato by *Agrobacterium tumefaciens*
   R.G.F. Visser, Wageningen, Netherlands

6. Transformation of tomato with *Agrobacterium tumefaciens*
   S. McCormick, Albany, USA

7. Regeneration and transformation of sugarbeet by *Agrobacterium tumefaciens*
   K. Lindsey, P. Gallois and C. Eady, Leicester, UK

8. Regeneration and transformation of apple (*Malus pumila* Mill.)
   D.J. James and A.M. Dandekar, Davis, USA

9. Transformation and regeneration of maize protoplasts
   C.A. Rhodes and D.W. Gray, Palo Alto, USA

10. Regeneration and transformation of barley protoplasts
    P. Lazzeri, A. Jähne and H. Lörz, Hamburg, Germany

11. *Agrobacterium*-mediated transformation of potato stem and tuber tissue, regeneration and PCR screening for transformation
    J.P. Spychalla and M.W. Bevan, Norwich, UK

12. Production of fertile transgenic wheat by microprojectile bombardment
    Dirk Becker and Horst Lörz, Hamburg, Germany

13. Transient gene expression and stable genetic transformation into conifer tissues by microprojectile bombardment
    Armand Séguin, Denis Lachance and Pierre J. Charest, Ontario, Canada

SECTION C: PROPAGATION & CONSERVATION OF GERMPLASM

1. Clonal propagation of orchids
   Y. Sagawa, Honolulu, USA

2. Clonal propagation of palms
   B. Tisserat, Pasadena, USA

3. Clonal propagation of conifers
   T.A. Thorpe and I.S. Harry, Alberta, Canada

4. Cytological techniques
   A. Karp, Long Ashton, UK

5. Restriction fragment analysis of somaclones
   R.H. Potter, As, Norway

6. Virus elimination and testing
   M.C. Coleman and W. Powell, Dundee, UK

7. Clonal propagation of *Citrus*
   T.S. Rangan, Pasadena, USA

8. Clonal propagation of eucalypts
   J.A. McComb, Perth, Australia

9. Cryopreservation of plant tissue cultures: the example of embryogenic
   tissue cultures from conifers
   P.J. Charest, J. Bongs and K. Klimaszewska, Ontario and New
   Brunswick, Canada

SECTION D: DIRECT GENE TRANSFER & PROTOPLAST FUSION

1. Gene transfer by particle bombardment
   T.M. Klein, S. Knowlton and R. Arentzen, Wilmington, USA

2. Transformation of pollen by particle bombardment
   D. Twell, T.M. Klein and S. McCormick, Albany, USA

3. Electrical fusion of protoplasts
   M.G.K. Jones, Murdoch, Australia

4. Cybrid production and selection
   E. Galun and D. Aviv, Rehovot, Israel

5. Fluorescence-activated analysis and sorting of protoplasts and somatic
   hybrids
   D.W. Galbraith, Lincoln, USA

6. RFLP analysis of organellar genomes in somatic hybrids
   E. Pehu, Helsinki, Finland

7. Isolation and uptake of plant nuclei
   P.K. Saxena and J. King, Saskatoon, Canada

8. *In situ* hybridization to plant metaphase chromosomes: Radioactive and
   non-radioactive detection of repetitive and low copy number genes
   J. Veuskens, S. Hinnisdaels and A. Mouras, Talence, France

9. Chemical fusion of protoplasts
   P. Anthony, R. Marchant, N.W. Blackhall, J.B. Power, M.R. Davey,
   Nottingham, UK

SECTION E: REPRODUCTIVE TISSUES

1. *In vitro* fertilisation of maize
   E. Kranz, Hamburg, Germany

2. Endosperm culture
   S. Stirn and H.-J. Jacobsen, Bonn, Germany

3. Endosperm culture
   B.M. Johri and P.S. Srivastava, Delhi, India

4. Hybrid embryo rescue
   A. Agnihotri, New Delhi, India

5. *In vitro* culture of *Brassica juncea* zygotic proembryos
   Chun-ming Liu, Zhi-hong Xu and Nam-Hai Chua,
   Singapore and New York, USA

*6. Production of haploids in *Brassica* spp. via microspore culture
   A.M.R. Ferrie and W.A. Keller, Saskatoon, Canada

SECTION F: MUTANT SELECTION

1. Use of chemical and physical mutagens *in vitro*
   P.J. Dix, Maynooth, Ireland

2. *In vitro* culture, mutant selection, genetic analysis and transformation of
   Physcomitrella patens
   David Cove, Leeds, UK

SECTION G: SECONDARY METABOLITES

1. Tropane alkaloid biosynthesis *in vitro*
   R. Robins, Colney, Norwich, UK

2. Anthocyanin biosynthesis *in vitro*
   A. Komamine and K. Kakegawa, Tsukuba City, Ibaraki, Japan

3. Biosynthesis of monoterpene indole alkaloids *in vitro*
   W.G.W. Kurz, K. Constabel, R. Tyler, Saskatchewan, Canada

SECTION H: TISSUE CULTURE TECHNIQUES FOR FUNDAMENTAL STUDIES

1. Establishment of photoautotrophic cell cultures
   W. Hüsemann, Münster, Germany

2. *Zinnia* mesophyll culture system to study xylogenesis
   M. Sugiyama and H. Fukuda, Sendai, Japan

---

* Included in Supplement 7.

3. Cell cycle studies: induction of synchrony in suspension cultures of
   Catharanthus roseus cells
   H. Kodama and A. Komamine

4. Thin Cell Layer (TCL) method to programme morphogenetic patterns
   K. Tran Thanh Van and C. Gendy, Paris, France

5. *In vitro* infection of *Arabidopsis* with nematodes
   Joke C. Klap and Peter C. Sijmons, Leiden and Wageningen,
   The Netherlands

*6. *Asparagus* cell cultures as a source of wound-inducible genes
   R.M. Darby and J. Draper, Leicester, U.K.

*7. Use of video cell tracking to identify embryogenic cultured cells
   Marcel A.J. Toonen and Sacco C. de Vries, Wageningen,
   The Netherlands

* Index

---

* Included in Supplement 7.

Preface

Plant tissue culture has a long history, dating back to the work of Gottlieb Haberlandt and others at the end of the 19th century, but the associated concepts and techniques have reached a level of usefulness and application which has never been greater. The technical innovations have given new insights into fundamental aspects of plant differentiation and development, and have paved the way to the identification of strategies for the genetic manipulation of plants. It is the aim of this manual to deliver a broad range of these techniques in a form which is accessible to students and research scientists of diverse backgrounds, including those with little or no previous experience. The themes of the manual aim to reflect those research areas which have been advanced by tissue culture technology.

As was the case for the sister volume *Plant Molecular Biology Manual*, the objective has been from the start to produce a manual which is at home on the laboratory bench. The plastic-covered, ring-bound format has proved to be most popular and is retained here. Equally, the emphasis has been on producing a collection of detailed step-by-step protocols, each supplemented with an introductory text and practical footnotes, to provide the next best thing to a supervisor at one's shoulder. Each author was chosen as one actively using the respective technique in his or her own laboratory, in order to give an authoritative account, with the most common difficulties or pitfalls highlighted for the benefit of the newcomer to the field.

The manual is published initially as a core text of basic techniques, to which will be added supplementary chapters and sections at regular intervals. One characteristic of plant tissue culture methodology is the fact that what works for one species quite possibly will not work for another (or even for a different genotype of the same species): each requires a carefully optimized protocol, whether it be for callus culture initiation, protoplast isolation and culture or plant regeneration following *Agrobacterium*-mediated transformation. With this in mind, we have decided on the one hand to generalize experimental approaches for some basic techniques, by reference to the use of model species. This should at least give the researcher a feel for the techniques, and indeed provide a starting point for species-specific optimization of protocols. On the other hand, we have provided detailed protocols for the manipulation of a number of specific crop species, which have been difficult to optimize and for which there may be particular interest. A detailed bibliography is provided in each chapter, so that access to the relevant literature is possible.

I would like to thank the authors both for their enthusiasm to contribute and for the rapidity with which they submitted their manuscripts. Thanks go in particular to Dr Mike Brewis of Kluwer for his valuable discussions during the embryonic stages of the manual and for his enormous help in taking on the administrative tasks.

Leicester, November 1990   K. LINDSEY

Section A:
Basic Techniques –
Cell & Tissue Culture of Model Species

Plant Tissue Culture
Manual

Plant Tissue Culture Manual **A1**: 1–24, 1991.
© 1991 *Kluwer Academic Publishers.*

Media preparation

OLUF L. GAMBORG
Tissue Culture for Crops Project, Biology Department, Colorado State University, Fort Collins, CO 80551, U.S.A.

Introduction

The degree of success in any technology employing cell, tissue or organ culture is related to relatively few major factors.

A significant factor is the choice of nutritional components and growth regulators. In the past two to three decades a large number of reports have appeared on modifications of about two dozen basic compositions [1–10].

It is not within the scope of this paper to review the nutritional requirements of cells, protoplasts, anthers, and meristems in culture in any detail. It is well known that particular media including the hormones used have given reproducible results and have provided efficient methods. However, the literature also contain reports on procedures which have not been reproduced. That is frequently a regeneration procedure. The difficulty in repeating the procedure is a reflection of the almost complete lack of understanding of the cell differentiation process which can lead to somatic embryogenesis and plant regeneration.

The following pages will outline a number of procedures for media preparation. The choice of media used as examples are suggested as a starting point for respective purposes. Some of the factors which may serve as a guide for choice of media and modes of preparation will be discussed.

Background information

A nutrient medium usually consists of inorganic salts, a carbon source, some vitamins and growth regulators. Other components added for specific purposes include organic nitrogen compounds, tricarboxylic acid compounds and plant extracts [12, 17]. Media compositions which are frequently used are listed in Table 1. The Murashige-Skoog (MS) (1962) [2] or Linsmaier and Skoog (LS) (1965) [3] salt compositions are the most widely used especially in plant regeneration procedures. The B5 (Gamborg *et al.*, 1968) [1], N6 [7] and Nitsch and Nitsch [5], and derivatives of these media have had wide applications for different plant species and for different culture objectives (See 8, 11 for more information).

An appreciation and a knowledge of the nutritional requirements and of the metabolic needs of the cultured cells and tissues is invaluable not only in a decision on the type of media to use but also in their preparation.

Table 1. Plant tissue culture media

Constituent	Conc. (mg liter^{-1}) in medium*				
	MS	B5	N6	NN	AA**
Macronutrients					
KNO_3	1900	2500	2830	950	
NH_4NO_3	1650			720	
$MgSO_4 \cdot 7H_2O$	370	250	185	185	250
KH_2PO_4	170		400	68	
$NaH_2PO_4 \cdot H_2O$		150			150
$CaCl_2 \cdot 2H_2O$	440	150	166	166	150
$(NH_4)_2 \cdot SO_4$		134	463		
KCl					2950
Micronutrients					
H_2BO_3	6.2	3	1.6	10	3.0
$MnSO_4 \cdot 4H_2O$	15.6	10	3.3	19.0	10.0
$ZnSO_4 \cdot 7H_2O$	8.6	2	1.5	10.0	2
$NaMoO_4 \cdot 2H_2O$	0.25	0.25	0.25	0.25	0.25
$CuSO_4 \cdot 5H_2O$	0.025	0.025	0.025	0.025	0.025
$CoCl_2 \cdot 6H_2O$	0.025	0.025		0.025	0.025
KI	0.83	0.75	0.8		
$FeSO_4 \cdot 7H_2O$	27.8		27.8		
Disodium EDTA	37.3		37.3		
EDTA sodium ferric salt		40		100	40
Glycine			40	5	
Sucrose	30×10^3	20×10^3	50×10^3	20×10^3	20×10^3
Vitamins					
Thiamine hydrochloride	0.5	10	1	0.5	10
Pyridoxine hydrochloride	0.5	1	0.5	0.5	1
Nicotinic acid	0.5	1	0.5	5.0	1
myo-inositol	100	100		100	100
pH	5.8	5.5	5.8	5.5	5.5

* Source: MS, Murashige and Skoog [2]; B5, Gamborg *et al.* [1]; N6, Chu [7]; NN, Nitsch and Nitsch [5]; AA, Toriyama *et al.* [15].
** The nitrogen source is: mg/l, L-glutamine 730; L-aspartate, 200; L-arginine, 176; Glycine, 7.5. Filter sterilize. Ref: 15.
NOTE: The LS medium used for cereals [18, 19] contain the MS mineral salts, Thiamine; 40 mg/L; and myoinositol; 100 mg/L.

Media composition

There is a large variety of media published in the literature (see ref 8 and 11). Some of the more commonly used media are listed in Table 1. Many media are modifications of the listed media. Included in the present paper is a description of the composition and the preparation of media for special purposes. The choice of media is dictated by the purpose of the tissue culture technology which will be employed and the species or type of plants [13].

The nutrient media for tissue culture contain the inorganic elements required by any growing plant. In addition they also contain a carbon or energy source, vitamins, growth regulators and sometimes organic supplements.

Inorganic salts

A relatively small number of mineral salts are used as components of media for plant tissue culture. For most purposes the medium should contain at least 30 mM of inorganic nitrogen and potassium. Ammonium can be used at 2–20 mM. However, the effect of ammonium salts can vary from inhibitory to essential depending upon the tissue and the purpose of the culture. The use of ammonium salts of malate or citrate makes it possible to use ammonium as the sole nitrogen source [14].

A concentration of 1 to 3 mM of calcium, sulfate, phosphorous and magnesium is usually adequate. The essential micronutrients and the amounts are shown in Table 1.

Carbon sources

Sucrose and glucose are the standard sources of sugars. Fructose is utilized much less readily. All other sugars are utilized only by a few variant cell lines. Most media contain m-inositol, which is beneficial as a component of cell wall metabolism.

Vitamins

Plants synthesize all vitamins, but cells in culture require thiamine and there are enhancing effects on growth and development by those listed in the media of Table 1.

Growth regulators

The four classes of compounds with growth regulator activity are auxins, cytokinins, gibberellins and abscisic acid.

Several of the compounds are listed in Table 2.

The auxins are required for the induction of cell division in cultured tissues. The

Table 2. Preparation of growth hormone stock solutions[a]

Compound	Common abbreviation	Mol wt	Quantity (mg for 50 ml of a 1 mM solution)	Prepn
Cytokinins				
Benzyladenine	BA	225.2	11.25	Dissolve in 2 to 5 ml of 0.5 N HCl; heat slightly; make to volume; adjust to pH 5.
Isopentenyl adenine	2-iP	203.2	10.15	
Kinetin	K	215	10.25	
Zeatin	Z	219	10.95	
Auxins				
2,4-Dichlorophenoxyacetic acid	2,4-D	221		Dissolve in 2 to 5 ml of ethanol; gradually add water; heat slightly; make to volume; adjust to pH 5.
2,4,5-Trichlorophenoxyacetic acid	2,4,5-T	255.5		
Naphthaleneacetic acid[b]	NAA	186.2		
Indolebutyric acid[b]	IBA	203.2		
Indoleacetic acid[b]	IAA	175.2		
Miscellaneous				
Gibberellic acid[b]	GA	364.4	17.3	Dissolve in 2 to 5 ml of 0.2 M KOH; adjust to pH 5.0
Picloram (9)	P	241.2	12.0	
Abscisic acid[b]	ABA	264	13.2	

[a] Store in refrigerator.
[b] Filter sterilize.

indole compounds and NAA also are used to induce root formation. They are often used in combination with cytokinins.

The cytokinins are adenine (aminopurine) derivatives. They have an essential role in differentiation and plant regeneration of most species.

The gibberellins are used in plant regeneration from meristems and after shoot primordia formation has occurred. The gibberellins are usually filter sterilized.

The role of abscisic acid in tissue culture is not well defined. The compound has been implicated in somatic embryogenesis. Abscisic acid should be filter sterilized.

Amino acids

The addition of amino acids may enhance growth of cells and facilitate differentiation towards plant regeneration. Plant cells can grow on L-glutamine as the only nitrogen source. A special medium in which all inorganic nitrogen has been replaced by amino acids has been used effectively for different purposes [15, 16]. See Preparation in Table D4.

Organic acids

The addition of acids of the tricarboxylic acid cycle intermediates such as malate or citrate is common in media for protoplasts. The compounds appear to alleviate any inhibitory effects of ammonium [14]. Cells can tolerate up to 10 mM of the potassium salts.

Complex organics

A variety of organic supplements can be used. The most commonly used are protein hydrolyzates and coconut milk.

The enzymic hydrolyzed protein such as K-Z-amine Type A is preferable, since the amino acids are intact. Acid hydrolysis destroys several amino acids.

Facilities and equipment

The facilities for the use of tissue culture operations are best divided into distinct areas. One area should be designated for media preparation. This area would have top-pan balances, stirrers, a pH meter, and a microwave oven or equivalent. The area will also house stock solutions. In the same area or adjacently would be the cold room or refrigerator and freezer, as well as space, or a separate room, for analytical balances.

The regularly used chemicals are also frequently stored in this area. Similarly clean glassware, containers or other supplies for the operations can be stored in the area.

The other major and separate area is the room where the autoclaves, the water demineralizer. and distilling of water are located and where the dish-washing is performed.

After autoclaving the media can be stored in or close to this area.

Specifics on major equipment

The most costly equipment is the autoclave. There are two general types. One is the horizontal and the other is the upright, model. The horizontal models have the larger capacity and require more space and may be more costly to operate. They may also have a shorter problem-free period. The upright models are more compact, and usually have a smaller capacity. They may be more reliable and economical.

In the preparation of sterile media it is essential to have access to a sterile laminar flow cabinet.

The other standard items include analytical balance, top-pan balance, pH meter, magnetic stirrers with hot plate, microwave oven or facility for melting agar gels and filter sterilizing apparatus and filter membranes.

The latter would require a vacuum pump or water suction.

There should be provision for distilled and highly purified demineralized water.

The containers are flasks, jars with stoppers and Petri dishes. Convenient and adequate containers are baby food jars or equivalent glass containers. A number of companies supply disposable sterile plastic boxes such as Magenta boxes for shoot cultures.

Materials for culture media

1. Mineral salts, carbon sources, vitamins, growth regulators as well as organic compounds should be of the highest grade available. Heat labile compounds should be filter sterilized. These include giberellic acids, thiamine, L-gluta-mine and indoleacetic acid.
2. Only the L-isomers of amino acids should be used. The most effective organic nitrogen compound is L-glutamine, which can also be used as the sole nitrogen source.
3. Protein hydrolyzates are available as acid and enzymatic hydrolyzates. The latter are preferable because no amino acids are destroyed during enzymatic hydrolysis.
4. Coconut milk from immature and ripe coconuts can be used. A hole is drilled through a germination pore. The liquid from several nuts are collected, heated to 80 °C with stirring, filtered and stored frozen.
5. There are several types of compounds available for producing gels. The standard agar can be substituted with Gel-Gro gellan gum (ICN Bio-

chemicals, Costa Mesa, CA) Other compounds are the SeaKems from the FMC Corporation, Rockland, Maine. Gelrite is a product of Kelco Co, San Diego, CA.

6. Water should be demineralized or distilled to high purity. Glass-distilled water is the most desirable.

Media preparation

Stock solutions are prepared as described below in the different sections. The chemicals are dissolved in distilled water. The salts should always be dissolved by adding one compound at a time. Precipitation is usually avoided by dissolving the inorganic nitrogen sources first. Precipitation can occur between the phosphate and the calcium sources, when the pH approaches 6. After the salts and other ingredients have been dissolved the pH is adjusted by using 0.5N HCl or 0.2N NaOH or KOH.

The kinds of stock solutions routinely made vary widely [12, 13, 19]. Each operation selects a system which is in line with needs and convenience. Stock solutions are prepared in 10 or 100 fold concentrations. The stocks can consist of groups of ingredients or nearly complete media. For example the inorganic salts, vitamins and perhaps sucrose can be prepared in 10 fold concentrations. After the ingredients are dissolved, the solution is distributed in for example Whirlpak bags (100 ml into 6-oz (ca 175 ml) bags or 400 ml into 18-oz (525 ml) bags). The bags are stored frozen. One bag of 100 ml is then used when needed to prepare 1 liter of complete medium. Other sealable plastic containers may also be used.

The medium is distributed in culture vessels before autoclaving, or the vessels and medium may be autoclaved separately. The autoclaving conditions are usually 120 °C for 15 to 20 min. After autoclaving the media are removed and cooled as quickly as feasible. Agar media are prepared in lots of for example 500 ml. The agar is melted using the microwave oven prior to autoclaving. After autoclaving the agar media are poured into sterile containers and cooled. The preferable temperature for media storage is about 10 °C.

Below are procedures for media preparation which can be used or adapted for most media. In subsequent sections are media for particular tissue culture experiments and which require a special procedure.

The examples are for MS and the B5 media. The composition of the media is listed in Table 1. A list of media compounds and their respective molecular weights are presented in Appendix A.

A Stock solutions

1. Stock solutions of MS-medium

MS-Major salts (10 ×)	g/liter
KNO_3	19.00
NH_4NO_3	16.50
$MgSO_4 \cdot 7H_2O$	3.70
$CaCl_2 \cdot 2H_2O$	4.40
KH_2PO_4	1.70

MS-Minor salts (100 ×)	mg/100 ml
H_3BO_3	620
$MnSO_4 \cdot 4H_2O$	2230
$ZnSO_4 \cdot 7H_2O$	860
$Na_2MoO_4 \cdot 2H_2O$	25
$CuSO_4 \cdot 5H_2O$	2.5
$CoCl_2 \cdot 6H_2O$	2.5

MS-Vitamins (100 ×) Store frozen	mg/100 ml
Nicotinic acid	50
Thiamine hydrochloride	50
Pyridoxine hydrochloride	10
myo-Inositol	10000

2. Stock solutions of B5 medium

B5-Major salts (10 ×)	g/liter
KNO_3	25.00
$(NH_4)_2 SO4$	1.50
$MgSO_4 \cdot 7H_2O$	2.50
$CaCl_2 \cdot 2H_2O$	1.50
$NaH_2PO_4 \cdot H_2O$	1.50

B5-Minor Salts (100 ×)	mg/100 ml
$MnSO_4 \cdot H_2O$	1000
H_3BO_3	300
$ZnSO_4 \cdot 7H_2O$	200
$Na_2MoO_4 \cdot 2H_2O$	25
$CuSO_4 \cdot 5H_2O$	2.5
$CoCl_2 \cdot 6H_2O$	2.5

B5-Vitamins (100 ×) Store frozen	mg/100 ml
Nicotinic acid	100
Thiamine hydrochloride	1000
Pyridoxine hydrochloride	100
myo-Inositol	10000

3. EDTA sodium ferric salt

The sodium ferric salt of EDTA can be obtained from a commercial source. A solution may also be prepared by dissolving 7.45 g of disodium EDTA and 5.57 g of $FeSO_4 \cdot 7H_2O$ in 1 liter of water. Five ml is used per liter of medium.

4. Stock solutions of growth regulators. See Table 2

B. Preparation of basic MS-medium	In 1 liter
Glass distilled water	400 ml
Macronutrients from MS-stock	100 ml
Micronutrients from MS-stock	10 ml
Vitamins from MS-stock	10 ml
EDTA sodium ferric from stock	5 ml
Sucrose	30 g

Add growth regulators and other ingredients, adjust the pH to 5.8 and make to 1 liter with distilled water.

C. Preparation of basic B5-medium   In 1 liter

Glass distilled water	400 ml
Macronutrients from B5-stock	100 ml
Micronutrients from B5-stock	10 ml
Vitamins from B5-stock	10 ml
EDTA Sodium Ferric from stock	5 ml
Sucrose	20 g

Add growth regulators and other ingredients, adjust the pH
to 5.5 and make to 1 liter with distilled water.

D. Media for special purposes

1. AlCl₃-containing medium

a. Stock solution.
The concentration of the solution will depend upon the desired final concentration in the medium.
An example. Dissolve 2.0 g of $AlCl_3$ in 100 ml distilled water.
Sterilize by filtration.

b. Medium preparation.
Prepare the medium to be used with the $AlCl_3$. Sterilize the medium and distribute in the containers. The $AlCl_3$ is then added aseptically to each of the containers. Using 0.1 ml per 10 ml medium in the vial will give a final $AlCl_3$ concentration of 200 mg/l. Cool the medium.

2. Polyethylene glycol (PEG) media.

a. Solution.
The solutions of the complete medium and the PEG solution are prepared separately. Prepare the complete medium in a known volume. Dissolve the desired amount of PEG in water so that the volume is : (1000 ml – medium volume).

b. Preparation.
Autoclave the two solutions separately. After autoclaving combine the two solutions. It is adviceable to leave a stirring bar in the larger of the two flasks during autoclaving.
After autoclaving use the stirrer and a hot plate during mixing of the medium and the PEG solution. Distribute the final medium to the culture containers.

3. Media for low pH testing (pH 4.2-4.5)

a. Solutions.
Prepare MS medium as described above.
Dissolve A. 5 mM citric acid + 6.6 mM Na_2HPO_4
B. 10 mM citric acid.

b. Procedure.
For pH 4.2. Use solution A and adjust the medium to pH 4.2 before autoclaving.
For pH 4.0. Use solution B and adjust the medium to 4.0 before autoclaving.
Note: Anticipate that the callus may alter the initial pH.
That appears to be a species-dependent characteristic.

4. The AA-medium

The AA refers to 'amino acids'. The medium was originally designed for selection of nitrate reductase mutants [16].

Since then it has had other useful purposes [15].

a. Solutions.
   The major salts are those of B5 or MS but omitting the nitrate and ammonium compounds.
   The micronutrients are the same as B5 or MS
   The vitamins are the same as for B5 or MS
   EDTA-Ferric sodium salt is the same as for other media

Additions:
   2.95 g per liter KCl
   Amino acids: L-Glutamine (6 mM)   876 mg/l
               L-Aspartic acid (2 mM)   266 mg/l
               L-Arginine (1 mM)   174 mg/l
               Glycine   (0.1mM)   7.5 mg/l

Note: The amino acids should be dissolved separately and filter sterilized before being added to the autoclaved medium.

Protoplast isolation and culture solutions

Procedures have been developed for the isolation of viable protoplasts and their culture back to cells and to plants of a wide range of species.

Equipment and Materials.

Most of the facilities and equipment required for protoplasts are identical to those for plant cell culture. The operations are performed under aseptic conditions.
The resource materials needed specifically for protoplasts include the following:
1. Conical test tubes, graduated, of capacity 15 ml and 40 ml with caps or capped with aluminium foil.
2. Miller disposable filter units and 0.45 and 0.20 µm membranes.
3. Plastipak or other disposable plastic syringes.
4. Pasteur pipettes with rubber bulbs of different bore sizes.
5. Stainless steel or nylon mesh of 44, 65 and 88 µm.
6. Agarose, Seachem or Gelrite (low melting point).
7. Enzymes:

Cellulases, Onozuka R10	Yakult Pharmaceuticals, Tokyo.
Onozuka RS	Yakult Pharmaceuticals, Tokyo
Macerozyme R10	Yakult Pharmaceuticals, Tokyo
Hemicellulase	Sigma Chemical Co. St. Louis
Pectolyase	Sigma Chemical Co. St. Louis

Isolation Solutions (IS)

Isolation solution A.

Ingredients	Amounts per 100 ml
Distilled water	40 ml
B5 Major salts (Stock × 10)	10 ml
B5 micronutrients (Stock × 100)	1 ml
B5 vitamins Stock × 100)	1 ml
$CaCl_2 \cdot 2H_2O$	73 mg
Sorbitol (0.3 M)	5.4 g
Glucose (0.2M)	3.6 g
MES Buffer (3 mM)	48.5 mg
Enzymes 0.1 to 2.0%	

Adjust to pH 5.6, make to volume with distilled water and filter sterilize.

Isolation solution B

Ingredients	Amounts per 100 ml
Distilled water	40 ml
Sorbitol (0.15 M)	2.73 g
Mannitol (0.15 M	2.73 g
Glucose (0.20 M)	3.6 g
$CaCl_2 \cdot 2H_2O$ (6 mM)	44.1 mg
KH_2PO_4 (0.7 mM)	9.5 mg
MES Buffer (3 mM)	65 mg
Enzymes (0.1 to 2.0%)	

Adjust to pH 5.6, make to volume with distilled water and filter sterilize.

Enzyme stability:
Enzymes obtained commercially are stable for one year or more when stored dry below 0 °C. Prior to weighing the enzymes, the vials should be placed in a desiccator at room temperature.

Protoplast culture medium

The compositions of media for culturing protoplasts vary, but they are becoming reasonably standardized. The variation usually is due to species requirements [20].

The following are stock solutions of groups of compounds which are commonly used. The absolute requirements for compounds other than those used for tissue culture have in most cases not been fully ascertained. However the observed beneficial effects of additional vitamins, L-glutamine and organic acids makes it advisable to include these compounds in the medium. The choice of growth regulators depends on the tissue source and the particular uses of the protoplasts.

Stock Solutions

Multivitamins (100 ×)

Ingredients	mg/100 ml
Ascorbic acid	200
Calcium panthothenate	50
Choline chloride	50
Folic acid	20
p-Aminobenzoic acid	2
Riboflavin	10
Biotin	0.5

Store frozen.

Organic acids (100 ×)

Ingredients	mg/100 ml
Sodium pyruvate	1000
Malic acid	2000
Citric acid	2000

Store frozen.

Protoplast culture medium

Murashige-Skoog-or B5-derived media.

Ingredients	Amounts per 100 ml
MS or B5 basic salts	
Vitamins (B5 Stock × 100)	1 ml
Multivitamins (Stock × 100)	1 ml
$CaCl_2 \cdot 2H_2O$	73.5 mg
L-Glutamine	25 mg
Casein hydrolyzate or Casamino acids	12.5 mg
Organic acids (stock × 100)	1 ml
Sucrose	15 mg
Sorbitol (0.2M)	3.6 gm
Glucose (0.2M)	3.6 gm

The medium is adjusted to pH 5.6, made to volume and filter sterilized.

Amino acid (AA) derived medium

Ingredients	Amounts per 100 ml
KCl	295 mg
NaH_2PO_4	15 mg
$CaCl_2 \cdot 2H_2O$	73.5
$MgSO_4 \cdot 7H_2O$	25 mg
Micronutrients (B5 stock × 100)	1 ml
Multivitamins (stock × 100)	1 ml
L-Glutamine	87.6 mg
L-Aspartic acid	26.6 mg
L-Arginine	17.6 mg
Glycine	0.75 mg
Sucrose	15 mg
Sorbitol (0.2M)	3.6 gm
Glucose (0.2M)	3.6 gm

The medium is adjusted to pH 5.5, made to volume and filter sterilized.

Acknowledgments

I would like to thank Fariha S. F. Sayied of TCCP, Colorado State University, Fort Collins, CO. for helpful information on the special media. The very able assistance of Susilowati Surachman, AARP, Bogor, Indonesia in the word processing preparation of the paper is gratefully acknowledged.

References

1. Gamborg, OL, RA Miller, and K Ojima (1968) Nutrient requirements of suspension cultures of soybean root cells. Exp. Cell Res. **50**:151–158.
2. Murashige, T, and F Skoog (1962) A revised medium for rapid growth and bioassays with tobacco tissue cultures. Physiol. Plant. **15**:473–497.
3. Linsmaier, EM, and F Skoog (1965) Organic growth factor requirements of tobacco tissue cultures. Physiol. Plant. **18**:100–127.
4. Schenk, RU and AC Hildebrandt (1972) Medium and techniques for induction and growth of monocotyledonous and dicotyledonous plant cell cultures. Can. J. Bot. **50**:199–204.
5. Nitsch, JP and C Nitsch (1969) Haploid plants from pollen grains. Science 163:85–87.
6. Phillips, GC and CB Collins (1980) Somatic embryogenesis from cell suspension cultures of red clover. Crop Sci. **20**:323–326.
7. Chu, CC (1978) The N6 medium and its applications to anther culture of cereal crops, p. 43–50. *In* Proceedings of the Symposium on plant Tissue Culture. Science Press, Beijing, China.
8. Street, HE and RD Shillito (1977) Nutrient media for plant organ, tissue and cell culture. p. 305–539. In M Rechcigl (ed.), CRC handbook in nutrient and food, vol. IV. CRC Press Inc., Boca Raton, Fla.
9. Keller, WA, and KC Armstrong (1978) High-frequency production of microspore-derived plants from *Brassica napus* anther cultures. Z. Pflanzenzucht **80**:100–108.
10. Kao, KN (1977) Chromosomal behavior in somatic hybrids of soybean-*Nicotiana glauca*. Mol. Gen. Genet. **150**:225–230.
11. Conger, BV (ed.) (1981) Cloning agricultural plants via *in vitro* techniques. CRC Press, Boca Raton, Fla.
12. Gamborg, Oluf L (1986) Protoplasts and plant regeneration in culture. In. Manual of Industrial Microbiology and Biotechnology. AL Demain and NA Solomon (eds.). Amer. Soc. Microbiol. Washington D. C. pp 263–273
13. Kyle, L. 1987. Plants from Test Tubes. An introduction to Micropropagation. Timber Press, Portlan Oregon.
14. Gamborg, OL and JP Shyluk (1970) The culture of plant cells using amino salts as the sole nitrogen source. Plant Physiol. 45, 598–600.
15. Toriyama, K, K Hinata, T Sasaki (1986) Haploid and diploid plant regeneration from protoplasts of anther callus in rice. Theor. Appl. Genet. 73, 16–19.
16. Muller, AJ. and R Grafe (1978) Isolation and characterization of cell lines of Nicotiana tabacum lacking nitrate reductase. Mol. Gen Genet. 161, 67–76.
17. Gamborg, OL and JP Shyluk (1981) Nutrition, media and characteristics of plant cell and tissue cultures. In: TA Thorpe (ed.). Plant Tissue Culture Methods and Applications in Agriculture. Academic Press, New York, pp 21–44.
18. Heyser, JW, TA Dykes, KJ DeMott and MW Nabors (1983) High frequency, long term regeneration of rice from callus cultures. Plant Science letters 29, 175–182.
19. Tissue Culture for Crops Project, Progress Report. 1987. Colorado State University, Fort Collins, Colorado, 80523.
20. Vasil, IK (ed.). Cell culture and somatic cell genetics in plants. Laboratory techniques. Academic Press, Inc., New York.

Appendix A: Molecular weight

Mineral salts

H_3BO_3	61.84
$CaCl_2 \cdot 2H_2O$	147.02
$Ca(NO_3)_2 \cdot 4H_2O$	236.15
$CoCl_2 \cdot 6H_2O$	237.93
$CuSO_4 \cdot 5H_2O$	249.68
EDTA sodium ferric salt (13% Fe)	366.85
KCl	74.56
KH_2PO_4	136.09
KI	166.01
KNO_3	101.1
KOH	56.1
K_2SO_4	174.1
$MgSO_4 \cdot 7H_2O$	246.5
$MnSO_4 \cdot H_2O$	169.01
$MnSO_4 \cdot 4H_2O$	223.09

Conversion $H_2O/4H_2O = 0.76$
$4H_2O/H_2O = 1.32$

NaCl	58.44
$NaH_2PO_4 \cdot 2H_2O$	137.98
NaOH	40.01
$Na_2MoO_4 \cdot 2H_2O$	241.95
Na_2SO_4	142.06
NH_4Cl	53.49
NH_4NO_3	80.09
$NH_4H_2PO_4$	115.03
$(NH_4)_2SO_4$	132.14
$ZnSO_4 \cdot 7H_2O$	287.55

Amino acids

Ala	89.09
Arg	174.2
Asn	132.12
Asp	133.1
Cys	121.16
Gln	146.2
Glu	147.13
Gly	75.1
His	155.16
Ile	131.17
Leu	131.17
Lys	146.19
Met	149.21
Phe	165.19
Pro	115.13
Ser	105.09
Thr	119.12
Trp	204.22
Tyr	181.19
Val	117.15

Sugars

Galactose	180.16
Glucose	180.16
Fructose	180.16
Mannitol	182.17
Sorbitol	182.17
Sucrose	342.30
Xylose	150.13
Ribose	150.13

Miscellaneous

Disodium citrate	258.08
Disodium succinate	162.2
Sodium malate	156.0
Sodium pyruvate	110.0
Morpholinoethanesulfonic acid	213.0
Adenine sulfate	184.2
Urea	60.1

Vitamins

myo-Inositol	180.16
Nicotinic acid (niacin)	123.11
Pyrodoxine hydrochloride	205.64
Thiamine hydrochloride	337.28
Calcium panthothenate	476.53
Ascorbic acid	176.12
Nictotinamide (niacinamide)	122.12
Choline chloride	139.63
p-Aminobenzoic acid	137.13
Folic acid	441.4
Riboflavin	576.4
Biotin	244.3

Appendix B Hoogland nutrient solution

Table 6. Hoogland nutient solution

Ingredients	Concn g/liter	Final molar concn
Macronutrients		
$Ca(NO_3)_2 \cdot 4H_2O$	0.94	4.0 mM
$MgSO_4 \cdot 7H_2O$	0.52	2.0 mM
KNO_3	0.66	6.0 mM
$NH_4H_2PO_4$	0.12	1.0 mM
Sequestrene 330 Fe[b]	0.07	
Micronutrients		
H_3BO_3	28	45 μM
$MnSO_4 \cdot H_2O$	34	20 μM
$CuSO_4 \cdot 5H_2O$	1.0	0.4 μM
$ZnSO_4 \cdot 7H_2O$	2.2	0.7 μM
$(NH_4)_6Mo_7O_{24} \cdot 4H_2O$	1.0	0.2 μM
H_2SO_4 (concentrated)	5 ml	

[a] A 0.1 ml volume of the micronutrient solution is mized with 1 liter of the macronutrients, and the pH is adjusted to 6.7.
[b] Geigy Agricultural Chemical Corp., Ardsley, N.Y.

The Hoogland nutrient solution is often used as a convenient fertilizer for most plants. The composition is also suitable for hydroponics culture.

Appendix C

There is increasing evidence for the importance of pH in plant cell culture growth and morphogenesis. The biological buffers are generally tolerated by plant cells and protoplasts. The compound are available from SIGMA Co, St. Louis, Mo and other suppliers.

Buffer	Biological buffers			
	M.W.	pKa (20°)	Range	0 °C (M)
MES	195.2	6.15	5.8–6.5	0.76
ADA	190.1	6.62	6.2–7.2	0.09
PIPES	302.2	6.80	6.4–7.2	slightly
ACES	182.2	6.88	6.4–7.4	0.22
BES	213.2	7.15	6.6–7.6	3.2
MOPS	209.3	7.20	6.5–7.9	3.09
TES	229.2	7.50	7.0–8.0	2.6
HEPES	238.3	7.55	7.0–8.0	2.25
HEPPS	252.3	8.00	7.6–8.6	1.58
Tricine	179.2	8.15	7.6–8.8	0.8
Glycine amide HCl	110.56	8.20	–	4.6
Tris	121.1	8.3	–	2.4
Bicine	163.2	8.35	7.8–8.8	1.1
Glycil glycine	132.13	8.40	–	1.1*
CHES	207.3	9.95	9.0–10.1	1.14
CAPS	221.3	10.40	9.7–11.1	0.47

* Molarity of Sat. Sol. MOC.

Plant Tissue Culture Manual **A2**: 1–19, 1991.
© 1991 *Kluwer Academic Publishers*.

The initiation and maintenance of callus cultures of carrot and tobacco

ROBERT D. HALL

Centre for Plant Breeding and Reproduction Research (CPRO), Wageningen, The Netherlands

Introduction

Terminology and characteristics. When an organ of a plant is damaged a wound repair response is induced to bring about the repair of the damaged portion. This response consists initially of the induction of division in the undamaged cells adjacent to the lesion, thus sealing off the wound. This is then followed by the hardening of this layer through the deposition of lignin, suberin, wax etc in order to regain the integrity of the protective outer barrier of the plant. If, however, wounding is followed by the aseptic culture of the damaged region on a defined medium, the initial cell division response can be stimulated and induced to continue indefinitely through the exogenous influence of the chemical constitution of the culture medium. Wounding may not be essential to the *in vitro* callussing response although it is generally stimulatory to the rate of callus formation. The result is a continually-dividing mass of generally poorly differentiated and disorganised plant cell aggregates termed a callus.

Callus is generally grown in Petri dishes, glass tubes or extra-wide necked Erlenmeyer flasks on medium solidified with agar or one of its replacements (agarose, Gelrite, etc). In morphological terms it can vary extensively, ranging from being very hard/compact, where the cells have extensive and strong cell to cell contact, to being 'friable' where the callus consists of small, disintegrating aggregates of poorly-associated cells and has a rather crumbly or creamy appearance. Friable callus is generally most sought-after as it is usually the fastest growing and most uniform type and is best suited for the initiation of cell suspension cultures (see appropriate chapter, this volume). Callus morphology is often explant-dependent but can usually be altered by the modification of the growth substance supplementation of the culture medium.

Due to their size and nature, callus cultures have an inherent degree of heterogeneity. As there is a unidirectional supply of nutrients (from the medium below) and gases and light (predominantly from above), chemical and physical gradients will be present within the callus mass. While, in some instances, this heterogeneity is a disadvantage (e.g. in the production of uniform biomass) it may also be an important influential factor in the developmental response of the callus in, for example, plant regeneration.

Applications. Callus cultures have a wide range of applications. For example, they are used directly as a source of cell material for biochemical studies [1], secondary metabolite production [2], or plant regeneration to obtain somaclonal variants (with or without the inclusion of a mutagenesis treatment) [3,4]. Most commonly however, callus cultures are involved as an intermediate step towards the initiation of cell suspension cultures or the production of plants from protoplasts.

Requirements to begin.

   i) *Laboratory equipment:* Basically a plant tissue culture laboratory requires facilities for media preparation and for the sterilization of media (autoclave, sterile filters) and glassware (autoclave and/or a 180 °C 'hot-box'); a sterile environment in which to carry out the aseptic manipulations (laminar flow cabinet or UV-sterilized transfer room or inoculation box) and finally, a constant environment room equipped for the control of light intensity, day-length and most importantly, temperature.

   ii) *Plant material:* The most commonly used starting materials are seeds (especially for monocots); young, aseptically-germinated seedlings and greenhouse-grown, young, healthy plants. The first two have the advantages that seed can generally withstand more severe sterilization conditions than plant tissues and thus it is easier to obtain sterile explants for culture. In addition, as all the plant material is at a very young stage there is a high potential for cell division within the explants. Low germination frequencies can however be problematic. Greenhouse-grown plants provide a larger source of explant material but sterilization can prove difficult both due to sensitivity to the chemicals used and to the external morphology of the plant (e.g. the presence of many hairs or waxy scales which can prevent good contact with the sterilization solution). If such greenhouse-grown plant material is to be used for callus initiation it must be entirely free from infection (including viruses), devoid of any signs of insect attack (which can lead to the internal contamination of the explant material) and be in a well maintained state. Unlike for *in vitro* grown plants, for which constant conditions are maintained, the physiological state of greenhouse-grown plants can, without care, vary considerably and this can have a profound influence upon the response of cultured explants.

   Essentially all organs can be used as explant sources. However, the degree of success with different tissues can vary extensively and calluses with differing morphologies (due to their origination from different cell types?) are frequently obtained. For dicotyledonous plants most commonly young leaves, petioles, stems and hypocotyls (from seedlings) are used. For monocots, meristematic regions are usually chosen (leaf bases, young inflorescences etc). Prior to culture all material must be surface sterilized, for which there is a range of chemicals to choose from (e.g. sodium hypochlorite, mercuric chloride etc, for overview see [5]). Following sterilization the material must be washed thoroughly to remove all traces of the sterilant, cut into suitably-sized pieces (explants) and plated onto solid medium. The choice of medium is determined by the species to be used and the aim of the experiment (i.e. biomass production, plant regeneration etc). Today, with the extensive work which has already taken place in this field, a suitable medium to begin with can usually be found in the literature (for a useful start see [6]). *In vitro* callus formation is often already visible after 4–7 d of culture although it can take much longer. Such callus can grow indefinitely *in vitro* if it is provided with a constant supply of the appropriate nutrients and plant growth substances. This thus requires that pieces of the growing cell mass be transferred (subcultured) to fresh medium at regular intervals.

How to proceed. This chapter is further divided into three parts. In the first two, examples are given of detailed experimental protocols for the initiation of callus cultures of two readily-available plant species suitable for use by researchers with little experience of plant cell culture. In the final part some advice is given as to how to tackle the two most commonly-occurring problems in callus initiation, those of contamination and the non-friability of cultures.

Procedures

The initiation and maintenance of Daucus carota *(carrot) callus cultures*

The following protocol for the production of carrot callus from tap-root material provides an excellent opportunity not only to practice the various techniques required for callus initiation but also allows one to see how different tissues within the explant respond differently to the *in vitro* conditions. Callus formation should be visible within 7 d and established, rapidly-growing, friable callus cultures can be obtained within 5–8 weeks [see Figure 1].

Steps in the procedure
[Steps 4–13, 16 & 18 under sterile conditions]

1. Choose as the explant source a healthy, undamaged carrot tap root 3–5 cm in diameter.
2. Clean the root thoroughly under running tap water.
3. Remove the uppermost 1 cm of the root (including all the remains of the shoot portion) and cut off all of the distal portion which is <2 cm thick and discard both. Cut the remainder transversely into *ca.* 7 cm lengths.
4. Under sterile conditions, remove the foil from a sterile 1 l beaker and pour in 500 ml of the well-mixed hypochlorite solution.
5. Submerge 2–3 of the carrot pieces in the solution and leave for 25 min.
6. Using sterile forceps, transfer the carrot pieces to a clean, sterile 1 l beaker and rinse with 100 ml sterile tap water.
7. Wash 3 times in 3 × 500 ml sterile tap water: first for 5 min, then for 10 min and finally for 15 min.
8. Transfer one of the sterile segments onto a sheet of sterile filter paper.
9. Remove and discard 1 cm thick slices from both ends using a sterile sharp knife.
10. Cut as much as possible of the remainder into 1 mm thick transverse slices and transfer these to a sterile Petri dish containing filter paper moistened with sterile distilled water.
11. Continue until 20–25 evenly-cut slices have been collected.
12. Place one slice on a suitable sterile cutting surface and using a sterile scalpel remove 2–3 (8 × 8 mm) explants so that each contains a segment of the cambial ring and the tissues on either side (phloem and xylem).
13. Transfer these explants directly to culture dishes, 5 per dish and seal all dishes with Parafilm.
14. Incubate in the dark or under low intensity light (*ca.* 25 μmol/m^2/sec) at 25 °C.
15. Examine all dishes after 7 d and 14 d and discard any which show signs of contamination.
16. After 21 d, select out the best-callussing explants and after cutting the callussing region into three evenly-sized pieces, transfer to fresh medium (5 per dish).

17. Seal all plates with Parafilm and incubate as above.
18. After 21 d, use the most friable callus for subculture as per step 16.
19. The cultures can hereafter be maintained by repeating step 18.

Equipment
a. Healthy, undamaged carrot roots.
b. Sterile, sharp, stainless steel knife suitable for flaming.
c. Sterile scalpel, long and short-handled forceps which, when not in use are stored with the ends in 70% EtOH. Before reuse the alcohol is removed by flaming.
d. Sterile Petri dishes.
e. 2 sterile 1 l beakers.
f. 500 ml 2% (w/v) sodium hypochlorite solution in water.
g. Sterile tap water and distilled water.
h. Sterile filter paper, coarse grade.
i. Sterile 9 cm Petri dishes each containing 20 ml culture medium.
j. Parafilm strips.

Medium

Murashige and Skoog medium	4.71 g/l*
Sucrose	30.00 g/l
2,4-D	1.0 mg/l
Agar (e.g. Oxoid No. 3)	8.00 g/l
pH prior to autoclaving	5.8

* The amount given here refers to the full MS medium (lacking agar and hormones) as supplied by Flow Labs. The corresponding amount of the equivalent medium as obtained from another supplier may vary slightly.

Notes (carrot callus)
1. If using purchased material choose carrots which have not been mechanically washed.
2. A coarse cloth or a fine, soft brush may be used.
4. Use a 2% (w/v) solution of sodium hypochlorite in tap water. This can be obtained by diluting most commercially — available domestic bleaches 1 : 4. If using domestic bleaches, those without a thickening agent should be used.
5. It is critical that the root segments remain submerged for the entire period. This can be ensured by weighing-down the pieces with the long-handled forceps or by covering the top of the sterilizing solution with 4 or 5 layers of sodden tissue paper.
7. Again as above, the root segments should be submerged. It is important that all of the hypochlorite solution, which is toxic, is removed otherwise the explants will not survive on culturing.
8. Or onto an empty Petri dish lid/base.
9. This is to remove the tissue which may have taken up the hypochlorite solution.
11. Use, when necessary the extra root pieces in order to obtain sufficient root slices.
12. It will be observed that a ring of the outermost tissue is bleached white by the sterilization protocol. This tissue is dead and should not form part of the explants.

13. It is preferable to retain the polarity of the explants by placing the root pole in contact with the culture medium. Set up *ca*. 15 dishes.
14. We generally observe better growth in the light.
15. Contamination will be apparent predominantly underneath the explant on the surface of the medium. No attempt should be made to 'rescue' non-contaminated explants from contaminated dishes. All contaminated dishes should be autoclaved before being discarded.
16. The callus pieces transferred should be *ca*. 5 × 5 × 2 mm.
18. When the callus is very friable it is better to place the small cell aggregates in 5 groups in each Petri dish rather than spreading them evenly over the surface of the medium.

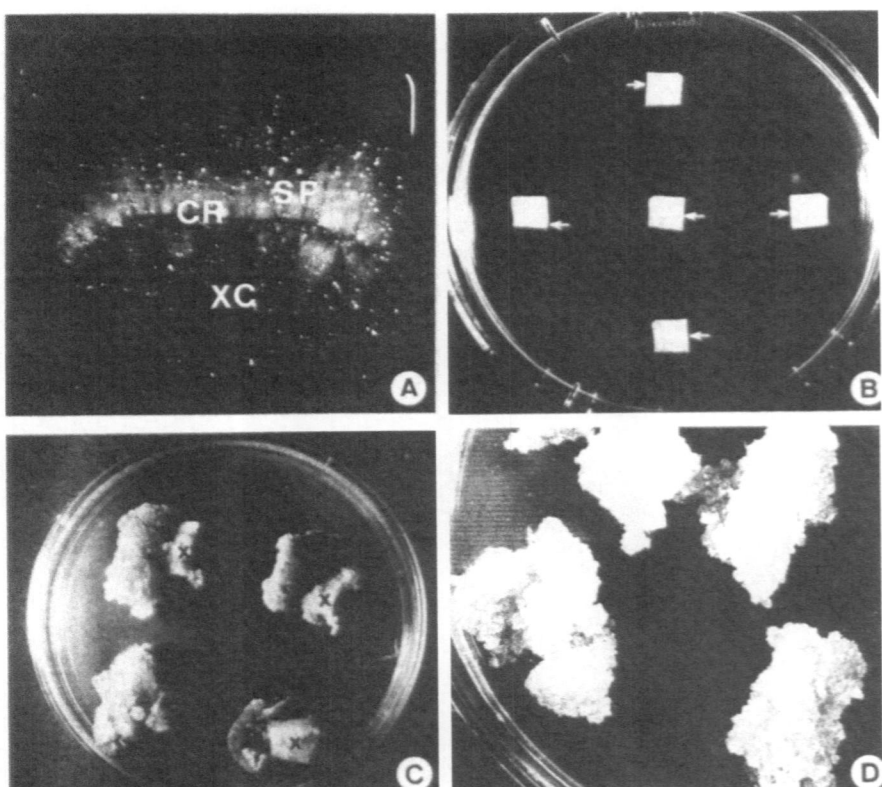

Fig. 1. [A] A transverse slice (8 mm × 8 mm) from a carrot tap root. SP – secondary phloem; CR – cambial ring; XC – xylem core. [B] Explants immediately after plating out in a 9 cm Petri dish (cambial ring arrowed). [C] Carrot root explants *ca.* 3 weeks after culture. Note that only the cambial and phloem tissues have callussed extensively while the xylem (X) tissue has not. [D] Established, friable callus cultures of carrot.

The initiation and maintenance of callus cultures from Nicotiana tabacum
(tobacco) pith tissue

Pith is that tissue found in the centre of the stems of most *Nicotiana* species. It consists entirely of parenchymatous cells — large, thin walled and highly vacuolate and represents one of the most uniform tissues which is readily available for *in vitro* culture. Indeed, the basal medium most commonly used for plant cell culture today, that of Murashige and Skoog, was developed using tobacco pith callus as the experimental system [7].

Callus formation from pith tissue is generally rapid — after 14 d the original explant will no longer be recognisable. The callus obtained is uniform and fast-growing and can be highly responsive to *in vitro* culture conditions, producing callus, or shoots, leafy structures or roots according to the type, level and balance of the growth substance supplements in the medium. As such, tobacco pith callus can be used as a reliable bioassay system for cytokinins and as an excellent system to demonstrate the influence of growth substances on plant morphogenesis (see [8]).

Steps in the procedure
[Steps 6—14 & 16—17 under sterile conditions]

1. Use healthy, rapidly growing, non-flowering plants of *N. tabacum* var. Maryland Mammoth or White Burley which are 0.5—1 m tall and which have been grown in the greenhouse.
2. Remove the upper 30 cm of the plant, cut off all the leaves and chop up the remaining stem immediately above and below each node.
3. Collect all the internode segments which have a diameter > 6 mm and cut them, if necessary, into 25 mm lengths. These should be stored in a Petri dish containing a disc of filter paper moistened with distilled water. Discard everything else.
4. Taking each internode in turn, dip both ends in molten paraffin wax (60—65 °C) to a depth of *ca.* 1 mm.
5. Allow the wax to harden fully.
6. Moving to the sterile work area, immerse the sealed internodes in 70% EtOH for 30 sec and then rinse briefly (10 sec) in distilled water.
7. Remove the foil from a sterile 250 ml beaker, pour in the well-mixed hypochlorite solution. Immerse the plant segments in this solution and leave for 25 min.
8. Using sterile, long-handled forceps transfer the internodes to a sterile 250 ml beaker and rinse with 100 ml sterile distilled water.
9. Wash the internodes three times with 3 × 200 ml sterile distilled water, first for 5 min, then 10 min and finally for 15 min.
10. Transfer all the internodes to a sterile Petri dish containing a filter paper disc moistened with sterile distilled water.
11. Taking the internodes in turn, transfer onto a sterile cutting surface (Petri dish lid/base, thick filter paper etc) and cut off and discard the waxed ends.

12. Place the internode on end and holding firmly with sterile forceps cut away all of the outer tissue by making 4 overlapping, longitudinal cuts to produce a rod of the central pith tissue *ca.* 10—12 mm long and with a square cross-section.
13. Slice the pith transversely into 1.5—2 mm sections and transfer directly to the culture plates, 5 sections per plate.
14. Repeat for the other internodes until *ca.* 15 dishes have been prepared.
15. Seal all dishes with Parafilm and incubate in the dark at 25 °C.
16. After 7 and 14 d examine all cultures and discard any dishes showing signs of contamination.
17. After 21 d, select the explants showing the best callussing response for subculture. Cut these up into 2 × 5 × 5 mm pieces and transfer onto fresh medium, 5 per plate.
18. Culture as per step 14.
19. Cultures can hereafter be routinely maintained by repeating steps 16 & 17.

Equipment
a. Healthy, greenhouse-grown tobacco plants *ca.* 0.5—1 m tall.
b. Sharp scalpel.
c. Sterile scalpel, long- and short handled forceps which, when not in use are stored with the ends in 70% EtOH. Before reuse the alcohol is removed by flaming.
d. Non-sterile 9 cm Petri dishes containing moistened filter paper.
e. Molten wax.
f. 250 ml beaker containing *ca.* 150 ml 70% (v/v) EtOH in water.
g. 2 sterile 250 ml beakers.
h. 200ml solution of 2% (w/v) sodium hypochlorite in water.
i. Sterile distilled water.
j. Sterile 9 cm Petri dishes.
k. Sterile 9 cm filter paper discs, coarse grade.
l. Sterile 9 cm Petri dishes each containing 20 ml culture medium.
m. Parafilm strips.

Medium

Murashige and Skoog medium	4.71 g/l*
Sucrose	30.00 g/l
1-Naphthaleneacetic acid (NAA)	2.00 mg/l
Kinetin	0.25 mg/l
Agar (e.g. Oxoid No. 3)	8.00 g/l
pH prior to autoclaving	5.8

* The amount given here refers to the full MS medium (lacking agar and hormones) as supplied by Flow Labs. The corresponding amount of the equivalent medium as obtained from another supplier may vary slightly.

Notes (tobacco)

1. No plant material showing any signs of (previous) disease or insect attack should be used.
2. Clean cuts are essential. Use a sharp scalpel.
4. Standard embedding wax can be used. As all waxed portions must later be discarded (as they will not be sterile) do not dip the segments too deeply.
6. Use a 2% (w/v) solution of sodium hypochlorite in tap water. This can be obtained by diluting most commercially – available domestic bleaches 1 : 4. If using domestic bleaches, those without a thickening agent should be used. Ensure that all segments are fully submerged, if necessary by covering the solution with several layers of sodden tissue paper. It is advisable to agitate the segments occasionally in order to release gas bubbles which may form on their surface.
9. It is critical that all of the hypochlorite is removed as it would prove toxic to the explanted tissue.
11. Take great care to remove all of the waxed portion. For certainty cut the internodes *ca*. 1 mm away from the limit of the wax.
16. Check particularly for contamination underneath the explant, i.e. between the tissue and the medium. All plates containing contaminated explants should be autoclaved before being discarded.
17. A sharp scalpel should be used.
19. A longer subculture period of 4–5 weeks may be used for stock cultures providing that they have not begun to show symptoms of stress (e.g. browning) by this time.

Troubleshooting

Even for experienced workers the initiation of callus cultures from a new species or even a new genotype frequently does not go according to plan. In such instances two basic problems generally predominate – culture contamination and the failure to obtain sufficient growth of a desired callus type (e.g. friable). However, with dedication such difficulties can usually be overcome and in this section some useful tips are given as to how one might proceed.

A. *Culture contamination*: It is most important to check regularly the source plants and discard immediately any which show signs of disease or damage. Keep the source plants isolated from older plants which are more likely to be harbourers of contaminants, maintain them in a healthy state and prevent insect attack (greenfly, leaf borers etc). However, if contamination still occurs with apparently clean and healthy plants a number of possible approaches can be taken. These can be grouped into two categories based either upon an extension or modification of the sterilization protocol to increase the likelihood of achieving complete decontamination or alternatively, on the inclusion of antimicrobial agents in the culture medium. The former is the best choice. However, if the contaminant is systemic (or internally seed-borne) then the latter is the only potential solution to the problem.

Protocol modifications:
a. If not already present, include in the protocol a brief submergence in 70% EtOH (v/v in water) prior to the sterilization step. For seed, up to 1 min is usual but for fresh tissue this step should not be longer than 30 sec. Material should be briefly rinsed in distilled water to remove the alcohol before transferring to the sterilizing solution.
b. Improve the wetting capacity of the sterilizing solution by including a drop of detergent (household or Tween 20/80).
c. If the plated explants show no signs of damage following the standard sterilization protocol (i.e bleaching, or rapid browning within the first 24h) increase the concentration of the sterilant (even up to 100% can be used for seeds) or switch to a more potent one (see [4]).
d. If it is suspected that poor sterilization is due to certain morphological characteristics of the plant material (extensive hairyness, uneven surfaces etc) choose the organs demonstrating these characteristics the least (often the youngest tissues) or alternatively:
e. Choose tissues which have not yet been exposed to the environment and should therefore, still be free from contaminants. For example, after surface 'sterilization' carefully dissect out the internal organs of young vegetative or floral buds.
d. Carry out a double sterilization where, following the first sterilization the outer layers of tissue are cleanly cut away and discarded and the remainder

resterilized. (This is clearly only possible with larger organs e.g. thick stems, root tubers etc.)

e. If the rough coat of seed is possibly the problem this can sometimes be removed prior to sterilization by coarse grinding or, if the seeds are large, by dissection. It may even be possible to dissect out the embryo. A softening treatment can also be employed – for sugarbeet we have found that a 15 min treatment with tap water at 55 °C enhances sterilization frequency considerably without having any apparent detrimental effect on germination. For grass seed (e.g. *Lolium*, Creemers-Molenaar, pers. comm.) a triple protocol of sterilization, then softening in sterile distilled water for 1 d followed by a second sterilization has proven very effective.

The use of antimicrobial agents:
Antibiotics should never be used to compensate for sloppy laboratory practise. Before considering their use one should be certain that the problem genuinely lies with a contamination of the explant source. There are a number of disadvantages to the use of such agents. Firstly, the fact that the source tissue is infected might suggest that it is probably not in the best physiological state for culturing. Secondly, the levels of antibiotics necessary to rid an explant of all viable microbes can prove equally phytotoxic thus rendering the procedure worthless. Even the most 'plant-friendly' compounds are growth inhibitory. The final difficulty is that antimicrobial agents are generally only toxic to a specific range of organisms. Contaminated explants can be infected with more than one (usually unknown) organism making it difficult to achieve a successful decontamination. However, if all else has failed there are a range of treatments to choose from involving either a single or a combination of antibiotics. As an indication, some treatments are presented here. However, before beginning it is strongly recommended to consult the specific literature on this subject [e.g. 9; 10 and the references therein].

Different compounds have different toxicity levels. It is therefore advisable to test compounds at a range of concentrations to avoid the risk of losing all the experimental material. As a general rule antibiotic solutions should be filter-sterilized and added to the medium after it has been autoclaved and allowed to cool to *ca.* 50 °C. A brief survey of the literature has revealed that the following treatments have proved valuable for various species: cefotaxime/carbenicillin (≤ 500 mg/l, alone or in combination), streptomycin (≤ 100 mg/l), kanamycin (≤ 200 mg/l), ampicillin (≤ 400 mg/l) and nystatin (≤ 20 mg/l).

B. *Non-friability of callus*: When beginning with a new species where one is unsure of the *in vitro* response it is always best to maximise the likelihood of obtaining suitably-friable callus right from the start of the culture process. This can be done by (1) using a range of genotypes (or at least several different plants) as the source of explant material; (2) choosing as many organs/tissues for culture as possible and (3) using a number of media containing a range of growth substance concentrations centered around a level determined from the literature. Furthermore, at each subculture routinely select only the least aggregated and most rapidly-growing callus for transfer to fresh medium. However, if the callus has already been established or the above procedure has consistently yielded only compact callus a number of manipulations can be attempted to improve culture friability. Unfortunately however, due to the varying response of different plant tissues no clear-cut guidelines can be given as to the most appropriate line of attack. The most commonly-used approach is to vary the growth substance supplementation of the medium both in quantitative and qualitative terms. Altering the auxin/cytokinin balance can also prove valuable. Transferring from light to dark or enhancing the far red portion of the incident light have also been suggested to be influencial in some systems. Finally, transferring the most friable callus to liquid culture (see relevant chapter, this volume) and selecting for the finest cell material produced therein for replating back onto agar medium can also be attempted.

References

[1] Treat, WJ, Engler, CR, Soltes, EJ, (1989). Comparison of photomixotrophic and heterotrophic callus and suspension cultures of *Pinus elliottii*. 1. Photosynthetic properties and ultrastructural evidence for coexistence of starch granules and secondary metabolites. Plant cell, tissue and organ culture 17, 205–224.

[2] Hall, RD, Yeoman, MM (1987). Variation in anthocyanin accumulation and its stability in cultures of *Catharanthus roseus* derived from single cells. Journal of experimental Botany 38, 1391–1398.

[3] Dix, PH (1986). Cell line selection. In: Yeoman, MM (Ed.) Plant cell culture technology, pp 143–201, Blackwell, Oxford.

[4] Reisch, B (1983). Genetic variability in regenerated plants. In: Evans, DA, Sharp, WR, Ammirato, PV, Yamada, Y, (Eds.) Handbook of plant cell culture 1, pp 748–769, Macmillan, New York.

[5] Yeoman, MM (1977). Tissue (Callus) culture techniques. In: Street, HE (Ed.) Plant tissue and cell culture, pp 31–58, Blackwell Sci. Publications, Oxford.

[6] George, EF, Puttock, DJM, George, HJ (1987). Plant culture media 1, Exegetics Ltd., Westbury, UK.

[7] Murashige, T, Skoog, F (1962). A revised medium for rapid growth and bioassays with tobacco tissue cultures. Physiologia Plantarum 15, 473–497.

[8] Reinert, J, Yeoman, MM (1982). Plant cell and tissue culture, a laboratory manual. Springer-Verlag, Berlin.

[9] Pollock, K, Barfield, DG, Shields, R (1983). The toxicity of antibiotics to plant cell cultures. Plant cell reports 2, 36–39.

[10] Leifert, C, Waites WM (1990). Contaminants of plant tissue cultures. IAPTC Newsletter #60, 1–13.

Plant Tissue Culture Manual **A3**: 1–21, 1991.
© 1991 *Kluwer Academic Publishers.*

The initiation and maintenance of plant cell suspension cultures

ROBERT D. HALL
Centre for Plant Breeding and Reproduction Research (CPRO), Wageningen, The Netherlands

Introduction

1. *Terminology*: Ideally, a plant cell suspension culture should consist of a population of single cells suspended, through continuous agitation, in a liquid nutrient medium. However, in only the rarest of instances, if ever, does such a culture exist. Increased cell dissociation means increased culture uniformity and consequently, most researchers strive to achieve as fine a cell suspension as possible. However, some degree of cell aggregation generally has to be tolerated and so-called 'fine' suspension cultures consist of micro- to submacroscopic colonies made up of *ca.* 5–200 cells. Furthermore, cultures consisting of larger aggregates (e.g. 0.5–1.0 mm in diameter) usually are more readily attainable, grow perfectly well and depending on the aim of the research are often sufficient to meet all requirements. Indeed, in certain instances a degree of cell aggregation has been argued to be greatly beneficial – e.g. in retaining the totipotent character of gramineous suspension cultures [1, 2] and in obtaining enhanced yields of desirable secondary metabolites from *in vitro* systems [3].

The most commonly-used cell suspensions are of the closed (or batch) type where the cells are grown in a fixed volume of liquid medium and which are routinely maintained through the transfer of a portion (usually *ca.* 10%) of a fully-grown culture to fresh medium at regular intervals. This is the type of cell system described in this chapter.

2. *Characteristics*: While cells from a suspension culture generally most resemble parenchymatous cells in having relatively large vacuoles, a thin layer of cytoplasm and thin, rounded cell walls [e.g. Figure 1.], this is by no means always the case. The species/genotype and medium composition used can influence *in vitro* cell morphology and different cell types (with different morphological/physiological properties) can co-exist within a single culture. The balance can however usually be manipulated to favour one specific type through selective subculture. This may be very important, as reported for example in both of the first two papers describing successful plant regeneration from maize suspension cell protoplasts [1,2]. Both groups, working independently, stressed the essentiality of selectively subculturing the small aggregates of densely-cytoplasmic, isodiametric cells and not the larger, highly vacuolate single cells for the maintenance of suitable stock cell suspension cultures.

The growth curve of a cell suspension culture [Figure 2] has a characteristic shape consisting of four essential stages – an initial lag phase, an exponential phase, a linear phase and ultimately a stationary phase. The duration of the different phases is dependent not only on the culture used (species/geno-type/cell line) but also on the medium and subculture regime. Inocula taken from a culture still in the linear growth phase will produce cultures with a shorter (or non-existent) lag phase than those initiated using cells taken from a stationary phase culture. The lag phase is also shortened when relatively large inocula are used although paradoxically, growth terminates earlier and overall biomass production is reduced. Conversely, a very small inoculum results in a greatly extended lag phase if indeed the culture grows at all. Furthermore, as can be seen from the experimentally-determined example in Figure 3 the relative time scales of different culture parameters are also quite different.

3. *Applications*: Plant cell suspension cultures have a wide range of applications and are generally used in preference to callus cultures due to their more rapid growth rate, greater homogeneity and the ease of experimental manipulation. The continuous agitation of such cultures mixes and aerates the medium and helps break up the cell aggregates. Consequently, a high proportion of the cells are in direct contact with the culture medium (and what is added to it) and thus the physical (light) and chemical (gases, nutrient) gradients present in callus cultures, are to a large extent eliminated. Cell suspensions have been success-fully used as a means of rapid and continuous production of uniform biomass for basic plant cell physiology studies [6], enzymological studies [7], and secondary metabolite production [8]. Through the manipulation of the medium/subculture regime, cells can be induced to divide synchronously thus providing a valuable tool for cell cycle studies [9]. Cell suspensions have also proven to be excellent starting materials for the isolation of protoplasts to be used in a wide range of applications (cell fusion, genetic manipulation etc.). Their use in embryogenesis studies is dealt with elsewhere in this volume.

4. *Requirements*: To a laboratory already equipped for growing callus cultures, essentially only three additions are required to enable the progression to sus-pension culture work; a supply of culture flasks (usually 100 ml or 250 ml wide-necked Erlenmeyer flasks are used), wide-mouthed glass pipettes for subculturing (we use 10 ml pipettes cleanly sawn off at the 9.75 ml gradation and a cheaper alternative, 20 cm × 3 mm bore glass tubing) and finally a culture shaker. Most commonly, a rotating platform shaker is used consisting of a table onto which flasks of varying sizes can be firmly fixed and which rotates in a horizontal plane at selected speeds of 50–150 rpm with an amplitude of *ca.* 1 cm. This must be situated in a constant environment room or have an independently-controlled culture chamber attached. A word of caution must be offered here. It is strongly advisable to have two smaller shakers rather than one large one, or at least, to have alternative shaking space available (in a nearby lab.) for emergencies. Culture shakers must function 24 h per day and

365(6) days per year. Eventually the motor will burn out and the bearings will wear away. Repairs take time and stationary suspension cultures frequently have very short 'shelf' lives of only a few hours.

5. *How to proceed*: This chapter is further divided into 4 parts. In the following 2 a couple of almost fool-proof methods are given for the initiation of suspension cultures of two readily-obtainable plants. Such examples should prove excellent starting points for the uninitiated. In addition, it is important to have a basic understanding of the experimental system; a section has thus been included detailing the basic growth analysis methods for plant cell suspension cultures. Finally, in the last section some brief advice is given concerning the progress towards working with other plant species and on how to overcome some of the possible difficulties which may be encountered.

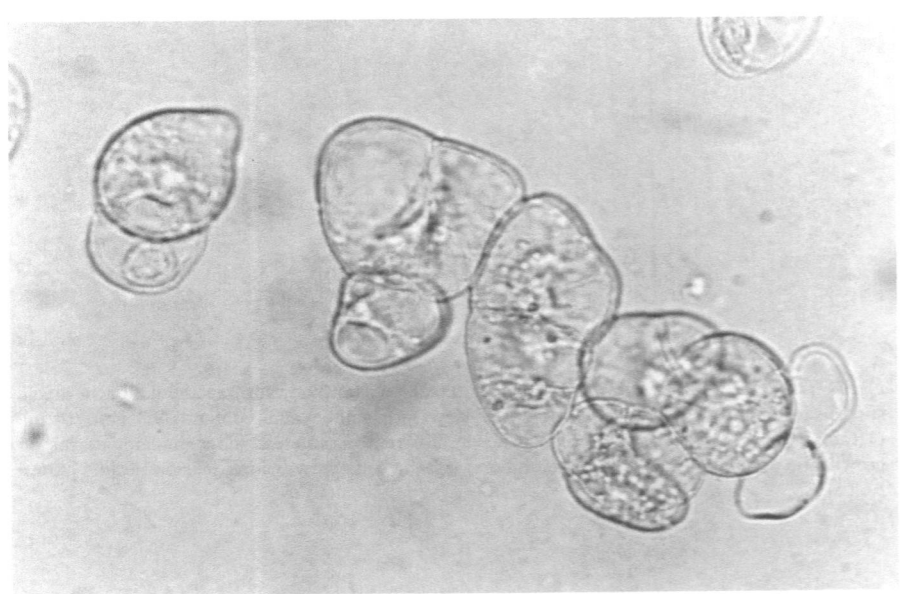

Fig. 1. The typical appearance of plant cells after culture in liquid medium.

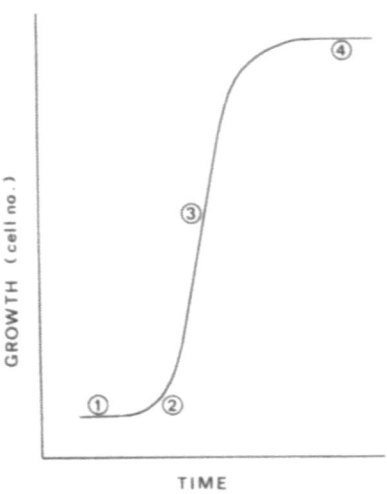

Fig. 2. Model growth curve for a plant cell suspension grown in a closed system. The four different growth phases are labelled; (1) Lag phase, (2) Exponential phase, (3) Linear phase, (4) Stationary phase. [After Ref. 4].

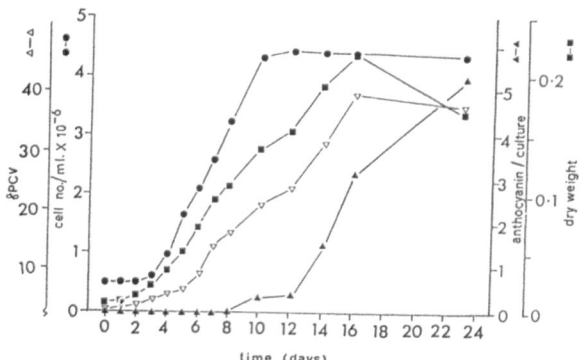

Fig. 3. The pattern of growth and secondary metabolite (anthocyanin) accumulation in a cell suspension culture of *Catharanthus roseus*. Dry wt (g); anthocyanin content (OD units/10 ml extraction solvent/culture). Note that while the different parameters show similar patterns of growth the individual time scales are different e.g. the cells grow before division begins. [After Ref. 5]

Procedures

Initiation of a carrot (Daucus carota) *cell suspension culture*

Carrot cell suspension cultures are one of the easiest types to establish. Beginning with friable callus, initiated as described in the previous chapter, a suitably fine suspension culture can be achieved rapidly, within 3–4 weeks. The basic protocol involves the inoculation of culture medium with, initially, a relatively large amount of cell material. Through the continuous agitation of the culture and the sustained division of the inoculum, single cells and small cell aggregates are released into the medium. These are selectively subcultured and after 1–2 further cycles the culture is established and typically, can subsequently be routinely maintained, using at each subculture, a smaller inoculum.

Steps in the procedure
[Steps 2–4, 7–10 under sterile conditions]

1. Use as the starting material friable, light-grown callus cultures, produced as described in the previous chapter, and which are still in their active growth phase (1–2 weeks following subculture).
2. Working in a sterile environment (laminar flow bench/UV – sterilized transfer room) quickly flame the neck of a culture flask and loosen but do not remove the aluminium foil cap.
3. Using forceps or a long-handled spoon spatula aseptically remove the outer periphery of each clump of callus; remove the foil cap from the flask, flame the neck and transfer *ca.* 2 g of this peripheral cell material to the culture flask.
4. Reflame the neck of the flask and reseal with the original or a new double foil cap.
5. Repeat for each flask after which all can then be firmly secured onto the rotary shaker under low intensity light (*ca.* 25 μmol/m²/sec).
6. After 7d examine all cultures and discard any which show signs of contamination.
7. After 10 d filter each culture through a sterile 250 μm nylon/stainless steel mesh and collect the filtrate in a sterile 100 ml Erlenmeyer flask.
8. Wait 10 min until the cells have settled and pipette off as much medium as possible. Use the remainder as the inoculum for a fresh culture.
9. After 10 d filter the culture through a sterile 100 μm nylon/stainless steel mesh. Repeat #8 but use only one half of the remainder as the inoculum.
10. After 10 d and every subsequent 10 d, subculture as for an established culture i.e. transfer directly, using a wide mouthed pipette, 1 ml of the well-mixed culture per 10 ml fresh medium.

Equipment:
a. Established, healthy friable callus cultures.
b. Autoclaved, 250 ml, wide-necked Erlenmeyer flasks containing 50 ml culture medium and sealed with a double layer of aluminium foil.
c. A supply of sterile aluminium foil sheets 10 × 10 cm.
d. Sterile forceps and a sterile spoon spatula which, when not in use, are stored with the ends in 70% EtOH. Before using, the alcohol is quickly removed by flaming.
e. A series of sterile 250 μm and 100 μm nylon or stainless steel filters + glass filter funnels.
f. Sterile 100 ml Erlenmeyer flasks.
g. Sterile, wide-mouthed glass pipettes.

Medium:

Murashige and Skoog medium	4.71 g/l*
Sucrose	30.00 g/l
2,4-D	2.21 mg/l
Kinetin	2.15 mg/l
pH prior to autoclaving	5.8.

* This amount corresponds to the full MS medium (without agar or hormones) as supplied by Flow Labs. The equivalent amount of medium supplied by other firms may vary slightly from this.
[Note: The hormone concentrations stated here are relatively high in order to 'guarantee' success with any carrot material. Much lower levels (e.g. 10% of the above values) may prove equally successful.]

Notes (carrot)
2. It is advisable, especially for inexperienced workers, to have an extra supply of sterile aluminium foil sheets available to use when the originals are accidentally torn on removal from the flasks.
3. If the callus is not characteristically 'crumbly' the outer, more friable regions can be gently scraped/sliced off for use as the inoculum. No callus showing signs of browning should be transferred.
5. Use speeds of 90–110 rpm with a throw amplitude of *ca.* 1 cm. Incubation at 25 °C is preferred.
6. Contamination is usually obvious: e.g. the medium appears cloudy and remains so after the culture has been stationary for 10 min; the cells appear browner and are not growing as well as the replicate flasks; a ring of slimy material has formed at the meniscus; fungal presence is visible as wafts of free hyphae in the medium, or as slimy, perfectly spherical balls among the cells. Contaminated cultures **must always be autoclaved** prior to removing the foil cap to discard the contents. [For possible causes/cures for contamination see final 'Troubleshooting' section.]
7. If an extensive amount of fine material has been released go directly to step 9.
8. To avoid cross-contamination use separate filters and flasks for each culture. Volume transferred should be < 15 ml/50 ml culture.
10. In the beginning it is advisable to check here that the initial cell density is in the $0.5-2.5 \times 10^5$ cells/ml range as recommended by Street [4] and adjust (if necessary) the subculture inoculum as required. Inoculum samples should be taken with care and be consistent. Continually agitate the suspension as the inoculum is removed, keep the pipette mouth free from the bottom of the flask and work with a reasonable speed to avoid clogging the pipette.

Initiation and maintenance of a cell suspension culture of the Madagascar
Periwinkle (Catharanthus roseus)

Some of the most rapidly-growing plant cell cultures on record are of *C. roseus*
suspension cultures. Growth rates of 5 g dry weight/l/day have been reported [10].
Many laboratories work with this species as it is the source of several highly valuable
secondary metabolites [11] which are used in anti-cancer therapy. Callus of *C. roseus*
[Figure 4] can be readily obtained from stem/petiole segments taken from young,
healthy plants when plated on MS medium supplemented with 30 g/l sucrose and
1 mg/l 2,4-D and 0.1 mg/l kinetin. Two types of callus can be obtained — a dry,
crumbly type and a wet, smooth creamy type. Both are excellent starting materials
for the initiation of cell suspension cultures.

Steps in the procedure
[Steps 1—4, 7—9 under sterile conditions]

1. Use as the starting material rapidly-growing friable callus (grown in the dark or
   in low intensity white light) which is midway through its growth cycle (7—10 d).
2. Working in a sterile environment, quickly flame and loosen the foil caps of the
   culture flasks.
3. Taking the flasks in turn, remove the foil cap, flame the neck of the open flask
   and using a spoon spatula inoculate each with 1—1.5 g of callus.
4. Flame the necks of the flasks and reseal with the original or new foil caps.
5. Secure all flasks firmly on the rotary shaker and incubate in the dark or in low
   intensity white light (25 μmol/m^2/sec).
6. Check all flasks after 7 d and discard any contaminated cultures.
7. After 10d:
   EITHER:
   [A] If the dry, crumbly callus was used, filter the culture through a sterile 100 μm
       nylon or stainless steel mesh and collect the filtrate in a sterile 100 ml flask.
       Mix the filtrate gently but well and transfer 20% (10 ml) to fresh medium
       using a sterile wide-mouthed glass pipette.
   OR:
   [B] If the creamy-type callus was used transfer 20% of the well-mixed culture
       directly to fresh medium using the sterile pipettes.
8. After 10 d check the extent of cell aggregation under the microscope. Further
   filtration is usually not required. Use a subculture ratio of 1 ml/10 ml fresh
   medium.
9. Maintenance as per step 8.

Equipment:
a. Established, friable callus cultures.
b. Autoclaved, 250 ml, wide-necked Erlenmeyer flasks containing 50 ml culture
   medium and sealed with a double layer of aluminium foil.
c. A supply of sterile aluminium foil sheets 10 × 10 cm.

d. Sterile spoon spatula which, when not in use, is stored with the end in 70% EtOH. Before using, the alcohol is quickly removed by flaming.
e. A series of sterile 100 μm nylon or stainless steel filters + glass filter funnels.
f. Sterile 100 ml Erlenmeyer flasks.
g. Sterile wide-mouthed glass pipettes.

Medium:

Murashige and Skoog medium	4.71 g/l*
Sucrose	30.00 g/l
2,4-D	1.0 mg/l
pH prior to autoclaving	5.8

* This amount corresponds to the full MS medium (without agar or hormones) as supplied by Flow Labs. The equivalent amount of medium supplied by other firms may vary slightly from this.

Notes (periwinkle)
3. Callus showing signs of browning should not be transferred.
4. Have a reserve supply of sterile foil sheets (10 × 10 cm) available to use as replacements for torn caps.
5. Use speeds of 110–140 rpm with a throw amplitude of *ca.* 1 cm. Incubation at 25 °C is preferred.
6. See note 6 of the previous section.
7. Use new filters and beakers for each culture to avoid cross-contamination. Inoculum samples should be taken with care — hold the pipette tip free from the flask bottom, move it around as the sample of mixed suspension is sucked up (to avoid taking just a sample of medium!) and both the taking and discharging of inocula should be performed with reasonable speed to avoid the clogging-up of the mouth of the pipette.
8. Check cell density as per note 10 of the previous section.

Fig. 4a.

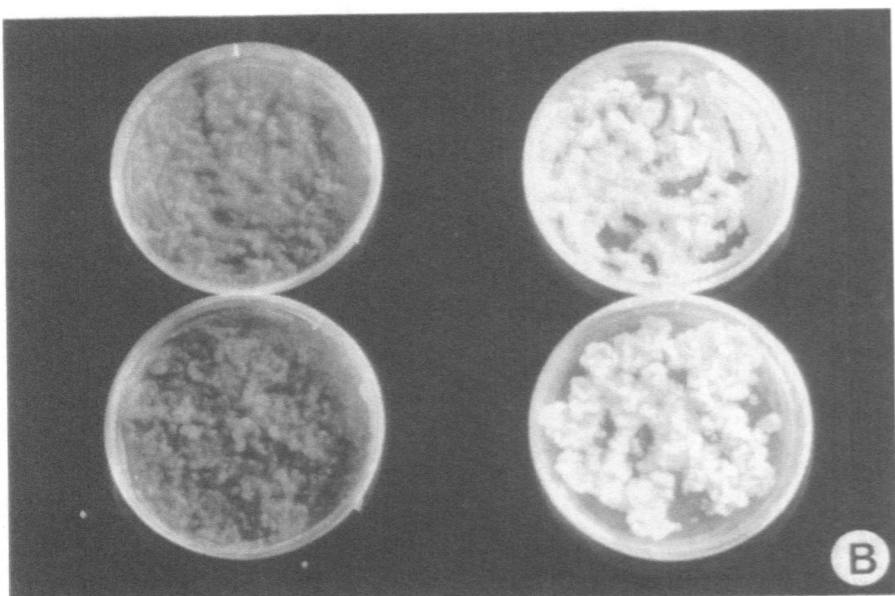

Fig. 4b.

PTCM-A3/11

Fig. 4c.

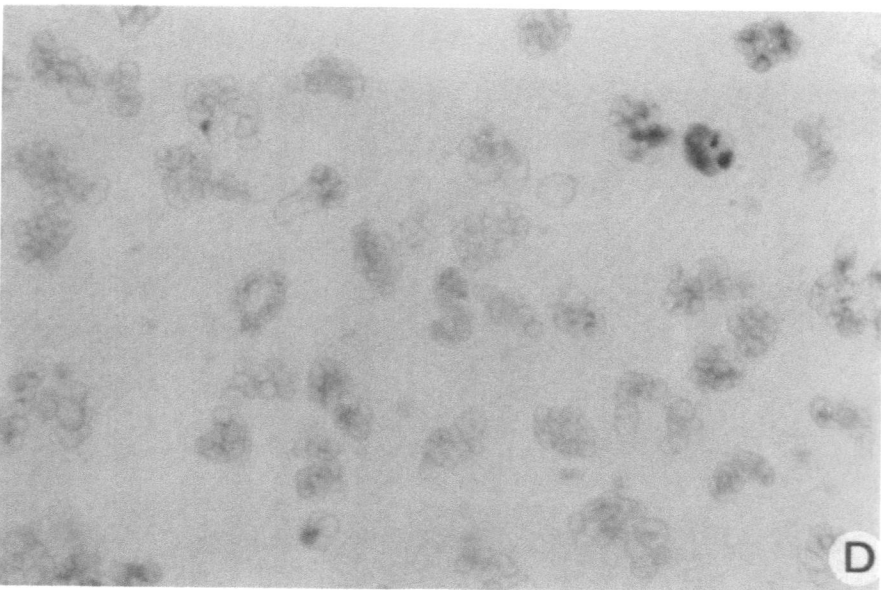

Fig. 4. [A] *Catharanthus roseus* (Madagascar periwinkle). [B] Typical 'creamy-type' callus cultures derived from [A]. Those on the left are pigmented after having been cultured in the light, those on the right were grown in darkness. [C] Established cell suspension cultures of *C. roseus*, light-grown (right) and dark-grown (left). [D] Sample from [C] viewed under the light microscope.

Methods for the determination of suspension culture growth and development

Once a suspension culture with the desired characteristics has been isolated it is strongly advisable to carry out some basic growth analyses in order to chart the development of the culture and determine the time scales of the different phases. Such information is important, for example, when the cultures are to be used for protoplast isolation or secondary metabolite production. However, even if such knowledge is not essential to the research aims it is always desirable to have a reasonable understanding of the basic experimental system in use.

Growth analysis is best carried out by initiating sufficient cultures to enable whole flasks to be harvested (sacrificed) on each measurement day. Replicate flasks must therefore also be set up to gain an estimate of interculture variation. However, if the cultures are fine and great care is taken at inoculation, 3 replicates per harvest time should be sufficient. Alternatively, a smaller number of cultures can be set up from which samples are taken on each measurement day. Where shaker space is limited this may be the only option. However, while this method has the advantage that all measurements are made using the same set of cultures there are also disadvantages. The risk of culture loss through contamination, arising from repeated sampling, is considerable. The samples taken must be of sufficient size to provide enough cell material for accurate measurement while not being so big that by the end of the experiment the culture volume has been drastically reduced (thus changing significantly the physical parameters of the closed culture system). This may prove to be a problem for cultures with a long culture cycle.

Measurements should be made every day for fast-growing cultures which are routinely transferred at weekly intervals. For slower-growing cultures measurements should be made daily for the first 6d and thereafter every second day.

Basic protocol
[Steps 2–6 under aseptic conditions]

1. Estimate how long the experiment is to run, how many harvest days there will be and how many replicates are needed.
2. Prepare in advance a sufficient number of cultures to provide an excess (50%!) of inoculum for the growth experiment.
3. To begin (0 d) mix a number of cultures of the age normally used for subculture in a large sterile conical flask to provide sufficient stock inoculum for all the flasks + 50% extra.
4. Label all the experimental flasks 1, 2, 3, 4, 5, ... etc.
5. Inoculate all flasks with uniform samples taken from the stock culture.
6. Flame, reseal and secure all flasks firmly on the culture shaker.
7. Immediately harvest the flasks for the 0d measurements beginning with # 1, 2, etc.
8. Carry out the measurements on the desired growth parameters [see next section].
9. Repeat 6 & 7 for each measurement day until the end of the experiment.
10. Repeat the experiment.

Notes (basic protocol)

3. The volume of the mixing flask must be at least 3× the volume of stock inoculum to enable proper and easy mixing. Excess inoculum is necessary to ensure that all inocula are uniform right up to the last one. If the stock volume becomes too low cells become stranded on the sides of the flask and the inoculum is effectively diluted.

5. Uniform inoculation is critical. The stock must be well mixed and continue to be mixed while the samples are taken. Sample taking should be practiced beforehand.

7. Harvesting in a predetermined sequence eliminates the temptation of subjectivity later on in the experiment.

To make the measurements

1. Mix each culture gently but thoroughly and remove a representative 2 ml sample to determine cell viability and cell number.

   Cell viability [After 13]:

   1a. Prepare a stock solution of fluorescein diacetate (FDA) in acetone (5 mg/ml).

   1b. To 5 ml fresh culture medium add drops of the FDA stock, mixing after each drop, until the medium begins to show a slight turbidity.

   1c. On a microscope slide with sample well, combine one drop of well-mixed cell suspension with one drop of the diluted FDA stock.

   1d. Apply a coverslip and leave for 10 min at room temperature.

   1e. For a randomly chosen field of view, count the total cell number under visible light.

   1f. Switch to UV light and count the fluorescent cells.

   1g. Repeat 1e. & 1f. until at least 500 viable (fluorescent) cells have been counted.

   1h. Calculate the proportion of the total cell number which is viable and express the result as a percentage.

   Cell number:

   EITHER: (When the suspension is fine)

   1i. Mix the sample thoroughly by pumping it into and out of a Pasteur pipette 5−10 times.

   1j. Immediately remove a sample with the pipette and place in a haemocyto-meter slide.

   1k. Count the number of cells above the entire grid ($= [A1]$).

   1l. Repeat 1i.−1k. a further 5 times ($= [A2] \rightarrow [A6]$).

   1m. Cell no. $_{(per\ ml\ culture)} =$

   $$= \frac{([A1] + \ldots [A6])}{6 \times \text{vol. above grid (ml)}} \times \text{dilution factor}$$

   OR: (When the aggregates are too large to count accurately the constituent cells)

   1o. Spin down the cells at $1000 \times g$ for 5 min and carefully pipette off the medium.

   1p. Resuspend the cells in 2 ml 10% (v/v) HCl.

   1q. Add 2 ml 10% (w/v) chromium trioxide in water and incubate at room temperature in the dark for 1−5 days.

   1r. Proceed as per 1i.−1m.

   1s. Cell no $_{(per\ ml\ culture)} =$

   $$= \frac{([A1] + \ldots [A6])}{6 \times \text{vol. above grid (ml)}} \times \frac{\text{Final sample volume}}{\text{Initial sample volume}}$$

2. Measurement of medium pH.

   2a. Allow the cells in the remainder of the culture to settle out and measure the pH of the medium with a standard pH meter.

3. Determination of the Packed Cell Volume (PCV).

   PCV is a non-destructive determination of the amount of culture biomass. If necessary it can be measured under sterile conditions and the cells reused. For fine cultures a change in the PCV is the most accurate estimation of culture growth. Increased culture aggregation decreases accuracy.

   3a. Mix the culture thoroughly and quickly transfer (pour or use a wide mouthed pipette) approximately 10 ml into a graduated conical centrifuge tube.

   3b. Centrifuge for 5 min at $500 \times$ g using a swing-out rotor.

   3c. Record the precise volume of the sample taken and the volume of the cell pellet.

   3d. $PCV = \dfrac{\text{volume of cell mass}}{\text{total volume of sample}} \times 100 \ (\%)$

4. Determination of the 'fresh' and dry weights.

   Culture fresh weight (probably better termed 'wet' weight) can only be determined following the separation of the cells from the medium. Herein lies a source of inaccuracy — overdrying small samples and underdrying large ones can sometimes cause problems. Consistent filtration is essential.

   4a. Combine the PCV sample with the remainder of the culture.

   4b. Record the weight of a disc of dry filter paper [A].

   4c. Place the filter paper in a Büchner funnel connected up to a side-arm flask, apply a vacuum and thoroughly wet the filter with distilled water.

   4d. When the water stops dripping down shut off the vacuum and remove and reweigh the, now moist, filter paper [B].

   4e. Replace the filter paper and filter the culture, washing all of the cells out of the flask with distilled water.

   4f. Weigh the cells + filter [C].

   4g. Place cells + filter in a 60 °C oven overnight.

   4h. Weigh the dried cells + filter [D].

   4i. $\text{Fresh weight}_{(culture)} = ([C] - [B]) \times \dfrac{\text{total culture volume}}{\text{culture volume} - 2 \text{ ml}}$

   4j. $\text{Dry weight}_{(culture)} = ([D] - [A]) \times \dfrac{\text{total culture volume}}{\text{culture volume} - 2 \text{ ml}}$

5. Methods for further measurements of parameters such as DNA content, mitotic index, protein synthesis etc can be found in [4].

Notes (Growth measurements)

1a. The acetone/FDA stock can be stored at 4 °C in the dark in a *tightly sealed* flask for several months.

1d. The combination of an intact plasmalemma and intracellular esterase activity in viable cells results in the production and accumulation of fluorescein, a dye which fluoresces yellow/green under UV light. FDA, in contrast, is colourless under UV light.

1f. Use the standard FITC filter combination – Excitation filter (450–490 nm) and barrier filter (*ca.* 510–520 nm). Weakly fluorescent cells should not be counted.

1i. This not only mixes the suspension but also helps to further break up the aggregates.

1k. Ideally there should be *ca.* 150–200 cells per grid. This will mean having to dilute samples from older cultures with a known quantity of distilled water prior to counting.

1m. Dilution factor = Final sample volume (after possible addition of water – see 1k.) divided by Initial sample volume. Results are usually expressed as cell no./ml or total cell number/culture.

1q. The chromium trioxide stock solution can be stored in the dark for several months. Great care must be taken as chromium trioxide is highly corrosive. The precise incubation time must be experimentally determined for each cell line. Set up a series of 15 replicate tubes containing suspension samples plus the digestion mixture. Count and discard three each day for 5 days. Choose the incubation time which yielded the highest cell counts.

2a. Note that it is usual that the pH of culture media is reduced following autoclaving, e.g. from 5.8–5.4. During cell culture the pH of the medium can vary dramatically (by as much as 3 units).

3a. Do not attempt to pour exactly 10 ml, or to get all the samples to exactly the same volume. This will lead to faulty sampling.

3c. In the early stages the pellet will be small and at an angle to the axis of the tube. An estimation by eye must be used to judge the actual pellet volume.

4a. Distilled water can be used to wash out the centrifuge tube completely.

4b. Use coarse-grade filter paper.

4c. A water vacuum pump is preferred.

4g. Do not use an oven with a convector fan as the cells, on drying, will be blown away.

4h. Reweigh immediately on removing from the oven or store the samples in a desiccator to prevent the very dry cells reabsorbing atmospheric moisture.

4i. & 4j. Includes correction factor for the initial 2 ml sample removed for cell no./viability determination.

[Avoiding the need for] Trouble shooting

Culture loss – some notes on good laboratory practice

The two 'model' examples described here both result in the obtention of cell suspension cultures relatively easily and quickly. However, in many instances a valued suspension cell line with the desired characteristics is only achieved after not inconsiderable time (months!) and effort. The loss of such a culture can be a major set-back requiring not only the instigation of the complete re-initiation process but also a subsequent recharacterisation of the system before proper experimentation can begin again. Following a few simple guidelines can avoid the need for this:

1. Where possible (and if the medium does not contain particularly labile supplements) prepare the flasks and medium 3–4 days prior to their requirement. In this way faulty sterilization can be recognised before subculture has occurred.
2. Always set up at each subculture 2–3 extra suspensions. At the next subculture these are left untouched. Culture loss through faults arising during the subculture process (non-sterile conditions due to contaminated pipettes, problems with the laminar flow cabinet etc.) can then be rectified by resubculturing using the (albeit a little older) reserve cultures.
3. If the luxury permits keep some cultures on a second shaker.
4. As a last resort, always maintain a stock of callus of the type best suited for the reinitiation of suspensions. Suspension culture cells, plated out and maintained on solid medium is often the best callus stock to keep.

Where to begin with a non-model species

It is not possible to recommend one basic protocol as different species often require different culture conditions. Furthermore, for a particular species the protocol necessary to produce an embryogenic cell suspension can contrast greatly with the one best employed to obtain a culture for e.g. rapid biomass production. The only pointer that can be given is the obvious one: before beginning, check the reference texts (e.g. [12] and such series as those edited by Bajaj (Biotechnology in Agriculture and Forestry; Springer-Verlag) and Evans *et al.*(Handbook of Plant Cell Culture; Macmillan, New York)) and carry out a literature survey of the more recent publications (a computer-assisted search of the international databases is invaluable here and is well worth the relatively small cost). Choose, to begin with, the method most closely related to your plant material/desired culture type/aims.

When setting up new suspension cultures the two most commonly encountered problems are those of contamination and non-friability of the cell material. Overcoming these problems is essential to the obtention of a suitable suspension culture.

1. *Contamination.* If the starting materials (callus, medium, etc) were aseptic, contamination can only have arisen through faulty transfer protocols. Applying extra care, and practice, will quickly overcome the problem. If the callus to be used is not totally free of micro-organisms – see the previous chapter for suitable decontamination protocols. Antibiotic treatments can be attempted with suspension cultures but are rarely successful.

2. *Friability.* Obtaining a suitable level of friability is likely to be the greatest problem encountered in suspension culture initiation. It is generally best to choose the most friable callus (see previous chapter) to obtain a fine suspension. Sometimes however, this does not always correlate. A number of possible manipulations can be attempted which have proved successful in certain cases:

Media manipulations:
2a. Increase the auxin/cytokinin ratio of the medium, or exclude cytokinin altogether.
2b. Switch to a more 'potent' auxin e.g. IAA → NAA or NAA → 2,4-D.
2c. Increase the shaker speed (although speeds of > 150 rpm are not recommended due to cell shearing problems).
2d. Reduce the pH of the culture medium by 0.5 (– 1) units.
2e. Include weak cellulase (0.1%)/pectinase (0.05%) concentrations in the culture medium to reduce cell-cell adhesion.

Alteration of physical parameters:
2f. Use Erlenmeyer flasks with internal baffles or temporarily include two 1 cm glass beads (marbles!) to help break up the aggregates.
2g. Briefly grind the callus using a sterile pestle & mortar or place for 5 sec in a Waring blender.
2h. Force the cells through a 200 μm sieve or repeatedly through a wide-bore syringe canula.

Modification of the subculture routine:
2i. Replating a poorly friable suspension back onto agar medium can yield colonies with enhanced friability. These can then be selectively used to reinitiate new liquid cultures. Repeatedly alternating cycles of growth on solid and then liquid media may be necessary to achieve a significant morphological effect.

2j. The degree of aggregation in an initially fine suspension culture can gradually increase with time. Routine reisolation of the finer material for selective subculture (by e.g. filtration) may be the only means to maintain fine suspensions of such cell lines on a long-term basis. Many researchers do this routinely (and often subconsciously) at each subculture through some deft work with the transfer pipettes.

References

[1] Shillito, RD, Carswell, GK, Johnston, CM, DiMaio, JJ, Harms, CT (1990). Regeneration of fertile plants from protoplasts of elite inbred maize. Bio/Technology 7, 581–587.

[2] Prioli, LM, Sondahl, MR (1990). Plant regeneration and recovery of fertile plants from protoplasts of maize (*Zea mays* L). Bio/Technology 7, 589–594.

[3] Lindsey, K, Yeoman, MM (1983). The relationship between growth rate, differentiation and alkaloid accumulation in cell cultures. Journal of experimental Botany 34, 1055–1065.

[4] Street, HE (1977). Cell (suspension) cultures – Techniques. In: Street, HE (ed.) Plant tissue and cell culture, pp 61–102, Blackwell Sci. Publns., Oxford.

[5] Hall, RD, Yeoman, MM (1986). Temporal and spatial heterogeneity in the accumulation of anthocyanin in cell cultures of *Catharanthus roseus* (L) G. Don. Journal of experimental Botany 37, 48–60.

[6] Fry, SC, (1990). Roles of the primary cell wall in morphogenesis. In: Nijkamp, HJJ, van der Plas, LHW, van Aartrijk, J (Eds.) Progress in plant cellular and molecular biology, pp 504–513. Kluwer Academic Publishers, Dordrecht.

[7] Herzbeck, H, Husemann, W (1985). Photosynthetic carbon metabolism in photoautotrophic cell suspension cultures of *Chenopodium rubrum* L. In: Neumann, KH, Barz, W, Reinhard, E (Eds.) Primary and secondary metabolism of plant cell cultures, pp 15–23, Springer-Verlag, Berlin.

[8] Hall, RD, Holden, MA, Yeoman, MM (1988). Immobilization of higher plant cells. In: Bajaj, YPS (Ed.) Biotechnology in agriculture and forestry 4, pp 136–156, Springer – Verlag, Berlin.

[9] Wang, AS, Phillips, RL (1984). Synchronization of suspension culture cells. In: Vasil, IK (Ed.) Cell culture and somatic cell genetics 1, pp 175–181, Academic Press, Orlando.

[10] Fowler, MW (1986). Industrial applications of plant cell cultures. In: Yeoman, MM (Ed.) Plant cell culture technology, pp 202–227, Blackwell Scientific Publishers, Oxford.

[11] Van der Heijden, R, Verpoorte, R, ten Hoopen, HJG (1989). Cell and tissue cultures of *Catharanthus roseus* (L) G.Don.: a literature survey. Plant cell, tissue and organ culture 18, 231–280.

[12] George, EF, Puttock, DJM, George, HJ (1987) Plant culture media 1, Exegetics Ltd., Westbury, UK.

[13] Widholm, J (1972). The use of fluorescein diacetate and phenosafranine for determining viability of cultured plant cells. Stain Technology 47, 189–194.

Plant Tissue Culture Manual **A4**: 1–13, 1991.

Shoot cultures and root cultures of tobacco

JENNIFER F TOPPING & KEITH LINDSEY
Leicester Biocentre, University of Leicester, Leicester LE1 7RH, UK

Introduction

Plant organ cultures are characterized by the maintenance of structural integrity, as a consequence of growth from defined meristems, and are thereby distinguished from callus and suspension cultures. Such morphological stability is in turn associated with improved genetic and metabolic stability compared with undifferentiated cultures, and has led to the use of organ cultures as a means of clonally propagating commercially important species [4], of producing specific secondary metabolites *in vitro*, though more specifically as genetically transformed 'hairy root' cultures [1], and of studying physiological properties of organs under relatively well defined environmental conditions [2].

Shoot cultures, then, typically comprise rootless aerial parts (apical and/or axillary buds, stems and leaves) growing on a defined medium supplemented with agar. In the case of tobacco shoots cultured on the media described in this article, however, some root development also occurs, within 2-3 weeks of shoot initiation. Most commonly shoot cultures are initiated from explant tissues containing meristems, such as apical buds, axillary buds or embryos, and are subcultured by the transplantation of excised buds onto fresh medium at intervals of approx. 4 weeks. The precise composition of the medium of course may vary with the nutritional requirements of different species, but are typically unsophisticated, comprising salts, vitamins and a carbon source (usually sucrose) and in some instances, growth substances (usually an auxin). Shoot cultures of tobacco (*Nicotiana tabacum* cv. Petit Havana SR1), described in detail in this article, have no requirement for growth substances. Although shoot cultures are usually provided with a carbon source, and so do not rely on photosynthetic carbon fixation for growth, full greening of the leaves requires a light supply, such as that provided by 'Warmwhite' or 'Coolwhite' fluorescent tubes at a photon flux density of between 20-200 $\mu mol/m^2/s$.

The culture vessels used to maintain shoot cultures can be of glass (such as Kilner jars) or plastic (such as those sold by Sigma). The major feature of the vessels which may be important in influencing the growth of shoot cultures of some species is gas permeability; an accumulation of ethylene or ethane within culture jars has been implicated in the inhibition of growth [5], resulting in stunted shoots. This problem can be eliminated by loosening lids or by covering jars with one half of a Petri dish, sealing either with e.g. Nescofilm or Parafilm and cutting slits in the film, or with gas-permeable 'micropore tape' (3M) to facilitate gas exchange (Fig. 1).

Shoot cultures of tobacco SR1 are an excellent source of leaf material as an explant tissue for inoculation in *Agrobacterium*-mediated gene transfer [3]. Furthermore, tobacco leaves may yield high numbers of protoplasts for use, for example, in transient assay and stable transformation studies (see [7]). The age, health and culture conditions of the shoot cultures are important factors in influencing transformation frequency and protoplast yield, features discussed in detail in the relevant references. By culturing shoots from the most apical cells only ('meristem culture'), it is possible to eliminate virus particles from infected plant material [6].

Root cultures are generally established by the transfer of aseptic root tips, isolated from germinated seeds, into a liquid culture medium. Subculture is carried out by excising the tips of the main axes or of laterals and transferring them to fresh medium, usually at intervals of 10–14 days.

Early work by White on the culture of tomato roots identified a requirement for B vitamins if prolonged growth is to be maintained [8]. As for shoot cultures, isolated roots further require inorganic salts and a carbon supply, but growth substances are usually unnecessary. Roots can be cultured in Erlenmeyer flasks or Petri dishes in the dark. The culture vessels can be shaken gently on a rotary shaker (20-50 r.p.m.) to eliminate gradients of oxygen and nutrients in the medium, but for many species (including tobacco) this is unnecessary for successful root growth.

Cultured roots provide a useful experimental system to study aspects of root biochemistry and physiology such as ion uptake, root exudation, nodulation and the synthesis of root-specific secondary metabolites. They can also act as a source of protoplasts for the study of tissue-specific trans-acting factors in the regulation of root-specific gene expression.

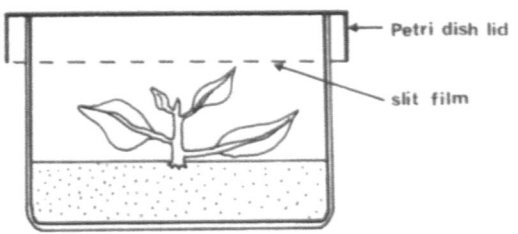

Fig. 1. Gas exchange can be facilitated by culturing shoots in a glass jar covered by a Petri dish lid, secured by plastic film which has been nicked to create slits.

Procedures

Sterilization of tobacco seeds

Steps in the procedure:
1. Place seeds in a beaker
2. Immerse seeds in 70% ethanol for 20s and agitate by gentle swirling
3. Tobacco seeds tend to float, so remove the ethanol using a sterile pipette. Keep the tip of the pipette close to the bottom of the beaker, to suck off the ethanol but leaving the seeds in the bottom of the beaker
4. Immerse seeds in a 5% (v/v) bleach (sodium or calcium hypochlorite) solution containing 0.05% (v/v) Tween 20 or 80, for 15 min. Agitate occasionally by swirling
5. Remove the sterilant using a sterile pipette as described in (3), and wash the seeds six times in sterile distilled water

Germination of the seeds

Steps in the procedure:

6. Transfer sterilized seeds to 9 cm Petri dishes containing solidified MS20 medium. Due to the small size of tobacco seeds, this is most easily achieved by suspending seeds in a small volume of sterile distilled water (approx. 50 seeds/ml) and transferring them to the Petri dish by pipette. Spread the seeds around the plate with a sterile glass rod, and remove excess water from the surface of the agar with a sterile pipette

7. Seal the Petri dish with Nescofilm or Parafilm, and culture at 24 °C in the dark to induce germination. The first signs of germination typically occur within about 3–5 days. If a dark incubator is not available, cover the Petri dishes with aluminium foil to exclude light.

Notes

2. Treatment with 70% (v/v) ethanol removes waxy substances from the surface of the seeds and kills some contaminating organisms present. Do not exceed the immersion time of about 20 s, as this will result in a loss of viability of the seed

3. Tween 20/80 acts as a surfactant to allow access of the sterilizing hypochlorite solution

5. Six washes in water are necessary to ensure the removal of all chlorine. Failure to do this will result in a loss of viability in a proportion of the seed

6. The germination of seeds on MS20 medium is preferable to the use of damp filter paper, as it allows the early identification of seeds which have not been sterilized successfully. These should be removed at the earliest opportunity to prevent contamination of the whole plate. If an unacceptably high proportion of the seeds are contaminated, modify the sterilization procedure either by increasing the concentration of the bleach solution to 10–20% (v/v) or by increasing the duration of the sterilization treatment (up to about 30 min).

Initiation and subculture of shoot cultures

Steps in the procedure :

1. When the germinated shoot is approx 1—2 weeks old, excise the shoot apex, cotyledons and emerging first leaves at the hypocotyl and transfer to MS30 (or MS20- see notes), embedding the cut surface in the agar medium. Culture at 24 °C in warm daylight fluorescent tubes (50-100 μmol/m^2/s) under long daylength conditions. Use culture vessels with loose or unsealed lids. After about 6 weeks, the shoots will be large enough to subculture.

2. To subculture the shoots, excise the apical buds together with the newest pair of leaves, or axillary buds and transfer to fresh MS30 (or MS20- see notes). Shoots should be subcultured routinely at intervals of approx 4 weeks.

Notes

1,2. We have found that, if the shoot cultures are to be used for leaf-disc transformation by *Agrobacterium tumefaciens*, optimum leaf production and transformation frequencies are obtained if the shoot cultures are maintained on MS30 medium. However, if the leaves are to be used for protoplast isolation and culture, we find that best results are obtained if the shoot cultures are maintained on MS20 medium. The reasons for this difference are not clear, but it may be that tissues grown in the presence of relatively high concentrations of sucrose accumulate relatively high levels of starch granules; and starch granules can cause protoplast bursting during the preparatory centrifugation steps, reducing protoplast yield.

If the shoot culture medium becomes contaminated during tissue manipulations, it may be possible to rescue the culture by removing the shoot apex, and transferring it to fresh medium. If the plant material itself becomes contaminated, it is usually best to dispose of the whole culture by autoclaving. However, if the culture is valuable, three approaches can be attempted to rescue it:

a) physically sterilize the shoot material by immersion for 10 min, in a solution of 5% (v/v) sodium hypochlorite containing 0.05% Tween 20/80; wash six times in sterile distilled water and transfer to fresh medium. This method is only useful if the plant material is contaminated on the surface rather than systemically;

b) transfer the shoot apex to fresh medium containing 200 mg/l cefotaxime or carbenicillin; these antibiotics will inhibit bacterial infection if the shoot apex is small (<0.5 cm) and is pushed well into the medium;

c) transfer the plantlet (i.e. shoot plus at least some roots) to soil; remove all agar, transfer to moist autoclaved compost in either a micropropagator or a plant pot covered with a transparent plastic 'clingfilm' to maintain a high humidity, and allow to harden off gradually over a 2 week period before transfer to greenhouse conditions; allow to flower and collect seed, from which new shoot cultures can be initiated.

Initiation and subculture of root cultures

Steps in the procedure;
1. Germinate sterile seeds
2. When the germinated radicle is approx. 1 week old, excise the apical 1 cm and carefully transfer it to liquid MS20 medium in a 9 cm Petri dish. Each dish should contain approx. 15 ml medium. A sterile microbiological loop is useful for the physical transfer of the root tips
3. Seal the Petri dish with Nescofilm or Parafilm, and incubate at 24 °C without shaking. If a dark incubator is not available, cover the Petri dishes with aluminium foil to exclude light
4. At intervals of 10-14 days, subculture the roots by excising the apical 1 cm and transferring it to fresh liquid MS20 medium

Notes
2. The use of a microbiological loop is useful for avoiding the physical damage to young root tissue which can occur with forceps

Solutions
— MS20 medium
  — 4.4 g/l Murashige and Skoog basal medium (Sigma)
  — 20 g/l sucrose
  — 8 g/l agar (Difco); not included in root culture medium
  — adjust pH to 5.8 with 1M KOH

— MS30 medium
  — 4.4 g/l Murashige and Skoog basal medium (Sigma)
  — 30 g/l sucrose
  — 8 gl agar (Difco)
  — adjust pH to 5.8 with 1M KOH

References

1. Flores, HE, Hoy, MW, Pickard, JJ (1987) Secondary metabolites from root cultures. Trends Biotechnol. 5: 64–69
2. Gautheret, RJ (1985) History of plant tissue culture; a personal account. In: Vasil, I.K. (ed) Cell Culture and Somatic Cell Genetics of Plants Vol. 2 pp. 2–59, Orlando, Academic Press
3. Horsch, RB, Fry, J, Hoffman, N, Neidermeyer, J, Rogers, S.G., Fraley, R.T. (1988) Leaf disc transformation. Plant Mol. Biol. Manual A5: 1–9 (eds. Gelvin, S.B., Schilperoort, R.A.)
4. Hussey, G (1986) Vegetative propagation of plants by tissue culture. In: Yeoman, M.M. (ed.) Plant Cell Culture Technology pp. 29–66. Oxford, Blackwell Scientific Publications
5. Jackson, MB, Abbott, AJ, Belcher, AR, Hall, KC (1987) Gas exchange in plant tissue cultures. In: Jackson, MB, Mantell, SH, Blake, J (eds.) Advances in the Chemical Manipulation of Plant Tissue Cultures pp 57–71, Bristol: The British Plant Growth Regulator Group
6. Kartha, K (1984) Elimination of viruses. In: Vasil, I.K. (ed) Cell Culture and Somatic Cell Genetics of Plants. Vol. 1 pp 577–585
7. Saul, MW, Shillito, RD, Negrutiu, I (1988) Direct DNA transfer to protoplasts with and without electroporation. Plant Mol. Biol. Manual A1: 1–16 (eds. Gelvin, SB, Schilperoort, RA)
8. White, PR (1937) Vitamin B1 in the nutrition of excised tomato roots. Plant Physiol. 12: 803–811

Plant Tissue Culture Manual **A5**: 1–15, 1991.
© 1991 *Kluwer Academic Publishers*.

Somatic embryogenesis in orchardgrass

MICHAEL E. HORN
Agrigenetics Co., 5649 E. Buckeye Road, Madison, WI 53716, U.S.A.

Introduction

If there exists a model monocotyledonous species with regard to somatic embryogenesis, then orchardgrass (*Dactylis glomerata* L.) must surely be nominated. As described by Conger and colleagues at the University of Tennessee [1], one is able to induce an embryogenic response from young leaf bases such that an abundance of mature, convertible somatic embryos are available within just a few weeks from the time of plating. The fact that one needs to start with selected *Dactylis* clones for this optimal embryogenic response [2] is not a trivial matter but does not detract from the important scientific possibilities these clones possess. Several studies on the origin and developmental pattern of the somatic embryos have been published [1–4] but the biochemical and molecular aspects of the embryogenesis phenomenon have only just begun to be examined [5, 6]. To those of us who have used the *Dactylis* system for our own purposes we recognize its potential and encourage its use as an experimental tool for the study of somatic embryogenesis in monocotyledonous plants.

The embryogenic response in *Dactylis*, as in most monocotyledonous species, is induced by exposure to auxin. Relatively high levels of the auxin are needed [1, 2, 4] and the recurrent selection of clones with high embryogenic potential appears not to change the auxin requirement quantitatively. Thus, scientists investigating the hormonal basis for induction of somatic embryogenesis have two distinct phenotypes with which to work: clones with low response regardless of the auxin level (original clones), and those with high response with high levels of auxin.

The classification of *Dactylis* as a model monocotyledonous system is further supported by reports of its use in protoplast regeneration and gene transfer experiments [7, 8]. In chapter B3, detailed procedures for gene transfer into protoplasts of *Dactylis* and subsequent regeneration of transformed plants will be described. Attempts to infect *Dactylis* leaves with *Agrobacterium* and achieve gene transfer have so far been in vain but the system lends itself well to such studies and should be used to test new ideas to achieve this goal.

The somatic embryogenesis response in *Dactylis* basal mesophyll sections has been examined at both the light and electron microscopic level [1]. When embryos were produced directly from mesophyll cells, i.e. with no callus phase, it appeared that the embryos arose from a single cell source located just beneath the epidermis. The single cell origin has not been unequivocally established, however the presence of a suspensor as well as other features suggested that, at most, only a few cells participated in the formation of the somatic embryos. Very few phenotypic variants such as albinos have been observed in plants derived from somatic embryos produced in this way.

Procedures

Procedure for producing somatic embryos from mesophyll sections

Steps in the procedure

1. Grow selected *Dactylis* clones in the greenhouse.
2. Cut out selected tillers by cutting down through root mass. Wash off dirt and take into a sterile hood.
3. Separate leaf sheath from meristem/root mass. Discard root mass.
4. Split leaves longitudinally along midrib and sterilize in 50% ethanol-H_2O solution for 2 min., rinse in 2–3 changes sterile distilled H_2O.
5. Put leaf into plastic Petri dish with ruler underneath dish. Cut leaf into 3 mm segments starting at base and extending out to 1.5 cm (Five segments)
6. Plate leaf segments onto Schenk and Hildebrandt (SH) basal medium [9] containing 30 µM 3,6-dichloro-*o*-anisic acid (DICAMBA) and solidified 2.5% Gelrite or 0.8% SeaPlaqueR agarose.
7. Place plates in a dark incubator at 25 °C
8. The appearance of somatic embryos is first visible after $1\frac{1}{2}$ to 2 weeks. After 3 weeks response from mother tissue should be at peak.

Note

1. Use supplemental lighting in the winter. Cut back dead tillers as needed. Cut off flower stalks if they appear. Propagate by separating root mass with sharp knife.

Solutions

Schenk and Hildebrandt salt base in mg/liter: KNO_3 2500, $MgSO_4 \cdot 7H_2O$ 400, $NH_4H_2PO_4$ 300, $CaCl_2 \cdot 2H_2O$ 200, $MnSO_4 \cdot H_2O$ 10, H_3BO_4 5, $ZnSO_4 \cdot 7H_2O$ 1, KI 1, $CuSO_4 \cdot 5H_2O$ 0.2, $NaMoO_4 \cdot 2H_2O$ 0.1, $CoCl_2 \cdot 6H_2O$ 0.1, $FeSO_4 \cdot 7H_2O$ 15, Na_2EDTA 20; Organics in mg/liter: inositol 10, thiamine · HCl 5, nicotinic acid 5, pyridoxine · HCl 0.5.

Description of embryogenic response

The most basal section of the leaf blade produces highly embryogenic callus almost exclusively. Sections farther from the base show less callus and a greater incidence of direct embryo formation (Fig. 1) but this also declines the farther distal the segment is from the meristem. Sections more distal than 1.8 cm from the base rarely show any embryogenic response. Callus may be subcultured onto the same medium (SH-30) repeatedly and somatic embryos will mature from the callus with time. The somatic embryos are quite pale when mature and thus are very easy to distinguish from the callus (Fig. 2). Both these embryos and somatic embryos derived directly from the mesophyll cells can be carefully removed and plated onto SH medium lacking phytohormones (SH-0) (Fig. 3). When incubated in the light $(30-60 \ \mu mol \ m^{-2} \ s^{-1})$ at 25 °C, 95–100% of these somatic embryos will convert into plantlets (Fig. 4). When plantlets reach the 6–12 leaf stage, they should be hardened off by opening the vessel closure for two days and then transferred to soil in the greenhouse [7].

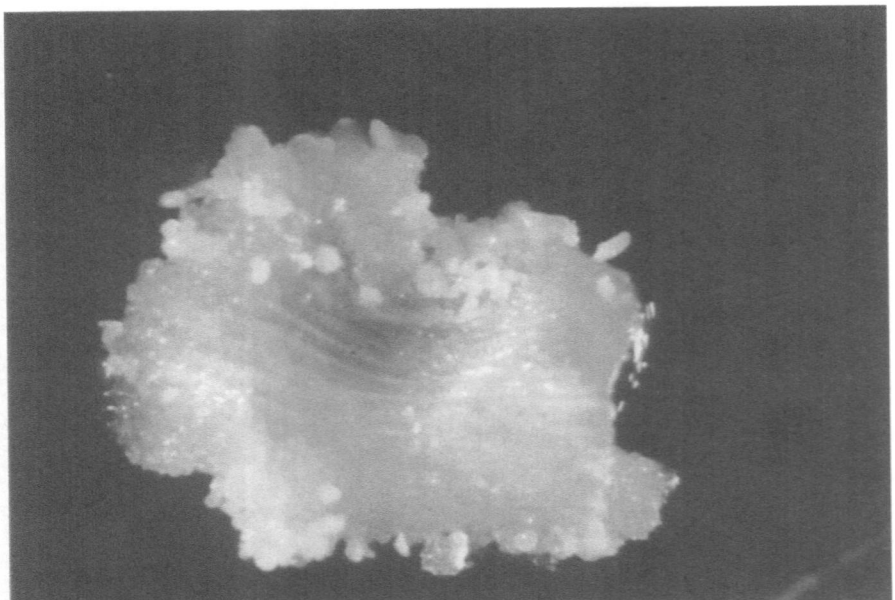

Fig. 1. Somatic embryos arising directly from *Dactylis* leaf segment. Photographed 18 days after plating segment taken 9 mm from leaf base onto SH-30 medium. (6 ×)

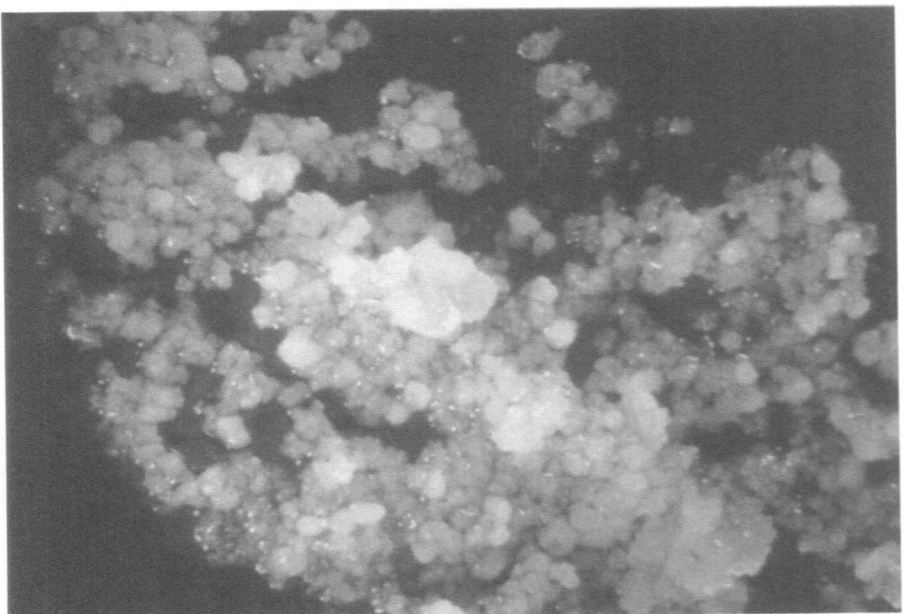

Fig. 2. Somatic embryos maturing from embryogenic suspension culture material plated onto SH-30 medium in the dark. Photograph taken about 11 days after plating. (6 ×)

PTCM-A5/7

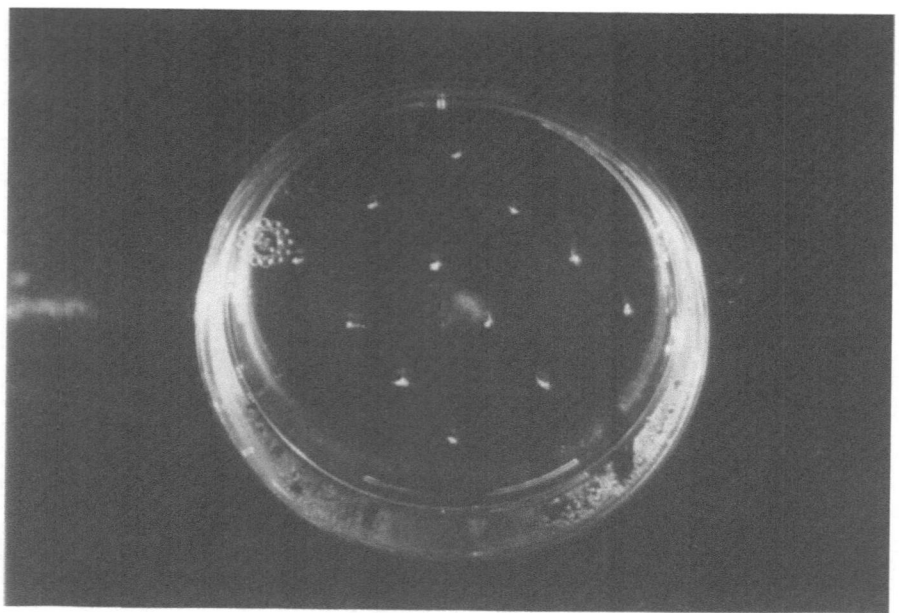

Fig. 3. Mature somatic embryos from *Dactylis* leaf segments plated onto SH-0 medium (no phytohormones).

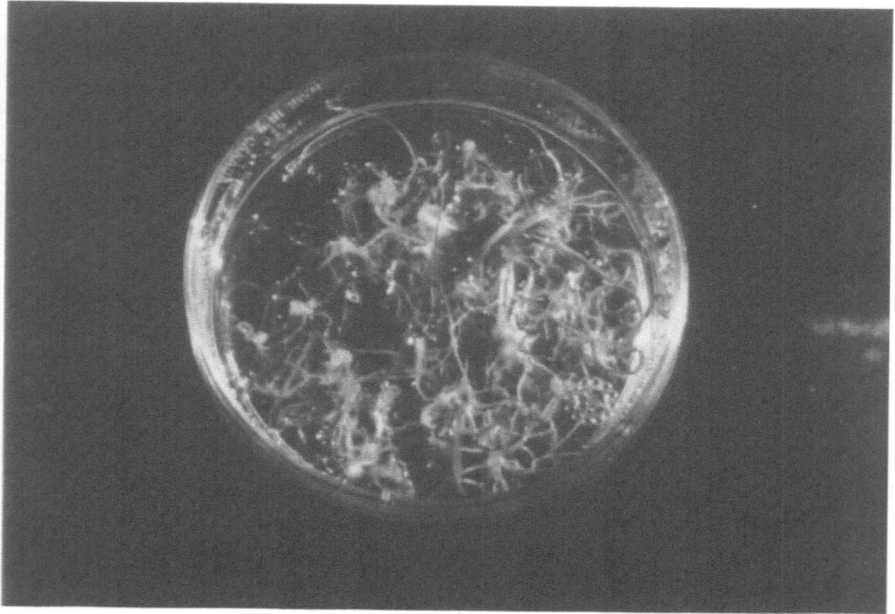

Fig. 4. Dactylis plantlets converted from plated somatic embryos of Fig. 3. Photograph taken 14 days after plating of embryos onto SH-0 medium and incubated in the light.

Procedure for establishment of embryogenic suspension cultures

Steps in the procedure

1. Inoculate 0.5 g of embryogenic callus into 50 ml of liquid SH medium containing 45 µM DICAMBA and 4 g/l casein hydrolysate [5] in a 125 ml baffled Delong flask (Bellco). Seal flask with a stainless steel cap with 'fingers' (Bellco). Parafilm is optional.
2. Grow cultures in darkness or subdued light on a gyratory shaker at 130 rpm.
3. After four weeks each flask is agitated by hand in a sterile hood and the larger clumps allowed to settle for about 30 s.
4. Take 10 ml aliquots of the supernatant by dipping the pipette just below the liquid surface. Each aliquot is used to inoculate another flask of the same medium (SH-45).
5. Repeat step 4 every three to four weeks. Examine flasks under an inverted microscope for progress toward a culture composed only of small proembryogenic masses (PEMs).
6. Somatic embryos can be obtained from embryogenic suspension cultures by plating the tissue onto solid SH-30 medium and incubating the plates in the dark as described above.
7. Remove mature somatic embryos and plate onto SH-0. Handle the resultant plantlets as described above.

Notes

1. Baffled flasks are very useful for establishing embryogenic suspension cultures in all monocotyledonous species tested. They improve flask aeration and the increased shearing action helps to break up the large clumps that typify young suspension cultures.
6. Depending on the size of the aggregates, the resultant callus may have to be subcultured to fresh SH-30 medium once before mature somatic embryos appear.

The usefulness of embryogenic suspension cultures

Embryogenic suspension cultures can be of tremendous usefulness for a number of goals. They have been absolutely essential as the source material for successful protoplast isolation, direct gene transfer, and subsequent plant regeneration. They are the best target for microparticle bombardment (see chapter by Klein *et al.*, this Volume) since they can be returned to liquid for the most stringent selection known for transformed cells. Embryogenic suspension cultures are the perfect material for freeze preservation. They are ideal for mutant selection studies due to their propensity for stringent selection. Finally, embryogenic suspension cultures are useful as the scale-up stage for rapid asexual propagation of select, unique genotypes [10].

Obtaining a good, established embryogenic suspension culture is essential for achieving any of the above goals. Initially the suspensions are composed largely of non-embryogenic cells. Most probably, these cells are derived from the suspensors or root apices of the mature embryos in the culture. Therefore, reducing the maturity of the aggregates such that they contain no mature embryos reduces the production of the non-embryogenic cells. The procedure described above for obtaining cultures composed exclusively of PEMs is only one way of achieving this desired goal. It succeeds because the smaller and somewhat less dense PEMs sink much slower than the larger aggregates that contain mature or nearly mature embryos. An established embryogenic suspension culture is also likely to be achieved by employing metal sieves to isolate and concentrate the few PEMs produced initially and then inoculating sieves to isolate and concentrate the few PEMs produced initially and then inoculating all of the PEMs into a single flask. We have used these sieves to achieve the same goal in rice [10] but we have not tested it in *Dactylis*.

Optimization of the Dactylis *somatic embryogenesis system*

Attempts at optimization of the *Dactylis* somatic embryogenesis system have been conducted by Conger and co-workers [2, 4, 5]. The particular auxin used for the induction of the embryogenic response is probably not critical. Both 2,4-D and DICAMBA have been used successfully [1, 2]. When compared, 10 μM 2,4-D gave a nearly equivalent response to that induced by 30 μM DICAMBA [2]. Clearly, only the first 2.0 cm or less of the leaf base is responsive to the auxin induction [2]. Moreover, if the entire leaf is plated on the solid SH-30 medium there is no embryogenic response at all, even from the basal area. In this case the distal part of the leaf blade turns brown after several days and this discoloration moves slowly down the blade toward the base (Horn, unpublished results). Why the more distal part of the blade prevents an embryogenic response in the basal part is unknown. The best results are obtained when the blade is cut into segments as described in the procedure above. There appeared to be no difference in the response when leaf segments were plated abaxially or adaxially onto the solid medium [1]. When leaf segments approximately 0.7–1 cm from the base were examined after three weeks, somatic embryos literally covered the entire surface (Fig. 5). It is unlikely that all of these embryo were derived from cells in the mesophyll. Rather the process of secondary embryogenesis is probably involved which results in somatic embryos arising from other somatic embryos (Horn, unpublished results). It is difficult to imagine how this step could possibly be improved.

PTCM-A5/11

The initiation of embryogenic suspension cultures and their subsequent growth has also been examined for possible optimization [4, 5]. The addition of casein hydrolysate was found to not only improve the growth rate of the suspension but also reduced the preponderance of non-embryogenic 'mucilagenous' cells. In fact, the authors claimed that casein hydrolysate was essential for the development of somatic embryos in suspension culture [5]. Such embryos were capable of converting directly to a plantlet stage while still in a thin layer of liquid. Why casein hydrolysate is critical for somatic embryo formation in liquid but not on solid medium is as yet unknown. Also unknown is precisely which component(s) of casein is responsible for its growth enhancing properties. Neither L-glutamine nor L-proline alone had a similar effect [4]. These studies also found that the DICAMBA level needed for optimal embryo production was 33% higher than that used with embryogenic callus on solid plates (40 vs. 30 µM).

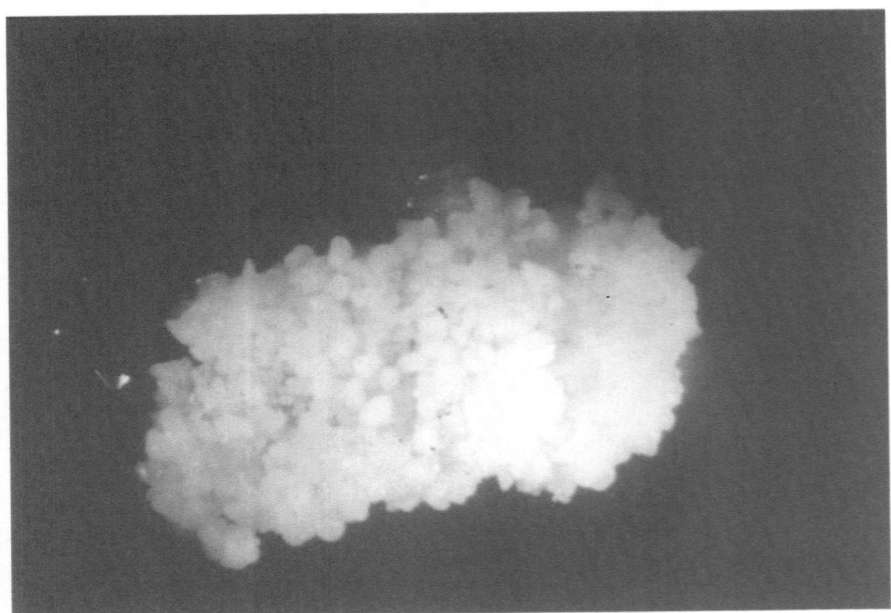

Fig. 5. Leaf segment from Fig. 1 29 days after plating onto SH-30 medium. (6 ×)

References

1. Conger BV, Hanning GE, Gray DJ, McDaniel JK (1983) Direct embryogenesis from mesophyll cells of orchardgrass. Science 221: 850–851
2. Hanning GE, Conger BV (1986) Factors influencing somatic embryogenesis from cultured leaf segments of *Dactylis glomerata*. J. Plant Physiol. 123: 23–29
3. McDaniel JK, Conger BV, Graham ET (1982) A histological study of tissue proliferation, embryogenesis, and organogenesis from tissue cultures of *Dactylis glomerata* L. Protoplasma 110: 121–128
4. Gray DJ, Conger BV, Hanning GE (1984) Somatic embryogenesis in suspension and suspension-derived callus cultures of *Dactylis glomerata*. Protoplasma 122: 196–202
5. Gray DJ, Conger BV (1985) Influence of dicamba and casein hydrolysate on somatic embryo number and culture quality in cell suspensions of *Dactylis glomerata* (Gramineae) Plant Cell Tissue Organ Culture 4: 123–133
6. Hahne G, Mayer JE, Lørz H (1988) Embryogenic and callus-specific proteins in somatic embryogenesis of the grass, *Dactylis glomerata* L. Plant Science 55: 267–279
7. Horn ME, Conger BV, Harms CT (1988) Plant regeneration from protoplasts of embryogenic suspension cultures of orchardgrass (*Dactylis glomerata* L.). Plant Cell Reports 7: 371–374
8. Horn ME, Shillito RD, Conger BV, Harms CT (1988) Transgenic plants of orchardgrass (*Dactylis glomerata* L.) from protoplasts. Plant Cell Reports 7: 469–472
9. Horn ME, Shillito RD, Conger BV, Harms CT (1989) Transgenic plants of orchardgrass (*Dactylis glomerata* L.) regenerated following direct gene transfer into protoplasts. IN: EUCARPIA Congress Proc., Genetic Manipulation in Plant Breeding, Helsingor/Denmark 1988, Plenum Press, NY
10. Schenk RU, Hildebrandt AC (1972) Medium and techniques for induction and growth of monocotyledonous and dicotyledonous plant cell cultures. Can. J. Bot. 50: 199–204
11. Horn, ME, Matsuno T, Rose BL, Brink KB. Rapid scale-up of rice embryogenic cell cultures for the production of clonal plants. In Preparation

Plant Tissue Culture Manual **A6**: 1–17, 1991.

Arabidopsis Regeneration and Transformation (Leaf & Cotyledon Explant System)

RENATE SCHMIDT[1] & LOTHAR WILLMITZER
Institut für Genbiologische Forschung Berlin GmbH, Ihnestraße 63, 1000 Berlin 33, Fed. Rep. of Germany; [1]present address: IPSR Cambridge Laboratory, John Innes Centre for Plant Science Research, Colney Lane, Norwich NR4 7UH, United Kingdom

Abbreviations

BAP: 6-Benzylaminopurine, CIM: Callus induction medium, 2,4-D: 2,4-Dichlorophenoxyacetic acid, DMSO: Dimethyl sulphoxide, GA_3: Gibberellic acid, G418: Geniticin, IAA: Indole 3-acetic acid, IBA: Indole 3-butyric acid, 2-IP: N-Isopentenylaminopurine, HPT: Hygromycin phosphotransferase, NAA: α-Naphthaleneacetic acid, NPT II: Neomycin phosphotransferase type II, SEM: Shoot elongation medium, SIM: Shoot induction medium, RIM: Root induction medium.

Introduction

Arabidopsis thaliana is a small cruciferous plant with a short generation time and large seed set after self-fertilization. Over 100 mutants have been identified and used for the establishment of an extensive genetic map. These advantages together with the small genome size and the amenability for tissue culture and transformation makes *Arabidopsis thaliana* ideally suited as a model system in plant molecular biology (for reviews see [4, 12, 16]).

The first transformation experiments of *Arabidopsis thaliana* were performed with wildtype strains of *Agrobacterium tumefaciens* resulting in crown gall formation [1].

Regeneration of fertile transformed plants from explants requires the use of disarmed *Agrobacterium* vectors and chimaeric antibiotic resistance genes as dominant selectable markers in leaf disc-type transformation experiments [8]. A major prerequisite for successful leaf disc transformation experiments is the availability of an efficient regeneration system for leaf explants. This had already been described in 1975 for *Arabidopsis thaliana* [14]. The initial leaf disc transformation experiments on *Arabidopsis thaliana* [9, 19] were much less efficient than those described for tobacco or petunia. Furthermore the regeneration of transformed tissue was not very efficient and took a long time, with shoot formation on callus tissue being observed 3 months after inoculation with *Agrobacterium* [9].

The analysis of the effects of several different parameters on transformation and regeneration efficiencies has led to an optimized protocol described in this chapter. The *Arabidopsis* ecotype and the *Agrobacterium* strain used have far more influence on transformation efficiency than the age of the explants or kind and duration of preincubation of explants. In a routine transformation experiment, with the protocol described in this chapter, 80–90% of the explants of the ecotype C24 form resistant calluses on selective medium. Approximately 50% of the explants of this ecotype will give rise to fertile transformed shoots and seed of the transformants can be obtained in sterile culture as early as 10 weeks after the start of the experiment [17].

The transformation protocol is most efficient for the *Arabidopsis thaliana* ecotype C24 but it is also amenable to other ecotypes – for example Landsberg erecta and Wassilewskija [17]. Differences in the regeneration and/or transformation frequencies for different ecotypes of *Arabidopsis thaliana* have also been observed for protoplasts or root explants [2, 20].

The selection conditions for hygromycin B as well as G418 and kanamycin have been established for this procedure allowing the use of vectors based on chimaeric HPT genes as well as NPT II genes as plant selectable markers [17]. Chimaeric genes encoding gentamycin acetyltransferases and dihydrofolate reductase have also been used successfully as selectable markers in *Arabidopsis thaliana* transformation experiments [7, 11].

Most of the *Agrobacterium* vectors tested gave very high transformation frequencies. No difference in efficiency could be shown for octopine or nopaline-type *Agrobacterium* strains and for binary or cointegrate vectors. However, the use of pGV3850Kan^r resulted in poor transformation rates and not all BIN19 derivatives tested were equally efficient [17 and unpublished results].

High regeneration rates for transformed calluses could be achieved by the successive use of two regeneration media. A change to an 2-IP-containing medium greatly enhances shoot formation of the transformed *Arabidopsis thaliana* explants and allows a more synchronous and rapid regeneration [17]. The usefulness of this hormone for rapid and efficient regeneration has already been demonstrated for *Arabidopsis* leaf and root explants [5, 20].

Cotyledons were also tested as an alternative explant source in the transformation experiments. They gave equally high transformation and regeneration rates supporting the previous report of their amenability in regeneration experiments [15].

Stable transformation of plants obtained with the described procedure was verified by Southern blot analysis of independent transformants [17]. Analysis of the progeny of transformants showed that the chimaeric resistance genes used as dominant selectable markers in the transformation experiments are stably inherited. In some of the transformants the selectable marker showed a behaviour consistent with a single dominant trait. The data for other transformants suggest that the resistance gene is present in more than one locus [17]. However, in a small number of cases more sensitive than resistant seedlings were found in the progeny of a transformant [17]. It is interesting to note that this phenomenon was also found in transformation experiments of *Arabidopsis thaliana* using direct gene transfer to protoplasts [3].

Procedures

Seed Sterilization and Growth of Plant Material

For the transformation protocol described in this chapter leaf or cotyledon explants are used. They are isolated from *Arabidopsis* plants grown from sterilized seeds in axenic culture.

Steps in the Procedure
1. Surface sterilize *Arabidopsis* seeds for 90 seconds with 70% ethanol and 10 minutes with 7% commercial bleach. After 3–5 careful washings with sterile water the seeds are allowed to dry completely. Sterilized dried seeds can either be used directly or stored at room temperature.
2. Plate 100–200 sterilized seeds on AM-medium in Petri dishes (94 × 16 mm). Expanded cotyledons can be isolated after germination directly from such a plate. If leaf explants are used for the transformation experiments transfer seedlings to glass containers at low density. Plant materials are grown in culture chambers (20 °C, 16 h photoperiod, 3000 lux, 50% relative humidity).

Solutions
— 70% Ethanol
— 7% Commercial bleach
— Sterile water
— AM-medium (see Table 1a)

Growth conditions for the feeder-layer cells
Cultivate tobacco suspension culture cells in liquid MS-medium supplemented with 2,4-D. Transfer suspension culture cells regularly to fresh culture medium.

Solutions
— Suspension culture cultivation medium: MS-medium without agar (see Table 1a) supplemented with 1 mg/l 2,4-D.
— 2,4-D stock: 1 mg/ml in aqueous ethanol.

Table 1a. Composition of the Tissue Culture Media

Medium	AM [17]	MS [17]	MG [18]
Macroelements			
NH_4NO_3	825	1650	1650
KNO_3	950	1900	1900
$CaCl_2 \cdot 2H_2O$	220	440	440
$MgSO_4 \cdot 7H_2O$	185	370	370
KH_2PO_4	85	170	170
Na_2EDTA	18.65	37.3	37.3
$FeSO_4 \cdot 7H_2O$	13.75	27.5	27.5
Microelements			
H_3BO_3	3.1	6.2	6.2
$MnSO_4 \cdot 4H_2O$	11.15	22.3	22.3
$ZnSO_4 \cdot 4H_2O$	4.3	8.6	8.6
KI	0.415	0.83	0.83
$Na_2MoO_4 \cdot 2H_2O$	0.125	0.25	0.25
$CuSO_4 \cdot 5H_2O$	0.0125	0.025	0.025
$CoCl_2 \cdot 6H_2O$	0.0125	0.025	0.025
Vitamins			
Nicotinic acid	0.5	1	1
Pyridoxine-HCl	0.5	1	1
Thiamine-HCl	5	10	10
Myo-Inositol	50	100	100
Sucrose	10000	20000	
Glucose			16000
Agar	8000	8000	8000
pH	5.8	5.8	5.8

All concentrations are given in mg/l. The microelements and vitamins can be made up as separate stock solutions (100x) and can be kept frozen. Na_2EDTA and $FeSO_4 \cdot 7H_2O$ are dissolved together by stirring the hot solution for 30 minutes. A $200 \times$ concentrated stock solution can be kept at 4 °C in the dark. The pH of the media is adjusted with KOH. Plant media are sterilized by autoclaving. All plant media used are based on the recipe of Murashige and Skoog [13] but are supplemented with vitamins according to Gamborg [6].

Isolation and Preincubation of Explants

The procedure given describes a preincubation of 2 days of the isolated explants on a tobacco feeder-layer prior to the cocultivation step.

Steps in the Procedure

1. Plate tobacco suspension culture cells on CIM-medium (94x16 mm Petri dish). Preferably collect the tobacco cells with a sterile fine meshed stainless steel sieve to avoid the transfer of too much liquid.
2. Take 2–4 weeks old plants and isolate the leaves under sterile conditions by cutting the leaf base — carefully avoiding tissue which contains axillary buds. Big leaves can be cut in 2 or more pieces. To isolate cotyledons start from seedlings which have expanded cotyledons. Cut them at the base thereby avoiding the shoot apex tissue. Try to prevent severe damage of the explants. For a routine transformation experiment isolate 50 explants. These can be cultivated on a single Petri dish.
3. Avoid desiccation of the explants by transferring them immediately after isolation to the CIM feeder-layer plates and put the abaxial side of the explants in contact with the tobacco suspension culture cells.
4. Incubate the explants for 2 days under low-light conditions.

Notes

1. The duration and type of the preincubation influences the transformation frequency. A longer preincubation enhances the transformation rate, but a preincubation period as long as 3–4 days also results in browning of the callus tissue, therefore, a preincubation time of 2 days seems to be optimal. The use of the tobacco feeder-layer during the preincubation is slightly advantageous but the effect on transformation rates is not very pronounced. However, the use of tobacco feeder-layer cells assures that the explants are not overgrown by the bacteria during the cocultivation. It is also possible to omit the preincubation completely, although this results in a lower transformation efficiency [17 and unpublished results].
2. The age of the plants (*Arabidopsis* ecotype C24) from which the explants are isolated does not significantly influence the transformation frequency, however, the explants should be isolated before the plants start to flower. If other ecotypes are used for transformation experiments it would be useful to optimize this parameter. There is a stronger influence of this parameter in the ecotype Wassilewskija (explants derived from younger plants give rise to significantly higher transformation rates than those from older plants) [17].
It is important to take tissue which does not contain axillary buds or the shoot apex, because the fast growth of these is normally not inhibited by antibiotics in the selective medium and therefore 'false positives' might be generated.

Solutions

— CIM (see Table Ib)
— 2,4-D stock: 1 mg/ml in aqueous ethanol.
— Kinetin stock: 1 mg/ml in HCl/H_2O. Dissolve kinetin in a drop of 1 M HCl, then add water to the final volume. Sterilize by filtration through 0.22 μm filters.

Table 1b. Composition of the Regeneration Media

	CIM [17]	SIM1 [9]	SIM2 [5]	SEM [10]	RIM [17]
MS-medium	+	+	+	+	−
AM-medium	−	−	−	−	+
Phytohormones (final concentration in the media)					
2,4-D (1 mg/l)	+	−	−	−	−
Kinetin (0.2 mg/l)	+	−	−	−	−
BAP (1 mg/l)	−	+	−	+	−
NAA (0.1 mg/l)	−	+	−	+	−
2-IP (7 mg/l)	−	−	+	−	−
IAA (0.05 mg/l)	−	−	+	−	−
GA_3 (0.1 mg/l)	−	−	−	+	−
IBA (1 mg/l)	−	−	−	−	(+)

References for the concentrations of the phytohormes in the different regeneration media are given in brackets. Add sterile phytohormone solutions to the plant media after autoclaving.

Growth of Agrobacterium

Steps in the Procedure

1. Bacteria are grown in 10 ml YEB-medium containing appropriate antibiotics at 28 °C for 1–2 days.
2. Spin the culture for 15 minutes at room temperature at $1500 \times g$. Discard the supernatant and resuspend the bacteria in 10 ml YEB-medium without antibiotics.
3. Repeat the centrifugation step. Discard the supernatant and resuspend the bacteria in 10 ml of a sterile $MgSO_4$ solution (10 mM) and transfer them to a 60x15 mm Petri dish. Perform all washing steps under sterile conditions.

Note

Different densities of the bacterial cultures (densities varied maximally by a factor of 10) did not result in major changes in the transformation efficiency (unpublished results). Freshly grown *Agrobacterium* cultures to be used for transformation experiments should be kept strictly at room temperature prior to transformation experiments. Storage in the $MgSO_4$ solution at room temperature for 1–2 hours does not affect transformation rates.

Solutions

- YEB-medium [22]: 0.5% beef extract, 0.1% yeast extract, 0.5% peptone, 0.5% sucrose, 2 mM $MgSO_4$.
- 10 mM $MgSO_4$

Agrobacterium growth medium and the $MgSO_4$ solution are sterilized by autoclaving.

Incubation with Agrobacterium

Steps in the Procedure

1. Transfer the pretreated explants to the *Agrobacterium* suspension and incubate them for 5–10 minutes.
2. Place the explants back on the CIM feeder-layer plates. It is not necessary to blot the explants dry before they are placed on the agar plates, however, try to avoid carrying over too much solution. Bring the abaxial side of the tissue in contact with the suspension culture cells.
3. Cocultivate the explants with *Agrobacterium* for 2 days under low-light conditions.

Selection and Regeneration of Transformed Tissue

The selection conditions for hygromycin B as well as G418 and kanamycin have been established for this procedure [17]. Explants on selective medium will first form green callus tissue, predominantly at the cut surfaces, proliferating callus tissue is easily detectable 1–2 weeks after *Agrobacterium* infection [21]. From these calluses shoots can be readily obtained.

To optimize transformation efficiencies for other *Arabidopsis thaliana* ecotypes advantage of an intron-containing marker gene (β-glucuronidase) can be taken. The use of this in transformation experiments allows reliable estimations of the transformation frequencies already a few days after the inoculation with *Agrobacterium* [21].

Steps in the Procedure

1. After cocultivation place the leaves directly on selective SIM1. To stop bacterial growth claforan (Roussel) is added to SIM1 to a final concentration of 500 mg/l. Add 20 µg/ml of hygromycin B to SIM1 to select for tissue which was transformed with a chimaeric HPT gene and G418 to a final concentration of 20 µg/ml in SIM1 if the explants were transformed with a chimaeric NPT II gene.
2. Transfer the explants to fresh medium at least once a week. Explants should develop green calluses on the selective SIM1 and the first calluses should start to form shoots about four weeks after the cocultivation.
3. As the first shoots are observed on some of the calluses, transfer all explants to selective SIM2 to promote the shooting-response. Within one week most of the calluses will then develop multiple shoots.
4. Transfer calluses with 3–4 mm long shoots to SEM to induce shoot elongation and keep them 1–2 weeks on this medium. Perform the shoot elongation step in glass containers. In this step there is no longer selection for the transformed cells but the medium still contains claforan.

Notes

1. In the case of extensive bacterial growth the leaves can be washed in liquid MS-medium. Normally this is not necessary if the cocultivation was done using a feeder-layer. After washing of the explants they should be blotted dry on sterile filter paper to remove excess liquid.

Kanamycin (final concentration: 25 µg/ml) can be used as an alternative to G418, however, the G418 selection is clearer because resistant calluses on kanamycin-containing medium have a brownish appearance [17 and unpublished results].

2. The regular transfer of the explants is extremely important, because claforan is unstable and the degradation products of the antibiotic are harmful to the explant tissue. However, media containing claforan can be stored at 4 °C for 1–2 weeks.

3. On SIM2 the regeneration of the transformed calluses is more rapid and efficient than on SIM1. Furthermore, regeneration of the callus tissue is more synchronous on SIM2. The successive use of the two different shoot induction media in the transformation experiments is crucial because a transfer from CIM to SIM2 directly after the cocultivation with *Agrobacterium* results in very poor transformation rates [17].

Solutions
- SIM1 (see Table Ib) supplemented with 500 mg/l claforan and 20 mg/l hygromycin or G418.
- SIM2 (see Table Ib) supplemented with 500 mg/l claforan and 20 mg/l hygromycin or G418.
- SEM (see Table Ib) supplemented with 500 mg/l claforan.
- BAP stock: 1 mg/ml in aqueous ethanol.
- NAA stock: 1 mg/ml in DMSO.
- GA_3 stock: 1 mg/ml in H_2O.
- 2-IP stock: 10 mg/ml in HCl/H_2O. Dissolve 2-IP in a drop of 1 M HCl and then add water to the final volume.
- IAA stock: 1 mg/ml in DMSO.
- Claforan stock: 250 mg/ml in H_2O.
- Hygromycin B stock: 20 mg/ml in H_2O.
- Kanamycin stock: 50 mg/ml in H_2O.
- G418 stock: 20 mg/ml in H_2O.

Antibiotic solutions and phytohormones dissolved in water are sterilized by filtration through 0.22 µm filters. Phytohormone and antibiotic stock solutions can be kept at 4 °C for at least 4 weeks, however, the claforan stock solution (storage at 4 °C) should be renewed weekly. Add phytohormones and antibiotics to the plant media after autoclaving.

Root Induction

Steps in the Procedure

1. Transfer elongated shoots to RIM in glass reaction tubes (16 cm long, 16 mm in diameter) which are closed with cotton-plugs. At this stage claforan is omitted from the medium. To allow root formation it is important to remove all callus tissue from the plantlets. Within 2 or 3 weeks the majority of plants will form roots. The plants are grown to maturity on this medium without an additional transfer.
2. The seeds can be removed from the plants when the capsules are yellowish and have already opened.

Note

In case of extensive *Agrobacterium* contamination claforan can be included in the root induction medium together with 0.1% activated charcoal.

IBA in RIM is not absolutely required to promote root formation. Rooting is not a prerequisite for seed setting of the plantlets in sterile culture, however, the probability of seed production is higher for rooted shoots than for non-rooted shoots. If 2–3 shoots derived from one transformed callus are transferred to RIM it is highly probable that seeds will be obtained from almost every regenerating callus. The plants are grown in tubes closed with cotton plugs in order to allow an efficient gas exchange. If the atmosphere is too humid the seed set will be significantly reduced. Ideally the medium should have dried out by the time the seed capsules have opened [17].

Solutions

– RIM (see Table Ib)
– IBA stock: 1 mg/l in DMSO.

Progeny Analysis of Transformants

To determine the segregation data for the dominant selectable marker genes in the progeny of the selfed transformants seeds are plated on selective germination medium. If a chimaeric HPT gene was used as selectable marker in the transformation experiment hygromycin is included in the medium if a chimaeric NPT II gene was used kanamycin is added.

Steps in the Procedure
1. Harvest the seeds of the regenerated plantlets and plate them on AM-medium containing 20 µg/ml hygromycin [17] or on MG-medium containing 50 µg/ml kanamycin [18].
2. Place the Petri dishes for 2—7 days at 4 °C in order to increase the germination frequency.
3. Transfer the Petri dishes to a culture chamber (20 °C, 16 h photoperiod, 3000 lux, 50% relative humidity). Under these conditions the sensitive plants will germinate but will not form true first leaves while resistant plants develop normal roots and shoots.

Note
The screening of kanamycin resistant versus sensitive seedlings is more reliable on tissue culture medium with glucose as a carbon source rather than sucrose. For the selection at seedling level use kanamycin (final concentration 50 µg/ml) rather than G418 (final concentration 20 µg/ml), because G418 selected seedlings may be considerable smaller than kanamycin selected seedlings.

Solutions
— AM-medium (see Table Ia) containing 20 mg/l hygromycin B.
— MG-medium (see Table Ia) containing 50 mg/ml kanamycin.
— Hygromycin stock: 20 mg/ml in H_2O.
— Kanamycin stock: 50 mg/ml in H_2O.
The media are sterilized by autoclaving and the antibiotics by filtration through 0.22 µm filters. Add the antibiotics to the media after autoclaving.

Acknowledgements

We thank Dr. Priska Stabel and Dr. Caroline Dean for critical reading of the manuscript.

References

1. Aerts M, Jacobs M, Hernalsteens JP, Van Montagu M, Schell J (1979) Induction and *in vitro* culture of *Arabidopsis thaliana* crown gall tumours. Plant Sci Lett 17: 43–50.
2. Damm B, Willmitzer L (1988) Regeneration of fertile plants from protoplasts of different *Arabidopsis thaliana* genotypes. Mol Gen Genet 213: 15–20.
3. Damm B, Schmidt R, Willmitzer L (1989) Efficient transformation of *Arabidopsis thaliana* using direct gene transfer to protoplasts. Mol Gen Genet 217: 6–12.
4. Estelle MA, Somerville CR (1986) The mutants of *Arabidopsis*. Trends in Genet 2: 89–93.
5. Feldmann KA, Marks MD (1986) Rapid and efficient regeneration of plants from explants of *Arabidopsis thaliana*. Plant Sci 47: 63–69.
6. Gamborg OL, Miller RA, Ojima K (1968) Nutrient requirements of suspension cultures of soybean root cells. Exp Cell Res 50: 151–158.
7. Hayford MB, Medford JI, Hoffman NL, Rogers SG, Klee HJ (1988) Development of a plant transformation selection system based on expression of genes encoding gentamicin acetyl-transferases. Plant Physiol 86: 1216–1222.
8. Horsch RB, Fry JE, Hoffmann NL, Eichholtz D, Rogers SG, Fraley RT (1985) A simple and general method for transferring genes into plants. Science 227: 1229–1231.
9. Lloyd AM, Barnason AR, Rogers SG, Byrne MC, Fraley RT, Horsch RB (1986) Trans-formation of *Arabidopsis thaliana* with *Agrobacterium tumefaciens*. Science 234: 464–466.
10. Lloyd A, Muskopf Y, Barnason A, Horsch R (1897) Transformation of *Arabidopsis thaliana* var. Columbia via leaf-piece cocultivation with an *Agrobacterium tumefaciens* strain containing a plasmid encoding for hygromycin resistance. In: Third International Meeting on *Arabidopsis*, Michigan State University, Abstract 23.
11. Masterson RV, Furtek DB, Grevelding C, Schell J (1989) A maize *Ds* transposable element containing a dihydrofolate reductase gene transposes in *Nicotiana tabacum* and *Arabidopsis thaliana*. Mol Gen Genet 219: 461–466.
12. Meyerowitz EM (1989) *Arabidopsis*, a useful weed. Cell 56: 263–269.
13. Murashige T, Skoog F (1962) A revised medium for rapid growth and bioassays with tobacco tissue cultures. Physiol Plant 15: 473–497.
14. Negrutiu I, Beeftink F, Jacobs M (1975) *Arabidopsis thaliana* as a model system in somatic cell genetics I. Cell and tissue culture. Plant Sci Lett 5: 293–304.
15. Patton DA, Meinke DW (1988) High-frequency plant regeneration from cultured cotyledons of *Arabidopsis thaliana*. Plant Cell Rep 7: 233–237.
16. Rédei GP (1975) *Arabidopsis* as a genetic tool. Ann Rev of Genet 9: 111–127.
17. Schmidt R, Willmitzer L (1988) High efficiency *Agrobacterium tumefaciens*-mediated trans-formation of *Arabidopsis thaliana* leaf and cotyledon explants. Plant Cell Rep 7: 583–586.
18. Schmidt R, Willmitzer L (1989) The maize autonomous element *Activator* (*Ac*) shows a minimal germinal excision frequency of 0.2%–0.5% in transgenic *Arabidopsis thaliana* plants. Mol Gen Genet 220: 17–24.
19. Sheikholeslam SN, Weeks DP (1987) Acetosyringone promotes high efficiency transforma-tion of *Arabidopsis thaliana* explants by *Agrobacterium tumefaciens*. Plant Mol Biol 8: 291–298.
20. Valvekens D, Van Montagu M, Van Lijsebettens M (1988) *Agrobacterium tumefaciens*-mediated transformation of *Arabidopsis thaliana* root explants by using kanamycin selection. Proc Natl Acad Sci USA 85: 5536–5540.
21. Vancanneyt G, Schmidt R, O'Connor-Sanchez A, Willmitzer L, Rocha-Sosa M (1990) Construction of an intron-containing marker gene: Splicing of the intron in transgenic plants and its use in monitoring early events in *Agrobacterium*-mediated plant transformation. Mol Gen Genet 220: 245–250.
22. Vervliet G, Holsters M, Teuchy H, Van Montagu M, Schell J (1975) Characterization of different plaque-forming and defective temperate phages in *Agrobacterium* strains. J Gen Virol 26: 33–48.

Plant Tissue Culture Manual A7: 1–20, 1991.
© 1991 *Kluwer Academic Publishers.*

Arabidopsis protoplast transformation and regeneration

BRIGITTE DAMM[1] & LOTHAR WILLMITZER[2]
[1] *Institut de Biologie Moléculaire des Plantes, 12, rue du Général Zimmer, 67084 Strasbourg, France*
[2] *Institut für Genbiologische Forschung Berlin GmbH, Ihnestraße 63, 1000 Berlin 33, FRG*

Introduction

Arabidopsis thaliana (L.) Heynh., a member of the Brassicaceae, has been used in classical plant genetics since the 1940s [14] because of the small size of the mature plant, its short generation time and the high number of offspring that is normally produced after self-pollination. During recent years *A. thaliana* has also been adopted as a model system for studies on the molecular genetic level due to its small genome size, the presence of a well defined genetic map and the availability of many mutants (for reviews see [1, 17, 18, 21]).

Several protocols have been developed for *Agrobacterium tumefaciens*-
-mediated gene transfer to *A. thaliana* involving either wounded tissue, i.e. explants [16, 22, 23, 25], or entire seeds [6] and resulting in the generation of transgenic plants. Only recently have protoplast-plant regeneration systems been established for this species [3, 7] which are prerequisites for the genetic manipulation of *A. thaliana* at the protoplast level. In our laboratory we have been using a highly efficient regeneration protocol starting from mesophyll protoplasts [3] for experiments on direct gene transfer to *Arabidopsis* as an alternative to DNA transfer methods involving *A. tumefaciens*. From the early 1980s to date, several methods for direct gene transfer to protoplasts (i.e. uptake of naked DNA mediated by Ca^{2+} and polyethylene glycol (PEG) [13], Ca^{2+} and polyvinyl alcohol (PVA) [11], electroporation [8, 24] or micro-injection [2]) have been developed for different plant species. For stable transformation of *Arabidopsis thaliana* via direct gene transfer to protoplasts we have chosen a combined Ca^{2+}/PEG procedure that has originally been developed to transform protoplasts of protoplasts of *Nicotiana tabacum* and *N. plumbaginifolia* [20]. Several modifications have been introduced to this protocol resulting in a highly efficient transformation system for *A. thaliana* [4].

In this chapter we describe the transformation and regeneration protocol for *A. thaliana* in detail, emphasizing the critical steps and their influence on the transformation efficiency (i.e. number of transformed calluses) as well as on the frequency of regeneration of fertile transgenic plants. The transformation and regeneration procedure presented is optimized for mesophyll protoplasts of *A. thaliana* genotype C24 and the use of the plasmids pGL2 and pRT103neo conferring resistance to hygromycin or to geneticin (G418) and kanamycin respectively. However, after several modifications the given protocol can be applied to mesophyll protoplasts of several ecotypes of *A. thaliana* such as Landsberg erecta, the Landsberg mutant ap-1 [12] and Estland resulting in the regeneration of hygromycin- as well as G418-resistant transformants [unpublished results]. On the basis of an efficient regeneration system [3] the major factors influencing the transformation frequency are the final PEG concentration and the presence of carrier DNA in the transformation experiment [4]. These latter factors as well as the regeneration conditions have to be varied for protoplasts of the different ecotypes and the optimum method determined empirically. Although the method presented here employs the genes for hygromycin phosphotransferase (HPT) and neomycin phosphotransferase II (NPT II) as selectable marker genes several other resistance markers, for example the gene coding for phosphinothricin acetyltransferase (PAT) which confers resistance against phosphinothricin, can probably also be used.

Procedures

Protoplast isolation from leaves of in vitro *grown plants of* Arabidopsis thaliana

The example given is for the isolation of protoplasts from sterile axenic grown plants of *A. thaliana* genotype C24. Three- to four-week-old plants that have not started to flower are taken for protoplast isolation.

Steps in the procedure
1. Take about 1–2 g of leaf material, corresponding to the leaves of 40 to 60 plants, under sterile conditions. Place the leaves bottom side down in a Petri dish (94 × 16 mm) and wet them with 4 to 7 ml of 0.5 M mannitol to prevent desiccation. Cut the leaves with a razor blade so that each leaf is cut once.
2. For plasmolysis the cut leaf material is transferred into two Petri dishes (94 × 16 mm) containing 12 ml of 0.5 M mannitol each. Plasmolysis is performed for 1–2 h in the dark at RT.
3. Remove the mannitol solution and add 12 ml enzyme solution per Petri dish and seal them with Parafilm. Incubation is performed in the dark for 16–20 h at 25 °C.
4. After incubation gently agitate the mixture and wait for a further half hour to complete digestion. Separate the protoplasts from undigested tissue by consecutive filtration of the mixture through a 125 μm stainless steel sieve on top of a 63 μm one. Wash each Petri dish with 6 ml of 0.2 M $CaCl_2$ in order to recover remaining protoplasts and add this solution to the filtrate by pouring it through the sieves.
5. Distribute the filtrate into four 12 ml centrifuge tubes. Centrifuge for 5 min at 60 × g in a swinging bucket rotor. The protoplasts collect in the pellet.
6. Carefully remove all the supernatant with a pipette. Resuspend each pellet in a solution containing 3 ml of 0.5 M mannitol and 6 ml of 0.2 M $CaCl_2$ and centrifuge for 5 min at 40 × g.
7. Repeat the resuspension and centrifugation step number 6 but take a 0.1 ml sample for determination of the protoplast density in a counting chamber (Fuchs-Rosenthall) before you centrifuge.
8. Resuspend the protoplast of each centrifuge tube after the last washing step in 5–7 ml of W5 medium and put the tubes on ice in the dark for at least 30 min. During that time the protoplasts stabilize.

Notes
1. Sterile plant cultures are established from seeds surface-sterilized with sodium hypochlorite (30 min, 5% [3]). After complete drying sterilized seeds can be stored at room temperature. For germination seeds are placed in a Petri dish (94 × 16 mm) containing AM medium and incubated for three to four days at 8 °C in the dark before they are transferred to germination conditions. About one week after the transfer the seedlings are placed into glass jars also containing AM medium and cultured for a further two or three weeks. Germination takes place in a culture chamber (16 h photoperiod, 3000 lux, 21 °C, 50% relative humidity). If the dried seeds are stored at 4 to 8 °C the stratification treatment can be omitted.

Only green, undamaged leaves should be taken for protoplast isolation.

4. Protoplasts can be released from leaf debris by sucking through a wide bore pipette. 1 g of leaf material should yield about 7–10 × 10⁶ protoplasts.

Solutions

In general all solutions and media are sterilized by filtration through 0.22 μm filters. Simple salt solutions can be sterilized by autoclaving.

— Enzyme solution [3]:   1.0% w/v Cellulase 'Onozuka' R-10 (Serva)
                              0.25% w/v Macerozyme R-10 (Serva)
                              8 mM $CaCl_2$
                              0.4 M mannitol
                              pH 5.5 (KOH)

— 0.5 M mannitol

— 0.2 M $CaCl_2$

— W5 solution [20]:   145 mM NaCl
                              125 mM $CaCl_2$
                              5 mM KCl
                              5 mM glucose
                              pH 5.6–6.0   — AM medium: see Table 1

Table 1. Composition of the media used. All of the media are derivatives of media in the indicated publications.

Table 1a: Inorganic salts (mg/l final concentration)[a]

	Medium				
	AM [19]	B5 [9]	MII [15]	SRM [19]	MG [19]
Macroelements					
KNO_3	950	2500	1900	1900	1900
$(NH_4)_2SO_4$		134			
NH_4NO_3	825		1650	1650	1650
$MgSO_4 \cdot 7H_2O$	165	250	370	370	370
$CaCl_2 \cdot 2H_2O$	220	150	440	440	440
KH_2PO_4	85		170	170	170
$NaH_2PO_4 \cdot H_2O$		150			
Microelements					
Na_2EDTA	18.65	37.3	37.3	37.3	37.3
$FeSO_4 \cdot 7H_2O$	13.9	27.8	27.8	27.8	27.8
$MnSO_4 \cdot H_2O$	11.15	10.0	22.3	22.3	22.3
H_3BO_3	3.1	3.0	6.2	6.2	6.2
$ZnSO_4 \cdot 7H_2O$	4.3	2.0	8.6	8.6	8.6
$Na_2MoO_4 \cdot 2H_2O$	0.125	0.25	0.25	0.25	0.25
KI	0.415	7.5	8.3	8.3	8.3
$CuSO_4 \cdot 5H_2O$	0.013	0.025	0.025	0.025	0.025
$CoCl_2 \cdot 6H_2O$	0.013	0.025	0.025	0.025	0.025

[a] Each macroelement is usually made up as 10 × concentrated stock, and the microelements together as 100 × concentrated stock solutions. Na_2EDTA and $FeSO_4 \cdot 7H_2O$ are dissolved together (gently warm to dissolve) as a 10 × concentrated stock.

Table 1b: Vitamins and other organics (mg/l final concentration)[b]

	Medium				
	AM [19]	B5 [9]	MII [15]	SRM [19]	MG [19]
m-Inositol	50	100	500	100	100
Glutamine			500		
Vitamins[b]					
Nicotinic acid	0.25	1.0	0.5	0.5	0.5
Pyridoxine HCl	0.25	1.0	0.5	0.5	0.5
Thiamine HCl	0.25	10.0	0.1	0.1	0.1
Glycine	1.0		2.0	2.0	2.0

[b] Vitamins are prepared separately as 10 × concentrated stocks.

Table 1c: Carbohydrates and phytohormones

	Medium				
	AM [19]	B5 [9][c]	MII [15]	SRM [19][d]	MG [19]
Carbohydrates (g/l)					
Sucrose	10		20.0	20.0	
Glucose		72.6	54.05		16.0
Phytohormones (mg/l)[e]					
2,4D		1.0	0.05		
IAA				0.05	
BAP		0.15			
Kinetin			1.0		
2-IP				7.0	
Final pH (KOH)	5.8	5.8	5.8	5.8	5.8
Agar (Difco)	0.8%			0.8%	0.8%

[c] Hormones according to Xuan and Menczel [26].
[d] Hormones according to Feldmann and Marks [5].
[e] 2,4-D = 2,4-dichlorophenoxyacetic acid (1 mg/ml dissolved in ethanol/H_2O); IAA = indole-3-acetic acid (1 mg/ml dissolved in ethanol/H_2O); BAP = 6-benzylaminopurine (1 mg/ml dissolved in HCl/H_2O); kinetin (1 mg/ml dissolved in HCl/H_2O); 2-IP = 2-isopentenyladenine (10 mg/ml dissolved in HCl/H_2O).

Stable transformation of protoplasts of A. thaliana *genotype C24*

For *A. thaliana* genotype C24 a simple transformation method involving the incubation of freshly isolated protoplasts in polyethylene glycol (PEG)/Ca^{2+} and the use of the plasmids pGL2 or pRT103neo results in transformation frequencies that are comparable to those obtained for *N. tabacum*, the model species for direct gene transfer [4, 20, 24]. The procedure described below is performed with plasmids that confer hygromycin or G418 and kanamycin resistance respectively. The vector pGL2 (J. Paszkowski, personal communication) contains the coding region of the hygromycin phosphotransferase (HPT) gene whereas pRT103neo (R. Töpfer, personal communication) carries the gene coding for the neomycin phosphotransferase II (NPT II). The expression of both marker genes is controlled by the 35S RNA gene promoter of cauliflower mosaic virus.

Steps in the procedure

1. Freshly isolated, purified protoplasts should have been incubated for at least 30 min on ice in W5 solution before the transformation experiment is started (see step number 8 in the previous section).
2. Centrifuge the protoplast at 40 × g for 5 min. Carefully remove all of the W5 solution. Resuspend the protoplasts in MaMg solution at a density of 1.6 × 10^6 protoplasts/ml and distribute 0.3 ml aliquots into 1.5 ml Eppendorf tubes.
3. Add 7 µl of the plasmid DNA solution and 14 µl of the carrier DNA solution and mix.
4. Thereafter add 0.32 ml of PEG-CMS solution and carefully mix the transformation sample by sucking it through a wide bore 1 ml plastic pipette. Incubate for 20–30 min at room temperature (RT).
5. Transfer the aliquots into 12 ml centrifuge tubes and gradually add 10 ml of W5 solution over a time period of about 10 min and then centrifuge for 5 min at 50 × g.
6. Resuspend the protoplasts in 2.5 ml 0.5 M mannitol and 0.5 ml W5 solution, and again centrifuge for 5 min at 50 × g. Carefully remove all of the supernatant and resuspend the protoplasts in 0.3 ml of 0.5 M mannitol for alginate embedding (see the following section).

Notes

1. If the protoplasts are not used immediately for transformation they can be stored for several hours (at least up to 4 h) in W5 solution on ice without loosing competence for transformation.
2. Transformation should take place without delay after the protoplasts have been transferred to the transformation buffer (MaMg solution). In contrast to the protocols for direct gene transfer to other species [10, 24, 27] where the application of a heat shock (i.e. incubation of the protoplasts for 5 min at 45 °C prior to transformation), has improved transformation efficiencies, this treatment cannot be applied in the case of *Arabidopsis* protoplasts as it destroys nearly all the cells.

The volumes indicated here can be altered. However, for stable transformation protoplast sample should not exceed 0.6 ml (corresponding to about 1 × 10^6 protoplasts) as we found that the use of bigger transformation aliquots led to a reduced viability of transformed protoplasts which had a negative influence on the transformation frequency [unpublished results]. Therefore one should take many smaller samples rather than a few big ones.

3. The concentration of the DNAs is worked out on the basis of the total number of protoplasts included in the transformation aliquot. For *A. thaliana* a concentration range of 20–50 μg of plasmid DNA and 50–100 μg of carrier DNA per 1.6×10^6 protoplasts was tested. The lowest amount of both DNAs was shown to be enough to obtain stable transformants at a high frequency [4]. Lower concentrations have not been tested but might lead to comparable transformation efficiencies. The carrier DNA could be omitted but this was shown to result in a four- to sixfold decrease in the transformation frequency [4]. If other, i.e. larger plasmids than pGL2 (4.5 kb) or pRT103neo (4.2 kb) were used for the transformation experiments equimolar amounts compared to the plasmids used in the given protocol should be taken. By doing so, vectors of up to 15 kb resulted in transformation frequencies comparable to those obtained by employing pGL2 or pRT103neo [unpublished results]. The plasmids can be taken in their circular form for the transformation experiment because for *A. thaliana* it could be shown that circular and linear plasmid DNA (20–50 μg/1.6×10^6 protoplasts) in combination with carrier DNA yielded similar transformation frequencies [4].

4. PEG is added to a final concentration of 20% which was
transformation of *A. thaliana* genotype C24 [4]. PEG 6000 (Merck) was taken to prepare PEG-CMS solution (see below) because protoplasts of the genotype C24 showed a better viability after the treatment with high molecular weight PEG compared to the treatment with low molecular weight compounds such as PEG 1500 [unpublished results].

5. The slow, stepwise dilution of the transformation mixture with W5 solution is very important for the recovery of mostly undamaged protoplasts after the PEG incubation.

Additional note

The transformation protocol described above can also be used in co-transformation experiments. This is particular of interest for the transfer of a non-selectable gene to the *Arabidopsis* genome. In case the non-selectable gene is linked to the selectable marker gene on the same plasmid molecule, the transformation is performed as described above. The regenerated transgenic plants carrying the selectable marker gene are then screened for the presence of the non-selected gene. In the case of a 15 kb plasmid containing both the gene for HPT as well as for β-glucuronidase (GUS) under the control of the 35 S promoter of cauliflower mosaic virus [pGSC1704gus; J. Leemans, personal communication] a co-transformation frequency for active copies of both genes of about 80% was obtained [unpublished results]. For the co-transformation of non-linked genes, simply mix the plasmid containing the non-selectable gene and the one carrying the selectable gene in molar ratios of about 3 : 1 to 10 : 1, perform the transformation experiment as described above, select for the presence of the resistance gene and screen the regenerated resistant plants for the presence of the second, non selected gene. If the plasmid carrying the non-selectable gene is used in excess co-transformation frequencies for active copies of both genes of up to 60% can be obtained [unpublished results].

Solutions

In general all solutions and media are sterilized by filtration through 0.22 μm filters. Simple salt solutions can be sterilized by autoclaving.

— MaMg solution [20]:　　15 mM $MgCl_2$
　　　　　　　　　　　　　0.5 M mannitol
　　　　　　　　　　　　　0.1% w/v MES
　　　　　　　　　　　　　pH 5.6 (KOH)

— Plasmid DNA solution: The DNA is sterilized by precipitation in ethanol followed by a wash in 70% ethanol and then dried in a sterile flow hood. The plasmids pGL2 and pRT103neo are dissolved at 1 μg/μl in sterile double distilled water.

— Carrier DNA solution: Calf thymus or salmon sperm DNA (Serva) is sheared to a size of about 10 kb, phenolysed and then sterilized in the same way as the plasmid DNA.

The carrier DNA is dissolved at 1 µg/µl in sterile double distilled water.

— PEG-CMS solution [20]: 40 g PEG 6000 (Merck) is dissolved in 100 ml of 0.1 M $Ca(NO_3)_2$ in 0.4 M mannitol (i.e. the final concentration of these two components will be lower due to the volume of the PEG). The pH is adjusted to 7–9 (KOH). It takes at least 2 h before the pH stabilizes in this solution. The filter-sterilized solution is stored in aliquots at −20 °C.

— W5 solution [20]:   145 mM NaCl
                           125 mM $CaCl_2$
                             5 mM KCl
                             5 mM glucose
                         pH 5.6–6.0

Embedding of transformed protoplasts

After the transformation protoplasts of A. *thaliana* have to be embedded in Na-alginate in order to induce divisions and to achieve the development of protoplast-derived microcolonies and calluses with a high frequency [3, 4]. The alginate gels produced via the method described below represent a culture system with many advantages, i.e. the cultures can be easily monitored through the microscope, the replacement of the culture media is simple and can be done without disturbing the development of the protoplast-derived cell colonies.

Steps in the procedure
1. Carefully mix 0.3 ml aliquots of the transformed protoplasts (see step 5 above) with equal volumes of Na-alginate solution by sucking the mixture through a wide bore 1 ml plastic pipette. Plate the mixture in small Petri dishes (60 × 15 mm) containing a layer of $CaCl_2$-agar. Carefully tilt the dishes in order to produce a thin gel (about 3 to 4 cm in diameter) of alginate-embedded protoplasts.
2. After 1 h incubation at room temperature 2 ml of solution 1 is added to the solidified alginate gels which are thereafter incubated for a further half to one hour. Carefully shake the Petri dishes in order to separate the gels from the $CaCl_2$-agar.
3. Remove the fully solidified gels with a broad spatula from the $CaCl_2$-agar and transfer them into small Petri dishes (60 × 15 mm) containing 2 ml of solution 2.
4. Seal the dishes with Parafilm and incubate the protoplast cultures in solution 2 for 2 d at 4–8 °C in the dark.

Notes
1. Take care that after the last washing step of the transformation (see step number 6 in the previous section) the supernatant is completely removed. Remaining Ca^{2+}-ions (present in MaMg and W5 solution) will result in premature polymerization of the protoplast/alginate mixture in the pipette and consequently in a massive destruction of protoplasts. Due to the viscosity of the alginate solution normally about 10% of the protoplasts were destroyed by shearing forces. However, this percentage can be much higher if the mixing is not done carefully enough.

Solutions
All solutions listed below are sterilized by autoclaving 121 °C, 20 min).
- Na-alginate solution: 2.4–2.8% w/v Na-alginate (Roth)
  0.4 M mannitol
- $CaCl_2$-agar: 20 mM $CaCl_2$
  0.4 M mannitol
  1% agar (Difco)
- Solution 1: 50 mM $CaCl_2$
  0.4 M mannitol
- Solution 2: 10 mM $CaCl_2$
  0.4 M mannitol

Protoplast culture and selection of transformed cell lines

Protoplast culture and the selection of transformed cell lines is performed in alginate gels. This culture method guarantees a continuous and controlled selection during the important early stages of protoplast development in transformation experiments. Consequently antibiotic-resistant and-sensitive protoplast-derived cell colonies can be clearly distinguished after 3 weeks of selection.

Steps in the procedure

1. After having kept the embedded protoplasts for 2 d at 4–8 °C in the dark replace solution 2 (see previous section) by 2 ml of protoplast culture medium B5. Seal the Petri dishes with Parafilm and incubate at 26 °C in the dark for 9 or 10 d without selection.
2. Start with the selection for transformants by exchanging protoplast culture medium B5 by 2 ml of the same medium but containing 20 µg/ml hygromycin B (Boehringer) in case the transformation experiment was performed with pGL2 or containing 10 µg/ml G418 (Sigma) in case the transformation experiment was carried out with pRT103neo. Incubate the Petri dishes for further 9–10 d in the dark.
3. After 3 weeks replace B5 medium containing 20 µg/ml hygromycin B or 10 µg/ml G418 by 2 ml MII medium supplemented with 20 µg/ml hygromycin or 20 µg/ml G418 respectively. Transfer the cultures to continuous dim light (700 lux).
4. Renew the selection medium every 10–14 d. Resistant and sensitive protoplast-derived colonies are easily distinguishable after 3 weeks of selection, and the resistant colonies can be scored by the naked eye after 8 weeks of culture to determine the transformation frequencies [4].
5. In order to induce morphogenesis, free the resistant colonies from the alginate matrix by picking them individuallly with a pair of forceps. The colonies have to be at least 1 mm in diameter before they can be transferred onto shoot regeneration medium (SRM). Smaller colonies will not survive the transfer to SRM medium.

Notes

2. A tight selection is only obtained when the antibiotics are applied as early as possible i.e. after the first cell divisions [4]. Routinely G418 rather than kanamycin was used to select for the presence of the NPT II gene as we found that only G418 selection allows a clear distinction between sensitive and resistant colonies after 8 weeks of culture [unpublished results]. Applying kanamycin as selective agent results in a high background of non-transformed protoplast-derived colonies.
4. The given protocol routinely yields relative transformation (RTF) of about 5% which correspond to absolute transformation frequencies (ATF) of 5×10^{-4} to 5×10^{-3} [4]. The RTF is described as the ratio between the number of resistant colonies in selected cultures and the developing colonies in non-selected cultures. The ATF is defined as the ratio between the number of resistant colonies and the initial number of protoplasts plated after transformation. Under the selective conditions described here no formation of macroscopically visible colonies in control cultures not treated with plasmid DNA was detectable eight weeks after the transformation experiment. Therefore plant regeneration from resistant colonies is routinely performed under non selective conditions (see below).

5. In order to ensure that all the resistant colonies survive the transfer to shoot regeneration medium (SRM) the larger colonies (1 mm in diameter) are continuously transferred whereas the smaller ones are kept in the alginate gels and cultured in MII medium without antibiotics until they have reached the critical size annd can also be plated. This prolonged liquid culture of the gels in MII medium is performed in a culture chamber with a photoperiod of 16 h, 3000 lux, 25 °C and 60% relative humidity.

Solutions
— B5 medium: see Table 1
— MII medium: see Table 1

Plant regeneration and seed setting

Plants are routinely regenerated from resistant colonies without selection as we observed that shoot formation occurred earlier and more frequently under non-selective compared to selective conditions [4]. Plant regeneration is performed on agar-solidified media in a culture chamber with a photoperiod of 16 h, 3000 lux, 25 °C and 60% relative humidity. The method described below is a modification of a previously published protocol [3].

Steps in the procedure

1. Plate antibiotic-resistant colonies individually with a pair of forceps on SRM medium to induce shoot formation. Shoot induction is carried out in Petri dishes.
2. Subculture the calluses every 14 d on the same medium. Subculturing will result in multiple shoot formation.
3. In order to induce shoot elongation transfer calluses with shoots to AM medium. Shoot elongation is performed in glass vessels.
4. After 1–3 weeks on AM medium separate the elongated shoots from the callus and transfer them into AM medium containing cotton-wool sealed test tubes for rooting and seed setting.
5. 4 to 8 weeks after the transfer into the test tubes rooted as well as non-rooted shoots have set seed that can be taken for progeny analysis.

Notes

1. No morphogenesis is achieved when the colonies are kept in the alginate gels. Only resistant colonies bigger than 1 mm should be transferred to SRM medium (see note number 5 in the previous section).
2. Calluses with developing shoots should not be kept for too long on SRM medium because the phytohormones will then hamper the regeneration of morphologically normal-looking plants from those shoots.
3. The developing shoots on the calluses have to be at least 3 mm before they are transferred on AM medium for shoot elongation.
4. Take care that no callus material remains on the detached shoot as this will hinder root formation. In order to obtain good seed setting with the regenerated shoots they should be transferred into cotton-wool sealed test tubes before they start to flower. A good gas exchange and a low humidity in the culture vessels is necessary to achieve a high fertility. Alternatively, to overcome the problem of reduced fertility due to a high humidity rooted transformants might be transferred to soil for seed setting. The agar is gently washed away and the plantlets potted. They should be kept in a humid atmosphere for about one week and can then be hardened off and grown in the greenhouse.
5. For storage let the seeds of the transformants dry completely and keep them at 4 or −20 °C to assure that they retain a high germination frequency over a long period of time.

Solutions

— SRM medium: See Table 1
— AM medium: See Table 1

Progeny analysis

Steps in the procedure
1. Collect seeds from selfed transformants (see step number 5 in the previous section). Plate them in Petri dishes containing AM medium supplemented with 20 µg/ml hygromycin B in the case of the progeny of hygromycin-resistant transformants. The seeds harvested from G418-resistant transformants were plated on MG medium containing 50 µg/ml kanamycin.
2. Incubate the Petri dishes for 3–4 d at 4–8 °C.
3. Transfer the Petri dishes to a culture chamber (16 h photoperiod, 3000 lux, 21 °C, 50% relative humidity).
4. After three weeks resistant and sensitive seedlings can be clearly distinguished.

Solutions
— AM medium: see Table 1
— MG medium: see Table 1

Acknowledgements

We want to express our thanks to T. Altmann, P.C. Morris, J. Siemens and R. Schmidt for careful reading of the manuscript.

References

1. Bowman JL, Yanofski MF, Meyerowitz EM (1988) *Arabidopsis thaliana*: A review. Oxford Surv Plant Mol Cell Biol 5: 57–87.
2. Crossway A, Oakes JV, Irvine JM, Ward B, Knauf VC, Shewmaker CK (1986) Integration of foreign DNA following microinjection of tobacco mesophyll protoplasts. Mol Gen Genet 202: 179–185.
3. Damm B, Willmitzer L (1988) Regeneration of fertile plants from protoplasts of different *Arabidopsis thaliana* genotypes. Mol Gen Genet 213: 15–20.
4. Damm B, Schmidt R, Willmitzer L (1989) Efficient transformation of *Arabidopsis thaliana* using direct gene transfer to protoplasts. Mol gen Genet 217: 6–12
5. Feldmann KA, Marks MD (1986) Rapid and efficient regeneration of explants of *Arabidopsis thaliana*. Plant Sci 47: 63–69
6. Feldmann KA, Marks MD (1987) *Agrobacterium*-mediated transformation of germinating seeds of *Arabidopsis thaliana*: A non-tissue culture approach. Mol Gen Genet 208: 1–9
7. Ford K (1990) Plant regeneration from *Arabidopsis thaliana* protoplasts. Plant Cell Rep 8: 534–537
8. Fromm ME, Taylor LP, Walbot V (1986) Stable transformation of maize after gene transfer by electroporation. Nature 319: 791–793
9. Gamborg OL, Miller RA, Ojima K (1968) Nutrient requirements of suspension cultures of soybean root cells. Exp Cell Res 50: 151–158
10. Guerche P, Charbonnier M, Jouanin L, Tourneur C, Paszkowski J, Pelletier G (1987) Direct gene transfer by electroporation in *Brassica napus*. Plant Sci 52: 111–116
11. Hain R, Stabel P, Czernilofski AP, Steinbiß HH, Herrera-Estrella L, Schell J (1985) Uptake, integration, expression and genetic transmission of a selectable chimeric gene by plant protoplasts. Mol Gen Genet 199: 161–168
12. Koornneef M, Van Eden J, Hanhart CJ, Stam P, Braaksma FJ, Feenstra WJ (1983) The linkage map of *Arabidopsis thaliana*. J Hered 74: 265–272
13. Krens FA, Molendijk L, Wullems GJ, Schilperoort RA (1982) *In vitro* transformation of plant protoplasts with Ti-plasmid DNA. Nature 296: 72–74
14. Laibach F (1943) *Arabidopsis thaliana* (L.) Heynh. als Objekt für genetische und entwicklungsphysiologische Untersuchungen. Bot Arch 44: 439–455
15. Li L-C, Kohlenbach HW (1982) Somatic embryogenesis in quite a direct way in cultures of mesophyll protoplasts of *Brassica napus* L. Plant Cell Rep 1: 209–212
16. Lloyd A, Barnason AR, Rogers SG, Byrne MC, Fraley RT, Horsch RB (1986) Transformation of *Arabidopsis thaliana* with *Agrobacterium tumefaciens*. Science 234: 464–468
17. Meyerowitz EM (1987) *Arabidopsis thaliana*. Ann Rev Genet 21: 93–111
18. Meyerowitz EM (1989) *Arabidopsis*, a useful weed. Cell 56: 262–269
19. Murashige T, Skoog F (1962) A revised medium for rapid growth and bioassays with tobacco tissue cultures. Physiol Plant 15: 473–497
20. Negrutiu I,, Shillito R, Potrykus I, Biasini G, Sala F (1987) Hybrid genes in the analysis of transformation conditions. I. setting up a simple method for direct gene transfer in plant protoplasts. Plant Mol Biol 8: 363–373
21. Redei GP (1975) *Arabidopsis* as a genetic tool. Ann Rev of Genet 9: 111–127
22. Schmidt R, Willmitzer L (1988) High efficiency *Agrobacterium tumefaciens*-mediated transformation of *Arabidopsis thaliana* leaf and cotyledon explants. Plant Cell Rep 7: 583–586
23. Sheikholeslam SN, Weeks DP (1987) Acetosyringone promotes high efficiency transformation of *Arabidopsis thaliana* explants by *Agrobacterium tumefaciens*. Plant Mol Biol 8: 291–298
24. Shillito RD, Saul M, Paszkowski J, Müller M, Potrykus I (1985) High efficiency direct gene transfer to plants. Bio/Technology 3: 1099–1103
25. Valvekens D, Van Montagu M, Van Lijsebettens M (1988) *Agrobacterium tumefaciens*-mediated transformation of *Arabidopsis thaliana* root explants by using kanamycin selection. Proc Natl Acad Sci USA 85: 5536–5540

26. Xuan LT, Menczel L (1980) Improved protoplast culture and plant regeneration from protoplast-derived callus in *Arabidopsis thaliana*. Z. Pflanzenphysiol 96: 77–80
27. Zhang HM, Yang H, Rech EL, Golds TJ, Davies AS, Mulligan BJ, Cocking EC, Davey MR (1988) Transgenic rice pants produced by electroporation-mediated plasmid uptake into protoplasts. Plant Cell Rep 7: 379–384

Plant Tissue Culture Manual **A8**: 1–17, 1992.
© 1992 *Kluwer Academic Publishers.*

Arabidopsis regeneration and transformation (Root Explant System)

DIRK VALVEKENS, MIEKE VAN LIJSEBETTENS & MARC VAN MONTAGU

Laboratorium voor Genetica, Universiteit Gent, K.L. Ledeganckstraat 35, B-9000 Gent, Belgium

Introduction

The crucifer *Arabidopsis thaliana* has become widely used as a model system for plant molecular biology [1]. In this chapter we describe a simple and highly reproducible procedure for regeneration and transformation of *Arabidopsis thaliana* (L.) Heynh. [1, 2]. For this method the explant source is the root system of the plant.

After a short treatment with 2,4-dichlorophenoxyacetic acid (2,4-D) and subsequent incubation on a medium with a high concentration of N^6-(2-isopentenyl)adenine (2ip), *Arabidopsis* roots give rise to a vigorous bush of shoots over their entire length after only three weeks (Fig. 1). These R1 shoots can then be transferred to a plant growth regulator-free medium to produce seeds three weeks later. The aseptic R2 seeds can immediately be germinated on a plant medium without any pretreatment.

The most widespread approach for creating transgenic plants is the *Agrobacterium* system [3, 4]. Therefore, we incorporated our regeneration method with *Agrobacterium tumefaciens* infection to develop an efficient and rapid transformation procedure using kanamycin (Km) selection. Towards the end of the 2,4-D treatment of the root explants, *Agrobacterium tumefaciens* which contain T-DNAs that carry an antibiotic resistance marker gene can be cocultivated with the root explants. Transgenic T1 shoots can then be selected on the 2iP medium by the addition of an antibiotic that can be inactivated by the protein product encoded by the antibiotic-resistant marker gene together with an antibiotic which prevents *A. tumefaciens* from overgrowing the root explants (Fig. 2). Using this strategy, transgenic T2 seeds can routinely be obtained within three months of tissue culture.

We have also described a rapid nondestructive screening procedure that allows transformed and untransformed plants to be distinguished (Fig. 3). This method is extremely valuable for analyzing the segregation of T-DNAs in relation to other markers among the offspring of genetic crosses.

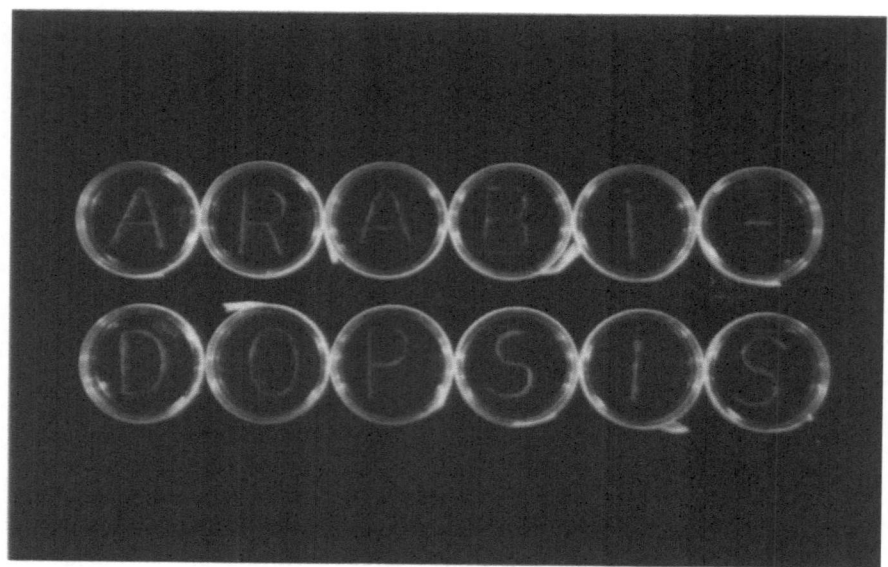

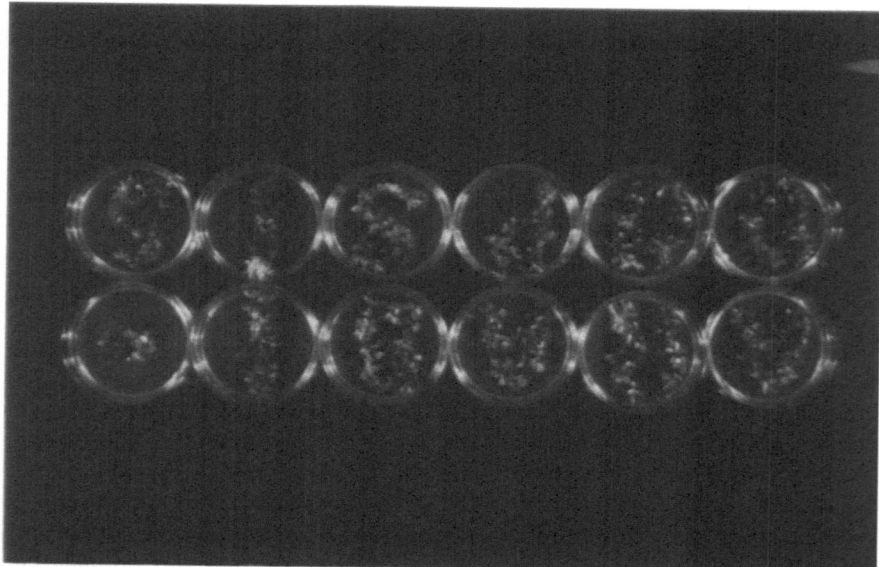

Fig. 1. 2,4-D-induced roots of *Arabidopsis thaliana* (a) give rise to a vigorous bush of shoots over their entire length after only three weeks of incubation on a medium with a high concentration of N^6-(2-isopentenyl)adenine (2ip) (b).

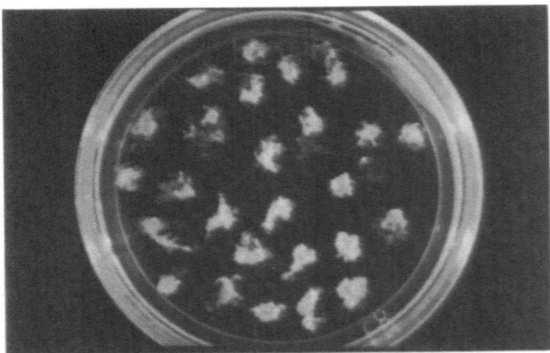

Fig. 2. After *Agrobacterium* infection transgenic T1 shoots are selected on a medium containing kanamycin and vancomycin.
Every C24 root explant gives rise to at least one transgenic shoot.

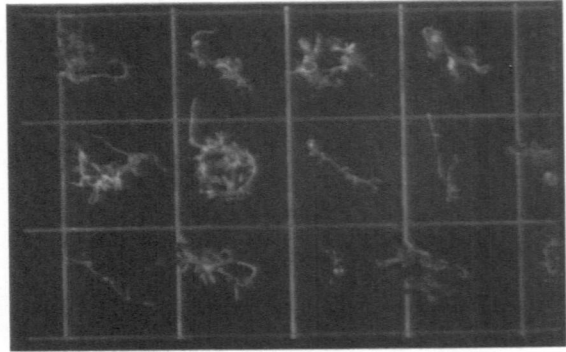

Fig. 3. Genetic test for marker gene expression in T2 progeny of transgenic lines.
Transgenic *Arabidopsis* T2 plants were identified by regenerating root segments on a kanamycin-containing medium. Eight out of the 12 root explants shown are resistant to kanamycin and produce numerous green shoots. The four remaining root explants originate from kanamycin-sensitive plants and do not produce any shoots.

Procedures

Seed sterilization and growth of seedlings
Steps in the procedure

1. Place seeds in 70% ethanol for two minutes (e.g. in a disposable, sterile 10 ml tube). Remove EtOH with a pipette.
2. Place seeds in 5% NaOCl/0.05% Tween 20 for 15 minutes. Shake regularly. Remove the sterilizing solution with a pipette.
3. Wash seeds in sterile, distilled water 5 times.
4. After the last wash, keep the seeds in 0.5–1.0 ml water. Take them up with a 2 ml pipette. Put drops with seeds on GM. Disperse the seeds homogeneously with the pipette tip.
5. Incubate the seeds on GM in a culture room for 7 days.
6. Transfer cotyledonary seedlings individually to fresh GM in 15 cm Petri dishes (Falcon® 1013) (15–30 seedlings/dish).

Notes
4. The seeds of some *Arabidopsis* ecotypes require a cold treatment prior to germination. In this case, the seeds can be stored dry for at least 7 days at 4 °C or can be incubated for 4 days at 4 °C on GM after sterilization.
5. The culture room should be operated with a 16 hour light/8 hour dark cycle and a temperature of 20 °C.
6. Sterilized seeds can also be immediately transferred individually to 15 cm Petri dishes; however, in our experience, germinated seedlings are easier to manipulate compared to seeds.

Regeneration from roots

For both regeneration and transformation *white* roots of aseptically grown *Arabidopsis* plants of any age can be used. After 8 weeks of growth, however, the roots will begin to turn green and should not be used.

Steps in the procedure

1. Pull plantlets *gently* out of the agar (GM) using forceps (the whole root system will come out easily).
2. Put the plantlets in a sterile Petri dish. Cut off root system from the rest of plantlet so that no green parts remain attached to the root.
3. Incubate the root system on 0.5/0.05 agar. Take care that the root is entirely in contact with the agar.
4. Incubate for 4 days in a growth room.
5. Transfer the roots to 0.15/5 agar. Numerous green shoots will completely cover the root system after 3 to 4 weeks.
6. Transfer individual shoots to GM in 15 cm Petri dishes (Falcon® 1013).
7. Incubate the growing shoots in the growth room for *in vitro* seed production.

Notes

1. Roots isolated from GM will quickly dessicate in a laminar-flow bench. Dessication can be reduced by quick handling and/or by keeping the roots submerged in some liquid plant medium (e.g. 0.5/0.05). For recovering *Arabidopsis* roots from agar several alternative methods can be used.
   - Remove the aerial parts of the plants with scalpel and forceps so that only the roots remain in the agar. Turn over the entire agar slab and recover the roots from the agar using forceps.
   - The use of 0.6% agarose instead of 0.8% agar causes the roots to grow on top rather than into the agar, leading to even more efficient recovery of root tissue [5].

7. To obtain seeds *in vitro* never put more than three shoots in each Petri dish. The lid of the Petri dish should always be absolutely free of condensation as high humidity inhibits anther dehiscence and hence prevents fertilization. The temperature should never exceed 22 °C. Rooted regenerants are obtained with 50% efficiency and can eventually also be transferred to soil to set seed. However, both rooted and unrooted regenerants will produce seeds *in vitro* with equal efficiencies.

Transformation of root explants

The root transformation procedure has been developed using the ecotype "collection number C24" [2]. For C24 an overall transformation efficiency of 80% (root explants giving rise to seed-producing T1 transformants) can be expected (Fig. 2). We have also successfully applied the method for transformation of Columbia and Landsberg *erecta* ecotypes. For Columbia, only 20% of the transformed root explants give rise to seed-producing regenerants. Transformation and complementation of an *Arabidopsis gl1* mutant, ecotype Columbia, has been reported [6], using slight modifications of our protocol. Landsberg *erecta* roots can be *regenerated* as efficiently as C24 roots. Also the transformation efficiency at the callus level (root explants producing green transformed calluses) is usually 100%. However, the regeneration of these transformed calluses is both retarded and less efficient in comparison with C24, suggesting that either the antibiotics or the agrobacteria are responsible for this negative effect. The overall transformation efficiency (i.e., root explants producing transformed T2 seeds) is usually about 20% over a 4-month period. The Bensheim and RLD ecotypes have also been successfully used for transformation [5, 7].

Using the root transformation procedure, "escapes" (isolates with 100% bleached seedlings in T2 populations germinated on GM K50) occur typically at a frequency of between 5% and 10%. When plant media are renewed more frequently than described below (e.g. weekly), it is advisable to use 60-70 mg/l kanamycin to prevent a higher "escape" frequency.

Although we exclusively used kanamycin selection in our transformation experiments, the successful use of hygromycin, chlorsulphuron, and methotrexate as selective agents for transformation of *Arabidopsis* root explants have recently been reported [8].

Steps in the procedure

1. Pull plantlets *gently* out of the agar (GM) using forceps (the whole root system will come out easily).
2. Put the plantlets in a sterile Petri dish. Cut off the roots from the rest of the plantlets so that no green parts remain attached to the roots.
3. Incubate the roots on 0.5/0.05 agar. Take care that the roots are entirely in contact with the agar.
4. Incubate in the culture room for 3 days.
5. Stack 3–5 2,4-D-induced roots in a sterile Petri dish. Cut the roots into 0.5 cm explants.
6. Transfer the root explants to a Petri dish containing 10–20 ml of liquid 0.5/0.05 medium.
7. Add 0.5–1.0 ml of an overnight grown *Agrobacterium* culture (28 °C; 200 rpm; Luria broth [9]). Shake gently for about 2 minutes.
8. Blot root explants briefly on a sterile filter paper to remove most of the liquid, and then place them on 0.5/0.05 agar.

9. Incubate up to 50 explants per 9 cm Petri dish in the growth room for 2 days to allow *Agrobacterium* infection. After 2 days of coculture the explants will be completely overgrown by agrobacteria.
10. Transfer the explants to 10—20 ml of liquid 0.5/0.05 medium. Shake vigorously to wash off agrobacteria. Blot root explants for a few seconds on sterile filter paper. Transfer explants to 0.15/5 V750 K50 agar, taking care that the root explants are in close contact with the agar.
11. Incubate for 2 weeks in a growth room (after approximately 2—3 weeks tiny, green calluses appear on the yellowish root explants).
12. Transfer the explants to 0.15/5 V500 K50 agar. Two weeks later the green calluses start to form shoots. These shoots are often vitreous (watery) at first. This "vitreous" effect is due to the presence of vancomycin. It is advisable to lower the vancomycin dose down gradually to 250 or even 0 mg/l after 4—5 weeks of tissue culture.
13. Transfer morphologically "normal" shoots to GM in 15 cm Petri dishes (Falcon® 1013) to allow further development. Among several other types of plant containers, this type of Petri dish was found to give the most efficient shoot regeneration and seed production. They can easily be kept free of condensation compared with other containers and are higher (2.5 cm) than standard 15 cm Petri dishes.

Notes

6. — A "root *explant*" is composed of multiple root *segments* and can be considered as a transverse cutting of about 0.5 cm through an entire root system and *not* as a cutting through only one branch of a root system. When performing infections or washings the explants sometimes fall apart, but then 5—10 root segments are taken together for further incubation. The root explants should not be longer than 0.5 cm. Histological examinations show that *Agrobacterium* often enters the vascular cylinder of root segments at the cut surface. Hence the more cut surfaces, the more entrance sites for the bacteria.
   — For convenience with *Agrobacterium* infections and washings of root explants, little autoclaved plastic baskets (3—4 cm high, approximately 6 cm in diameter) are used which have a 100 μm nylon mesh in the bottom (Fig. 4). The root explants are put in these little sieves to be infected or washed. In this way the explants can be removed from the mesh very efficiently rather than "fishing" them out of the liquid medium. This reduces the time needed for infections and washings by at least a factor of 2.
7. Only *Agrobacterium* cultures that were grown in nonselective LB medium should be used.
11. During vancomycin counter-selection bacterial overgrowth of the root explants can occur. This problem arises when a fluid layer is present on top of the agar. Therefore always allow the 0.15/5 vancomycin/kanamycin media to dry completely (30 minutes) in the laminar-flow cabinet before closing them.
13. *In vitro* seed set of regenerants is positively influenced by low humidity, so condensation of the lid of the Petri dish should by all means be prevented. Condensation arises when the bottom of a plant container has a higher temperature than the top. This situation often occurs in culture room racks by the heat produced by the illumination source. Further, one should be careful to culture shoots at low density (maximum 3 per 15 cm Falcon® 1013 Petri dish) and at low temperature (approximately 20 °C). Temperatures above 22 °C should be avoided.

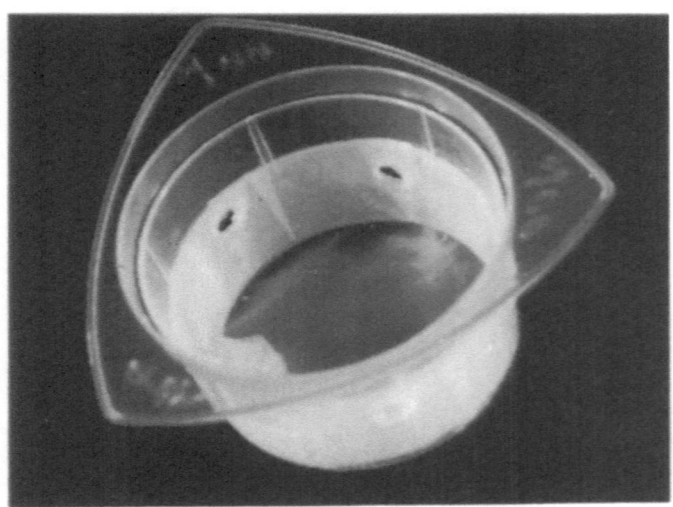

Fig. 4. Home-made basket containing a 100 μm mesh, which can be used for *Agrobacterium* infection and washing of *Arabidopsis* root explants.

Genetic tests for marker gene activity in T2 plants

The best criterion for demonstrating stable genetic transformation of a plant is the expression and Mendelian segregation of the selectable marker gene in the T2 progeny of T1 transformants. We describe both a selection and a screening procedure for transformed T2 seedlings.

Selection of transformed T2 plants
Steps in the procedure

1. Place T2 seeds on GM K50 agar in Petri dishes. Seal with Urgopore® tape.
2. Put the Petri dishes in the dark at 4 °C (refrigerator) for at least 4 days (stratification).
3. Incubate the germinating seedlings for 2 weeks in the growth room. Sensitive seedlings do not form roots nor leaves and develop white cotyledons; transformed seedlings are phenotypically normal.

Note
2. This step is not necessary if seeds were preserved for more than one week at 4 °C.

Screening for transformed T2 plants
Steps in the procedure

1. Remove young plants (4–6 leaves) from GM without damaging roots. Put plant in a sterile Petri dish.
2. Cut off half of the root system transversely.
3. Incubate the root explant on 0.5/0.05 agar. Put plantlet back onto GM. For this assay we use Falcon® 1013 (15 × 2.5 cm) Petri dishes. They contain a grid so that root explants and plantlets can be put in corresponding squares on either media.
4. After 4 days, transfer the root explants to 0.15/5 agar supplemented with 50 mg/l kanamycin in Falcon® 1013 Petri dishes (use corresponding squares).
5. About 4 days later, resistant roots can be distinguished from roots of sensitive plantlets with a high degree of accuracy. Roots of resistant plants bear many, long, straight root hairs whereas sensitive plants have at best a few shrunken ones. Another 5 days later, resistant roots turn green and begin to form shoots, whereas sensitive roots turn yellow and die. Greening of regenerating root tissue is a 100%-safe criterion for the determination of kanamycin resistance (Fig. 3).

Plant media

— GM (germination medium)
  1 × Murashige and Skoog salt mixture (Flow Laboratories)
  10 g/l sucrose
  100 mg/l inositol
  1.0 mg/l thiamine (stock 1.0 mg/ml)
  0.5 mg/l pyridoxine (stock 0.5 mg/ml)
  0.5 mg/l nicotinic acid (stock 0.5 mg/ml)
  0.5 g/l 2-(N-morpholino)ethanesulfonic acid (MES) (adjusted to pH 5.7 with 1 N KOH)
  8 g/l Difco Bacto agar
— 0.5/0.05 (callus-inducing medium)
  1 × Gamborg's B5 medium (without 2,4-D, kinetin, and sucrose) (Flow Laboratories)
  20 g/l glucose
  0.5 g/l MES (pH 5.7 adjusted with 1 N KOH)
  8 g/l Difco Bacto agar (for solid 0.5/0.05 medium)
  0.5 mg/l 2,4-D (stock 10 mg/ml)
  0.05 mg/l kinetin (stock 5 mg/ml)
— 0.15/5 (shoot-inducing medium)
  1 × Gamborg's B5 medium (without 2,4-D, kinetin and sucrose) (Flow Laboratories)
  20 g/l glucose
  0.5 g/l MES (pH 5.7)
  8 g/l Difco Bacto agar

5 mg/l 2ip (stock 20 mg/ml)

0.15 mg/l indole-3-acetic acid (IAA) (stock 1.5 mg/ml)

— 0.15/5 V750 K50. As 0.15/5, but supplemented with 750 mg/l vancomycin (Vancocin HCl; Eli Lilly) (stock 200 mg/ml) and 50 mg/l kanamycin sulphate (Sigma) (stock 50 mg/ml)

— 0.15/5 V500 K50. As 0.15/5, but supplemented with 500 mg/l vancomycin and 50 mg/l kanamycin

— GM K50. As GM, but supplemented with 50 mg/l kanamycin

Notes

— All media should be autoclaved for 15 minutes at 121 °C. Vitamins are added and the pH adjusted to 5.7 before autoclaving. Plant growth regulators and antibiotics are added after autoclaving and cooling of the media to 35 °C.

— Plant growth regulators are dissolved in dimethylsulfoxide (DMSO) and added to the autoclaved media without any further treatment. Antibiotics are dissolved in water and subsequently filter-sterilized using a 0.22 μm Millipore filter.

— The choice of the agar is extremely important for efficient regeneration. Several agars that we have tested proved to contain inhibiting agents or simply to have a bad structure, resulting both in slower regeneration and a severely reduced regeneration efficiency.

— Falcon® Optilux 1005 (10 cm × 2.0 cm) Petri dishes are used for pouring 0.5/0.05, 0.15/5, 0.15/5 V750 K50, and 0.15/5 V500 K50 agars. Falcon® 1013 (15 cm × 2.5 cm) Petri dishes are used for pouring the GM and GM K50.

— It is very important to allow the media to solidify completely in a laminar flow and preferentially to wait 15 minutes further before closing the Petri dishes in order to prevent the formation of a liquid layer on top of the medium.

— All Petri dishes were sealed with Urgopore® (Chenove, France) medical gas-permeable tape. Many other types of tape were found to have a negative effect on the regeneration efficiency.

References

1. Meyerowitz, E.M. (1989). *Arabidopsis*, a useful weed. *Cell* 56, 263–269.
2. Valvekens, D., Van Montagu, M., and Van Lijsebettens, M. (1988). *Agrobacterium tumefaciens* -mediated transformation of *Arabidopsis* root explants using kanamycin selection. *Proc. Natl. Acad. Sci. USA* 85, 5536–5540.
3. Gheysen, G., Herman, L., Breyne, P., Van Montagu, M., and Depicker, A. (1989). *Agrobacterium tumefaciens* as a tool for the genetic transformation of plants. In *Genetic transformation and expression*, L.O. Butler, C. Harwood, and B.E.B. Moseley (Eds.). Andover, Intercept, 151–174.
4. Hooykaas, P.J.J. (1989). Transformation of plant cells via *Agrobacterium*. *Plant Mol. Biol.* 13, 327–336.
5. Chaudhury, A.M., and Signer, E.R. (1989). Non-destructive transformation of *Arabidopsis*. *Plant Mol. Biol. Reporter* 7, 258–265.
6. Herman, P.L., and Marks, M.D. (1989). Trichome development in *Arabidopsis thaliana*. II. Isolation and complementation of the *GLABROUS1* gene. *Plant Cell* 1, 1051–1055.
7. Márton, L. and Browse, J. (1991). Facile transformation of *Arabidopsis*. *Plant Cell Reports* 10, 235–239.
8. Honma, M.A., Waddell, C.S., and Baker, B. (1990). Development of an *Ac/Ds* transposon tagging system in *Arabidopsis*. Abstract presented at the Fourth International Conference on *Arabidopsis* Research, Vienna (Österreich), p. 12.
9. Miller, J.H. (1972). *Experiments in Molecular genetics*. New York, Cold Spring Harbor Laboratory.

Plant Tissue Culture Manual **A9**: 1–32, 1992.

Somatic Embryogenesis in Carrot *(Daucus carota)*

ABRAHAM D. KRIKORIAN & DAVID L. SMITH
Department of Biochemistry and Cell Biology, State University of New York at Stony Brook, Stony Brook, New York 11794–5215, USA

Introduction

The culture of carrot cells in liquid suspension dates from 1953 and the recognition of their totipotency from 1956 [10]. By 1962 it was feasible to maintain in the laboratory, routinely, liquid cultures, heterogeneous as to their unit size, but in which large numbers of embryos readily developed from suspended cell clusters and single cells [8]. By this time, the role of synergistic combinations of the growth-promoting complex as it occurs in coconut water with auxins such as naphthaleneacetic acid (NAA) and 2,4-dichloro-phenoxyacetic acid (2,4-D) had become well-known [11, 23]. Moreover, the advantages to be gained in some otherwise morphogenetically recalcitrant cell cultures, of sequential treatments with different growth-promoting complexes and systems, became appreciated [23]. By these general means it was shown that a number of umbelliferous plants (family Apiaceae) and species or cultivars from other families, could yield cells and somatic embryos which in turn could give rise to whole plants. But when the main sequence of embryogenic development of carrot cells became known [24], it was found that its outcome could be greatly altered by the environmental conditions and the identity and mode of application of the growth and morphogenetic stimuli [11, 23].

As it became clearer that the innate totipotency of somatic (normally diploid) cells is a presumptively general property, the following problems emerged:

(a) cell cultures of some plants (natural species or cultivars) may prove to be so recalcitrant that somatic embryos either do not form or do so too infrequently for this to be of practical use;

(b) unlike the normal course of zygotic embryogenic development, which is very closely controlled in the environment of the ovule and the embryo sac, that of somatic embryogenesis is much more responsive to external conditions [23, 24]. In the outcome, therefore, somatic embryos may differ in varying degrees from their zygotic counterparts and are much more variable, even within a given *in vitro* culture, in both size and form. Therefore, it is a real problem for an investigator to adjust, and readjust, the environmental conditions and the composition of the ambient medium so that embryogenic cell cultures may resemble more closely the development of zygotes and also develop along their assigned course as predictably.

(c) since the various methods for isolating plant protoplasts are now well established [5], it should be possible to begin the somatic embryogenic process with naked somatic protoplasts, prepared from embryogenic carrot cells in culture. These should, in theory at least, resemble zygotes. However, to the present, it has proven unexpectedly difficult to promote, reliably, high fidelity somatic embryogenesis in free protoplast cultures of carrot (and any other species for that matter) even when prepared from highly embryogenic units [13, 24], that as such, will develop in high yield to well-formed somatic embryos and plantlets.

Totipotency in cultured cells

Somatic embryogenesis in cultured plant cells relates to the origin in small proembryonic clusters, cells or initials that give rise to shoot and root growing points. After this has occurred, embryogenic and subsequent development is subject to all the environmental factors and interactions that beset whole plants.

Some definitions

In the early period, somatic embryos were frequently referred to as "embryoids". This was to distinguish them from zygotic embryos. Use of the word "embryoid" has diminished over the years and its complete abandonment would constitute no great loss. We refer to embryogenically competent cells or cell clusters as preglobular stage proembryos (PGSPs). This follows Halperin who first used the phrase in connection with somatic embryogenesis [6]. The term is fully justified since, if PGSPs are generated in the manner to be detailed in this section, they are, indeed, precursors to globular stage embryos. Others have preferred to call them embryogenic cell clusters, proembryonic masses (PEMs), or proembryonic globules and argue that the term PGSP begs the issue whether a single cell in the cluster gives rise to embryos, or whether a group of cells yields the next (globular) stage. The fact is that until the culture methods become standardized and better understood, these issues will, for us, remain moot points [3, 4, 8, 24]. Whatever term is used to designate these units (be they single cells or cell clusters of varying size), the terms are justified only if the next stage can be demonstrated as occurring. If something is a called a PGSP, it should only be called a PGSP if it can progress to a globular stage embryo. Similarly, the term globular stage embryo is preferable to the earlier and more common designation, proembryo. If it progresses to the heart stage, then one is justified in calling it a globular stage somatic embryo. Similarly, a heart-shape stage progresses to a torpedo stage and that, in turn, to the cotyledonary stage. Aberrant or abnormal embryos may be defined precisely in morphological terms if that is possible. In this laboratory, the term "neomorph" has been used to designate abnormal embryonal forms. This word was invented by Waris [26] to describe severely deformed seedlings that had been achieved *in vitro* by the use of anti-metabolites and amino acids. For us, neomorphs represent terminal products of inaccurate delivery or processing of various signals in the embryogenic or developmental process. A key feature of neomorph status, as it were, is that gene mutation is not involved. The term "terminal product" seems justified since the aberrant form (neomorph) does not progress directly to a plantlet. They invariably do not survive to become plantlets because their growing points are absent or ill-formed. On the other hand, explanted tissues from neomorphs can be induced to de-differentiate and, provided the environment is such that embryogenic signals are correct, cells from the neomorph can yield somatic embryos. These, in turn, can make plantlets. All this suggests that neomorphosis can be used as a tool to dissect the epigenetic and physiological aspects of developmental pathways.

Somatic embryogenesis protocols for carrot

The classic literature [1, 2, 6, 8, 10] reveals that the culture procedures were developed in a largely empirical fashion and leave something to be desired in terms of reliability. The fact remains, moreover, that despite the use of carrot to study somatic embryogenesis for a third of a century [10], and its being viewed by many as a routine and well-established procedure (so much so that it is frequently cited as the "model system"), there are fewer principles that can be grasped in terms of why a given procedure "works" than one might hope for. Because it was the first case where somatic embryogenesis was encountered *in vitro*, an extensive literature indeed exists [1, 2, 3, 4, 7].

Nevertheless, it will be helpful to recapitulate in some detail how embryogenic carrot cultures may be established using these nominally long-available procedures. But, since many laboratories have used various protocols, it is hoped that investigators will see opportunities to probe further both the "older" and "newer" methods to be presented and thus through comparative studies be able to gain better understanding of the full range of controls that must be in place in developing and growing plants.

Initiation of primary cultures

Source of explants

Cells derived from carrot root secondary phloem explants are embryogenically competent [23, 24]. Similarly, zygotic embryos from developing or mature carrot "seeds" (botanically they are "mericarps" or fruits) also yield vigorous cultures capable of organizing. Nevertheless, a real problem in all of this is to have materials which are not only embryogenic but which maintain over time a high level of response to the stimuli which allow them to organize. It may come as a surprise to some investigators seeking to work with carrot that considerable work must be done to enable one to select clones that respond in a predictable and desirable way. In this laboratory, over the years, selection procedures have been applied to cultures (primarily suspensions) initiated from the secondary phloem and even the cambium of cultivated carrot root [24]. They have also been applied to cultures started from explanted tissues of aseptically germinated carrot seedlings such as hypocotyls. Whereas only minor differences in morphogenetic response have been detected between cultures derived from cambium or secondary phloem, differences are detectable in cultures derived from seedlings of several cultivars. For instance, although Gold Pak, Royal Chantenay, Imperator Long, Scarlet Nantes, Spartan Delight, Hi Pak etc. all respond and produce somatic embryos, they have different requirements [24] and whether one seeks to study one or the other cultivar will depend on one's objectives. This is an important point and underscores the view that "model" systems are dependent on objectives.

"Seeds" or mericarps

Two types of carrot "seed" have generally been used for establishing embryo-genic or totipotent carrot cultures: (1) *Daucus carota* var. *carota* L. (the Wild Carrot or Queen-Anne's-Lace) and *Daucus carota* var. *sativus* (the cultivated carrot) [17]. In the former case, the plants grow wild in open fields, meadows and roadsides, and "seeds" may be readily harvested. It is essential to establish on the spot, however, that the mericarps contain embryos. A common pest of carrot is the "tarnished plant bug" (*Lygus* spp.). This insect is able to selectively suck out the developing embryo in a "seed" without leaving signs of obvious damage. A mericarp presumed to be healthy may contain no embryo [14]!

"Seed" sterilization

The two following procedures provide, step-wise, instructions for the surface decontamination or disinfestation of carrot "seeds"- assuming that no micro-organisms are present within the boundaries of the fruit wall.

Procedure 1 is the easiest and quickest method if the "seeds" are to be plated out for germination and if filtering pans with autoclavable (stainless steel) sieves are available. (In the USA, Cellector Tissue Sieves by E-C Apparatus 3831 Tyrone Blvd, North St Petersburg, FL 33709 have been satisfactory. See Fig. 1 on p. 17 for the appearance of these.)

Procedure 2 is the preferred method if zygotic embryos are to be dissected from the "seeds". Each procedure will disinfest approximately 500 "seeds" (there some 1000 per 10 ml glass beaker).

Procedure 1

Steps in the procedure

1. Sterilize one 250 ml Erlenmeyer flask containing 50 ml of glass distilled water (or equivalent), two 500 ml Erlenmeyer flasks containing 250 ml each of water, filtering pans with sieves comprising either a # 60, # 50, or # 40 mesh screen (see Table 1, p. 12) supported on a 400 ml glass beaker (protect filtering pan and beaker with aluminum foil), and one 100 mm diameter glass Petri dish per batch of 100 ''seeds'' being disinfested.
2. Add 50 ml of commercial bleach (e.g. Clorox which contains about 5.25 percent w/v sodium hypochlorite) and a drop of a surfactant such as Tween 20 into the 250 ml Erlenmeyer flask containing 50 ml of water.
3. Carefully pour the desired number of ''seeds'' into the 250 ml flask directly from the seed package, or from a small (10 ml) beaker and cover. Five ml is about 500 ''seeds''.
4. Briskly swirl every 15 minutes for 1.5 hours.
5. Pour ''seeds'' onto the filtering pan sieve and allow to drain.
6. Hold filtering pan over a one liter beaker and slowly pour the contents of both 500 ml flasks over the ''seeds'' to rinse off the hypochlorite.
7. Turn the filtering pan over and gently tap out the ''seeds'' into the bottom of the glass Petri dishes, or, alternatively use a sterile scoop to transfer the ''seeds'' into the Petri dishes.

Procedure 2

Steps in the procedure

1. Sterilize seven 250 ml Erlenmeyer flasks, one containing 50 ml, the others 100 ml each of distilled water and one glass Petri dish (100 mm diameter) per batch of 100 "seeds".
2. Add 50 ml of commercial bleach and 1 drop of a surfactant to the 250 ml flask containing 50 ml of water.
3. Carefully pour the desired number of "seeds" into the flask directly from the seed package, or from a small beaker and cover.
4. Briskly swirl every 15 minutes for 1.5 hours.
5. Carefully pour out the solution after flaming the neck of the flask. Some "seeds" will fall out but not many if you're careful. Again flame the neck of the flask.
6. Pour in 100 ml of water from the other flasks one at a time repeating step 5 after each addition. The last addition of water remains in the flask to allow the "seeds" to soak.
7. Soak the "seeds" for at least two hours, or as long as overnight. Approximately six hours is optimum.
8. Drain off the water.
9. Using a sterile scoop, place approximately 50 "seeds" into a sterile Petri dish for dissection under a stereoscopic microscope, for plating or germination.
10. Embryos may be excised from the "seeds" with two pairs of fine point forceps (e.g. Dumont # 3) or a single pair of forceps and a scalpel. This can be done by nicking the pointed end of the "seed" and pushing the embryo out from the other end. With practice, this procedure can be carried out readily. The embryos are then placed on an appropriate culture medium. For culturing either "seeds", embryos or seedlings, any of a number of culture vessels may be used.

Initiation of primary cultures

Media used in initiation of seedlings as a source of primary explants

The media most commonly used for germination of seeds and culture of embryos are one-half strength basal Murashige and Skoog salts (abbreviated in this laboratory as $B_{MS1/2}$), "vitamins" (micro-organics) [12] and iron chelate [15], or one-half strength modified White's salts (abbreviated as $B_{W1/2}$), iron, and "vitamins" [16], each supplemented with 0.5% w/v sucrose. B_{MS} is usually adjusted to pH 5.4; B_W to pH 6.2. B_W can profit from addition of 200 mg/l of casein hydrolysate (enzymatic digest or acid hydrolyzed). Normally B_{MS} is supplemented with 3% w/v sucrose; B_W with 2%. The low amount of sucrose (0.5%) used in the germination of seedlings has been shown to give a more uniform response. Indeed, this is added to permit disclosure of any microbial contaminants. Coconut water, abbreviated CW, (10% by volume) has frequently been used as an additive to B_W. It is also sometimes used in conjunction with full strength Murashige and Skoog salts, iron and vitamins (B_{MS}), especially with excised embryos. Coconut water seems to produce a more vigorous seedling.

Media utilized in initiation of callus

If callus is desired from the "seed", embryo or seedling, growth regulators such as NAA or 2,4-D (either at 2 mg/l, i.e. 10.8 μM or 9.0 μM respectively) may be added to the full strength salts. Seedlings are allowed to germinate in darkness at about 22 °C on basal medium. After about 10 days, when seedlings have reached a height of about 8 to 10 cm, including the root, hypocotyl and two (or three) etiolated cotyledons, they are suitable for culture. (That is not to imply that earlier stages or light-grown ones cannot be used). The seedlings are usually cut into sections approximately one cm long, and are then gently wounded with a scalpel along their length, and then placed in or on the culture medium. The sections may be placed on semi-solid medium in culture vessels or Petri dishes or directly into liquid medium.

Normally, an inoculum should comprise the tissue from a single seedling. This implies that a clonal population is being sought. The disadvantage of a culture derived from a single source is that it takes a longer period to obtain a large enough population of cells to work with but it provides the only means whereby one can establish specifics of performance of a given clone.

Since carrot suffers from inbreeding depression, there is little opportunity at present to work with isogenic stocks. By using clonal lines, one can, hopefully, come to a better understanding of control processes in embryo development. The vast majority of embryogenic cultures of carrot (and other species as well) have not, however, been clonally derived and hence represent mixtures of cells that may have very different characteristics.

Initiation of embryogenic cultures is best carried out in darkness, although dim light will be satisfactory in some cases. This will depend on the clone in question [24]. Within a week, or two at the most, signs of globular stage embryo production should be apparent. These will be discernible by their "pearly" white appearance. Their numbers and place of production on the primary explant will vary depending on how a culture is initiated. If a petiole or hypocotyl explant is used, for example, the cut or wounded ends will show the somatic embryogenic cells first.

Recognition of these small, pearly white and globular structures (globular stage embryos) is crucial to achieving an embryogenically active culture. These globular stage somatic embryos must be removed and multiplied via subculture under appropriate conditions to get a dense or vigorous culture, be it in suspension or on semi-solid medium. In addition to globular stage embryos giving rise to other globular embryos (proembryonic globules), they may shed, especially in liquid, embryogenic cell clusters (so-called proembryogenic masses or preglobular stage proembryos, PGSPs).

It is right at the beginning that choices in medium must be made, i.e. the exact auxin concentration to be used, whether or not to include coconut water or a chemically identified cytokinin in its stead, or whether to subculture in the same medium or to switch to another one, etc. Empirical evidence shows: somewhere in the culture protocol the use of basal medium plus an auxin is beneficial in yielding cells which can undergo embryogenesis more readily (apparent when the cultures are later evaluated for their embryogenic potential; this turns out to be "sooner rather than later"). Probably, the use of auxin at the outset is best, whether on semi-solid or in liquid medium. But if auxin is used at the outset, the medium must eventually be changed to include coconut water or a cytokinin, because with auxin by itself, the suspensions do not grow well for long, and indeed, are rarely maintainable with the right qualitative characteristics for more than three or four subcultures. That is, the suspension grown in medium supplemented only with auxin predominantly yields relatively large globular embryos, and there are few, if any, small isodiametric cells or tight clusters comprised of only a few (4–6) cells. These latter cell types are the ones that are most embryogenically responsive [24, 25].

Actively embryogenic suspensions have also been obtained over the years in this laboratory by placing primary explants of seedling sections directly into liquid medium with an auxin and coconut water. This can save at least a couple of weeks in establishing suspensions.

The basal salts medium of Murashige and Skoog [12] (i.e. a high nitrogen medium) promotes rapid proliferation of cells and is well suited to the initiation of growth from explants; it is also useful later as a medium in which totipotent cells may be placed to test their embryogenic potential. The B_{MS} may be inappropriate, however, for some cell lines and in these cases it is not possible to maintain the required level of competent cells that can develop into somatic embryos when placed in an auxin-free medium. Rather, the cells tend to grow into amorphous callus masses with no externally visible organization. Modified White's medium (B_W) [16] seems to be better suited for most routine and sustainable subcultures of embryogenic suspensions provided the appropriate growth regulators (e.g. CW, 2,4-D or NAA and casein hydrolysate) are added.

Procedures used for subculturing cells in suspension and evaluating them for their embryogenic potential

After the primary explants placed on semi-solid medium have callused and localized areas have produced globular stage embryos, the pearly white sectors may be transferred or subcultured onto semi-solid medium, or this embryogenic callus may be placed in liquid where, in time, it will produce a suspension. Seedling segments directly placed in small flasks usually produce a suspension in four to five weeks when the medium contains a cytokinin and auxin, and in approximately six weeks when it does not. Suspensions can be achieved earlier by combining contents of several flasks or increasing the number of units comprising an inoculum in a culture vessel but, as suggested above, it is better to obtain clonal lines for studies of development.

Routine subculturing of material grown in flasks may be done in the following way

Ideally, suspensions should be separated into fractions of known size by sieving (filtration) of the suspensions through a series of graded stainless steel or nylon sieves. Embryogenically competent suspensions are readily grown and multiplied as small cell clusters in various media, and, as stated above, these media may comprise any of several formulations comprising inorganic salts, sucrose, "vitamins", coconut water and an auxin, usually either 2,4-D or NAA. Lowering or removing the auxin from a totipotent cell culture-maintenance medium such as that of White and transferral to a hormone-free (H-F) one that utilizes the salts and vitamins of Murashige and Skoog [12] leads to the production of later stage somatic embryos [1, 2, 6].If coconut water is used, it and the auxins are usually removed, although the removal of auxin is by far the more important step. Presence of auxin inhibits progression of globular stage embryos to later stages of development.

Somatic embryo formation depends on a number of things including the line of carrot employed. In certain lines there is only a very low level of response, with unorganized clumps also being present in the suspension. This also tends to be the case when the auxin concentration is only lowered but not removed completely. Frequently, in a suspension that has had growth regulators removed via washing in H-F medium or subculture to a H-F medium, the newly formed embryos grow very rapidly and pass quickly through the embryo stages into plantlets. Another noteworthy fact is that with the removal of hormones, the cell culture tends to become very sparse and usually is lost after two passages on medium with no hormones. (More information about the use of H-F media is described later). This is probably due to the fact that somatic embryos do not (under those circumstances) retain their ability to slough off the new cells that would enable the maintenance of the suspension even as somatic embryos were being produced. Carrot suspensions in the presence of hormones normally comprise embryogenic cell clusters and the extent to which these can be recognized by a skilled investigator as embryogenic cells, and that may be equated with certain early stages of zygotic embryogenesis such as the globular stage embryo stage or the preglobular proembryo stage, is a moot point.

Be that as it may, development of cells and clusters into somatic embryos and plantlets can proceed readily with cells obtained by either of two procedures. 1) The fraction that has passed through a crude filter such as that provided by a single or double layer of cheesecloth stretched over a 400 ml beaker and supported by a taught string may be used, or 2) further procedures may be adopted using the cheesecloth filtrate first, to yield a more refined fraction. As an example of the latter: the cheesecloth filtrate can be poured through sieves of different mesh sizes such as # 100, # 200, # 400 (see Table 1).

Table 1. Some examples of screen sizes as used in the protocols presented. Note: The nominal dimensions of mesh (wires/inch) and pore size do not correspond exactly to US Bureau of Standards Sieve sizes but values are close.

Mesh (#)	Pore Size (μm)	US Standard Sieve (μm)
20	860	841
30	520	545
40	380	420
50	280	297
60	230	250
80	190	177
100	140	149
200	73.7	74
300	45.7	not available
400	38.1	37
500	25.4	25

The resulting material may be collected in two different ways: (1) by allowing the filtrate to stand so that the cells settle, followed by pouring off the supernatant; or (2) after pouring material through a series of filtering pans, the material can be collected directly off the smaller mesh by transferring the residue on the mesh into a beaker with a "rinse" medium (comprising the medium into which transfer or platings/plantings are to be made), thus producing a suspension of cells and/or clusters in a particular size range, e.g. 74–140 μm. In either case, the material finally collected is poured into a capped centrifuge tube and centrifuged at low speed, approximately 100–300 rpm ($2-15 \times g$) for about 10 minutes. The supernatant is poured off and the remaining cells and clusters may be further washed in the centrifuge tube with the rinse medium, re-centrifuged, etc., or it may be put into a culture medium directly without carrying out additional washes. The suspension in the centrifuge tube is brought up to volume using the same medium that is to be used as the test medium. The final volume to be achieved can be estimated from experience by judging (by use of graduated test tubes) the density of the cells. This may also be quantified by counting an aliquot of cells.

All this means that when a culture is active and hormone is provided in the maintenance phase, few if any late stage somatic embryos are produced. Accordingly, it is a relatively simple matter to maintain early stage embryonic suspensions via subculture. Hormone-maintained suspensions are poured through cheesecloth or sieves to achieve filtration. Inocula can be used from those fractions or the fraction which does best in a particular setting. A volume consisting of some 1 to 10 ml (some 1 g fresh weight at the higher volume) may be used to inoculate a flask which accommodates 225 ml of liquid medium. Since the auxins generally prevent further development or progression of the somatic embryogenic process, it follows that if 2,4-D or NAA is subsequently removed, numerous somatic embryos will form. Also, it is at the time of transferral to H-F medium that the largest number of somatic embryos are produced. These are, however, inevitably of different sizes and hence must be separated by sieving if they are to be suitable for detailed investigation.

Isolation and fractionation of embryos

Embryogenically competent suspensions of carrot containing mixed populations of materials, whether they be competent cells or cell clusters, or somatic embryos ranging from freely suspended preglobular stage proembryos or globular stage embryos to fully formed somatic embryos can be separated as follows. Fig. 1 (p. 17) provides an abbreviated schematic version of some of the key steps in the process.

Steps in the procedure

1. Flame the neck of the culture flask and allow to cool for about 30 seconds.
2. While the neck of the flask cools, remove an aluminum foil protective cover from a 250 ml glass beaker and place a # 20 (864 μm pore size) filtering pan over the beaker.
3. Pour the contents of the flask through the # 20 screen into the beaker. The largest somatic embryos and plantlets are retained on this size sieve.
4. Wash the sieve with a rinse medium consisting of the salts and vitamins and sucrose to flush through any smaller embryos remaining in the larger mass. Usually these larger embryos retained on the # 20 screen are discarded. A small amount of the embryo suspension from the original culture flask may be put aside in a beaker instead of passing the entire contents of the flask through the # 20 sieve. This material may be used later as a source of inoculum for a new maintenance culture flask.
5. After a few minutes, most of the embryos in the filtrate settle to the bottom of the beaker. Some of the upper filtrate can be decanted so that the volume of the filtrate does not become too large. This might occur because rinse medium is added following each sieving. This decanting step can be done at the same step throughout the procedure.
6. Repeat the same procedure of sieving by pouring the filtrate from the # 20 sieve through a # 40 sieve (380 μm) which also rests on a 250 ml beaker. The material retained on the # 40 sieve is washed with rinse medium.
7. Collect the filtrate in the beaker, cover with foil and set aside while the # 40 sieve with retained embryos is flipped over and placed on a sterile 400 ml glass beaker.
8. Pour rinse medium through this sieve to rinse off these embryos into the 400 ml beaker.
9. Pour the embryos then into a 35 ml conical centrifuge tube and label # 20–40. A small aliquot of this embryo suspension can be put into a small Petri dish (e.g. a Falcon plastic 1006, 50 mm diameter) for observation under an inverted microscope.
10. Pass the filtrate in the beaker from the # 40 sieve then through a # 60 sieve in a similar manner and rinse with fresh medium.
11. This filtrate is again set aside and the sieve with retained embryos again flipped over and the embryos collected in a 400 ml beaker, then poured into a centrifuge tube and labelled # 40–60.

12. Repeat this procedure with a # 80 (190 μm) sieve and then a # 100 (140 μm) sieve with embryos collected from each sieve, yielding # 60–80 and # 80–100 fraction.

13. Pour the filtrate that passes through the smallest sieve, the # 100 sieve, into several centrifuge tubes and label < # 100.

By this process embryos have now been collected in the centrifuge tubes in the following ranges: # 20–40, 380–860 μm; # 40–60, 380–230 μm; # 60–80, 230–190 μm; # 80–100, 190–140 μm; < # 100, < 140 μm.

Sieves of intermediate size can also be used, i.e. # 30, 50, 70, etc. to achieve whatever size ranges are desired. Fewer sieves, of course, can be used to achieve larger size ranges.

The < # 100 fraction is comprised of small embryos at the heart stage, as well as smaller units and cells. Thus, this fraction usually requires centrifugation to concentrate the embryos and cells. This is done by centrifugation at about 300 rpm for 10 minutes. The fraction of the larger sizes generally settles quickly and completely, and thus does not require centrifugation. The supernatant is then carefully poured off all the tubes and the embryos are re-suspended in fresh medium without hormones or coconut water. The amount of medium to be added is estimated by determining the density of the embryos. The embryo concentration at this point can be quantified by withdrawing an aliquot, say one ml, pipetting it (make sure it has an appropriately wide mouth) into a small Petri dish and counting the number of embryonic units. Care must be taken to ensure that a well-mixed sample is withdrawn, particularly in the case of the larger embryo sizes, since they rapidly sink to the bottom of the tube or pipette. When the count is made, the density can be adjusted further by adding more medium or units and a second count carried out, if necessary.

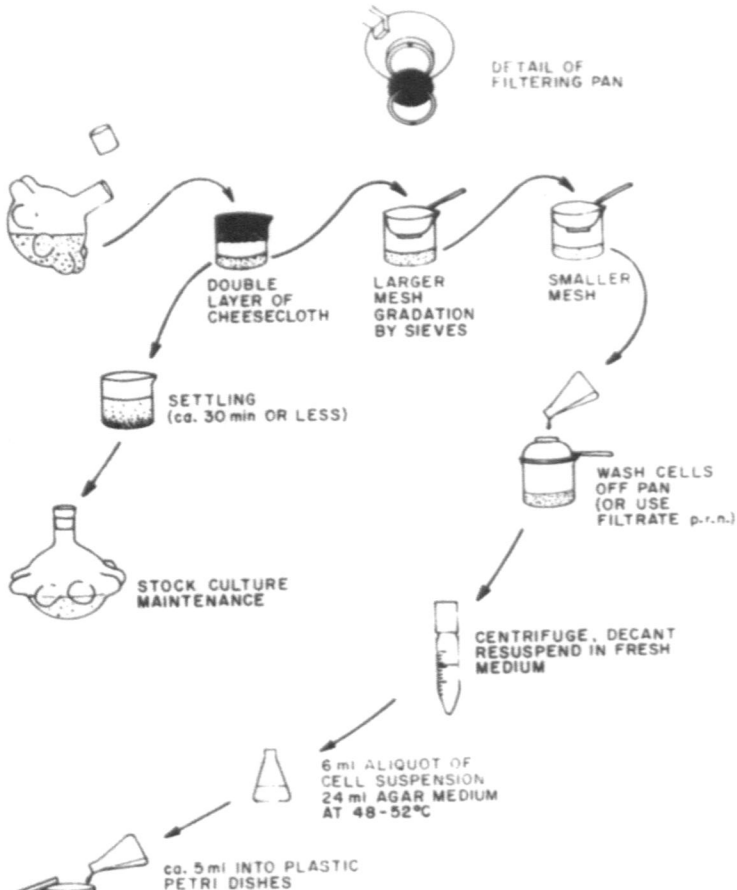

Fig. 1. Diagrammatic representation of procedures followed in filtration of cultured cells grown in suspension to uniform unit sizes for experimentation such as plating in an agar medium. In this illustration, a so-called "nipple flask" [24] has been used to culture the cells; an Erlenmeyer flask serves equally well. Similarly, materials grown on semi-solid media may be suspended in nutrient medium and subjected to separation procedures. p.r.n. means 'as needed'.

Culture of somatic embryos

The somatic embryos are now ready for experimentation in various media and cultures vessels, or for biochemical analysis etc. A typical experiment in semi-solid medium in small plastic disposable Petri dishes may be performed by mixing 6 ml of somatic embryos taken from a centrifuge tube with 24 ml of medium to which autoclaved agar-containing medium has been added. This is done by thoroughly mixing the embryos in the centrifuge tube by swirling and withdrawing 6 ml by pipette. The 6 ml are then introduced into 24 ml of agar medium (the concentration of agar to be used will have to be determined empirically!) which has been cooled to approximately 45 °C. (As noted above, B_{MS} supplemented with "vitamins", and sucrose will sustain the continued growth of these somatic embryos.) Other additions may be tested as desired. The agar must be swirled gently but thoroughly to ensure complete mixing; care must be taken to exclude air bubbles during mixing and final distribution of the tube contents into the final culture vessel. If Falcon # 1006 Petri dishes or an equivalent are to be used, 5 ml per dish is an appropriate amount. It is important to obtain uniform distribution of embryos in the dish. This may be done by pouring the agar-embryo mixture so that approximately two-thirds of the Petri dish is covered; then the dish is quickly tilted so that the remainder of the bottom of the dish is covered. With practice this results in an even distribution of somatic embryos. The snap-tight lids are then attached, and the agar is allowed to harden. Once the agar has hardened, the dishes can be stacked and manipulated.

A count of "X" somatic embryos per one ml in the centrifuge tube is equivalent to "X" embryos per dish in agar medium. Diluting 6 ml of somatic embryos with 24 ml agar medium is equivalent to each ml in the centrifuge tube being diluted 5-fold. Since 5 ml of agar-somatic embryo medium are poured into each Petri dish, each dish contains one ml of the centrifuged embryo suspension. This procedure, even though time-consuming, permits somatic embryos at varying stages of development to be analyzed, grown and manipulated or further evaluated.

Embryogenic suspensions may be passed, successively, through sieves with smaller pore size. Fractions that pass through a # 500 screen (pore size < 25 μm) can be grown but may require extra supplements and care.

A summary evaluation of some key observations and problems of somatic embryogenesis using cell cultures of carrot or other umbellifers according to the "classic" or "conventional model" system described above is outlined briefly as follows:

(a) in the first instance, there has been (and for the foreseeable future will remain) the outstanding question as to where the embryogenic cells come from [25]. One is interested to know whether cells which can develop into somatic embryos are induced to be embryogenically responsive by a specific *in vitro* protocol or regimen, or whether cells which for one reason or another are already embryogenically competent and therefore need only be selected for *in vitro* via permissive media, and cell populations or stocks built up from these for further manipulation. The point here is that in the first instance one is dealing with induction, followed by further manipulation of the circumstances or conditions of expression; in the second, it is a matter of selecting predetermined embryogenic cells or units, building up stocks or populations and manipulating these. While an understanding of the modulation of expression of later stages in somatic embryogenesis is important, it is perhaps a more fundamental question to ask what makes cells embryogenically competent in the first place? Or, conversely, what prevents somatic cells from expressing their presumed ability to become somatic embryos?

Whatever the correct answer to the above question may be, a number of points have emerged from study of somatic carrot embryogenesis in the now "classic" system.

(b) The parallelism with the zygotic embryos is closer if:

(i) their development stems from very small units.(A culture that is heterogeneous as to size fosters the normal development of its smaller ones; a culture rendered uniformly composed of small units by sieving profits by the addition to it of some medium that has been "conditioned" by the growth in it of a large heterogeneous crops of somatic embryos.)

(ii) if the total osmotic concentration of the medium (achieved e.g. by sorbitol) is high and comparable with fluids of embryo sacs which may be in the order of 12–14 atmospheres. (This keeps the somatic embryos small and favors organized growth in contrast to random proliferation.)

(c) the origin and development of small, globular, proembryonic units of carrot in suspension cultures is then fostered by:

(i) a period of dark induction after the cell cultures have been grown and multiplied in continuous light.

(ii) the use, in the ambient medium, of some liquid "pre-conditioned" by the growth of somatic embryos in densely inoculated cultures heterogeneous as to their unit size.

(iii) the duration of the dark induction and the degree of "conditioning" of the medium for somatic embryo formation interact and so promote the embryonic development of the smallest clusters (174 μm which consist of small free cells and cell clusters).

(iv) by withholding reduced carbon (sucrose), while maintaining high osmotic values by use of sorbitol or an equivalent osmoticum in the medium, proembryonic cell cultures of carrot may be maintained in a "poised" state (i.e. their further development is arrested) until these conditions are reversed.

(v) the cells of the pro-embryonic globules that develop under these conditions into somatic embryos are small, with a few small vacuoles and a cytoplasm which is notably free of unusual inclusions but is rich in microtubules [24, 25].

However, despite all the above it has been very clear for a long time that much more needs to be done to prescribe fully the growth factor requirements and the external conditions which are most conducive to somatic embryogenesis under *in vitro* conditions [3, 7].

Some recent alternatives or modifications to the "classic" embryogenic carrot system protocols

It has long been suspected that the mineral element composition of culture media influences virtually every aspect of *in vitro culture* including initiation, maintenance and development of somatic embryos. Most often though, these suspicions have been ignored and emphasis has been placed on the extent of somatic embryo formation, and this only when exogenously added growth regulators were used to initiate and maintain the cultures. Also, considerable attention has been given to developing protocols that can yield somatic embryos, which in turn, develop into mature plants phenotypically identical to the plant from which the original explant was derived. Moreover, the require-ments for, or the ability to respond to, certain mineral elements and other media components has widely been thought to be controlled by the exogenously added growth regulator(s) and that these requirements then direct the developmental fate of those cells. Hence, there has been little incentive to investigate inorganic components in isolation from the more usual organic, hormonal additives.

The procedures to be described below show that the classic or model carrot system, as described above, generally thought in the past only to be achievable in a reliable way via added growth regulator(s), is controllable with high confidence without the use of any added growth regulator(s).

Hormone-free (H-F) procedures

Using mericarps

1. Surface sterilize "seeds" of a carrot cultivar such as "Scarlet Nantes" (see seed sterilization procedures 1 or 2 above on p. 4 and following) and place them directly onto semi-solid or in liquid medium (see Fig. 2 on p. 25 for a schematic representation of the responses in cultures to be described).
   All tests are carried out in 15 mm × 100 mm plastic Petri dishes containing 40 ml of various media made semi-solid by the addition of 0.7% agar or in 125 ml Erlenmeyer flasks containing 25 ml of medium.
2. Inoculate onto each dish either five "seeds" (number 1 in the scheme) or five "seeds" each devoid of their zygotic embryo after germination (number 3) but still retaining their endosperm. (Mericarps rendered free of their embryos by dissection are a poor tissue source but will work). Culture conditions may include either continuous darkness, or light at 22 °C [18]. Under these conditions, radicles emerge at approximately day 5 (number 2).
3. Physically separate the seedling (number 4) from the "seed" after day 7–10 (number 3). At day 7–10, a small non-proliferating callus is observable on nearly all fruits at the site of embryo exit if the mericarp remains on the culture medium (number 3a). Proembryos are observable as small 0.5 mm to 1.0 mm diameter, whitish globules at day 30–35 (number 3b).

At this time the proembryos can be left undisturbed (3b to 3c) or can easily be separated from the mericarp (3b'-5 or 3b"-3h") with the same results. As illustrated in Fig. 2 (3b'–5), normal development from the proembryo (3b') through the heart stage (3c'), torpedo stage (3d') stage, cotyledonary stage (3e') and the plantlet is possible on certain basal media.

Figure 2 (3b"-3h") also illustrates that normal development up to the torpedo stage (3d") can result in abnormal cotyledonary stage embryos (3e^{III}-3^{VI}) which in turn can give rise to a secondary somatic embryogenesis system. Proembryos, depicted in 3f", arise from the axis of pre-existing cotyledonary stage embryos. The secondary proembryos give rise to both somatic embryos and adventitious shoots (3g") that again give rise to more proembryos (3h") in a cyclical manner (3h"-3b").

For example, certain media foster embryo growth that follows pathway 3b"-3e^{V} thus resulting in "neomorphs", i.e. embryonic forms which do not go on to further developmental stages unless they are cultured on an appropriate basal medium prior to the torpedo stage of their development (horizontal arrows 3c" to 3c').

The mineral element composition of the basal medium quantitatively affects zygotic embryo "germination", and the number of mericarps giving rise to proembryos. It furthermore qualitatively affects the size of the proembryos, morphology of the cotyledonary stage embryos and the continued growth of embryos into plantlets or the initiation of cyclical secondary embryo formation.

Table 2 provides a base or skeleton formulation of salts to which nitrate or ammonium may be added for experimentation. In this laboratory, the base formulation is referred to as "DS-5a salts". For tissue maintenance as PGSPs, add 1 mM NH_4Cl (54 mg/l) and buffer with 25 mM MES (free acid) (5000 mg/l) titrated with potassium hydroxide to pH 4 to 4.2. Alternatively, add 10 mg/l of casein hydrolysate (acid hydrolyzed). For continuation of the embryogenic process beyond the PGSP stage, add 5 mM NH_4Cl (270 mg/l) or 5 mM NH_4OH and buffer with 25 mM $MES.H_2O$ free acid (5000 mg/l) at pH 6.0. The former may be referred to as "maintenance medium" and the latter to as "embryo growth medium". Further points concerning other additions to the medium such as activated charcoal will be covered below.

Table 2. DS-5a "Salts" or Base Formulation. Note: nitrogen source and buffering to be carried out as needed. (See text for details).

	mM	mg/l
$Ca(H_2PO_4).H_2O$	1.0	252.10
$MgSO_4.7H_2O$	1.0	246.49
NaCl	0.5	29.23
H_3BO_3	0.1	6.18
$MnSO_4.H_2O$	0.1	16.9
$ZnSO_4.H_2O$	0.01	2.88
KI	0.005	0.83
$Na_2MoO_4.2H_2O$	0.0001	0.024
$CuSO_4.5H_2O$	0.0001	0.025
$CoCl_2.6H_2O$	0.0001	0.024
$Na_2FeEDTA.2H_2O$ $FeSO_4.7H_2O$	0.1	*
thiamine HCl	0.0037	1.0
sucrose	58.4	20,000.00

* 27.8 mg/l $FeSO_4.7H_2O$; 37.23 mg/l $Na_2EDTA.2H_2O$ or 33.6 m/l Na_2EDTA.
See [15 and 16] for details of chelated iron solutions.

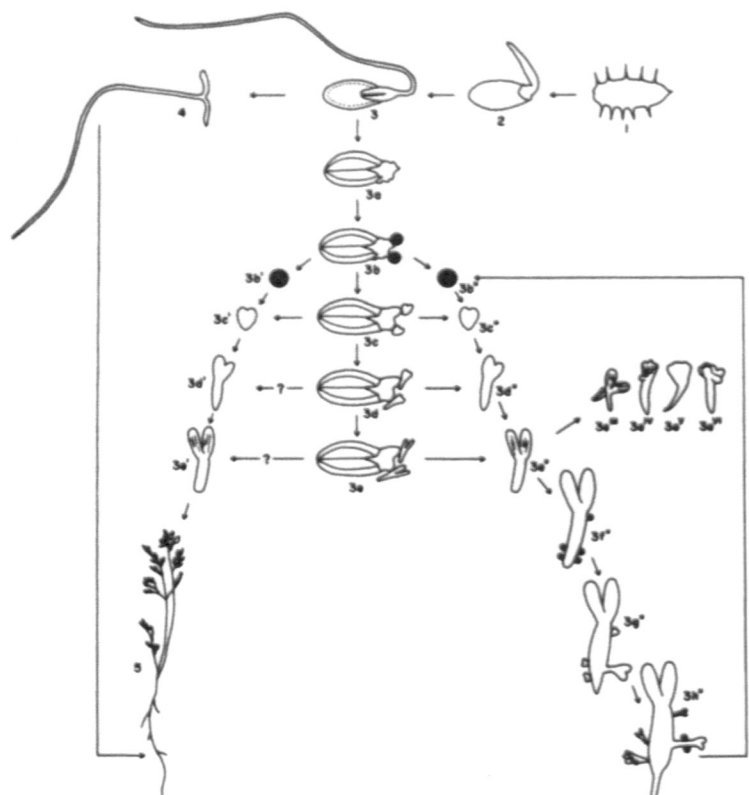

Fig. 2. Somatic embryogenesis of carrot from mericarp tissue in the absence of exogenous hormone. See text for details. The two question marks indicate that these steps have not been investigated although there is no reason to believe that they would not proceed in a predictable fashion.

Using zygotic embryos

Early work suggested that mericarps were a dramatically better source of embryogenically competent cells than zygotic embryos. Moreover, only wounded zygotic embryos were able to yield somatic embryos and only a few at that [18].

1. Attempts to increase somatic embryo production from wounded zygotic embryos were realized, however. Experiments showed that H-F nutrient medium, containing 1 mM NH_4^+ as the sole nitrogen source, fostered production of somatic embryos from wounded zygotic embryos to the same high degree found with mericarps. Intact embryos or seedling hypocotyls, on the other hand, virtually never produce somatic embryos [19].
2. Somatic embryo formation from wounded zygotic embryos occurs mainly from the cotyledons, but root tips and hypocotyls (also with shoot tips present) of mature zygotic embryos are responsive as well [19].
3. On media containing unreduced nitrogen, somatic embryo formation leads to the generation of vigorous cultures entirely comprising somatic embryos at various stages of development which in turn proliferate still other somatic embryos [19].
4. Growth on nutrient medium containing 1 mM NH_4^+ prevents the initially formed somatic embryos from developing into later stages, but does not prevent cell multiplication. This apparent inability to continue development results in cultures consisting entirely of preglobular stage proembryos.
5. This medium containing 1 mM NH_4^+ as the sole nitrogen source apparently does not induce somatic proembryo formation. Instead, the medium is best thought of as permissive to the expression of certain embryogenically determined cells within zygotic embryos, especially those in the cotyledon [20].
6. Histological examination has confirmed that preglobular stage proembryos from wounded zygotic embryos are maintained on 1 mM NH_4^+-containing medium.
7. The first-formed somatic embryos continue development into later embryo stages, without continued secondary embryo proliferation, if the medium pH is maintained above 4.5. (Tested at pH 4.5, 5.0, 5.5 and 6.0) [21].
8. The establishment of cultures consisting entirely of preglobular stage proembryos is a process, not an event. The first-formed somatic embryos multiply in the beginning as globular stage embryos, only when the pH of the medium is allowed to fall during the culture period. During each successive culture period, the volume per tissue mass made up of preglobular stage proembryos increases. A total of 4 to 6 transfers of the entire tissue mass after initiation of somatic embryos is required to establish a culture consisting of preglobular stage proembryos.
9. Establishment of preglobular stage proembryos can be hastened by repeated mashing or wounding of the first-formed globular stage embryos at the time transfers are made.

Some specifics in the protocols

Maintaining preglobular stage proembryos on semi-solid medium

1. Preglobular stage proembryos are maintained and multiplied on semi-solid hormone-free (H-F) medium initially at pH 4.3 (cf. Table 2, p. 24 with an appropriate nitrogen source) by subculturing at 2–3 week intervals. To assure a good crop of preglobular stage proembryos, subcultures should be carried out every 2 weeks. Each subculture is carried out by moving about one-half of the tissue mass, so that four tissue masses, each of approximately 5 mm, diameter are cultured per 100×15 mm diameter plastic Petri dish. Each dish contains between 40–50 ml of medium.
2. At the end of a 3 week growth period, the masses referred to above should have grown to about 1 cm diameter. If all the tissue from one dish is collected and dispersed in liquid, there should be about 1 ml settled cell volume. Settled cell volume is determined by allowing the cells to settle for 10 min in either 30 ml or 15 ml plastic pre-sterilized graduated tubes. The tubes are then tapped gently on a surface 3 times before measuring the volume.

Continuation of somatic embryogenesis from PGSPs into later stages

1. PGSPs larger than 140 μm in diameter often yield multiple or twinned somatic embryos. PGSPs should therefore be sieved through a # 100 sieve (140 μm pore size) and collected in a beaker (cf. Fig. 1, p. 17). This can then be poured through a # 200 sieve (74 μm pore opening), and the cell clusters which are retained are backwashed onto the # 200 sieve. Thus, one is using cell clusters with a diameter between 74 and 140 μm. This yields a relatively uniform population of units and hence the response to experimental treatments will be more uniform.
2. PGSPs so collected may be washed once with "embryo growth" medium but this is not absolutely necessary.
3. The tradeoff in the procedure that seeks to assure a more uniform response and single somatic embryos is, however, that there will be a considerable reduction in the usable volume of PGSPs.
   This can be offset, of course, by maintaining larger stocks of PGSPs in readiness for experimentation.
4. For experiments in either liquid or on semi-solid medium, the final settled cell volume per volume of medium should be around 0.1 ml settled cell volume to 50 ml medium. To enable subsequent calculation of the percent progression of distributed or suspended units into later embryonic stages, an aliquot of the cell clusters should be removed and counted in a haemocytometer or other appropriate cell counting device. For example, if you have 0.5 ml settled cell volume, adjust the total volume to 5 ml; then pipette 1 ml out for each 50 ml culture vessel. For semi-solid medium,

drop-by-drop application and spreading out of the PGSPs over the solidi-fied surface by some appropriate means needs to be carried out. In the case of liquid medium, inoculation is carried out directly by means of a pipette.

Semi-solid medium

A mixture of washed agar (National Formulary of the U.S.A. grade has been very satisfactory), 0.4% w/v plus 0.1% gellan gum (Kelco Gelrite) permits an appropriate level of gelation for semi-solid medium preparation for "main-tenance medium". For "embryo growth medium" 1.2% washed agar is satis-factory. See [19] for agar washing procedures.

The use of activated charcoal

Activated charcoal provides a major benefit in the progression of PGSPs to later embryo stages [20, 21]. Powdered activated charcoal, maintained and stored so that activation level is assured [9], can be used around 1% w/v for liquid cultures or 0.5% for semi-solid agar media. These percentages are approximate figures, however, since different charcoals will behave differently and will have to be tested in a given setting [9, 21] to ascertain what the response level generated will be. The objective will necessarily determine what is used and how. Having determined an appropriate kind of charcoal to use, the matter becomes one of assuring even distribution. Various procedures are used by different laboratories to assure even distribution of particles and to prevent settling before gelling occurs. One convenient way is to pour semi-solid Petri dishes in two or three layers, allowing each layer to solidify before the next is added. The final or "finishing layer" should be the smallest in volume.

Charcoal-impregnated filter papers (# 508, Schleicher and Schuell, Keene, New Hampshire), placed on a semi-solid surface of a Petri dish have been used successfully [20, 21] and provide the added advantage of enabling the handling of tissues conveniently since tissues resting on the papers can be transferred and manipulated with relative ease.

These papers have to be washed, however, and autoclaved twice to assure sterility [21].

Use of low external pH to replace synthetic auxin like 2,4-D in maintaining and multiplying embryogenic cells of carrot

Integration of the knowledge gained from the use of some of the classic methods described above to initiate embryogenic cultures of carrot have led to the view that cultures that contain embryogenic cells generated by any means can be, in time, sustained virtually exclusively at the pre-globular proembryogenic level by exposing them to low pH in medium devoid of the initiating/maintaining auxin. This procedure may be likened to a "jump-start" wherein a hormone like

2,4-D is used to initiate an embryogenic culture (see p. 9 and following, above), but low pH is then employed to clean-up the culture, i.e. enrich it with embryogenic cells, and to maintain it predominantly in the PGSP-replicative condition [22]. This approach would seem to have major advantages since numerous plant parts do not yield demonstrably embryogenic cultures in media devoid of growth regulator like 2,4-D. In the carrot system, for example, seedling hypocotyl does not yield embryogenic cultures like explanted wounded zygotic embryos, or mericarps [19]. Therefore, depending on one's objectives, one can initiate embryogenic cultures using hormone or whatever means possible, and then transferral to hormone-free [H-F] medium at low pH may be effected. In this way, the advantages of both procedures may be capitalized upon.

Ultimately, one cannot emphasize too strongly that objectives should determine strategies adopted and that somatic embryo-generation procedures should be closely tied to the experiments in question.

Acknowledgments

The feasibility of the investigations referred to here arose from long continued support from the National Aeronautics and Space Administration. This help is gratefully acknowledged.

References

1. Ammirato PV (1983) Embryogenesis. In: Evans DA, Sharp WR, Ammirato PV, Yamada Y (eds) Handbook of Plant Cell Culture. Vol I, Techniques for propagation and breeding, pp 82–123. New York: Macmillan.
2. Ammirato, PV (1984) Induction, maintenance, and manipulation of development in embryogenic cell suspension cultures. In: Vasil, IK (ed) Cell Culture and Somatic Cell Genetics of Plants. Vol. I, Laboratory procedures and their applications, pp 139–151. Orlando: Academic Press.
3. Carman J (1990) Embryogenic cells in plant tissue cultures: Occurrence and behavior. *In Vitro* Cellular and Developmental Biology 26: 746–753.
4. Choi JH, Sung ZR (1989) Induction, commitment, and progression of plant embryogenesis. In: Kung S-d, Arntzen CJ (eds) Plant Biotechnology, pp 141–159. Boston, London: Butterworths.
5. Fitter MS, Krikorian AD (1982) Plant Protoplasts. Some guidelines for their preparation and manipulation in culture.San Diego: Behring Diagnostics.
6. Halperin W (1966) Alternative morphogenetic events in cell suspensions. American Journal of Botany 53: 443–453.
7. Komamine A, Matsumoto M, Tsukahara M, Fujiwara A, Kawahara R, Ito M, Smith J, Nomura K, and Fujimura T (1990) Mechanisms of somatic embryogenesis in cell cultures-Physiology, biochemistry and molecular biology. In: Nijkamp, HJJ, van der Plas, LHW, van Aartrijk, J (eds) Progress in Plant Cellular and Molecular Biology, pp 307–313. Dordrecht and Boston: Kluwer Academic Pubs.
8. Krikorian AD (1982) Cloning higher plants from aseptically cultured cells. Biological Reviews 57: 151–218.
9. Krikorian AD (1988) Plant tissue culture: Perceptions and realities. Proceedings of the Indian Academy of Sciences (Plant Science) 98: 425–464.
10. Krikorian AD (1989) Introduction to: Growth and organized development of cultured cells by Steward, FC, Mapes, MO, Mears, K, Amer. J. Bot. 45: 705–708.1958. In: Janick, J (ed) Classical Papers in Horticultural Science, pp 40–55. W.H. Freeman, New York.
11. Krikorian AD, Kelly K, Smith, DL (1987) Hormones in plant tissue culture and propagation. In: Davies PJ (ed) Plant Hormones and their Role in Plant Growth and development. pp. 592–613. Martinus Nijhoff/Dr. W. Junk, Dordrecht, Netherlands.
12. Murashige T, Skoog F (1962) A revised medium for rapid growth and bioassays with tobacco tissue. Physiologia Plantarum 15: 473–497.
13. Nomura K, Komamine A (1986) Somatic embryogenesis in cultured carrot cells. Development, Growth and Differentiation 28: 511–517.
14. Scott DR, Walz, AJ, Manis, HC (1966) The effect of Lygus spp. on carrot seed production in Idaho (Hemiptera: Miridae). University of Idaho Research Bulletin no. 69: 1–12.
15. Singh M Krikorian AD (1980) Chelated iron in culture media. Annals of Botany 46: 807–809.
16. Singh, M, Krikorian AD (1980) White's standard nutrient solution. Annals of Botany 47: 133–139.
17. Small E (1978) A numerical taxonomic analysis of the *Daucus carota* complex. Canadian Journal of Botany 56: 248–276.
18. Smith DL, Krikorian AD (1988) Production of somatic embryos from carrot tissues in hormone-free medium. Plant Science 58: 103–110.
19. Smith DL, Krikorian AD (1989) Release of somatic embryogenic potential from excised zygotic embryos of carrot and maintenance of proembryonic cultures in hormone-free medium. American Journal of Botany 76: 1832–1843.
20. Smith DL, Krikorian AD (1990) Somatic proembryo production from excised, wounded zygotic carrot embryos on hormone-free medium: evaluation of the effects of pH, ethylene and activated charcoal. Plant Cell Reports 9: 34–37.
21. Smith DL, Krikorian AD (1990) Somatic embryogenesis of carrot in hormone-free medium: External pH control over morphogenesis. American Journal of Botany 77: 1634–1647.

22. Smith DL, Krikorian AD (1990) Low external pH replaces 2,4-D in maintaining and multiplying 2,4-D-initiated embryogenic cells of carrot. Physiologia Plantarum 80: 329–336.
23. Steward, FC, Krikorian AD (1971) Plants, Chemicals and Growth. Academic Press, New York.
24. Steward FC, Israel HW, Mott RL, Wilson, HJ, Krikorian AD (1975) Observations on growth and morphogenesis in cultured cells of carrot (*Daucus carota* L.). Philosophical Transactions of the Royal Society of London B 273: 33–53.
25. Street HE (1976) Experimental embryogenesis- The totipotency of cultured plant cells. In: Graham CF, Wareing, PF (eds) The Developmental Biology of Plants and Animals, pp 73–91. Philadelphia: W.B. Saunders.
26. Waris H (1959) Neomorphosis in seed plants induced by amino acids. I. *Oenanthe aquatica*. Physiologia Plantarum 12: 753–766.

Plant Tissue Culture Manual **A10**: 1–28, 1992.
© 1992 *Kluwer Academic Publishers.*

Low density cultures: microdroplets and single cell nurse cultures

G. SPANGENBERG[1] & H.-U. KOOP[2]

[1] *Institute for Plant Sciences, Swiss Federal Institute of Technology, CH-8092 Zürich, Switzerland*
[2] *Institute of Botany, University of Munich, D-8000 Munich 19, FRG*

Introduction

The development of low density cell culture systems is mainly based on two assumptions. Firstly, although plant cell (protoplast) cultures are commonly referred to as relatively homogeneous cell populations the results of closer analysis have often shown the existence of a high degree of cellular hetero-geneity. Thus, plant cell cultures do not consist of a large number of uniform cells, but rather are a population of cells showing a range of variation in individual genotype, phenotype and age [5]. This range of variation can be further increased by the use of mutagenic agents, genomic mixing and recombi-nation achieved through protoplast fusion and genetic transformation. Secondly, most chemically defined plant cell culture media will not support cell division at low cell densities, commonly less than about 9,000–15,000 cells/ml: the *critical inoculum density* [41], and plant cell culture under optimal conditions deals with population densities in the range of 10^4–10^6 cells/ml.

Thus, the need for the development of methods allowing "density independent" plant cell growth, in order to improve the possibility of recovering rare variants, becomes evident.

In this context, the use of conditioned media [14, 25, 2], nurse cell layers [4, 10] or growing cells in very low culture volumes in multiwell dishes or hanging droplets [15, 11, 6, 26] have all been successfully used for growing clones from single cells or protoplasts.

In addition, all these techniques have been applied for *in vitro* selection thus dissecting the "natural" heterogeneity of the original population of plant cells in more homogeneous subpopulations [25, 10], or for the "density independent" culture of the extended range of – more or less rare – variants produced by mutagenic agents [44], by protoplast fusion [12, 22, 27] and through genetic transformation [6].

An extension of the techniques for culturing small numbers of cells, following the strategy of reducing the volume of culture medium, namely using *micro-cultures* [13, 43, 15, 11] was achieved with the success of *ab initio* microculture in chemically defined and unconditioned culture medium of individually selected single plant cells in *nanodroplets* (10–50 nl) of culture medium pro-tected from evaporation by a layer or droplets of mineral oil [16, 18, 36]. This was based on a reduction of the volume of culture medium to a similar cell population density as used in mass culture, i.e., if a population density in the range of 10^4–10^5 cells/ml is required for *en masse* plant cell or protoplast culture, then *one single* cell or protoplast should be cultured in a nanodroplet of 100–10 nl (microculture or microdroplet culture), respectively.

A similar development for establishing techniques in order to culture small numbers of "target" cells following the principle of "nurse culture" [24, 7, 30, 42] led finally to a method for the nurse culture of individual cells [8, 31].

Our aim is not to provide a detailed review of tissue culture methods dealing with the culture of small populations of plant cells, but rather to emphasise two experimental protocols for the individual culture of single "target" cells, namely 1) the individual selection and exclusive culture of defined single plant cells or protoplasts in nanodroplets of unconditioned culture medium injected into separate microdroplets of mineral oil: *microculture* or *microdroplet culture*, and 2) the individual culture of "target" plant cells grown close to, but physically separated from a second plant cell population (feeder cells), supporting their growth: *single cell nurse culture*.

We also describe a computerized hydraulic system which allows a) the microscopical selection of defined plant cells from a population and their individual inoculation, applied in both experimental protocols outlined above, b) the preparation of the *microculture chamber* in the case of the microdroplet culture and c) the positioning of micro-tools used in different micro-manipulation steps, e.g. positioning of microelectrodes for one-to-one micro-(electro)fusion of preselected pairs of protoplasts, or positioning of micro-injection needles and holding capillaries.

A. Microdroplet Culture

Individual cell culture represents *a priori* a useful tool for *in vitro* selection of different cell types and studies on their physiology, for the analysis of cell differentiation programs and cell-to-cell interactions, as well as a culture cloning step after the performance of genetic manipulations at the single cell level: *single cell engineering* [32, 20, 38].

By using a microculture chamber [18] and a computer-aided instrumental setup [32, 20] the microscopical selection of single plant cells or protoplasts and their individual culture at the same population density as in mass culture can be performed.

This microdroplet culture system has been so far successfully used for the individual selection, and culture in microculture chambers, of mesophyll proto-plasts of *Nicotiana tabacum* [18], hypocotyl protoplasts of *Brassica napus* [36] and protonemata protoplasts of *Funaria hygrometrica* [23] and *Physcomitrella patens* [1] and microspores of barley [3], as well as for the analysis of conditioning effects at the single cell level in different cell types of *B. napus* [33].

Single cell engineering applications, based in part on this micromanipulation instrumental setup or on the culture in microdroplets of the micromanipulated plant cells or protoplasts, have been reported for a number of systems, such as: intranuclear microinjection of protoplasts and karyoplasts of *B. napus* [37], microfusion of preselected pairs of protoplasts of *N. tabacum* [17, 19], cell reconstitution by protoplast-subprotoplast microfusion in *B. napus* [35]; protoplast microfusion in *F. hygrometrica* [35], *in vitro* fertilization via microfusion of isolated gamete cells in *Zea mays* [23]; defined somatic hybridization via protoplast microfusion in *N. tabacum* [39]; organelle transfer [9] and defined cybridization via protoplast-cytoplast microfusion in *N. tabacum* [40].

Procedures

1. Major Equipment: Instrumental Setup for Microdroplet Culture

As already mentioned, the microdroplet culture is mainly based on a microprocessor-aided instrumental setup (Fig. 1) [32, 20].

   This instrumental setup which is assembled in an air-flow cabinet in order to fulfill the sterility requirements, consists basically of an optical and operational-positioning unit (a); a mineral oil filled hydraulic system (b and c) and a control unit (d) (Fig. 1A, B). In more detail, these units are:

a) An inverted microscope (IM 35 Zeiss microscope with Nomarski optics, C. Zeiss, Oberkochen, FRG) with a microprocessor-controlled programmable microscope stage (EK8b-S4 microscope stage and control unit MCC 13 JS RS232, Gebr. Märzhäuser OHG, Wetzlar, FRG, or from Lang-Elektronik, D-6338 Hüttenberg, FRG) driven by adjustable stepmotors in x-, y- and z-axis. This allows the positioning of a microculture chamber and a selection chamber while selecting single defined cells under microscopical control (optionally based on a TV-screen image) from the population of cells kept in the selection chamber, and their individual inoculation into microdroplets of the microculture chamber (see section 3) (Fig. 1A-C).
   Alternatively, the insert of the microscope stage can be replaced by a number of different inserts allowing for example the positioning of up to six coverslips for the automatic preparation of microculture chambers (see section 2.3) or the use of standard culture dishes (Petri dishes, slide chambers, microtiter plates) in various combinations.

b) A hand-pulled selection microcapillary, fixed to a holding device driven by the z-axis stepmotor, is connected via a teflon tubing filled with mineral oil to:

c) A microprocessor-controlled nanoliter-pump (modified diluter Microlab M, Hamilton, Bonaduz, Switzerland), which can be programmed for volume, speed and direction of the pumping, thus allowing delivery and withdrawal of volumes in the nanoliter-range (Fig. 1E) and, finally:

d) A microprocessor (Apple II, Apollo-IIASKF/64K, AAA Electronic GmbH, Freiburg, FRG with interface CCS 7710, California Computer System, CA 94086, USA or from Pete & Pam Micro Computers, Haslingden, Rossendale, Lancs BB4 5HU, U.K.; and IBS-AP2, IBS Computertechnik, Bielefeld, FRG) controlling the positioning electronics of the x-, y- and z-axis stepmotors of the programmable microscope stage as well as the nanoliter-pump (Fig. 1B). Alternatively other computers can be used. For software see section 2.3.8.
   With these elements, this instrumental setup can be programmed to perform the different working steps required, for example:

A) *Fully-automatic preparation of a microculture chamber:* The setup adjusts under the optical axis of the inverted microscope to a given position in x- and y-axis, and the microcapillary supported in the z-axis-device is lowered so that the tip of the microcapillary contacts the coverslip, placed on the microscope stage at the pre-fixed position. The nanoliter-pump is activated to deliver a pre-established volume with a given speed, and after delivery of the microdroplet the microcapillary is lifted and adjusted to the next position following a pre-determined array in x- and y-axis. Thus all working steps required for the preparation of microculture chambers for microdroplet culture can be performed with minimal intervention by the experimenter.

B) *Selection of defined cells and their transfer into microdroplets:* While activating the same functions, the position of a particular cell in the selection chamber can be chosen under microscopical control by the experimenter with the aid of a joy-stick, manually controlling the positioning system in x-, y- and z-axis. The selection microcapillary is lowered over the chosen cell and the selected cell withdrawn in a volume in the nl-range after foot-switch activation of the nanoliter-pump. Then the selection microcapillary is automatically lifted, the microscope stage moved so that a microdroplet at a pre-established position of the microculture chamber is now under optical control, the microcapillary is lowered into the nanodroplet of culture medium covered by a microdroplet of mineral oil and by a second activation of the nanoliter-pump, the chosen cell can be delivered into the desired microdroplet.

If access to all of these major pieces of equipment is not possible, successful microdroplet culture, namely preparation of microculture chambers as well as selection of defined plant cells into microdroplets, can still be performed, but obviously with more intervention and expertise required from the experimenter.

In this case, the positioning system in x-, y- and z-axis, namely the automatic microscope stage with the three stepmotors and the corresponding control-electronics can be omitted and replaced by a standard manually moveable micro-scope stage (simply adapted for two 24 × 40 mm coverslips, namely for the selection and microculture chambers) (Fig. 2C).

In addition, the selection microcapillary will then also be manually controlled while delivering the nanodroplets of culture medium into the oil microdroplets (see section 2.3) and selecting cells from the selection chamber under microscopical control and individually transferring them into previously (also mainly manually) prepared microculture chambers (see section 3). For the exclusive application in microdroplet culture, this "less automatic" version of the instrumental setup [18]

Fig. 1. Major equipment for single cell culture.
A) Setup for microprocessor-controlled selection and transfer into microculture of defined plant cells.
B) Components of the selection and microculture setup: (1) inverted microscope, (2) nanoliter-pump, (3) microprocessor and (4) 1 μl microdroplets-dispensing device.
C) Microprocessor-controlled microscope stage with x-, y-, z-stepmotors (1, 2 and 3) with support for microtools (4) e.g. for microelectrodes for microfusion (5).
D) Detailed view of 1 μl microdroplets-dispensing device from B(4) showing tubing connection to a selection microcapillary.
E) Detailed view of microprocessor-controlled nanoliter-pump from B(2).

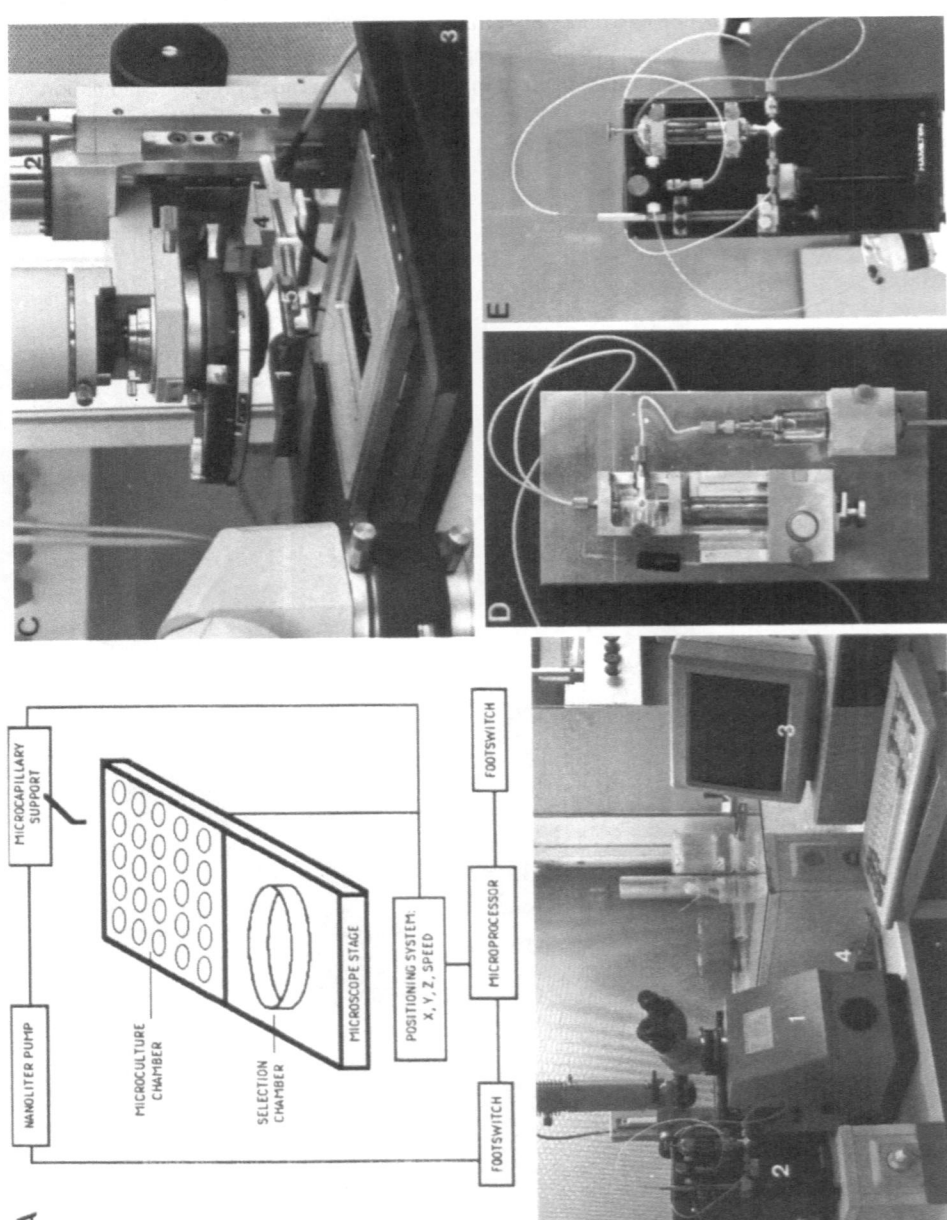

still allows a skillful experimenter — after a reasonable training period — a comparably fast (e.g. selection and transfer of 200 cells/h) and sometimes more flexible performance. Thus, for microdroplet culture alone, in the authors' experience, this simplified version is equally suited compared to the full computer-aided setup. However, if more precise positioning of micro-tools for particular micromanipulation purposes is envisaged, e.g. positioning of micro-electrodes for micro(electro)fusion, then the full automatic setup will provide a more accurate and efficient working basis.

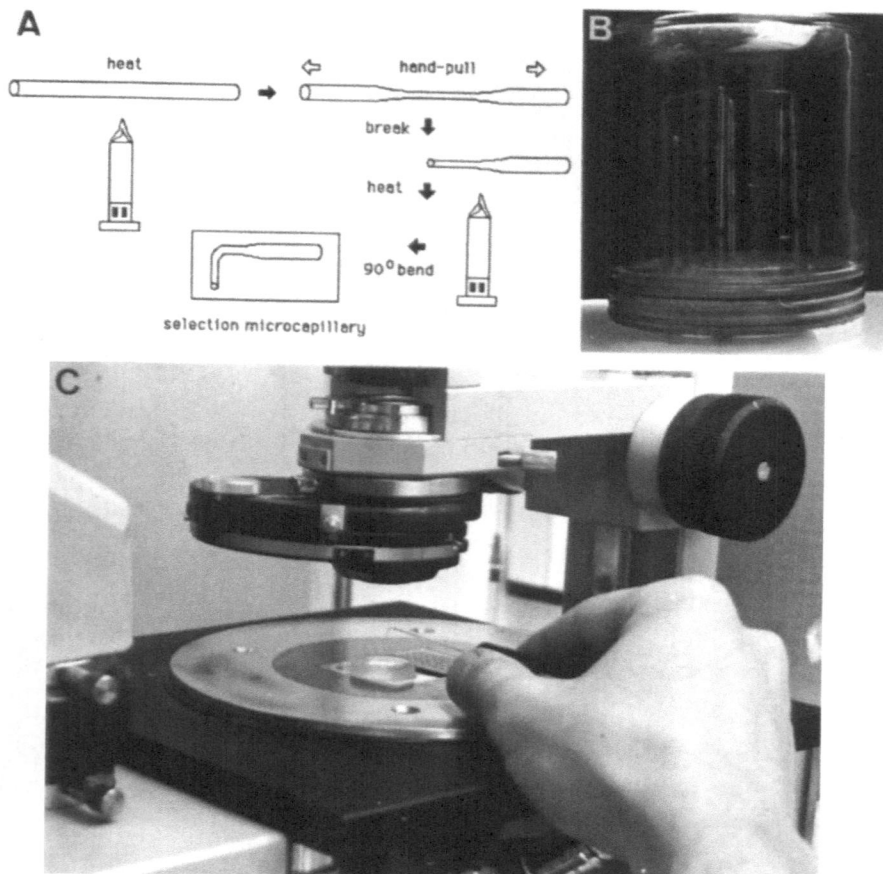

Fig. 2. Selection microcapillary and manual selection of defined plant cells.
A) Preparation of selection microcapillaries for single cell culture.
B) Ready-to-use drawn selection microcapillaries.
C) Manual selection of defined plant cells from selection chamber into microdroplets.

2. Preparation of Selection Microcapillaries, Selection Chambers and Microculture Chambers

In the following sections a detailed description of the different components and tools required as accessories for performing microdroplet culture is presented; important steps and practical considerations regarding the procedure and minor equipment are stated in the notes.

2.1. Preparation and maintenance of selection microcapillaries

The selection microcapillaries are hand-drawn from 50 µl disposable micropipettes (Blaubrand, 50 µl, green color code, No. 708733, Brand, IntraMark) over a small flame.

Steps in the procedure

1) Heat the center of the micropipette until soft, remove it from the heat, wait 1—2 s and then quickly pull the two ends apart, so that the 127 mm long micropipettes are elongated by an approximately 3—4 cm pulled central region (Fig. 2A).

2) Break the pulled micropipettes with the aid of a diamond pencil in the middle of the narrowed portion and bevel the rough broken tips with fine polishing-paper.

3) Bend the drawn-out portions of both extended capillary-halves to a right angle by holding the capillary-half over a small flame and heat just before the expanded portion and wait until the drawn-out portion bends down due to gravity (Fig. 2A, B).

Notes

2.1.1 It is recommended to pull micropipettes with a series of different opening diameters. On average, for selection of cells the inner diameter of the open tip should be approximately 2-3 fold the diameter of the cell to be selected, and thus will vary if selection of large vacuolated protoplasts, karyoplasts, cytoplasts, gamete cells, etc. is intended.

2.1.2 For the delivery of liquids (e.g. mineral oil; 2.5 M sucrose or culture medium), the choice of the appropriate microcapillary will depend mainly on the viscosity and volume of the solution to be handled. For viscous liquids (e.g. mineral oil and 2.5 M sucrose), which are normally used for the delivery of 1 µl microdroplets (see section 2.3), larger opening-capillaries (inner diameter: 600—800 µm) are preferred. For the delivery of nanodroplets (e.g. 10—20 nl) of standard plant cell culture media into mineral oil microdroplets, narrower opening-capillaries (inner diameter: 300—500 µm) are recommended.

2.1.3 Drawn-out selection microcapillaries should be carefully checked (eventually, requiring the use of a stereomicroscope), keeping in mind the following criteria: a) a correct bend at 90°, b) 10—15 mm length of the bent terminal portion, c) 45—60 mm length of the not-drawn portion, d) evenly broken, smooth and thin-walled tips. After strictly discarding those not fulfilling the requirements from a) to d), they can be sorted in two or three categories depending on the opening size of their tips for different experimental purposes (see 2.1.1 and 2.1.2) and finally, autoclaved (Fig. 2B).

2.1.4 Pulled selection microcapillaries can be re-used as long as they fulfill the sterility and other requirements (see notes 2.1.1—2.1.3). The chances for re-use in subsequent experiments are increased if the same capillary is used for the same experimental purpose (e.g. a capillary appropriate for delivering mineral oil droplets should be used exclusively for dispensing oil and not sucrose solution). As long as the microcapillary is sterile, it can be replaced (and kept under axenic conditions) if, with the same delivery-setup, another experimental purpose (and thus another microcapillary, e.g. for selecting protoplasts, is required) is to be accomplished. If the microcapillaries do not work properly any longer, it is recommended to discard them, instead of attempting a time-consuming, cumbersome washing procedure.

2.2 Preparation of selection chambers

Steps in the procedure

1) Prepare selection chambers by glueing a polycarbonate ring (5 mm high, 24 mm outer diameter) onto a 24 × 40 mm coverslip (coverglass AL No. 9161040, Menzel, FRG) with silicone glue (Fugendichtung Praktikus, Praktikus-Chemie, D-4048 Grevenbroich 5, FRG) (Fig. 2C). The rings are made out of 5 mm high transverse sections from a 24 mm diameter polycarbonate cylinder.
2) Bevel the sections to a ring leaving inside an empty volume with the shape of an inverted truncated cone (13 mm inner lower diameter, 22 mm inner upper diameter).
3) Smear the wider ring-base with the silicone glue and smoothly press onto the center of the coverslip and slightly rotate while pressing, in order to allow for an even distribution of the glue and a tight sealing-contact between the poly-carbonate ring and the coverslip.

Notes

2.2.1 After hardening of the glue, the selection chambers can be autoclaved. Care should be taken to allow possible cytotoxic solvents (e.g. acetic acid) to evaporate from the silicone glue, so a 2–3 days air drying of the selection chambers is recommended before autoclaving.

2.2.2 Other types of selection chambers are conceivable, however, they should allow for good optical quality (e.g. recommended coverglass), should be easily autoclavable (e.g. polycarbonate, autoclavable silicone glue), should have dimensions providing 1–2 ml content and should be compatible with following considerations: manipulation of reasonable populations of plant cells at a time, with population densities compatible with easy identification of "target" cells to be selected and facilitating the withdrawal of individual cells in volumes of 10–50 nl; provide an easy access to the "target" cell with the selection microcapillary considering a working distance of 3–4 cm between the microscope stage and the condensor (Zeiss IV Z/7(0,63) with removed front lens) of the inverted microscope; have a convenient surface area/volume ratio under comparatively high evaporation conditions (e.g. air-flow cabinet during 1–3 h), etc. Standard poylstyrene Petri dishes can also be used as selection chambers, where optimal optical conditions are not critical.

2.2.3 Non-leaking selection chambers can be re-used (after washing with a tissue culture compatible detergent and re-autoclaving). Otherwise, the polycarbonate ring can be removed and newly glued onto another coverslip.

2.2.4 When preparing selection chambers, a small amount of silicone glue remains at the inner edge of the ring on the glass surface. This remaining silicone can be helpful for removing mineral oil, eventually clogging the tip of the selection microcapillary, while transferring cells into microdroplet culture (see section 3).

2.3 Preparation of microculture chambers

Steps in the procedure

1) Prepare the standard microculture chambers [18, 32] from 24 × 40 mm cover-slips (microcover glasses Thomas No. 6663-F82, A. Thomas Co., USA, marked e.g. by cutting off one corner), where 50 1.0 µl microdroplets of 2.5 M sucrose are positioned onto the coverslip in an array of five rows with 10 microdroplets per row (Fig. 3A, B).

2) Immerse the coverslips carrying the sucrose-droplets for 1–2 s in a 2% dimethyl-dichlorosilane solution in 1,1,1 trichloro-ethane (Repel Silane, No. 1850-252, LKB, Sweden), drain and leave for 10–15 min. in a fume-hood to evaporate the solvent.

3) Wash with warm tap-water until the sucrose-droplets are completely removed, rinse with ethanol and double distilled water and carefully dry with optical-quality paper.

4) Each coverslip is then placed, with the side where the sucrose droplets had been positioned facing upward, in a two-compartment 6 cm diameter culture dish (organ tissue culture dish Falcon, No. 3037, Becton-Dickinson & Co., Cockeys-ville, USA) and UV-sterilized for 15–20 mins. The coverslips are now handled under sterile conditions for positioning 50 1,0 µl droplets of autoclaved mineral oil (paraffin oil for spectroscopy Merck, No. 7161, Merck, Darmstadt, FRG) exactly in the center of the non-siliconized circular areas previously occupied by the sucrose-droplets.

5) Finally with the aid of the fully automatic instrumental setup, inject nanodroplets of culture medium (normally in the range of 15–100 nl) with the microcapillary centered in each mineral oil droplet previously placed on the coverslip (Fig. 3C).

6) The prepared microculture chamber is transferred back to the two-compartment culture dish, 1.5–2 ml of autoclaved 0.2 M mannitol are pipetted into the outer compartment of the culture dish, thus serving as moist chamber, and finally the latter is sealed with Parafilm (American Can Company, Greenwich, CT 06830, USA) (Fig. 3B).

Notes

Notes 2.3.1 to 2.3.8 deal with minor equipment required for and some technical aspects concerning the preparation of microculture chambers, notes 2.3.9 to 2.3.14 refer to the different components of the microculture chamber *per se*.

2.3.1 The 50 1.0 µl droplets of sucrose and of mineral oil are distributed routinely, while using the full automatic setup with the microprocessor controlled nanoliter-pump (see section 1), 3,350 µm apart from each other in rows of 10 with a distance of 3,400 µm between the rows, when the standard microculture chamber is prepared. However, the volume and number of droplets and their distribution in x- and y-axis as well as the distance between droplets and rows can be freely chosen and programmed.

2.3.2 If no programmable microscope stage/positioning system is available (see section 1), still a very accurate distribution of the 1.0 µl microdroplets (of sucrose solution or mineral oil) can be achieved by manually applying them onto the coverslip after placing it on a plate with a dotted array (5 rows per 10 dots each, every dot marked approximately 3 mm from each other on a mm-squared paper). The application of the 1.0 µl droplets of sucrose solution or mineral oil can then be performed with the aid of a simplified dispenser-device (described under note 2.3.3 and illustrated on Fig. 1D).

2.3.3 This simplified 1.0 µl microdroplets-dispensing arrangement consists of a 10 ml syringe (luer-lok, Ultra-Asept for Braun-Perfusor, No. 872 855/0, B. Braun Melsungen AG, D-3508 Melsungen, FRG) used as oil reservoir, a 50-teeth mechanical device (dispensor-device Hamilton PB 600-1, No. 83700, Hamilton Bonaduz AG, CH-7402 Bonaduz, Switzerland) driving the piston of a 50 µl syringe (Hamilton type 705N, No. 80500 Hamilton) and allowing the stepwise (max. 50 steps, 1 µl each) emptying of its contents, and the connections (B. Braun Melsungen, FRG) required to link via a three-way-cock (Edwards-Bentley model K75B, Bentley Lab. Europe, 5404 AA Uden, Holland) both syringes to the selection microcapillary (see description on section 2.1). All mechanical elements are anchored to a stable plate and all required connections are achieved by a clear flexible teflon tubing (Tefzel tubing 1.8 mm outer diameter; 0.8 mm inner diameter, No. 19-7435-01, Pharmacia, Sweden). The 10 ml syringe oil-reservoir is attached to one of the openings of the three-way cock to provide auxiliary negative pressure (e.g. for loading the selection microcapillary and in part the tubing system with sucrose solution or mineral oil) or positive pressure (e.g. while pulling out the piston of the 50 µl syringe linked to the 50-teeth mechanical device). The terminal connection of the teflon tubing to the selection microcapillary is achieved by a plastic screw-type fitting (tubing connector for 1.8 mm o.d. tubing, No. 19-7476-01, Pharmacia, Sweden), thus allowing an easy exchange as well as tight fitting of selection microcapillaries, as well as representing a holding element in the case of manual selection. The entire arrangement functions as an hydraulic system filled with autoclaved mineral oil (Fig. 1D).

2.3.4 The positioning of the culture medium nanodroplets into each of the 50 1.0 µl mineral oil microdroplets loaded onto the coverslip in a specified array is done under optical control (the use of a 2.5× objective is recommended, as it allows for the observation of four mineral oil microdroplets on the image field and thus provides an easy orientation in x- and y-axis). The delivery of the nanodroplets of culture medium is performed with the aid of the nanoliter-pump (mentioned in section 1), independently of whether the PC-controlled microscope stage/positioning system or a manual control of the positioning of the microscope stage and selection microcapillary is used. The modifications performed on the commercially-available dispenser (Microlab M, Hamilton, Bonaduz) for adapting it to the purpose of microdroplet culture are described under note 2.3.5 and illustrated on Fig. 1E.

2.3.5 The nanoliter-pump for microdroplet culture is a slightly modified dilutor MircolabM (Hamilton, Switzerland). On the front plate of the commercially available dispenser a 1 µl or, alternatively, a 5 µl syringe (Hamilton, type 70001N, No. 80100, and 7005N, No. 200305, respectively, Hamilton Bonaduz) is installed and mechanically linked to the movable dispensing axis-device driven by a stepmotor (with 1000 steps working range) of the dilutor. Thus, per step minimal volumes of 1 nl (1 µl/1000 steps) or 5 nl (5 µl/1000 steps), respectively, can be delivered or withdrawn with this arrangement in the range of 1 nl up to 5 µl (by varying the number of driven steps). In addition, the arrangement consists of a 10 ml syringe (luer-lok, Ultra-Asept for Braun-Perfusor, No. 872 855/0, B. Braun Melsungen AG, D-3508 Melsungen, FRG) connected to the selection microcapillary and to the 1 µl (or 5 µl) syringe via a three-way cock and the required clear flexible teflon tubing for linking all these components. Further modifications (not shown) can include three outlet valves and tubings for separate delivery of sucrose, oil and culture medium, respectively and a separation unit in the sucrose outlet to prevent mixing of oil and sucrose in the tubing system. An oil reservoir is attached to one of the openings of the central three-way valve of the nanoliter-pump to provide oil supply while delivering (from the 1 µl or 5 µl syringes) volumes in excess to the total syringe content (1 or 5 µl, respectively). The 10 ml syringe, also anchored to the front plate of the nanoliter pump, acts as a wider-volume range reservoir providing auxiliary negative pressure (e.g. for loading the selection microcapillary and in part the teflon tubing with e.g. culture medium) or positive pressure (e.g. for flushing the hydraulic system free of disturbing air-bubbles). The entire arrangement acts as an hydraulic system and is filled with autoclaved mineral oil (Fig. 1E).

2.3.6 The ready-to-use microculture chamber should already contain a fraction of the final volume of culture medium/microdroplet which will, by the end, after individual transfer of the selected cell into the microdroplet (see section 3) complete the expected volume for achieving an optimal volume of culture medium/cell ratio (e.g. nearly equivalent to an optimal population density in *en masse* culture). Routinely, at this initial step, volumes in the range of 10—30 nl of the culture medium are injected in each oil microdroplet and directly dispersed onto the unsiliconized area of the coverslip forming a flat central nanodroplet of culture medium, evenly covered by the oil microdroplet (Fig. 3C-D).

2.3.7 Independently of the mode of positioning control in x- and y-axis (via the PC-controlled microscope stage or a manually slidable standard stage) as well as for the z-axis (manually holding the selection microcapillary at the screw-type fitting or fixing it to a support which can be vertically displaced by a PC-controlled stepmotor anchored under the condensor of the inverted microscope), the reproducible delivery of defined nanodroplets of culture medium into the oil microdroplets can only be achieved under microscopical control (see note 2.3.4) with accuracy using the nanoliter-pump (described under note 2.3.5). In the case of the PC-controlled positioning system, by pressing once the corresponding foot-switch the first position (center of an oil microdroplet) is automatically placed under the optical axis and the z-axis stepmotor vertically driving the selection microcapillary lowered (and a final adjustment by the experimenter of the coordinates in all axis via a joystick is allowed). These finally adjusted coordinates can then be saved by re-pressing the same foot-switch. By pressing then once the foot-switch corresponding to the nanoliter-pump control, the programmed volume (e.g. 10—30 nl) is delivered at the first chosen position. From now on, alternating the positioning system and the nanoliter-pump will be automatically activated repeating the delivery of culture medium into the oil-microdroplets following the pre-programmed positioning pattern (freely selectable from a corresponding menu) as well as a selectable time factor (for allowing the reproducible delivery of different volumes or of different culture media with distinct viscosities into the oil microdroplets) in a stand-alone function without further intervention by the experimenter. This procedure for preparation of microculture chambers can be performed on parallel with up to six coverslips on the programmable microscope stage. In the case of the manually driven standard microscope stage and manual holding of the selection microcapillary, a single foot-switch activating the nanoliter-pump has to be pressed by the experimenter every time the tip of the microcapillary is immersed into each oil microdroplet and serially completing the 50 positions on a standard microculture chamber. This procedure can be performed with e.g. two coverslips placed on the standard microscope stage at a time.

2.3.8 Required software for both systems outlined before (see note 2.3.7): a) exclusive control of the nanoliter-pump, and b) coordinated control of positioning system and nanoliter-pump, allowing a flexible arrangement of delivered and withdrawn volumes and of positions and arrays of droplets can be obtained upon request for Apple II and IBM-compatible PCs from the authors. For IBM-PCs XT or AT (or compatibles) equipped with a standard games part and two serial parts, extensive software (developed by the second author) is available. In addition, similar software for Apple II, IBM and Epson systems for the same experimental purposes can be obtained upon request from E. Kranz (Department of Botany, University of Hamburg, D-2000 Hamburg, FRG), U. Lagercranz (Department of Plant Breeding, Swedish University of Agricultural Sciences, S-750 07 Uppsala, Sweden) and F.L. Olsen (Department of Physiology, Carlsberg Laboratory, DK-2500 Copenhagen Valby, Denmark), respectively.

2.3.9 The rationale behind the preparation of the microculture chamber [18] is to protect the nanodroplets of culture medium from evaporation by covering them with oil microdroplets separately (in order to guarantee "individual" culture) positioned onto a coverslip (in order to achieve good optical quality). To avoid merging of the oil microdroplets on the surface of the coverslip, the latter is siliconized between the oil droplets, which is achieved as indicated on Fig. 3.

2.3.10 For defining the array of unsiliconized circular areas on the coverslip (Fig. 3A) a viscous solution, namely 2.5 M sucrose (filter-sterilized, not autoclaved: otherwise there is a loss of viscosity) is recommended over the 2.0 M described in the original protocol [18]. In order to be able to discriminate between the 2.5 M sucrose solution and the oil (contained in the hydraulic system used for dispensing 1 μl droplets) it is recommended to include in the sucrose solution a vital dye (e.g. cochenille red or food colorant E124). While delivering the sucrose droplets, care should be taken to avoid oil coming out of the microcapillary, as it would be dissolved by the Repel-Silane solvent, the glass surface below the sucrose droplet would then be partially siliconized and thus the nanodroplet of culture medium will not be properly dispersed onto the coverslip.

2.3.11 It is recommended for the preparation of the oil microdroplets and for filling the hydraulic system to use good quality paraffin oil (.e.g. paraffin oil for spectroscopy, Merck No. 7161, E. Merck, Darmstadt, FRG). For the microdroplets preparation, some activated charcoal (activated charcoal Merck No. 2514, particle size: 1.5 mm) for adsorbing possible cytotoxic contaminants from the oil, can be added to the oil. The oil can then be saturated with water (by adding double distilled water to the oil-activated charcoal suspension and autoclaving it; re-autoclave it at least once a week). The oil may have to be discarded after multiple re-autoclaving.

2.3.12 The use of standard coverslips (e.g. Menzel AL No. 9161040, 24 × 40 mm, type 1, Menzel, FRG) has proven to be as good as more expensive microcover glasses of better − lead free − quality (e.g. microcover glasses Thomas No. 6663-F82, A. Thomas Co., USA) for most experimental purposes. However, if problems are encountered in microculture of particular "sensitive" plant cell types, attention should be paid to quality requirements of all the different components of the microculture chamber, as cells cultured in these "micro-environments" may react more sensitively than if mass-cultured.

2.3.13 Immediately after preparation of the microculture chambers (after the nanodroplets of culture medium have been injected into the oil microdroplets), these should be kept before use in a moist chamber (two-compartment Falcon dish, as described in section 2.3). Only freshly prepared micro-culture chambers (not more than 2−4 days after injection of the culture medium nanodroplets) should be used. Coverslips with sucrose droplets (but not siliconized) can be prepared well in advance and be stored (it is even recommended to let them air-dry for 2−3 days before siliconization if freshly prepared 2.5 M sucrose solution is used). Microculture chambers in preparation should not be stored after the oil microdroplets have been positioned onto the coverslip (this makes the delivery of the nanodroplets of culture medium more difficult).

2.3.14 Regarding the culture medium to be used for microculture, the medium used for mass culture of the cells or protoplasts of interest has − in general − proven to be satisfactory. It is however recommended, before starting microculture experiments, to optimize the composition of the culture medium and culture conditions (macro- and micronutrients, hormones, vitamins, cell density, buffering system, pH, osmolality, etc.) using a population of protoplasts (e.g. using the MDA-technique; [29]) until achieving a reproducible and efficient plant cell (protoplast) mass culture protocol. If so, optimization of the microculture conditions would be restricted to eventually checking particular parameters affecting the performance of the individually cultured cells (e.g. volume of nanodroplets), however, in general, no "microdroplet culture-specific" requirements to the culture medium are necessary.

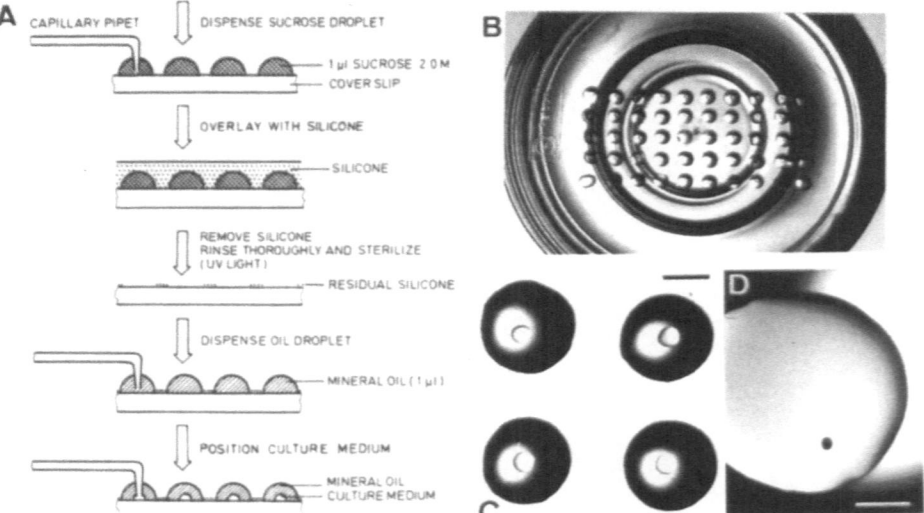

Fig. 3. Microculture chamber for individual culture of defined plant cells.
A) Preparation of microculture chambers for single cell microdroplet culture.
B) Microculture chamber in a moist chamber (two-compartment dish).
C) Detailed view of part of a microculture chamber. Each 1 µl droplet of mineral oil contains 30 nl of culture medium. Bar: 1 mm.
D) Individually selected mesophyll protoplast of *N. tabacum* in a nanodroplet of culture medium. Bar: 500: µm.

3. Selection of defined Plant Cells and Individual Culture in Microdroplets

As mentioned before, the instrumental setup for microdroplet culture (see section 1, Fig. 1) in both its variations (namely, manual positioning or full automatic control) has been used by the authors for the selection of defined protoplasts out of heterogeneous cell populations, their transfer into microdroplets on microculture chambers, their individual culture and further growth until plant regeneration for isolated protoplats and/or fusion products of following plant species: N. tabacum [18, 19, 39, 40], B. napus [34, 32], F. hygrometrica [23] and P. patens [1]. Details on the performance of protoplasts from the above mentioned species are provided in the original references. Only some relevant aspects, for achieving successful microculture of protoplasts from these species are described in the following, illustrating some considerations for microdroplet culture of more general interest when dealing with other species. Some special recommendations are presented in the notes.

Once the technical elements of the instrumental setup for microdroplet culture, the microculture chambers and other accessories are established, the reader interested in performing microdroplet culture of protoplasts or plant cells from a system efficiently working in his own hands at a mass culture scale, will be faced first with two questions: 1) which initial volume should the nanodroplet of culture medium have? and 2) which type of cell or protoplast should be selected?

Indeed these two parameters play an essential role on the success of the individual microculture of defined protoplasts (establishment of cell clones, finally leading to plant regeneration) using the experimental setup described here.

As regards to the initial volume of the nanodroplet of culture medium into which a single protoplast has been selected, volumes in the range of 30–50 nl have proven adequate for tobacco mesophyll protoplasts and rapeseed hypocotyl protoplasts. Larger nanodroplets of 125 nl culture medium have proven advantageous for the microdroplet culture of protonemata protoplasts of F. hygrometrica.

Concerning which type of cell should be selected, obviously, the experimental purpose followed, e.g. cell cloning for in vitro cell line selection for increased yields of secondary metabolites, or microisolation of heteroplasmic cells after protoplast mass-fusion, etc., would be a key factor upon this decision.

However, the experimenter interested in microdroplet culture will necessarily identify discrete morphologically distinguishable cell (or protoplast) types (e.g. according to cell size, distribution and number of organelles, degree of vacuolization, vacuolar accumulation of anthocyanins, evident presence of cytoplasmic strands, etc.) in the apparently homogeneous population of cells one is dealing with, and thus learn to look at populations of plant cells as collections of individuals. In this context, it is recommended to invest particular time and attention in comparing the behaviour (e.g. viability, rates of cell division and microcallusing, etc.) in microdroplet culture of selected subpopulations of discrete individual-types.

In addition, since the experimental setup described allows on average the manipulation of 100 to 200 cells/hour, the experimenter is restricted to the use of subpopulations limited in number selected out from plant cells (or protoplasts) populations which are heterogeneous. This inherent variation as well as the variability present between different protoplast preparations, due at least in part to a series of external factors affecting protoplast culture [28], may account for much of the variation observed in different experiments while performing microdroplet culture with apparently the "same" protoplast system. Nevertheless, in the authors' experience, plant regeneration from microcultured protoplasts of *F. hygrometrica* can be reproducibly achieved with overall frequencies in the range of 40—60% [23], for tobacco protoplasts on average between 3—60% [18, 39, 40] and for rapeseed hypocotyl protoplasts with frequencies of 0. 1—0.2% [32, 38]. However, cell division frequencies from microcultured protoplatsts can significantly vary — within the ranges described — from one microculture chamber to another.

For achieving sustained cell division in microculture it is in some cases (e.g. for tobacco and rapeseed protoplasts) required to add fresh culture medium to the microdroplets. Optimization of a "feeding"-system for microcultured protoplasts is recommended, as this factor may even prove to be a critical step [36]. The optimal "feeding" schedule should be determined considering following factors: a) starting time for the addition of fresh culture medium (normally not required before first cell divisions have been observed), b) frequency of the addition of culture medium (e.g. every 3—4 days has proven convenient for tobacco and rapeseed protoplasts), c) volume of the added culture medium and dilution rate (e.g. 10—30 nl in the case of tobacco protoplasts, and exclusively in those microdroplets where protoplasts have divided, in order to avoid dilution of the population density of microcultured protoplasts being "laggards" in cell division), d) composition, e.g. regarding hormones and osmoticum used, of the added culture medium.

In this way, microcolonies (consisting of ca. 20—100 cells) derived from *ab initio* microcultured protoplasts can be obtained under oil microdroplets, after 20—30 days of individual culture, with frequencies of up to 90% (on average, however, 20—30%) of the cases for tobacco and between 1—2% for rapeseed hypocotyl protoplasts. In the case of *F. hygrometrica* protoplasts, protonema filaments will be formed from the individually selected protoplasts after 10—15 days in individual culture without requiring addition of fresh culture medium to the microdroplets containing individual protoplasts in nanodroplets of initially 125 nl culture medium (having a size which will directly allow — with the aid of the tip of a needle — to transfer them out of the oil microdroplets directly to standard culture conditions for further growth and plant regeneration (see: [23]) in up to 60% of the cases.

In a species-dependent manner, the timepoint for transferring the microcultured protoplast-derived microcolonies out of the oil microdroplets and further culture steps have proven to be critical in particular cases. In the case of tobacco protoplasts — after 20—30 days in microculture — microcalluses can be individually transferred onto wells of multiwell dishes containing 1 ml of agarose-solidified morphogenesis medium and efficiently regenerated to plants. However, for microcultured rapeseed hypocotyl protoplasts, the microcolonies obtained — after up to 1 month in micro-culture — will necessarily require for further growth intermediate culture steps, including first, the transfer onto 1 µl wells of polycarbonate microdishes containing 500 nl of liquid culture medium and later, up to three additional microculture steps with a progressive increase of the culture volume (2—5 fold from one step to the next) per microcallus for finally establishing callus clones with potential for plant regeneration [32].

Notes

3.1 In most cases, conditions optimized for microdroplet culture for one particular genotype of one species allow for acceptable performance (e.g. viability, cell division, microcallus formation and finally plant regeneration) of protoplasts from other genotypes of the same species or intraspecific protoplast microfusants. This holds true: a) for tobacco protoplast microculture in initial 55 nl droplets of culture medium PNT [39] for *N. tabacum* cv. Xanthi, cv. Petit Havana SR1, cv. Badischer Burley, tobacco cms analogs with cytoplasms of *N. debneyi*, *N. bigelovii* and *N. suaveolens* and some of their pairwise fusants; b) for rapeseed hypocotyl protoplast microculture in initial 60 nl droplets of culture medium PBN-7 [34, 36, 32], for *B. napus* cv. Bronowski, cv. Tower and rapeseed cms analog with cytoplasm of *Raphanus sativus* and some of their pairwise fusants; and c) for protonemata protoplasts of the moss *F. hygrometrica* in microdroplets of initial 125 nl of culture medium Fha [23] for wild type and four different auxin-resistant development mutants and some of their pairwise microfusants. In most cases tested, if genotype-dependent differences in behaviour in microdroplet culture were evident, a similar trend was detectable in mass culture. Nevertheless, genotype dependent "microdroplet culture specific" differences cannot be generally ruled out.

3.2 If no significant differences in performance of individually selected protoplasts in microculture can be detected in a particular range of initial volume of the nanodroplets of culture medium, e.g. for *F. hygrometrica* protoplasts in the volume range of 30—125 nl, then the larger volumes are recom-mended, as: a) a faster delivery of the protoplasts is possible during the selection and transfer into microculture, b) further "feeding"-steps, where fresh culture medium is added/exchanged to the microdroplets, can be limited to a minimum or fully avoided, thus reducing tedious working steps, risk of contamination, osmotic shock of the microcultured cells, etc.

3.3 When transferring a defined cell from the selection chamber to a microdroplet of the microculture chamber the optimal total initial volume of the nanodroplet of culture medium will be divided into a portion already present under the oil microdroplet in the ready-to-use microculture chamber and the residual volume required for withdrawing the cell from the selection chamber. A halving of the desired total volume of the nanodroplet of culture medium for fulfilling both purposes is recommended, if possible. In general, the portion of the microdroplet volume of culture medium already present under the oil microdroplet (*recipient volume*) should not be less than 5—10 nl (in order to avoid "burying" the selected cell in oil while transferring it) — if possible — and the same holds true for the withdrawing volume from the selection chamber (*transfer volume*) while selecting the cell (in order to facilitate its delivery into microculture, as minor elasticities in the mineral oil-filled hydraulic system cannot be completely excluded). If larger volumes (more than 10—30 nl) are tolerable as initial total nanodroplet volume, then the recipient volume can be increased and the transfer volume can be split into two working steps of the nanoliter pump: one for withdrawing only culture medium (without the cell) from the selection chamber and the second for withdrawing the cell to be selected and transferred. In this way, the cell to be transferred will be more terminal to the tip of the selection microcapillary and thus rapidly delivered into the microdroplet while activating the nanoliter-pump and a risk of selecting the "target" cell with an "undesired" cell from the selection chamber — due to large negative pressure and withdrawing volume — will be reduced.

3.4 Indicated overall plant regeneration frequencies for the microdroplet culture of protoplasts of tobacco, rapeseed and *F. hygrometrica* are calculated as % of the individually selected protoplasts transferred into microdroplets which formed microcolonies under microculture conditions, these latter being transferred out of the oil microdroplets and separately plated onto further culture steps and finally regenerating to plants.

B. Single cell nurse culture

The feasibility, in principle, to culture individual higher plant cells by exploiting the supportive effect of "nurse cells" was initially demonstrated by Muir *et al.* [24] in their classical paper, in which it was shown that a higher plant cell line can be derived from an individual cell. There, an individual cell was placed on a sheet of filter paper on top of a nurse callus serving as a physical but not a diffusion barrier between the individual and the nurse cells. Here, we describe the simultaneous nurse culture of many single cells in a culture system, which uses an agarose layer containing a number of pits for inoculation with pre-selected single cells and a stainless steel basket with a nylon sieve or dialysis membrane bottom as a container for nurse cells in liquid medium [8]. With this culture system [8] a number of physiological parameters relevant to single cell nurse culture have been analysed [31].

In contrast to nanodroplets single cell nurse culture does not provide completely defined culture conditions. However, the system may offer a number of advantages, which can be summarized as follows:

A) *Transfer of single cells*
Cells are routinely cultured in a total volume of 4 ml (see below) of culture medium. Thus, it is not critical to use a defined and very small volume for withdrawal and transfer of single cells from selection chambers. Therefore, a nanoliter pump is not mandatory for this procedure, although still highly recommendable for selection and transfer of individual cells.

B) *Renewal of culture medium*
Culture medium is simultaneously replenished for a whole set of target cells using standard pipettes. Thus it is not necessary to "feed" fresh medium into individual microculture chambers as might be the case for nanodroplets. Single cell nurse culture therefore provides a culture system for individual cells which is much easier to handle.

C) *Same culture dish during regeneration from single cells or protoplasts to macroscopically visible colonies*
Whereas specific "feeding" protocols (compare section 3) and a precise step-up procedure regarding the total culture volume at each developmental stage of a microcolony may be required when using nanodroplets, in single cell nurse culture, the cell and the colonies derived thereof are maintained at the same sites in the same culture dish until macrocolonies of 1 to 2 mm diameter have been achieved, sufficient in size for transfer to standard culture conditions. This again contributes to the ease of handling of this procedure.

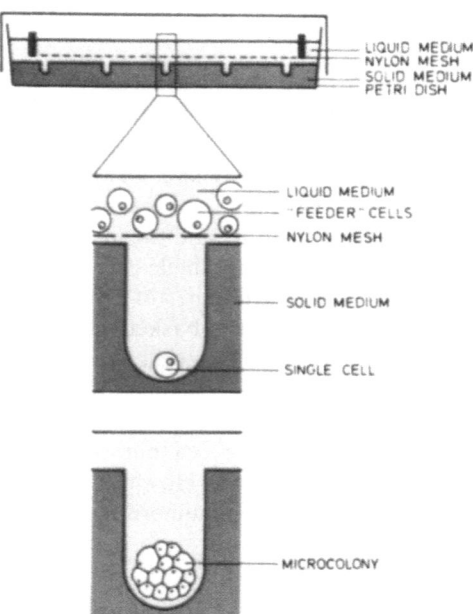

LIQUID MEDIUM
NYLON MESH
SOLID MEDIUM
PETRI DISH

LIQUID MEDIUM
"FEEDER" CELLS
NYLON MESH

SOLID MEDIUM

SINGLE CELL

MICROCOLONY

Fig. 4. Schematical representation of the single cell nurse culture system. The upper part shows the total setup, a section with a single protoplast and the feeder layer is depicted in the central part, the lower part demonstrates a single microcolony after removal of the feeder layer.

Procedures

5. Major equipment

Basically, the same hardware and software as described for microdroplet culture is used for selection and transfer of individual cells in single cell nurse culture. Since this culture system uses however standard Petri-dishes (3.5 cm diameter) rather than coverslips, a stage insert containing at least one position for a Petri-dish is required.

6. Making a single cell nurse culture dish

A special casting tool (Fig. 5B) is required for preparing a single cell nurse culture dish. It is machined by inserting the 25 pins (diameter 1 mm, lengt 7 mm) of a standard computer plug (Fig. 5A, male DB 25P, ITT Cannon, Santa Ana, CA, USA) in five rows of five pins into a brass holder. The distance between pins in a row and between rows is 4 mm. The brass holder is adjusted to provide a distance of 0.5 mm between the lower ends of the pins and the inner surface of the bottom of the Petri-dish, when the holder is inserted into the Petri-dish (Fig. 5C). The casting tool has a handle for easier insertion into and removal from the Petri-dishes. It is sterilized by autoclaving prior to its usage. For casting of single cell nurse culture dishes, 2 ml of liquified 2% agarose (LMP agarose, BRL, Bethesda) made up with the respective culture medium, autoclaved and stored in a water bath at 45 °C, are pipetted into a Petri-dish, and the casting tool is inserted. After 10 to 15 min at room temperature, 1 ml of liquid culture medium is added on top of the solidified agarose layer, and the casting device is carefully removed.

Notes

6.1 The dimensions of the system are critical: if the agarose layer is thinner, the pits are not deep enough, and the slightest turbulence will remove the single cells from the pits.

6.2 Single cell nurse culture dishes can be stored in a refrigerator for at least one week. During storage the density of the medium tends to increase causing the cells inoculated into the agarose pits to float rather then sediment into the pit. The liquid layer should then be replaced with fresh culture medium prior to inoculation with single cells.

7. Making a basket for feeder cells

Two stainless steel rings (30 mm outer diameter; 6,5 mm height) are used for preparing a basket for feeder cells (Fig. 5D). The outer diameter of the lower part of the upper ring is adjusted to slightly less than the inner diameter of the lower ring, such that when sliding the lower ring onto the lower part of the upper ring, a dialysis membrane (or nylon sieve) is safely held in place between the two rings. When the bottom of the basket, providing a physical but not a diffusion barrier is attached, the parts of the membrane, protruding between the two rings are trimmed off using small scissors. Whereas nylon sieve of 5 μm pore size is also suitable, a single layer of dialysis tubing (43 mm width, Sigma #9527), providing much better optical conditions, is recommended [31].

Notes

7.1 Stainless steel rings and barrier membranes should be carefully cleaned by boiling in a suitable laboratory detergent and subsequent rinses with bidistilled water.

7.2 When using dialysis membrane, it should not be allowed to dry during the whole procedure in order to prevent formation of folds.

7.3 Assembled baskets are autoclaved and stored in bidistilled water and excess water is drained off prior to inserting the basket into the Petri dish.

8. *Assembling a nurse culture dish after inoculation of the pits in the agarose layer with single cells*

After inoculation with single cells, a feeder basket is inserted into the culture dish, carefully avoiding entrapment of air between the agarose layer and the basket, and feeder cells, suspended in one ml of culture medium are pipetted into the basket.

9. *Culture procedure*

As with culture of individual cells in nanodroplets, protocols for single cell nurse culture have to be adjusted to the type of cells used in a particular experiment. For tobacco leaf protoplasts the following procedure is used:

1) use the culture medium as described earlier [8, 31];
2) adjust the density of nurse cells to 5×10^4 cells per ml of the total volume (4 ml);
3) once a week, and starting after one week of culture, remove one eight to one quarter of the liquid phase outside and inside of the basket and replace with fresh culture medium;
4) gradually reduce the osmolality after approximately three weeks of culture.

Following this procedure, colonies become macroscopically visible (300 µm diameter) after three to four weeks of culture and feeder baskets can be removed at this time. Routinely, about 50% to 90% of the single cells develop into colonies, and subsequently, into plants.

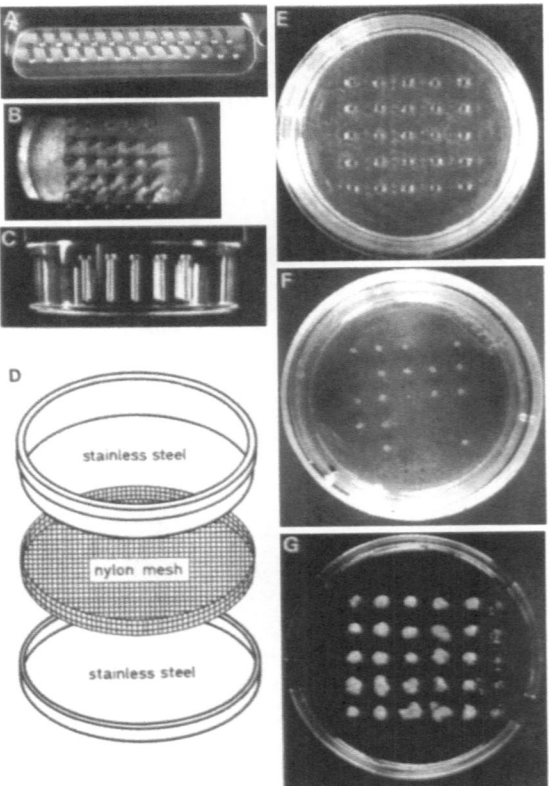

Fig. 5. Single cell nurse culture system.

Fig. 5A-E: preparation of single cell nurse culture dish. Fig. 5F: microcolonies derived from single protoplasts, 5G: test for contamination of single cells with feeder cells during single cell nurse culture.

Fig. 5A: computer plug containing 25 gold plated pins, Fig. 5B: pins removed from the plug and mounted into a brass stamp, Fig. 5C: stamp with pins inserted into the bottom of a standard plastic Petri-dish (3.5 cm diameter) for molding of an agarose layer, Fig. 5D: schematical representation of a custom made basket for taking up of the feeder cells, Fig. 5E: Petri-dish with agarose layer containing pits for inoculation with single cells, Fig. 5F: microcolonies in a single cell nurse culture dish, derived from leaf protoplasts of *N. tabacum* after four weeks of culture and after removal of the feeder layer, Fig. 5G: growth of 25 colonies produced from single cell nurse culture of leaf protoplasts of a kanamycin resistant line of *N. tabacum* and 5 colonies produced from feeder protoplasts derived from a non-resistant wildtype; all the colonies originating from culture of single protoplasts show normal growth on selection medium. Illustrations are not to scale.

References

1. Abel WO, Knebel W, Koop HU, Marienfeld JR, Quader H, Reski R, Schnepf E, Spörlein B (1989) A cytokinin-sensitive mutant of the moss, *Physcomitrella patens*, defective in chloroplast division. Protoplasma 152: 1–13.
2. Bariaud-Fontanel A, Tabata M (1988) Somaclonal variation in the berberine-producing capability of a culture strain of *Thalictrum minus*. Plant Cell Rep 7: 206–209.
3. Bolik M, Koop HU (1991) Identification of embryogenic microscopes of barley (Hordeum vulgare L.) by individual selection and culture and their potential for transformation by microinjection. Protoplamsma (*in press*).
4. Cella R, Galun E (1980) Utilization of irradiated carrot cell suspensions as feeder layer for cultured *Nicotiana* cells and protoplasts. Plant Sci Lett 19: 243–252.
5. Collin HA, Dix PJ (1990) Culture systems and selection procedures. In: Dix P (ed) Plant Cell Line Selection, pp 3–17. Weinheim: VCH Verlagsgesellschaft.
6. Crossway A, Oakes JV, Irivne JM, Ward B, Knauf VC, Shewmaker CK (1986) Integration of foreign DNA following microinjection of tobacco mesophyll protoplasts. Mol Gen Genet 202: 179–185.
7. De Ropp RS (1955) The growth and behaviours in vitro of isolated plant cells. Proc Roy Soc B 144: 86–93.
8. Eigel L, Koop HU (1989) Nurse culture of individual cells: Regeneration of colonies from single protoplasts of *Nicotiana tabacum, Brassica napus* and *Hordeum vulgare*. J Plant Physiol 134: 577–581.
9. Eigel L. Koop HU (1990) Organelle transfer by protoplast-subprotoplast microfusion: transfer of very low numbers of wild type chloroplasts leads to variegated regenerants. Abstract In: Proceedings VII IAPTC Meeting, Amsterdam 1990, pp 209.
10. Ellis BE (1985) Characterization of clonal cultures of *Anchusa officinalis* derived from single cells of known productivity. J Plant Physiol 119: 149–158.
11. Gleba YY (1978) Microdroplet culture: tobacco plants from single mesophyll protoplasts. Naturwissenschaften 65: 158–159.
12. Gleba YY, Hoffmann F (1980) "*Arabidobrassica*": A novel plant obtained by protoplast fusion. Planta 149: 112–117.
13. Jones LE, Hildebrandt AC, Riker AJ, Wu JH (1960) Growth of somatic tobacco cells in microculture. Am J Bot 47: 468–475.
14. Kao KN, Michayluk M (1975) Nutritional requirement for growth of *Vicia hajastana* cells and protoplasts at a very low population density in liquid media. Planta 126: 105–110.
15. Kao KN (1977) Chromosomal behaviour in somatic hybrids of soybean and *Nicotiana glauca*. Mol Gen Genet 150: 225–230.
16. Koop HU, Weber G. Schweiger HG (1983a) Individual culture of selected single cells and protoplasts of higher plants in microdroplets of defined media. Z Pflanzenphysiol 112: 21–34.
17. Koop HU, Dirk J, Wolff D, Schweiger HG (1983b) Somatic hybridization of two selected single cells. Cell Biol Int Rep 7: 1123–1128.
18. Koop HU, Schweiger HG (1985a) Regeneration of plants from individually cultivated protoplasts using an improved microculture system. J Plant Physiol 121: 245–257.
19. Koop HU, Schweiger HG (1985b) Regeneration of plants after electrofusion of selected pairs of protoplasts. Eur J Cell Biol 39: 46–49.
20. Koop HU, Spangenberg G (1989) Electric field-induced fusion and cell reconstitution with preselected single protoplasts and subprotoplasts of higher plants. In: Neumann E, Sowers A, Jordan C (eds) Electroporation and Electrofusion in Cell Biology, pp 355–366. New York, London: Plenum Press.
21. Kranz E, Bautor L, Lörz H (1990) In vitro fertilization of single, isolated gametes, transmission of cytoplasmic organelles and cell reconstitution of maize (*Zea mays L.*). In: Proceedings VII IAPTC Meeting, Amsterdam 1990, pp 252–257.
22. Krumbiegel G, Schieder O (1981) Comparison of somatic and sexual incompatibility between *Datura innoxia* and *Atropa belladonna*. Planta 53: 466–470.

23. Mejia A, Spangenberg G, Koop HU, Bopp M (1988) Microculture and electrofusion of defined protoplasts of the moss *Funaria hygrometrica*. Botanica Acta 101: 166–171.
24. Muir WH, Hildebrandt AC, Riker AJ (1954) Plant tissue cultures produced from single isolated cells. Science 119: 877–878.
25. Ogino T, Hiraoka, N, Tabata M (1978) Selection of high nicotine-producing cell lines of tobacco callus by single-cell cloning. Phytochemistry 17: 1907–1910.
26. Oksman-Caldentey K, Strauss A (1986) Somaclonal variation of scopolamine content in protoplast-derived cell culture clones of *Hyoscyamus muticus*. Planta Medica 1: 6–12.
27. Patnaik G, Cocking EC, Hasmill J, Pental D (1982) A simple procedure for the manual isolation and identification of plant heterokaryons. Plant Sci Lett 24: 105–110.
28. Potrykus I, Shillito RD (1986) Protoplasts: Isolation, culture, plant regeneration. In: Weissbach A, Weissbach H (eds). Methods in Enzymology, Plant Molecular Biology 118 pp 549–578. Orlando: Academic Press.
29. Potrykus I, Harms CT, Lörz H (1979) Multiple-drop-array (MDA) technique for the large-scale testing of culture media variations in hanging microdrop cultures of single cell systems. I. The technique. Plant Sci Lett 14: 231–235.
30. Reinert L (1956) Dissociation of cultures from *Picea glauca* into small tissue fragments and single cells. Science 123: 457–458.
31. Schäffler E, Koop HU (1990) Single cell nurse culture of tobacco protoplasts: physiological analysis of conditioning factors. J Plant Physiol 137: 95–101.
32. Schweiger HG, Dirk J, Koop HU, Kranz E, Neuhaus G, Spangenberg G, Wolff D (1987) Individual selection, culture and manipulation of higher plant cells. Theor Appl Genet 73: 769–783.
33. Spangenberg G, Koop HU, Schweiger HG (1985) Different types of protoplasts from *Brassica napus*: analysis of conditioning effects at the single cell level. Eur J Cell Biol 39: 41–45.
34. Spangenberg G (1986) Manipulation individueller Zellen der Nutzpflanze *Brassica napus* mit Hilfe von Elektrofusion, Zellrekonstruktion und Mikroinjektion. PhD thesis, Universität Heidelberg.
35. Spangenberg G, Schweiger HG (1986) Controlled electrofusion of different types of proto-plasts including cell reconstitution in *Brassica napus*. Eur J Cell Biol 41: 51–56.
36. Spangenberg G. Koop HU, Schweiger HG (1986a) Microculture of single protoplasts of *Brassica napus*. Physiol Plant 66: 1–8.
37. Spangenberg G, Neuhaus G, Schweiger HG (1986b) Expression of foreign genes in a higher plant cell after electrofusion-mediated cell reconstitution of a microinjected karyoplast and a cytoplast. Eur J Cell Biol 42: 236–238.
38. Spangenberg G, Neuhaus G, Potrykus I (1990a) Micromanipulation of higher plant cells. In: Dix P (ed) Plant Cell Line Selection, pp 87–109. Weinheim: VCH Verlagsgesellschaft.
39. Spangenberg G, Osusky M, Oliveira MM, Freydl E, Nagel J, Pais MS, Potrykus I (1990b) Somatic hybridization by microfusion of defined protoplast pairs in *Nicotiana*: morphological, genetic, and molecular characterization. Theor Appl Genet 80: 577–587.
40. Spangenberg G, Freydl E, Osusky M, Nagel J, Potrykus I (1990c) Organelle transfer by microfusion of defined protoplast-cytoplast pairs. Theor Appl Genet (in press).
41. Stuart R, Street HE (1971) Studies on the growth in culture of plant cells. X. Further studies on the conditioning of culture media by suspensions of *Acer pseudoplatanus L.* J Exp Bot 22: 96–106.
42. Torrey JG (1957) Cell division in isolated single cells in vitro. Proc Natl Acad Sci USA 43: 887–891.
43. Vasil V, Hildebrandt AC (1965) Differentiation of tobacco plants from single, isolated cells in microcultures. Science 150: 889–892.
44. Weber G, Lark KG (1979) An efficient plating system for rapid isolation of mutants from plant cell suspension. Theor Appl Genet 55: 81–86.

Plant Tissue Culture Manual **A11**: 1–11, 1992.

Tobacco Protoplast Isolation, Culture and Regeneration

I. NEGRUTIU

Plantengenetica, Paardenstraat 65, B-1620 St. Genesius Rode, Belgium.

Introduction

Plant protoplasts have been in constant use for more than two decades, and have become one of the most versatile analytical tools in plant biology. Depending upon the species and culture conditions, the protoplasts may have the potential to:

(1) regenerate a cell wall;
(2) dedifferentiate;
(3) divide mitotically and proliferate clonally;
(4) differentiate into shoots, roots or embryos and produce a complete plant [1, 3].

They can be isolated in large quantities from a variety of tissues or organs. Freed from the cellulosic cell walls, the plasma membrane becomes accessible for investigation or purification. The relatively homogenous population of individualized wall-less cells can be subjected to various experimental treatments.

These characteristics of plant protoplasts make them the material of choice for the following applications:

(1) Transient gene expression experiments with considerable implications in (a) elucidating tissue-specific regulation and *cis*- and *trans*-regulatory interactions among foreign and/or endogenous genes, (b) analysing translational processes independent of transcriptional events.
(2) Molecular cytogenetics, with particular emphasis on *in situ* hybridization for gene localization on metaphase chromosomes (for example root meristem protoplasts).
(3) Flow-cytometry experiments, comprising cell-cycle analysis and chromosome sorting.
(4) Cloning of large DNA inserts (megabase cloning) in DNA preparations from agarose embedded protoplasts.
(5) DNA analysis and long-range mapping via pulsed-field gel electrophoresis.

One major requirement in performing these types of experiments is the production of high yields of robust protoplasts. Tobacco is one of the pioneering exprimental species in this area, for which have been established simple and reproducible protoplast isolation and culture procedures. The isolation technique described here has been shown to work also on a variety of other plant species.

Materials

Nicotiana tabacum

Haploid or diploid *in vitro* cuttings, grown on basal MS medium (Murashige and Skoog [2], obtainable, e.g. from Sigma; see [4]), long days, of maximum light intensity of 2500 lux, 25 °C. Low protoplast yield and survival are observed when plants are grown under continuous light.

Equipment

- forceps, scalpels and blades (sterile)
- beaker 100 ml, sterile
- capped plastic centrifuge tubes
- rack for centrifuge tubes
- incubator
- 500–100 μm sieve batteries with beaker
- counting chamber, type 'Thoma'
- Parafilm
- sterile pipettes 1 ml, 5 ml, 10 ml
- Petri dishes 9 cm, 6 cm.

Protoplast isolation and culture

Steps in the procedure

1. Fully expanded leaves are sliced in the enzyme solution (10 ml per Petri dish) and incubated overnight at 20 °C.
2. Next morning shake the dish gently to free the protoplasts and sieve them through sterile 50–100 µm sieves and add half volume ML0.6.
3. Transfer to 15 ml centrifuge tubes and spin at 700 rpm for 5 min.
4. Remove the band of floating protoplasts with a Pasteur pipette. The band will be easy to handle if sucked as a mixture of air and protoplast suspension. Dilute with 3–5 volumes of W5 salts solution.
5. Spin at 700 rpm for 5 min. Remove the supernatant with a pipette or vacuum pump. Shake the pellet gently and resuspend it in 5–10 ml W5 salts solution. Count the density in a counting chamber. Measure the volume exactly.
6. Dilute with culture medium and distribute into Petri dishes at 5×10^4 protoplasts per ml, 10 ml per 9 cm dish. Incubate for 72 h in the dark. Later the cultures are maintained in low light/dark cycles.
7. To embed protoplasts in agarose, prepare fresh medium by mixing liquid K3M and autoclaved agarose powder (calculate agarose amount so that the final concentration is 0.6%) by melting in a microwave oven. Surplus medium can be remelted a couple of times. Perform this one hour before use and keep the medium at 37 °C until use. When embedding in agarose, make sure that the temperature of the medium is below 37 °C.

Diluton and planting of dividing colonies

Steps in the procedure

8. After 7–10 days in culture, SR1 protoplasts have undergone 4–5 division cycles. They can be diluted to reduce the osmoticum concentration and adjust the hormonal balance.
9. Dilute by a factor 10 with AG medium. The same conditions are used when selecting for resistance to antibiotics, herbicides, etc. in transformation experiments. In those cases in which agarose was included in K3M, sectors can be sliced in the original dish and transferred with a spatula to a new container. If more than 12 ml of culture medium is to be used, shake the dilution mix for appropriate aeration.
10. After 2–3 weeks in culture, colonies visible to the naked eye are obtained. Refresh the medium by removing the old one. The osmotic pressure in the fresh AG medium should be 0.2M. If agarose sectors undergo dilution steps, break them into smaller pieces to reduce local colony density.
11. 2–3 mm microcalluses can be picked manually or through wide-mouth pipettes and spread on top of agar plates containing MR_1 medium. During two successive transfers onto MR_1, callus is produced and regeneration of multiple shoot structures should occur. If the regeneration frequency is too low, reduce the BAP concentration to 0.25 mg/l (MR025).

12. Rooting and elongation of the shoots is obtained on half strength basal MS medium [2].
13. Rooted plantlets can be transferred to the greenhouse and/or maintained in culture as cuttings [4].

Note

Short-term cultures, such as those for transient gene expression experiments, are maintained in liquid media. Embedding in agarose (Sea-Plaque grade) is recommended for recalcitrant species or whenever optimal division rates and plating efficiencies are essential.

Troubleshooting guide

Many protoplasts burst (seen as cell "ghosts" sticking at the bottom of the Petri dish), and vacuolated cells may frequently exhibit low division rates. This problem can be reduced by the following method.

Add $<0.25\%$ PEG (from a standard 40% stock solution) to the protoplasts and/or preincubate at a high density ($2.5-5 \times 10^5$ protoplasts per ml) for 24–48 h. Dilute afterwards to the standard 5×10^4 protoplast per ml. Furthermore, survival and division activity of the protoplasts are usually better when using a longer enzyme incubation time (overnight) and, thus, less concentrated enzyme solutions. The Ca^{2+} concentration during digestion is kept to a minimum (5 mM).

Adding NAA and BAP (1–5 mg/l each) to the enzyme can activate the protoplasts during initial stages of culture.

Solutions

Enzyme solution

Universal enzyme (EG, Yu Yu Gleba, Ukrainian Biotechnology Centre for Agriculture, Kiev)

- 1% Driselase (Fluka)
- 0.3% Cellulase R10 ('Onozuka', Yakult Pharmac.)
- 0.2% Macerozyme R10 (as above)
- 0.1% Cellulysin (Sigma)
- $CaCl_2 \cdot 2H_2O$ 5mM
- Sucrose 0.5M
- pH $\leqslant$ 5

A 50–75% strength of the enzyme mix is usually sufficient for a complete overnight digestion in the case of leaf protoplasts.

Washing solutions

W5 salt solution
154mM NaCl
125mM $CaCl_2 \cdot 2H_2O$
 5mM KCl
 5mM glucose
pH 5.8–6.0

Floating solution (ML 0.6)
0.6M Sucrose
15mM $CaCl_2 \cdot 2H_2O$
0.1% MES
pH 5.6

Culture media

MS salts stock solutions [2]
Microelements MS (1000 ×) g/l
H_3BO_3	6.20
$MnSO_4 \cdot 4H_2O$	22.30
$ZnSO_4 \cdot 7H_2O$	8.60
$Na_2MoO_4 \cdot 2H_2O$	0.25
$CuSO_4 \cdot 5H_2O$	0.025
$CoCl_2 \cdot 6H_2O$	0.025

Macroelements MS (20 ×) g/l
KNO_3	38.00
NH_4NO_3	33.00
$MgSO_4 \cdot 7H_2O$	7.40
$CaCl_2 \cdot 2H_2O$	8.80
KH_2PO_4	3.40

Morel vitamins [5] stock solution (500 ×) mg/l
thiamine · HCl	168.5
myo-inositol	9000.0
pyridoxine · HCl	102.5
Ca pantothenate	238.0
biotin	1.22
nicotinic acid	61.5

K3M basal (protoplast culture)
Per liter:
– macroelements K3 (10 ×)	100 ml
– microelements K3 (1000 ×) [2]	1 ml
– Fe-EDTA (100 ×)	5 ml
– Morel vitamins (500 ×) [5]	2 ml
– thiamine	100 mg
– glucose	0.45 M
– Sea-Plaque agarose	0.6% (optional)
– pH 5.5	

Macroelements K3 (g/l)

NH_4NO_3	0.6
KNO_3	1.9
$CaCl_2 \cdot 2H_2O$	0.6
$MgSO_4 \cdot 7H_2O$	0.3
KH_2PO_4	0.17
KCl	0.3

AG medium (dilution of protoplast-derived colonies) [1]

Per liter:
- Macroelements (10 ×)   100 ml
- Microelements, Morel vitamins, Fe-EDTA as in K_3M
- Sucrose   30 g
- Mannitol   50 g
- NAA   0.1 mg
- BAP   1 mg
- pH   5.7

Macroelements (g/l)

NH_4NO_3	1.01
$CaCl_2 \cdot 2H_2O$	0.44
$MgSO_4 \cdot 7H_2O$	0.74
KH_2PO_4	0.136
$(NH_4)_2$ succinate	0.1

MR_1 medium (callus formation and shoot regeneration)

Per liter:
- macroelements MS (20 ×) [2] 50 ml
- microelements MS (1000 ×) [2] 1 ml
- Fe-EDTA (100 ×)   10 ml
- Morel vitamins (500 ×) [5]   2 ml
- thiamine-HCl   100 mg
- sucrose   30 g
- BAP   (0.25–) 1 mg
- agar   0.6%
- pH 5.8

Enzyme solutions and K_3M medium are sterilized by filtration and stored frozen at − 20 °C. All other solutions are autoclaved. Autoclave Sea-Plaque agarose as dry powder.

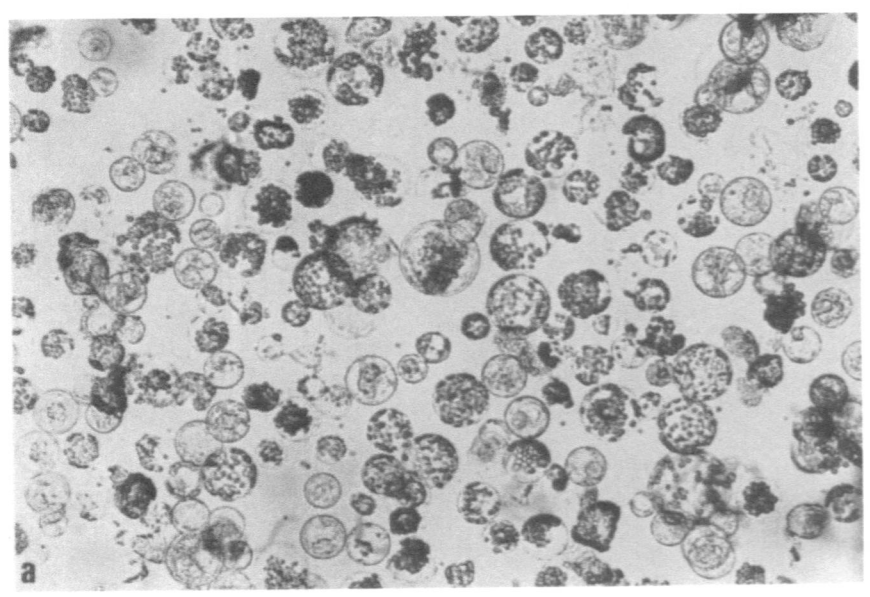

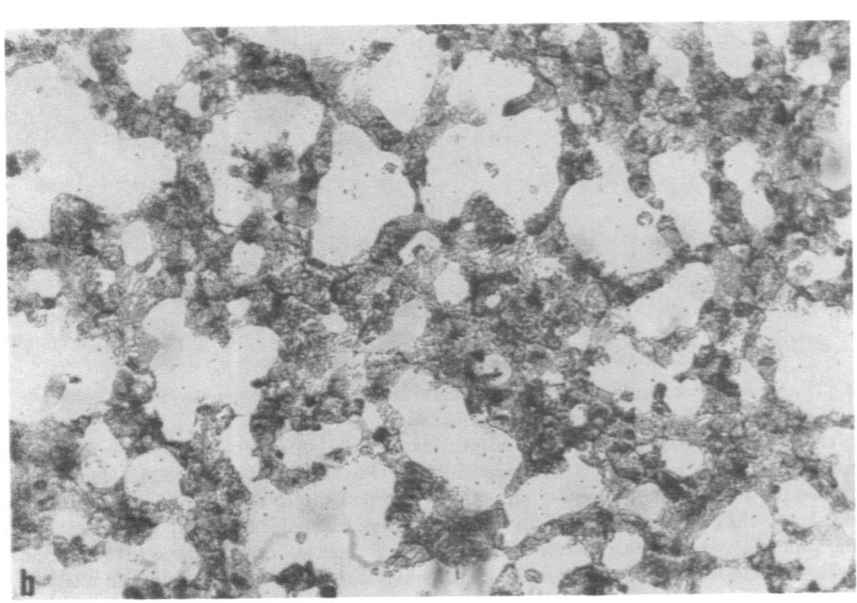

PTCM-A11/9

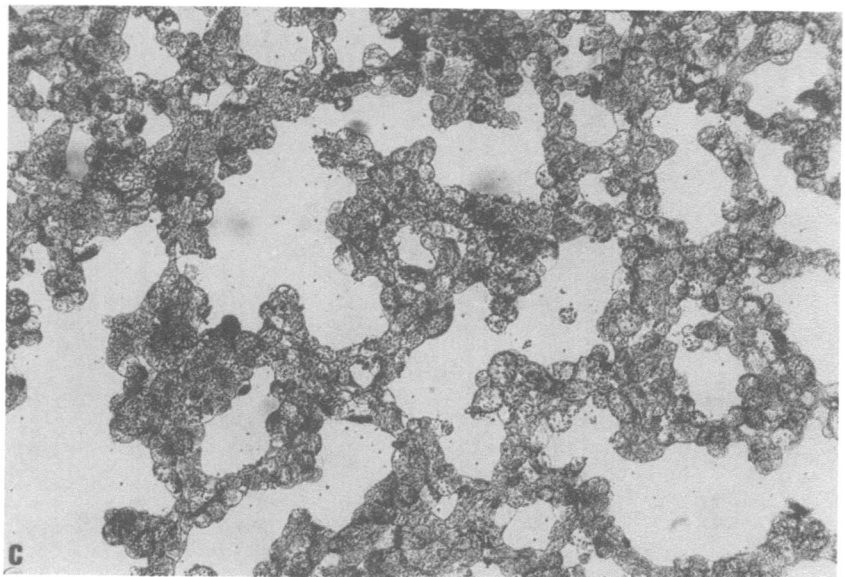

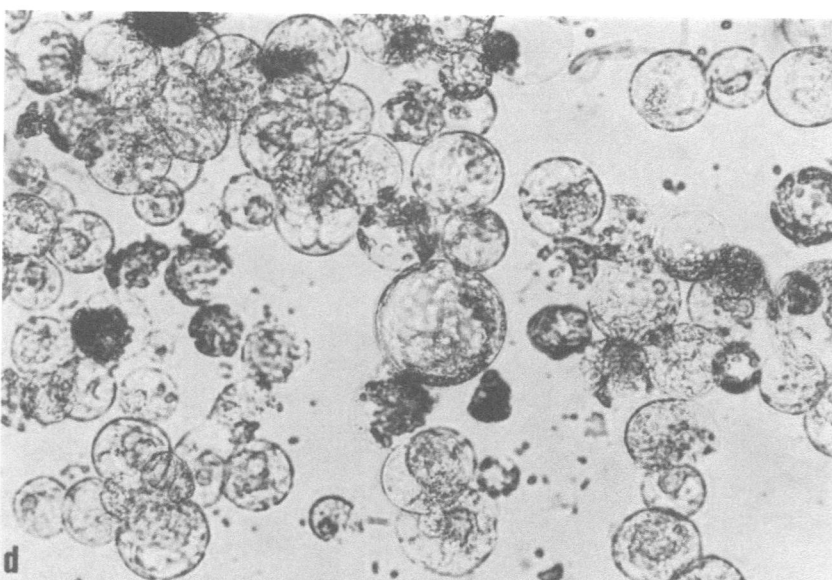

Fig. 1. A PEG test on freshly isolated protoplasts of *Nicotiana tabacum* to assess their physiological condition.

(a) A mixture of mesophyll- and cell suspensions-derived protoplasts just after isolation (the former contain chloroplasts and are vacuolated). (b) PEG 4000 is applied at 26% final concentration for 10 min. The protoplasts agglutinate and undergo extreme modifications of shape, demonstrating a high level of elasticity. (c) The PEG is gradually diluted with W5 salts solution, the protoplasts recovering spherical shapes but still showing agglutination. (d) PEG-treated protoplasts in culture medium exhibiting high survival rates (few protoplasts died after treatment).

References

1. Caboche M (1980) Nutritional requirements of protoplast-derived, haploid tobacco cells grown at low densities in liquid medium. Planta 149, 7–18.
2. Murashige T, Skoog F (1962) A revised medium for rapid growth and bioassays with tobacco tissue culture. Physiol Plant 15, 473–497.
3. Potrykus I, Shillito RD (1986) Protoplasts: isolation, culture, plant regeneration. Meth Enzym 118, 549–578.
4. Topping JF, Lindsey K (1991) Shoot cultures and root cultures of tobacco. This volume, A4/1–14.
5. Morel G, Wetmore RH (1951) Fern tissue culture. Amer J Bot 38, 141–143.

Section B:
Tissue Culture & Transformation of Crop Species

Plant Tissue Culture
Manual

Crop Species

Sect. B

Plant Tissue Culture Manual **B1**: 1–16, 1991.
© 1991 *Kluwer Academic Publishers.*

Embryogenic callus, cell suspension and protoplast cultures of cereals

INDRA K. VASIL & VIMLA VASIL
Laboratory of Plant Cell and Molecular Biology, Department of Vegetable Crops, University of Florida, Gainesville, FL 32611-0514, USA

Introduction

Cereals constitute the most important source of calories and protein for man since more than 52% of our food is derived from grains such as wheat, rice, maize, barley, millets, etc. Cereals, therefore, are an obvious and important target for genetic manipulation by modern biotechnological methods which require efficient regeneration of plants from cultured tissues and cells, and genetic transformation, two important and interacting components of plant biotechnology. In addition, owing to the serious problems still being faced in *Agrobacterium*-mediated transformation of cereals, the development of a reliable protoplast regeneration system for direct DNA delivery is also a necessity.

Prior to 1980, there were only a few scattered reports of plant regeneration from tissue cultures of cereals [1]. In the few instances where plant regeneration was possible, it was found to be inefficient, sporadic, of short-term nature, and restricted to one or two genotypes. However, rapid progress has been made during the past ten years so that efficient, reliable and long-term regeneration of plants is now possible from tissue cultures of all of the important species of cereals as well as grasses [2–4]. This was made possible because some of the early pioneering studies recognized the importance of culturing immature tissue/organ explants containing undifferentiated meristematic cells at precisely defined stages of development in high 2,4-dichlorophenoxyacetic acid (2,4-D) media for the establishment of embryogenic cultures. Such cultures have proven invaluable for the development of regenerable cell suspension cultures which are still the only source of totipotent protoplasts for this important group of crop species [5–6]. Embryogenic suspension cultures as well as protoplasts isolated from them have been used successfully to obtain transgenic cereals [7–9].

Embryogenic Callus Cultures

Immature embryos are the most suitable explant for the establishment of embryogenic cultures [3, 4, 10]. The developmental stage and the physiological status of the explant are two of the most significant factors which regulate the response in vitro. The genotype of the explant is also considered by some to be an important factor, although its effect can be largely overcome by physiological and developmental factors. Best response is obtained from embryos in which morphological development has been nearly completed, and the deposition of starch in the scutellar cells has just begun. Culture of such immature embryos in simple nutrient media supplemented with a strong auxin, such as 2,4-D, is generally sufficient to induce the formation of embryogenic callus in most species. Callus is formed from several of the outermost layers of the scutellar cells in the coleorhizal half of the embryo. Segments of young inflorescences [10, 11], or the bases of young leaves [12], are also suitable for the initiation of embryogenic cultures in many species [11]. Irrespective of the explant used, both embryogenic and non-embryogenic calluses are formed. The former is white, off-white or pale-yellow in color, and is compact, organized and often nodular in appearance (Type I callus). It comprises small, richly cytoplasmic cells containing a large nucleus, many amyloplasts, lipid bodies, and small vesicles but no large vacuoles. The non-embryogenic callus is soft and translucent, grows more rapidly, but consists mostly of large and vacuolated cells. Early identification of embryogenic sectors, their physical separation from non-embryogenic callus, and selective subculture on fresh media every 2–4 weeks are critical for long term maintenance of the cultures and their embryogenic potential. Somatic embryos and plantlets are often formed on the callus induction medium itself, as the effective amount of 2,4-D in the medium is reduced gradually during culture, although transfer of the callus to a medium with low levels of 2,4-D (generally 10% of the amount used for callus induction) gives better results.

A second type of embryogenic callus, described as Type II callus, is formed at a very low frequency (often in only 0.1% of the cultured embryos) in a few inbred lines and hybrids of maize [13, 14]. Inclusion of L-proline in the medium is considered beneficial. Type II callus is relatively soft, friable, translucent, and fast growing, but contains the typical embryogenic cells found in Type I callus. It forms somatic embryos in abundance, can be maintained for long periods of time by subculture, and is essential for the establishment of embryogenic cell suspension cultures. Type I callus of maize can not be maintained in culture for more than 4-6 months and will not form cell suspensions.

There is increasing evidence that embryogenic cultures are genetically more stable than non-embryogenic cultures [1, 5]. In addition, there is a strong selection in favor of normal cells during the formation of somatic embryos so that most of the plants recovered from embryogenic cultures are genetically similar to the explant material [5, 15]. Such cultures are, therefore, ideal for the rapid clonal multiplication of new hybrids, cultivars, etc., and for genetic transformation.

Embryogenic Cell Suspension Cultures

Since 1980, embryogenic callus cultures have been used to establish fast growing and finely dispersed suspension cultures in a few species, including important crops like wheat, maize, rice, sugarcane, etc. [7–9, 13, 14, 16–18]. As stated earlier, embryogenic suspensions of maize can be established only from Type II callus. Nevertheless, well established cell suspension cultures of gramineous species are still rare and rather difficult to obtain and maintain. Fortunately, such valuable cell lines, once established, can be preserved for long term use by cryopreservation [19, 20]. Plating of the suspensions on agar media is required for the formation of somatic embryos and for plant regeneration.

Embryogenic Protoplast Cultures

In spite of intensive attempts for more than 15 years it is still not possible to regenerate plants from, or even induce sustained cell divisions in, mesophyll protoplasts of gramineous species [21]. This is particularly surprising in view of the fact that since 1970 mesophyll protoplasts of dicotyledonous species have been used extensively for plant regeneration. Fortunately, suspension cultures have proven to be an excellent source of dividing and totipotent protoplasts, and plants have been obtained from protoplasts of almost all species of the Gramineae in which regenerable suspension cultures have become available [1, 5, 19, 21–23]. Further improvements of this technology are needed as in many instances protoplast plating efficiencies are still low, and recovery of mature and fertile plants is either difficult or not possible.

Applications

Protoplasts isolated from suspension cultures have been used for somatic hybridization [24], particularly for the production of cytoplasmic male sterile cybrids of rice [25], which can be very useful in hybrid seed production. Protoplasts of several gramineous species, including wheat, maize and rice, have also proven to be valuable for the evaluation of various methods of DNA delivery and selectable markers, the expression of introduced genes, the elucidation of promoter and intron function, etc. [26–31]. Fertile transgenic plants of rice have been produced following DNA delivery into protoplasts [7, 9]. Transgenic maize plants, carrying herbicide resistant genes, were obtained by the direct delivery of DNA into intact embryogenic suspension culture cells by the microprojectile bombardment method [8, 32].

    It should be a matter of satisfaction and pride to those working with cereal crops that within a decade they have brought the technology to a stage where practical applications are imminent. The protocols described in the following pages are generic in nature, and should be applicable to most species of cereals and grasses, with appropriate adjustments made for species, genotypes and cell lines of interest.

Acknowledgements

We thank the large number of graduate students and post-doctoral associates who over the years have greatly helped in the development of efficient protocols used for cereal and grass species in our laboratory. Support for this research has been provided by the Monsanto Co. (St. Louis, MO), and by a collaborative project between the Institute of Food and Agricultural Sciences (University of Florida) and the Gas Research Institute (Chicago, IL).

Procedures

Culture of immature embryos [10, 11, 13, 14, 33]

Steps in the procedure

1. Obtain inflorescences bearing young caryopses 10—15 days after pollination.
2. Remove the immature caryopses and sterilize by immersion in 70—90% ethanol for 30—60 seconds, followed by 10—20 minutes in 10—20% commercial bleach containing a wetting agent (e.g., 0.05% Tween 80). Wash the caryopses thoroughly with sterile distilled water at least three times.
3. Carefully dissect the embryos from the caryopses and place them so that the embryo axis is in contact with the medium and the scutellum is fully exposed. Place 10—20 embryos in each 10 cm Petri dish containing 20—25 ml of medium (5—7 embryos in 6 cm Petri dishes with 10 ml medium). The Murashige and Skoog [34] medium is most commonly used, supplemented with 2—5 mg/l of 2,4-D, 3% sucrose and solidified with 8 g/l agar or 2 g/l Gelrite. In some species, picloram and dicamba have been found to be equally suitable. Supplementing the media with casein hydrolysate, coconut water, proline, etc. have proven to be useful in many instances. In some species, such as maize, higher levels of sucrose (6—12%) are recommended.
   Whole mature seeds have also been used successfully to establish embryogenic callus cultures in a few species. However, the efficiency of callus formation in these instances is generally poor and unreliable.
4. The culture dishes should be sealed with Parafilm, and incubated in the dark at 28 °C.
5. Visible callus is formed within 7—15 days. At the end of 3—4 weeks, several types of callus may be seen. It is important at this stage to physically separate the embryogenic callus from the non-embryogenic callus and subculture it on fresh nutrient medium. In maize, Type II embryogenic callus is formed either directly on the scutellum of the cultured embryos, or appears later as very small sectors during the subculture of Type I callus. Once formed, it can be maintained by subculture for long periods of time.
6. A subculture routine of 2—4 weeks is recommended, but care should be taken at the time of each subculture to exclude non-embryogenic segments. Longer subculture periods may reduce regeneration ability because of the formation of increased amount of non-embryogenic callus.
7. Transfer portions of embryogenic callus to media containing about 10% of the original level of 2,4-D to induce further organization and development of somatic embryos. In many species this may not be necessary as the effective levels of 2,4-D in the medium fall substantially during the period of subculture to allow the development of somatic embryos. In some species, addition of abscisic acid to the medium promotes normal development and maturation of embryos.
8. Germinating somatic embryos should be placed on half-strength MS medium to obtain plantlets, which can be transferred to soil after the formation of a well

developed shoot and root system, and grown to maturity. In most instances somatic embryo-derived plants can be established in soil without difficulty. When problems arise, the plantlets can be hardened in a high humidity chamber for about a week before transfer to the greenhouse.

Culture of bases of young leaves [12, 35—37]

Steps in the procedure
1. Grass leaves grow from a basal meristem. Therefore, the youngest part of the leaf is at the base and the oldest at the tip. Young unexpanded leaves are obtained either from seedlings grown aseptically in vitro or from young shoots which have been sterilized by treatment with 95% ethanol (60 sec) and 5—10% commercial bleach (10—20 min), followed by three rinses in sterile distilled water.
2. The outermost leaves are carefully removed and discarded. The inner 4—5 youngest leaves are cut into 1—2 mm transverse segments and placed on the nutrient medium used for immature embryos. Polarity does not seem to be important here.
3. Formation of callus and regeneration of plants is similar to that described for immature embryo culture. Both temporal and spatial gradients in response are seen. The optimal response is from the basal portions of the leaf, generally not extending beyond a few centimeters. Similarly, the youngest 2—5 leaves are best for culture.

Culture of young inflorescences [10, 38]

Steps in the procedure
1. The developmental stage of the inflorescence is critical for obtaining the best response. Young, unemerged, premeiotic inflorescences, 1—2 cm in length, are sterilized while still enclosed in protective leaves. The leaves are then cut open, the inflorescence is removed and divided into 1—2 mm thick segments and placed in culture. Callus formation and plant regeneration are similar to the descriptions above.

Cultured explants of many cereal and grass species, such as sorghum and sugar-cane, release copious amounts of phenolic and other substances resulting in strong pigmentation of the surrounding medium. Their presence is quite harmful and either inhibits or prevents cell proliferation and growth. Although addition of substances such as activated charcoal can be beneficial, the best results are obtained by frequent transfer (every 1—3 days) of the explants to fresh media until discoloration of the medium stops.

Establishment of embryogenic suspension cultures [13, 14, 16, 17, 39, 40]

Steps in the procedure

1. Slice or tease apart Type I embryogenic callus into small pieces, taking care to remove any sectors of non-embryogenic callus. Place 1.0—1.5 gm of callus pieces in 20—25 ml liquid MS [34] medium (1—2.5 mg/l 2,4-D, 3% sucrose, with or without 5—10% coconut water) in 250 ml Erlenmeyer flasks. If sufficient amount of callus is not available, suspensions can be initiated in 50 or 125 ml flasks with 5—15 ml of medium. The cultures are placed on a gyratory shaker at 120 rpm in the dark at 28 °C.

2. In some species, such as *Sorghum*, the culture medium rapidly becomes dark in color owing to the release of phenolic and other compounds which are harmful to the cells. The flasks should be removed from the shaker every two to three days and allowed to rest for a minute to permit the callus pieces and cell groups to settle to the bottom. The supernatant medium is then completely drained off and replaced with an equal volume of fresh medium. This removes not only the harmful substances but also cellular debris, and large, vacuolated and non-dividing cells which are continuously sloughed off from the surface of the callus pieces. This procedure should be repeated, as often as needed, until the medium stays clear, and the callus pieces and cell masses start to become friable and dissociate into small groups of richly cytoplasmic embryogenic cells. Frequent microscopic monitoring of the cultures is important. During this period, embryogenic suspensions in many species become very viscous and mucilaginous. This is transient, does not generally affect their embryogenic competence, and may even be needed for the establishment of the suspension.

3. The growth of the suspensions becomes more rapid with the release of small groups of embryogenic cells into the medium, allowing subculture of the suspension by 1:1 dilution with fresh medium every four to seven days. The cultures are removed from the shaker, and allowed to settle for 15—30 seconds. Cell groups from the middle portion of the cultures are removed with a spring-loaded pipette fitted with a wide mouth tip and transferred to fresh medium. Large, non-dividing and vacuolated cells at the top of the medium, as well as larger tissue pieces at the bottom of the culture, are thus excluded. Filtration or sieving of the suspension is neither required nor recommended. However, in those cases where small clumps of tissue continue to be present, it might be necessary to sieve the suspension through 500 μm or even 1000 μm sieves. Established suspensions are fast growing (cell doubling times of 30—50 hr), finely dispersed (in some species the suspensions may contain 0.5—1.00 mm diameter aggregates of cells), and do not contain any organized structures or meristems. They can be subcultured every 3—7 days by 1:1 to 1:5 dilution ratios. Depending on the species used, it may take 3—6 months, or even longer, to reach this stage. Finely dispersed and fast growing embryogenic suspensions can be obtained in maize within 4—6 weeks by placing Type II callus in liquid medium. In this case, the difficulties faced with the use of Type I callus are not encountered.

4. Two procedures can be followed to induce the formation of somatic embryos and

plants. An aliquot of 7–9 day old suspensions is first plated on MS agar medium containing 2.5 mg/l 2,4-D. Once a continuous lawn of tissue has been formed, portions of it can be transferred to low 2,4-D (0.1–0.25 mg/l) media to induce the formation of embryos and plants. Alternatively, the suspension can be overlaid directly on MS medium with low levels of 2,4-D.

Culture of embryogenic protoplasts [19, 22, 23, 41–43]

Steps in the procedure

1. Best results are obtained when cells from a rapidly growing, finely dispersed embryogenic suspension culture are used during the exponential phase of growth. Approximately 1.5 ml of the settled cells are harvested 2–7 days after subculture and placed in 15 ml of a filter-sterilized enzyme solution in a Petri dish. The amount and combinations of enzymes used vary from species to species, and from cell line to cell line. However, the following may be used as a good starting point: 1–3% cellulase Onozuka RS, 1% pectinase Serva, dissolved in a buffer solution consisting of 7 mM $CaCl_2 \cdot 2H_2O$, 0.7 mM $NaH_2PO_4 \cdot H_2O$, 0.5 M mannitol, and 3 mM MES at pH 5.6 (osmolality adjusted to 650 m osm/kg H_2O). The cells are incubated at room temperature on a gyratory shaker at 50 rpm. Protoplast yields of 3–5 × 10^6/ml can be obtained within 4–6 hr. In some species, use of small amounts of Driselase and Pectolyase Y23 is necessary to obtain good protoplast yields. Incubation in the enzyme mixture for longer than 4–6 hours should be avoided. Commercially available enzymes can be used directly, without any additional purification.

2. The protoplast-enzyme mixture is filtered through a layer of Miracloth, and then successively through 100 μm and 25 μm stainless steel filters. The protoplasts are collected by centrifugation at 100 × g for 3–4 min, and washed three times with the wash solution (the enzyme solution without the enzymes, but containing 0.5–0.6 M mannitol).

3. Protoplasts are cultured in shallow layers (1.5–3.0 ml medium) in 6 cm Petri dishes at a density of 1–3 × 10^5/ml, either in liquid or 0.3–1.2% low-melting-point Seaplaque agarose (FMC) solidified Kao and Michayluk's [44] medium as modified by Vasil and Vasil [41], containing various concentrations of 2,4-D (there are some recent reports of good plating efficiencies obtained even in MS medium). In some species addition of cytokinins, like zeatin (0.1–0.2 mg/l), has been found to be beneficial. The cultures are incubated in the dark at 28 °C. Fresh nutrient medium with reduced osmoticum (0.3 M glucose) is added after 10–14 days. Protocolonies become visible within four weeks and can be transferred to fresh medium with 3% sucrose or 2% each of glucose and sucrose. The use of feeder layers or floating sectors of agarose-plated protoplasts in liquid media (agarose bead culture) have been found to be particularly useful in maize [19, 43] and other species where plating efficiencies may be rather low, and in the selection of transformed cells.

4. Protoplast-derived embryogenic callus is transferred to regeneration media as described earlier to induce the formation of somatic embryos and plants, in much the same manner as that obtained from the culture of immature embryos, inflorescences, leaves and suspension cultures.

References

1. Vasil IK (1987) Developing cell and tissue culture systems for the improvement of cereal and grass crops. J Pl Physiol 128: 193–218.
2. Bright SWJ, Jones MGK, eds (1985) Cereal Tissue and Cell Culture. Amsterdam: Martinus Nijhoff/Dr W Junk.
3. Vasil IK, Vasil V (1986) Regeneration in cereal and other grass species. In: Vasil IK (ed) Cell Culture and Somatic Cell Genetics of Plants, Vol 3, Plant Regeneration and Genetic Variability, pp 121–150. Orlando: Academic Press.
4. Morrish, F, Vasil V, Vasil IK (1987) Developmental morphogenesis and genetic manipulation in tissue and cell cultures of the Gramineae. Adv Genet 24: 431–499.
5. Vasil IK (1988) Progress in the regeneration and genetic manipulation of cereal crops. Bio/Technology 6: 397–402.
6. Potrykus, I (1990) Gene transfer to cereals: an assessment. Bio/Technology 8: 535–542.
7. Shimamoto K, Terada R, Izawa T, Fujimoto H (1989) Fertile transgenic rice plants regenerated from transformed protoplasts. Nature 338: 274–277.
8. Gordon-Kamm WJ, Spencer TM, Mangano ML, Adams TR, Daines RJ, Start WG, O'Brien JV, Chambers SA, Adams Jr WR, Willetts NG, Rice TB, Mackey CJ, Krueger RW, Kausch AP, Lemaux PG (1990) Transformation of maize cells and regeneration of fertile transgenic plants. The Pl Cell 2: 603–618.
9. Datta SK, Peterhans A, Datta K, Potrykus I (1990) Genetically engineered fertile indica-rice recovered from protoplasts. Bio/Technology 8: 736–740.
10. Vasil V, Vasil IK (1981) Somatic embryogenesis and plant regeneration from tissue cultures of *Pennisetum americanum* and *P. americanum* × *P. purpureum* hybrid. Amer J Bot 68: 864–872.
11. Vasil V, Vasil IK (1984) Induction and maintenance of embryogenic callus cultures of Gramineae. In: Vasil IK (ed) Cell Culture and Somatic Cell Genetics of Plants, Vol 1, Laboratory Procedures and Their Applications, pp 36–42. Orlando: Academic Press.
12. Wernicke W, Brettell R (1980) Somatic embryogenesis from *Sorghum bicolor* leaves. Nature 287: 138–139.
13. Armstrong CL, Green CE (1985) Establishment and maintenance of friable, embryogenic maize callus and the involvement of proline. Planta 164: 207–214.
14. Vasil V, Vasil IK (1986) Plant regeneration from friable embryogenic callus and cell suspension cultures of *Zea mays*. J Pl Physiol 124: 399–408.
15. Swedlund B, Vasil IK (1985) Cytogenetic characterization of embryogenic callus and regenerated plants of *Pennisetum americanum* (L) K Schum. Theoret Appl Genet 69: 575–581.
16. Vasil V, Vasil IK (1981) Somatic embryogenesis and plant regeneration from suspension cultures of pearl millet (*Pennisetum americanum*). Ann Bot 47: 669–678.
17. Vasil V, Vasil IK (1984) Isolation and maintenance of embryogenic cell suspension cultures of Gramineae. In: Vasil IK (ed) Cell Culture and Somatic Cell Genetics of Plants, Vol 1, Laboratory Procedures and Their Applications, pp 152–158. Orlando: Academic Press.
18. Redway, F, Vasil V, Vasil IK (1990) Characterization and regeneration of wheat (*Triticum aestivum*) embryogenic cell suspension cultures. Pl Cell Rep 8: 714–717.
19. Shillito RD, Carswell, GK, Johnson CM, DiMaio JJ, Harms CT (1989) Regeneration of fertile plants from protoplasts of elite inbred maize. Bio/Technology 7: 581–587.
20. Gnanaprasagam S, Vasil IK (1990) Plant regeneration from cryopreserved embryogenic suspension cultures of a commercial sugarcane hybrid (*Saccharum spp*). Pl Cell Rep 9: 419–423.
21. Vasil IK (1983) Isolation and culture of protoplasts of grasses. Int Rev Cytol Supp 16: 79–88.
22. Yamada Y, Yang Z, Tang D (1986) Plant regeneration from protoplast derived callus of rice (*Oryza sativa* L). Pl Cell Rep 5: 85–88.
23. Vasil V, Redway F, Vasil IK (1990) Regeneration of plants from embryogenic suspension culture protoplasts of wheat (*Triticum aestivum* L). Bio/Technology 8: 429–434.

24. Tabaeizadeh, Z, Ferl RJ, Vasil IK (1986) Somatic hybridization in the Gramineae: *Saccharum officinarum* L (sugarcane) + *Pennisetum americanum* (L) K Schum (pearl millet). Proc Nat Acad Sci USA 83: 5616–5619.

25. Kyozuka J, Kaneda T, Shimamoto K (1989) Production of cytoplasmic male sterile rice (*Oryza sativa* L) by cell fusion. Bio/Technology 7: 1171–1174.

26. Fromm ME, Taylor LP, Walbot V (1986) Stable transformation of maize after electroporation. Nature 319: 791–793.

27. Hauptmann RM, Vasil V, Ozias-Akins P, Tabaeizadeh Z, Rogers SG, Fraley RT, Horsch RB, Vasil IK (1988) Evaluation of selectable markers for obtaining stable transformants in the Gramineae. Pl Physiol 86: 602–606.

28. Vasil V, Hauptmann RM, Morrish FM, Vasil IK (1988) Comparative analysis of free DNA delivery and expression into protoplasts of *Panicum maximum* Jacq (Guinea grass) by electroporation and polyethylene glycol. Pl Cell Rep 7: 499–503.

29. Callis, J, Fromm M, Walbot V (1987) Introns increase gene expression in cultured maize cells. Genes Develop 1: 1183–1200.

30. Vasil V, Clancy M, Ferl RJ, Vasil IK, Hannah LC (1989) Increased gene expression by the first intron of maize *shrunken-1* locus in grass species. Pl Physiol 91: 1575–1579.

31. Klein TM, Kornstein L, Sanford JC, Fromm ME (1989) Genetic transformation of maize cells by particle bombardment. Pl Physiol 91: 440–444.

32. Fromm ME, Morrish F, Armstrong C, Williams R, Thomas J, Klein TM (1990) Inheritance and expression of chimeric genes in the progeny of transgenic maize plants. Bio/Technology 8: 833–839.

33. Redway FA, Vasil V, Lu D, Vasil IK (1990) Identification of callus types for long-term maintenance and regeneration from commercial cultivars of wheat (*Triticum aestivum* L). Theoret Appl Genet 79: 609–617.

34. Murashige T, Skoog F (1962) A revised medium for rapid growth and bioassays with tobacco tissue cultures. Physiol Plant 15: 473–497.

35. Haydu Z, Vasil IK (1981) Somatic embryogenesis and plant regeneration from leaf tissues and anthers of *Pennisetum purpureum*. Theoret Appl Genet 59: 269-273.

36. Lu C, Vasil IK (1981) Somatic embryogenesis and plant regeneration from leaf tissues of *Panicum maximum* Jacq. Theoret Appl Genet 59: 275–280.

37. Ho W, Vasil IK (1983) Somatic embryogenesis in sugarcane (*Saccharum officinarum* L). I. The morphology and physiology of callus formation and the ontogeny of somatic embryos. Protoplasma 118: 169–180.

38. Botti C, Vasil IK (1984) The ontogeny of somatic embryos of *Pennisetum americanum* (L) K Schum. II. In immature inflorescences. Canad J Bot 62: 1629–1635.

39. Vasil V, Vasil IK (1982) Characterization of an embryogenic cell suspension culture derived from inflorescences of *Pennisetum americanum* (pearl millet; Gramineae). Amer J Bot 69: 1441–1449.

40. Green CE, Armstrong CL, Anderson PC (1983) Somatic cell genetic systems in corn. In: Downey K, Voellmy RW, Ahmad F, Schulz J (eds), Advances in Gene Technology: Molecular Genetics of Plants and Animals, pp 147–157. New York: Academic Press.

41. Vasil V, Vasil IK (1980) Isolation and culture of cereal protoplasts. II. Embryogenesis and plantlet formation from protoplasts of *Pennisetum americanum*. Theoret Appl Genet 56: 97–99.

42. Srinivasan C, Vasil IK (1986) Plant regeneration from protoplasts of sugarcane. J Pl Physiol 126: 4–48.

43. Rhodes CA, Lowe KS, Ruby KL (1988) Plant regeneration from protoplasts isolated from embryogenic maize cell cultures. Bio/Technology 6: 56–60.

44. Kao, KN, Michayluk MR (1975) Nutritional requirements for growth of *Vicia hajastana* cells and protoplasts at a very low population density in liquid media. Planta 126: 105–110.

Plant Tissue Culture Manual **B2**: 1–17, 1991.
© 1991 *Kluwer Academic Publishers.*

Transformation and regeneration of rice protoplasts

JUNKO KYOZUKA and KO SHIMAMOTO
Plantech Research Institute, 1000 Kamoshida, Midori-ku Yokohama, 227 Japan

Introduction

The production of transgenic plants has provided new insights into plant biology and has become an important tool for the improvement of crop species (for review, see [2, 15]). Until recently, however, most of the studies using transgenic plants employed dicotyledonous species because of the relative ease of transformation. Monocotyledonous plants are not generally susceptible to *Agrobacterium tumefaciens* which is routinely used for transformation of dicotyledonous plants. Furthermore, lack of efficient regeneration of fertile plants from protoplasts of monocot species has made it difficult to use the alternative, direct DNA transfer method.

Genes from monocotyledonous species are not always expressed correctly in transgenic dicotyledonous plants [3, 6, 20]. This suggests that some differences may exist in DNA sequences or cellular factors influencing gene expression between monocots and dicots. Thus, it is desirable to use transgenic monocot plants to study expression of structual or regulatory genes derived from monocot species. In addition to basic studies using transgenic plants, transformation of monocot plants, particularly those in the Gramineae, is urgently needed for crop improvement because this group contains the major crops such as rice, corn, barley and wheat.

In graminaceous species, rice is exceptional in that routine and efficient regeneration of plants from protoplasts is possible (for review, see [9]). Recently, successful transformation of rice using protoplasts has been achieved. Several research groups reported the production of transformed rice callus [19], plants [18, 21, 22] and fertile transgenic plants [16] by a direct gene transfer method using protoplasts. Moreover, tissue specific expression of the CaMV35S promoter [17] and maize *Adh1* promoter [10] in transgenic rice plants was examined. In addition, transient assay using rice protoplasts is becoming a useful tool for identifying DNA sequences that regulate expression of genes not only from rice [12] but also from other graminaceous species [10, 11]. Therefore, rice is considered to be a model plant among graminaceous species for studies in regulation in expression of monocot genes and for application of genetic engineering approaches in crop improvement.

In this chapter we will describe a detailed procedure to produce transgenic rice plants including methods for the production of embryogenic suspension

cultures, protoplast isolation and electroporation, selection of stably transformed callus and plant regeneration. Co-transformation of unselectable genes
and transient expression assay with protoplasts will be also described.

Procedures

Procedures described here have been developed for generating transgenic rice plants, however, these techniques should be in principle applicable to other cereal species. Two methods have been developed to introduce foreign DNA into rice protoplasts, namely a polyethylene glycol method [21] and electroporation [16, 18, 22]. At present, it is unclear from the available data as to which method is better for direct DNA uptake in rice cells because a number of other factors influence the efficiency in generation of transgenic plants. After having examined various parameters influencing the efficiency in production of transgenic rice plants, however, we have developed a protocol for rice transformation by using electroporation in our laboratory. The scheme of the protocol is illustrated in Fig. 1 and compositions of culture media used are listed in Table 1.

Protoplasts are isolated from embryogenic suspension cultures [7] for production of stably transformed plants or the 'Oc' cell line [1] for transient assay. Embryogenic protoplasts can be obtained from suspension culture in less than 2 months after the initiation of callus from mature dry seed. Electroporated protoplasts are cultured by the nurse culture method (Fig. 2) [7] and the presence of nurse cells has been found to increase their plating efficiency considerably.

Stably transformed calluses are selected with hygromycin B (Hm). we found that the Hm^r (hph) is a more effective marker for rice cells although the Km^r (nptII) gene has been also used as a selectable marker in rice transformation [18, 22]. Effective selection of transformed cells is not always easy with the Km^r marker and albino or sterile plants are often obtained from Km^r rice callus. Selection by hygromycin B is started after 10—14 days of culture and performed by two steps for 1—2 weeks each. Hm^r calluses thus selected are further grown on hygromycin-free medium for one week before the transfer onto the regeneration medium. Shoots arise from transgenic callus within 6—8 weeks after protoplast isolation. Regenerated plantlets are transferred to plastic boxes and they are transferred to pots when they become approximately 15 cm in height and grown to maturity in greenhouse.

Transient expression with protoplasts is a convenient assay to assess quickly expression capacities of various vector constructs. It is in practice more useful than stable transformation [10] because more than 10^6 protoplasts are used in one treatment and the level of expression in each cell is averaged. Thus the problem of clonal variation observed in transgenic plants can be avoided. However, it should be noted that some highly regulated plant promoters do not always express in protoplasts in a correctly regulated manner.

Table 1. Compositions of culture media

	MSC	R2S	R2P	R2SA	R2A	R2R
Major elements (mg/l)						
KNO_3	1900	4000	4000	4000	4000	4000
NH_4NO_3	1650	–	–	–	–	–
$(NH4)_2SO_4$	–	335	335	335	335	335
$MgSO_4 \cdot 7H_2O$	370	250	250	250	250	250
$CaCl_2 \cdot 2H_2O$	440	150	150	150	150	150
$NaH_2PO \cdot H_2O$	–	273	273	273	273	273
KH_2PO_4	170	–	–	–	–	–
Iron and minor elements (mg/l)						
Na_2EDTA	37.3	7.5	7.5	7.5	7.5	7.5
$FeSO_4 \cdot 7H_2O$	27.8	5.5	5.5	5.5	5.5	5.5
$MnSO_4 \cdot 4H_2O$	22.3	1.6	1.6	1.6	1.6	1.6
$ZnSO_4 \cdot 7H_2O$	8.6	2.2	2.2	2.2	2.2	2.2
$CuSO_4 \cdot 5H_2O$	0.025	0.125	0.125	0.125	0.125	0.125
$CoCl_2 \cdot 6H_2O$	0.025	–	–	–	–	–
KI	0.83	–	–	–	–	–
H_3BO_3	6.2	3.0	3.0	3.0	3.0	3.0
$NaMoO_4 \cdot 2H_2O$	0.25	0.125	0.125	0.125	0.125	0.125
Organic components (mg/l)						
m-Inositol	100	100	100	100	100	100
Nicotinic acid	0.5	0.5	0.5	0.5	0.5	0.5
Pyridoxine HCl	0.5	0.5	0.5	0.5	0.5	0.5
Thiamine HCl	0.5	0.5	0.5	0.5	0.5	0.5
Glycine	2.0	2.0	2.0	2.0	2.0	2.0
Sucrose (g/l)	20	30	136	60	60	20
Sorbitol (g/l)	–	–	–	–	–	30
2,4-D (mg/l)	2.0	2.0	2.0	2.0	2.0	–
Agarose (type I, Sigma) (g/l)	8.0	0	0	2.5	5.0	10.0
pH[a]	5.6	5.6	5.6	5.6	5.6	5.6

[a] pH was adjusted before autoclaving for MSC, R2S, R2SA, R2A and R2R. R2P is filter-sterilized.

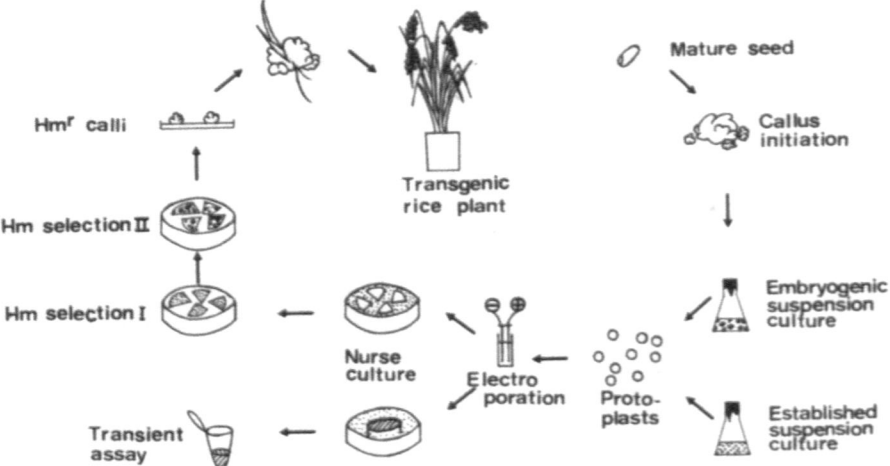

Fig. 1.

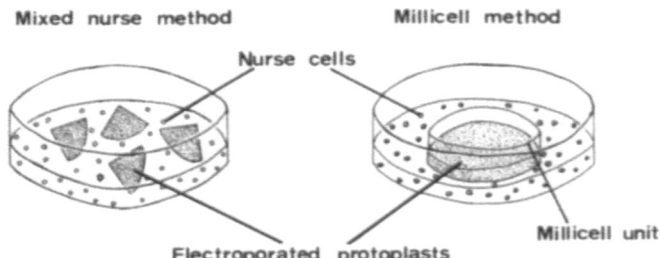

Fig. 2.

Production of embryogenic suspension culture

Embryogenic suspension cultures are produced from mature-seed-derived callus, and therefore the culture can be initiated at any time of a year. The ease with which embryogenic suspension culture can be generated varies among cultivars. Generally, japonica cultivars respond better than indica cultivars. Callus of indica cultivars tends to become brown and die in liquid medium. The method described below can be applied to most of the japonica cultivars and some indica cultivars [8]. For recalcitrant cultivars, some modifications in medium composition or culture conditions will be requirted.

Frequency of plant regeneration from protoplast-derived callus is high (30—80%) because embryogenic suspension cultures are used as the source of protoplasts. The 'quality' of embryogenic suspension cultures from which protoplasts are isolated, is the most important factor for successful plant regeneration from protoplasts.

Steps in the procedure
1. Sterilize dehusked mature seeds in 70% ethanol for 30 s and in 1.2% sodium hypochlorite for 45—60 min with vigorous shaking. Wash seeds twice in sterilized distilled water.
2. Place seeds on MS medium (Murashige and Skoog [13]) for callus induction (MSC) and incubate at 30 °C. After 2—3 weeks callus develops around the scutellum of the embryo.
3. Transfer 5—10 small blocks of callus each of which is 2—5 mm in diameter in a 125 ml flask containing 15 ml of suspension medium (R2S).
4. Shake suspension cultures on a gyrotary shaker (80 rpm) under light (ca. 3000 lx) at 30 °C for 4—6 weeks. The cultures are subcultured every 7 days. Subculture at an uniform interval is an important factor to keep suspension cultures suitable for isolation of a large number of competent protoplasts. At the start of culture, cells grow slowly, then after 2—3 subcultures they start to grow vigorously and each callus in coarse suspension cultures becomes yellowish in color. In such suspension cultures, cell debris or vacuolated cells are rarely observed and highly viable protoplasts are easily obtained from them.

Protoplast isolation and electroporation

Protoplasts are isolated from suspension cultures at 3–4 days after the subculture. The yield of protoplasts is $4–5 \times 10^6$/g fresh weight cells. Isolated protoplasts are able to divide in 3–4 days and form colonies with an efficiency of 1–10% depending on the genotype, quality of the suspension and other factors. Considering the fact that embryogenic protoplasts of rice and 'Oc' cell line protoplasts are small (10–20 µm in diameter), electroporation was carried out using a long pulse (10–20 ms), generated by a large capacitor (800–1200 µF) and a moderate voltage (300–500 V/cm). Under these conditions, the plating efficiency (0.1–1.2%) is 10–20% of that of non-electroporated control protoplasts when grown by the nurse culture method [16]. The conditions for electroporation shoud be determined by taking into account the efficiency for uptake of DNA into protoplasts and the survival rate of protoplasts after the electroporation. Our studies with a number of japonica cultivars show that 'competence' or efficiency of DNA uptake by rice protoplasts is variable between cultivars and sometimes between different suspension cultures derived from a cultivar.

Steps in the procedure

1. Place a sample of a 3–5 days-old suspension culture in a 10 cm Petri dish and pipette out the culture medium. Then add 20 ml of the enzyme solution [7].
2. Incubate the Petri dish in the dark at 30 °C for 3–5 hr without shaking.
3. After the incubation, gently pipette enzyme-protoplast mixture a few times and pour it into a glass funnel with a fine nylon mesh (pore size 20 µm). Add $4\times$ volume of KMC [4] solution to the filtered protoplast suspension.
4. Centrifuge protoplast suspension at 800 rpm for 10 min.
5. Dispose of the supernatant and resuspend pelleted protoplasts in KMC solution, then count the number of protoplasts.
6. Wash the protoplasts twice in EP 3 buffer [16] by centrifugation (800 rpm, for 5 min each).
7. Resuspend protoplasts in EP 3 buffer at the desired cell density (4–8 × 10^6 protoplasts/ml).
8. Transfer aliquots (500 µl) of protoplast suspension to an 1.5 ml Eppendorf tube and add the plasmid DNA solution (15–25 µl depending on its concentration) to the protoplast suspension.
9. Incubate the protoplast-DNA mixture in ice for 20 min.
10. Transfer the protoplast-DNA mixture to a plastic cuvette precooled in ice and electroporate by a capacitor-discharge system (X-cell 450, Promega).
11. Incubate the electroporated protoplasts in an Eppendorf tube for 20 min more on ice.

Transient assay

The protocol described here is for a transient assay using protoplasts and the GUS reporter system [5]. The *E. coli* β-glucuronidase (GUS) gene is a highly sensitive reporter and its enzymatic activity is easy to quantify in crude extracts of rice cells.

Usually we use the 'Oc' cell line for the transient assay because the 'competence' for DNA uptake of 'Oc' protoplasts is higher than that of embryogenic protoplasts. For the transient assay highly viable protoplasts should be used to obtain high expression levels of various vectors. Also to accurately measure the level of expression of a vector construct, it is important to optimize transformation conditions that give reproducible results.

Steps in the procedure
1. Culture electroporated-protoplasts by the nurse culture method using the Millicell (Millipore) (Fig. 2b) for 48 hr at 30 °C in the dark without shaking. In the Millicell method [7], the electroporated-protoplast suspension (0.5 ml) and 1 ml of R2P medium are placed inside the Millicell. Nurse cells and 4.5 ml of R2P medium are added outside the Millicell.
2. After 48 hr of incubation, transfer the protoplast suspension to a 1.5 ml Eppendorf tube and fill up with 0.4 M mannitol.
3. Collect protoplasts by centrifugation ($10\,000 \times g$, for 3 min).
4. Resuspend the pellet in 100 µl GUS Extraction Buffer [5].
5. Sonicate cells for 1 min using a setting of duty cycle (10%) and micro tip limit (1) by a Sonifer 450 (Branson).
6. Centrifuge at $10\,000 \times g$ for 5 min at 4 °C.
7. Measure the protein content in the extract using 5 µl of the extract.
8. Add 50 µl of the extract to 150 µl GUS assay buffer, mix thoroughly and incubate at 37 °C.
9. At regular time intervals, transfer 50 µl aliquots of the reaction mixture into Eppendorf tubes containing 950 µl stop buffer. Normally we take 3–4 time points to obtain accurate measurement of the enzymatic activity.
10. Determine the activity by measuring MU (methylumbelliferone) concentration with a spectrofluorimeter, excitation at 365 nm, emission at 455 nm.

Selection of stably transformed calluses and regeneration

The frequency of generation of hygromycin resistant calluses varies depending on 1) types of Hm^r vectors used, 2) quality of protoplasts, and 3) conditions of electroporation. Transformed rice cells can be clearly selected at 30 µg/ml of hygromycin B and approximately 1% of protoplast-derived calluses are Hm^r.

Little background colony growth is observed in control experiments in which no Hm^r plasmid is added to protoplasts in selection I (Fig. 1). Even if some untransformed cells survive the selection I they do not continue growth in the second cycle of selection with the same concentration of Hm (30 µg/ml) (selection II). Frequency of plant regeneration from transformed callus is comparable to that from non-transformed callus.

Steps in the procedure
1. Culture electroporated protoplasts by the mixed nurse culture method (Fig. 2a) with gentle shaking (ca. 30 rpm) for 10 days in the dark at 30 °C.
2. After 10 days of the culture, transfer agarose blocks a new culture dish containing 5 ml of R2P medium, remove nurse cells by repeated washes with the culture medium and add 5 ml of R2P medium.
3. Add hygromycin B to make a final concentration at 30 µg/ml.
4. Culture the cells in the dark with gentle shaking for more 10 days.
5. Transfer agarose blocks onto a new plate containing 6 ml of R2SA medium containing 30 µg/ml hygromycin B, and incubate plates for 10 days at 30 °C under light.
6. Pick up Hm^r microcalluses individually using forceps and place them on R2A medium without hygromycin B for further growth.
7. Transfer colonies about 5 mm in diameter onto the regeneration medium (R2R).
8. Within 2–8 weeks shoots and roots emerge. Transfer plantlets (ca. 2 cm height) to a plastic box containing 30 ml of hormone-free R2R medium with 0.8% agarose (Type I, Sigma).
9. Transfer regenerated plants of 10–15 cm in height to soil in a growth room (30 °C, 3000 lx), and then transfer them to greenhouse and grow until maturity.

Co-transformation

Introduction of non-selectable genes into rice cells can be performed using co-transformation with the Hm^r selectable marker gene (Fig. 3). In co-transformation, plasmid DNA of the non-selectable gene and plasmid DNA carrying the selectable marker gene are mixed with protoplasts and electroporated as described above. The addition of the non-selectable gene does not influence the plating efficiency or the frequency of Hm^r colonies. The efficiency of co-transformation may depend on the ratio in the concentration of selectable marker to non-selectable gene. In our protocol the Hm^r plasmid and various GUS plasmids are mixed in 1 : 1 ratio (30–50 µg/ml each) and the efficiency of co-transformation (frequency of GUS-positive transformants in Hm^r transformants) is 30–50%. One of the advantages of this method is that construction of a plasmid carrying both the selectable marker and the non-selectable gene is not required, thus, it is convenient to generate transgenic plants carrying multiple genes.

Solutions

– Enzyme for protoplast isolation
  - 4% Cellulase RS (Kinki Yakult, Japan)
  - 1% Macerozyme R10 (Kinki Yakult, Japan)
  - 0.4 M mannitol
  pH 5.6

– KMC solution
  - 0.35 M KCl
  - 0.245 M $MgCl_2$
  - 0.254 M $CaCl_2$
  pH 6.0

– EP 3 buffer
  - 70 mM KCl
  - 5 mM $MgCl_2$
  - 0.1% MES
  - 0.4 M mannitol
  pH 5.8

– GUS extraction buffer
  - 50 mM NaH_2PO_4, pH 7.0
  - 10 mM beta-mercaptoethanol
  - 10 mM Na_2EDTA
  - 0.1% sodium lauryl sarcosine
  - 0.1% Triton X-100

– GUS assay buffer
  - 1 mM MUG (4-Methyl umbelliferyl beta-D-glucuronide) in extraction buffer

– stop buffer
  - 0.2 M Na_2CO_3

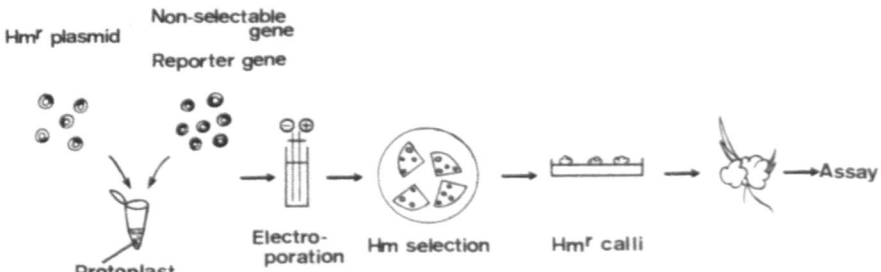

Fig. 3.

References

1. Baba A, Hasazawa S, Shono K (1986) Cultivation of rice protoplasts and their transformation mediated by *Agrobacterium* spheroplasts. Plant Cell Physiol 27: 463–472
2. Benfey PN, Chua N-H (1989) Regulated genes in transgenic plants. Science 244: 174–181
3. Ellis JG, Llewellyn DJ, Dennis ES, Peacock WJ (1987) Maize *Adh-1* promoter sequences control anaerobic regulation: addition of upstream promoter elements from constitutive genes is necessary for expression in tobacco. EMBO J 6: 11–16
4. Harms CT, Potrykus I (1978) Fractionation of polant protoplast types by iso-osmotic density gradient centrifugation. Theor Appl Genet 53: 57–63
5. Jefferson RA, Kavanagh TA, Bevan MW (1987) GUS fusions: β-glucuronidase as a sensitive and versatile gene fusion marker in higher plants. EMBO J 6: 3901–3907
6. Keith B, Chua N-H (1986) Monocot and dicot pre-mRNAs processed with different efficiencies in transgenic tobacco. EMBO J 5: 2419–2425
7. Kyozuka J, Hayashi Y, Shimamoto K (1987) High frequency plant regeneration from rice protoplasts by novel nurse culture methods. Mol Gen Genet 206: 408–413
8. Kyozuka J, Otoo E, Shimamoto K (1988) Plant regeneration from protoplasts of indica rice: genotypic differences in culture response. Theor Appl Genet 76: 887–890
9. Kyozuka J, Shimamoto K, Ogura H (1989) Regeneration of plants from rice protoplasts. In Biotechnology in Agriculture and Forestry, Vol. 8, Plant protoplasts and Genetic Engineering I (ed. by Bajaj YSP) pp. 109–123, Springer-Verlag Berlin, Heidelberg, New York, Tokyo
10. Kyozuka J, Izawa T, Nakajima M, Shimamoto K (1990) Effect of the promoter and the first intron of maize *Adh1* on foreign gene expression in rice. Maydica 35: 1–5
11. Marcotte WR, Christopher CB, Quatrano RS (1988à) Regulation of a wheat promoter by abscisic acid in rice protoplasts. Nature 335: 454–457
12. Mundy J, Yamaguchi-Shinozaki K, Chua NH (1990) Nuclear proteins bind conserved elements in the abscisic acid-responsive promoter of rice *rab* gene. Proc Natl Acad Sci USA 87: 1406–1410
13. Murashige T, Skoog F (1962) A revised medium for rapid growth and bioassays with tobacco tissue cultures. Physiol Plant 15: 473–497
14. Ohira K, Ojima K, Fujiwara A (1973) Studies on the nutrition of rice celll culture 1. A simple, defined medium for rapid growth in suspension culture. Plant Cell Physiol 14: 1113–1121
15. Schell JS (1987) Transgenic plants as tools to study the molecular organization of plant genes. Science 237: 1176–1182
16. Shimamoto K, Terada R, Izawa T, Fujimoto H (1989) Fertile transgenic rice plants regenerated from transformed protoplasts. Nature 338: 274–276
17. Terada R, Shimamoto K (1990) Expression of CaMV35S-GUS gene in transgenic rice plants. Mol Gen Genet 220: 389–392
18. Toriyama K, Arimoto Y, Uchimiya H, Hinata K (1988) Transgenic rice plants after direct gene transfer into protoplasts. Bio/Technology 6: 1072–1074
19. Uchimiya H, Fushimi T, Hashimoto H, Harada H, Shono K, Sugawara Y (1986) Expression of a foreign gene in callus derived from DNA-treated protoplasts of rice (*Oryza sativa* L.) Mol Gen Genet 204: 204–207
20. Wesley BB, Christensen AH, Klein T, Fromm M, Quail PH (1989) Photoregulation of a phytochrome gene promoter from oat transformed into rice by particle bombardment. Proc Natl Acad Sci USA 86: 9692–9696
21. Zhang W, Wu R (1988) Efficient regeneration of transgenic plants from rice protoplasts and correctly regulated expression of the foreign gene in plants. Theor Appl genet 76: 835–840
22. Zhang HM, Yang H, Reach EL, Golds TJ, Davis AS, Mulligan BJ, Cocking EC, Davey MR (1988) Transgenic rice plants produced by electroporation-mediated plasmid uptake into protoplasts. Plant Cell Rep 7: 379–384

Plant Tissue Culture Manual **B3**: 1–15, 1991.
© 1991 *Kluwer Academic Publishers.*

Transformation and regeneration of orchardgrass protoplasts

MICHAEL E. HORN
Agrigenetics Co., 5649 E. Buckeye Road, Madison, WI 53716, U.S.A.

Introduction

The idea of using embryogenic suspension cultures as a source material for protoplasts was first popularized by I.K. Vasil and co-workers. The reasons for the success of embryogenic suspension cells giving totipotent protoplasts are, in hindsight, very clear. First, the cells in these suspension cultures (indeed, in *all* suspension cultures) are dividing relatively rapidly. The resultant protoplasts are already programmed to divide and, in fact, do so after only three to four days following plating. Second, as embryogenic cultures, every cell in every aggregate should have the ability to regenerate into somatic embryos which can convert into plantlets. Since non-embryogenic cultures rarely are able to produce plantlets, it seems unlikely such cultures could give protoplasts capable of doing much more than reconstituting a cell wall and going through a division or two. Starting with meristems of some type as the protoplast source material would seem to satisfy the above criteria but they are hard to isolate *en masse* and actually contain only a few percent of cells dividing. This latter point is also true of embryogenic callus as the plating efficiency is always extremely poor with embryogenic callus-derived protoplasts. Thus, currently the only source from which to reliably obtain large quantities of totipotent monocotyledonous protoplasts is from embryogenic suspension cultures.

Possessing a suspension culture with embryogenic cell aggregates is usually not adequate for successful protoplast isolation and plant regeneration. There is a strong correlation between the quality of the starting embryogenic suspension culture and subsequent plating efficiency. In Section 1, a procedure is described by which one may develop in orchardgrass (*Dactylis glomerata* L.) an established embryogenic suspension culture, i.e. one that is composed exclusively of small proembryogenic masses or PEMs. The presence of nonembryogenic cells should be avoided when starting protoplast isolation experiments. This is true not only in *Dactylis* but with all monocotyledonous species and those dicotyledonous species that regenerate through somatic embryogenesis.

Orchardgrass has not been the only member of the Gramineae to be transformed by direct gene transfer into protoplasts followed by regeneration [1, 2]. As examples, both rice [3] and maize [4] have been reported to be transformed in this fashion. Recently, fertile transformed maize plants were recovered after embryogenic suspension cells were bombarded with microparticles coated with the foreign DNA [5; and see chapter by Klein *et al.*, this volume]. Although particle bombardment may offer some advantages as a transformation method compared to direct gene transfer into protoplasts, the future method of choice still remains to be chosen.

Procedure

Procedure for the isolation and culture of Dactylis *protoplasts*

Steps in the procedure
1. Collect cellular aggregates from rapidly growing suspension cultures by vacuum filtration on a 0.22 μm or 0.45 μm Nalgene membrane filter.
2. Weigh out 0.5 g aliquots of cells in sterile Petri dishes.
3. Add 12.5 ml of filter sterilized protoplast enzyme mixture (EM). Seal plates with Parafilm.
4. Place plates onto orbital shaker and agitate slowly (approx. 50 rpm) for 4 hours at room temperature in dim light.
5. When wall digestion is determined to be complete the entire mixture is poured through a sterile stainless steel sieve having pores 94 μm in average diameter (Fig. 1). The eluant is collected in two 12 ml centrifuge tubes per digestion plate.
6. Centrifuge the eluant at 100 × g for 10 min. Pour off and discard the supernatant.
7. Wash the sediment three times with 10 ml filter-sterilized protoplast culture medium (KM-8p) per centrifuge tube.
8. Combine and resuspend the two sediments in 2 ml KM-8p and pour through a sterile stainless steel sieve having pores 20 μm in diameter (Fig. 2). Collect the protoplast preparation in a fresh centrifuge tube.

Notes
1. This allows the sterile collection of conditioned culture medium for use in the protoplast culture medium following protoplast isolation.
4. The actual time of incubation depends on the size of the proembryogenic masses, or PEMs, and the activity of the cellulase enzyme. Optimal incubation time should be determined empirically by the researcher.
7. Because of the very high density of embryogenic cells/protoplasts it is difficult to wash these protoplasts by flotation. To do so requires an osmolality of approx. 700 mOs/kg H_2O. Washing via sedimentation results in some small undigested aggregates (< 94 μm) to also pellet but these are almost entirely eliminated by the 20 μm sieving of step 8.

Factors involved in protoplast isolation

Yields of protoplasts have been very high in *Dactylis*, ranging up to more than 60×10^6 protoplasts per gram fresh weight of starting suspension material. If the protoplasts appear to clump too much then the amount of Ca^{2+} in the KM-8p protoplast culture medium should be varied. The age and quality of the wall degrading enzyme(s) can have significant effects on the yield and subsequent regrowth of the protoplasts. If protoplasts yields are low or if the protoplasts die quickly, another enzyme lot should be tested.

Solutions

Enzyme mixture (Em) [11]

	mg/l
$CaCl_2 \cdot 2H_2O$	1 029
$NaH_2PO_4 \cdot H_2O$	96.6
MES	585.6
Onozuka RS Cellulase	20 000
glucose	81 090

Adjust osmolality to 550 mos/kg H_2O

[MES = morpholinoethane sulphonic acid]

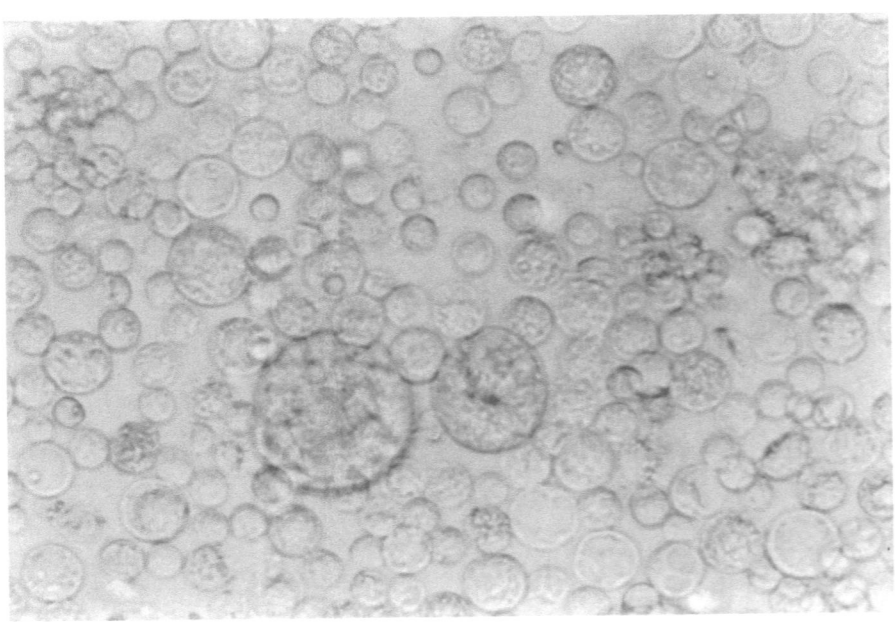

Fig. 1. Dactylis protoplast preparation following sieving through 94 μm pore stainless steel screen. (320 ×)

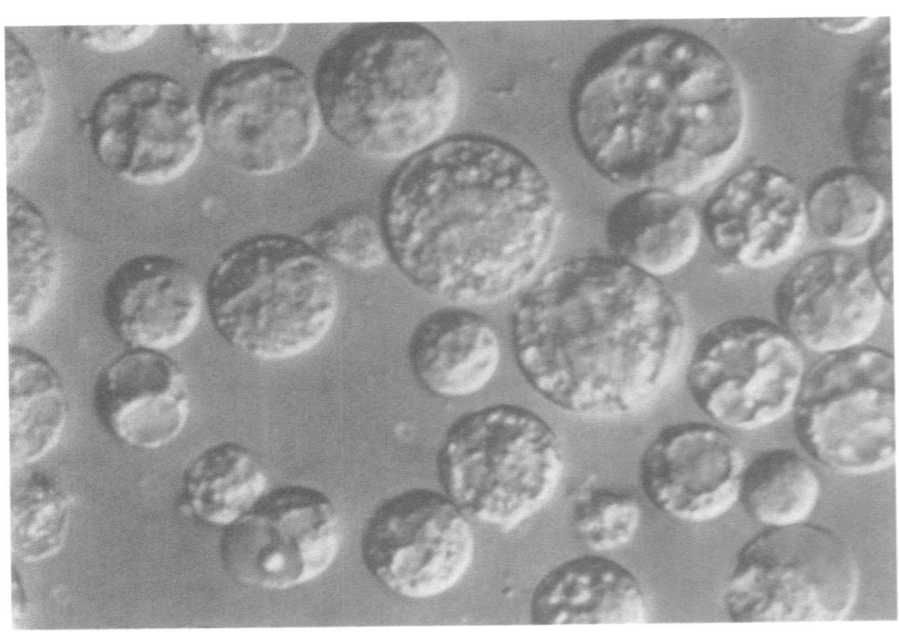

Fig. 2. Dactylis protoplast preparation following sieving through 20 μm pore stainless steel screen. (400 ×)

PTCM-B3/5

Procedure for culturing of Dactylis *protoplasts*

Steps in the procedure

1. Using a cell counter, determine the density of protoplasts in the protoplast preparation.
2. Dilute preparation to a concentration of 1.25-2.5×10^6 protoplasts per milliliter. Mix by inverting the tube several times.
3. Dispense 0.6 ml of protoplasts into a sterile Petri dish (6 cm diameter) and mix with 2.4 ml of KM-8p medium containing 1.5% (w/v) SeaPlaqueR agarose. This gives a final protoplast density of 250,000–500,000/ml.
4. Swirl Petri dish gently until protoplasts are evenly distributed around the plate.
5. Seal plates with Parafilm and incubate in the dark at 28 °C. Cell divisions should begin within four days after plating. Subsequent divisions should occur every three to four days (Fig. 3).
6. After 14 days the agarose can be sliced into five wedges [bead culture [6]] and each wedge floated on 20 ml SH-45 embryogenic suspension culture medium (see Horn, chapter A5, this volume) in a Petri dish.
7. Agitate the plates at 50 rpm on a platform shaker in darkness or dim light. Proto-colonies will grow until they emerge from the agarose at which time fragments will break off and start new suspension cultures (Fig. 4).
8. New suspension cultures can be plated onto SH-30 medium solidified with 2.5% Gelrite and incubated in the dark for somatic embryo production.
9. Somatic embryos can be converted to plantlets by transfer to SH-0 medium (no phytohormones) and incubated in the light.

Notes

2. Pipets with a small bore orifice should never be used with protoplasts. Use a wide bore pipet when pipetting is necessary.
3. The KM-8p with agarose should have the agarose melted and maintained at 45 °C until needed.
8. We have observed the protoplast-derived suspension cultures take longer to produce somatic embryos when plated onto SH-30 medium. In many cases a subculture onto fresh SH-30 medium is required for embryos to fully mature.

From protoplasts to plants

The addition of 30-40% (final concentration) conditioned medium in the KM-8p protoplast culture medium appears to give a significant boost to the growth rate of young microcolonies [1]. Note the conditioned medium should be added to the KM-8p medium before adding and melting the agarose. The reason for the growth enhancement properties of conditioned medium is still obscure. One possibility is that while glucose is the only carbohydrate that allowed cell divisions, it is not the carbohydrate of choice once new cell walls are formed. Cell wall-bound invertase allows the metabolism of the preferred sucrose once again. Thus, having some sucrose (from conditioned medium) in the protoplast culture medium may allow more rapid growth as soon as new cell walls are formed when compared to medium containing no sucrose whatsoever.

PTCM-B3/7

Once one is able to obtain plants from protoplasts, direct gene transfer experiments can proceed. It is, of course, more useful if the protoplast-derived plants are fertile. Sterility has been a problem using the protoplast approach but it is my opinion that sterility is a function not of the protoplast step but of the long time needed to reach the most critical step, that of the established embryogenic suspension culture. Research needs to be performed which results in a shortened time period from explant to established culture. We have conducted such research in rice with remarkable success [7].

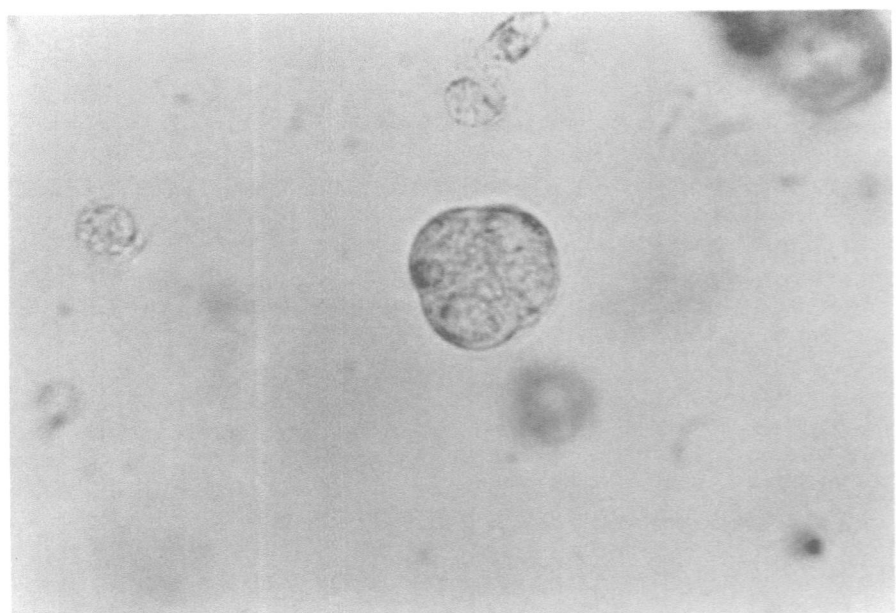

Fig. 3. *Dactylis* suspension-derived proto-colony after second division. Photograph taken six days following protoplast isolation and plating in KM-8p medium. (400 ×)

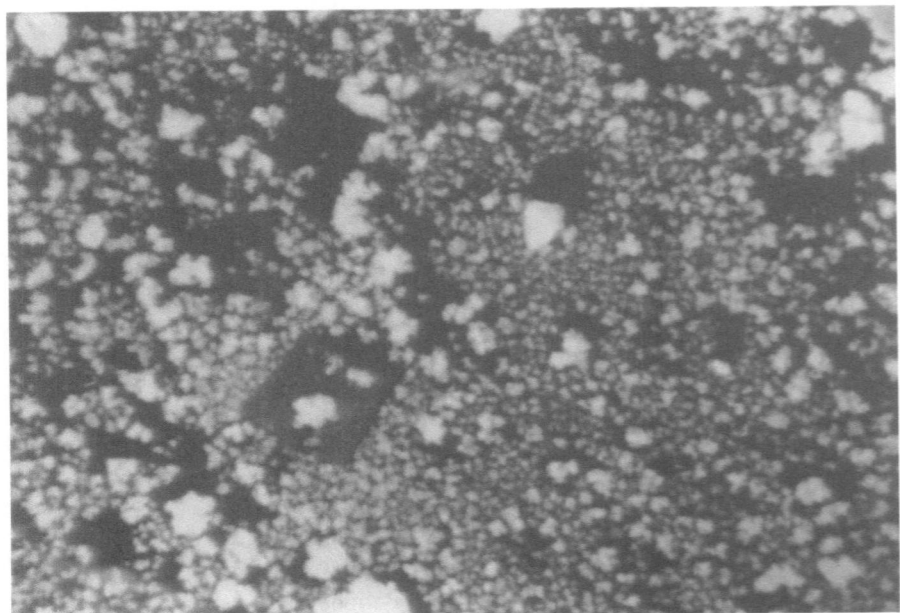

Fig. 4. New embryogenic suspension culture resulting from cells breaking off from original proto-colony still visible in agarose block. Colony growth and suspension reformation occurred during selection in 20 μg/ml hygromycin. (6 ×)

Procedure for direct gene transfer into Dactylis *protoplasts via electroporation*

Steps in the procedure [from ref. (8)]

1. After the protoplasts are filtered through the 20 μm stainless steel sieve they are pelleted by centrifugation at 60 × g for 5 min.
2. Resuspend the protoplasts to a density of $1.25-2.5 \times 10^6$/ml using ~0.4 M mannitol (550 mOs/kg H_2O) containing 6 mM $MgCl_2$, 1 g/l morpholinoethane-sulphonic acid (MES) as a buffer (pH 5.6).
3. Take a 0.34 ml aliquot of the protoplast suspension and test its resistance in a 1 ml electroporation chamber (e.g. 'DIA-LOG' Elektroporator, DIA-LOG GmbH., D-4000 Dusseldorf 13, FRG). The resistance should be between 1.0 and 1.1 kΩ which can be obtained by adding 3-5% (v/v) 0.3 M $MgCl_2$ to the protoplast suspension and retesting.
4. Place the tube containing the protoplast suspension into a 45 °C water bath for five minutes.
5. Distribute 0.7 ml aliquots of the protoplast suspension into 5 ml polycarbonate tubes.
6. Add 60 μl of the DNA solution containing 10 μg plasmid DNA and 50 μg calf thymus carrier DNA. Mix gently.
7. Add 0.38 ml of PEG Solution 1. Mix gently but completely. Wait ten minutes.
8. Transfer the samples to the electroporator chamber and pulse three times at 10 second intervals with pulses of initial field strengths of 3 kV/cm and an exponential decay constant of 10 μs. Other treatments should include 4 and 5 kV/cm.
9. Pour every three samples into a Petri dish and wait for 10 min. before slowly adding 3 ml KM-8p protoplast culture medium containing 40% (v/v) condi-tioned suspension medium.
10. Let agarose solidify, wrap plates in Parafilm, and incubate them in the dark at 28 °C as described above.

Note

3. The 0.3 M $MgCl_2$ solution must be filter sterilized but because of its viscosity this is difficult to do. Heating the solution to 50 °C before filter sterilizing can ameliorate this problem.

Procedure for direct gene transfer into Dactylis *protoplasts via the PEG protocol [from ref. (8)]*

Steps in the procedure

1. Sediment and resuspend the protoplasts in the 0.4M mannitol solution as described above for electroporation at a density of 1.25-2.5×10^6/ml.
2. Distribute the protoplasts into 0.3 ml aliquots in polycarbonate tubes.
3. Add 30 µl of DNA solution containing 4 µg plasmid DNA and 10 µg calf thymus carrier DNA. Mix gently and incubate at room temp. with occasional shaking.
4. Add 0.3 ml of PEG Solution 2. Incubate for 30 minutes at room temperature with occasional gentle mixing.
5. Slowly add 1 ml of KM-8p protoplast culture medium with no agarose. Three minutes later slowly add 3 ml KM-8p. Three minutes later add 6 ml KM-8p.
6. Wait five minutes then centrifuge at 60 × g for five minutes.
7. Resuspend protoplasts in 0.3 ml KM-8p without agarose, wait ten minutes, then gently mix with KM-8p with agarose as described above for electroporation.

Solutions

KM-8p [10] in mg/liter: NH_4NO_3 600, KNO_3 1900, $CaCl_2 \cdot H_2O$ 600, $MgSO_4 \cdot 7H_2O$ 300, KH_2PO_4 170, KCl 300, Fe·EDTA 28, KI 0.75, H_3BO_3 3, $MnSO_4 \cdot H_2O$ 10, $ZnSO_4 \cdot 7H_2O$ 2, $Na_2MoO_4 \cdot 2H_2O$ 0.25, $CuSO_4 \cdot 5H_2O$ 0.025, $CoCl_2 \cdot 6H_2O$ 0.025, glucose to 550 mOs/kg H_2O, inositol 100, nicotinamide 1, pyridoxine·HCl 1, thiamine·HCl 1, D-calcium pantothenate 1, folic acid 0.4, p—aminobenzoic acid 0.02, biotin 0.01, choline chloride 1, riboflavin 0.20, ascorbic acid 2, vitamin A 0.01, vitamin D_3 0.01, vitamin B_{12} 0.02, 2,4-dichloro-phenoxyacetic acid 0.2, zeatin 0.5, naphthaleneacetic acid 1.0, sodium pyruvate 20, citric acid 40, malic acid 40, fumaric acid 40, fructose 250, ribose 250, xylose 250, mannose 250, rhamnose 250, cellobiose 250, sorbitol 250, mannitol 250, casein hydrolysate 250, coconut water (filtered, heat cured) 20 ml/liter, pH 5.6, filter sterilized.

PEG Solution 1: 24% (w/v) PEG 6000 in 0.4 M mannitol, 0.1% MES (pH 5.6) and 30 mM $MgCl_2$ (electrical resistance of this solution in the electroporation chamber should be between 1.2 and 1.3 kΩ. Use $MgCl_2$ to change resistance.).

PEG Solution 2: 40% (w/v) PEG 4000 in 0.4 M mannitol, 0.1 M $Ca(NO_3)_2$, pH 8.0 with KOH. The pH takes 3-4 hours to stabilize.

Concerning direct gene transfer techniques

Both electroporation and the PEG protocol have been successful in giving transformed proto-colonies and transformed plants [2]. However, the PEG method has tended to give far more independent transformation events due to its gentle nature relative to electroporation. We routinely found 50% mortality of *Dactylis* protoplasts following electroporation while the PEG treatment itself resulted in virtually no cell death. Another consideration that might be important to some scientists is that no special equipment is needed for the PEG protocol. Both methods require empirical testing for optimal results. For example, the exact concentration of $MgCl_2$ needed in the mannitol solutions appears to be critical and may vary between laboratories and certainly varies between plant species. The plasmid should be linearized by cutting the DNA with a restriction enzyme in some non-essential location on the sequence. Linear DNA seems to result in a higher transformation frequency than does circular DNA.

Selection for antibiotic resistance following direct gene transfer

Selection for the stable, viable, transformed proto-calluses should begin at between 10-14 days following plating. This corresponds very well with the beginning of the 'bead culture' phase so that the antibiotic can be introduced to the proto-colonies in the liquid culture medium supporting each agarose wedge. Selection begun earlier than ten days after plating reduced the number of resistant proto-colonies which survived. Selection started later than 14 days allowed for growth of some non-resistant colonies although eventually their growth would falter and they were recognized as 'escapes'. Our experiments utilized a hygromycin[R] gene bordered by a 35S promoter region and a 35S termination region constructed by Rothstein *et al.* [9]. Hygromycin at 20 µg/ml when applied no later than 14 days after plating allowed very few false positives. Kanamycin up to 200 µg/ml did not show acceptable selection stringency in *Dactylis* embryogenic suspension cultures. Selection using the antibiotic G418 may allow transformation with the kanamycin[R] gene but this has not been tested. Selection should continue throughout the regeneration process. Embryo conversion on SH-0 medium is very sensitive to antibiotics and the hygromycin level at this step should be reduced to 0.1-0.5 µg/ml.

References

1. Horn ME, Shillito RD, Conger BV, Harms CT (1988) Transgenic plants of orchardgrass (*Dactylis glomerata* L.) from protoplasts. Plant Cell Reports 7: 469-472
2. Horn ME, Shillito RD, Conger BV, Harms CT (1989) Transgenic plants of orchardgrass (*Dactylis glomerata* L.) regenerated following direct gene transfer into protoplasts. IN: EUCARPIA Congress Proc., Genetic Manipulation in Plant Breeding, Helsingor/Denmark 1988, Plenum Press, NY
3. Shimamoto K, Terada R, Izawa T, Fujimoto H (1989) Fertile transgenic rice plants regenerated from transformed protoplasts. Nature 338: 274-276
4. Rhodes CA, Pierce DA, Mettler IJ, Mascarenhas D, Detmer JJ (1988) Genetically transformed maize plants from protoplasts. Science 240: 204-207
5. Gordon-Kamm WJ, Spencer TM, Mangano ML, Adams TR, Daines RJ, Start WG, Chambers SA, Adams WR Jr, Willetts NG, Rice TB, Mackey CJ, Krueger RW, Kausch AP, Lemaux PG (1990) Transformation of maize cells and regeneration of fertile transgenic plants. The Plant Cell 2: 603-618
6. Shillito RD, Paszkowski J, Potrykus I (1983) Agarose plating and a bead type culture technique enable and stimulate development of protoplast-derived colonies in a number of plant species. Plant Cell Reports 2: 244-247
7. Horn, ME, Matsuno T, Rose BL, Brink KB. Rapid scale-up of rice embryogenic cell cultures for the production of clonal plants. In Preparation
8. Shillito RD, Saul MW (1988) Protoplast isolation and transformation. In: Plant Molecular Biology, A Practical Application, IRL Press, United Kingdom, pp 161-186
9. Rothstein SJ, Lahners KN, Lotstein RJ, Carozzi NB, Jayne SM, Rice DA (1987) Promoter cassettes, antibiotic resistence genes, and vectors for plant transformation. Gene 53: 153-161
10. Kao KN, Michayluk MR (1972) Nutritional requirements for growth of *Vicia hajastana* cells and protoplasts at a very low population density in liquid media. Planta 126: 105-110
11. Horn ME, Conger BV, Harms CT (1988) Plant regeneration from protoplasts of embryonic suspension cultures of Orchardgrass (*Dactylis glomerata* L.). Plant Cell Reports 7: 371–374

Plant Tissue Culture Manual **B4**: 1–24, 1991.

Transformation and regeneration of oilseed rape protoplasts

DOMINIQUE ROUAN* AND PHILIPPE GUERCHE
Laboratoire de Biologie cellulaire, INRA, Route de St. Cyr, F.78026 Versailles cedex, France

* Present address:
Plant Genetic Systems, J. Plateaustraat 22, B. 9000 Gent, Belgium

Introduction

Genetic transformation of higher plants is one of the new tools offered by the development of biotechnology and it is already helping to understand some of the mechanisms of regulation and expression of the genes. It is likely that the transfer of genes carrying agronomically interesting characters in major plant crops will rapidly have an influence on the breeding programs of these plants.

For this reason, soon after the first success of genetic transformation achieved in model species like tobacco and petunia [2, 15, 16] different laboratories have tried to transpose this technology to crop species.

Rapeseed is cultivated for oil and animal feed production on an increasing scale throughout the world [7]. Transgenic rapeseed plants were first obtained after regeneration of hairy roots induced by stem infection with wild-type *Agrobacterium rhizogenes* [11, 17, 26]. Soon after this, antibiotic- or herbicide-resistant plants were created using *Agrobacterium tumefaciens* [4, 6, 9, 22, 23, 31, 32; Mauvais (Dupont, Inc), personal communication; Primard, personal communication], microinjection of microspore derived embryos [24] and direct gene transfer of protoplasts [10, 12]. Some recent experiments have attempted to alter the protein composition of the seeds [14, 32] and others in progress seek to obtain virus resistant plants (Yot, personal communication).

Nevertheless, the transposition of the transformation techniques initially developed on some Solanaceous species to rapeseed led to numerous problems as shown by analysis of the different studies mentioned above. While it seems possible to obtain regeneration from primary explants (hypocotyl, stem thin-layer explant, leaf petiole, etc) with a high efficiency both on spring and winter rapeseed varieties, the overall *Agrobacterium* transformation process appears to be more problematic. Only spring varieties (mainly Westar) were successfully transformed, showing that this technique is highly dependent on the genotype. However recently, De Block *et al.* [6] have described obtaining transgenic rapeseed plants from a winter variety.

It also appeared that each step of the transformation procedure can present limiting factors. The regeneration potential of the tissues greatly decreased after infection with *Agrobacterium tumefaciens* and decontamination. Many authors mentioned problems of expression of the selectable marker gene, and difficulties in selecting the transformed buds. None of the various associations of promoters (pNos, p19S, p35S) with selectable marker gene sequences (kanamycin (Km), hygromycin (Hyg), phosphinothricin (PPT), methotrexate or chloramphenicol resistance) (for review, see [19]) used in transformation experiments enabled a clearcut selection of the transformed tissue, showing that many factors controlling the transformation efficiency are still unknown.

We developed a reasonably efficient rapeseed regeneration procedure in our laboratory [12, 29]. Taking advantage of the fact that protoplasts are easily obtained in large numbers and with a high regeneration ability, we tried to develop direct gene transfer by electroporation in such material. Electroporation [25] consists of subjecting cells to a pulse of high electric field for a very short time. This electric shock creates a reversibly permeable area in the membrane of the cell, which will allow the transfer of nucleic acid into the cell. Among the different kinds of electric shocks which may be used for electroporation, we chose the discharge of a capacitor within the protoplast-DNA suspension [8, 20, 35].

1. Transient expression assay for the optimization of rapeseed protoplast electroporation

Transient expression is a convenient way to determine optimum conditions for DNA transfer in plant protoplasts, most often by assays of choramphenicol acetyl transferase (CAT) [8] or β 1–4 glucuronidase (GUS) [18] activity. However, especially for rapeseed, CAT is not recommended for this purpose because of the high endogenous CAT activity of rapeseed protoplasts [1], but also because of the presence of an inhibitor factor which lowers the introduced CAT activity [5; Rouan *et al.*, in preparation]. GUS appears to be the most efficient reporter gene for this species. In our experiments, no significant endogenous GUS-like activity was ever detected.

Transient expression allows an efficient optimization of parameters leading to stable transformations, however only when a high level of protoplast viability has been conserved. We carried out several preliminary electroporation experiments with the pCH 1 plasmid carrying the 35S promoter with its duplicated enhancer and the GUS coding sequence (Horlow *et al.*, in preparation), using various capacitances and field strength values. For each assay, the viability was measured by observation of the morphological appearance of protoplasts under an optical microscope just before harvesting them for the GUS activity assay.

The mean results obtained with increasing field strengths using a 64 µF capacitor are shown in Fig. 1. The maximum GUS activity is obtained at between 20 to 50% of protoplast viability, while lower field strengths gave less activity but higher viability.

In our case, various conditions of DNA transfer revealed detectable GUS activity with 60 to 70% viability using either a long pulse duration (350 ms) with a 64 µF capacitor charged at a low voltage (225 V/cm), or a shorter duration (90 ms) with a 16 µF capacitor charged with a higher voltage (300 V/cm).

2. Stable transformation experiments

Electrical parameters

The electrical conditions defined for transient expression appear also to be reliable for stable transformation, especially with respect to protoplast viability (60–70%) which seems to be the critical point. When protoplast viability was not affected after the electroporation treatment, the protoplasts were not sufficiently permeabilized to give transformed cells. However, when protoplast viability was less than 50-60%, the resulting capability of the surviving protoplasts to undergo division and subsequently to yield microcolonies in the selective medium was seriously compromised because of the resulting decrease in the cell density.

Using the electrical conditions described above, the frequency of transformation reached $2.09 \pm 1.15 \times 10^{-4}$ as the mean of five different experiments (which represent 104 ± 53 resistant colonies per 10^6 surviving protoplasts), whatever the selective agent used.

Culture technique and colony selection

The selective agent must be applied when the protoplast-derived cells have undergone their first division. At this moment, the micro-colony density is critical in achieving selection of the resistant colonies because the sensitive colonies can prevent the development of the resistant ones by releasing toxic substances into the medium. One of the best solutions to this problem would be to cultivate colonies at a very low density (i.e. at 1 colony/ml), as is possible for tobacco cell culture [3] without losing any regeneration capacity. This cannot yet be achieved for rapeseed cells. An alternative was to use agarose-embedded protoplast culture ('bead type culture'), which allows the selective medium to be renewed often without affecting the colonies' growth [34]. For antibiotic resistance experiments, we used paromomycin, an analog of kanamycin, which at least in tobacco, allows a clear-cut selection of resistant colonies at higher density levels than with kanamycin [13].

In our experiments, antibiotic-resistant rapeseed colonies were selected using paromomycin and the bead-type culture technique. Some reconstruction experiments were done to show that these culture and selection conditions were critical for efficient selection of transformed micro-colonies. Mixtures at 4 different ratios (20%, 2%, 0.2%, 0.02% of resistant protoplasts) of antibiotic-sensitive (from a Brutor plant) and antibiotic resistant protoplasts (from the plant line PG20; [12]), were cultivated in 3 different ways:

– in a first experiment, the protoplast mixture was cultivated 11 days in liquid medium B and diluted 1:1 with medium C containing 20 mg/l of paromomycin. In this case, even when the resistant protoplasts represent 20% of the initial protoplast population, it was not possible to recover any resistant colonies;
– in a second experiment (Table 1), the protoplasts were cultured in liquid

Table 1. Percentage of recovered kanamycin-resistant colonies after bulk culture of different proportions of sensitive (Brutor), and resistant (PG20) protoplasts, in liquid medium (L) or in agarose (A)

Plants[a]				
– PG20	10000	1000	100	10
– Brutor	40000	50000	50000	50000
R(%) L	2.7	2.2	1.3	0
R(%) A	100	25	12	9

[a] Number of protoplasts in bulk culture

$$R(\%) = \frac{\text{Number of recovered colonies}}{\text{Theoretical number of colonies}} \times 100$$

medium or embedded in a culture medium containing agarose. At 11 days, the colonies were diluted 1:1 in medium C, and 2 days later, diluted again 1:1 in medium C containing the antibiotic. We defined the percentage of colony recovery (R) as the percentage of the resistant colonies observed relative to the theoretical number of resistant colonies (which is the product of the number of resistant protoplasts initially present in the mixture multiplied by the plating efficiency). This ratio is higher when colonies are more dilute in the liquid medium, however the best R value is obtained when the colonies are cultivated in agarose medium and diluted once again before adding the selective agent. It is important to note that successive dilutions and the selection process reduce the growth rate of the colonies. This fact also antagonizes the regeneration frequency of these colonies. Therefore, it is important to find a compromise between these different factors.

Resistant colonies can be obtained on medium containing kanamycin, though with a lower efficiency than with paromomycin (such colonies are less green and appear later than on paromomycin medium).

Marker genes and selection

We used various marker genes like Km^r, Hyg^r and PPT^r with 3 different promoters (19S, 35S and NOS). All of these were introduced into diverse plasmids (pABDI [27]; pHP23b [28]; pGL2 [30] and pIB 16.1 [36]. Though their transformation and expression efficiency could not be directly compared, all of them gave similar results in terms of the relative transformation frequency (RTF) as shown in Table 2.

Table 2. Transformation frequencies obtained with stable transformation of rapeseed protoplasts in various plasmid/promoter systems: pGL2 (HYGr) and plB16.1 (PPTr, Kmr). Electrical conditions were the same in all cases: 300 V/cm, 16 µF and 3 pulses at 5 seconds intervals.

Selection	Concentration (µg/ml)	P.E. (%)	Number of resistant colonies/0.4×10^6 pps	ATF ($\times 10^{-4}$)	RTF ($\times 10^{-4}$)
Paromomycin	20	47.5	47	1.2	2.55
Hygromycin	25	19	15	0.375	2
Phosphinothricin	5	47	74	1.85	3.8

P.E.: plating efficiency in percent of control. ATF: absolute transformation frequency. RTF: relative transformation frequency.

The main differences were qualitative. Table 3 shows that with PPT^r and Hyg^r selection, resistant colonies grew faster than with Km^r (or paromomycin) selection. This is of great importance because the colonies can be removed earlier from the selection medium and be regenerated with a higher efficiency

Table 3. Effect of gene marker type on the growth kinetics and physical appearance of the resistant colonies. Electrical conditions: 300 V/cm, 16 µF, 3 pulses at 5 seconds intervals.

Selection	Number of recovered colonies[a]			Physical appearance of the colonies[c]
	7 days[b]	14 days[b]	21 days[b]	
Paromomycin	27	49	24	+
Hygromycin	43	50	7	+ +
Phosphinothricin	29	57	14	+ + +

[a] In percent of total number of colonies recovered in this experiment.
[b] Time after selection medium applied.
[c] Subjective analysis on the basis of green color and size.

than 'older' colonies. Colonies obtained through PPT^r selection seem to be more 'competent' for regeneration than the colonies selected on hygromycin or kanamycin.

It was possible to select Km^r colonies on lower concentration of paromomycin (10 to 15 µg/ml). This permits a more rapid growth of transformed cells by delaying the toxicity induced by the sensitive dying ones. Such lower doses of paromomycin did not allow growth of the control colonies. This kind of selection at low dose of antibiotic was not easy to apply with hygromycin, because sensitive cells died very slowly or continued to grow at doses of less than 20–25 µg/ml, making selection ambiguous.

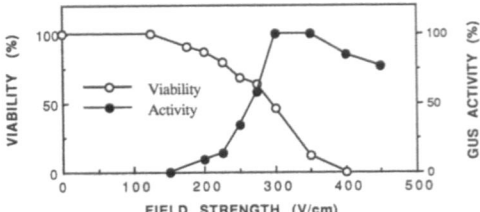

Fig. 1. Effect of field strength on protoplast viability and GUS transient expression. Viability is expressed as a percent of the control. GUS activity is expressed as the percent of the highest value (319 pmole 4 · Mu/min/mg of protein). The capacity was 64 μF, 3 pulses.

Procedures

Plant Material

Steps in the procedure

1. Soak the seeds (*Brassica napus*, var: Brutor) in 70% (v/v) ethanol for 1min. Surface-sterilize for 30 to 40 minutes in a commercial sodium hypochlorite solution (12 CHL) containing 0.1% (v/v) Tween 80. Rinse three times in sterile distilled water. Sow the seeds for germination on medium A (see note a).
2. Maintain shoot tip cultures *in vitro* (see Chapter by Topping and Lindsey, this volume) on medium A under approximately 2000 lux with a 16h photoperiod at 26 °C by repeated subcultures at three-to-four week intervals (see note b).

Protoplast Isolation

Steps in the procedure
1. Place 12 to 16 sterile fully expanded (1 to 2 cm^2) and scarified leaves located below the rapidly growing leaves of the apex area, on the surface of 20 ml of the maceration solution in a 250 ml flask (see note c).
2. Let them macerate overnight with gentle agitation at room temperature.
3. Filter 20 ml of the protoplast suspension through a 40 μm sieve into a 30 ml tube containing 10 ml of washing buffer (2.5% (w/v) KCl, 0.2% (w/v) CaCl2), gently mix, and centrifuge at 70 × g for 10 min.
4. Re-suspend the pellet in 20 ml of MKCl buffer [13], gently mix and centrifuge at 70 × g for 10 min.
5. Re-suspend the pellet in MKCl buffer to a final concentration of 2 × 10^6 protoplasts/ml (counted in a Malassez-type hemocytometer). Keep protoplasts on ice (see note d).

Electroporation Procedure

Steps in the procedure
1. Add 10 µg of the plasmid and 50 µg of sonicated calf thymus DNA (approximately 1 kb size), or 60 µg of the plasmid alone, to 1 ml of the protoplast suspension in the electroporation chamber.
2. After homogenization, measure the conductivity of the solution in the chamber using an alternating current multimeter operating at 1 kHz (Hoelzle and Chelius, FRG) (see note e).
3. Apply 3 electric pulses at 5 sec intervals into the cuvette at room temperature. Transfer contents to three 10 cm Petri dishes (tissue culture treated, Greiner, FRG) for stable transformation or into only one dish for transient expression.
4. After 10 minutes at room temperature, gently add 10 ml of medium B containing 0.6% (w/v) melted agarose (LSM type, LITEX, Denmark) at 37 °C for stable transformation or 10 ml of liquid medium B for transient expression.

Cell Culture and Selection

Steps in the procedure

1. Incubate protoplasts in the dark at 26 °C in a plastic box, without sealing the Petri dishes. After 7 to 9 days, when the majority of protoplast-derived cells have divided once, cut the agarose disk in two parts. Place each half in a new Petri dish containing 5 ml of liquid medium C, seal with Parafilm and transfer culture to light (2000 lux, 16h photoperiod at 26 °C).

2. Two days later, cut each half in two parts and transfer each resulting quarter onto a Petri dish with 2.5 ml of the previous liquid medium plus 5 ml of fresh medium C. Add antibiotic or herbicide to the Petri dishes (which contain 7.5 ml liquid medium + 2.5 ml solid medium with cells in it) to obtain the following final concentrations:
   - Paromomycin sulfate (Substantia laboratories) 20 mg/l
   - Hygromycin B (Sigma) 20 mg/l
   - D/L Phosphinothricin (Hoechst) 5 mg/l.

3. Cut the agarose fragment into small pieces (about 0.25 cm^2 each).

4. Renew all the liquid selection medium every 6 days by discarding the old liquid and adding the same volume of fresh selective medium (7.5 ml).

5. Three to four weeks after protoplast isolation, pick up any green colonies greater than 1 mm in diameter from the agarose.

6. Transfer 'resistant' colonies into the following medium for 4 to 7 days: 2 ml of fresh medium D mixed with 2 ml of 'conditioned' medium C (removed from a culture plate with no added selection agent) in a 6 × 15 mm petri dish.

7. Transfer colonies onto shoot regeneration medium E (10-15 colonies on 25 ml medium in a 10 cm Petri dish).

8. After buds appear (2–4 weeks of culture on medium E), transfer to medium F to enhance shoot elongation.

9. Two weeks later, transfer the plantlets to medium G for rooting.

Equipment
- General tissue culture materials
- Jouan centrifuge or equivalent
- 40 μm sieve
- Hemocytometer (Malassez type)
- Phase-contrast microscope
- Tissue-culture, surface-treated, and normal Petri dishes (94 × 16 mm)
- Electroporation equipment
   - an ISCO 490 power supply coupled with one to ten 16 μF (600 V) capacitors connected in parallel.
   - a memory monitoring oscilloscope
   - a plexiglass electroporation chamber (1 cm^3) where two stainless steel electrodes (1 cm^2) placed on two opposite sides of the interior of the chamber are connected to the electrical system.
   - a conductivimeter (Hoelzle and Chelius)

Solutions

— Digestion solution: [12]

Cellulase R10	0.2% (w/v)
Pectolyase Y23	0.1% (w/v)
$CaCl_2 \ 2H_2O$	0.9 g/l
NaH_2PO_4	0.08 g/l
Mannitol	100 g/l
NAA	1 mg/l
BAP	1 mg/l

The medium (pH 5.8) is filtered through a 0.2 μm sieve.

— MKCl buffer [13]

Mannitol	90 g/l
KCl	0.37 g/l
MOPS	0.042 g/l

$MgCl_2$ was added to a final concentration of 1 mM.
The buffer, pH 7.2, was autoclaved at 110 °C for 20 min.
Conductivity: 800 μS/cm.

— Culture media: See Table 4

— DNA solutions:
  — Plasmids were extracted according to [21] and were re-suspended in sterile MKCl buffer at a final concentration of 1μg/μl and kept on ice until electroporation.
  — Calf Thymus DNA (Sigma) was sonicated (average of 1 kb long fragments), ethanol-precipitated and resuspended in sterile MKCl buffer to a final concentration of 1 μg/μl.

Table 4. Compositions (mg/l) of the different media used for plant regeneration from mesophyll protoplasts of rapeseed. [12, 29].

	A	B	C	D	E	F	G
NH_4NO_3	1.650		200	200	1.650	1.650	825
KNO_3	1.900	2.500	1.250	1.250	1.900	1.900	950
$(NH_4)_2SO_4$		134	67	67			
NaH_2PO_4		150	75	75			
KH_2PO_4	170		35	35	170	170	85
$CaCl_2, 2H_2O$	440	750	525	525	440	440	220
$MgSO_4, 7H_2O$	370	250	250	250	370	370	185
H_3BO_4	12.4	3	3	12.4	12.4	6.2	6.2
$MnSO_4, 4H_2O$	33.6	10	10	33.6	33.6	22.3	22.3
$ZnSO_4, 7H_2O$	21	2	2	21	21	8.6	8.6
KI	1.66	0.75	0.75	1.66	1.66	0.83	0.83
$Na_2MoO_4, 2H_2O$	0.5	0.25	0.25	0.5	0.5	0.25	0.25
$CuSO_4, 5H_2O$	0.05	0.025	0.025	0.05	0.05	0.025	0.025
$CoCl_2, 6H_2O$	0.05	0.025	0.025	0.05	0.05	0.025	0.025
$FeSO_4, 7H_2O$	27.8	27.8	27.8	27.8	27.8	27.8	27.8
Na_2 EDTA	37.3	37.3	37.3	37.3	37.3	37.3	37.3
Inositol	100	100	100	100	100	100	100
Nicotinic acid	0.5	1	1	1	0.5	1	0.5
Pyridoxine HCl	0.5	1	1	1	0.5	1	0.5
Thiamine		10	10	10		10	
Glycine	2				2		2
Glucose	10000	20000			10000		10000
Sucrose	10000		20000	20000	10000	10000	
D-mannitol		70000	40000		10000		
NAA		1	0.2		1	0.1	0.01
BA		1	1		0.5	0.5	
2,4-D		0.25		1			
Adenine sulfate				30			
IPA					0.5		
GA_3					0.02	0.02	
Agar (Biomar)	8000					8000	8000
Agarose (Litex)		600			10000		
pH (KOH)	5.8	5.8	5.8	5.8	5.9	5.8	5.8

NAA: naphthalene-acetic acid; BA: 6-benzyl aminopurine; 2,4-D: 2,4 dichloro-phenoxy-acetic acid; IPA: N^6-(Δ^2 iso-pentenyl)-adenine; GA_3: gibberellic acid.

GUS assay [18]

Steps in the procedure

1. Protoplasts were harvested 24 hours after electroporation (1x106 protoplasts were sufficient for one assay). Wash the protoplasts twice in washing buffer by centrifugation at 70 X g, and resuspend them in 100 µl of the GUS extraction buffer.

2. Break up the cells by at least two cycles of freezing and thawing, and vortex for 45 sec. Centrifuge at 12000 X g for 5 minutes. Recover the supernatant for GUS assay.

3. Determine protein concentration in crude extracts by Bradford assay (Biorad) with BSA as standard.

4. Mix an equal amount of protein (25 µg) per sample, with the GUS reaction buffer to a final volume of 100 µl. Incubate at 37 °C. (For each sample, pre-incubate the GUS reaction buffer at 37 °C).

5. Transfer 20 µl of the reaction mixture into 1 ml of stop buffer at time 0, 10, 20 and 30 minutes (see note f).

6. Measure the fluorescence emission at 455 nm in a fluorescence spectro-photometer under an excitation wavelength of 365 nm. Calibrate the apparatus with 100 and 10 nM of freshly prepared 4-Methyl umbelliferone (4.Mu) in the lysis buffer.

7. Express GUS activity in pmoles of 4.Mu formed per min and per mg of protein.

— Equipment:
  — General molecular biology equipment
  — Fluorimeter (Hitachi F. 2000 fluorescence spectrophotometer) — Quartz cuvettes
  — Water bath at 37 °C

— Buffers:
  — GUS extraction buffer. For 100 ml:

— Na_2HPO_4	0.2 M	15.25 ml
— NaH_2PO_4	0.2 M	9.75 ml
— Na_2EDTA	0.5 M, pH 8.0	0.2 ml
— Triton X100 (Sigma)		0.1 ml
— Sarkosyl (Sigma)		0.1 g
— Dithiothreitol	1 M	1 ml

  — GUS assay buffer: 1 mM 4-methyl umbelliferone glucuronide (MUG, M.9130, Sigma) in extraction buffer. For 50 ml: add 22 mg of MUG to 50 ml of GUS extraction buffer. Store at 4 °C for a maximum of 2 weeks.

  — Stop buffer 0.2M Na_2CO3. For 100 ml: dissolve 2.12 g in H_2O.
  — 1 mM 4-Methyl umbelliferone (4.Mu, M.1508, Sigma). For 100 ml: dissolve 19.82 mg in Stop buffer and store in the dark at 4 °C.

Notes

a. It was sometimes difficult to completely sterilize rapeseed tissues, as endogenous bacteria could appear during shoot tip subculture. This kind of infection is not often visible with the naked eye, thus it may be necessary to check by putting a tissue sample of each in vitro plant on top of solidified LB medium which allows bacteria to grow quickly.

b. The nature of the agar (Biomar) used is important for the leaf quality of the plants grown in vitro. For certain batches, it was necessary to mix agar with agarose to a final concentration of 6 g/l, the proportion depending on the type of agar.

c. At a given field intensity, the permeability of the membrane depends on the diameter of the protoplasts and other physiological factors [33]. In order to obtain optimum permeation of the protoplast population, the homogeneity of the cell preparation remains an important factor. For these reasons, well-expanded leaves which are as similar as possible gave the best results.

d. Most of the electroporation protocols published so far have used highly conductive mediums containing either Mg^{++} or Ca^{++}. For example, Fromm et al. [8], who gave electroporation shocks with approximately the same voltage as us, used a very low resistivity buffer (between 11 to 18 $\Omega \times$ cm) and very high capacitance (1000 µF). By increasing the resistivity of the buffer (900 $\Omega \times$ cm) and lowering the capacitance, it was possible to decrease the heating of the protoplast solution during the electric shock (by Joule effect) and to retain a good transformation efficiency. Mg^{++} ions were added only at a 1 mM concentration in order to stabilize the protoplast membrane. Neumann et al. [25], demonstrated that high $MgCl_2$ concentration could facilitate adsorption of DNA on the plasmalemma at the expense of its transfer.

e. Checking pulses on an oscilloscope should not be necessary if it is possible to measure the conductivity of the MKCl buffer-protoplast mixture between the electrodes. Pulse length (t_0) is determined by the time within which the initial peak voltage E_0 decays to a field strength approaching 0. This decay time can be calculated empirically to be equal to 5-fold the product of the resistance R(Ω) (which for a 1 cm3 chamber is equal to the resistivity ($\Omega \times$ cm)) times the capacitance value C (F) of the capacitance: $t_0 = 5 \times R \times C$, and so the pulse's decay time can be calculated if both these simple values are known.

f. In some protoplast preparations, endogenous GUS-like activity could interfere with the introduced activity. Then the 4.Mu production would not be linear over time. This could be counteracted by assaying short reaction times, such as 15 to 30 minutes. After this time, non-linear reaction appears. Nevertheless, endogenous GUS activity was negligible in our experiments, but seems to depend on the protoplast preparation and/or extraction procedure.

Acknowledgments

We would like to thank Christine Horlow for the generous gift of the pCH 1 plasmid, Georges Pelletier, Yves Chupeau, Ian Small, Tony Lough, Marie Hélène Montané and Eric Huttner for critical reading of the manuscript and helpful discussions. We thank Marie France Commeau and Jo Anne Marie for corrections to the English. D.R is grateful to Jean Pierre Bourgin for his hospitality and assistance during her stay in his laboratory. D.R was supported by a grant from Le Ministère de la Recherche et de la Technologie.

References

1. Balazs E, Bonneville JM (1987) Chloramphenicol acetyl transferase activity in *Brassica* spp. Plant Sci 50, 65–68.
2. Bevan MW, Flavell RB (1983) A chimaeric antibiotic resistance gene as a selectable marker for plant cell transformation. Nature 304, 184–187.
3. Caboche M (1980) Nutritional requirement of protoplast-derived, haploid tobacco cells grown at low cell densities in liquid medium. Planta 149, 7–18.
4. Charest PJ, Holbrook LA, Gabard J, Iyer VN, Miki BL (1988) *Agrobacterium*-mediated transformation of thin cell layer explants from Brassica napus L. Theor Appl Genet 75, 438–445.
5. Charest P, Iyer VN, Miki BL (1989) Factors affecting the use of chloramphenicol acetyltransferase as a marker for *Brassica* genetic transformation. Plant Cell Reports 7, 628–631.
6. De Block M, De Brouwer D, Tenning P (1989) Transformation of *Brassica napus* and *Brassica oleracea* using *Agrobacterium tumefaciens* and the expression of the *bar* and *neo* genes in the transgenic plants. Plant Physiol 91, 694–701.
7. Food and Agriculture Organisation (1988) Yearbook Production, 42, 161.
8. Fromm M, Taylor LP, Walbot V (1985) Expression of genes transferred into monocot and dicot plant cells by electroporation. Proc Natl Acad Sci USA 82, 5824–5828.
9. Fry J, Barnason A, Horsch RB (1987) Transformation of *Brassica napus* with *Agrobacterium tumefaciens* based vectors. Plant Cell Rep 6, 321–325.
10. Golz C, Köhler F, Sacristan MD, Schieder O (1990) Transformation of *Brassica* species via direct gene transfer. Abstracts VIIth international congress on plant tissue and cell culture, Amsterdam, June 24–29, A2–57.
11. Guerche P, Jouanin L, Tepfer D, Pelletier G (1987a) Genetic transformation of oilseed rape (*Brassica napus*) by the Ri T-DNA of *Agrobacterium rhizogenes* and analysis of inheritance of the transformed phenotype. Mol Gen Genet 206, 382–386.
12. Guerche P, Charbonnier M, Jouanin L, Tourneur C, Paszkowski J, Pelletier G (1987b) Direct gene transfer by electroporation in *Brassica napus*. Plant Sci 52, 111–116.
13. Guerche P, Bellini C, Le Moullec JM, Caboche M (1987c) Use of a transient expression assay for the optimisation of direct gene transfer into tobacco mesophyll protoplasts by electroporation. Biochimie 69, 621–628.
14. Guerche P, De Almeida ERP, Schwarztein MA, Gander E, Krebbers E, Pelletier G (1990) Expression of the 2S albumin from *Bertholletia excelsa* in *Brassica napus*. Mol Gen Genet 221, 306–314.
15. Herrera-Estrella L, Depicker A, Van Montagu M, Schell J (1983) Expression of chimaeric genes transferred into plant cells using a Ti-plasmid-derived vector. Nature 303, 209–213.
16. Horsch RB, Fraley RT, Rogers SG, Sanders PR, Lloyd A, Hoffmann N (1983) Inheritance of functional foreign genes in plants. Science 223, 496–498.
17. Hrouda M, Dusbabkova J, Necasek J (1988) Detection of Ri T-DNA in transformed oilseed rape regenerated from hairy roots. Biol. Plant. 30, 234–236.
18. Jefferson RA, Kavanagh TA, Bevan MW (1987) GUS fusions: β Glucuronase as a sensitive and versatile gene marker in higher plants. EMBO J. 6, 3901–3907.
19. Klee HJ, Rogers SG (1989) Plant gene vectors and genetic transformation: plant transformation systems based on the use of *Agrobacterium tumefaciens*. Cell Culture and Somatic Cell Genetics of Plants, Vol 6: Molecular Biology of Plant Nuclear Genes. Schell J and Vasil I.K Eds, Academic Press Inc, 1–23.
20. Langridge WHR, Li BJ, Szalay AA (1985) Electric field mediated stable transformation of carrot protoplasts with naked DNA. Plant Cell Reports 4, 355–359.
21. Maniatis, T, Fritsch, EF, Sambrook, J (1982) Molecular Cloning: a Laboratory Manual. Cold Spring Harbor, New York.
22. Moloney MM, Walker JM, Sharma KK (1989) High efficiency transformation of *Brassica napus* using *Agrobacterium* vectors. Plant Cell Reports 8, 238–242.

23. Misra S (1990) Transformation of *Brassica napus* L with a 'disarmed' octopine plasmid of *Agrobacterium tumefaciens*: molecular analysis and inheritance of the transformed phenotype. J. Exp Bot 41, 224, 269–275.

24. Neuhaus G, Spangenberg G, Mittelsten Scheid, Schweiger HG, (1987) Transgenic rapeseed plants obtained by the microinjection of DNA into microspore-derived embryoids. Theor Appl Genet 75, 30–36.

25. Neumann E, Schaefer-Ridder M, Wang Y, Hofschneider PH (1982) Gene transfer into mouse lyoma cells by electroporation in high electric fields. EMBO J 1, 841–845.

26. Ooms G, Bains A, Burrell M, Karp A, Twell D, Wilcox E (1985) Genetic Manipulation in cultivars of oilseed rape (*Brassica napus*) using *Agrobacterium*. Theor Appl Genet 71, 325, 329.

27. Paszkowski J, Shillito R.D, Saul M, Mandak V, Hohn T, Hohn B, Potrykus I (1984) Direct gene transfer to plants. EMBO J 3, 2717–2722.

28. Paszkowski J, Baur M, Bogucki A, Potrykus I (1988) Gene targeting in plants. EMBO J. 7, 4021–4026.

29. Pelletier G, Primard C, Vedel F, Chetrit P, Remy R, Rousselle P, Renard M (1983) Intergeneric cytoplasmic hybridization in *Cruciferæ* by protoplast fusion. Mol Gen Genet 191, 244–250.

30. Pietrzak M, Shillito RD, Hohn T, Potrykus I (1986) Expression in plants of two bacterial antibiotic resistance genes after protoplast transformation with a new plant expression vector. Nucleic Acid Res 14, 5857–5868.

31. Pua EC, Mehra-Palta A, Nagy F, Chua N-H (1987) Transgenic plants of *Brassica napus* L. Bio/technology 5, 815–817.

32. Radke SE, Andrews BM, Moloney M.M, Crouch ML, Kridl JC, Knauf V.C (1988) Transformation of *Brassica napus* L. using *Agrobacterium tumefaciens*: developmentally regulated expression of the reintroduced napin gene. Theor Appl Genet 75, 685–694.

33. Rouan D, Montané MH, Yot P, Alibert G, Tessié (1990) A reliable method for efficient DNA uptake by *Brassica* mesophyll protoplasts with electropulsing: relationship between protoplasts size and critical field strength. Abstracts VIIth international congress on plant tissue and cell culture, Amsterdam, June 24–29, A2-121.

34. Shillito RD, Paszkowski J, Potrykus I (1983) Agarose plating and a bead type culture technique enable and stimulate development of protoplast-derived colonies in a number of plant species. Plant Cell Rep 2, 244–247.

35. Shillito RD, Saul MW, Paszkowski J, Müller M, Potrykus I (1985) High efficiency direct gene transfer to plants. Biotechnology 3, 1099–1103.

36. Wohlleben W, Arnold W, Broer I, Hillemann D, Strauch E, Pühler A (1988) Nucleotide sequence of the phosphinothricin N-acetyltransferase gene from *Streptomyces viridochromogenes* Tü494 and its expression in *Nicotiana tabacum*. Gene 70, 25–37.

Plant Tissue Culture Manual **B5**: 1–9, 1991.
© 1991 *Kluwer Academic Publishers.*

Regeneration and transformation of potato by *Agrobacterium tumefaciens*

RICHARD G. F. VISSER

Department of Plant Breeding (IvP), Agricultural University Wageningen, P.O. Box 386, 6700 AJ Wageningen, The Netherlands

Introduction

Potato is one of the major crop species and which is quite amenable to genetic transformation through *Agrobacterium*. Because a wide variety of cell culture techniques is available, all of the requirements for efficient transformation with *Agrobacterium* have been satisfied.

Transformation of the potato has been mainly achieved in tetraploid genotypes, especially cv. Desiree [2, 7, 9, 10, 11, 12, 14]. Several transformation methods exist which rely on the cocultivation of various tissue or organ explants, e.g. leaf and stem segments or tuber discs, with engineered bacteria.

In this chapter attention will be focused on the use of stem and leaf explants. Transformation of tuber explants will be dealt with in another chapter of this manual [2].

The use of stem and leaf explants has several advantages over the use of tuber explants:

- vigorously growing *in vitro* material is available in all times of the year;
- transformation of potato genotypes which give little or no tubers is possible;
- (haploid or) diploid transformants can be obtained from explants derived from haploid or diploid genotypes. This is not possible with potato tuber discs since the ploidy level of most cells in tubers is much higher than that of the rest of the plant and in general only the tetraploid cells in a tuber grow out to form a plant [5].

A major disadvantage of using leaf and stem explants in transformation experiments is that because of a callus phase somaclonal variation can occur. However, this phenomenon can be kept in check by making sure that the callus phase is as short as possible.

In general the transformation procedure of leaf and stem explants is one based upon the leaf disc transformation procedure as originally described by Horsch *et al.* [4]. The choice of the explant type is not an easy one. A number of researchers have claimed that stem segments are the most optimal for use [7, 12, 13], whereas others found that leaf segments are better [1, 3]. However, before even attempting transformation of either of these explant types optimal regeneration conditions should be worked out, both for stem and leaf explants, of the particular genotype of interest. Because every genotype has its own special requirements as far as hormone composition and concentration is concerned, it is virtually impossible to give a regeneration and transformation protocol that works well with every genotype. In spite of this I have made an attempt to point out what the basic components of a callus- and shoot induction medium should be, as follows (Table 1).

Table 1. General composition of callus- and shoot induction medium

---

A. Callus induction medium
  − MS salts
  − sucrose: 10 g/l−30 g/l
  − cytokinins: 0.5 mg/l−2.25 mg/l BA*
  − auxins   : 0.01 mg/l−0.2 mg/l NAA* or
                : 0.2 mg/l−2 mg/l   2.4-D*
  optional
  − gibberellins
  − vitamins
  − glutamine
    pH 5.8

B. Shoot induction medium
  − MS salts
  − sucrose   : 10 g/l−30 g/l
  − cytokinins : 0.25 mg/l−2.25 mg/l BA*
  − gibberellins: 0.1 mg/l−10 mg/l GA_3*
  − NO auxins
    pH 5.8

---

* BA = benzyladenine; NAA = naphthaleneacetic acid; GA_3 = gibberellic acid; 2.4-D = 2.4-dichlorophenoxyacetic acid.

The importance of having a good regeneration procedure for the genotype of interest becomes clear when one realizes that after transformation with *Agrobacterium* the regeneration efficiency can drop 5- to 100-fold.

The transformation protocol described below has led to transformants in different diploid (6) and tetraploid (3) potato genotypes. In Figure 1 four different stages during potato transformation are shown.

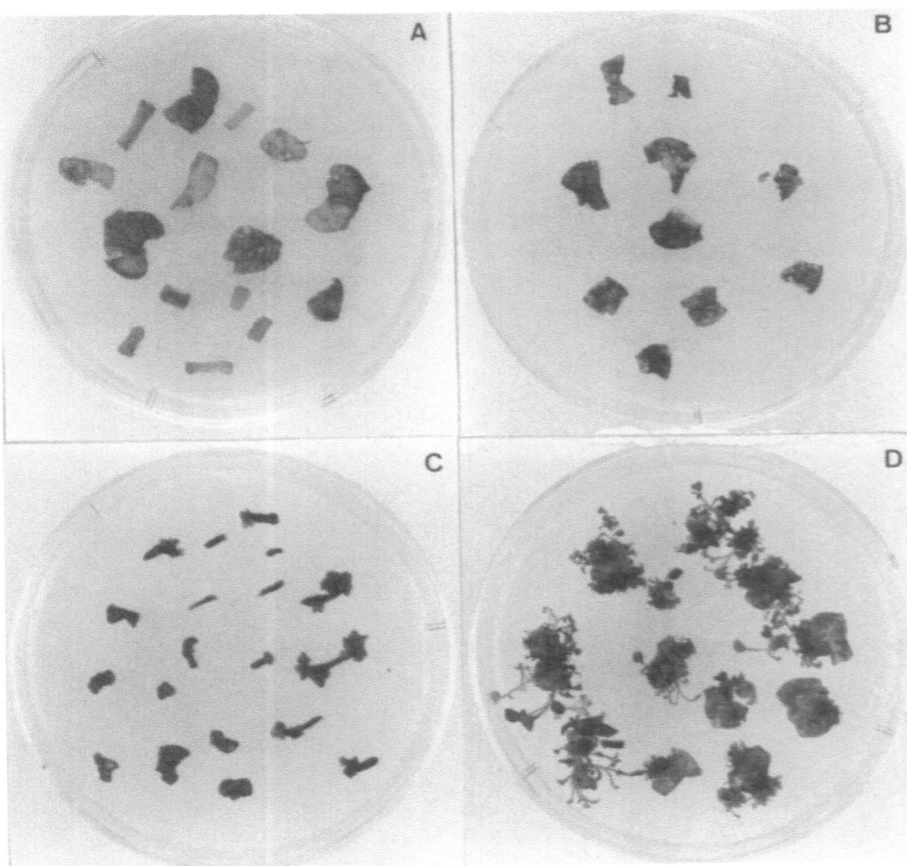

Fig. 1. Four stages during potato transformation
a. Leaf and stem explants on MC plates shortly after inoculation with *Agrobacterium tumefaciens*.
b. Leaf and stem explants two weeks after incubation with *A. tumefaciens*, plated on selective medium. Both explant types have turned brown.
c. Callus formation on explants three to four weeks after incubation with *A. tumefaciens* on M400 selective medium.
d. Shoot formation on callus developed on leaf explants after approximately 5–6 weeks on M400 selective medium.

Procedures

Setting up plant material
1. Sterilize axillary buds in bleach solution for 10 minutes, followed by 3–5 washes in sterile water.
2. Place the axillary buds on MS10 or MS30 medium and grow under moderate light (2000 lux, 16 hr) at 25 °C.
3. Propagate *in vitro* plants by subculturing every three to four weeks on MS10 or MS30 medium.
4. Prepare explants for transformation by cutting leaves with a sharp scalpel in two pieces, cut stem segments of about 1 cm devoid of axillary buds.
5. Preculture the cut explants for 1 to 3 days on MC plates (M300 standard Petri plates overlayed with 1.5 to 2 ml of M100 medium and then covered with a piece of sterile filter paper, e.g. Whatman no. 1) at 25 °C with moderate light. Twenty to twenty-five explants are placed on one plate.

Transformation
6. Grow *A. tumefaciens* culture in LB medium with appropriate antibiotics to select for the vector. Prepare the culture for inoculation of explants by taking the overnight culture and diluting it 1 to 10 in liquid MS10 medium.
7. After preculturing the explants, inoculate by immersion in the diluted *A. tumefaciens* culture in MS10. Soak for 10 min to ensure good inoculation. Blot explants dry for a short moment (5–20 sec) on a sterile filter paper to drain off excess fluid. Take care that they do not dry out.
8. Transfer the explants back to the MC plates. Make sure that the explants are in good contact with the medium. Incubate for 1 to 2 days at 25 °C (no intense light).
9. Place explants (about 8–14 per plate) on M400 medium with cefotaxime (or carbenicillin), but *without* the *plant selectable antibiotic* at 25 °C. Tape the plates with a foil which allows a good gas exchange.
10. After 3 to 5 days transfer the explants to M400 selection medium containing cefotaxime (carbenicillin) and a plant selectable antibiotic (e.g. kanamycin or hygromycin). Transfer explants every 2 weeks to fresh selection medium.
11. Remove green calluses ($\pm$ 3 mm) which appear on the explants on selection medium within 3 to 6 weeks and transfer to plates with M13 medium.
12. Remove calluses with shoots and transfer only independent shoots, preferably one shoot per callus, to MS30 selection medium containing antibiotics.
13. When roots appear on MS30 selection medium shoots are considered transformed. Optional: this can be further checked by performing other tests, e.g. NPT-II or β-glucuronidase tests.
14. Transfer shoots which have rooted on MS30 selection medium to fresh MS30 selection medium.
15. When small roots have appeared transfer plantlets to soil. Wash the agar from

base and shoots and transfer in sterile soil in 5 cm pots. Place the pots in a humidity chamber.

16. After approximately 14 days transfer the plantlets to bigger pots ($\emptyset \pm 15$ cm) and grow under standard plant growth conditions.

Notes

2. Transfer about 10 axillary buds per glass jar ($\emptyset$ 8 cm).

3. Ensure that the plant material is as uniform as possible and make sure that enough plant material is at hand. Young and vigorously growing plant material is essential for transformation.

4. Explants can be prepared the day before starting preculturing and be kept in M387 'float' medium which seems to have a positive effect on transformation with certain genotypes.

6. It is necessary to start cultures 2 days before inoculation. Dilute a good growing culture the day before the transformation experiment actually is going to take place. The addition of acetosyringone (10 mg/l) to the overnight bacterial culture in some cases results in a better response of the genotype to the transformation procedure.

9. An inexpensive source of cefotaxime is Claforan and of carbenicillin is Geopen, both available from pharmacies. Working concentration for cefotaxime is 200 mg/l and for carbenicillin 400 mg/l.

10. Use a selection concentration of kanamycin (e.g. 50 mg/l) or hygromycin (e.g. 10 mg/l) well above the concentration on which non-transformed tissue is able to grow.
The addition of $AgNO_3$ (10 mg/l) to the medium can lead to a more effective callus- and shoot induction with certain genotypes.

11. When using monoploid or diploid genotypes fast selection and transfer of calluses to M13 selection medium will result in a relatively higher production of mono- and/or diploid transformants. Furthermore, somaclonal variation, due to a short callus phase, will in most cases also be lower.

12. Isolation of one shoot from each individual callus, and the shoot itself should be free of callus, ensures that each plantlet is really an independent transformant. Use a concentration of the antibiotic from which you know that it prevents rooting of non-transformed shoots.

13. When rooting is a problem the addition of indole acetic acid (IAA) or indole butyric acid (IBA) at a concentration of ± 0.1 mg/l to the medium results in a better root formation of the transformants. Rooting in the presence of hygromycin always seems to be a problem.

16. Before transfer to larger pots, roots can be harvested and stored to perform cytological studies to determine the chromosome number of the transformants.

Additional notes

A number of general points have to be taken into account when attempting to transform different genotypes. First, the choice of explant source. Either stem or leaf explants will perform better in transformation experiments, depending on the vector used in combination with the *Agrobacterium* strain.

Second, preculturing and/or preconditioning. Nurse cultures, for example with an exponentially growing tobacco/petunia/potato cell line, can greatly facilitate the process and increase the frequency of transformation. This alternative to the described protocol is however more time-consuming and also genotype dependent.

Third, condition of *in vitro* plant material. With some genotypes there is a strong seasonal influence on the transformation efficiency [13] for no apparent reason other than the season. The best time to start and optimize transformation with any genotype, therefore, is spring time.

Fourth, appropriate controls such as plates without selective antibiotic and explants without inoculation both on non- and selective plates have to be included in the transformation experiments.

By using the outlined protocol a tight selection on positive transformants is carried out. However, in almost all transformation experiments and with all genotypes some shoots will be selected that, although growing in the presence of the selectable antibiotic, do not express and/or contain the foreign DNA. The reason(s) for these escapes as they are called are not clear but may include DNA methylation or loss of DNA during plant development. A good test is to select for callus inducing ability and root forming ability of a given shoot on antibiotic containing medium.

Solutions
— Bleach solution
  — 2% Na-hypochlorite
  — 0.1% SDS

— MS10 medium
  — 4.7 g/l MS medium (Flow labs.)
  — 10 g/l sucrose
  — 8 g/l agar
  pH 5.8

— MS30 medium
  — 4.7 g/l MS medium (Flow labs.)
  — 30 g/l sucrose
  — 8 g/l agar
  pH 5.8

— MS30 selection medium
  — MS30 medium
  — 200 mg/l cefotaxime or 400 mg/l carbenicillin
  — 50 mg/l kanamycin

— LB (Luria Broth)
  — 1% Bacto tryptone
  — 0.5% Yeast extract
  — 1% NaCl
  pH 7.0

— M13 (shoot induction) medium
  — MS10 or MS30 medium
  — 0.25 mg/l BA
  — 0.1 mg/l GA3
  — 200 mg/l cefotaxime or 400 mg/l carbenicillin
  — 50 mg/l kanamycin or 10 mg/l hygromycin
  — 8 g/l agar
  pH 5.8

— M100 medium
  — 4.7 g/l MS medium (Flow labs.)
  — 30 g/l sucrose
  — 0.5 mg/l thiamine-HCl
  — 0.5 mg/l pyridoxine-HCl

- 1 mg/l nicotinic acid
- 29.8 mg/l $FeSO_4 \cdot 7H_2O$
- 1 mg/l 2.4-D
- 0.5 mg/l kinetin
- 2 g/l casein hydrolysate
pH 6.5

- M300 medium
  - MS30 medium
  - 2 mg/l NAA
  - 1 mg/l BA
  - 8 g/l agar
  pH 5.8

- M387 medium
  - 4.7 g/l MS medium (Flow labs.)
  - 80 mg/l NH_4NO_3
  - 14.7 mg/l $CaCl_2$
  - 10 mg/l NAA
- 10 mg/l BA
  pH 5.8

- M400 (selection) medium
  - MS10
  - 2 mg/l zeatin
  - 0.01 mg/l NAA
  - 0.1 mg/l GA3
  - 200 mg/l cefotaxime or 400 mg/l carbenicillin
  - 50 mg/l kanamycin or 10 mg/l hygromycin
  - 8 g/l agar
  pH 5.8

References

1. An G, Watson BD, Chiang CC (1986) Transformation of tobacco, tomato, potato and *Arabidopsis thaliana* using a binary Ti-vector system. Plant Physiol 81: 301–305.
2. Bevan M (1991) Regeneration and transformation of potato by *Agrobacterium tumefaciens* (tuber explant system). Plant Tissue Culture Manual (K. Lindsey, ed.), in preparation.
3. De Block M (1988) Genotype-independent leaf disc transformation of potato (*Solanum tuberosum*) using *Agrobacterium tumefaciens*. Theor Appl Genet 76: 767–774.
4. Horsch RB, Fry JE, Hoffmann NL, Wallroth M, Eichholz D, Rogers SG, Fraley RT (1985) A simple and general method for transferring genes into plants. Science 227: 1229–1231.
5. Jacobsen E (1987) Genetic diversity in protoplasts- and cell-derived plants of potato. In: YPS Bajaj. Biotechnology in agriculture and forestry. Vol. 3: Potato pp 358–374.
6. Knapp S, Coupland G, Uhrig H, Starlinger P, Salamini F (1988) Transposition of the maize transposable element Ac in *Solanum tuberosum*. Mol Gen Genet 213: 285–290.
7. Ooms G, Burrell MM, Bevam MW, Hille J, Karp A (1987) Genetic transformation in two potato cultivars with T-DNA from disarmed *Agrobacterium*. Theor Appl Genet 73: 744–750.
8. Shahin EA, Simpson RB (1986) Gene transfer system for potato. Hort Science 21: 1199–1201.
9. Sheerman S, Bevan MW (1988) A rapid transformation method for *Solanum tuberosum* using binary *Agrobacterium tumefaciens* vectors. Plant Cell Rep 7: 13–16.
10. Stiekema WJ, Heidekamp F, Louwerse JD, Verhoeven HA, Dijkhuis P (1988) Introduction of foreign genes into potato cultivars Bintje and Desiree using an *Agrobacterium tumefaciens* binary vector. Plant Cell Rep 7: 47–50.
11. Tavazza R, Tavazza M, Ordas RJ, Ancora G, Benvenuto E (1988) Genetic transformation of potato (*Solanum tuberosum*): an efficient method to obtain transgenic plants. Plant Science 59: 175–181.
12. Twell D, Ooms G (1987) The 5′ flanking DNA of a patatin gene directs tuber specific expression of a chimearic gene in potato. Plant Mol Biol 9: 365–375.
13. Visser RGF, Jacobsen E, Hesseling-Meinders A, Schans MJ, Witholt B, Feenstra WJ (1989) Transformation of homozygous diploid potato with an *Agrobacterium tumefaciens* binary vector system by adventitious shoot regeneration on leaf and stem segments. Plant Mol Biol 12: 329–337.
14. Wenzler H, Mignery G, May G, Park W (1989) A rapid and efficient transformation method for the production of large numbers of transgenic potato plants. Plant Science 63: 79–85.

Plant Tissue Culture Manual **B6**: 1–9, 1991.
© 1991 *Kluwer Academic Publishers.*

Transformation of tomato with *Agrobacterium tumefaciens*

SHEILA McCORMICK
*Plant Gene Expression Center, USDA/ARS–UC-Berkeley, 800 Buchanan St., Albany, CA 94710
USA*

Introduction

Agrobacterium-mediated transformation and regeneration of a variety of tomato (*Lycopersicon esculentum*) cultivars was first reported by McCormick *et al.* [12]; other procedures for *L. esculentum* transformation include those of Fillatti *et al.* [5] and Bird *et al.* [2]. Chyi *et al.* [4] published a protocol for transformation of *L. esculentum* cv. VF36 × *L. pennellii* (LA716) interspecific hybrids, and procedures for transformation of a derivative of an interspecific hybrid between *L. esculentum* and *L. peruvianum* [9, 11] are available. Since 1986, numerous groups have used these procedures or modifications thereof to obtain transgenic tomatoes [e.g., 3, 7, 14, 15] for analysis of gene expression. However, tomato is still considered more difficult to transform than species such as *Petunia hybrida* and *Nicotiana tabacum*, and can show widely different success rates, possibly depending on cultivar, *Agrobacterium* strain, antibiotic selection, and/or the personnel performing experiments. This paper describes a detailed protocol with trouble-shooting comments that should serve as a basis for initiating a tomato transformation program, or for improving an existing program.

Protocol

1. Sterilize dry seeds in 20% household bleach + 0.1% Tween-20 for 15 minutes, followed by 3 or more rinses in sterile H_2O.
   Alternatively, a 5 minute treatment with 50% household bleach, followed by rinses with sterile water can be used with seeds directly extracted from fruit.

2. Germinate the sterilized seeds on 1/2 × MS0 in sterile magenta boxes (obtainable from Sigma). Cover bottom of box with a dense monolayer of seeds (100 or so); grow at 26 °C under 16L/8D light conditions for 10–14 days.

3. Cotyledons can be used for transformation when there are no or minimal true leaves present on seedlings. The tops of the seedlings (approx. the top half of hypocotyl and attached cotyledons) are cut off in batches and floated in MS0 liquid for cutting. Cutting the cotyledon explants is conveniently done in a 150 × 15 mm Petri dish. Approximately one third of the seedlings in a magenta box are removed and cut at a time; this ensures that the cotyledons don't dry out while waiting to be cut and placed on medium. Cotyledons are cut near the proximal (wide) end (an additional cut surface near the distal end can be made, but one is usually sufficient). Attempts should be made to cut the cotyledons when they are submerged in the MS0. The cotyledons are placed upside down on 100 × 15 mm Petri dishes containing D1 medium. 500 ml of medium should yield 20 plates. The cotyledons are placed very densely (sides touching) on the plate (50–75 cotyledons per plate). Alternatively, cotyledons are placed upside down on MS0 plates (with hormones appropriate for callus formation) with a 1 ml layer of *Nicotiana tabacum* feeder cells under a 3MM filter paper disc, cut to fit precisely, as described in McCormick *et al.* [12].

4. 5 ml of a 20-fold diluted *Agrobacterium* 2 day 'overnight' culture is poured over the cotyledons on the surface of the plate, incubated for 1–2 hrs, then the excess sucked off with a pipetteman (usually you can remove 3–4 ml of solution from the plate surface). The plates can be swirled occasionally during the 1–2 hr incubation to ensure that the cut surfaces get wet. The cotyledons can be straightened out (but still upside down) so they are not clumped to one side on the plates. Then the plates are taped shut with two pieces of time tape and incubated in a growth room (26 °C, 16L/8D) for 2 days. For leaf transformation, or for cotyledons on feeder plates, it is possible to treat with *Agrobacterium* as above; alternatively, the cotyledons or leaf pieces can be swirled in a dish until the edges are wet, and then blotted dry and placed on the feeders for co-cultivation, as described in McCormick *et al.* [12].

5. After two days co-cultivation, the cotyledons are transferred to deep plates (100 × 20 or 25 mm) of D1 medium + 500 µg/ml carbenicillin + 100 µg/ml kanamycin. Ten to twelve plates can be poured from 500 ml of medium. The cotyledons don't need to be blotted dry. Cotyledons should be placed upside down on the selection plates, the plates sealed with tape, and returned to the growth room for incubation. Callus, green bumps, or shoots usually will be seen at the cut edges within 10 days, and certainly after 3 weeks incubation, if the transformation was successful.

6. Cotyledons are transferred at 3 weeks to D2 medium (deep plates) + 500 µg/ml carbenicillin + 100 µg/ml kanamycin, for further shoot organogenesis. At the first transfer to D2 plates, the entire explant should be transferred; at subsequent 3 week transfers, the callus/shoot should be excised away from the dying cotyledon.

7. After another three weeks, often shoots with true meristems will have arisen, and can be cut off cleanly from the callus/cotyledon and transferred to rooting medium (MS0 + carbenicillin + kanamycin, but no hormones). Shoots will usually root within 7–10 days.

8. Shoots can then be transferred into very wet soil in a 2 inch pot within a magenta box, with the lid tightly closed. Any residual agar should be rinsed from the roots before transplanting. These magenta boxes are incubated under low light conditions (1200 ftc) (incandescent + fluorescent) and the lid of the magenta box slowly tilted off over a period of 5 days. The transformants are left within the magenta box (without the lid) until they are a reasonable size. They are then removed from the magenta box, and eventually potted up and transferred to the greenhouse.

9. R1 seed is scored for kanamycin resistance by sterilization and germination on 1/2 MS0 + 100 mg/l kanamycin in magenta boxes. Resistant seedlings will be taller, will have branched roots and reduced or no anthocyanin pigmentation in the hypocotyls. Sensitive seedlings will germinate but will be short and stunted, will have stubby unbranched roots, and often have enhanced anthocyanin pigmentation in the hypocotyls. Scoring is most reliable at approximately 2 weeks after seed are set out, but can be done up to a month after germination.

Notes

1. The above protocol was optimized for the cultivar VF36. VF36 seeds are not commercially available, and thus have to be bulked up by the researcher. In addition to the cultivars cited in McCormick et al. [12], other cultivars that give a reasonable frequency of transformants with this protocol are New Yorker and the interspecific cross L. esculentum cv. VF36 X L. pennellii (LA716). Cultivars that did not respond favorably to this protocol include Vendor and VFNT cherry; the protocol of Fillatti et al. [5] reportedly is successful with VFNT cherry [15].
Before any new cultivar is used for transformation, it is strongly recommended that an optimized regeneration protocol be developed, based on protocols from existing literature or by empirical tests of different hormone grids.

2. The cotyledons should be dark green and the seedlings compact, not pale green and spindly. This can be assured by dense platings of the seeds, 50 ml volumes of 1/2 X MS0 in the magenta boxes for germination, and use within 10–14 days after seed germination. Some cultivars or experiments will show anthocyanin pigmentation in the hypocotyls, or on the abaxial surface of the cotyledons during culture (a probable stress response). However, there is no particular correlation between successful transformation and any of these features.
Hypocotyls can also be used for transformation, although they are not as efficient in generating transgenic shoots, and the shoots take longer to develop. For these reasons, hypocotyls are not worth the extra effort, unless the plant material is limiting.
Leaf pieces from greenhouse-grown 5–6 weeks old plants (optimally), or young healthy leaves from older plants can be used if required, although success with leaf transformation is more of an 'art' than with cotyledons. For certain experimental protocols (such as re-transformation of transformants) leaf transformation is more expedient than waiting for F1 seed from the transformants. Batches of medium can be made and autoclaved in 500 ml aliquots and stored for months at room

temperature, then melted in the microwave and hormones and antibiotics added before pouring plates. Carbenicillin (250 mg/ml), kanamycin (50 mg/ml) and zeatin (5 mg/ml) are stored as stocks at -20 °C, melted and added to stored medium just before pouring plates. Poured plates with kan/carb/zeatin can be stored for at least 1–2 months at 4 °C.

3. Cotyledons can be pre-callused, or pre-incubated for 1–2 days before addition of *Agrobacterium*. Select for transformation only the cotyledons which swell in size. This is extra work, but may be worth trying if the cut edges of the cotyledons are turning brown. Usually 2 plates per construct per transformation date is sufficient to obtain at least 10–20 independent transformants, although these transformants might be transferred to rooting medium and to soil over a period of several weeks. A typical transformation (performed on 6/13/89) with two plates of cotyledons yielded 26 independent transformants that were transferred to soil during the period of 8/23/89 to 9/25/89.

For various reasons (receptivity of cotyledons, *Agrobacterium* overnight not at right cell density, 'bad' day) experiments occasionally do not give a good yield of positives within the first three week incubation. It is more effective to do multiple 2 plate transformations on different days, than to do a large scale transformation on a particular day (if 2 plates give unsatisfactory efficiencies, doing 8 plates with the same material will not help matters).

For cotyledon transformations, acetosyringone (final concentration of 375 µM) can substitute for *N. tabacum* feeder layers, and saves two days time. Feeder layers are recommended for leaf transformations, although acetosyringone can be used. Feeders are prepared and used as in McCormick *et al*. [12]. A *N. tabacum* suspension culture is grown in liquid MSO medium containing 2 mg/l p-chlorophenoxyacetic acid. The suspension is subcultured weekly.

4. Although it seems reasonable to place the cotyledons on the plates right-side up, so that the cut edge is in contact with the medium, experience suggests that better regeneration is attained (and few to no escapes obtained) if the cotyledons are placed upside down for selection/regeneration. Parafilm can also be used to seal plates, but more condensation occurs on the plate lids. Time tape closure is fine if a reasonably sterile growth room is used to incubate the plates. However, if there is a problem with fungal contamination, use parafilm.

Agrobacterium (2 days) overnights are started from glycerol cultures stored at −80 °C, and grown in 5 ml cultures in LB, with no selection in 250 ml Erlenmeyer flasks at 29 °C, shaking at 250 rpm. These are diluted 1:20 into liquid MSO, and acetosyringone added (final concentration of 375 µM). If problems with *Agrobacterium*-overrun occur, or if no transformants are obtained, a more careful study of bacterial titer might be warranted; however, it is usually not necessary.

Acetosyringone is stored in EtOH at −20 °C, stock 0.0148M. (Conveniently, then 25 µl is added per ml of diluted *Agrobacterium* to give a final concentration of 375 µM). If the *Agrobacterium* culture has to be grown under selection, then the culture should be spun down and resuspended in 5 ml LB without antibiotics, and treated as above.

This protocol has been successful with both cointegrate [6, 16] and binary vectors [1, 10], although problems with a particular binary construct (pMON505) were noted [12].

5. Typical experiments should give at least 50–60% of the cotyledons regenerating callus, green bump structures, or shoots within the first 3 weeks, which should result in 10–20 independent transformants in soil within 9 weeks. Different cultivars and/or *Agrobacterium* strain combinations may take longer or shorter to achieve these numbers of transformants. If there are no visible calluses or green bumps during the first three week incubation, it is not worth transferring these cotyledons to another 3 weeks incubation on D1 selection plates; it is better to attempt a new transformation. Other selectable markers that can be used with tomato include hygromycin [13] and gentamycin [8]. Some *Agrobacterium* strains carry resistance to carbenicillin; claforan can be used to control the bacteria, but the cotyledons may have to be transferred at more frequent intervals than every three weeks (because claforan is light-sensitive). In this case, transfers should be made so that the cotyledons are exposed to 1 mg/l zeatin for three weeks, and then transferred to 0.1 mg/l zeatin plates.

6. A convenient method of keeping track of independent events is to assign a letter (a, b, c, ... z, a^2, b^2, ..) to each cotyledon that is to be transferred.

Occasionally, shoots that appear abnormal, or shoots with unusual leaf shapes will arise from the cut edges of the cotyledons — if deemed necessary (because of lack of an abundance of normal shoots), these abnormal shoots can be transferred several times and usually will eventually 'grow out' of the problem and a normal meristem will arise.

It is usually not worth the effort to do more than 3 serial transfers onto D2 selection plates, because you reach a point of diminishing returns. If something hasn't regenerated normal shoots by then, it probably won't. However, if a particular experiment looked promising at the 3 week point, but it is difficult to obtain shoots, it might be worthwhile to try a different hormone regime to stimulate shooting; for example, maintaining selection, but switching from 0.1 mg zeatin/l plates to 1 mg/l benzyladenine (BA) + 0.1 mg/l indole acetic acid (IAA) for a 3 weeks interval.

7. Any shoots that arise during the second or subsequent transfers (D2 + selection plates) should be transferred to rooting medium as soon as they look big enough; it is not necessary to only do transfers at three week intervals if something looks ready to go to the next step.

   Plantlets should have 2 to 3 good roots and an obvious meristem before transfer to soil. Some transgenic plants may not root (for physiological reasons) on selection plates, but remain green and healthy, and continue to grow. Such plants can be dipped in Rootone and potted into soil as above; they are often *bonafide* transformants.

   During the shoot initiation and shoot elongation steps (D1 and D2 media) it is important to maintain kanamycin selection at 100 µg/ml, rather than reducing the level and hoping thereby to obtain more shoots. Shoots that arise at 25–50 µg/ml kanamycin are likely to be escapes.

8. Plants should be acclimated to larger pots and allowed to become sturdy plants under the 1200 ftc light conditions before transfer to the harsher conditions of the greenhouse.

9. Only one obvious, probable, somaclonal mutation (e.g. chlorophyll defects) was noticed in the process of screening R1 seed from over 100 independent transformants that were generated using this transformation procedure (McCormick, unpublished). Yoder *et al.* [15] reported a mutation rate of over 20% in VF36 transformants — possibly, differences in the transformation regime can account for this difference.

Media

- *MSO:*
  - 4.3 g/l MS salts (Gibco)
  - 3% (w/v) sucrose
  - 1× Gamborg's B5 vitamins (Gibco or Sigma).
  - 0.8% (w/v) agar. for solid medium
    - adjust pH to 5.8 with 1M KOH

- *½xMSO:*
  - 2.3 g/l MS salts (Gibco)
  - 3% (w/v) sucrose
  - 1× Gamborg's B5 vitamins (Gibco or Sigma)
  - 0.8% (w/v) agar
    - adjust pH to 5.8 with IM KOH

- *D1 medium:*
  - 4.3 g/l MS salts (Gibco)
  - 3% (w/v) glucose
  - 1× Gamborg's B5 vitamins (Gibco or Sigma)
  - 1 mg/l zeatin
  - 0.8% (w/v) agar
    - adjust pH to 5.8 with 1M KOH

- *D2 medium:*
  - 4.3 g/l MS salts (Gibco)
  - 3% (w/v) glucose
  - 1× Gamborg's B5 vitamins (Gibco or Sigma)
  - 0.1 mg/l zeatin
  - 0.8% (w/v) agar
    - adjust pH to 5.8 with IM KOH

- *LB (Luria—Bertani) medium:*
  - 10 g/l bacto-tryptone
  - 5 g/l bacto-yeast extract
  - 10 g/l NaCl
    - adjust pH to 7.5 with 1M NaOH

References

1. Bevan M (1984) Binary *Agrobacterium* vectors for plant transformation. Nucleic Acids Res 12: 8711–8721.
2. Bird CR, Smith CJS, Ray JA, Moureau P, Bevan MW, Bird AS, Hughes S, Morris PC, Grierson D, Schuch W (1988) The tomato polygalacturonase gene and ripening-specific expression in transgenic plants. Plant Molec Biol 11: 651–662.
3. Boylan MT, Quail PH (1989) Oat phytochrome is biologically active in transgenic tomatoes. Plant Cell 1: 765–773.
4. Chyi Y-S, Jorgensen RA, Goldstein D, Tanksley SD, Loaiza-Figueroa, F (1986) Locations and stability of *Agrobacterium*-mediated T-DNA insertions in the *Lycopersicon* genome. Mol. Gen. Genet. 204: 64–69.
5. Fillatti JJ, Kiser J, Rose R, Comai L (1987) Efficient transfer of a glyphosate tolerance gene into tomato using a binary *Agrobacterium tumefaciens* vector. Bio/technology 5: 726–730.
6. Fraley RT, Rogers SG, Horsch RB, Eichholtz DA, Flick JS, Fink CL, Hoffmann NL, Sanders PR (1985) The SEV system: a new disarmed Ti plasmid vector system for plant transformation. Bio/technology 3: 629–635.
7. Hamilton AJ, Lycett GW, Grierson D (1990) Antisense gene that inhibits synthesis of the hormone ethylene in transgenic plants. Nature 346: 284–287.
8. Hayford MB, Medford JI, Hoffmann NL, Rogers SG, Klee HJ (1988) Development of a plant transformation selection system based on expression of genes encoding gentamicin acetyl-transferases. Plant Physiol 86: 1216–1222.
9. Horsch RB, Fry JE, Hoffmann NL, Wallroth M, Eichholtz D, Rogers SG, Fraley RT (1985) A simple and general method for transferring genes into plants. Science 227: 1229–1231.
10. Koncz C, Schell J (1986) The promoter of T_L-DNA gene 5 controls the tissue-specific expression of chimaeric genes carried by a novel type of *Agrobacterium* binary vector. Molec Gen Genet 204: 383–396.
11. Koornneef M, Jongsma M, Weide R, Zabel P, Hille J (1987) Transformation of tomato. In: Tomato Biotechnology. Nevins DJ, Jones RA (eds), Alan R. Liss, New York, pp 169–178.
12. McCormick S, Niedermeyer J, Fry J, Barnason A, Horsch R, Fraley R (1986) Leaf disc transformation of cultivated tomato (*L.esculentum*) using *Agrobacterium tumefaciens*. Plant Cell Rep 5: 81–84.
13. Rogers SG, Klee HJ, Horsch RB, Fraley RT (1987) Improved vectors for plant transformation: expression cassette vectors and new selectable markers. Methods Enzymology 153: 253–277.
14. Twell D, Yamaguchi J, McCormick S (1990) Pollen-specific gene expression in transgenic plants: coordinate regulation of two different tomato gene promoters during microsporogenesis. Development 109: 705–713.
15. Yoder JI, Palys J, Alpert K, Lassner M (1988) *Ac* transposition in transgenic tomato plants. Mol Gen Genet 213: 291–296.
16. Zambryski P, Joos H, Genetello C, Leemans J, Van Montagu M, Schell J (1983) Ti plasmid vector for the introduction of DNA into plant cells without alteration of their normal regeneration capacity. EMBO J 2: 2143–2150.

Plant Tissue Culture Manual **B7**: 1–13, 1991.

Regeneration and transformation of sugarbeet by *Agrobacterium tumefaciens*

KEITH LINDSEY[1], PATRICK GALLOIS[2] & COLIN EADY[3]

[1] *Leicester Biocentre, University of Leicester, Leicester, LE1 7RH, UK*
[2] *Laboratoire de Physiologie Vegetale, Universite de Perpignan, Perpignan 66025, France*
[3] *Dept of Botany, University of Nottingham, Nottingham, NG7 2RD, UK*

Introduction

Sugarbeet, *Beta vulgaris* L., is of significant commercial importance as the major sucrose producer in temperate climates and world-wide currently contributes about 35% of the total sucrose yield [1]. Sugarcane, a monocot, is limited to warmer climates, and so the two species can be considered complementary in their global distribution. Sugarbeet has also attracted further attention, by virtue of the potential to accumulate novel or valuable specific metabolites in the storage tissues: in other words, to use beets, as 'green bioreactors' for the synthesis of products other than sucrose. Sugarbeet is a natural cross-pollinator, and so modern cultivars are highly heterozygous. It is also a biennial species, and both these features contribute to a slow rate of progress in the generation of new varieties by conventional breeding. A genetic engineering strategy for sugarbeet therefore would be expected to aid the breeder in introducing specific traits directly into 'elite' (commercially valuable) genotypes and also would open up the possibility of using sugarbeet to produce metabolites novel to that species.

Progress in devising protocols for regeneration and transformation of sugarbeet has been relatively slow, and the species has acquired a reputation for recalcitrance *in vitro*. Organogenesis has now, however, been reported from callus, [9], [11], [12], leaf tissue, [3], [9], [10], hypocotyls [4], suspension culture cells [13] and protoplasts [2], [6]. Regeneration of transgenic sugarbeet plants may now be possible via a direct gene transfer route, since both the regeneration of whole plants from protoplasts [6] and the stable transformation of protoplasts following electroporation [8] have been reported. In this article, however, we describe a protocol for the regeneration and transformation of sugarbeet using *Agrobacterium tumefaciens*-mediated gene transfer and a kanamycin-resistance selection system [7]. No other methods have yet been published. Highly morphogenetic tissues, derived from the shoot bases of aseptic shoot cultures, are inoculated with *Agrobacterium* harbouring vectors carrying the selectable *npt-II* gene encoding neomycin phosphotransferase, which confers resistance to the aminoglycoside antibiotics including kanamycin and G418. Following coincubation, inoculated tissues are cultured on shoot-inducing medium in the presence of cefotaxime, to kill off the bacteria, and kanamycin, to prevent shoot initiation by non-transformed cells. Putatively transformed shoots are transferred to root-inducing medium (also containing cefotaxime and kanamycin) and plantlets are transferred to sterile compost in a micropropagator.

Procedures

Establishment of shoot cultures

Steps in the procedure
Sterilization of seeds:
1. Wrap seeds in muslin and soak overnight in running tap water, to imbibe.
2. Shake off excess water and place seeds in a beaker.
3. Immerse seeds in 70% (v/v) ethanol for 30 s.
4. Pour off ethanol, and immerse seeds in a 10% (v/v) bleach (e.g. sodium or calcium hypochlorite) solution containing 0.05% (v/v) Tween 20 or 80, for 20 min.
5. Pour off sterilant and wash seeds six times in sterile distilled water.

Germination of seeds and initiation of shoot cultures:
6. Transfer sterilized seeds to 9 cm Petri dishes containing MS20 medium, and culture at 24 ± 1 °C in the dark, to induce germination.
7. When the shoots are approx. 5 cm long, excise the shoot apex, leaves and cotyledons at the hypocotyl, and transfer to MSsh (shoot culture) medium. Culture at 24 ± 1 °C in warm daylight fluorescent tubes (160–180 µmol $m^{-2} s^{-1}$) under long daylength conditions. Use culture vessels with loose or unsealed lids.
8. At intervals of 4–6 weeks, subculture the shoots by excising the buds found on the now swollen basal tissue (Fig. 1), and transfer to fresh MSsh medium. This swollen shoot base tissue is the explant source used for inoculation with *Agrobacterium*.

Notes
3. Because of their irregular and corky surface, sugarbeet seeds are not always easy to sterilize. Treatment with 70% (v/v) ethanol removes waxy substances from the surface of the seeds and kills some contaminating microorganisms present. Do not exceed an immersion time of about 30 s, otherwise a proportion of the seeds may become inviable.
4. Tween 20/80 acts as a surfactant to allow access of the sterilizing hypochlorite solution. If problems are encountered in achieving sterility, try 15–20% bleach solution, or longer sterilization times.
5. Six washes in water are necessary to ensure the removal of all chlorine.
6. The germination of seeds on MS20 medium is preferable to the use of damp filter paper, as it allows early identification of seeds which have not been sterilized successfully. These should be removed at the earliest opportunity to prevent contamination of the whole plate.
7. If sugarbeet shoot cultures are maintained in fully sealed cultures vessels, growth is stunted. This is presumed to be due to the accumulation within the vessel of growth-inhibitory gases (such as ethylene). This problem is avoided if lids are loosened.

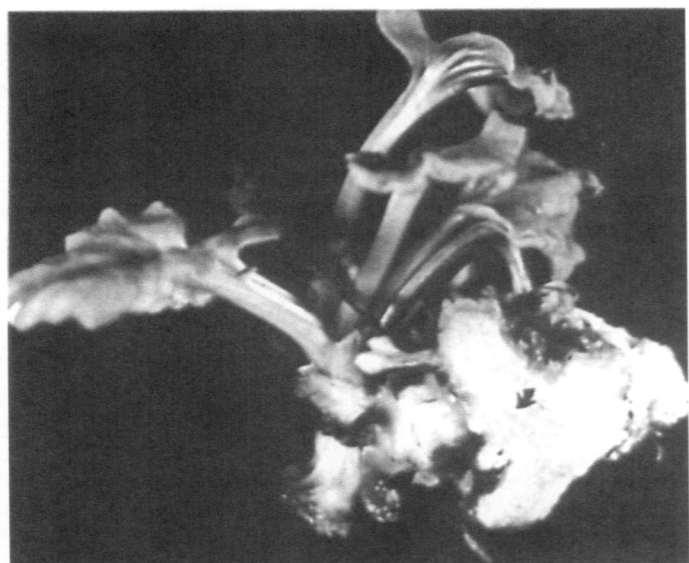

Fig. 1. Cultured shoot of sugarbeet, showing swollen shoot base tissue which is used as explant for inoculation by *Agrobacterium*.

Transformation

Steps in the procedure

1. Grow up *A. tumefaciens* strain LBA4404 harbouring desired plasmids overnight in 25 ml LB medium containing appropriate antibiotics, on a rotary shaker (200 r.p.m.) at 29 °C, to an optical density at 600 nm of 1.1 (approx 2×10^8 bacteria ml^{-1}).

2. Using a sterile scalpel and forceps, cut shoot base tissues into slices (approx 1 cm $\times$ 1 cm $\times$ 2 mm) and inoculate by immersing in 10 ml of the bacterial suspension for about 30 s.

3. Remove excess liquid from the tissue by gently blotting on sterilized filter paper.

4. Transfer tissue slices to 9 cm Petri dishes containing MS30 medium, seal the dishes with Nescofilm, and allow co-cultivation to proceed at 24 $\pm$ 1 °C in the light for 48 h. Culture 5 or 6 tissue slices per Petri dish.

5. Transfer inoculated tissues to Petri dishes containing shooting/selection medium, ensuring they are well embedded. Regeneration of non-transformed shoots is inhibited by 100 mg l^{-1} kanamycin. To eliminate agrobacteria, the medium is supplemented with 200 mg l^{-1} cefotaxime and the Petri dish lids are not sealed for 48 h — the resultant drying of the cultures helps to kill off the bacteria. After this period seal the dishes with Nescofilm and incubate at 24 $\pm$ 1 °C in the light for several weeks, during which time green shoots appear. To maintain the selection pressure, transfer explants to fresh shooting/selection medium at intervals of 2–3 weeks.

6. Carefully excise green shoots (2–4 cm long) from the explant tissue, ensuring no explant is carried over, and transfer to 'polypots' containing rooting medium supplemented with 100 mg l^{-1} kanamycin and 200 mg l^{-1} cefotaxime. Also ensure that only one shoot is transferred from a given region of explant tissue, to eliminate the possibility of propagating separately the products of a single transformation event.

7. Once shoots have rooted, remove them from the polypots, rinse the agar from the roots and plant out in sterile compost in a micropropagator. Maintain the plantlets at 25 °C and at relatively high humidity (by keeping the micropropagator vents closed) for 1 week. After this period transfer the plantlets to greenhouse conditions (18–20 °C) and open the micropropagator vents. After a further week, transfer the plantlets to plant pots (Fig. 2).

Notes

1. For the development of the transformation system, we have used *A. tumefaciens* LBA4404 harbouring binary vectors: either 1) pBin19CAT (M Bevan, AFRC IPSR, John Innes Institute, Norwich, UK) carrying the screenable chloramphenicol acetyltransferase (CAT) gene and the selectable npt-II gene, or 2) pBI121 (R Jefferson, M Bevan, AFRC IPSR John Innes Institute, Norwich, UK), carrying the β-glucuronidase (GUS) gene and the npt-II gene. *Agrobacterium* harbouring either of these plasmids is cultured in LB medium supplemented with 50 mg l^{-1} kanamycin and 25 mg l^{-1} streptomycin. The bacterial cultures may develop a 'stringy' appearance, but this is normal. See various chapters in Kluwer's Plant Molecular Biology Manual, including [5], for details of growing bacterial strains.

2. Shoot cultures which are 6—8 weeks old should be used, having relatively large amounts of swollen basal tissue for inoculation.
3. Ensure excess bacterial suspension is removed from the tissues, to prevent overgrowth of the explants at a later date. If explants show evidence of bacterial overgrowth, dispose of them.
4. The use of *Nicotiana* nurse cultures during the co-cultivation period has been found to increase the transformation frequency for some species (see [5]), but had no detectable effect with sugarbeet [7].
5. We have found that 100 mg l^{-1} kanamycin is efficient at inhibiting shoot initiation in non-inoculated control cultures, but some escapes may be observed in inoculated cultures. The reasons for this are unclear: it may be that transformed cells afford a degree of cross-protection to non-transformed cells; transient expression of transferred npt-II genes may protect non-transformed cells, allowing them to begin regeneration; or agrobacteria may degrade antibiotics in the medium so reducing the local concentration. If screenable as well as selectable genes are transferred, GUS or CAT assays (or more laborious npt-II assays) can be carried out at the shooting stage to provide further information on foreign gene expression.

   There is anecdotal evidence that different sugarbeet cultivars are differentially sensitive to kanamycin.

   We have found that G418 may be too toxic a selection agent for sugarbeet. No transformants were regenerated if G418 (50 mg l^{-1}) was incorporated into the shooting medium.
6. Root development is slow (first detectable 4—6 weeks after shoot transfer). Shoots may develop roots in the absence of rooting medium, i.e. simply on transfer to sterile compost. However, by culturing in root medium containing 100 mg l^{-1} kanamycin, a further round of selection for transformants is provided: non-transformed shoots fail to root in kanamycin.

Solutions

MS20 medium
- 4.4 g l^{-1} Murashige and Skoog basal medium (Sigma)
- 20 g l^{-1} sucrose
- 8 g l^{-1} agar (Difco)
- adjust pH to 5.8 with 1M KOH

MS30 medium
- 4.4 g l^{-1} Murashige and Skoog basal medium (Sigma)
- 30 g l^{-1} sucrose
- 8 g l^{-1} agar (Difco)
- adjust pH to 5.8 with 1M KOH

MSsh medium
- MS30 supplemented with —
- 0.25 mg l^{-1} benzylamino purine (BAP)
- 8 g l^{-1} agar (Difco)
- adjust pH to 5.8 with 1M KOH

LB (Luria—Bertani) medium
- 10 g l^{-1} bacto-tryptone
- 5 g l^{-1} bacto-yeast extract
- 10 g l^{-1} NaCl
- adjust pH to 7.5 with 1M NaOH

Shooting/selection medium
— MS30 medium supplemented with : —
— 1 mg l^{-1} BAP
— 100 mg l^{-1} kanamycin sulphate (Sigma), (see 'Stock solutions' below)
— 200 mg l^{-1} cefotaxime ('Claforan', Roussel), (see 'Stock solutions' below)
— adjust pH to 5.8 with 1M KOH

Rooting Medium
— MS30 medium supplemented with
— 5 mg l^{-1} napthalene acetic acid
— 100 mg 1^{-1} kanamycin sulphate (Sigma)
— 100 mg l^{-1} cefotaxime ('Claforan', Roussel)
— adjust pH to 5.8 with 1M KOH

Stock solutions
1. Dissolve NAA in a small volume of absolute ethanol and make up to 1 mg ml^{-1} with distilled water. Store at 4 °C in the dark
2. Dissolve BAP in a small volume of 1M HCl and make up to 1 mg ml^{-1} with water. Store at 4 °C in the dark
3. Make up 100 mg ml^{-1} stock solutions of cefotaxime and kanamycin in distilled water. Add to autoclaved media after filter sterilzation, e.g. using a 0.2 μm sieve and syringe. Store antibiotic stocks in the dark at −20 °C

Acknowledgements

The protocols described in this article were established with financial support from Danisco A/S, Copenhagen, awarded to Dr Mike Jones of the Department of Biochemistry and Physiology, AFRC Institute of Arable Crops Research at Harpenden.

Fig. 2. Sugarbeet regenerants.

References

1. Atanassov, AI (1986) Sugar beet. In: Handbook of Plant Cell Culture Vol 4 Eds. DA Evans, WR Sharp, P V Ammirato, pp 652–680
2. Bhat, S, Ford–Lloyd, BV, Callow, JA (1986) Isolation and culture of mesophyll protoplasts of garden, fodder and sugar beets using a nurse culture system: callus formation and embryogenesis. J. Plant Physiol. 124: 419–423
3. Freytag, AH, Anand, SC, Rao–Ardelli, AP, Owens, LD (1988) An improved method for adventitious shoot formation and callus induction in *Beta vulgaris* L. *in vitro*. Plant Cell Rep. 7: 30–34
4. Hooker, MP, Nabors, MW (1977) Callus initiation, growth and organogenesis in sugarbeet (*Beta vulgaris* L.). Z. Pflanzenphysiol. 84: 237–246
5. Horsch, RB, Fry, J, Hoffman, N, Neidermeyer, J, Rogers, SG, Fraley, RT (1988) Leaf disc transformation. Plant Molec. Biol. Manual A5: 1–9 (eds. Gelvin, SB, Schilperoort, RA).
6. Krens, FA, Jamar, D, Rouwendal, GJA, Hall, RD (1990) Transfer of cytoplasm from new *Beta* CMS sources to sugar beet by asymmetric fusion. 1. Shoot regeneration from mesophyll protoplasts and characterization of regenerated plants. Theor. Appl. Genet. 79: 390–396
7. Lindsey, K, Gallois, P (1990) Transformation of sugarbeet (*Beta vulgaris*) by *Agrobacterium tumefaciens*. J. Exp. Bot. 41: 529–536
8. Lindsey, K, Jones, MGK (1989) Stable transformation of sugarbeet protoplasts by electroporation. Plant Cell Rep. 8: 71–74
9. Ritchie, GA, Short, KC, Davey, MR (1989) *In vitro* shoot regeneration from callus, leaf axils and petioles of sugar beet (*Beta vulgaris* L.) J. Exp. Bot. 40: 277–283
10. Rogozinska, J, Goska, M (1978) Induction of differentiation and plant formation in isolated sugar beet leaves. Bull. Acad. Polon. Sci. 26: 343–345
11. Saunders, JW, Daub, ME (1984) Shoot regeneration from hormone-autonomous callus from shoot cultures of several sugar beet (*Beta vulgaris* L.) genotypes. Plant Sci. Lett. 34: 219–223
12. Tetu, T, Sangwan, RS, Sangwan–Norreel, BS (1987) Hormonal control of organogenesis and somatic embryogenesis in *Beta vulgaris* callus. J. Exp. Bot. 38: 506–517
13. Van Geyt, JPC, Jacobs, M (1985) Suspension culture of sugarbeet (*Beta vulgaris* L.) Induction and habituation of dedifferentiated and self-regenerating cell lines. Plant Cell Rep. 4: 66–69

Plant Tissue Culture Manual **B8**: 1–18, 1991.
© 1991 *Kluwer Academic Publishers.*

Regeneration and Transformation of Apple (*Malus pumila Mill.*)

DAVID J. JAMES[1] and ABHAYA M. DANDEKAR[2]

Horticulture Research International East Malling, West Malling, Kent, ME19 6BJ, U.K.[1]
Department of Pomology, University of California, Davis, CA 95616, USA[2]

Introduction

In recent years the formulation of high efficiency, reproducible in vitro regeneration protocols for apple [1–4] has paved the way for the introduction of novel genetic information (genes) via non-sexual methods. There are now an array of transformation procedures that can be adopted; *Agrobacterium*-mediated transformation [5], direct or forced DNA uptake into protoplasts [6], and the use of DNA-coated microprojectiles and their introduction into regenerable plant tissue [7]. *Agrobacterium*-mediated transformation has so far been the most successful and reproducible method for a number of crops including apple [8] and this paper will be concerned only with this aspect.

The introduction of binary plasmid vectors in disarmed strains of *Agrobacterium* [9] combined with an adaptation of the leaf-disc transformation procedure [10] has permitted the production of several transgenic apple clones of the variety Greensleeves with apparently normal phenotypes [11]. In addition the use of efficient micropropagation procedures for apple [12] has ensured a continuous supply of sterile, juvenile and fast-growing tissue for further transformation experiments.

The transgenic apple plants produced so far have been derived from transformation experiments using the binary plasmid vector pBIN6 [9]. The construction of pBIN6 has been described in detail by Bevan [9] and contains both a scorable marker gene *nos* (from pTiT37), encoding the enzyme nopaline synthase (NOS) and a selectable marker gene encoding a bacterial kanamycin phosphotransferase, APH(3′)II. The APH(3′)II coding region was expressed by making a chimeric construct utilizing the 5′ and 3′ regulatory regions from the nopaline synthase gene [9]. A sucessful transformation event could be identified by the ability of the tissue to grow on the antibiotic kanamycin and produce nopaline. Although the leaf-disc transformation procedure we have reported was successful in producing transgenic apple plants, the frequency of transformation was low with less than 1 per cent of explants producing transgenic plants [13]. Current work is aimed at identifying factors limiting transformation, selection and regeneration of transformed apple plants. We shall point out some of our unpublished findings on apple that may be of value in increasing the frequency. We have considered and implemented a number of findings relating to the factors controlling virulence of the bacteria prior to and during infection of host plant cells (see e.g. [14]).

Our aim then must be to make the transformation of apple and other trees a simple, reproducible and high frequency procedure where clonal fidelity is retained and somaclonal variation [15] excluded.

Current procedures for producing transgenic apple plants

The procedures may conveniently be considered under five headings:

1) Propagation and tissue culture of cultivar(s) of interest for the supply of source plant tissue prior to inoculation and transformation.
2) The genetic background of the bacteria used in transformation experiments and factors controlling their virulence.
3) The selection of transformed cells that lead to the regeneration of transgenic plants.
4) All factors pertaining to the regeneration of transformed plants after the antibiotic selection.
5) Cloning and analysis of putative transformants.

Methods

1. Propagation and tissue culture prior to inoculation – Shoot proliferation, rooting and organogenesis

a) Shoot proliferation
Micropropagation systems offer the possibility of using a range of tissues for transformation but leaves have been the organ of choice because of their abundance and ease of manipulation for infection with *Agrobacterium*. The procedure for the micropropagation of all the cultivars that we have used for transformation work does not deviate significantly from published methods. The variety Greensleeves is particularly useful for transformation work since it is vigorous, roots well, requires no alterations to the standard MS media and is not subject to 'vitrification' during the shoot multiplication stage (Viss, pers. comm.). We use a standard medium referred to as A17 and details for its preparation are given below. The addition of the antibiotic cefotaxime is not usually necessary with the parent cultures but is often necessary during the cloning of putative transformants (see below). For Greensleeves and related cultivars axillary shoot cultures should be transferred to fresh media at approximately monthly intervals. Other cultivars such as Spartan or M.9 require more frequent subculturing and will require the presence of phenolic compounds such as phloroglucinol [12] if long term maintenance is anticipated.

b) Root induction and root emergence
The two-phase method for rooting apple cultivars *in vitro* was published some time ago [12]. The induction phase requires a 2–4 day period on an auxin (usually IBA)-containing medium (R13), followed by a transfer to a hormone-free medium with half strength Murashige and Skoog (MS) macroelements (R37). Roots appear within 10–14 days, without callus formation, under these conditions. This procedure has been consistently successful in rooting a wide

variety of apple cultivars *in vitro*. We have not found that incubating the cultures in the dark during the root induction phase was of any greater benefit than culturing them in the light. It has been shown however that phenolic compounds need to be added for some cultivars [12] although the mechanism(s) of auxin-phenolic synergism remains unknown.

Culture vessels for micropropagation and regeneration

Apple shoot cultures have been successfully grown in a wide variety of glass or plastic vessels. We now routinely use standard, plastic 20 ml Coulter Counter vials (also known as disposable blood dilution vessels, American Scientific Products, McGaw Park, Il USA) that are sterile, cheap and disposable. They can also be used for shoot multiplication, root induction and root growth. Ten ml of media will support the growth of shoot cultures for at least a month and rooted shoots for 6 to 8 weeks before necrotic symptoms appear. If these are not available 'Magenta GA7' boxes (Magenta, Chicago, Illinois) or 100 and 250 ml conical flasks with aluminium foil closures may be used for shoot proliferation and 25 ml flat bottomed glass culture tubes for root elongation. The 'snap-on' type of closure present on Coulter vessels markedly reduces evaporation and we have noticed that this will give increased multiplication rates compared to glass vessels with more loosely fitting closures.

Much of the published work on the regeneration and transformation of apple has been carried out in 5 × 5 multiwell dishes (RepliDishes, Sterilin, U.K.). The wells are square and up to 3ml of media per well can be added (see Fig. 1 in [16]). These dishes have the advantage of separating explants from each other so that should contamination occur during the lengthy incubation time required for regeneration and transformation not all explants will be lost, as is often the case when using non-compartmented Petri dishes. The major disadvantage of this container is cost and many 9 cm plastic Petri dishes can be bought for the cost of a single multiwell dish.

Choice and excision of explant

Almost all of our work on apple regeneration has employed 7 mm leaf discs excised from rooted cultures using a cork borer. The cultures are approximately 4 weeks old and have expanded leaves of sufficient size to allow the operator to comfortably excise an entire disc. The use of other devices such as punches we have found less efficient since discs tend to become lodged in the apparatus. Although a cork borer may feel difficult to handle at first, particularly for producing large numbers (i.e. 500 or more per day), we have found that most operators quickly adapt to its use. Other investigators have successfully used leaf pieces and segments rather than discs since they quite rightly claim the method is faster. Although this is correct when the sole aim is merely to produce regenerants (e.g. for studies on somaclonal variation) segments are definitely inferior to discs for transformation purposes. We have found that segments can expand their surface area considerably in the weeks following infection, co-cultivation, selection and regeneration. This leads to very inefficient contact with the selection media and greatly increases the possibility of regenerating shoots escaping kanamycin selection. Although discs can also expand they do so to a far lesser extent. However the most overwhelming argument in favour of using discs is the ability to quantify effects and variables within the experimental procedure. For example the use of leaf discs has been used in *Petunia* to quantify the efficiency of transformation using a GUS construct in a disarmed binary vector [17].

There are reports that preculturing plant tissues prior to bacterial inoculation increases the frequency of transformants that can be produced [19] and there are suggestions that cells actively undergoing division are more easily infected and transformed than older tissues [20]. In apple however when leaf discs were exposed to regeneration media for time periods of 0–8 days prior to inoculation any preculture treatment approximately halved the number of explants that eventually produced kanamycin resistant callus (James, unpublished). These results need confirmation with biochemical and molecular analyses but if validated suggest that wounding factors may be lost by preculture, at least in apple.

Other preculture effects on regeneration ability have been noted by Kobayashi [18] who showed that one week of etiolation at the end of the subculture period and before excising the tissue was the most effective method of inducing adventitious shoot formation in *Malus prunifolia* Borkh.

c) Organogenesis – Regeneration from leaf and stem explants
Leaves grown under strictly controlled tissue culture conditions still exhibit marked differences in both their transformation potential with wild type *Agrobacterium* and in their regeneration ability. Significant improvements in performance of both transformation and regeneration can be obtained by using leaves of the 'correct' age and stage of leaf development [13]. For this reason we regularly use rooted plantlets rather than those from micropropagating cultures where inter-shoot competition for light and nutrients will maximise variations in the response of excised tissues. Other work has also shown that discs from rooted shoots performed better than discs taken from leaves of unrooted plants [1]. Other more recent work using the same cultivar has confirmed this (Viss, pers. comm. 1990). Good regeneration levels can be obtained in several apple cultivars using media based on full strength M&S salts containing BAP and NAA as the cytokinin and auxin respectively. The concentrations of these compounds may need to be optimised for individual apple cultivars but the range 1–2 mg/l BAP and 0.1–0.5 mg/l NAA seem to be adequate for many. Much higher levels e.g. 10 mg/l BAP: 3 mg/l NAA (Cheng, pers. comm.) may also be effective although the ratios of one to the other are

similar. Below are detailed two regeneration media that have been effective for regeneration from leaf explants of apple.

After discs have been removed, the remainder of the leaf can also be used for regeneration and transformation since it provides an 'identical' wounded surface. The pieces are difficult to handle however and have a tendency to become necrotic. The use of the petiole may be more promising although in apple it is very short and has a tendency to curl upwards over a long time period, thus restricting contact with the media, a distinct disadvantage during the selection of transformed cells.

2) The genetic background of the bacteria and genetic determinants of virulence

The ability of *Agrobacterium* to transform plants is in part a function of the virulence genes in the bacteria. The end result of the combined action of these genes is the transfer of T-DNA into the plant cell. Genetic determinants that regulate virulence have been a subject of intensive study during the past few years and the subject of excellent reviews (see e.g. [14]). The *vir* genes that determine virulence lie within both the Ti-plasmid and chromosomal DNA. The plasmid complement of *vir* genes is critical in host recognition, stimulation of virulence or virulence induction and processing and transfer of T-DNA. This should emphasise the need to use strains of *Agrobacterium* that are known to naturally infect the target host plant or its close family relatives [21, 35, 36]. For example we [21] have recently shown that the *Agrobacterium* strains A281 and C58 are highly virulent on the apple cultivar Greensleeves and most other apple cultivars (Uratsu unpublished). Virulence of the A281 strain could be further enhanced if the strain contained extra copies of the Ti-encoded *vir* genes A, B and G. Wild type strains are very useful for matching strain with host plant but are of little use in producing regenerated plants with a normal phenotype. Disarmed strains of agrobacteria are still relatively few in number and therefore we have restricted our selection to wild type counterparts of available disarmed strains.

Induction of virulence in Agrobacterium

The expression of individual genes present in the Ti-plasmid-encoded virulence region are essential for virulence and are regulated by a two component system involving the virulence genes *vir*A and *vir*G [22, 33]. Some physiological and biochemical factors have been identified and shown to be important for optimal virulence induction. The most important of these is the phenolic compound acetosyringone (AS) used at low pH (5.2) [23]. Although the growth of *Agrobacterium* is inhibited at low pH, the addition of glycine betaine has been shown to improve growth and virulence induction at pH 5.2 [34]. Currently *Agrobacterium* cultures to be used in transformation experiments are grown for 24–36 h in YEP medium at 24–28 °C. The cultures are centrifuged and resuspended in a simplified induction medium (SIM) to an OD of 0.5 (550 nm) and allowed to grow at 26 °C for 5 h to induce virulence. The SIM medium we use is a modification of that previously reported by Alt-Moerbe et al. [24]. These authors and other groups have shown that at temperatures below 28 °C, the presence of 2–3% sucrose and acidic pH (5.0–5.5) were essential for expression of the vir genes and that such expression was stimulated by 100 μm AS. Furthermore there were important lag phases for the appearance of the gene products for each of the vir genes which, depending on the gene, varied from 2 to several hours duration; the induction of expression of virD and virB only occurred in SIM media. Usually before transforming plant cells bacterial cells are resuspended in MS medium at pH 5.5 with sucrose. However MS media has a low buffering capacity and a drop in the pH below 5.2 (a distinct possibility after autoclaving) could drastically reduce viability and virulence.

3) The selection of transformed cells that lead to the regeneration of transformed plants

a) The length of the co-cultivation period
The co-cultivation period immediately after inoculation is 1–3 days and is the time during which transformation occurs. From an empirical standpoint the time required is something of a compromise between allowing enough time for transformation to occur and not letting agrobacteria overgrow the plant tissue. For apple explants we have found 3 days to be slightly better than 1 or 2 as assessed by the number of kanamycin resistant calluses formed in a 3 month period following infection.

Feeder layers of plant cells, usually in the form of cell suspensions, are often used during the co-cultivation period to enhance transformation and restrict the growth of agrobacteria [10]. We have found in apple that neither potato or apple endosperm cell suspensions affect the frequency of transformed callus or shoots that subsequently develop (unpublished data). We retain two ingredients from the SIM medium i.e., acetosyringone and betaine phosphate during this period but do not attempt to replace the MS salts with SIM during this period. The discs will not survive 3 days on an agar-based SIM media!

b) Antibiotic selection – concentration effects
In apple we find a marked decrease in the frequency of regeneration in the presence of more than 20 mg/l kanamycin. One possible explanation for a low number of transformants in an otherwise highly efficient regeneration system is that the inhibitory action of the antibiotic on the non-transformed cells in the tissue somehow interferes with the process of regeneration of the transformed cells. This has already been claimed in sunflower [26] and may be operative in apple. Ways of obviating this are to lower the contact time, the concentration or both. Although there is a greater risk of permitting 'escapes' to regenerate it may allow some true transformants to develop that would otherwise not have done so at higher concentrations of the antibiotic. Although our transgenic apple clones were initially produced at 60 mg/l we are currently using 25 mg/l since we know that this permits no root formation in control untransformed

Greensleeves material that normally gives very high rooting levels [11]. The use of the different selection strategies and their effects on the regeneration of apple transformants have been discussed elsewhere [11, 16]. Although we have obtained transgenic apple plants with selection periods as short as 5 days this only slightly reduces the time taken for transgenic shoots to regenerate compared to shoots regenerated after much longer periods of exposure to the antibiotic. At the same time short selection times often produce large numbers of 'escapes' which take considerable time to screen for the scorable marker and even longer to screen for kanamycin resistance since several subcultures are required to produce enough shoots to test for their ability to root on kanamycin (see below). We have now compromised on using 21 days exposure to kanamycin as a suitable time for screening out untransformed cells without subjecting the tissues to overlong exposures to the antibiotic should it interfere with the regeneration process. There are reports that other antibiotics related to kanamycin and equally effective at selection are less inhibitory to the regeneration process *viz* paromomycin [27] and phleomycin [28].

4) Factors pertaining to the regeneration of transformed plants after the selection period

If we assume that the majority of cells within a leaf explant are not transformed immediately after the transformation process they will be either dead or dying after exposure to the antibiotic. The necrotic products could then have an inhibitory effect on growth of nearby viable transformed cells. De Block [29] was able to regenerate transformants of potato cultivars at a much higher frequency by including silver nitrate at 10 mg/l in the regeneration medium. Silver nitrate has long been known to inhibit the production of ethylene during senescence and other age-related phenomenon [30–31]. More recently it has been shown to stimulate shoot regeneration in tissue cultures of both monocots and dicots [32]. Higher regeneration frequencies using altered tissue culture media for apple have already been mentioned in the context of measuring adventitious bud formation *per se*. Media manipulations during the antibiotic selection phase have so far not been reported in apple but it is likely that frequencies of regeneration from apple leaf explants can be much higher than those already recorded. Recent unpublished work by Viss (pers. comm. 1990) showed that leaf explants of the apple cultivar Greensleeves could regenerate on average more than 100 shoots per explant using the diphenyl urea thidiazuron (TDZ) combined with other media changes. It needs to be determined how many of these were true *de novo* regeneration events and how many were 'condensed' axillary shoots laid down on initial *de novo* primordia. This question can only be resolved by a detailed histological analysis based on the time course of regeneration. Whatever the answer to this question observations such as this and others already mentioned serve to show that the role of media manipulations in increasing the efficiency of transformation may yet be paramount.

5) Cloning and analyses of putative transformants

Not all apple shoots that regenerate after kanamycin selection have proven to be transgenic and 'escapes' are screened out with a suitable scorable marker such as NOS or GUS. Moreover shoots that have proven positive for the scorable marker activity do not always go on to produce micropropagating cultures. The reasons for this are unknown. Unfortunately matters are further complicated by the knowledge that the absence of expression of the scorable marker does not always mean that a clone is not transformed. In fact our first kanamycin-resistant clone was negative for NOS activity but has since been shown to be contain multiple copies of the APH(3')II gene [11] and [unpublished]. Once a clone is established however it can take several months to produce sufficient material for rooting, biochemical and molecular analyses. To produce sufficient apple shoots for the rooting assay is critical since regenerates can show a reduced rooting capacity in the absence of kanamycin compared to the control parent line. Consequently we routinely produce a minimum of 30 shoots for rooting assays on kanamycin before subjecting the clone to molecular analyses. One of our clones, B, has consistently rooted at a significantly lower level than other transgenic clones irrespective of the presence of kanamycin in the rooting media [11].

One of the major problems in plant cell transformation work is the inability to identify quickly the effects of the several variables that are being examined in the assessment of transformation efficiency. For this recourse to a quantitative measure of the efficiency of transformation such as the GUS assay [17, 26] may permit a more rapid assessment of the role of the variables discussed above.

Procedures

Apple Leaf Disc Transformation

A) Preparations for Experiment:

1. Petri dishes containing MS20 (full strength in 2% w/v sucrose).
2. Regeneration plates for co-cultivation (BN51 or C81).
3. MS20 or SIM media (100 ml) liquid.
4. YEP medium + kanamycin (50 mg/l) in liquid medium.
5. 9 cm filter papers cut to fit 9 cm Petri dishes autoclaved and ready for blotting.
6. Prepare co-cultivation media and dispense into 9 cm sterile Petri dishes. This comprises regeneration media, acetosyringone and betaine phosphate.

B) Growth and culture of agrobacteria prior to infection

Steps in the procedure
1. Use one loopful of bacteria from selection plate kept at 4 °C.
2. Resuspend in 10 ml YEP or LB media containing 50 mg/l kanamycin in a sterile, 30 ml screw top centrifuge tube with conical base so that the bacteria are more efficiently pelleted.
3. Grow overnight at 28 °C with agitation at 100 rpm.
4. Check growth to OD of 0.5 at 550 nm.
5. Next morning spin 15 min @ 2500 rpm.
6. Pour off supernatant.
7. Resuspend pellet in either MS20 or SIM media.
8. Shake for 5 h at 25 °C.
9. Pour agrobacterial suspension into Petri dishes.

C) Cutting and infecting leaf discs

Steps in the procedure
1. Remove expanded leaves from rooted plants and float in sterile distilled water in 9 cm Petri dishes.
2. Cut discs in dishes under sterile water. Prepare batches of 10–20.
3. Place in agrobacteria suspensions for 10–20 min.
4. Blot onto sterile, Whatman filter paper.
5. Transfer discs to Petri dishes containing co-cultivation media covered with sterile 9cm Whatman filter papers previously cut to fit.
6. For large scale experiments we routinely cut about 800 discs and inoculate Petri dishes with about 24/25 discs per plate.

D) Co-cultivation

Steps in the procedure
1. Seal all Petri dishes with Nesco Film or ParaFilm.
2. Incubate cultures at 25 °C dark for at least 3 days.
3. The use of a cell suspension layer is optional but we have found it of no use in our work on Greensleeves.

E) Selection for transformation events

Steps in the procedure
1. After 3 days transfer discs to Petri dishes with regeneration medium (BN51 or C81) containing filter-sterilised cefotaxime (200 mg/l) and kanamycin sulphate (25, or 50 mg/l).
2. Maintain cultures at 25 °C in the dark for 21 days. Check regularly for contamination.

F) Scoring and testing regenerates.

Steps in the procedure
1. Depending on kanamycin concentration shoots will appear in 2–9 months.
2. Carefully remove disc or explant with regenerated shoot and place in a small test tube (Coulter vessels are ideal) containing 10 ml of A17 with 200 mg/l cefotaxime.
3. Subculture at monthly intervals until sufficient shoot material for physiological, biochemical and molecular analyses is obtained.
4. Use scorable marker (e.g. NOS, GUS, CAT) to screen large numbers of putative transformants.
5. Perform rooting assays when a large number (minimum 30) of shoots is available. Root on R13, R37 in both presence AND absence of 50 or 25 mg/l kanamycin. Some regenerants may root at a decreased frequency compared to the parent culture after regeneration irrespective of whether they are transformed.
6. Perform molecular analyses on shoots giving positive expression of scorable marker.

Media

A. *Shoot multiplication medium — A17 — 1 litre*

1. Sorbitol or sucrose as carbon source	30.00 g (3%)
2. M & S (Gibco) macroelements	4.31 g
3. Add MS Inositol from stock (100 mg/ml)	1.0 ml
4. Add MS Vitamins from stock × 1000	1.0 ml
5. Add BA from stock (2 mg/ml)	0.5 ml
6. Add IBA from stock (1 mg/ml)	0.1 ml
7. Make up to 1 litre with deionised/distilled water	
8. Adjust pH to 5.8 with 1M and 0.1M KOH	
9. Add Difco-Bacto agar ('Certified')	8.0 g
10. Autoclave 20 min, 15 psi	

Filter-sterilisable optional extras to be added after autoclaving: (use Acro discs, 0.22 μm)

11. Kanamycin sulphate	25–50.0 mg/ml
12. Cefotaxime from stock (frozen) 100 mg/l	2.0 ml

B. *Root induction media — R13 — 1 litre*

1. Sorbitol or sucrose as carbon source	30.00 g (3%)
2. M & S (Gibco) macroelements	4.31 g
3. Add MS Inositol from stock (100 mg/ml)	1.0 ml
4. Add MS Vitamins from stock × 1000	1.0 ml
5. Add IBA (1000 mg/ml stock)	3.0 ml
6. Add IBA from stock (1mg/ml)	0.1 ml
7. Make up to 1 litre with deionised/distilled water	
8. Adjust pH to 5.8 with 1M and 0.1M KOH	
9. Add 8 g Difco-Bacto agar ('Certified')	
10. Autoclave 20 min, 15 psi	

Filter-sterilisable optional extras to be added after autoclaving: (use Acro discs, 0.22 μm)

11. Kanamycin sulphate (stock, frozen) 50 mg/ml	1.0 ml
12. Cefotaxime (stock, frozen) 100 mg/ml	2.0 ml

C. *Root emergence media — R37 — 1 litre*

1. Sorbitol or sucrose as carbon source	30.00 g (3%)
2. M & S (Gibco), half strength macroelements — 2.15 g	
3. Add MS Inositol from stock (100 mg/ml)	1.0 ml
4. Add MS Vitamins from stock × 1000	1.0 ml
5. Add IBA (1000 mg/ml stock)	3.0 ml

        6. Add IBA from stock (1 mg/ml)   0.1 ml
        7. Add deionised/distilled water   1.0 litre
        8. Adjust pH to 5.8 with 1M and 0.1M KOH
        9. Add Difco-Bacto agar ('Certified')   8.0 g
       10. Autoclave 20 min, 15 psi

Filter-sterilisable optional extras to be added after autoclaving:

       11. Kanamycin sulphate (stock, frozen) 50 mg/ml   1.0 ml
       12. Cefotaxime (stock, frozen) 100 mg/ml   2.0 ml

   D. *Regeneration media — BN51 — 1 litre*

        1. Sorbitol or sucrose as carbon source   30.00 g (3%)
        2. M & S (Gibco) macroelements   4.31 g
        3. Add MS Inositol from stock (100 mg/ml)   1.0 ml
        4. Add MS Vitamins from stock × 1000   1.0 ml
        5. Add BA from stock (2 mg/ml)   2.5 ml
        6. Add NAA from stock (1 mg/ml)   1.0 ml
        7. Make up to 1 litre with deionised/distilled water
        8. Adjust pH to 5.8 with 1M and 0.1M KOH
        9. Add Difco-Bacto agar ('Certified')   8.0 g
       10. Autoclave 20 min, 15 psi
       11. If necessary add filter sterilised cefotaxime from stock
           (frozen — 100 mg/ml)   2.0 ml

   E. *Regeneration media — C81 — 1 litre*

        1. Sorbitol or sucrose as carbon source   30.00 g (3%)
        2. M & S (Gibco) macroelements   4.31 g
        3. Add MS Inositol from stock (100 mg/ml)   1.0 ml
        4. Add MS Vitamins from stock × 1000   1.0 ml
        5. Add BA from stock (2 mg/ml)   1.0 ml
        6. Add NAA from stock (1 mg/ml)   0.5 ml
        7. Make up to 1 litre with deionised/distilled water
        8. Adjust pH to 5.8 with 1M and 0.1M KOH
        9. Add Difco-Bacto agar ('Certified')   8.0 g
       10. Autoclave 20 min, 15 psi
       11. If necessary add filter sterilised cefotaxime from stock
           (frozen — 100 mg/l)   2.0 ml

   F. *SIM media (Simplified induction media) [24]*

       Per 100 ml
   a) Sucrose 2% w/v (2 g in 100 ml)

b) 0.1 mM acetosyringone (stock solution: 0.1M in ethanol)
c) 20 mM sodium citrate, pH 5.2 (100 ml)
d) Add betaine phosphate at a final concentration of 1mM. Prepare a 1M solution, filter sterilise and maintain at 4 °C

G. Co-cultivation media

1. Sorbitol or sucrose as carbon source	30.00 g (3%)
2. M & S (Gibco) macroelements	4.31 g
3. Add MS Inositol from stock (100 mg/ml)	1.0 ml
4. Add MS Vitamins from stock × 1000	1.0 ml
5. Add BA from stock (2 mg/ml)	1.0 ml
6. Add NAA from stock (1 mg/ml)	0.5 ml
7. Make up to 1 litre with deionised/distilled water	
8. Adjust pH to 5.8 with 1M and 0.1M KOH	
9. Add Difco-Bacto agar ('Certified')	8.0 g
10. Autoclave 20 min, 15 psi	
11. Add Betaine phosphate to 1mM and acetosyringone to 0.1 mM by filter sterilisation.	
12. Add sterile filter paper to surface of solidified agar plates (9 cm)	

Maintenance of disarmed strains on solid media
Per litre

---

a) Luria Broth

Yeast extract (Difco)	5.0 g
Tryptone (Difco)	10.0 g
NaCl	10.0 g
Adjust pH to 7.2	
BactoAgar (Difco)	15.0 g
Kanamycin	50.0 mg/l
Streptomycin (include occassionally)	0.5 g/l

Filter-sterilise antibiotics

b) YEP MEDIUM

Yeast extract (Difco)	5.0 g
Peptone (Difco)	10.0 g
NaCl	10.0 g
Adjust pH to 7.2	
BactoAgar (Difco)	15.0 g
Kanamycin	50.0 mg/l
Streptomycin (include occasionally)	0.5 g/l

References

1. James DJ, Passey AJ, Rugini E (1988) Factors affecting high frequency plant regeneration from apple leaf tissues cultured in vitro. J Plant Physiol 132: 148–154.
2. Fasolo F, Zimmerman RH, Fordham I (1989) Adventitious shoot formation on excised leaves of *in vitro* grown shoots of apple cultivars, Plant Cell Tissue Organ Culture 16: 75–87.
3. Predieri S, Fasolo Fabbri Malavasi, F (1989) High frequency shoot regeneration from leaves of the apple rootstock M.26 (*Malus pumila* Mill.) Plant Cell Tissue Organ Cult 17: 133–142.
4. Welander M (1988) Plant regeneration from leaf and stem segments of shoots raised in vitro from mature apple trees J Plant Physiol 132: 738–744.
5. Zambryski P, Tempe J, Schell J (1989) Transfer and function of T-DNA genes from *Agrobacterium* Ti- and Ri-plasmids in plants. Cell 56: 193–201.
6. Jones H, Jones, MGK (1989) Direct gene transfer into plant protoplasts. Plants Today 2: 175–178.
7. Finer JJ and McMullen MD (1990), Transformation of cotton (*Gossypium hirsutum* L.) via particle bombardment. Plant Cell Rep 8: 586–589.
8. Gasser CS and Fraley RT (1989) Genetic engineering plants for crop improvement. Science 244: 1293–1299.
9. Bevan MW (1984) Binary *Agrobacterium* vectors for plant transformation, Nucleic Acids Res 12: 8711–8721.
10. Horsch RB, Fry JE, Hoffman NL, Eicholtz D, Rogers SG, Fraley RT (1985) A simple and general method of transferring genes into plants. Science 227: 1229–1231.
11. James DJ, Passey AJ, Barbara DJ, Bevan MW (1989) Genetic transformation of apple (*Malus pumila* Mill.) using a disarmed Ti-binary vector, Plant Cell Rep., 7: 658–666.
12. James DJ, Thurbon IJ (1979) Rapid *in vitro* rooting of the apple rootstock M.9. J Hort Sci 54: 309–311
13. James DJ, Passey AJ, Barbara DJ (1990) Regeneration and transformation of apple and strawberry using disarmed Ti-binary vectors. in: 'Genetic Engineering of Crop Plants', 49th Nottingham Easter School, University of Nottingham Sutton Bonington, G Lycett and D Grierson, eds., Butterworth. pp.239–248.
14. Zambryski PC (1988) Basic Processes underlying *Agrobacterium* mediated DNA transfer to plant cells. Ann Rev Genet 22: 1–30.
15. Larkin PJ, Scowcroft WR (1983) Somaclonal variation and crop improvement, in: 'Genetic Engineering of Plants: An Agricultural Perspective,' T Kosuge, CP Meredith, and A Hollaender, eds., pp. 289–314, Plenum Press, New York.
16. James DJ, Passey AJ, Barbara DJ (1990) Agrobacteria-mediated transformation of the cultivated strawberry (Fragaria × anannassa Duch.) using disarmed Ti-binary vectors. Plant Science 69: 79–94.
17. Janssen B and Gardner RC (1989), Localized transient expression of GUS in leaf discs following co-cultivation with *Agrobacterium*. Plant Molec Biol 14: 61–72.
18. Kobayashi A (1989) Studies on cell culture and plant regeneration in apple. III. Influence of etiolation on adventitious shoot formation from leaf discs, Bull. Fruit Trees Res Stn C, 16: 15–21.
19. McHughen A, Jordan M, Feist G (1989) A preculture period prior to *Agrobacterium* inoculation increases production of transgenic plants. J Plant Physiol, 135: 245–248.
20. Chyi Y-S, Phillips GC (1987) High efficiency *Agrobacterium* mediated transformation of Lycopersicon based on conditions favourable for regeneration, Plant Cell Reports, 6: 105–108.
21. Dandekar AD, Uratsu SL, Matsuta N (1990) Factors influencing virulence in agrobacteria-mediated transformation of apple. In: *In vitro* Culture and Horticultural Breeding, Acta Hort 280: 483–493.
22. Rogowsky PM, Close TJ, Chimera JA, Shaw JJ, Kado CI (1987) Regulation of the vir genes of *Agrobacterium tumefaciens* plasmid pTiC58. J Bact 169: 5101–5112.

23. Stachel SE, Messens E, Van Montague M, Zambryski PC (1985) Identification of the signal molecules produced by wounded plant cells that activate T-DNA transfer in *Agrobacterium tumefaciens*. Nature 318: 624–629.

24. Alt-Moerbe J, Kuhlmann H, Schroder J (1989) Differences in induction of Ti plasmid virulence genes virG and virD and continued control of *virD* expression by four external factors, Molecular Plant-Microbe Interactions 2: 301–308.

25. Jefferson RA, Kavanagh TA, Bevan MW, (1987) GUS fusions: β-glucuronidase as a sensitive and versatile gene fusion marker in higher plants, The EMBO J 6: 3901–3907.

26. Everett NP, Robinson KEP, Mascarenhas D,(1987) Genetic engineering of sunflower (*Helianthus annuus* L.), Bio/technology, 5: 1201–1204.

27. Chupeau M-C, Bellini C, Guerche P, Maisonneuve B, Vastra G, Chupeau, Y.,(1989) Transgenic plants of lettuce obtained through electroporation of protoplasts. Bio/Technology 7: 503–508.

28. Perez, P, Tiraby G, Kallerhoff J, Perret J (1989) Phleomycin resistance as a dominant selectable marker for plant cell transformation, Plant Mol Biol 13: 365–373.

29. De Block M (1988) Genotype independent leaf disc transformation of potato (*Solanum tuberosum*) using *Agrobacterium tumefaciens*, Theor Appl Genet 76: 767–774.

30. Aharoni N, Anderson JD, Lieberman M (1979) Production and action of ethylene in senescing leaf discs. Effects of indoleacetic acid, kinetin, silver ion, and carbon dioxide. Plant Physiol 64: 805–809.

31. Beyer EM (1976) A potent inhibitor of ethylene action in plants. Plant Physiol 58: 268–271.

32. Purnhauser L, Medgyesy P, Czako M, Dix PJ, Marton L (1987) Stimulation of shoot regeneration in *Triticum aestivum* and *Nicotiana plumbaginifolia* Viv. tissue cultures using the ethylene inhibitor AgNO3. Plant Cell Rep. 6: 1–4.

33. Winans S, Kerstettner R, Nester E,(1988) Transcriptional regulation of the *virA* and *virG* genes of *Agrobacterium tumefaciens*. J Bact 170: 4047–4054.

34. Vernade D, Herrera-Estrella a, Wang K, Van Montagu M,(1988) Glycine betaine allows enhanced induction of the *Agrobacterium tumefaciens* vir genes by acetosyringone at low pH. J Bact 170: 5822–5829.

35. Dandekar, AM, Gupta, PK, Durzan, DJ, and Knauf, V (1987) Transformation and foreign gene expression in micropropagated Douglas-fir (*Pseudotsuga menziesii*), Bio/Technology 5, 587–590.

36. Dandekar, AM, Martin, LA, and McGranahan, G.H. (1988) Genetic transformation and foreign gene expression in walnut tissue, J. Am. Soc. Hort. Sci. **113**, 945–949.

Plant Tissue Culture Manual **B9**: 1–13, 1992.

Transformation and regeneration of maize protoplasts

CAROL A. RHODES & DAVID W. GRAY
Plant Biotechnology Department, Sandoz Crop Protection Corporation

Introduction

Several laboratories have successfully recovered mature plants from maize protoplasts [9, 11, 12, 15, 16]. In all cases, embryogenic suspension cultures were the source of competent protoplasts (those capable of division and plant regeneration). The limiting step has been to produce such suspensions repeatedly. Friable, embryogenic callus cultures were used as initial tissue for suspension cultures. The only commonality among the methods used to derive suspensions in these laboratories was that the donor suspensions seemed to require a minimum period (weeks to months) of growth in liquid culture before yielding competent protoplasts.

Genotype alone is not limiting, as success has been achieved with diverse genotypes, including a tropical inbred [11], a supersweet hybrid-derived haploid line [16], elite field corn lines [15] as well as a line derived from genetic mixtures [9]. This does not imply that any genotype will work as readily as any other, but rather that success is not dependent on one specific genotype.

Another difficulty is the frequency of partial or complete sterility in plants regenerated from protoplasts. This frequency seems to depend on the donor culture and may be related to the length of time *in vitro*. Frequency and vigor of plant regeneration directly from suspension cells often declines with increasing time in culture. These declines of regeneration vigor may be at least partly responsible for sterile plants regenerated from protoplasts. Recovery of plants in which there are problems with fertility may indicate genetic changes have occurred *in vitro* and/or regeneration protocols need to be improved for this trait.

Maize protoplasts appear capable of growing and dividing under a variety of culture conditions. Variations of both MS-based and N6-based media have been used successfully, in either liquid or solidified form. Addition of acetyl-salicylic acid improved plating efficiency in one case [2]. In several other cases, nurse cells have been shown to increase plating efficiency when used effectively [5, 7, 12, 16]. Choice of feeder cells is critical; among other desirable traits, cells must be growing rapidly to be effective.

Transformation of protoplasts can be accomplished by several means of direct DNA uptake. Two methods are used most often: electroporation or polyethylene glycol (PEG)-mediated uptake. Both methods have been used successfully with cereal protoplasts to produce stably transformed cells [1, 3,

6, 10]. Methods which combine both PEG and electroporation have been shown to increase transformation efficiency in some [14] but not all [18] systems. We describe here our procedure for electroporation [13].

Several genes have been used to select and identify stably transformed cells. These include genes conferring resistance to antibiotics, herbicides or other growth-inhibiting compounds. Examples of such genes are those encoding the enzymes neomycin phosphotransferase (NPT II), hygromycin phosphotransferase (HPT), phosphinothricin acetyltransferase (PAT), and acetolactate synthase (ALS). Critical factors for any of these systems are timing of transfers, concentration of selective agent, and density of selected cells. Expression of the inserted gene must be sufficient to confer resistance to the selective agent. An example of one selection scheme using NPT II as the selectable marker gene is given here. Kanamycin has not always worked well in all cell systems; each case must be optimized and evaluated individually.

Procedures

Protoplast isolation

Steps in the procedure

1. Maize cell lines are maintained in liquid culture at 5 to 10 grams fresh weight per 50 ml of medium. Transfer cells twice weekly to fresh N6ap medium [12], N6 supplemented with 790 mg/l asparagine, 1.38 g/l proline, and 1 mg/l 2,4-dichlorophenoxyacetic acid (2,4-D).
2. Transfer suspension to a 50 ml graduated centrifuge tube (Corning #25330) and spin at $100 \times g$ (Hermle Z320 centrifuge) for 5 minutes.
3. Pour off the supernatant. The packed volume of cells is approximately equal to grams fresh weight. Add 10 ml of Digestion Buffer for each gram of tissue. Generally, 5 grams of tissue is sufficient for a high yield of protoplasts (5 to 20×10^6).
4. Transfer the mixture to a 250 ml flask and place on a rotary shaker at 100 rpm for $1\frac{1}{2}$ to 2 hours in the dark.
5. After $1\frac{1}{2}$ hours, check a sample of the digest under a microscope for quality and approximate yield of protoplasts. Digest up to 1 hour longer if yield is below 1×10^6 per g.
6. Separate undigested cells from protoplasts by filtering the digest through a 149 μm polypropylene mesh screen (Spectrum Medial Industries #146432) and then through a 30 μm nylon mesh screen (SMI #146506). Rinse screens and collection beakers each time with 10 ml of Mannitol Buffer.
7. Distribute the protoplast suspension equally among four clear round-bottom centrifuge tubes (Nalge #3118, 50 ml polycarbonate). Add mannitol buffer to fill the tubes and centrifuge at $100 \times g$ for 5 minutes.
8. Carefully pipette off the supernatant without disturbing the protoplast pellet. Discard the supernatant and add 15 ml of fresh Mannitol Buffer. Combine two tubes into one and centrifuge again at $100 \times g$ for 5 minutes.
9. Repeat step 8 so that there is only one remaining tube with protoplasts. Spin and discard the supernatant. To prepare the protoplasts for electroporation, add 5 ml of Protoplast Culture Medium with the pH adjusted to 8.0. Estimate the number of viable protoplasts by staining a sample with 0.01% fluorescein diacetate [17].
10. Adjust the protoplast concentration to 3 to 4×10^6 per ml for transformation.

Direct uptake of DNA via electroporation

Steps in the procedure

1. Just prior to electroporation, protoplasts may be heat treated by submersing the tube in a 42 °C water bath for 3 minutes. This step should be followed as soon as possible by all subsequent steps to the dilution of protoplasts (step 5).
2. Transfer 0.5 ml aliquots of protoplasts into wells of a 24 well microtiter plate (Corning #25820). The number of wells loaded is determined by the number of protoplasts and the size of the experiment. A typical experiment is as follows:

Treatments	DNA (μg)	Voltage (V)	Capacitance (μF)
a) control	none	none	. . . .
b) control	none	225	1200
c) pZO3	30	200	1200
d) pZO3	30	225	1200
e) pZO3	30	250	1200

3. Add 75 μl of 2M KCl (final concentration 150 mM), 0.4 ml Protoplast Culture Medium pH 8.0, and plasmid DNA (10 to 50 μg per well) to each of the wells with protoplasts. Circular or linear DNA can be used.
4. Electroporate each well immediately after adding DNA and mixing with protoplasts (Hoeffer Scientific Instruments, Progenetor™). Optimum voltages may vary with protoplast size.
5. After treatments are electroporated, transfer protoplasts to 60 × 15 mm Petri plates and dilute with 2 volumes of Protoplast Culture Medium (final volume, 3 ml).

Note
4. Alternative electroporation apparatus: Bio-Rad Gene Pulser® with samples placed in cuvettes.

Growth of electroporated protoplasts

Steps in the procedure

1. Prepare feeder plates by suspending 0.5 g of BMS (Black Mexican Sweet suspension culture) cells in 1 ml of Protoplast Culture Medium per plate. Choice of feeder cells is important [7, 12].
2. Pipette 1.5 ml of feeder suspension mixture onto solidified Protoplast Culture Medium (100 × 15 mm Petri plates) and spread the mixture over the entire plate. Place a filter (Millipore #AABG 047 SO or HAEP 047 SW) over the feeder layer. Make 5 plates for each treatment.
3. Transfer 0.6 ml of protoplasts to each of the feeder plates and carefully pipette protoplasts over the surface of the filters.
4. Place finished plates in the dark or subdued light at 28 °C.

Selection of stable transformants

Steps in the procedure

1. One week after electroporation, transfer filters with protoplasts to plates with fresh culture medium at reduced osmolarity (N6ap with 0.1 M mannitol) containing 100 mg/l kanamycin. These plates may also contain feeder cells, made as before. A kanamycin-resistant feeder culture may be used, but is not essential.
2. Transfer filters after another week to N6ap supplemented with 1 mg/l 2,4-D and 100 mg/l kanamycin (no added mannitol) without any feeder cells.
3. Repeat weekly or biweekly transfers as needed, until rapidly growing colonies are distinctive among any background growth. This should take a total of 4 to 5 weeks after electroporation.
4. Callus can be assayed for presence of the introduced gene via enzyme assay [8] and DNA analysis.
5. Regenerate shoots by transferring callus to N6 or MS medium with reduced (0.05 to 0.25 mg/l) or no 2,4-D. Other hormonal and nutritional regimes which have been suggested include the use of cytokinins for a brief period [15] or increased sucrose concentrations [4].
6. Transfer shoots to clear jars or boxes (Magenta GA7) containing MS medium without any added hormones. It is important that shoots be grown under bright light all the way to maturity. Transfer to sterile soil mix after roots have grown out. Each genotype or cell culture may regenerate best under particular conditions which must be determined empirically.

Solutions
Mannitol Buffer

> 80 mM $CaCl_2$
> 0.5% w/v MES
> 0.30 M mannitol
> adjust pH to 6.0

Digestion Buffer

> To 100 ml of mannitol buffer, add:
>> 2 g Cellulase RS (Yakult Pharmaceutical Co.)
>> 100 mg Pectolyase Y23 (Seishin Pharmaceutical Co.)
> Filter sterilize. Stable at $-20\ ^\circ C$.

N6 medium

100X Iron Stock

Na_2EDTA	3.72 g
$FeSO_4 \cdot 7H_2O$	2.78 g

Boil the EDTA vigorously for 1-2 min in 200 ml H_2O. Add to the $FeSO_4$ dissolved in 200 ml H_2O and bring volume to 1 litre with H_2O.

10X N6 Major Salts

KNO_3	28.30 g
$MgSO_4$	0.90 g
$CaCl_2$	1.25 g
$(NH_4)_2SO_4$	4.63 g
KH_2PO_4	4.00 g

Add salts in order to a beaker with ca. 500 ml H_2O. Stir until dissolved and bring volume to 1 litre with H_2O.

100X N6 Minor Salts

$MnSO_4.H_2O$	0.33 g
H_3BO_3	0.16 g
$ZnSO_4.7H_2O$	0.15 g
KI	0.08 g

Add salts in order to a beaker with ca. 500 ml H_2O. Stir until dissolved and bring volume to 1 litre with H_2O.

100X N6 Vitamin Stock

Thiamine-HCl	20 mg
Nicotinic acid	10 mg
Pyridoxine-HCl	10 mg
Glycine	40 mg
Casein hydrolysate	2 g

Dissolve in H_2O and bring to a final volume of 200 ml.

N6ap 1D Medium

Sucrose	30 g
L-asparagine	0.79 g
Proline	1.38 g
100X Iron stock	10 ml
10X N6 Major salts	100 ml
100X N6 Minor salts	10 ml
100X N6 Vitamins	10 ml
Inositol	2 ml of a 100 mg/ml stock
2,4-D	1 ml of a 1 mg/ml stock

Add sucrose, asparagine and proline to 800 ml H_2O and start mixing. Add remaining stock solutions. Bring volume to 1 litre. Adjust pH to 5.8 after all compounds are dissolved. For plates, add 7 g agar. Autoclave in volumes not greater than 1 litre for 20 min.

Composition of N6 medium

I. *Salts*   *Final Concentration*

Major elements	*mg/l*	*mM*
$(NH_4)_2SO_4$	463	3.5
KNO_3	2830	28.0
$CaCl_2$	125	1.13
$MgSO_4$	90	0.75
KH_2PO_4	400	2.94
Na_2 EDTA	37.2	0.20 (Na)
$FeSO_4\ 7H_2O$	27.8	0.10 (Fe)

Minor elements	*mg/l*	*μM*
H_3BO_3	1.6	25.8
$MnSO_4\ 1H_2O$	3.3	19.5
$ZnSO_4 7H_2O$	1.5	5.2
KI	0.8	5.0

Organic constituents	*mg/l*
inositol*	200
thiamine HCl	1.0
glycine	2.0
pyridoxine	0.5
nicotinic acid	0.5
cassein hydrolysate*	200
sucrose	30 g/l

Protoplast Culture Medium

N6ap with 2 mg/L 2,4-D
0.30 M mannitol
pH 5.8

* not in original formula as given by Chu *et al.*, 1975, Scient. Sinica 18:659.

References

1. Armstrong CL, Petersen WL, Buchholz WG, Bowen BJ, Sulc SL (1990) Factors affecting PEG-mediated stable transformation of maize protoplasts. Plant Cell Rep. 9: 335–339.
2. Carswell GK, Johnson CM, Shillito RD, Harms CT (1989) O-acetyl-salicylic acid promotes colony formation from protoplasts of an elite maize inbred. Plant Cell Rep. 8: 282–284.
3. Fromm ME, Taylor LP, Walbot V (1986) Stable transformation of maize after gene transfer by electroporation. Nature 319: 791–793.
4. Hodges TK, Kamo KK, Imbrie CW, Becwar MR (1986) Genotype specificity of somatic embryogenesis and regeneration in maize. Bio/Tech. 4: 219–223.
5. Kuang VK, Shamina ZB, Butenko RG (1983) Use of nurse tissue culture to obtain clones from cultured cells and protoplasts of corn. Fiziologiya Rastenii 30: 803–812.
6. Lörz H, Baker B, Schell J (1985) Gene transfer to cereal cells mediated by protoplast transformation. Mol. Gen. Genet. 199: 178–182.
7. Lyznik LA, Kamo KK, Grimes HD, Ryan R, Chang K, Hodges TK (1989) Stable transformation of maize: the impact of feeder cells on protoplast growth and transformation efficiency. Plant Cell Rep. 8: 292–295.
8. McDonnell RE, Clark RD, Smith WA, Hinchee MA (1987) A simplified method for the detection of neomycin phosphotransferase II activity in transformed plant tissues. Plant Mol. Biol. Rep. 5: 380–386.
9. Morocz S, Donn G, Nemeth J, Dudits D (1990) Plant regeneration from haploid and diploid Zea mays (L.) protoplast cultures. International Plant Tissue Culture Congress, Amsterdam. Abstract A1–102.
10. Potrykus I, Saul M, Petruska J, Paszkowski J, Shillito R (1985) Direct gene transfer to cells of a graminaceous monocot. Mol. Gen. Genet. 199: 183–188.
11. Prioli LM, Sondahl MR (1989) Plant regeneration and recovery of fertile plants from protoplasts of maize (Zea mays L). Bio/Tech. 7: 589–594.
12. Rhodes CA, Lowe KS, Ruby KL (1988) Plant regeneration from protoplasts isolated from embryogenic maize cell cultures. Bio/Tech. 6: 56–60.
13. Rhodes CA, Pierce DA, Mettler IJ, Mascarenhas D, Detmer JJ (1988) Genetically transformed maize plants from protoplasts. Science 240: 204–207.
14. Shillito R, Saul M, Paszkowski J, Muller M, Potrykus I (1985) High efficiency direct gene transfer to plants. Bio/Tech. 3: 1099–1103.
15. Shillito RD, Carswell GK, Johnson CM, DiMaio JJ, Harms CT (1989) Regeneration of fertile plants from protoplasts of elite inbred maize. Bio/Tech. 7: 581–587.
16. Sun CS, Prioli LM, Sondahl MR (1989) Regeneration of haploid and dihaploid plants from protoplasts of supersweet (sh2sh2) corn. Plant Cell Rep. 8: 313–316.
17. Widholm JM (1972) The use of fluorescein diacetate and phenosafranine for determining viability of cultured plant cells. Stain Tech. 47: 189–194.
18. Yang H, Zhang HM, Davey MR, Mulligan BJ, Cocking EC (1988) Production of kanamycin resistant rice tissues following DNA uptake into protoplasts. Plant Cell Rep. 7: 421–425.

Plant Tissue Culture Manual **B10**: 1–11, 1992.
© 1992 *Kluwer Academic Publishers.*

Culture, regeneration and transformation of barley protoplasts

[1] PAUL A. LAZZERI, ALWINE JÄHNE & HORST LÖRZ
*Institut für Allgemeine Botanik der Universität Hamburg (AMP II), Ohnhorststraße 18, D-2000
Hamburg 52, Germany.*
[1] *Current address: AFRC IACR, Rothamsted Experimental Station, Harpenden, Herts. AL5 2JQ,
U.K.*

Introduction

For plant genetic engineering to benefit agriculture it must be possible to transform the world's major crop species. While most of the important dicot crops have been shown to be susceptible to *Agrobacterium*-mediated transformation [2], monocot species, including the Poaceae, are resistant to *Agrobacterium* infection, so alternative transformation methods are required for the transformation of cereal and grass crops [13, 19]. Various different transformation methods have been proposed and tested in poaceous species, but at present only two, direct DNA uptake into protoplasts [18], and particle bombardment [8], have yielded transformed plants whose progeny clearly inherit the transferred genes [20, 3]. Both methods have particular advantages and disadvantages. Particle bombardment should, in theory, be tissue and genotype independent, as DNA-bearing particles can be delivered to most plant cells. In practice, however, successful application requires large numbers of regenerable cells which can subsequently be selected to identify transformants. In poaceaous species, embryogenic suspensions are at present the only source of such cells and such suspensions are generally difficult to produce [16].

Protoplast transformation has the advantages that very many cells can be handled with ease, allowing many independent transformants to be produced, that the selection of transformants is relatively simple and, that with PEG-mediated transformation, no specialised equipment is needed. The major disadvantage is the difficulty of regenerating from poaceous protoplasts. This is a long-standing problem, but in recent years there has been real progress and plants have been recovered from protoplasts of most of the major poaceous crops [22, 12].

In barley, dividing **protoplasts** were first obtained in 1976, using callus as the source material [9], but the resulting calluses were not morphogenic. Regeneration from barley protoplast cultures was first reported in 1988 [17], when albino plantlets were recovered from suspension-derived protoplasts. Subsequently, green plantlets were regenerated from cell suspension protoplasts, but these did not survive transfer to soil [14]. Recently, two groups have reported the regeneration of green plants from protoplast cultures [23, 5]; in the latter work fertile plants which set seed were recovered.

There are now several reports of transient gene expression in barley protoplasts, derived either from plant tissues [21, 15] or from suspensions [6, 1] (see

[7] for review of transformation in barley). In most cases plasmid constructs containing the CaMV 35S promoter have been used, driving either the NPT [21, 6], GUS [1] or CAT [15] genes as scorable markers. To date, however, there is only a single report of stable transformation of barley, via PEG-mediated DNA uptake into suspension protoplasts [11]. In this work, a plasmid containing both the NPT and GUS genes was used, with both under the control of CaMV 35S promoters. Transformed colonies were selected on media containing G418 sulphate. All antibiotic-resistant colonies had NPT activity and contained copies of the NPT gene, while around half the colonies showed GUS activity, although most contained the GUS gene.

In the following pages we present protocols for the isolation, culture, regeneration and transformation of barley suspension protoplasts. These protocols have been developed with suspensions of the cultivars Dissa [11, 14] or Igri [4, 5], but with very little modification have proved suitable for other *H. vulgare* cultivars, and for *H. marinum* and *H. murinum* (unpublished data).

Procedures

Protoplast isolation

The protocol is for the isolation of protoplasts from barley cell suspensions. Such suspensions are generally established from immature embryo or anther cultures [14, 17, 4]. A strict regime for suspension maintenance is essential, as the 'health/quality' of the protoplasts (and hence of the donor suspension) is the single most important factor deciding the success and reproducibility of subsequent manipulations.

Steps in the procedure

1. Maintain suspensions in L1D2 medium by twice-weekly subculture (see notes). At subculture, transfer 1–3 g cells to a 9 cm plastic Petri dish and add 7–10 ml enzyme solution per gramme cells. Seal dish.
2. Place dishes on a rotary shaker set at 30 rpm, at 25 °C. Check protoplast yield hourly; digestion time is usually 2–3 hrs.
3. Sieve digestion mixture through 100, 50 and 25 μm sieves and wash through with LW (washing) solution.
4. Using a wide-mouth pipette, distribute the protoplast suspension in centrifuge tubes and spin for 5 min at 50 × g.
5. Pour supernatants off pellets, then immediately disperse them in residual liquid by gentle shaking. Wash protoplasts by adding 10 ml LW solution to the tubes and then centrifuge as in step 4 and disperse pellets.
6. Resuspend protoplasts in LW and pool the whole isolation in a known volume (20–50 ml) of the solution in a flask. Swirl gently to evenly distribute the protoplasts and take a sample to count the density in a haemocytometer.
7. Either continue immediately with the culture procedure, or store pooled protoplasts at 4–7 °C until use.

Notes
1. Suspensions are cultured in 190 ml plastic vessels (Greiner 967161), shaken at 100 rpm, in dim light, at 25 °C. At subculture (3 1/2 d intervals), the medium is pipetted off and cells are harvested, leaving 2.5–3 g fresh weight per vessel. Ten ml of fresh L1D2 medium are then added to the vessel, and the harvested cells used for protoplast isolation. The standard enzyme solution contains mannitol as osmoticum; however, enzyme activity is increased when KCl is used as osmoticum. This reduces digestion time and increases yield from hard-to-digest suspensions, but more care must be taken to avoid 'overdigestion', resulting in damage to protoplasts.
2. Faster shaking increases damage to protoplasts; for sensitive suspensions it may be advisable to shake for the first hour only. Digestion is faster at higher temperatures (27–30 °C), but spontaneous protoplast fusion is also increased. Isolation times vary with the physiological status of the suspension. Complete digestion is never seen – only small aggregates and the outer layers of larger aggregates are digested. When to terminate digestion must be learned by experience and will always be a compromise between maximum yield and minimum damage to protoplasts from 'overdigestion'.
4. For less dense protoplasts, using LW washing solution with KCl instead of mannitol will improve sedimentation. For most bench-top centrifuges, speeds between 500 and 1000 rpm are suitable.
5. If pellets are not dispersed quickly, protoplasts will form aggregates which are later difficult to separate.
7. Protoplasts may be stored up to 24 h at 4–7 °C and still give good plating efficiencies, although there is obviously some loss of viability.

Protoplast culture is most efficient in agarose medium, although protoplasts will divide in liquid medium. For protoplasts from young, newly-established cell suspensions feeder cultures [10] may be useful, but in most cases adequate plating efficiencies are obtained without feeders, by using appropriate plating densities.

Steps in the procedure

1. Distribute aliquots of the protoplast suspension among centrifuge tubes such that individual tubes contain $1-4 \times 10^6$ protoplasts, depending on the final plating density required (see notes). Centrifuge at $50 \times g$ for 5 min, pour off supernatants and disperse pellets.
2. Suspend protoplasts in a 1 : 1 mixture of x2 L1D0.5 protoplast medium and 2% SeaPlaque agarose, at 42 °C, using a wide-mouth pipette (see notes). Carefully pipette 2.5 ml aliquots of the suspended protoplasts into 6 cm plastic Petri dishes (tissue culture quality).
3. Allow agarose to solidify, seal dishes with Nescofilm and incubate at 25 °C, under dim light or in darkness.
4. Monitor culture development; after 15 d calculate plating efficiency (number of microcalluses formed/number of protoplasts originally plated).
5. After 20 d, add 0.5 ml of L1D2 suspension medium to the surface of the agarose in culture dishes (making cuts through the gel to allow the liquid to penetrate). Return dishes to incubator.
6. After 30 d, transfer agarose segments containing calluses to the surface of solid regeneration (L3D0.5B1) or induction (L3D2.5) medium. Incubate dishes at 25 °C, under continuous light (ca. 1000 lux).
7. Subculture calluses at 25–30 d intervals on L3D0.5B1 medium, selecting for compact, yellow embryogenic tissue. Transfer calluses with distinct somatic embryos or shoots to hormone-free (L3-) medium in larger vessels and culture under higher light conditions (16 h, 6000 lux).
8. Transfer rooted plantlets to a peat/soil mix and grow on under greenhouse conditions.

Notes

1. Higher protoplasts plating densities, up to values as high as 1.5×10^6/ml, generally give higher plating efficiencies, presumably due to cross-feeding effects. The disadvantage of such high-density cultures is that the developing microcalluses quickly exhaust the medium, so that such cultures must be fed early and frequently with liquid medium. In most cases, plating densities between 0.5 and 0.75×10^6/ml are suitable.
2. For example, for three dishes, each with protoplasts at density 0.5×10^6/ml, a volume of protoplast suspension containing 3.75×10^6 protoplasts would be spun down and the pellet resuspended in 7.5 ml of medium/agarose mixture, 2.5 ml of which would be pipetted into each of the three dishes. Final agarose concentration is therefore 1%.
4. Plating efficiency is calculated by finding the mean number of microcalluses in a known area of the culture dish (e.g. the field of view for a $\times 10$ objective of an inverted microscope) and using this figure to calculate the total number of microcalluses present. Plating efficiency is expressed as the percentage of protoplasts plated which give rise to viable microcalluses. Some workers estimate the frequency of surviving protoplasts after a few days of culture and calculate the frequency of

microcallus formation from these cells, or base counts on the number of cells showing at least one division: both these methods yield inflated plating efficiency values, but are not true measure of the 'useable' protoplasts in a preparation.

6. In some cases, protoplast microcalluses may form embryogenic tissue when cultured directly on regeneration medium, while in other cases embryogenesis may have to be 're-induced' by culture on medium with a higher 2,4-D content.
7. Frequently, somatic embryos do not appear during the first passage on regeneration medium; selective subculturing enriches for embryogenic tissue and allows small groups of embryogenic cells to proliferate and subsequently to regenerate.
8. Plantlets with healthy shoot systems which do not form roots *in vitro* may sometimes be rooted by the use of a commercial rooting powder, directly in a peat/soil mix. As with all *in vitro* regenerants, plantlets must be gradually hardened after transfer to soil, by careful regulation of air humidity during the first days.

Transformation

Procedures are given both for stable transformation and transient gene expression in barley protoplasts. Where methods differ alternatives are given.

Steps in the procedure

1. Take centrifuge tubes containing dispersed pellets of $1-4 \times 10^6$ protoplasts (see *Culture and Regeneration*, step 1).
2. Add 1 ml LW solution to (untransformed) control tube, 0.2 ml C100M solution to the mock transformation tube and to the transformation tubes.
3. Add 50 µl plasmid DNA solution (concn. 1 µg/µl) to all transformation tubes, 50 µl 'control' DNA (e.g. salmon sperm DNA) to the mock transformation tube. Shake tubes gently.
4. Add 0.6 ml C100M/PEG 1500 solution, dropwise, with gentle shaking, to the transformation and mock transformation tubes. Allow transformation mixtures to stand for 10 min, with gentle shaking at intervals.
5. Fill all tubes to 10 ml with LW solution, taking care that the washing solution dilutes the viscous PEG mixtures rather than floating above them.
6. Centrifuge tubes at $50 \times g$ for 7 min, discard supernatants and disperse pellets.
7. For stable transformation, embed and culture protoplasts as in steps 2, 3 and 4 in *Culture and Regeneration* above, with the exception that a 2.5% agarose solution is used. For transient expression, suspend protoplasts in 1 ml liquid L1D0.5G protoplast medium, at density $1-2 \times 10^6$/ml, in tissue-culture quality plastic Petri dishes. Incubate cultures at 25 °C, in dim light.

Notes

1. Up to 4×10^6 protoplasts may be conveniently handled in a 12.5 ml centrifuge tube; for larger numbers larger tubes or replicate tubes can be used.
2. Two containing a selectable or scorable marker gene is replaced by other plasmid DNA or by salmon sperm or calf thymus DNA (sheared to approximately the size of the transforming plasmid by passage through an 18G needle). A further control, that of the addition of plasmid DNA, without a transformation treatment, may also be made.
3. DNA concentration is an important factor — both transient and stable transformation efficiencies are improved by high DNA: protoplast ratios. To avoid having to use large amounts of plasmid the volume of the transformation mixture is kept as low as possible. Routinely, 50 µg DNA are used for pellets of $1-3 \times 10^6$ protoplasts. With the volumes specified this gives a DNA concentration

of 0.125 µg/µl before the addition of PEG solution and a concentration of 0.05 µg/µl after PEG addition. Two further factors, namely DNA configuration and the use of carrier DNA can also affect transformation efficiency: in transient expression experiments linearised plasmid molecules and the addition of carrier DNA (e.g. sheared salmon sperm or calf thymus DNA at two- or threefold plasmid concentration) may increase expression levels, but it is not yet clear whether there are similar effects on stable transformation.

4. With a 40% w/v PEG solution, the final PEG concentration is ca. 25%. Higher PEG concentrations (30–35%) may be useful in transient expression experiments, but these levels significantly reduce protoplast plating efficiencies, which is disadvantageous in stable trdansformation experiments. Similarly, higher molecular weight PEGs (4000, 6000 or 8000) are generally more active than PEG 1500, but also do more damage to protoplasts.

7. The harder gel used for embedding transformed protoplast facilitates the subsequent transfer of agarose segments. For transient expression studies, protoplasts are cultured in liquid for ease of harvesting. Glucose is used instead of maltose in the medium as they float in the latter sugar and are therefore difficult to collect by centrifugation.

Harvesting and Selection

Steps in the procedure

1a. For transient expression studies, protoplasts (cells) are harvested after 36–48 h culture: scrape adhering cells from the floor of the Petri dish with a spatula and pipette the suspended cells into a 2.2 ml Eppendorf tube. Rinse the dish with 1 ml LW solution to collect any cells remaining, pipetting these into the Eppendorf tube.

2a. Centrifuge samples at 10000 rpm, at 4 °C for 15 min. Discard supernatants and store Eppendorf tubes containing pellets at −70 °C until analysis.

1b. For stable transformation, selection pressure is applied after ca. 12 d culture: assess protoplast plating efficiency, by counting microcalluses as described in note 4 under *Culture and Regeneration*.

2b. Cut agarose in dishes into four segments and transfer these to a 9 cm Petri dish. Pipette 5 ml of L1D2 suspension medium into the dish and add 0.18 ml of sterile G418 sulphate solution of concentration 1 mg/ml (final antibiotic concentration therefore 0.18 mg in total agarose and medium volume of 7.5 ml = 25 mg/l G418). Microcalluses from the untreated control are cultured with antibiotic selection, while the mock-transformed microcalluses are cultured without selection.

3b. Place dishes on a rotary shaker at 30 rpm, at a temperature of 25 °C, under dim light. After 10 d, remove old medium (using a narrow-mouth pipette, so as not to take any free-swimming microcallus) and replace with 5 ml of fresh medium, with or without 25 mg/l G418, as appropriate. Return dishes to shaker.

4b. After 20 d liquid selection culture, pick any calluses > 1 mm in diameter and transfer them to the surface of solid L2D2.5/G25 medium (containing 25 mg/l G418). Calluses showing sustained growth after two 25 d passages on this medium are almost certainly transformants this can be verified by assays for activity of marker genes, or by DNA analysis to demonstrate integration of the transferred gene(s).

5b. For regeneration from transformed callus, follow the steps 6 to 8 under *Culture and Regeneration*, with the modification that media should contain G418, although the concentration may be reduced to 10–15 mg/l, as organised tissues are generally more sensitive than callus to antibiotics.

Notes

1b. The timing of selection is critical: when too early the trauma will be too great for small, sensitive microcalluses and no resistant colonies will be recovered; when too late many large microcalluses will grow through the selection pressure and will subsequently be very difficult to select, even with very high antibiotic concentrations. Plating efficiency also affects selection: very high microcallus densities require higher antibiotic levels for control. As a guide, selection pressure can generally be applied when the majority of microcalluses contain 8 to 20 cells.

2b. Antibiotic concentration is again a critical factor which may be varied to suit the particular conditions. If selection is applied at the right time (see above), good discrimination can be achieved with 25 mg/l G418 or 50 mg/l hygromycin, but for late selection double these concentrations may be required for control. If the untreated control microcalluses are selected and the mock-transformed calluses left unselected then both the efficiency of the selection pressure and the viability of treated protoplasts may be assessed.

3b. Under the liquid selection conditions, microcallus growth is greatly improved by the slow shaking.

Media and Solutions

Media
All media sterilised by filtration through 0.22 μm membranes.

	L1	L2	L3
macrosalts (mg/l)			
NH_4NO_3	750	1500	200
KNO_3	1750	1750	1750
KH_2PO_4	200	200	200
$MgSO_4 \cdot 7H_2O$	350	350	350
$CaCl_2 \cdot 2H_2O$	450	450	450
microsalts (mg/l)			
$MnSO_4 \cdot H_2O$	15	15	15
H_3BO_4	5	5	5
$ZnSO_4$	7.5	7.5	7.5
KI	0.75	0.75	0.75
$Na_2MoO_4 \cdot H_2O$	0.25	0.25	0.25
$CuSO_4 \cdot 5H_2O$	0.025	0.025	0.025
$CoCl_2 \cdot 6H_2O$	0.025	0.025	0.025
iron (mg/l)			
Na_2EDTA	37	37	37
$FeSO_4 \cdot 7H_2O$	28	28	28
vitamins (mg/l)			
ascorbic acid	1	1	1
biotin	0.005	–	–
Ca-pantothenate	0.5	0.5	0.5
choline chloride	0.5	–	–
folic acid	0.2	–	–
myo-inositol	100	200	100
nicotinic acid	1	1	1
p-aminobenzoic acid	1	–	–
pyridoxine-HCl	1	1	1
riboflavin	0.1	–	–
thiamine-HCl	10	10	10
amino acids (mg/l)			
glutamine	750	750	750
proline	150	150	150
asparagine	100	100	100
sugars (g/l)			
maltose	50	30	30
pH	5.7	5.7	5.7
osmolarity (mOsm)	~300	~200	~200

- L1D2: L1 medium as above, with 2 mg/l 2,4-D.
- L1D0.5 protoplast medium: L1 medium, with 0.5 mg/l 2,4-D and 180 g/l maltose ($\sim$700 mOsm). Made at $\times 2$ concentration and mixed 1:1 with double-concentrated SeaPlaque agarose (see below).
  L1D0.5G protoplast medium: L1 medium, with 0.5 mg/l 2,4-D, and with 90 g/l glucose instead of maltose. Used as liquid medium, $\times 1$ concentration.
- L3D0.5B1: L3 medium, with 0.5 mg/l 2,4-D and 1 mg/l BAP, solidified with 0.4% (final concentration) Sigma 1A agarose.
- L3D2.5: L3 medium, with 2 mg/l 2,4-D, and 0.4% Sigma 1A agarose.
- L3 – : L3 medium, without hormones, with 0.4% Sigma 1A agarose.
- L2D2.5/G25: L2 medium, with 2 mg/l 2,4-D and 25 mg/l G418 sulphate (Gibco), with 0.4% Sigma 1A agarose.

Solutions

(Unless stated otherwise, all solutions sterilised by filtration.)

- SeaPlaque agarose: made as $\times 2$ solution (2% for regeneration, 2.5% for transformation) in double-distilled water. Autoclave to sterilise.
- Sigma 1A agarose: $\times 2$ solution; 0.8% in double-distilled water. Autoclave to sterilise.
- LW (washing) solution: macro- and microsalts and amino acids as L1 medium, 100 g/l mannitol, pH 5.7, $\sim$700 mOsm. Autoclave to sterilise.
- Enzyme solution: 1% Onozuka RS cellulase (Yakult), 0.5% macerozyme (Serva), 0.05% pectolyase Y23 (Seishin), dissolved in LW solution (stir at least 2 hr), pH 5.7, $\sim$725 mOsm.
- C100M: 15 g/l $CaCl_2 \cdot 2H_2O$ ($\sim$100 mM), 1 g/l MES, 70 g/l mannitol, pH 5.7, $\sim$650 mOsm.
- C100M/PEG 1500:40% w/v Merck PEG 1500, dissolved in C100M solution, pH 7.0.
- DNA solution: sterile (95% alcohol) plasmid DNA, dissolved in autoclaved 10:1 Tris:EDTA buffer, concentration 1 µg/µl.
- G418 stock: 1 mg/ml G418 sulphate (Gibco) in water.

References

1. Baratz S, Breiman A (1991) Parameters affecting transient expression of β-glucuronidase in protoplasts derived from *Hordeum bulbosum* cell suspensions (submitted).
2. Gasser CS, Fraley RT (1989) Genetically engineering plants for crop improvement. Science 244: 1293–1299.
3. Gordon-Kamm WJ, Spencer TM, Mangona ML, Adams TR, Daines RJ, Start WG, O'Brien JV, Chambers SA, Adams WR, Willets NG, Rice TB, Mackey CJ, Krueger RW, Kausch AP, Lemaux PG (1990). Transformation of maize cells and regeneration of fertile transgenic plants. The Plant Cell 2: 603–618.
4. Jähne A, Lazzeri PA, Jäger-Gussen M, Lörz H (1991) Plant regeneration from embryogenic suspensions derived from anther culture of barley. Theor Appl Genet 82: 74–80.
5. Jähne A, Lazzeri PA, Lörz H. (1991) Regeneration of fertile plants from protoplasts derived from embryogenic cell suspensions of barley (*Hordeum vulgare* L.) Plant Cell Rep (in press).
6. Junker B, Zimny J, Lührs R, Lörz H (1987) Transient expression of chimaeric genes in dividing and nondividing cereal protoplats after PEG-induced DNA uptake. Plant Cell Rep 6: 329–332.
7. Karp A and Lazzeri PA (1991) Regeneration, stability and transformation of barley. In: Barley: Genetics, Molecular Biology and Biotechnology, ed. PR Shewry. C.A.B. International, Wallingford (in press).
8. Klein TM, Wolf ED, Wu T, Sanford JC (1987) High velocity microprojectiles for delivering nucleic acids into living cells. Nature 327: 70–73.
9. Koblitz H (1976) Isolation and cultivation of protoplasts from callus cultures of barley. Biochem Physiol Pflanz 170: 287–293.
10. Kyozuka J, Mayasmi Y, Shimamoto K (1987) High frequency of plant regeneration from rice protoplasts by novel nurse culture methods. Mol Gen Genet 206: 408–13.
11. Lazzeri PA, Brettsschneider R, Lührs R, Lörz H (1991) Stable transformation of barley via PEG-induced direct DNA uptake into protoplasts. Theor Appl Genet 81: 437–444.
12. Lazzeri PA, Kollmorgen J, Lörz H (1990) *In vitro* technology. In Reproductive Versatility in the Grasses ed. GP Chapman. Cambridge University Press, pp 182–219.
13. Lazzeri PA and Lörz H (1988) *In vitro* Genetic Manipulation of Cereals and Grasses. Adv Cell Cult 8: 291–325.
14. Lazzeri PA and Lörz H (1990) Regenerable suspension and protoplast cultures of barley and stable transformation via DNA uptake into protoplasts. In: Genetic Engineering of Crop Plants eds. GW Lycett and D Grierson. Butterworths. London, pp 231–237.
15. Lee BT, Murdoch K, Topping J, Kreis M, Jones MGK (1989) Transient expression in aleurone protoplasts isolated from developing caryopses of barley and wheat. Plant Mol Biol 13: 21–19.
16. Lörz H, Göbel E, Brown PTH (1988) Advances in tissue culture and progress towards genetic transformation of cereals. Plant Breed 100: 1–25.
17. Lührs R and Lörz H (1988) Initiation of morphogenic cell- suspension and protoplast cultures of barley. Planta 175: 71–81.
18. Paszkowski J, Shillito RD, Saul MW, Mandak V, Hohn B, Potrykus I (1985) Direct gene transfer to plants. EMBO J 3: 2717–2722.
19. Potrykus I (1990) Gene transfer to cereals: an assessment. Bio/technology 8: 535–542.
20. Shimamoto K, Terada R, Izawa T, Fujimoto H (1989) Fertile transgenic rice plants regenerated from transformed protoplasts. Nature 338: 274–276.
21. Teeri TH, Patel GH, Aspegren K, Kauppinen V (1989) Chloroplast targeting of *neomycin phosphotransferase* II with a pea transit peptide in electroporated barley mesophyll protoplasts. Plant Cell Rep 8: 187–190.
22. Vasil I (1988) Progress in the regeneration and genetic manipulation of cereal crops. Bio/technology 6: 397–402.
23. Yan Q, Zhang X, Shi J, Li J (1990) Green plant regeneration from protoplasts of barley (*Hordeum vulgare* L.). Kexue Tongbao 35: 1581–1583.

Plant Tissue Culture Manual **B11**: 1–18, 1993.

Agrobacterium-mediated transformation of potato stem and tuber tissue, regeneration and PCR screening for transformation

JAMES P. SPYCHALLA & MICHAEL W. BEVAN
Molecular Genetics Department, Cambridge Laboratory, Centre for Plant Science Research, Colney Lane, Norwich, NR4 7UJ, U.K.

Introduction

Potato (*Solanum tuberosum* L.) callus tissue culture was first reported by Steward and Caplin in 1951 [25]. Potato has since proven to be very amenable to a wide range of tissue culture and regeneration schemes (reviewed in [16]). The first reports of *Agrobacterium*-mediated transformation and regeneration of potato utilised shoot cultures [19–21]. Since then several methods which transform either leaf tissue [1, 5, 7, 12, 13, 22, 23, 27–31], stem internodes [18, 29–31] or tuber discs [6, 9, 24, 26] have been reported (reviewed in [17]).

We transform potato with the tuber disc method of Sheerman and Bevan [24]. The tuber disc method is facile but requires tubers recently harvested from the field or glasshouse. We also transform potato with the *in vitro*-grown stem method of Newell *et al.* [18]. The *in vitro* grown stem method is more labour intensive but avoids any seasonal production problems of potato tubers. Both methods begin to yield rooted transgenic plants in 6–8 weeks.

The choice of cultivar, type and physiological status of explant tissue, transformation vector, and the *Agrobacterium* strain utilised are major variables in any attempted transformation. In our laboratory we use the binary plant vector pBIN400, which is a derivative of the pBIN19 vector [4]. In pBIN400 a ColE1 origin of replication has been added to the plasmid and a spectinomycin resistance gene from the bacterial transposon Tn7 has replaced the kanamycin resistance gene of pBIN19 for selection in *Agrobacterium*. The mutant NPT II gene found in pBIN19 [32] still functions adequately as a selectable marker for potato transformation in pBIN400. Transformation of the potato cultivar Desiree is accomplished with the *Agrobacterium* strain T37 containing a modified nopaline-type avirulent helper plasmid pTiT37-SE (Km^r). The disarmed pTiT37-SE was a gift from the Monsanto Company constructed according to the method of Fraley *et al.* [8]. The efficiencies of potato transformation using other combinations of cultivar, source of explant, vector and *Agrobacterium* strain are likely to differ. We utilise a two step screen for successful potato transformation. Regenerated shoots are first screened for the ability to form roots on kanamycin containing medium and secondly for the presence of PCR amplifiable T-DNA. In this chapter we detail the steps of the transformation, regeneration and screening procedures from our laboratory.

Transformation of potato tuber discs

Procedure (based upon [24]

A. Preparation of materials

Steps in the procedure

Preparation of Agrobacterium
1. Maintain the desired T37 :: pTiT37-SE :: pBIN400 *Agrobacterium* on minimal medium plates (0.07 mM $CaCl_2$, 0.02 mM $FeSO_4$, 33 mM KH_2PO_4, 46 mM K_2HPO_4, 0.8 mM $MgSO_4$, 0.01 mM $MnCl_2$, 19 mM NH_4Cl, 11 mM glucose, 3 g/l Difco BiTek agar, 50 μg/ml kanamycin and 100 μg/ml spectinomycin).
2. Day 1. Inoculate a 10 ml culture of Luria broth containing spectinomycin (100 μg/ml) and kanamycin (50 μg/ml) with a toothpick smear from a fresh plate of the *Agrobacterium*. Grow the culture overnight in a shaker at 30 °C.
3. Day 2. Inoculate a 100 ml culture of Luria broth containing spectinomycin (100 μg/ml) and kanamycin (50 μg/ml) with 1 ml of the previous overnight culture and again grow overnight in a shaker at 30 °C.
4. Day 3. Harvest the *Agrobacterium* from 50 ml of the overnight culture by centrifugation in a 100 ml vessel. Resuspend the *Agrobacterium* in 100 ml of MS30 medium (Table 1). The *Agrobacterium* suspension is now ready for the cocultivation step.

Preparation of tobacco feeder plates
5. Day 2. A tobacco cell suspension culture is continuously propagated on a weekly basis in fresh MS-Tobacco media (Table 1). Pipette one ml of 3–5 day old tobacco cell suspension culture onto 9 cm Petri plates of 3C5ZR medium (Table 1) and cover with a disc of # 1 Whatman filter paper. The feeder plates are now ready for the cocultivation step.

Surface sterilization and harvesting of tuber discs
6. Day 3. Wash the Desiree tubers (2–6 cm diameter) in water.
7. To facilitate surface sterilization, peel the potatoes to remove the periderm, and rinse the tubers in water.
8. Place the tubers in a large 1 l screw top container, and soak for 1 min in 70% ethanol.
9. Remove the ethanol and soak for 10 min in 0.7% sodium hypochlorite plus 0.05% Tween-20. All steps beyond this point must be performed in a sterile flow hood.
10. Rinse the tubers in sterile water 3 times for 30 min each rinse, and return the tubers to sterile water. The tubers may now remain in water several hours without harm.

Table 1. Composition of media used in potato stem and tuber disc transformations (mg/L)

	MS30	MS-Potato and MS-Select[1]	MS-Tobacco	1/10th P1	P1 and P2[2]	3C5ZR and 3C5ZR-Select[3]
Inorganic Salts						
$CaCl_2 \cdot 2H_2O$	440					
$CoCl_2 \cdot 6H_2O$	0.025					
$CuSO_4 \cdot 5H_2O$	0.025					
FeNa·EDTA	36.7					
H_3BO_4	6.2	as per MS30	as per MS30	use 1/10th of the inorganic salts listed in MS30	as per MS30	as per MS30
KH_2PO_4	170					
KI	0.83					
KNO_3	1900					
$MgSO_4 \cdot 7H_2O$	370					
$MnSO_4 \cdot 4H_2O$	22.3					
$Na_2MoO_4 \cdot 2H_2O$	0.25					
NH_4NO_3	1650					
$ZnSO_4 \cdot 7H_2O$	8.6					
Organics						
sucrose	30000	20000	30000	1000	30000	30000
inositol		100	100	10	100	
glycine		2.0		0.2	2.0	
Vitamins[4]						
folic acid					0.5	
nicotinic acid		0.5	1.0	0.05	0.5	0.5
thiamine·HCl		0.1	10	0.01	0.5	0.1
pyridoxine·HCl		0.5	1.0	0.05	0.5	0.5
biotin					0.05	
Antibiotics[5]						
carbenicillin		500[1]			500	500[3]
kanamycin		100[1]			100	100[3]
Phytohormones[6]						
2,4-D			1.0			
6-BAP			0.2		3.0[2]	
NAA					0.01[2]	
GA_3					0.3[2]	
IAA-aspartic acid						0.87
zeatin riboside						1.75
Gel Agent						
Gel-Rite		2000				
Difco Bacto-Agar				8000	8000	8000
Final pH (KOH)	5.9	5.8	5.9	5.6	5.6	5.9

[1] MS-Potato does not contain antibiotics, while MS-Select contains both carbenicillin and kanamycin.

[2] P1 and P2 differ solely in phytohormone composition. P1 contains only BAP and NAA, while P2 contains only GA_3.

[3] C5ZR does not contain antibiotics, while 3C5ZR-Select contains both carbenicillin and kanamycin.

[4] Vitamins are dissolved in H_2O as $1000 \times$ concentrated stocks, stored at $-20\,°C$, and added to the medium prior to autoclaving.

[5] Antibiotics are dissolved in H_2O as $500 \times$ concentrated stocks, filter sterilized (0.22 μm) and stored at $-20\,°C$.

[6] Phytohormones are dissolved in fresh DMSO as $10,000 \times$ concentrated stocks, filter sterilized (0.22 μm) and stored at $-20\,°C$. Abbreviations; 2,4-D = 2,4-dichlorophenoxyacetic acid; 6-BAP = 6-benzylaminopurine; NAA = naphthaleneacetic acid; GA_3 = gibberellic acid A_3; IAA-aspartic acid = indole-3-acetyl-L-aspartic acid.

11. Remove as many cores as possible from each tuber with a sterile 1 cm cork borer.
12. Submerge each core immediately into MS30 medium in 13.5 cm Petri plate.
13. Remove and discard the outermost 5 mm from the ends of each core.
14. Slice the remainder of the core into 2 mm discs and keep the discs submerged in MS30 medium.

B. Transformation and regeneration

Steps in the procedure

Cocultivation
1. Day 3. Remove the MS30 medium from the tuber discs and replace it with the *Agrobacterium* suspension, swirl the plates well and allow to stand for 30 min.
2. Remove each individual disc and briefly touch it against the side of the large Petri plate to remove excess bacterial suspension.
3. Place 10 discs on each tobacco feeder plate, and seal the feeder plates with 3M micropore surgical tape.
4. Cocultivate the *Agrobacterium* and tuber discs for two days at 25 °C with a 16 hour day-length and irradiance of 30–100 μmol m^{-2} s^{-1}.

Regeneration of transformed shoots
5. Day 5. Transfer the tuber discs without blotting to Petri plates containing 3C5ZR-Select medium (Table 1).
6. Place 10 discs per plate, seal with 3M micropore surgical tape and incubate the tuber discs at 20 °C, with a 16 hour daylength and irradiance of 30–100 μmol m^{-2} s^{-1}. The carbenicillin in the medium selects against the *Agrobacterium* and the kanamycin selects for transformed plant cell growth.
7. Day 26 and 47. Transfer the tuber discs to fresh 3C5ZR-Select medium.

Screening for transformed plantlets by use of the rooting assay
8. Day 26+. With no visible callus formation small green bumps will appear on the upper surface and sides of the tuber disc in as little as 4 weeks. After 6 weeks there are often shoots greater than 5 mm in height with several shoots per disc (Fig. 1A). Conversely some discs regenerate no shoots. Resist the temptation to harvest the young shoots until they have reached a height of at least 1 cm. A larger shoot is more likely to survive harvest and the rooting assay. Sever the base of the shoots from the tuber discs and place 20–30 shoots into a Magenta GA-7 pot containing MS-Select medium (Table 1). Expression of the NPT II gene in transformed plants will confer an enhanced ability to form roots under kanamycin selection.
9. Day 40+. The shoots will demonstrate a range of responses on the MS-Select medium. In this primary rooting assay some shoots will grow rapidly and form vigorous *de novo* roots. *De novo* roots are those which arise form the lower stem just above cut surface, not from axillary meristems. Other shoots will grow

rapidly and develop roots from axillary meristems. The formation of roots from axillary meristems of wild type tissues is insensitive to kanamycin. Non-transformed shoots will also form *de novo* roots, but these roots tend to be very short and fewer in number. The base of non-transformed shoots also tends to develop a deep red colour. In reality the primary rooting assay is a poor screen. When each shoot reaches a height of about 6–8 cm subculture the uppermost 1.5 cm of the shoot which contains the apical tip to MS-Select medium (Table 1). In the secondary rooting screen transformed shoots will generate typically 6–8 vigorous *de novo* roots from the base of the stem section (Fig. 1C). The uppermost section of a shoot containing the apical tip always forms the most vigorous roots. When the plants again reach a height of 6–8 cm a single leaf can also be removed from the shoot and screened by the PCR for transformation. Tuber discs produce shoots at an average rate of 70%. Approximately 8% of these regenerated shoots will form roots under kanamycin selection in the secondary rooting screen. The net result is that tuber discs transform and regenerate shoots which root under kanamycin selection at the rate of 5.6%.

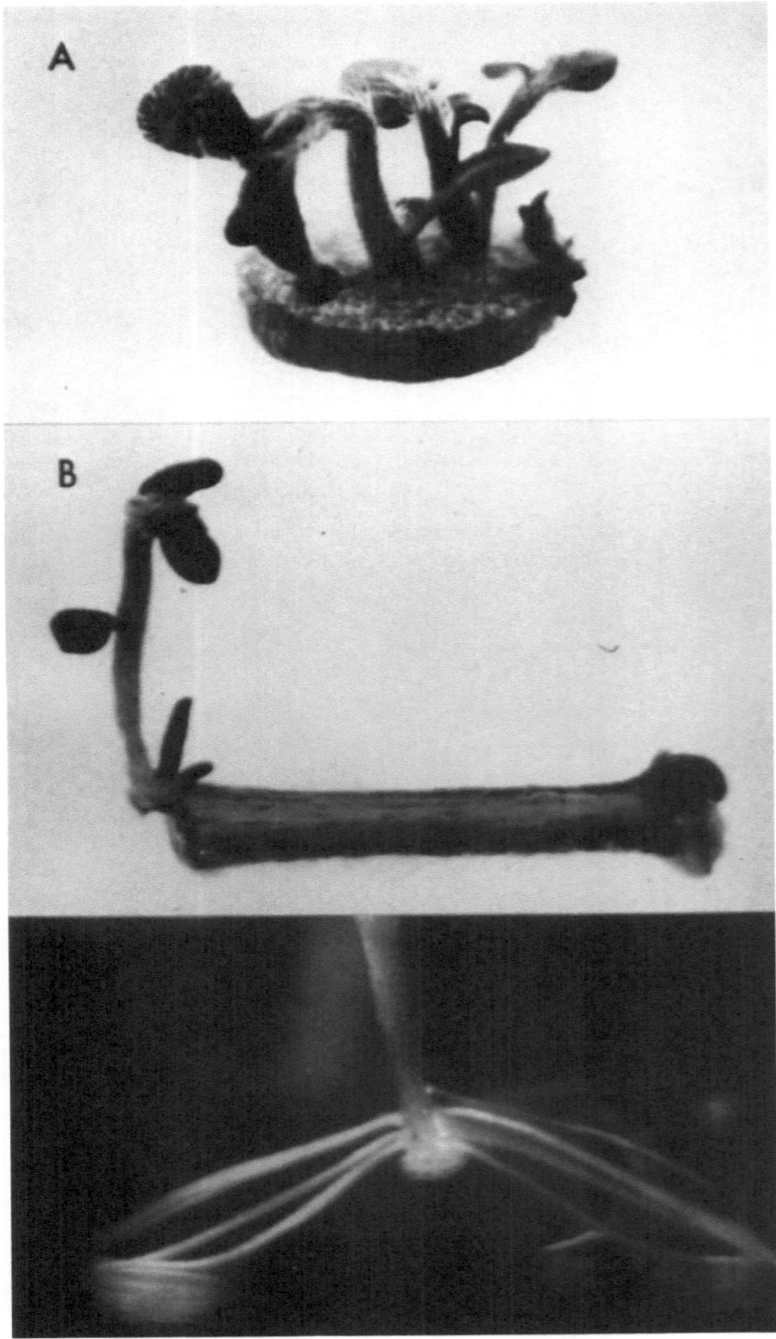

Fig. 1. A) Shoots regenerated from a potato tuber disc after 7 weeks. B) Shoot regenerated from a potato stem segment after 7 weeks. C) *De novo* root formation from a transformed shoot under kanamycin selection.

Transformation of potato stems

Procedure (based upon [18]

A. Preparation of materials

Steps in the procedure

In vitro propagation of stock plants
1. Remove about 200 non-sterile stem sections approximately 3 cm in length, each containing an axillary meristem from glasshouse-grown Desiree plants. The stem sections are surface sterilized exactly as the tubers were sterilized in section II.
2. After surface sterilization place each stem section into a test tube of MS-Potato medium (Table 1) and grow at 20 °C with a 16 hour daylength and irradiance of 30–100 μmol m^{-2} s^{-1}. Fungal contamination of the explants can be severe.
3. After 2–3 weeks subculture any newly arisen non-contaminated shoots to Magenta GA-7 pots of MS-Potato medium. An important factor for successful stem transformation is the vigour of the *in vitro* grown stock plants. Vigourous stock plants can be maintained indefinitely by subculturing every 2–3 weeks in MS-Potato medium (Table 1). Subculture only the uppermost 1.5 cm of the shoots which contains the apical meristem. Place 5 apical shoot explants in each Magenta pot. Root growth, leaf size and stem thickness are much greater than shoots arising from axillary meristems. Occasionally a faint bacterial contamination is observed in the MS-Potato medium near the roots. Never attempt to subculture or transform plants with traces of bacterial contamination.

Preparation of Agrobacterium
4. Day 1. From a fresh plate of the desired T37::pTiT37-SE::pBIN400 *Agrobacterium* streak new minimal medium plates extensively so that the *Agrobacterium* will grow over the entire surface of the plate. Incubate the plates at 20 °C. The *Agrobacterium* will be ready for use in the cocultivation step in 3 days.

Harvesting of in vitro *stem tissue*
5. Day 4. Remove the cover from the Magenta pot containing the 2 week old Desiree plants. The apical tip is usually subcultured at this time.
6. Remove all the leaves from the stems by cutting the petiole within a few mm of the main stem.
7. Remove the stem and place into MS30 medium in a 13.5 cm Petri plate.
8. Cut the stem into internodal sections approximately 1 cm in length while submerged under MS30 medium.

B. Transformation and regeneration

Steps in the procedure

Cocultivation
1. Day 4. With forceps remove 20 internodal stem sections from the MS30 medium and place onto sterile # 1 Whatman filter paper.
2. Gently hold each stem section and smear the cut surfaces with *Agrobacterium* from the confluent plates.
3. Immediately place the inoculated stem section onto the surface of well-dried 1/10th P1 medium (Table 1) in a 9 cm Petri plate. Place twenty stem sections on each Petri plate.
4. Pipette 1.5 ml of 3—5 day old tobacco stem cell suspension culture onto the stem segments.
5. Wet a # 1 Whatman filter paper disc with water, drain excess water from the paper disc and cover the stem segments.
6. Seal the plates with 3M micropore surgical tape and cocultivate the *Agrobacterium* and stem sections at 20 °C with a 16 hour daylength and irradiance of 30—100 μmol m^{-2} s^{-1} for two days.

Regeneration of transformed shoots
7. Day 6. Transfer the stem sections to Petri plates of P1 callus forming medium (Table 1). Place 20 stem sections per plate, seal with 3M micropore surgical tape and incubate the stem sections at 20 °C with a 16 hour daylength and irradiance of 30—100 μmol m^{-2} s^{-1}. The P1 medium promotes callus formation, the carbenicillin selects against the *Agrobacterium* and the kanamycin selects for transformed plant cell growth. The ends of healthy stem sections swell and turn red after a week.
8. Day 34. Transfer the stem segments to Petri plates of P2 shoot inducing medium (Table 1).
9. Day 62. Transfer any remaining stem segments which still appear to have a possibility of shooting to fresh P2 medium.

Screening for transformed plants by use of the rooting assay
10. Day 48+. A small amount of callus forms on most of the stem sections, and about 14% of the stem sections produce shoots (Fig. 1B). The harvest of regenerated shoots and the root-forming screen for transformation are now identical to the procedures described for tuber discs. Approximately 26% of the regenerated shoots will form roots under kanamycin selection. The net result is that *in vitro*-grown stem sections transform and regenerate shoots which root under kanamycin selection at the rate of 3.6%.

Solutions
- Minimal medium plates (for *Agrobacterium*)
    - 0.07 mM $CaCl_2 \cdot 2H_2O$
    - 0.02 mM $FeSO_4$
    - 33 mM KH_2PO_4
    - 46 mM K_2HPO_4
    - 0.8 mM $MgSO_4$
    - 0.01 mM $MnCl_2$
    - 19 mM NH_4Cl
    - 11 mM glucose
    - 3 g/l Difco BiTek agar

Autoclave to sterilize, and add filter-sterilized antibiotics:
    - 50 μg/ml kanamycin
    - 100 μg/l spectinomycin

- Luria broth
    - 1% w/v Bacto tryptone
    - 0.5% w/v Yeast extract
    - 1% w/v NaCl

pH 7.0

PCR screening for transformation

Procedure

Steps in the Procedure

DNA extraction (based upon [14])
1. Pinch approximately 10 mg of potato leaf directly into a 1.5 ml microfuge tube.
2. Grind the tissue with a small pestle into 100 µl of DNA extraction buffer.
3. Add 100 µl of $CHCl_3$, vortex at high speed for 10 s, and heat for 30 min at 65 °C.
4. Spin in a microfuge for 5 min and transfer the upper aqueous layer to a fresh tube.
5. Precipitate the nucleic acids with an equal volume of isopropanol and pellet by centrifugation.
6. Wash the pellet once in 70% ethanol and resuspend the pellet in TE buffer. Extreme care must be used with all samples to avoid cross contamination.

The PCR Reaction (based upon [15])
One particular gene construct which we have inserted into potato plants is a 0.56 kb EcoRI fragment of a potato tuber UDP-glucose pyrophosphorylase cDNA oriented in the antisense direction and fused to a 3.5 kb patatin promoter [10] (Fig. 2). Two deoxynucleotide primers 500 bp apart corresponding to a portion of the UDP-glucose pyrophosphorylase cDNA [11] and the NOS terminator [3] were used to screen for the presence of the T-DNA. The UDP-glucose pyrophosphorylase primer (P1) sequence was 5′ ● TTT ● CTC ● CAG ● TCA ● ATG ● TGT ● TGC ● 3′. The NOS terminator primer (P2) sequence was 5′ ● TAG ● ATG ● ACA ● CCG ● CGC ● GCG ● ATA ● 3′. In order to check for contamination of samples with pBIN plasmid DNA, the PCR was performed with two different primers 460 bp apart corresponding to a portion of the vector beyond the right border of the T-DNA [3] and to a portion of the NPT II protein coding region [2]. The non-T-DNA primer (P3) sequence was 5′ ● CGC ● TCT ● TTT ● CTC ● TTA ● GGT ● TTA ● 3′. The NPT II primer (P4) sequence was 5′ ● TTG ● TGC ● CCA ● GTC ● ATA ● GCC ● GAA ● 3′. The 25 µl PCR mixtures contained 50 mM KCl, 10 mM Tris ● HCl pH 8.4, 1.5 mM $MgCl_2$, 2.5 µg gelatin, 200 µM each dNTP, 4 µM each primer, 1 µl of potato DNA extract or 100 pg plasmid DNA, and 1.25 units of AmpliTaq (Perkin Elmer Cetus, Norwalk, CT). The reaction mixture was overlaid with paraffin oil. The reaction temperature conditions were 4 min at 94 °C, followed by 30 cycles of 1 min at 94 °C, 1 min at 55 °C and 1 min at 72 °C, followed by a final extension for 7 min at 72 °C. 15 µl (plant samples) or 3 µl (plasmid samples) of the PCR products were loaded to a 1.5% agarose gel, electrophoresed and stained with ethidium bromide to screen for the presence of the appropriate sized DNAs. Approximately 40% of the regenerated plants which form vigorous *de novo* roots under kanamycin selection tested positive for the presence of the desired T-DNA with the verification primers P1 and P2 (Fig. 3,

lane 3) and negative for plasmid contamination with primers P3 and P4 (Fig. 3, lane 4).

Solutions

DNA extraction buffer
— 0.14 M sorbitol
— 0.22 M Tris·HCl pH 8.0
— 0.022 M EDTA
— 0.8 NaCl
— 0.8% w/v hexadecyltrimethylammonium bromide
— 1% w/v N-lauroylsarcosine

TE buffer
— 10 mM Tris · HCl pH 8.0
.— 10 mM EDTA

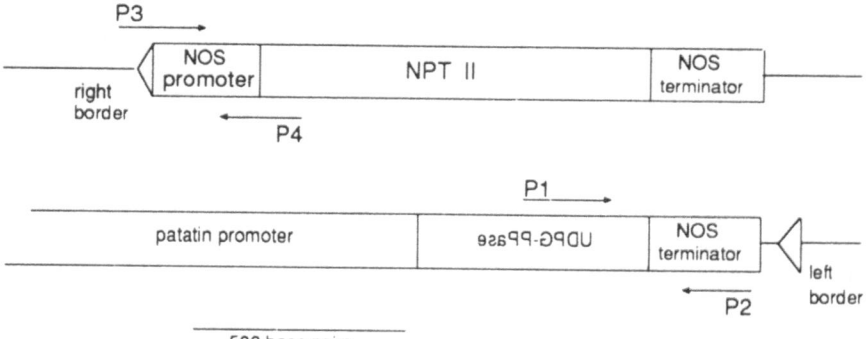

Fig. 2. A schematic of a patatin promoter-antisense UDP-glucose pyrophosphorylase fusion inserted into pBIN 400, showing the location of the primer sequences (P1-P4) used in the PCR screen for transformation (see text for details).

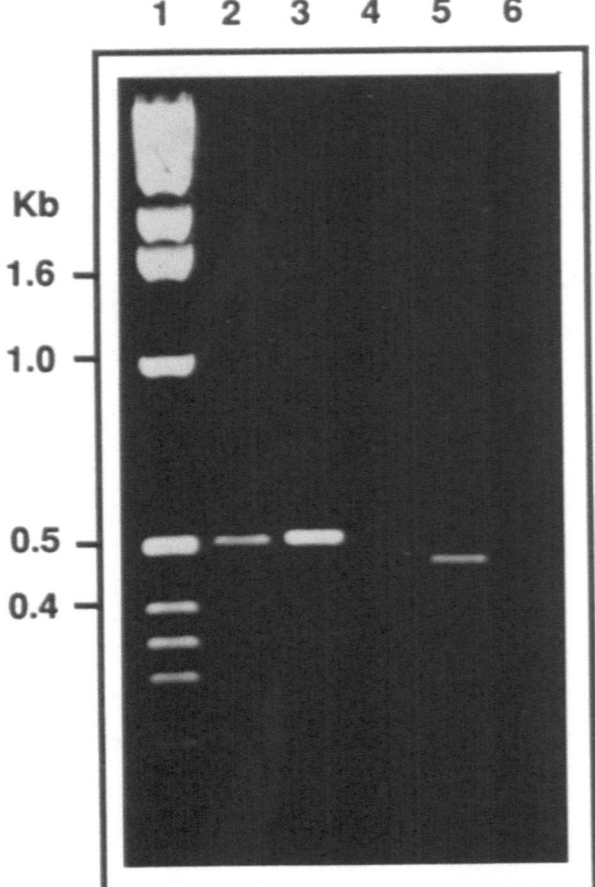

Fig. 3. A 1.5% agarose gel of ethidium bromide stained PCR products. Lane 1, molecular weight markers; lane 2, PCR positive plasmid DNA using verification primers P1 and P2, 3 μl; lane 3, PCR positive DNA from transformed plant using verification primers P1 and P2, 15 μl; lane 4, PCR negative DNA from transformed plant using contamination primers P3 and P4, 15 μl; lane 5, PCR positive plasmid DNA using contamination primers P3 and P4, 3 μl; lane 6, PCR negative DNA from wild type plant using verification primers P1 and P2, 15 μl.

References

1. An, G, Watson, BD, Chiang, CC (1986) Transformation of tobacco, potato, and *Arabadopsis thaliana* using a binary Ti vector system. Plant Physiol 81: 301–305.
2. Beck, E, Ludwig, G, Auerswald, EA, Reiss, B, Schaller, H (1982) Nucleotide sequence and exact localization of the neomycin phosphotransferase gene from transposon Tn5. Gene 19: 327–336.
3. Bevan, M, Barnes, WM, Dell-Chilton, M (1983) Structure and transcription of the nopaline synthase gene region of T-DNA. NAR 11: 369–385.
4. Bevan, M (1984) Binary *Agrobacterium* vectors for plant transformation. NAR 12: 8711–8721.
5. De Block, M (1988) Genotype-independent leaf disc transformation of potato (*Solanum tuberosum*) using *Agrobacterium tumefaciens*. Theor Appl Genet 76: 767–774.
6. Eckes, P, Rosahl, S, Schell, J, Willmitzer, L (1986) Isolation and characterization of a light-inducible, organ-specific gene from potato and analysis of its expression after tagging and transfer into tobacco and potato shoots. Mol Gen Gen 205: 14–22.
7. Fladung, M (1990) Transformation of diploid and tetraploid potato clones with the *rol C* gene of *Agrobacterium rhizogenes* and characterization of transgenic plants. Plant Breeding 104: 295–304.
8. Fraley, RT, Rogers, SG, Horsch, RB, Eichholtz, DA, Flick, JS, Fink, CL, Hoffman, NL, Sanders, PR (1985) The SEV system: A new disarmed Ti plasmid vector system for plant transformation. Bio/Technology 3: 629–635.
9. Ishida, BK, Snyder Jr., GW, Belknap, WR (1989) The use of in vitro-grown microtuber discs in *Agrobacterium*-mediated transformation of Russet Burbank and Lemhi Russet potatoes. Plant Cell Reports 8: 325–328.
10. Jefferson, R, Goldsbrough, A, Bevan, M (1990) Transcriptional regulation of a patatin-1 gene in potato. Plant Mol Biol 14: 995–1006.
11. Katsube, T, Kazuta, Y, Mori, H, Nakano, K, Tanizawa, K, Fukui, T (1990) UDP-glucose pyrophosphorylase from potato tuber: cDNA cloning and sequencing. J Biochem 108: 321–326.
12. Keil, M, Sánchez-Serrano, JJ, Willmitzer, L (1989) Both wound-inducible and tuber-specific expression are mediated by the promoter of a single member of the potato proteinase inhibitor II gene family. EMBO 8: 1323–1330.
13. Knapp, S, Coupland, G, Uhrig, H, Starlinger, P, Salamini, F (1988) Transposition of the maize transposable element *Ac* in *Solanum tuberosum*. Mol Gen Genet 213: 285–290.
14. Lassner, MW, Peterson, P, Yoder, JI (1989) Simultaneous amplification of multiple DNA fragments by polymerase chain reaction in the analysis of transgenic plants and their progeny. Plant Mol Biol Rep 7: 116–128.
15. McGarvey, R, Kaper, JM (1991) A simple and rapid method for screening transgenic plants using the PCR. BioTechniques 11: 428–432.
16. Miller, SA, Lipschutz, L (1984) Potato. In: Ammirato, PV, Evans, DA, Sharp, WR, Yamada, Y (Eds.) Handbook of Plant Cell Culture: Crop Species, Vol 3, pp 291–326, Macmillan Publishing Company, New York.
17. Mitten, DH, Horn, M, Burrell, MM, Blundy, KS (1990) Strategies for potato transformation and regeneration. In: Vayda, ME, Park, WD (Eds.) Molecular and Cellular Biology of the Potato, pp 181–191, C.A.B. International, Wallingford.
18. Newell, CA, Rozman, R, Hinchee, MA, Lawson, EC, Haley, L, Sanders, P, Kaniewski, W, Tumer, NE, Horsch, RB, Fraley, RT (1991) *Agrobacterium*-mediated transformation of *Solanum tuberosum* L. cv. "Russet Burbank". Plant Cell Reports 10: 30–34.
19. Ooms, G, Karp, A, Roberts, J (1983) From tumour to tuber; tumour cell characteristics and chromosome numbers of crown gall-derived tetraploid potato plants (*Solanum tuberosum* cv. "Maris Bard"). Theor Appl Genet 66: 169–172.
20. Ooms, G, Karp, A, Burrell, MM, Twell, D, Roberts, J (1985) Genetic modifications of potato development using Ri T-DNA. Theor Appl Genet 70: 440–446.

21. Ooms, G, Burell, MM, Karp, A, Bevan, M, Hille, J (1987) Genetic transformation in two potato cultivars with T-DNA from disarmed *Agrobacterium*. Theor Appl Genet 73: 744–750.

22. Rocha-Sosa, M, Sonnewald, U, Frommer, W, Stratmann, M, Schell, J, Willmitzer, L (1989) Both developmental and metabolic signals activate the promoter of a class I patatin gene. EMBO 8: 23–29.

23. Shahin, EA, Simpson, RB (1986) Gene transfer system for potato. HortScience 21: 1199–1201.

24. Sheerman, S, Bevan, MW (1988) A rapid transformation method for *Solanum tuberosum* using binary *Agrobacterium tumefaciens* vectors. Plant Cell Reports 7: 13–16.

25. Steward, FC, Caplin, SM (1951) A tissue culture from potato tuber: The synergistic action of 2,4-D and of coconut milk. Science 113: 518–520.

26. Stiekema, WJ, Heidekamp, F, Louwerse, JD, Verhoeven, HA, Dijkhuis, P (1988) Introduction of foreign genes into potato cultivars Bintje and Désirée using an *Agrobacterium tumefaciens* binary vector. Plant Cell Reports 7: 47–50.

27. Tavazza, R, Tavazza, M, Ordas, RJ, Ancora, G, Benvenuto, E (1988) Genetic transformation of potato (*Solanum tuberosum*): An efficient method to obtain transgenic plants. Plant Science 59: 175–181.

28. Visser, RGF, Jacobsen, E, Witholt, B, Feenstra, WJ (1989) Efficient transformation of potato (*Solanum tuberosum* L.) using a binary vector in *Agrobacterium rhizogenes*. Theor Appl Genet 78: 594–600.

29. Visser, RGF, Jacobsen, E, Hesseling-Meinders, A, Schans, MJ, Witholt, B, Feenstra, WJ (1989) Transformation of homozygous diploid potato with an *Agrobacterium tumefaciens* binary vector system by adventious shoot regeneration on leaf and stem segments. Plant Molecular Biology 12: 329–337.

30. Visser, RGF (1991) Regeneration and transformation of potato by *Agrobacterium tumefaciens*. In: Lindsay, K (Ed.) Plant Tissue Culture Manual: Fundamentals and Applications, Vol B5, pp 1–9, Kluwer Academic Publishers, Dordrecht.

31. Wenzler, H, Mignery, G, May, G, Park, W (1989) A rapid and efficient transformation method for the production of large numbers of transgenic potato plants. Plant Science 63: 79–85.

32. Yenofsky, RL, Fine, M, Pellow, JW (1990) A mutant neomycin phosphotransferase II gene reduces the resistance of transformants to antibiotic selection pressure. Proc Natl Acad Sci USA 87: 3435–3439.

Plant Tissue Culture Manual **B12**, 1–20, 1996.

Production of fertile transgenic wheat by microprojectile bombardment

DIRK BECKER & HORST LÖRZ

Institut für Allgemeine Botanik der Universität Hamburg (AMPII), Ohnhorststraße 18, 22609 Hamburg, Germany

Introduction

The development of plant transformation techniques during the past decade has made it possible to improve plants by introduction of cloned genes. For most dicotyledonous species, the *Agrobacterium*-mediated transformation system can be used to generate many transformants while for most of the monocotyledonous species, this transformation system still needs to be improved. However, in the last two years two reports were published describing the regeneration of transgenic fertile rice by using *Agrobacterium tumefaciens* as vector [4, 11]. Although this is a very recent development, the future prospects also for other cereals like maize, wheat and barley will surely be investigated in the near future. Of the various approaches to gene transfer, three transformation methods have led predominantly to the production of transgenic plants:
– protoplast based direct gene transfer,
– tissue electroporation,
– microprojectile-mediated gene transfer.
In the past considerable progress has been made in establishing efficient *in vitro* culture systems for most cereals. However, thus far embryogenic suspension cultures are the only reliable source for totipotent protoplasts. Nevertheless, it is very difficult and time-consuming to start and maintain these cultures. Furthermore, regeneration capability has been observed to decline gradually during cultivation in cereal suspension cultures. The direct DNA transfer into isolated protoplasts, induced by polyethylene glycol (PEG) or electric pulses is a successful and routinely used method to obtain transformed cell lines, but the regeneration of transgenic plants still remains difficult. Only in rice and maize it has been possible until now to obtain fertile, transgenic plants by protoplast transformation [21, 7, 8]. The fundamental problem of this transformation method is the continous loss of embryogenic capacity of the suspension cultures during long term culture [13], occurrence of somaclonal variation, and expenditure of labour and energy. As an alternative, microprojectile-mediated gene transfer [20] or tissue electroporation [6] have the potential to overcome these limitations. The essence of microprojectile systems for plant genetic transformation is to use high velocity particles to penetrate cell walls and to introduce DNA into intact cells thus circumventing the host range limitation of *Agrobacterium* and the problems of plant regeneration from protoplasts.

The transfer of DNA into cells and tissues with embryogenic capacity takes place with high efficiency. The choice of appropriate target cells is of major importance as there are only few tissues and cells capable of plant regeneration. Using embryogenic suspension cells and embryogenic callus cultures, successful transformation and regeneration of cereals, such as maize, rice, wheat and oat [10, 9, 3, 23, 22] could be achieved. However, the morphogenetic competence of cells is significantly reduced during long term maintenance and the phenomenon of somaclonal variation limits the suitability of these cells for transformation.

These limitations could be overcome by directly targeting tissues or cells which can be obtained easily and manipulated *in vitro*. In cereals, scutellar tissue of immature embryos, immature inflorescences or microspores are suitable primary explants for bombardment or tissue electroporation as it was reported in maize, wheat, barley, rice, tritordeum and triticale [16, 26, 24, 2, 19, 5, 13, 25, 1, 27]. The time necessary for preparation of the target cells is comparatively low and the risk of somaclonal variation is negligible as the period in tissue culture is reduced to a few weeks. Another advantage of microprojectile bombardment of primary explants is that even genotypes which are recalcitrant in protoplast culture can be transformed easily [5].

In the following pages we will present protocols for the isolation and transformation of scutellar tissue of immature embryos and the culture and selection procedure used to obtain fertile transgenic wheat plants.

Procedures

The primary requirement for an optimal target is that the tissue or cells receiving exogenous DNA are culturable *in vitro*, actively dividing and capable of giving rise to fertile plants. In our experiments we used scutellar tissue of immature embryos of the winter wheat genotype "Florida" and the spring type line "Veery" as a target for particle bombardment. The *in vitro* culture system used, allowed plant regeneration at high frequencies. Optimization of the transformation parameters was performed by transient transformation experiments using the plasmid pDB1 [2], containing the *gus* gene under control of the actin1-promoter from rice [17] and the selectable marker *bar* driven by the CaMV 35S-promoter. The aim of these experiments was to enhance transient transformation by minimizing tissue damage, which is correlated with a reduced regeneration capability.

For the selection of putative transgenic plants we have developed an *in vitro* selection system based on the resistant gene *bar* which confers resistance against the herbicide BASTA. The advantage of herbicide selection is not only the possibility to select plants *in vitro*, but also to identify transgenic plants at each time point of development by spraying plants with a herbicide solution.

Putative transgenic regenerants, transferred into soil, were analyzed for the presence of GUS-activity histochemically and for PAT-activity by spraying plants with a BASTA herbicide solution.

Isolation of immature embryos and preparation for bombardment

Steps in the procedure
1. Obtain spikes bearing immature caryopses 12–14 days after pollination.
2. Remove the immature caryopses and sterilize for 1 min in 70% ethanol and for 20 min in 1% NaOCl, 0.5 % Mucasol.
3. Wash three times with sterile distilled water.
4. Dissect the embryos in a sterile environment and place scutellum-side up on modified L3D2 [13] callus induction medium (without amino acids). For particle bombardment, place 20–30 embryos in the center of a 9 cm Petri dish.
5. Seal the culture dishes with Parafilm and incubate in the dark at 26 °C.
6. The isolated embryos can be bombarded directly after isolation or after a one to two days preculture.

Notes
1. The developmental stage of the immature embryos is the most significant factor which regulates the response *in vitro*. Best response (high capability of somatic embryogenesis and plant regeneration) is obtained from embryos in which the morphological development has been nearly completed, and the deposition of starch in the scutellar tissue has just begun.
6. A preculture treatment prior to bombardment depends on the genotype used. For example, in our experiments scutellar tissue of the winter wheat genotype "Florida" showed no significant differences in culture response without a preculture prior to bombardment, whereas in the case of immature embryos from the spring type line "Veery" only a preculture of one or, preferably, two days gave the same frequencies of somatic embryogenesis as non-bombarded controls. Using longer preculture treatments, we never obtained transgenic plants.

Preparation of DNA coated gold particles

Steps in the procedure
1. Place 40 mg gold particles (0.4–1.2 μm) per 1 ml of 96 % ethanol in a microtube. Sonicate the particle suspension for 30 sec using a standard tip to destroy particle aggregates. Centrifuge the microtube for 1 min at 4200 × g in a microfuge, remove the supernatant and add 1 ml 96 % ethanol. Resuspend the particles briefly and repeat three times.
2. Wash particles three times in 1 ml sterile distilled water as described previously. After the last centrifugation step, resuspend particles in 1 ml sterile distilled water. Aliquot 50 μl of the final suspension into microtubes, while vortexing the suspension continually. Store aliquots at −20 °C.
3. For precipitation of plasmid DNA, add to a 50 μl aliquot, 5 μl of plasmid DNA (1 μg/μl) and vortex briefly.
4. Add 50 μl of a 2.5 M solution of calcium chloride and 20 μl of a 0.1 M spermidine (free base) solution. The suspension should be vortexed while adding each solution. Incubate on ice for 15 min.
5. Spin the microcarriers down in a microfuge at 4200 x g for 5 sec. and remove the supernatant. Wash particles with 250 μl of absolute ethanol by vortexing for 1 min, centrifuge in the same manner and remove the supernatant.
6. Resuspend particles in 240 μl of absolute ethanol. Particles attached to the microtube wall can be scraped and resuspended thoroughly with a pipet.

Notes
1. Gold particles with an average size of 0.4–1.2 μm are available from Heraeus, Karlsruhe, Germany.
6. DNA coated microprojectiles should be used for bombardment immediately after preparation.

Bombardment of scutellar tissue

The particle gun employed in these experiments was a DuPont PDS 1000/He gun. The bombardment parameters used are summarized in Table 1.

Steps in the procedure

1. Clean the PDS 1000/He particle delivery system, the sample chamber and all the other material used for bombardment with 70 % ethanol. Allow time for drying in a sterile environment.
2. Seat a macrocarrier into the macrocarrier holder. Pipet 3.5 μl of the freshly prepared DNA-microcarrier suspension in the centre of the macrocarrier. The particle suspension must briefly be vortexed each time. Dry for 2 min.
3. Place a stopping screen in the stopping screen support and the macrocarrier holder containing the macrocarrier on the top rung of the fixed nest. Fix the macrocarrier holder with the macrocarrier cover lid. Place in position.
4. Place a rupture disk of the desired burst pressure in the recess of the rupture disk retaining cap and screw the rupture disk retaining cap onto the gas acceleration tube.
5. Put a Petri dish containing the immature embryos on the Petri dish holder and place in position.
6. Close the sample chamber door and evacuate to a partial vacuum of 27 inch Hg.
7. Press the FIRE switch to allow pressure to build in the acceleration tube.
8. After the rupture disk has burst release the vacuum in the sample chamber.

Table 1. Successfully used parameters for particle bombardment of scutellar tissue of wheat

Distance between	
– rupture disk and macrocarrier	2.5 cm
– macrocarrier and stopping screen	0.8 cm
– stopping screen and target cells	5.5 cm
gas pressure	900–1550 psi
partial vacuum	27 inch Hg
particles	gold, 0.4–1.2 μm
particle amount per bombardment	29 μg

An average number of 100 transient transformation events per embryo could be observed using a plasmid construct containing the *gus* marker gene driven by the actin-1 promoter [17]. The number of transient events was reduced, using greater distances between the stopping screen and the target cells and/or lower partial vacuum. No significant differences in transient transformation numbers were observed using helium gas pressures between 900–1550 psi. Lower or higher gas pressures gave reduced transient numbers and, in the case of higher gas pressures, higher tissue damage. The important step in minimizing tissue damage was the reduction of the amount of particles used per bombardment. We observed a direct correlation between the particle amount (between 29 and 116 μg per bombardment) used for bombardment and the capability of bombarded tissue to develop somatic embryos in high frequency. On the other hand, there were no significant differences in the number of transient transformation events. Using 29 μg particles per bombardment, the same rate of somatic embryogenesis and plant regeneration could be observed as in non-bombarded controls.

Transient expression assay

Following bombardment culture embryos for two days at 26 °C in the dark. For histochemical detection of GUS-activity [14, 15], overlay bombarded embryos with x-Gluc staining buffer. Incubate for 12–18 h at 37 °C. Count the number of blue spots.

Culture, selection and plant regeneration

Steps in the procedure

1. Spread the embryos over the culture plate, one day after bombardment. Developing coleoptiles from the embryo must be cut off in the following days.
2. Subculture developing calluses 2 weeks after transformation for additional 14 days on L3D2/B3 selection medium. At this time point, the development of embryoid-like structures are visible on the surface of the developing calluses.
3. Transfer callus showing the development of somatic embryos after 14 days on L3D0.1Z10/B3 regeneration medium for shoot development. Culture under light conditions (3000 lx for 16 h) at 26 °C.
4. Subculture after additional 14 days on L3D0.1/B3 regeneration medium for root development. Subculture at 14–21 d intervals on the same medium.
5. Transfer rooted plantlets with a leaf length of 1.5–2.0 cm to half strenght MS-medium for additional 14 days.
6. Transfer rooted plants to a peat/soil mix and grow under greenhouse conditions to maturity.

Notes

1. The further development of the coleoptile can be observed in embryos of an older developmental stage. The induction of somatic embryogenesis is in principle possible, but it takes place at lower frequency. The development of somatic embryos on the scutellum is inhibited when the coleoptile develops.
2. For the selection of transgenic plants, we use the herbicide BASTA. The solution contains 20% phosphinothricin, the active component of the herbicide.
3. and 4. The same selection pressure used during the callus induction phase was also used during plant regeneration.
5. Plantlets with healthy shoot systems which do not form roots *in vitro* may sometimes be rooted by placing a sterile droplet of a 1 mg/ml solution of IBA (Indole-3-butyric acid) to the leaf bases.
6. The plantlets must be hardened in a high humidity chamber for about 1 week before transfer to the greenhouse.

Identification of transgenic regenerants

After the transfer of putative transgenic regenerants into soil, all plants are analyzed for enzyme activity of both introduced genes. GUS-activity is monitored histochemically and PAT-activity indirectly by spraying plants with a BASTA-solution which is toxic for non transformed plants.

Steps in the procedure

A. Detection of GUS-activity in leaf segments
1. Harvest leaf pieces of 1 cm in length and incubate them in GUS-assay buffer for 12 to 16 h at 37 °C.
2. Extract the chlorophyll by incubating leaf pieces in a 3 : 1 solution of ethanol : glacial acetic acid for 1 h at 70 °C.
3. GUS-activity is visible under the microscope.

B. Detection of PAT-activity
4. 14 days after transfer of plants into soil, spray whole plants or only single leaves with an aqueous solution of the herbicide BASTA (150 mg/l PPT, 0.1% Tween 20).
5. Examine plants one week after herbicide application for necrosis. Resistant plants do not show necrosis or only partial necrosis at the leaf tips, whereas sensitive plants do not survive herbicide treatment. As a negative control, use in each experiment non transformed regenerants of the same developmental stage.

Notes
3. The staining intensity depends on the expression level and the developmental stage of the leafs. Younger leaves normally show a stronger reaction than older ones.
5. The lethal dose of PPT depends on the developmental stage of the plants. Older plants tolerate higher concentrations than younger ones. In order to get clear results of the status of older plants it is necessary to use PPT concentrations between 200 and 250 mg/l.

Media and solutions

Media

Prepare culture media double concentrated and sterilise it through 0.22 μm membranes.

		L3	MS
Macrosalts (mg/l)			
	NH_4NO_3	200	1650
	KNO_3	1750	1900
	KH_2PO_4	200	170
	$MgSO_4 \cdot H_2O$	350	370
	$CaCl_2 \cdot H_2O$	450	440
Microsalts (mg/l)			
	$MnSO_4 \cdot H_2O$	15	22.3
	H_3BO_4	5	6.2
	$ZnSO_4$	7.5	8.6
	KI	0.75	0.83
	$Na_2MoO_4 \cdot H_2O$	0.25	0.25
	$CuSO_4 \cdot 5H_2O$	0.025	0.025
	$CoCl_2 \cdot 6\,H_2O$	0.025	0.025
iron (mg/l)			
	Na_2EDTA	37	37.2
	$FeSO_4 \cdot 7H_2O$	28	27.8
vitamins (mg/l)			
ascorbic acid		1	–
Ca-pantothenate		0.5	–
myo-inositol		100	100
nicotinic acid		1	0.5
pyridoxin-HCl		1	0.5
thiamine-HCl		10	0.4
glycine		–	2
sugar (g/l)			
maltose		50	–
sucrose		–	30

pH is adjusted to 5.7 with KOH
L3D2: L3 medium with 2 mg/l 2,4-D
L3D2/B3: L3 medium with 2 mg/l 2,4-D and 3 mg/l phosphinothricin (PPT, active component of the herbicide BASTA).
L3D0.1Z10/B3: L3 medium with 0.1 mg/l 2,4-D, 10 mg/l zeatin and 3 mg/l PPT.
L3D0.1/B3: L3 medium with 0.1 mg/l 2,4-D and 3 mg/l PPT.

Solutions

GUS assay buffer
- 5 mM potassium ferricyanide
- 5 mM potassium ferrocyanide
- 0.05% (w/v) 5-bromo-4-chloro-3-indolyl-β-D-glucuronic acid
- 0.06% (v/v) Triton X-100
- 0.2 M sodium phosphate buffer, pH 7.0

Sterilise by filtration. Store aliquots at -20 °C.

- Sigma 1-A agarose: prepare double-concentrated (1.6% for all media) in distilled water. Autoclave befor using.
- PPT stock solution: 20 mg/l PPT (use the herbicide BASTA) in sterile distilled water.
- DNA solution: dissolve DNA in sterile 10 : 1 mM Tris/HCl:EDTA buffer, pH 8.0. Final concentration 1 $\mu g/\mu l$.
- Spermidine solution: 0.1 M in sterile distilled water. Store at -80 °C.
- $CaCl_2$ solution: 2.5 M in sterile distilled water. Store at -20 °C

References

1. Barcelo P, Hagel C, Becker D, Martin A, Lörz H (1994) Transgenic cereal (tritordeum) plants obtained at high efficiency by microprojectile bombardment of inflorescence tissue. The Plant Journal 5(4): 583–592.
2. Becker D, Brettschneider R, Lörz H (1994) Fertile transgenic wheat from microprojectile bombardment of scutellar tissue. The Plant Journal 5(2): 299–307.
3. Cao J, Duan X, McElroy D, Wu R (1992) Regeneration of herbicide resistant transgenic rice plants following microprojectile mediated transformation of suspension culture cells. Plant Cell Rep 11: 586–591.
4. Chan MT, Chang HH, Ho SL, Tong WF, Yu SM (1993) *Agrobacterium*-mediated production of transgenic rice plants expressing a chimeric alpha-amylase promoter/beta-glucuronidase gene. Plant Mol Biol 22: 491–506.
5. Christou P, Ford TL, Kofron M (1991) Production of transgenic rice (*Oryza sativa* L.) plants from agronomically important indica and japonica varieties via electrical discharge particle acceleration of exogenous DNA into immature zygotic embryos. Bio/Technol 9: 957–962.
6. D'Halluin K, Bonne E, Bossut M, De Beuckeleer M, Leemans J (1993) Transgenic maize plants by tissue electroporation. Plant Cell 4: 1495–1505.
7. Datta SK, Peterhans A, Datta K, Potrykus I (1990) Genetically engineered fertile indica-rice recovered from protoplasts. Bio/Technol 8: 736–740.
8. Donn G, Eckes P, Müller H (1992) Genübertragung auf Nutzpflanzen. BioEngineering 8: 40–46.
9. Fromm ME, Morrish F, Amstrong C, Williams R, Thomas J, Klein TM (1990) Inheritance and expression of chimeric genes in progeny of transgenic maize plants. Bio/Technol 8: 833–839.
10. Gordon-Kamm WJ, Spencer TM, Mangano ML, Adams TR, Daines RJ, Start WG, O'Brian JV, Chambers SA, Adams JWR, Willetts NG, Rice TB, Mackey CJ, Krueger W, Kausch AP, Lemaux PG (1990) Transformation of maize cells and regeneration of fertile transgenic plants. Plant Cell 2: 603–618.
11. Hiei Y, Ohta S, Komari T, Kumashiro T (1994) Efficient transformation of rice (Oryza sativa L.) mediated by Agrobacterium and sequence analysis of the boundaries of the T-DNA. The Plant Journal 6(2): 271–282.
12. Jähne A, Becker D, Brettschneider R, Lörz H (1994) Regeneration of transgenic microspore-derived, fertile barley. Theor Appl Genet 89: 525–533.
13. Jähne A, Lazzeri PA, Jäger-Gussen M, Lörz H (1991a) Plant regeneration from embryogenic cell suspensions derived from anther cultures of barley (*Hordeum vulgare* L.). Theor Appl Genet 82: 74–80.
14. Jefferson RA (1987a) Assaying chimeric genes in plants: The GUS gene fusion system. Plant Mol Biol Rep 5: 387–405.
15. Jefferson RA, Kavanagh TA, Bevan MW (1987b) GUS fusions: β-glucuronidase as a sensitive and versatile gene fusion marker in plants. EMBO J 6: 3901–3907.
16. Koziel GM, Beland GL, Bowman C, Carozzi NB, Crenshaw R, Crossland L, Dawson J, Desai N, Hill M, Kadwell S, Launis K, Lewis K, Maddox D, McPherson K, Meghji MR, Merlin E, Rhodes R, Warren GW, Wright M, Evola SV (1993) Field performance of elite transgenic maize plants expressing an insecticidal protein derived from *Bacillus thuringiensis*. Bio/Technol 11: 194–200.
17. McElroy D, Zhang W, Cao J, Wu R (1990) Isolation of an efficient actin promoter for use in rice transformation. Plant Cell 2: 163–171.
18. Murashige T, Skoog F (1962) A revised medium for rapid growth and bioassays with tobacco tissue cultures. Physiol Plant 15: 473–497.

19. Nehra NS, Chibbar RN, Leung N, Caswell K, Mallard C, Steinhauer L, Baga M, Kartha K (1994) Self-fertile transgenic wheat plants regenerated from isolated scutellar tissues following microprojectile bombardment with two distinct gene constructs. The Plant Journal 5(2): 285–297.

20. Sanford JC, Klein TM, Wolf ED, Allen N (1987) Delivery of substances into cells and tissues using a particle bombardment process. J Part Sci Technol 5: 27–37.

21. Shimamoto K, Terada R, Izawa T, Fujimoto H (1989) Fertile transgenic rice plants regenerated from transformed protoplasts. Nature 338: 2734–276.

22. Somers DA, Rines HW, Gu W, Kaeppler HF, Bushnell W R (1992) Fertile, transgenic oat plants. Bio/Technol 10: 1589–1594.

23. Vasil V, Castillo AM, Fromm ME, Vasil IK (1992) Herbicide resistant fertile transgenic wheat plants obtained by microprojectile bombardment of regenerable embryogenic callus. Bio/Technol 10: 667–674.

24. Vasil V, Srivastava V, Castillo AM, Fromm ME, Vasil IK (1993) Rapid production of transgenic wheat plants by direct bombardment of cultured immature embryos. Bio/Technol 11: 1553–1558.

25. Wan Y, Lemaux PG (1994) Generation of large numbers of independently transformed fertile barley plants. Plant Physiol 104: 37–48.

26. Weeks JT, Anderson OD, Blechl AE (1993). Rapid production of multiple independent lines of fertile transgenic wheat (*Triticum aestivum*). Plant Physiol 102: 1077–1084.

27. Zimny J, Becker D, Brettschneider R, Lörz H (1995). Fertile, transgenic Triticale (x *Triticosecale* Wittmack). Mol Breeding 1, No.2: 155–164.

Plant Tissue Culture Manual **B13**, 1–46, 1996.
© *1996 Kluwer Academic Publishers.*

Transient gene expression and stable genetic transformation into conifer tissues by microprojectile bombardment

ARMAND SÉGUIN, DENIS LACHANCE & PIERRE J. CHAREST

Molecular Genetics and Tissue Culture Group; Petawawa National Forestry Institute; Box 2000, Chalk River, Ontario K0J 1J0, Canada; E-mail: aseguinpnfi.forestry.ca

Introduction

Genetic tranformation technologies are essential to programs of molecular biology and genetic engineering. Although significant efforts in conifer molecular biology have been initiated since the late 1980s, so far only a limited number of genes have been cloned [1]. Conifers are by far the most difficult plant group for this type of study because of their large genomes and lengthy life cycles. Furthermore, progress has been hindered by the present inefficiencies in gene transfer methods and tissue culture protocols for certain species such as pines. Nevertheless, there is a significant body of literature on gene transfer in several conifer species using *Agrobacterium*-mediated transformation and other protocols of direct DNA transfer.

For example, specific strains of *Agrobacterium* that induce tumors on seedlings have been identified and, in some cases, the transfer and integration of T-DNA encoded genes has been confirmed (see Table 1 and ref. [2, 3, 4] for a complete list). The only success for transgenic tree regeneration was reported with *Larix decidua* [5] but the number of trees produced was limited to less than a dozen [6]. There is no report of *Agrobacterium* transformation using a conifer tissue culture system.

Table 1. Agrobacterium-mediated DNA transformation of conifers

Genus	Observations (results)	Reference
Abies	Tumours obtained on seedlings; opines detected	[7, 8]
Larix	Roots observed on in vitro *grown stem*	[9, 10, 11]
	Transgenic trees regenerated from inoculated seedlings	[5, 6]
Libocedrus	Tumours obtained on seedlings	[12]
Picea	Tumours obtained on seedlings; opines detected	[7, 13, 14, 15, 16]
Pinus	Tumours obtained on plants and seedlings; NPT II transgene expression, Southern blot analysis	[8, 11, 12, 13, 16, 17, 18, 19, 20]
Pseudotsuga	Tumours obtained on seedlings; opines detected; NPT II transgene expression; Southern blot analysis	[8, 12, 13, 21, 22]
Taxus	Tumours obtained on shoot segments; opines detected; Southern blot analysis	[23]
Tsuga	Tumours obtained on seedlings	[8]

NPT II = neomycin phosphotransferase.

With direct DNA transfer methods, transient gene expression of the delivered genes has been obtained with electroporation, polyethylene glycol DNA delivery, silicon carbide-mediated DNA delivery and microprojectile bombardment (Table 2). This last method has emerged as simple and promising for the stable genetic transformation of conifers and other recalcitrant species, and has been used to regenerate transgenic plants of black spruce (*Picea mariana*), white spruce (*Picea glauca*), and tamarack (*Larix laricina*) [24, 25]. Furthermore, microprojectile DNA delivery has been an invaluable tool for studying expression and regulation in conifers of various genes from both angiosperms and gymnosperms (Table 2 and ref. [2]). It provides a tool to bypass the long life cycle of conifers by allowing gene delivery and expression in mature tree tissues such as flowers, pollen, differentiating wood, and needles [26].

We shall describe the protocols employed in our laboratory for direct DNA transfer in black spruce using microprojectile-mediated DNA delivery and we shall also describe the gene expression assay procedures used (β-glucuronidase,

Table 2. Direct DNA transformation in conifers

Genus	Method used and observations	Reference
Larix	Transient expression of GUS and CAT genes by electroporation of protoplasts	[27]
	Transient expression of GUS gene by microprojectile bombardment of somatic embryogenic tissues	[28, 29]
	Stable transformation by microprojectile bombardment of somatic embryogenic tissues	[25]
Picea	Transient expression of GUS and CAT genes by electroporation or PEG-mediated expression of protoplasts	[30, 31, 32, 33]
	Transient expression of GUS gene by silicon carbide-mediated DNA delivery	[34]
	Transient expression of GUS gene by microprojectile bombardment of somatic embryogenic tissues	[28, 35, 36, 37, 38, 39, 40]
	Stable transformation by microprojectile bombardment of somatic embryogenic tissues	[24, 25, 41, 42]
Pinus	Transient expression of Luc or CAT genes by electroporation of protoplasts	[32, 43]
	Transient expression of GUS gene by microprojectile bombardment of cell suspension, cotyledons, and differentiating wood	[43, 44, 45, 46]
Pseudotsuga	Transient expression of GUS gene by microprojectile bombardment of cotyledons	[47]

CAT= chloramphenicol acetyltransferase, GUS=β-glucuronidase, Luc= firefly luciferase, PEG= polyethylene glycol.

neomycin phosphotransferase, and luciferase assays). The microprojectile DNA delivery method can be used with any plant tissue but this section will only cover protocols for somatic embryogenic tissues and for pollen.

With conifers, somatic embryogenesis is an ideal tissue culture system for gene transfer experiments because it can be induced readily from tree tissues (immature and mature zygotic embryos, cotyledons, and needles from young seedlings) and plants can be regenerated from tissue culture lines [48, 49]. Two stages of somatic embryogenesis were used for transient gene expression and stable genetic transformation; mature somatic embryos and embryonal masses. In transient gene expression studies, our laboratory has tested over 35 different chimeric gene constructs for level of expression and tissue specificity. Several factors affecting the level of expression of introduced reporter genes using particle bombardment in *Picea* embryogenic tissues have been described previously. For instance, the choice of tissue line and the time kept in culture will result in variation in transient gene expression of the reporter gene [35, 37, 38]. Moreover, the type of vector used and the strength of the promoter driving the reporter gene have also been shown to be important [35, 36, 38].

For stable gene transfer, we have obtained transgenic tissue culture lines using kanamycin selection for black spruce, white spruce, and tamarack. From these, transgenic trees have been regenerated in black spruce and tamarack and, depending on the effort invested, unlimited numbers of transgenic trees may be obtained. The procedure yields trangenic tissues at low frequency, but can be repeated consistently. Some of the transgenic tissue cultures have been maintained for 3 years without loss in the level of foreign gene expression. Our laboratory is improving transgenic embryogenic line recovery by investigating (a) the use of other selective agents (e.g., geneticin), (b) by designing vectors carrying scaffold or matrix attachment regions [50] and; (c) by altering the physiology of the tissue culture lines. We have already tested selection using hygromycin and methotrexate resistance genes as markers but with no success.

It is difficult to estimate the frequency of stable transformant recovery in relation to the level of transient gene expression obtained. According to several authors, a large variation can be observed in the estimation of the conversion rates from transiently expressing cells to stably transformed cells when using microprojectile-DNA delivery in plant cells as detected by GUS histochemical staining. The data can vary from a conversion rate of approximately 1% [51] to 5% of the cells that transiently expressed a foreign gene and then stably integrated it [52]. For the conifer species with which we regenerated transgenic lines, the ratio of cells showing transient gene expression to the number of lines stably transformed is lower than 0.1%.

For pollen, microprojectile bombardment has been used successfully to achieve transient gene expression in tobacco [53, 54], lily [54, 55], and maize [56] for the study of tissue-specific gene expression. Pollen grains have also been proposed as target material for gene transfer in plants [57, 58]. Because pollen grains from conifers are usually relatively large in diameter (about 50–100 μm for spruce), are easy to collect in large quantities, and can be preserved for an extended period of time, we have investigated the genetic transformation of pollen using microprojectile bombardment technology [40, 59]. Comparison of the germination frequency between bombarded pollen expressing GUS activity and non-bombarded pollen showed that pollen vigour does not seem affected by the bombardment procedure [40]. The frequency of gene transfer into this tissue without optimization as indicated by transient gene expression is in the order of 5–8% [59]. These data are encouraging from the perspective of using microprojectile bombardment of pollen for the stable transformation of conifer germlines.

Protocols for gene delivery in pollen and embryogenic tissues of black spruce

The protocols that follow were used to optimize gene delivery, as indicated by transient gene expression, and to assess the strength of different gene constructs driven by various promoters. Although optimization of gene transfer for transient gene expression is often considered a first step in determining the parameters needed for stable transformation, conditions for microprojectile bombardment used for stable transformation differed from those for transient gene expression. We describe here protocols for black spruce and we give some indications of the conditions that can differ for white spruce and tamarack. Furthermore, these protocols can be applied to other spruce and larch species using different tissue culture media.

Procedures

i– Tissue culture protocols

The tissue culture protocols for black spruce embryogenic cultures including initiation of embryogenic cultures, maintenance, maturation of somatic embryos, germination, and transfer to soil have been described in Lelu *et al.* [49] and in Cheliak and Klimaszewska [60]. Furthermore, chapter C3 (pp. 1–16) of this manual, by Thorpe and Harry [61], covers somatic embryogenesis of conifers. The embryogenic tissue culture lines of black spruce used in this procedure were induced from mature seeds.

Embryonal masses

Embryogenic cell suspension cultures are established by transferring approximately 5 g of embryonal masses from gelled 1/2 LM medium to 40 ml liquid 1/2 LM medium. The suspensions are maintained by weekly transfer of embryogenic suspension into fresh medium (ratio 1 : 1) for a total volume of 40 ml in 250 ml Erlenmeyer flasks. These flasks are kept on a gyratory shaker at 120 rpm under indirect light with a 16h photoperiod. Cell suspension cultures of embryonal masses of black spruce are preferable for transformation although embryonal masses maintained on solid media can be broken up in liquid media just prior to bombardment (see note). Actively dividing four-day-old suspensions are used for transformation experiments. In some cases, pretreatment of embryogenic tissues before bombardment in liquid medium with increasing osmoticum can result in enhanced expression of the reporter gene [40]. Augmentation of the osmoticum medium may induce plasmolysis of the cells, making them less likely to release protoplasm after the penetration of the microprojectiles into the cells. Similar results, using particle bombardment, have been obtained with white spruce embryogenic material and suspension-cultured cells of tobacco [62] and maize [63] using mannitol, sorbitol, or raffinose as the osmoticum.

Somatic embryos

Somatic embryos are produced by placing embryonal masses on 1/2LM maturation media. Mature cotyledonary somatic embryos (size: 2–3 mm) are produced after 4–8 weeks at which stage they are used for targets for microprojectile bombardment.

Notes
1. Higher levels of transient gene expression are observed with cell suspension cultures because they are more easily and uniformly spread than callus on solid media.

ii– Microprojectile bombardment and reporter gene assays

A– Microprojectile preparation

Steps in the procedure
1. In a 1.5 ml microfuge tube, place 60 mg of gold microprojectiles.
2. Add 1 ml absolute ethanol and vortex for 1–2 minutes. Allow tube to stand for 1 hr with brief periodic vortexing.
3. Briefly centrifuge the tube and remove supernatant. Wash the microprojectiles two times with 1 ml sterile distilled water by repeating resuspension and centrifugation. Finally, resuspend in sterile distilled water to a final volume of 500 μl.
4. Aliquot 25 μl of gold microprojectile suspension into 0.5 ml microfuge tubes.
5. Add 10 μg of vector DNA at a concentration of 1 μg/μl in water or TE and mix thoroughly.
6. While vortexing the DNA/gold microprojectile mixture, add 50 μl of 2.5 M $CaCl_2$ and 20 μl of 0.1M spermidine (free base). Continue vortexing for 30 sec. then let stand for 10 minutes.
7. Briefly centrifuge the tube and remove supernatant.
8. Add 200 μl of absolute ethanol and vortex. Briefly spin down microprojectiles and remove ethanol.
9. Resuspend the microprojectiles in 50 μl of absolute ethanol. 5 μl of this suspension (1 μg of DNA) will be used for each bombardment.
10. Use the DNA/ microprojectile preparations as soon as possible.

Notes
1. Gold microprojectiles with a diameter of 1.6 μm are from Bio-Rad Laboratories, Richmond, CA.
3. The gold microprojectiles may be stored at room temperature for up to a month.
5. Plasmid DNA was isolated by alkaline lysis [64] and subsequently purified on Qiagen™ anion exchange resin according to the protocol provided by the manufacturer (Qiagen, Chatsworth, CA) or by CsCl gradient [65]. The optimal concentration of plasmid DNA to be added to the gold microprojectiles for maximum transient transformation can be established by a dose-response curve. However, increasing gene delivery with higher DNA quantities will reach a plateau and eventually lead to a decrease in transformation efficiency, presumably due to inappropriate conditions for DNA precipitation on the gold microprojectiles. Aggregation of the gold microprojectiles is often observed at high DNA concentrations and this reduces efficiency of cell penetration and could cause cell injuries.
6. Plasmid DNA is adsorbed onto the gold microprojectiles using the procedure described originally by Klein [52].

B– Tissue bombardment

DNA transfer is carried out using the Biolistic™ Particle Delivery system PDS-1000/He System (referred to as the "gene gun" in the following pages; Bio-Rad Laboratories, Richmond, CA) following the manufacturer's recommendations and as described in Kikkert [66] (See note). The various settings and parameters used are given in Table 3.

Table 3. Potential settings of the PDS-1000/He apparatus and parameters used for transient expression

PDS-1000/He parts	Tissue bombarded			
	Settings	Pollen	Embryonal masses	Somatic embryos
Rupture Disc pressure	400	O		
(PSI)	650			
	900		X	
	1100	×	×	×
	1300			
	1550			
	1800		O	
	2000			
	2200			
Gap distance (cm)	0.32			
	0.64	×		
	0.95	O	×, O	×
Fixed nest assembly	3	×		
(macrocarrier flight distance,	8	O	O	
mm)	13		×	×
Sample holder distance	6			×
(cm)	9	×, O	×, O	
	12			

×=settings used for black spruce
O=settings used for white spruce

Steps in the procedure
1. Sterilize stop screens, kapton discs, and disc holders in ethanol (70%) and then allow to air dry.
2. Position the kapton disc within the disc holder.
3. Pipet 5 µl of the DNA preparation from the tube (while vortexing) and deposit in the center of the kapton disc. Let dry under a laminar flow until ethanol has evaporated.
4. Position a rupture disc in the rupture disc holder and screw tightly into position.

5. Assemble the fixed nest as per the manufacturers' instructions and position in the gene gun housing.
6. Place the target tissue on the sample holder beneath the nest assembly and close the gene gun door.
7. Evacuate chamber with a vacuum of 675 mm Hg (26.5 inch Hg) and fire the gene gun.
8. Release vacuum slowly and remove the bombarded sample.
9. Repeat steps 2–8 for each sample.

Notes

1. Our laboratory compared the PDS-1000 gunpowder system with the helium system. Also, the tungsten microprojectiles were tested in both systems. The helium system and the gold microprojectiles consistently gave better results. The stop screens, kapton discs, and disc holders should be resterilized before each use.

C– Reporter gene assays

Standard protocols to detect expression of the β-glucuronidase gene [67], neomycin phosphotransferase II gene, and luciferase gene [68] are suitable for enzymatic assays in conifer tissues. Small modifications have been made to prevent interference by endogenous biochemicals such as the addition of methanol for the fluorescent GUS assay and increase in the concentration of phosphate buffer in the luciferase assay for black spruce.

β-glucuronidase assay:

Histochemical staining

Steps in the procedure
1. Completely submerge tissue to be assayed in GUS histochemical buffer.
2. Incubate material in the dark at 37 °C for 24 hours.
3. Count the number of cell clusters (GUS expression units) in embryonal masses or of pollen grains with a blue colouration, indicating GUS gene expression.

Note
1. The GUS histochemical buffer can be vacuum infiltrated for 5 min for increased sensitivity.

Fluorometric assay

Steps in the procedure
1. Harvest tissue and place in 1.5 ml microfuge tube.
2. Grind or sonicate tissue in 200 μl of GUS extraction buffer (GEB).
3. Pellet the cell debris by centrifugation.
4. Take 100 μl sample from the clarified extract and combine with 100 μl GEB containing 2 mM MUG (4-methyl umbelliferyl β-D-glucuronide) and 40% methanol in a new microfuge tube.
5. After incubation at 37 °C for 1, 2, and 4 hours, take out aliquots of 20 μl and stop the reaction by adding to the aliquot to 1980 μl of 0.2 M Na_2CO_3.
6. Determine the fluorescence for each sample and calculate GUS activity as described in Jefferson [69] by comparing with a standard curve using MU (4-methyl umbelliferone) as a standard.

Notes
4. As described in Kosugi [70], we found that addition of methanol in the reaction buffer is essential to remove endogenous fluorescence of spruce and larch, tissues probably due to high phenolic content.
6. In our laboratory, the fluorescence was determined in a TKO-100 fluorometer (Hoeffer Scientific Instrument, San Francisco, CA). A dose-response curve can be obtained with both the histochemical staining and the fluorogenic assays. In several publications using microprojectile bombardment technology, the quantitative determination of GUS gene activity is done by assessing the number of discrete areas of blue histochemical staining. Determination of the transient GUS activity in pmole of 4-methylumbelliferyl β-D-glucuronide (MUG) per minute per mg of protein extracted facilitate the evaluation of the efficiency of the method used.

Neomycin phosphotransferase assay

The neomycin phosphotransferase gene product is assayed using an ELISA (enzyme linked immunosorbent assay) kit from 5 Prime → 3 Prime, Inc. (Boulder, CO, USA). The steps are as indicated by the manufacturer and the extraction buffer used is the one described for plant tissues (0.25M Tris-Cl, pH 7.8, 1.0 mM phenylmethylsulfonylfluoride). Alternatively, the fluorometric GUS extraction buffer can be substituted when both GUS and NPT II are being used to quantify relative gene expression of two different promoters.

Firefly luciferase assay

This assay is done according to the luciferase assay kit from Promega Biotech (Fisher Scientific, Ottawa, Ontario, Canada). Modifications to the extraction buffer and procedure were required due to luciferase enzyme inhibition with black spruce tissues.

Steps in the procedure
1. Collect the bombarded tissues 36 hours after bombardment and add 100 μl of cell culture lysis buffer.
2. Freeze the mixture in liquid nitrogen and grind to produce an uniform powder or, in an Eppendorf tube, grind directly the sample in cell culture lysis buffer.
3. Centrifuge cells and add 20 μl of supernatant to 100 μl of beetle luciferin (470 μM) in reaction buffer.
4. Count immediately the number of photons emitted by using a scintillation counter (specially equipped) or a luminometer. Express the data as 10^6 photon events/mg of protein.

Protein assay

Protein concentration of the extracts is evaluated with the Bio-Rad Bradford Protein assay kit or the Bio-Rad *DC* protein assay kit (Bio-Rad, Mississauga, Ont., Canada) following manufacturer's protocols.

iii– Gene transfer for transient gene expression into conifer pollen

A– Pollen collection and storage

Steps in the procedure
1. At the first sign of dehiscence, collect and bag sections of twigs containing several microstrobili (male flower cones) directly from the field.
2. Release the pollen by shaking bags and collect by pouring into scintillation vials.
3. Store black spruce and white spruce pollen desiccated at −20 °C.

Notes
1. The male flower can also be forced to dehiscence under controlled laboratory conditions.
3. The pollen can be stored for up to a year with marginal loss in germination rate.

B– Pollen bombardment

Steps in the procedure
1. Weigh out the desired amount of pollen.
2. Make a suspension of pollen in sterile distilled water. 5 ml of the suspension should contain the desired amount pollen per bombardment.
3. Vacuum filter 5 ml of suspension onto a nylon membrane using a sintered glass filter assembly to ensure uniform distribution.
4. Place the nylon filter in the center of a 9 cm Petri dish containing solidified pollen germination media or three layers of Whatman filter papers soaked with 1.5 ml of the same liquid media.
5. Prepare DNA-coated gold microprojectiles for transformation (see section ii A).
6. Bombard pollen within 30 minutes of plating.
7. Incubate the pollen at 24 °C until it germinates.
8. Assay the pollen for expression of the transferred gene(s).

Notes
3. Nylon membrane (e.g., for nucleic acid hybridization) is preferred as the pollen can be easily washed or scraped off and recovered if desired. Do not use a Buchner funnel as the pollen tends to be unevenly distributed on the filter. Vigorous stirring of the suspension is required to maintain homogeneity. For transient assays, various amounts of pollen (1 to 5 mg) are used.
7. 16–24 hours is sufficient for black spruce.

iv– Gene transfer for transient gene expression in embryogenic tissues

A– Embryonal masses

Steps in the procedure
1. Vacuum filter 0.1–0.5 g black spruce cell suspension culture onto paper filter discs (5.5 cm diam, Whatman # 2) using a Buchner funnel.
2. Position the filter in the center of a 9 cm diameter Petri dish containing solidified 1/2LM maintenance media.
3. Prepare vector DNA for transformation.
4. Bombard the target tissue using the PDS1000/He.
5. Incubate the tissue at 24 °C for 48hrs in the dark.
6. Assay tissue for expression of the introduced genes.

B– Somatic embryos

Steps in the procedure

1. Position a 3 cm × 3 cm piece of nylon mesh (200 μm pore size) in the center of a 9 cm diameter Petri dish containing solidified 1/2LM maintenance media. Carefully position (disperse horizontally) 30 to 35 mature somatic embryos on the nylon mesh.
2. Prepare vector DNA for transformation.
3. Bombard the target somatic embryos using PDS1000/He.
4. Incubate the bombarded somatic embryos at 24 °C in the dark for 48 hours.
5. Assay the somatic embryos for the expression of the introduced genes.

Notes

1. The embryos are placed on a fine mesh to prevent them from becoming embedded in the media and also for ease of handling.

v– Gene transfer for stable transformation of embryogenic tissues

A– Determination of the antibiotic concentration for optimal selection.

The antibiotic concentration to be used for selection of tranformed tissues is determined by establishing toxicity curves and by identifying a concentration that will cause inhibition of growth but not complete killing. This was essential for spruce and larch embryogenic cultures selected for kanamycin resistance.

Embryonal masses

Steps in the procedure

1. Dilute a 4-day-old cell suspension culture in 1/2 LM maintenance media to a concentration of 20 mg/ml.
2. Vacuum filter 5 ml of diluted suspension (100 mg of embryonal masses) onto paper filter disc (5.5 cm diam. Whatman #2) using a Buchner funnel.
3. Transfer the filter bearing the cells on a Petri dish containing solidified 1/2 LM maintenance media.
4. Weigh the filter paper and the cells after one week of incubation and transfer them on fresh 1/2 LM media containing a given concentration of kanamycin (8 Petri dishes with each of the following concentrations: 0, 10, 15, 20, 25, and 30 μg/ml).
5. At weekly intervals, the filter paper discs from each Petri dish are weighed and subcultured to fresh media each 2 weeks.
6. The lowest concentration that inhibits growth and results in cell mortality after 4 weeks on selection should be chosen (viability is assessed by vital staining or growth).

Somatic embryos

Steps in the procedure
1. Place somatic embryos (16) in a Petri dish containing 1/2 LM gelled media with a given concentration of kanamycin (0, 5, 7.5, 10, 12.5 and 15 μg/ml).
2. Incubate in the dark at 24 °C for 8 weeks with a subculture after 4 weeks.
3. Identify minimal (threshold) antibiotic concentration that inhibits secondary embryogenic growth but does not kill all regrowth.

B– Stable transformation using embryonal masses as target material.

Steps in the procedure
1. Tissue preparation and bombardment procedures are the same as those used for transient gene expression.
2. After bombardment, incubate tissues on a filter paper placed on gelled 1/2 LM medium in the dark for 7 days at 24 °C.
3. Then transfer the bombarded tissues on the filter paper to maintenance media containing the selective agent (25 μg/ml kanamycin for black spruce and tamarack). Return to previous incubation conditions for 6–8 weeks. Subculture to fresh media after 4 weeks.
4. Identify putatively transformed embryonal masses by screening any growing tissue for expression (enzymatic assay) or presence of the introduced gene by polymerase chain reaction (PCR) [71]. After 8 weeks, remove all tissues from the selective media and place on regular maintenance media and continue to monitor for any growth.
5. Bulk up transformed embryonal masses then pass through liquid media containing 500–750 μg/l of kanamycin for two weeks to eliminate or reduce chimerism.
6. Maintain transgenic lines on selection free media or, preferably, cryopreserve the tissues [72, this manual] to avoid loss of regeneration capacity or change in gene expression.
7. Place tissues on maturation media to produce somatic embryos that will germinate to regenerate transgenic trees. For this, follow the established procedure described in Lelu [49] and Thorpe and Harry [61, this manual].

C– Stable transformation using mature somatic embryos as target material.

Steps in the procedure
1. Tissue preparation and tissue bombardment procedures are the same as those used in transient expression.
2. Incubate the bombarded somatic embryos on a filter paper deposited on gelled media at 24 °C in the dark for 7 days.
3. After this time, spread embryos out over the surface of fresh maintenance media containing the selective agent (10 μg/ml Km- no mesh required) and return to the previous incubation conditions for 6–8 weeks.
4. Identify putatively transformed callus formed through secondary somatic embryogenesis by screening all secondary embryogenic growth for expression (enzymatic assay) or presence of the introduced genes (PCR).
5. Bulk up the transformed callus and then briefly pass (2 weeks) through liquid media containing 500–750μg/l kanamycin to eliminate or reduce chimerism.
6. Maintain the transgenic lines on selection free media or preferrably cryopreserve the transgenic lines.
7. Tissues can be matured to produce somatic embryos from which transformed plantlets can be obtained.

Note
4. Secondary somatic embryogenesis is defined as the re-induction of embryonal masses from somatic embryos using the same conditions as for the induction of embryonal masses from zygotic embryos. The tissue culture medium used for induction of somatic embryogenesis is the same as used for maintenance.

Solutions:

For 1 litre of half strength Litvay's medium (1/2LM; [41])

1/2 LM 5X frozen stock	200 ml
Casein hydrolysate (casamino acids)	1 g
Sucrose (1% final)	10 g
2,4-D (1mg/ml stock solution)	2.2 ml
6-benzylaminopurine (BA, 0.5mg/ml stock solution)	2.2 ml
Add distilled H_2O to 1 l	
pH	5.7

Autoclave and add 20 ml of filter sterilized glutamine (25 mg/ml stock solution) to cooled medium.

If solid medium is required, add gelrite (4g/l) before adjusting pH and before autoclaving.

Maturation medium (1 litre)
same as 1/2LM but replace 2,4-D and BA with abscisic acid

Abscisic acid (10 mM stock solution)	2.0 ml
Sucrose (6% final)	60 g

For 2 litres of Litvay's 5X stock (1/2 LM 5X frozen stock)

NH_4NO_3	8.21 g
KNO_3	9.5 g
$MgSO_4.7H_2O$	9.25 g
KH_2PO_4 (monobasic)	1.7 g
$CaCl_2.2H_2O$	0.11 g
LM micro nutrient stock (100X)	100 ml
LM vitamin stock (100X)	100 ml
Myo-inositol	1 g
Fe diethylene triamine pentaacetate	0.4 g

Add distilled H_2O to 2000 ml and store frozen in 100 ml aliquots for further use.

For 1 litre of LM micronutrient stock 100X

KI	0.415 g
H_3BO_3	3.1 g
$MnSO_4.H_2O$	2.1 g
$ZnSO_4.7H_2O$	4.3 g
$Na_2MoO_4.2H_2O$	0.125 g
$CuSO_4.5H_2O$	0.05 g
$CoCl_2.6H_2O$	0.013 g

Add distilled H_2O to 1000 ml and store frozen in 100 ml aliquots for further use.

For 1 litre of lm vitamin stock 100X
Nicotinic acid   0.05 g
Pyridoxine HCl   0.01 g
Thiamine HCl   0.01 g
Add distilled H_2O to 1000ml and store frozen in 100 ml aliquots for further use.

GUS histochemical buffer
0.5 mg/ml X-GLUC (5-bromo-4-chloro-3-indolyl glucuronide)
100 mM sodium phosphate pH 7.0
0.5 mM ferrocyanide
0.5 mM ferricyanide
0.5% (V/V) Triton X-100
1 mM EDTA

GUS extraction buffer (GEB)
50 mM NaHPO4 pH 7.0
10 mM β-mercaptoethanol
0.5 M Na_2EDTA pH 8.0
10% Triton X-100

Luciferase assay cell lysis buffer
100 mM Tris-phosphate pH 7.8
2 mM DTT
2 mM DACT
10% glycerol
1% Triton

Luciferase assay reaction buffer
20 mM Tricine
1.07 mM $((MgC)_3)_4MG(OH)_2$
267 mM $MgSO_4$
0.1 mM EDTA
33 mM DTT
270 μM Coenzyme A
530 μM ATP pH7.8

Pollen germination medium (Brewbaker [73])
8.0 mM H_3BO_3
1.3 mM $Ca(NO_3)_2$
1.0 mM KNO_3
1.5 mM $MgSO_4$
5% (W/V) sucrose
pH of 5.2
0.6% agar

References

1. Charest PJ, Rutledge R (1993) Is genetic engineering a viable option for tree improvement? In: Proceedings of the twenty-fourth meeting of the Canadian Tree Improvement Association, pp 27–40, CTIA proceedings.
2. Ellis DD (1992) Transformation in *Picea*. In: Bajaj YPS (Ed.) Biotechnology in Agriculture and Forestry, Springer-Verlag, Berlin and New York (in press).
3. Manders G, Davey MR, Power JB (1992) New Genes for Old Trees. J Exp Bot 43: 1181–1190.
4. van Wordragen MF, Dons HJM (1992) *Agrobacterium tumefaciens*-mediated transformation of recalcitrant crops. Plant Mol Biol Rep 10: 12–36.
5. Huang Y, Diner AM, Karnosky DF (1991) *Agrobacterium rhizogenes*-mediated genetic transformation and regeneration of a conifer: *Larix decidua*. In Vitro Cell Dev Biol 27P: 201–207.
6. Karnosky DF, Podila GK, Tsai CJ, Chiang VL, Shin D-I (1994). Progress in production of transgenic trees with value-added genes: Results with larch and aspen. In: Biological Sciences Symposium, pp 157–160, Tappi Press, Minneapolis, Minnesota.
7. Clapham DH, Ekberg I (1986) Induction of tumours by various strains of *Agrobacterium tumefaciens* on *Abies nordmanniana* and *Picea abies*. Scand J For Res 1: 435–437.
8. Morris JW, Castle LA, Morris RO (1989) Efficacy of different *Agrobacterium tumefaciens* strains in transformation of pinaceous gymnosperms. Physiol Mol Plant Pathol 34: 451–461.
9. Diner AM, Karnosky DF (1987) Differential responses of two conifers to *in vitro* inoculation with *Agrobacterium rhizogenes*. Eur J For Path 17: 211–216.
10. Karnosky DF, Diner AM, Barnes WM (1988) A model system for gene transfer in conifers: european larch and *Agrobacterium*. In: Somatic cell genetics of woody plants, Ahuja MR (Ed.), pp 55–63, Kluwer Academic Publishers, Dordrecht, The Netherlands.
11. McAfee BJ, White EE, Pelcher LE, Lapp MS (1993) Root induction in pine (*Pinus*) and larch (*Larix*) spp. using *Agrobacterium rhizogenes*. Plant Cell Tiss Org Cult 34: 53–62.
12. Stomp A-M, Loopstra CA, Chilton WS, Sederoff RR, Moore LW (1990) Extended host range of *Agrobacterium tumefaciens* in the genus *Pinus*. Plant Physiol 92: 1226–1232.
13. Ellis DD, Roberts D, Sutton B, Lazaroff W, Webb D, Flinn B (1989) Transformation of white spruce and other conifer species by *Agrobacterium tumefaciens*. Plant Cell Rep 8: 16–20.
14. Clapham D, Ekberg I, Eriksson G, Hood EE, Norell L (1990) Within-population variation in susceptibility to *Agrobacterium tumefaciens* A281 in *Picea abies* (L.) Karst. Theor Appl Genet 79: 654–656.
15. Hood EE, Clapham DH, Ekberg I, Johannson T (1990) T-DNA presence and opine production in turmors of *Picea abies* (L.) Karst induced by *Agrobacterium tumafaciens* A281. Plant Mol Biol 14: 111–117.
16. Magnussen D, Clapham D, Grönroos R, von Arnold S (1994) Induction of hairy and normal roots on *Picea abies*, *Pinus sylvestris* and *Pinus cortorta* by *Agrobacterium rhizogenes*. Scand J For Res 9: 46–51.
17. Sederoff R, Stomp A-M, Chilton WS, Moore LW (1986) Gene transfer into loblolly pine by *Agrobacterium tumefaciens*. Bio/Technol 4: 647–650.
18. Gupta PK, Dandekar AM, Durzan DJ (1988) Somatic proembryo formation and transient expression of a luciferase gene in Douglas-fir and loblolly pine protoplasts. Plant Sci 58: 85–92.
19. Loopstra CA, Stomp A-M, Sederoff RR (1990) *Agrobacterium*-mediated DNA transfer in sugar pine. Plant Mol Biol 15: 1–9.
20. Bergmann BA, Stomp A-M (1992) Effect of host plant genotype and growth rate on *Agrobacterium tumefaciens*-mediated gall formation in *Pinus radiata*. Phytopathol 82: 1457–1462.
21. Dandekar AM, Gupta PK, Durzan DJ, Knauf V (1987) Transformation and foreign gene expression in micropropagated Douglas-fir (*Pseudotsuga menziesii*). Bio/Technol 5: 587–590.

22. Morris JW, Morris RO (1990) Identification of an *Agrobacterium tumefaciens* virulence gene inducer from the pinaceous gymnosperm *Pseudotsuga menziesii*. Proc Nat Acad Sci USA 87: 3614–3618.

23. Han KH, Fleming P, Walker K, Loper M, Chilton WS, Mocek U, Gordon MP, Floss HG (1994) Genetic transformation of mature *Taxus* – an approach to genetically control the in vitro production of the anticancer drug, taxol. Plant Sci 95: 187–196.

24. Ellis DD, McCabe DE, McInnis S, Ramachandran R, Russell DR, Wallace KM, Martinell BJ, Roberts DR, Raffa KF, McCown BH (1993) Stable transformation of *Picea glauca* by particle acceleration. Bio/Technol 11: 84–89.

25. Charest PJ, Lachance D, Devantier Y, Klimaszewska KK (1994) Transient gene expression and stable genetic transformation in *Picea mariana* (black spruce) and *Larix laricina* (tamarack). Rev Invest Agraria: Serie Syst Rec For 4: 187–192.

26. Charest PJ, Calero N, Lachance D, Mitsumune M, Yoo BY (1993) The use of microprojetile DNA delivery to bypass the long life cycle of tree species in gene expression studies. In: Current topics in Botanical Research vol. 1, Menon J (Ed.), Council of Scientific Research Integration, pp 151–163, India.

27. Charest PJ, Devantier Y, Ward C, Schaffer U, Klimaszewska KK (1991) Transient expression of foreign chimeric genes in the gymnosperm hybrid larch following electroporation. Can J Bot 69: 1731–1736.

28. Duchesne LC and Charest PJ (1992) Effect of promoter sequence on transient expression of the β-glucuronidase gene in embryogenic calli of *Larix* x *eurolepsis* and *Picea mariana* following microprojection. Can J Bot 70: 175–180.

29. Duchesne LC, Lelu MA, von Aderkas P, Charest PJ (1993) Microprojectile-mediated DNA delivery in haploid and diploid embryogenic cells of *Larix* spp. Can J For Res 23: 312–316.

30. Bekkaoui F, Pilon M, Laine E, Raju DSS, Crosby WL, Dunstan DI (1988) Transient gene expression in electroporated *Picea glauca* protoplasts. Plant Cell Rep 7: 481–484.

31. Bekkaoui F, Datla RSS, Pilon M, Tautorus TE, Crosby WL, Dunstan DI (1990) The effects of promoter on transient expression in conifer cell lines. Theor Appl Genet 79: 353–359.

32. Tautorus TE, Bekkaoui F, Pilon M, Datla RSS, Crosby WL, Fowke LC, Dunstan DI (1989) Factors affecting transient gene expression in electroporated black spruce (*Picea mariana*) and jack pine (*Pinus banksiana*) protoplasts. Theor Appl Genet 78: 531–536.

33. Wilson SM, Thorpe TA, Moloney MM (1989) PEG-mediated expression of GUS and CAT genes in protoplasts from embryogenic suspension cultures of *Picea glauca*. Plant Cell Rep 7: 704–707.

34. Charest PJ, Lachance D, Jones C, Devantier Y (1993) Microprojectile and silicon carbide mediated DNA delivery in conifers and recovery of transgenic black spruce. In Vitro Cell Devel Biol 29A: 87.

35. Duchesne LC, Charest PJ (1991) Transient expression of the β-Glucuronidase gene in embryogenic callus of *Picea mariana* following microprojection. Plant Cell Rep 10: 191–194.

36. Ellis DD, McCabe D, Russel D, Martinell B, McCown BH (1991) Expression of inducible angiosperm promoters in a gymnosperm, *Picea glauca* (white spruce). Plant Mol Biol 17: 19–27.

37. Newton RJ, Yibrah HS, Dong N, Clapham DH, von Arnold S (1992) Expression of an abscisic acid responsive promoter in *Picea abies* (L.) Karst. following bombardment from an electric discharge particle accelerator. Plant Cell Rep 11: 188–191.

38. Charest PJ, Calero N, Lachance D, Datla RSS, Duchesne LC, Tsang EWT (1993) Microprojectile-DNA delivery in conifer species – factors affecting assessment of transient gene expression using the β-Glucuronidase reporter gene. Plant Cell Rep 12: 189–193.

39. Bommineni VR, Datla RSS, Tsang EWT (1994) Expression of *gus* in somatic embryo cultures of black spruce after microprojectile bombardment. J Exp Bot 45: 491–495.

40. Li Y-h, Tremblay FM, Séguin A (1994) Transient transformation of pollen and embryogenic tissues of white spruce (*Picea glauca* (Moench.) Voss) resulting from microprojectile bombardment. Plant Cell Rep 13: 661–665.

41. Robertson D, Weissinger AK, Ackley R, Glover S, Sederoff RR (1992) Genetic transformation of norway spruce (*Picea abies* (L) Karst) using somatic embryo explants by microprojectile bombardment. Plant Mol Biol 19: 925–935.

42. Bommineni VR, Chibbar RN, Datla RSS, Tsang EWT (1993) Transformation of white spruce (*Picea glauca*) somatic embryos by microprojectile bombardment. Plant Cell Rep 13: 17–23.

43. Campbell MA, Kinlaw CS, Neale DB (1992) Expression of luciferase and β-glucuronidase in *Pinus radiata* suspension cells using electroporation and particle bombardment. Can J For Res 22: 2014–2018.

44. Stomp A-M, Weissinger A, Sederoff RR (1991) Transient expression from microprojectile-mediated DNA transfer in *Pinus taeda*. Plant Cell Rep 10: 187–190.

45. Loopstra CA, Weissinger AK, Sederoff RR (1992) Transient gene expression in differentiating pine wood using microprojectile bombardment. Can J For Res 22: 993–996.

46. Walter C, Smith DR, Connett MB, Grace L, White DWR (1994) A biolistic approach for the transfer and expression of a *gusA* reporter gene in embryogenic cultures of *Pinus radiata*. Plant Cell Rep 14: 69–74.

47. Goldfarb B, Strauss SH, Howe GT, Zaerr JB (1991) Transient gene expression of microprojectile-introduced DNA in Douglas-fir cotyledons. Plant Cell Rep 10: 517–521.

48. Tautorus TE, Fowke LC, Dunstan DI (1991) Somatic embryogenesis in conifers. Can J Bot 69: 1873–1899.

49. Lelu MA, Klimaszewska KK, Jones C, Ward C, von Aderkas P, Charest PJ (1993). A laboratory guide to somatic embryogenesis in spruce and larch. No. PI-X-111. Petawawa National Forestry Institute.

50. Allen GC, Hall GE, Childs LC, Weissinger AK, Spiker S, Thompson WF (1993) Scaffold attachment regions increase reporter gene expression in stably transformed plant cells. Plant Cell 5: 603–613.

51. Gordon-Kamm WJ, Spencer TM, Mangano ML, Adams TR, Daines RJ, Start WG, O'Brien JV, Chambers SA, Adams WRJ, Willetts NG, Rice TB, Mackey CJ, Krueger RW, Kausch AP, Lemaux PG (1990) Transformation of maize cells and regeneration of fertile transgenic plants. Plant Cell 2: 603–618.

52. Klein TM, Harper EC, Svab Z, Sanford JC, Fromm ME, Maliga P (1988) Stable genetic transformation of intact *Nicotiana* cells by the particle bombardment process. Proc Natl Acad Sci USA 85: 8502–8505.

53. Twell D, Klein TM, Fromm ME, McCormick S (1989) Transient expression of chimeric genes delivered into pollen by microprojectile bombardment. Plant Physiol 91: 1270–1274.

54. Nishihara M, Ito M, Tanaka I, Kyo M, Ono K, Irifune K, Morikawa H (1993) Expression of the β-glucuronidase gene in pollen of lily (*Lilium longiflorum*), tobacco (*Nicotiana tabacum*), *Nicotiana rustica*, and peony (*Paeonia lactiflora*) by particle bombardment. Plant Physiol 102: 357–361.

55. van der Leede-Plegt LM, van de Ven BCE, Bino RJ, van der Salm TPM, van Tunen AJ (1992) Introduction and differential use of various promoters in pollen grains of *Nicotiana glutinosa* and *Lilium longiflorum*. Plant Cell Rep 11: 20–24.

56. Hamilton D, Roy M, Rueda J, Sindhu R, Sanford J, Mascarenhas J (1992) Dissection of a pollen-specific promoter from maize by transient transformation assays. Plant Mol Biol 18: 211–218.

57. Negrutiu I, Heberle-Bors E, Potrykus I (1986) Attempts to transform for kanamycin-resistance in mature pollen of tobacco. In: Mulcahy DL, Bergamini-Mulcahy G, Ottaviano E (Eds.) Biotechnology and Ecology of Pollen, pp 65–70, Springer-Verlag, New York.

58. Twell D, Klein TM, McCormick S (1991) Transformation of pollen by particle bombardment. In: Lindsey K (Ed.) Plant Tissue Culture Manual, pp 1–14, Kluwer Academic, Dordrecht.

59. Hay I, Lachance D, von Aderkas P, Charest PJ (1994) Transient chimeric gene expression in pollen of five conifer species following microparticle bombardment. Can J For Res 24: 2417–2423.

60. Cheliak WM, Klimaszewska KK (1991) Genetic variation in somatic embryogenesis response in open-pollinated families of black spruce. Theor Appl Genet 82: 185–190.

61. Thorpe TA, Harry IS (1991) Clonal propagation of conifers. In: Lindsey K (Ed.) Plant Tissue Culture Manual, C3: pp 1–16, Kluwer Academic Publishers, Dordrecht.

62. Russell JA, Roy MK, Sanford JC (1992) Major improvements in biolistic transformation of suspension-cultured tobacco cells. In Vitro Cell Dev Biol-Plant 28P: 97–105.

63. Vain P, McMullen MD, Finer JJ (1993) Osmotic treatment enhances particle bombardment-mediated transient and stable transformation of maize. Plant Cell Rep 12: 84–88.

64. Birnboim HC, Doly J (1979) A rapid alkaline extraction procedure for screening recombinant plasmid DNA. Nucleic Acids Res 7: 1513–1517.

65. Sambrook J, Fritsch EF, Maniatis T (1989) Molecular Cloning: A Laboratory Manual (2nd ed.), Cold Spring Harbor, NY: Cold Spring Habor Laboratory.

66. Kikkert JR (1993) The Biolistic PDS-1000/He device. Plant Cell Tissue Organ Cult 33: 221–226.

67. Jefferson RA (1987) Assaying chimeric genes in plants: The GUS gene fusion system. Plant Mol Biol Rep 5: 387–405.

68. Howell SH, Ow DW, Schneider M (1989) Use of the firefly luciferase gene as a reporter of gene expression in plants. In: Gelvin SB, Schilperoort RA (Eds.) Plant Molecular Biology Manual, pp 1–11, Kluwer Academic Publishers, Dordrecht.

69. Jefferson RA, Kavanagh TA, Bevan MW (1987) GUS fusion: β-Glucuronidase as a sensitive and versatile gene fusion marker in higher plants. EMBO J 6: 3901–3907.

70. Kosugi S, Ohashi Y, Nakajima K, Arai Y (1990) An improved assay for β-glucuronidase in transformed cells: methanol almost completely suppresses a putative endogenous β-glucuronidase activity. Plant Science 70: 133–140.

71. Saiki RK, Gelfand DH, Stoffel S, Scharf SJ, Higuchi R, Horn GT, Mullis KB, Erlich HA (1988) Primer-directed enzymatic amplification of DNA with a thermostable DNA polymerase. Science 239: 487–491.

72. Charest PJ, Bonga J, Klimaszewska K (1996) Cryopreservation of plant tissue cultures: the example of embryogenic tissues from conifers. In: Plant Tissue Culture Manual, Lindsey K (Ed.), C9: pp 1–27 Kluwer Academic Publishers, Dordrecht.

73. Brewbaker JL, Kwack BH (1963) The essential role of calcium ion in pollen germination and pollen tube growth. Am J Bot 50: 859–865.

Section C:
Propagation & Conservation of Germplasm

Germplasm

Sect. C

Plant Tissue Culture
Manual

Plant Tissue Culture Manual **C1**: 1–7, 1991.
© 1991 *Kluwer Academic Publishers*.

Clonal Propagation of Orchids

YONEO SAGAWA

University of Hawaii, Honolulu, HI 96822 USA

Introduction

Ever since Knudson [3] developed a nonsymbiotic in vitro culture method for germination of orchid seeds, tissue (in vitro) culture has played an important and highly significant role in the cultivation, propagation, breeding and preservation of orchid species and hybrids. The purpose of this chapter is to present the procedures in detail for clonal propagation of orchids as successfully practiced in our laboratory.

Initially in vitro techniques were used for germination of orchid seeds shed from mature fruits of species and hybrids. However, by the middle of this century the practice of harvesting immature (green) fruits and aseptically culturing the contents was used to overcome some barriers of incompatibility and produce some unique hybrids.

Since Morel [4] and Wimber [9] demonstrated the applicability of tissue culture for rapid clonal propagation of orchids, this technology has been successfully applied for commercial production of selected orchid plants. Large quantities of quality plants, establishment of new hybrids and multiplication and preservation of germplasm are some benefits which have been derived from application of this technology.

Although orchids are considered by many to be easy to propagate via in vitro culture owing to the number of papers which have been published on this topic, since the family Orchidaceae incorporates over 450 genera and 15,000 species and is extremely heterozygous, a diversity of response to tissue culture is inevitable depending on genotype, species and genus. This diversity is poorly documented in the literature.

Furthermore, since the time from initiation of tissue culture to maturity of plant may range from one to several years, data on uniformity, stability and fidelity of end product are scarce. Therefore, in this chapter, general procedures are detailed with some notations on precautions to alleviate variations.

Culture medium

The formulations for media to successfully grow orchids in vitro are relatively simple. Many are presented in an appendix of a book written by Arditti [1]. Two commonly used basic media are Knudson C [3] and Vacin and Went [8] in which growth regulators or organic additives are incorporated. In our laboratory a modified Vacin and Went medium (Sagawa and Kunisaki [7]) with 15% coconut water is preferred since it is better buffered. If the medium is prepared properly, very little adjustment of pH (4.8–5.0) is necessary. Coconut water (15% v/v) is added to the medium for initiation and multiplication while homogenized green banana (50–100 g/l) is added to the medium for rooting. Supplemental sucrose is either reduced or deleted in media for initiation and multiplication of monopodial orchids and restored in media for shoot promotion and rooting. If coconut water is not readily available other hormones may be used [1].

Explants

Explants are excised from vegetative shoots, young inflorescences, leaves and roots. Our preference for source of explant is terminal and axillary buds from active vegetative shoots from healthy plants. Apical bud explants include the apical meristem with 3 to 4 nodes with attached young leaves. Axillary bud explants consist of the cube with axillary bud attached to node. Root explants consist of 3 to 5 mm root tips.

Disinfestation

Disinfestation is accomplished in either 2 or 3 steps using household bleach (clorox) in descending concentrations (20, 15, 10%) and time (20, 15, 10 min) Tween 20 (2–3 drops/100 ml), a wetting agent, is usually added to the initial disinfesting solution. Sterile deionized water is used to rinse after each treatment. If highly contaminated, shoots are rinsed in running tap water for approximately 30 min to 2 hours. More drastic treatment would be rinsing shoots in 70% ethanol for one minute followed by rinsing with sterile deionized water prior to disinfestation with household bleach.

Culture conditions

Aseptic cultures are maintained under continuous or interrupted light from cool white or power groove fluorescent lamps (approximately 100 fc) at approximately 25 °C to 28 °C. Light may be cycled to provide a 16 hour day. Liquid cultures are placed on vertical or horizontal shakers. Some genera such as *Phalaenopsis* and *Paphiopedilum* are initiated on solid media since explants bruise easily and soon become necrotic when agitated on shakers.

Procedures

The growth habit of orchids is either monopodial or sympodial. Monopodial orchids have a single upright indeterminate axis of growth; inflorescences are axillary; e.g. genera of subtribe Sarcanthinae including *Ascocentrum, Doritis, Vanda, Phalaenopsis*. Sympodial orchids have two axes of growth — a horizontal axis which is indeterminate and a vertical axis which is determinate and terminated by an inflorescence; e.g. *Cattleya, Dendrobium, Cymbidium, Epidendrum, Miltonia, Oncidium*. Recognition of basic growth habit is important for selection of the appropriate procedure for initiation and multiplication.

Initiation of sympodial orchid

The example given is for initiation of cultures of *Dendrobium* sp., a sympodial orchid, from shoot tip and axillary bud explants from vegetative shoot. The protocol was developed in our laboratory and has proven to be suitable for most sympodial orchids.

Steps in the procedure
1. Select a young vegetative shoot without expanded leaves. Sever the shoot from rhizome with a razor blade.
2. Remove all papery scale-like leaves from the shoot and rinse in running water for a minimum of 30 min.
3. Remove in sequence all exposed scale-like leaves with forceps until the first visible axillary bud is exposed. Make sure that each leaf is torn off to the node. It is better to tear leaf off rather than to cut it.
4. Disinfest in 15% household bleach with Tween 20 for 15 min; shake occasionally.
5. Rinse in sterile deionized water.
6. Remove all but 2 or 3 leaves which cover the apical meristem with sterile forceps.
7. Disinfest with 10% household bleach for 10 min; shake occasionally.
8. Rinse in sterile deionized water.
9. For apical shoot explants cut transversely across the shoot tip to obtain a piece with an apical meristem and 2 to 3 nodes.
10. For axillary shoot explants isolate the cube with an axillary bud attached to the node by slicing on each side of the bud, across young leaves and below the node.
11. Culture both apical and axillary bud explants in sterile liquid media.
12. Place cultures in liquid medium on rotary shaker.
13. If liquid or solid medium discolors, change the medium daily or as necessary until the medium remains clear.
14. The first sign of growth appears within 4 to 6 weeks after culture.

Initiation of monopodial orchid

1. Sever 10 to 15 cm of shoot from a stock plant.
2. Trim expanded leaves flush to the stem axis.
3. Cut the stem into segments with two to three nodes.
4. Disinfest stem segments in 20% v/v commercial bleach with Tween 20 for 10 min; shake occasionally.
5. In a sterile transfer hood, tear the leaf base from the stem to expose the axillary bud, using sterile forceps.
6. Disinfest the stem in 5% commercial bleach for 10 min; shake occasionally.
7. Rinse in sterile deionized water.
8. For apical shoot explants cut transversely across shoot tip to obtain a piece including the apical meristem and 2 to 3 nodes.
9. For axillary shoot explants isolate the cube which includes a bud attached to a node.
10. Transfer the explant into liquid medium without supplemental sucrose and place on a gyrotary shaker (approx. 100 rpm).
11. If the liquid medium discolors, change the medium daily or as necessary until the liquid remains clear.
12. The first sign of growth appears within 6 to 8 weeks after culture.

Initiation from young inflorescence

1. Carefully collect young inflorescences which are still enclosed in bracts.
2. Disinfest in 10% commercial bleach with Tween 20 for 15 min; shake occasionally.
3. In a transfer hood, remove bracts by tearing with sterile forceps.
4. Disinfest in 5% commercial bleach for 10 min; shake occasionally.
5. Rinse in sterile deionized water.
6. Culture the inflorescence in liquid medium.

Multiplication and rooting

1. For multiplication of liquid cultures, subculture clusters into separate flasks with the same medium; protocorm-like bodies or shoots break apart with time.
2. For multiplication of cultures on solid medium, separate protocorm-like bodies or shoots and subculture into flasks with the same medium.
3. For rooting, transfer shoots to solid agar medium with sucrose. Homogenized green banana (50–100 g/l) is used as an additive to promote rooting.

Transfer to greenhouse

1. When plantlets have roots approximately 1 to 2 cm in length, flush plants out of the container with water, rinse agar off and plant in media appropriate for growing orchid seedlings.

2. Place pots with plants under low light and high humidity for a few weeks; do not overwater.
3. When active root growth begins, move pots to higher light conditions.

Solutions and media

Use deionized water for preparation of stock solutions and media. Use sterile deionized water for preparation of disinfesting solution and rinsing during disinfestation. The concentration of Tween 20 wetting agent used is 2 to 3 drops per 100 ml.

Modified Vacin and Went medium

	Amount/liter
$Ca_3(PO_4)_2$	0.20 g
KNO_3	0.525
KH_2PO_4	0.25
$MgSO_4 \cdot 7H_2O$	0.25
$(NH_4)_2SO_4$	0.50
$MnSO_4 \cdot 4H_2O*$	0.0068
Sucrose	20.00
Agar*	8.00
Water*	845.00 ml
Coconut Water*	150.00
$FeSO_4 \cdot 7H_2O$	27.8 mg
$Na_2 \cdot EDTA$	37.3
pH 4.8–5.0	

* Modified constituents

If cultures are incubated in a greenhouse or at temperatures above 32 °C, agar may be increased up to 15 g/l to maintain medium solidity. Fresh liquid from immature coconuts is filtered through 2 : 1 mixture of Diatomaceous Earth (Sigma D5509) and Cellulose (Sigma 9004-34-6) and stored in a freezer. Coconut water is defrosted in a microwave oven just prior to use.

Modified Vacin and Went medium can also be prepared by use of stock solutions as follows:

Stock Solution A:
KNO_3	5.25 g/l
KH_2PO_4	2.50
$(NH_4)_2SO_4$	5.00
$MnSO_4 \cdot 4H_2O$	0.068

Stock Solution B:
$MgSO_4 \cdot 7H_2O$	2.50 g/l

Stock Solution F:
Na_2EDTA	7.450 g/l
$FeSO_4 \cdot 7H_2O$	5.570

1. To 250 ml deionized water add 5 ml stock solution F.
2. Add 100 ml stock solution A and 100 ml stock solution B.
3. In a small vial or beaker thoroughly dissolve 0.20 g $Ca_3(PO_4)_2$ with 1N HCl. Add to mixture.
4. Add 150 ml coconut water.
5. Add 20.0 g sucrose.
6. Bring the volume to 1000 ml.
7. Adjust pH to 4.8 to 5.0.
8. Add 8.0 g agar.
9. Boil the mixture to dissolve the agar, on a hotplate, in an autoclave or in a microwave oven.
10. Dispense into containers; cap containers.
11. Autoclave for 10 min. at 15 psi and 121° C.
12. Cool. If slants are desired, slant containers at this time.

Other natural complex: green banana

1. Harvest, wash and store mature green bananas in a freezer.
2. Just prior to use, defrost the banana at room temperature or in a microwave oven and peel. Fresh green bananas are very difficult to peel.
3. Weigh amount of banana to be added (50–100 g/l).
4. Homogenize in a blender for 30 to 60 sec at high speed and add to the medium.

Trouble shooting

– The best sources of explants are active vegetative shoots without expanded leaves from healthy plants or young inflorescences. Leaves and roots are less desirable due to a higher mutation rate. Dormant buds from rhizomes of sympodial orchids such as cattleya rarely respond.
– Seedlings *in vitro* are poor sources of explants since orchids are highly heterozygous and the phenotype of the end product cannot be predicted.
– Sucrose is generally reduced or deleted from initiation and multiplication media of monopodial orchids.
– Browning and necrosis of explants upon culture or subculture may occur. The causes and methods for prevention are as yet unknown.
– Some genotypes are highly mutable. The cause and methods for prevention are as yet unknown.

References

1. Arditti J (1977) Clonal propagation of orchids by means of tissue culture – A manual. In: Orchid Biology Reviews and Perspectives, I (J Arditti, ed.) pp. 203–293 Cornell Univ. Press, Ithaca, NY
2. Goh CJ (1990) Orchids, Monopodials. In: Handbook of Plant Cell Culture 5 (Ammirato PV, Evans DA, Sharp WP, Bajaj YPS, eds.) pp. 598–637 McGraw-Hill Inc. NY
3. Knudson L (1946) A new nutrient solution for germination of orchid seed. Amer. Orchid Soc. Bull. 15: 214–217.
4. Morel GM (1960) Producing virus-free Cymbidiums. Amer. Orchid Soc. Bull. 29: 495–497.
5. Rao AN (1977) Tissue culture in the orchid industry. In: Applied and Fundamental Aspects of Plant Cell, Tissue and Organ Culture (Reinert J & Bajaj YPS, eds.) pp. 44–69. Springer-Verlag, NY
6. Sagawa Y (1990) Orchids, Other Considerations. In: Handbook of Plant Cell Culture 5 (Ammirato PV, Evans DA, Sharp WP, Bajaj YPS, eds.) pp. 638–653 McGraw-Hill Inc. NY
7. Sagawa Y & Kunisaki JT (1984) Clonal propagation: Orchids. In: Cell Culture and Somatic Cell Genetics of Plants I (Vasil IK, ed.) pp. 61–67 Academic Press NY
8. Vacin E & Went F (1949) Some pH changes in nutrient solution. Botan. Gaz. 110: 605–613.
9. Wimber DE (1963) Clonal multiplication of Cymbidiums through tissue culture of shoot meristem. Amer. Orchid Soc. Bull. 32: 105–107.

Plant Tissue Culture Manual **C2**: 1–14, 1991.
© 1991 *Kluwer Academic Publishers.*

Clonal propagation of palms

BRENT TISSERAT
U.S. Department of Agriculture, Agricultural Research Service, Fruit and Vegetable Chemistry Laboratory, 263 S. Chester Avenue, Pasadena, CA 91106, USA

Introduction

Palm products, historically and currently, are important sources of economic revenues for sub-tropical and tropical countries (e.g. palm oil accounts for 15% of the world's total vegetable oil) [20]. The only means of vegetative propagation in palms, being monocotyledons, is restricted to vegetative lateral bud outgrowths (offshoots) which are produced during the juvenile phase of their life-cycle (e.g. *Phoenix dactylifera* L., date palm). Only a limited number of vegetative lateral bud outgrowths are produced in the life of a tree. When the palm tree reaches the adult phase of their life-cycle, lateral buds become entirely reproductive in nature (i.e. inflorescences). Several economically important palms (e.g. coconut palm, *Cocos nucifera* L. and oil palm, *Elaeis guineensis* L.) do not produce any offshoots during their juvenility. Micropropagation of palms appears to provide an ideal method to obtain numerous clonal palms. Production of desirable clonal palms through tissue culture has been suggested as a means to substantially increase fruit product yields/tree [1, 3, 5, 7, 8, 13, 15, 17, 24, 29, 31].

The Arecaceae (palm) family is large and morphologically diverse and a variety of media, culture techniques and explant choices have been employed to clone palms [4, 5, 8, 11, 13, 17–31, 37, 38]. However, a general similarity appears to exist regarding the response of palm tissues to *in vitro* conditions [31, 38]. For example, date palm and oil palm somatic embryogenesis and culture conditions appear to be quite similar [30, 38]. It remains to be determined through extensive comparative survey experiments if standardization of media, techniques and explant selections is possible.

Currently, at the time of writing (August, 1990) the state of cloning in palms is uncertain [2, 4, 6, 7, 9, 10, 14, 16, 21, 26, 28, 29, 31]. The two commonly 'accepted' methods to clone palms are: 1) somatic embryogenesis *via* callus, and 2) lateral bud proliferation. Date palms may be propagated from lateral bud branching *in vitro* [25, 37]. The apparent inability of cultured oil palm and coconut palm shoot tips to readily produce vegetative lateral bud outgrowths suggests that clonal propagation of these palms may only be achieved by somatic embryogenesis. However, it should be noted that sporadic reports of vegetative branching from cultured normally non-branching palms (i.e. non-date palms) including coconut [12], *Metroxylon sp.* [38] and adventitious budding in oil palm [23] have appeared in the literature.

Such reports suggest that a method for the vegetative propagation of normally non-branching palms *in vitro* through offshoots is feasible.

A few grams of callus capable of prolific somatic embryogenesis may be an unlimited source of somatic 'clonal' embryoids and plantlets [11, 18, 19, 22, 23, 27, 30–36]. However, acceptance of somatic embyrogenesis as a means to clonally propagate palms is premature [4, 6, 7, 9, 10, 14, 16, 19, 21, 26, 28, 29, 31]. The genetic stability of embryogenic callus is unknown. Comparison of the composition, quality and yield of fruit produced from 'cloned' trees with fruit from the mother tree would provide the most conclusive information to assess the viability of employing the somatic embryogenesis process. However, the long-lived nature of palms in which a sequential juvenile growth phase (usually in the first 3–5 years of life) precedes the adult stage of development has precluded such an analysis, to date. Indirect means to determine the clonal nature of plantlets including chromosome counts of root tip cells, e.g. oil palm [9] and date palm [26] and isozyme analysis, e.g. date palm [26, 32], have been employed. The importance of resolving the clonal status of palm plantlets produced in tissue culture can not be over emphasized.

Although several hundred somatic embryos may be produced within a single culture tube during a culture passage of 8 weeks, few germinate to become plantlets [32–34]. Similarly, only a few axillary buds are produced from a cultured tip during a single culture passage [37]. A procedure is provided to accelerate the production rates of somatic embryos and plantlets obtained from embryogenic callus or lateral bud outgrowths using an automated culture system (ACS). Using the ACS, the original culture has access to a much larger growing area and more media than can be provided within a culture tube. The necessity for artificial division and manual routine reculturing which are standard procedures in traditional micropropagation is much reduced resulting therefore in enhanced culture growth [39, 40].

Procedures

Preparation, culture and plant generation using the asexual embryogenesis process from clonal tissues of P. dactylifera *(date palm)*

The protocol given for plantlet formation from somatic callus has been previously reported [31, 33, 38]. Offshoots are employed as the source material due to their ease of obtainment. The same procedure may also be employed to culture shoot tips from the mature, solitary-trunk, adult palm trees *in vitro*.

Steps in the procedure

1. Remove offshoots from the parent tree by severing the vascular connection with either a sledge hammer and chisel or a chain saw. Dissect offshoots using a hatchet and serrated knife. Remove leaves acropetally, exposing lateral buds at the axil and each leaf. Shoot tips are obtained after peeling off all mature leaves. Lateral buds in various stages of differentiation occur within the same source. Store buds and tips in a cold antioxidant solution (150 mg/l citric acid and 100 mg/l ascorbic acid). Keep explants in refrigerator at 0 °C until conducting the surface sterilization procedure.
2. Once all buds and tips are obtained from offshoots, trim away the outermost leaves of the buds and tips to obtain explants that are about 0.5 cm^2.
3. Sterilize explants by wrapping them in cheesecloth to prevent loss in handling procedures and place in a 25- × 150-mm culture tube. Sterilize in 2.6% solutions of sodium hypochlorite solution (containing 1 drop of Tween-20 per 100 ml solution) for 15 min. Agitating the culture tube periodically to dislodge air bubbles from tissues. Pour off bleach solution and rinse three times with sterile water. Remove explant package and transfer aseptically to the sterile petri dish (15- × 150-mm dia.).
4. Remove additional leaves one at a time from shoot tip and bud explants to obtain a 1–3-mm^2 culture. An additional 10-sec dip in the bleach solution of this explant before planting reduces contamination.
5. Plant explant on the surface of the 'callus medium' described in Tables 1 and 2. Orient apical end aerially upwards.
6. Reculture explants at 8-week intervals. Nodular yellow-brown callus initiation becomes evident after 2–3 culture passages.
7. Eventually the original bud or tip structure becomes obliterated through the production of yellow-white friable and nodular callus. Subculture 1-cm^2 pieces to 'plantlet germination medium' (Table 2). Asexual embryos and green plantlets usually will become apparent within 2 to 4 weeks in culture.
8. To obtain greater multiplication rates, transfer 1 cm^2 piece of callus to ACS culture chamber.
9. To obtain free-living plantlet follow procedure outlined in 'Transferring somatic plantlets obtained from clonal tissues to a field environment'.

Solutions

Media may be sterilized by autoclaving.

Table 1. Composition of inorganic salt formulation.

Ingredients	mg/l
Inorganic salts:	
Macronutrients:	
$MgSO_4 \cdot 7H_2O$	370
KH_2PO_4	170
$NaH_2PO_4 \cdot H_2O$	170
KNO_3	1900
NH_4NO_3	1650
$CaCl_2 \cdot 2H_2O$	440
Micronutrients:	
H_3BO_3	6.2
$MnSO_4 \cdot H_2O$	15.6
$ZnSO_4 \cdot 7H_2O$	8.6
$NaMoO_4 \cdot 2H_2O$	0.25
$CuSO_4 \cdot 5H_2O$	0.025
$CoCl_2 \cdot 6H_2O$	0.025
KI	0.83
$FeSO_4 \cdot 7H_2O$	27.8
Na_2EDTA	37.3

Table 2. Composition of date palm tissue culture media. The same inorganic basal salts are employed in each medium.

Ingredients	Callus (mg/l)	Plantlet germination (mg/l)	Shoot tip (mg/l)	Shooting (mg/l)	Rooting (mg/l)
Vitamins:					
m-inositol	100	100	100	100	100
thiamine · RCl	0.4	0.4	0.4	0.4	0.4
Carbohydrate source:					
Sucrose	30000	30000	30000	30000	30000
Complex addenda:					
Agar	8000	8000	8000	8000	8000
Charcoal, activated	3000	3000	3000		
Phytohormones[a]					
2,4-D	100		10		
2iP	3			10	
NAA				0.1	0.1
pH	5.8	5.8	5.8	5.8	5.8

[a] 2,4-D = 2,4-dichlorophenoxyacetic acid; 2iP = N^6(2-isopentenyl)adenine; NAA = 1-naphtha leneacetic acid.

Preparation, culture and plant generation using axillary bud branching process from clonal tissues of P. dactylifera *(date palm)*

The protocol given is for somatic plantlet formation has been previously reported [31, 33, 38]. Offshoots are employed as the source parent clonal material due to the ease of obtainment.

Steps in the procedure
1. Follow procedure steps 1—4 outlined in 'Preparation, culture and plant generation using asexual embyrogenesis process from clonal tissues of *P. dactylifera* (date palm)' to obtain surface-sterile explants.
2. Plant explant on the surface of the 'shoot tip medium' described in Table 1. Orient apical end aerially upwards.
3. Cultures will initiate leaves and enlarge considerably within the next 4 to 6 weeks. Reculture explants at 8-week intervals.
4. To induce axillary buds, reculture tips to 'shooting medium' to induce axillary proliferations. For enhanced axillary bud formation follow steps outlined in 'Enhanced mass propagation of date palm clonal plantlets using the automated culture system (ACS) technology'.
5. Root isolated buds or bud clumps by methods outline in 'Transferring plantlets to free-living conditions'.

Enhanced mass propagation of date palm clonal plantlets using the automated culture system (ACS) technology

This protocol has been previous reported [39, 40]. Commercial automated systems can also be procured (De Novo, Automated Micropropagation Systems, 736 N. Bradish Avenue, San Dimas, CA 91773).

Steps in the procedure
1. Culture 1–3 cm^2 clumps of embryogenic callus on 'plantlet medium'; culture a single shooting bud cultures on 'shooting medium'.
2. Within 2–8 weeks after culture establishment a dramatic increase in size and growth will occur. Eventually after 16–32 weeks in culture plantlets as large as 10–15 cm in length can be removed and transferred to free living conditions. Growth rates of 2 to 5 times that obtained in agar medium are plausible. In some cases, depending on the cultured species, even higher growth rates are obtainable (Fig. 1).
3. Remove large plantlets every 8–16 weeks, as desired, and reculture these plantlets to free-living conditions as described in 'Transferring somatic plantlets obtained from clonal tissues to field environment' section. Replace media every 8–16 weeks. Media showing excessive browning should be replaced frequently.

ACS Components
— Culture chamber, 1 quart (0.95 l) glass mason jar filled to a depth of 1.5 cm with 5 mm diameter solid glass beads.
— Peristaltic pump, variable speed type with a 100 rpm motor capable of delivering a flow range of 5–80 ml per minute using tubing size 16. To be equipped with Lexan polycarbonate heads and cold-rolled steel rotor (Cole-Parmer Instrument Company, Chicago, Ill).
— Silicone tubing, dimensions: size 16, 3.1 mm inside diameter (i.d.); 3.2 mm hose barb size; and 7 mm outer diameter (o.d.).
— Media reservoir, 1 liter capacity pyrex glass media storage bottles (101 mm o.d. × 225 mm height (h) × 45 mm screw cap size fitted with linerless polypropylene screw caps (Bellco Glass, Inc., Vineland, NJ).

Construction of ACS
— Centrally position a No. 7 (30 mm dia.) 'Twistit' rubber stopper containing 3 predrilled countersunk holes (American Sci. Products, McPaw Park, Ill) in the center of the mason lid of the culture chamber by drilling a 2.86 cm diameter hole using a key-hole drill. Insert two 6 mm o.d. × 120 mm length (l.) glass tubes with right angle bends 60 mm from one end into two holes. Connect 3 cm l. of 6 mm o.d. silicon tubing attached to an air filter (ACRO 37 TF 37, Gelman 4464) to these tubes. Insert an inlet line tube (6 mm o.d. × 260 mm l.) into the third stopper hole with a right angle bend 50 mm from end in atmosphere. Make sure that this tubing touches the bottom of the chamber. This is important to ensure

good evacuation of medium following culture soaking time. Connect a 120 cm length of silicone tubing to this outlet line and connect to culture reservoir.

— Drill a 2.86 cm diameter hole using a key-hole drill in the twist top closure of the culture media reservoir to accommodate the 'Twistit' stopper. Insert 2 glass tubings as previously described.

— Diagram for the construction of the relays employed in the interface system is shown.

— Add 950 ml to each media reservoir bottle. Use pinch clamp closures to seal all silicone connections for medium reservoirs during autoclaving. Autoclave at 15 lbs/in^2 for 15 min at 121 °C.

— A variety of microcomputers or microprocessor-controller systems and even some multi-channel timers can be used for this system.

— Adjust the input flow of medium into the chamber to allow the culture to be only half submerged in solution. Thereafter, program the computer or controller for this time. Program entry of medium into the chamber every 4 hours.

Fig. 1. Growth responses of date palm callus culture grown for 16 weeks in agar tubes versus that obtained using the ACS.

Transferring somatic plantlets obtained from clonal tissues to a field environment

Steps in the procedure

1. Reculture the plantlet on 'rooting medium' with the primary root trimmed to 1 to 2 cm long to enhance the formation for adventitious roots (Table 2). Position the somatic embryos and plantlets during the first culture transfer so that the root portion is embedded in the agar medium and the shoot and leaves grow upwards. Continue the reculturing every 8 weeks for 2 or 3 culture passages until plantlets reach a length of 10–15 cm with 2 or 3 leaves and have a well-developed adventitious root system.

2. Transfer plantlets to free-living conditions as follows: Remove plantlets carefully from agar medium without damaging the root systems and soak in distilled water for 15 min to avoid dehydration and to remove excess media that may be adhering to them. Rinse plantlets three times with distilled water, spray them with 0.5% benolate containing active ingredient benomyl (DuPont, Wilmington, DE) fungicide solution and transfer plantlets to soil medium, which consists of peat moss and vermiculite in a 1 : 1 v/v ratio. Pot Plantlets in either 7.6 cm diameter plastic or jiffy peat containers and enclose within a transparent tent composed of two interlocking clear polystyrene tumblers.

3. Spray foliage weekly with 0.5% benolate to minimize fungal growth. Water pots every other day with distilled water and once a week with one-fourth strength Hoagland's solution during the first 2 months of plantlet development. Incubate plantlets initially in an environmentally controlled chamber under 800 fc light intensity, 16-hr photoperiod at 28 °C for 2 weeks, then transfer to a shaded greenhouse. Gradually acclimate plantlets to the greenhouse humidity conditions by punching holes in the plastic cover. After 2 months remove covers and treat the plant thereafter as a palm seedling.

Disclaimer

References

1. Anonymous (1981) Yield increase expected from cloned palms. Ceres 14: 10–11.
2. Al-Jiboury AAM, Salman RM, Omar MS (1988) Transfer of the in vitro regenerated date palms to the soil. Date Palm J 6: 390–400.
3. Smith RW (1975) Progress in the vegetative propagation of coconut, using tissue culture techniques. Fourth session of the FAO technical working party on coconut production, protection and processing, Kingston, 14–25 September 1975, FAO UN, Rome, pp 1–6.
4. Branton RL, Blake J (1983) A lovely clone of coconuts. New Scientist 98: 554–557.
5. Choo WK, Yew WC, Corley RHV (1981) Tissue culture of palms–A review. In AN Rao, ed, Tissue Culture of Economically Important Plants, Commit on Sci and Tech in Dev Count and Asian Network for Biol Sci, Singapore, pp 138–144.
6. Corley RHV (1977) First clonal oil palms planted in the field. Planter 53: 331–332.
7. Corley RHV (1981) Vegetative Propagation. In WR Stanton, M Flach, eds, Sago, The Equatorial Swamp as a Natural Research, Proc Second Int Sago Symp, Martinus Nijhoff, The Hauge-Boston-London, pp 98–109.
8. Corley RHV, Barrett JN, Jones LH (1976) Vegetative propagation of oil palm via tissue culture. Malay Int Agric Oil Palm Conf 1976: 1–7.
9. Corley RHV, Wong CY, Wooi KC, Jones LH (1980) Early results from the first oil palm clone trials. Malay Int Agric Oil Palm Conf 1980: 1–19.
10. Corley RHV, Lee CH, Law IM, Wong CY (1986) Abnormal flower development in oil palm clones. Planter 62: 233–240.
11. Daikh H, Demarly Y (1987) Resultats preliminaires sur l'obtention d'embryons somatiques et la realisation de semences artificielles de palmier dattier (*Phoenix dactylifera* L.). Fruits 42: 593–596.
12. Fisher JB, Tsai JH (1979) A branched coconut seedling in tissue culture. Principes 23: 128–131.
13. Jones LH (1983) The oil palm and its clonal propagation by tissue culture. Biologist 30: 181–188.
14. Jones LH (1984) Novel Palm Oils from cloned Palms. J Amer Oil Chem Soc 61: 1717–1719.
15. Jones LH (1988) Commercial development of oil palm clones. Fett Wiss Tech 90: 58–61.
16. Jones LH, Barfield D, Barrett J, Flook A, Pollock K, Robinson P (1982) Cytology of oil palm cultures and regenerated plants. In A Fujiwara, ed, Plant Tissue Culture 1982, Proc 5th Int Congr Plant Tissue and Cell Cult, Tokyo, 11–16 July 1982, Maruzen Com, Tokyo, pp 727–728.
17. Kovoor A (1981) Palm tissue culture: State of the art its application to the coconut. FAO Plant Prod and Prot Paper, FAO UN, Rome, pp 1–69.
18. Lioret C (1980) Vegetative propagation of oil palm by somatic embryogenesis. Malay Int Agric Oil Palm Conf 1980: 1–10.
19. Noiret JM, Gascon JP, Pannetier C (1985) Oil palm production by in vitro culture. Oleagineux 40: 365–372.
20. Paranjothy, K (1984) Oil Palm. In DA Evans, WR Sharp, PV Ammirato, Y Yamada, eds, Applications of Plant Tissue Culture Methods for Crop Improvement, MacMillan Press, New York 3: 591–605.
21. Paranjothy, K (1986) Recent developments in cell and tissue culture of oil bearing palms. PORIM Occasional Paper No 19: 1–12.
22. Pannetier C, Buffard-Morel J (1982) First results of somatic embryo production from leaf tissue of coconut, *Cocos nucifera* L. Oleagineux 27: 349–354.
23. Raju CR, Prakash Kumar P, Chandramohan M, Iyer RD (1984) Coconut plantlets from leaf tissue cultures. J Plant Crops 12: 75–91.
24. Reynolds JF (1982) Vegetative propagation of palm trees. In JM Bonga, DJ Durzan, eds, Tissue Culture in Forestry, Martinus Nijhoff/Dr W Junk, The Hauge, pp 182–207.
25. Rhiss A, Poulain, C, Poulain G, Beauchesne (1979) La culture *in vitro* appliquee a la multiplication vegetative du palmier-dattier (*Phoenix dactylifera* L.). Fruits 34: 551–554.

26. Salman RM, Al-Jiboury AAM, Al-Quadhy WK, Omar MS. 1988. Isozyme and chromosomal analyses of tissue culture derived date palm. Date palm J 6: 401–411.
27. Sharma DR, Deepak S, Chowdhury JB (1986) Regeneration of plantlets from somatic tissues of the date palm, *Phoenix dactylifera* Linn. Ind J Exp Biol 24: 763–766.
28. Soh AC (1987) Abnormal oil palm clones. Possible causes and implications: further discussions. Planter 63: 59–65.
29. Soh AC, Wong G, Tan CC (1988) Clonal propagation of oil palm; current experiences and their implications to breeding and cloning. Newsletter ISOPB 5: 4–7.
30. Thomas V, Rao PS (1985) In vitro propagation of oil palm (*Elaeis guineensis* Jacq var. Tenera through somatic embryogenesis in leaf-derived callus. Curr Sci 54: 184-185.
31. Tisserat B (1981) Date palm tissue culture. Adv Agric Tech, West Region, Ser No 17, USDA, ARS, pp 1–50.
32. Tisserat B (1981) Production of free-living date palms through tissue culture. Date Palm J 1: 43–54.
33. Tisserat B (1982) Development of new technology to aid in the cultivation and crop improvement of date palms. In YM Makki, ed, Proc First Symp Date Palms, Al-Hassa, 23–25 March, 1982, King Faisal University, Al-Hassa, pp 126–139.
34. Tisserat B (1982) Factors involved in the production of plantlets from date palm callus cultures. Euphytica 31: 201–214.
35. Tisserat B (1984) Clonal propagation: palms. In IK Vasil, ed, Cell Culture and Somatic Cell Genetics of Plants. Academic Press, New York 1: 74–80.
36. Tisserat B (1984) Palm Tissue Culture. In DA Evans, WR Sharp, PV Ammirato, Y Yamada, eds, Applications of Plant Tissue Culture Methods for Crop Improvement, MacMillan Press, New York 2: 505–545.
37. Tisserat B (1984) Propagation of date palms by shoot tip cultures. HortScience 19: 230–231.
38. Tisserat B (1988) Palm tissue culture. ARS-55, USDA, ARS, pp 1–60.
39. Tisserat B (1990) Micromachines to automate plant tissue culture. Meth Mol Biol 6: 563–569.
40. Tisserat B, Vandercook CE (1985) Development of an automated plant culture system. Plant Cell Tissue Org Cult 5: 107–117.

Plant Tissue Culture Manual **C3**: 1–16, 1991.
© 1991 *Kluwer Academic Publishers*.

Clonal propagation of conifers

TREVOR A. THORPE & INDRA S. HARRY
Plant Physiology Research Group, Department of Biological Sciences, University of Calgary, Calgary, Alberta T2N 1N4, Canada

Introduction

Conifers are the best known and most economically important of the gymnosperms, which make up approximately 60% of the forested areas of the world. There are about 50 genera and 300–500 species of conifers [24]. Pines (*Pinus* spp), spruces (*Picea* spp), firs (*Abies* spp), Douglas fir (*Pseudotsuga*), larch or tamarack (*Larix* spp), hemlock (*Tsuga* spp), bald cypress (*Taxodium* spp), redwood (*Sequoia* spp), arbor vitae (*Thuja* spp) and juniper (*Juniperus* spp), all belong to this group. Conifers are evolutionary a very old group of plants from the Permian period, i.e., 180–205 million years ago. They grow best in temperate zones, where they form huge forests in North America, Europe, and Asia. Very few of these softwoods are strictly tropical in distribution and they usually occur at high altitudes in these regions. Although not native to Australia and New Zealand, extensive stands of certain conifers can be found there.

Interest in conifer regeneration is very high at this time, as it is generally recognized that the forests are being harvested at a faster rate than they are being regenerated, either naturally or artificially, due to increased demand for wood and wood products [11]. Thus, the possibility of a shortage exists. In addition, certain diseases and pests, as well as fires threaten the very existence of some species. Therefore, there is an urgent need for large numbers of improved, fast-growing trees [37].

The potential benefits of using clonal planting stock over seedlings for reforestation have long been recognized, as at least a 10% increase in gain can be expected [27]. However, for the maximum possible genetic gain for forest improvement, both sexual reproduction and vegetative multiplication must be used [23]. The former is important for both introducing new genes to prevent inbreeding and for achieving genetic gain for those characteristics controlled by additive gene effects; the latter allows for the multiplication of elite full-sib families or individuals in a family, that show significant gain due to non-additive gene effects. The traditional methods for vegetative propagation of forest trees are rooted cuttings or rooted needle fascicles for the pine species, and grafting. However, for the majority of conifers it is not possible to use these methods [37, 40]. Micropropagation is potentially useful, but its promise has remained largely unfulfilled.

Clonal propagation

Clonal or vegetative propagation of plants by tissue culture methods of micropropagation can be achieved by three approaches, namely by (a) enhancing axillary bud breaking, (b) production of adventitious buds, and (c) somatic embryogenesis. The first two approaches lead to plantlet formation via organogenesis through the production of unipolar shoots, which must then be rooted in a multi-staged process. However, the first method is of little value with conifers [37]. In contrast, somatic embryogenesis leads to the formation of a bipolar embryo, through steps that are often similar to zygotic embryogenesis. The potential for forming large numbers of plantlets *in vitro* increases in the above order, but unfortunately so does the difficulty in producing plantlets. The governing principles for the micropropagation of woody species have been outlined recently [14, 39, 40]. Using the latter two approaches, it has been possible during the last 20 years to produce plantlets in over 50 gymnosperms [40]. However, large-scale regeneration is not possible at present for the majority of these.

Micropropagation via adventitious budding

This is a multi-staged process consisting of at least four distinct stages, namely (a) bud induction on the explant, (b) shoot development and multiplication, (c) rooting of developed shoots, and (d) hardening of plantlets. In some cases, the last two stages are combined, particularly when rooting is carried out *ex vitro*. The optimum requirements for each stage must be empirically determined, although the broad principles to be used have been established [39].

The formation of adventitious buds involves an interplay between the inoculum, the medium and the culture environmental conditions [39]. In conifers, buds are usually induced directly on the explant, and the callus stage is bypassed. In general, the more juvenile the tissue, the better it will respond to *in vitro* treatments leading to *de novo* organogenesis. Thus, mature isolated embryos are frequently the explants of choice. In addition, other juvenile explants such as cotyledons and epicotyls are also frequently used, as well as lateral bud explants from adolescent and mature trees.

Various mineral salt formulations have been used for *in vitro* culture. However, full strength mineral salts are not always optimum, and different formulations may work better at different stages [39, 40]. In addition, a carbon-energy source (usually sucrose), vitamins, reduced nitrogen (normally amino acids) and phytohormones are needed. Many factors of the culture environment influence growth and differentiation. These include (a) the physical form of the medium, (b) pH, (c) humidity, (d) gaseous atmosphere, (e) light, and (f) temperature. Most success has been achieved with agar-solidified medium. The effects of these factors have been recently discussed [6].

The organogenic process involves the induction of localized meristematic tissue by phytohormone treatment, leading to primordium differentiation and shoot development, the latter often in the absence of a phytohormone. Generally, N^6-benzyladenine (BA) at a concentration of up to 25 μM is the cytokinin of choice and usually the only phytohormone required, although mixed cytokinins have proved beneficial in some cases, e.g., in black and white spruces [35]. In conifers, the addition of auxin or other phytohormones tends to enhance callus formation and reduce organogenesis [4].

Indirect adventitious budding can also occur via callus formation. Although, this is fairly common among angiosperms, it is rare in subcultured conifer callus, which can be induced from a variety of plant organs and tissues. Only one case of plantlet regeneration (*Pinus eldarica*) from long-term subcultured conifer callus has been reported [17]. It has recently been possible to use subculturable meristemic nodule callus (not a true callus) in *Pinus radiata* for plantlet regeneration [2]. In addition to the rapid loss of regeneration capacity in conifer callus, the possibility of producing abnormal plants is increased [31].

The second stage of plantlet formation in conifers involves the development of the bud primordia into shoots with primary needles and their subsequent multiplication [39]. The approaches used are designed to achieve the above and at the same time lead to minimum callus formation. Generally, the formation of true shoot apices with juvenile leaf primordia in conifers requires transfer onto a medium with altered nutritional and/or phytohormonal levels, and often with the inclusion of activated charcoal [37, 39]. In a few cases, shoot primordia are formed in the initial culture and a change in medium is needed for stem elongation. For most conifers, no phytohormones are needed at this stage, but several cycles of culture may be required to produce rootable shoots, e.g., in black, and white spruce [35].

Remultiplication of radiata pine shoots can be carried out by a topping method, in which the shoot apex and apical shaft of the primary needles (tops) are severed from each shoot [1]. New shoots arise from the axillary meristems in the basal portions. These are ready for rooting in 3 months, but also can be remultiplied. The use of low levels of cytokinin may also enhance axillary bud development without the need to remove the apical meristem, e.g. in eastern white cedar [22]. Several other culture variables must be examined for multiplication [39]. These include the salt formulation used, its concentration, the gelling agent and its concentration, sucrose concentration, temperature and light intensity.

The rooting and hardening phases of plantlet formation may be carried out separately or together, depending on whether rooting is carried out *in vitro* or *ex vitro*. A main advantage of using the latter approach is that callus rarely forms at the base of the shoots, thereby ensuring a continuous vascular connection between the shoot and root. Rooting in agar is often more synchronous, although the quality of the roots may not be as good as those produced *in vitro* in soilless substrates or under *ex vitro* conditions [30]. Generally, auxin is required for rooting with indole butyric acid (IBA) often being best. This treatment can be given as a pulse treatment (hrs to days) or may be continuous.

Sterilized commercial rooting powders have also been effective. In some cases so called 'auxin synergists' or 'rooting cofactors' have enhanced the process [25]. Finally, for *in vitro* rooting the salts and the sucrose concentration must often be reduced to half to one-third that of the previous stages.

Micropropagation via somatic embryogenesis

Plantlet regeneration via this route is preferred, whenever possible because (a) of the difficulty and time expended with rooting of shoots derived from adventitious budding, (b) somatic embryogenesis provides an effective method for rapid propagation of large numbers of plants, and (c) embryogenic suspensions obtained from embryogenic callus can serve as a source of embryogenic protoplasts that can be used for genetic engineering of trees [40]. Several other advantages of this method have been noted: (a) it represents a method of obtaining true rejuvenation from mature trees, (b) costs are reduced when compared to plantlet production via adventitious budding, (c) problems associated with field performance of plantlets will be eradicated, and (d) large numbers of somatic embryos can be stored relatively easily in liquid cultures [9]. Embryogenic tissue and cells can also be cryopreserved for future use [26].

While the number of species that can be propagated through this method has been increasing, the percentage recovery of plantlets still remains relatively low. In Norway spruce, for example, it has been estimated that less than 1% of induced embryos develop into plantlets [3]. Clearly, for the adoption of this technology, a higher conversion rate is necessary. However, good progress is being made in all aspects of somatic embryogenesis in conifers [13].

Success in somatic embryogenesis in conifers is largely limited to the use of immature zygotic embryos as explants, although a few reports on the use of mature embryos and excised seedling parts exist [13]. The process generally requires the induction of so-called embryogenic callus (not a true callus) in the presence of auxin, e.g., 2,4-dichlorophenoxy acetic acid or naphthalene acetic acid. Once embryogenic callus is formed it appears to be stable and can be maintained indefinitely by subculture. In contrast, the process of somatic polyembryogenesis in some conifers is said to occur without a callus phase [18]. The production of mature cotyledonary stage embryos usually requires transfer to auxin-free medium and the addition of abscisic acid. Further growth of plantlets in the greenhouse or in the field has been rather problematic to date, as the plantlets often become dormant. It appears that partial dehydration of the plantlets may overcome this problem [33].

Micropropagation of mature conifers

As is apparent from the above, at present micropropagation is largely limited to embryonic and seedling explants. The inability to easily manipulate explants taken from mature conifers is one of the major problems limiting a wider application of this technology [37, 40]. Since tree species tend to be outbreeders, the use of juvenile explants means that unproven material is being used for regeneration. This problem is compounded by the fact that most tree breeders prefer to make selections at half the rotation age, an age unfortunately at which most conifers are recalcitrant *in vitro*. Nevertheless, some success in using explants from mature trees has been achieved. Perhaps the best example is the micropropagation of *Pinus pinaster* from 11-year-old fascicle shoots [12]. Redwood micropropagation was also achieved through the use of current year basal sprouts [8, 32]. Such sprouts are juvenile.

In conifers, since juvenile sprouts are not readily available, special 'rejuvenating' treatments have been applied before or during cloning [5, 16]. These include repeated spraying of selected branches with cytokinin (BA being the most commonly used), use of pruned hedges, serial grafting i.e., repeated grafting of selected scions from mature trees onto seedling root stocks, serial rooting of mature cuttings, and repeated subculture of shoot apices on a cytokinin-containing medium. Modification of the culture medium, e.g., the addition of NH_4NO_3, allowed for the culture of vegetative buds of 17 to 20 year-old Douglas fir [15].

Problems and prospects

A major problem, as indicated above, is the difficulty of micropropagating mature proven elite material. In addition, plagiotropic growth habits are not uncommon with some conifers, thus minimizing the value of the plantlets. It is often difficult to decontaminate explants taken from mature specimens in the field. Other problems such as the production of auto-inhibitory volatiles and vitrification are not unique to conifers *in vitro*. One severe problem with hardwoods, namely the secretion of large amounts of phenolics by explants, does not normally occur with softwoods. In general, hardwoods have performed very well in the field. In contrast, however, much less information exists on conifers and the data that is available is somewhat equivocal [40].

There are both short- and long-term prospects for clonal propagation of conifers. In the short-term, micropropagation will be involved with the multiplication of elite genotypes and of rare and endangered plant species, where a large number of plants can be produced from a few seeds in a relatively short time [40]. Also, characteristics such as wood quality, disease and pollution resistance, drought and frost tolerance and morphology will be selected for and produced on a commercial scale [38]. Other advantages, such as the propagation of selected ecotypes for specific sites, the maintenance of genetic diversity (from different clones) in plantations, and the use of hedged orchards which are cheaper to manage than seed orchards, will also come into play [10]. Longer term goals include increasing genetic gain from subsequent generations in breeding programs, producing commercial stock for reforestation, plantations, and orchards, and the utilization of techniques such as callus cultures and somatic embryogenesis, once they are optimized [40]. The potential for easy scale-up and artificial seed technology will then be feasible. Micropropagation technology is thus destined to play an increasingly important role in forest tree improvement [38, 40], and with time will also be integrated into the forest nursery operations [19].

Procedures

Micropropagation of Pinus canariensis *via organogenesis (Figs. 1—4) [29].*

Steps in the procedure

1. Sterilization, imbibition, and stratification:

Scarify seeds for 5 min in conc. H_2SO_4; rinse thoroughly, and discard floating seeds. Imbibe remaining seeds under running tap water for 48 hr. Sterilize for 30 min with 50% commercial bleach (v/v) plus 0.05% Tween 20 followed by 10% H_2O_2 for 10 min; rinse thoroughly with sterile distilled water after each sterilant. Stratify for 2 days at 4—5 °C. Remove seed coat (using sterile pliers) and re-sterilize megagametophytes with 15% bleach (15 min) and 10% H_2O_2 (5 min); again, rinse carefully after each sterilant.

2. Selection of explant:

Germinate excised embryos on 1% sucrose agar (Difco Bacto agar at 0.8%), under 16 h photoperiod, 26 °C, and 30—40 μmol m^{-2} s^{-1}; harvest cotyledons after 3—4 days.

3. Selection of induction medium:

Try several formulations at full and half-strength macro salts containing 5—10 μM BA. Based on overall appearance, percent explants responding and adventitious shoot production, MCM medium [7] was chosen for bud induction.

4. Selection of induction treatments:

— Determine the optimum BA concentration, time of exposure and sucrose concentration. For *P. canariensis,* use MCM medium with 10 μM BA and 3% sucrose for 14 days.

— Also, determine whether or not other phytohormones have an effect on induction and subsequent quality of shoots e.g., BA/2iP (each at 5 μM) produces adventitious shoots of *P. canariensis* which elongate faster than those generated by BA alone.

5. Selection of bud development medium:

— Test several formulations (usually at half-strength) to obtain best elongation medium.

— Determine optimum sucrose and activated charcoal (AC) concentrations; buds should elongate faster than control, and needles should be non-vitreous.

— Section explants to allow faster bud elongation; when shoots reach 5 mm remove them from explant and cultivate singly.

— Medium selected: 1/2 MCM, 2% sucrose and 0.5% AC (Fig. 2).

6. Selection of shoot elongation and maintenance medium:

— Test shoot growth on various salt formulations (half-strength); check for rate of elongation, needle appearance, and vitrification problems. For *P. canariensis,* best elongation medium: 1/2SH [36]. Vitrification problems are eradicated by culturing affected shoots on medium containing 0.5% Gelrite ^{R} for one passage.

7. Re-multiplication:

Elongate shoots on 1/2SH until they have reached $\pm$1.5 cm. Remove shoot apex and allow elongation of axillary buds; culture axillary shoots individually when they have reached lengths of 5—10 mm (Fig. 3).

8. Rooting and acclimatization:
— Use shoots greater than 1 cm for rooting.
— Prepare rooting substrate by autoclaving peat/vermiculite (1 : 1) moistened with 1/4 MCM containing 1% sucrose.
— Pulse shoots in IBA 1 mM, pH 5.0 for 4 hr.; plant pulsed shoots in prepared substrates, and seal jars with Parafilm^R.
— After roots appear (4—6 wk), harden shoots for 1—2 wk at 20 °C, and transfer to greenhouse conditions (Fig. 4). Place plantlets in individual pots with peat/sand (1 : 1).

Notes

1. Not all species require these elaborate pre-germination treatments. The degree of contamination, size of seeds, viability, desired explant, etc., all determine how seeds should be treated before use.
2. Whole excised embryos or isolated parts of germinating seedlings can be used. However, if embryos are large, only a few cotyledons will be in contact with the induction medium.
3. In general, Difco Bacto agar can be used with most species. If persistent browning occurs, which cannot be traced to salt formulation, light etc., try a different brand or a more purified agar, or even a different gelling agent, such as gellan gum.
4. As a general rule, BA is the most effective cytokinin for adventitious shoot production. Others seem to provide qualitative differences. For *P. canariensis*, 2iP and kinetin have an affect on elongation and may have an effect on rooting.
5. Rooting is very problematic with conifers. Several points should be considered: the quality of adventitious shoots, method(s) of application of auxin used and type and application of rooting co-factors all have an effect on rooting [25, 30, 39].

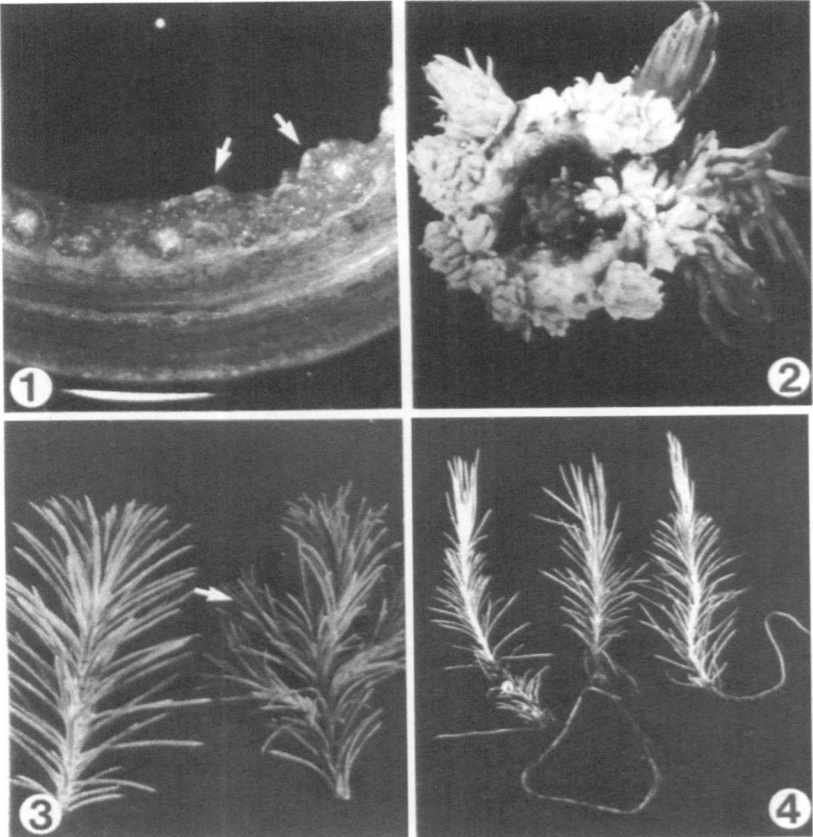

Fig. 1. Developing bud primordia on 14-day-old cotyledonary explant of *Pinus canariensis.*

Fig. 2. Cotyledon of *Pinus canariensis* after 5 weeks, showing elongating buds.

Fig. 3. A typical adventitious shoot of *Pinus canariensis* (left), and one (right) which has been treated to allow axillary shoot development (arrow).

Fig. 4. Shoots of *Pinus canariensis* pulsed for 4 hrs in IBA and rooted in peat-vermiculite.

Somatic embryogenesis and plantlet regeneration of Picea glauca, (Figs. 5, 6) [20, 28].

Steps in the procedure

1. Selection of explant:

   Collect immature cones weekly over a period of time, e.g., in western Canada, from early July to mid-August. Immature embryonic explants should be 1.5–2 mm long, and should be at the early cotyledonary stage. Cones can be stored in brown paper bags for up to 2 months at 4 °C [20, 28].

2. Seed sterilization:

   Remove seeds from cones and sterilize using 20% bleach (v/v) and 0.05% (v/v) Tween 20 for 5 min, followed by 2 min in 70% ethanol. Rinse thoroughly with sterile distilled water before dissection.

3. Induction medium:

   – Use AE medium [41] supplemented with 10 μM 2,4-D or picloram and 5 μM BA, 1% sucrose, 500, 100 and 100 mg I^{-1} of casein hydrolysate, glutamine and myo-inositol, respectively; gel medium with 0.7–0.8% Difco Bacto agar. Place excised immature embryos on media and incubate in the dark at ±25 °C until appearance of embryogenic callus [20, 28].

   – Mucilaginous, white embryogenic callus containing proembryos, is visible after 2–4 weeks; remove and subculture (multiply) this callus on the induction medium (Fig. 5).

4. Maintenance medium:

   Use induction medium gelled with 0.4% Gelrite^R or reduce the level of hormones; subculture every 3–4 weeks.

5. Development of somatic embryos:

   – Use basal medium with no growth regulators or 2,4-D at 0.5 μM; culture in the light conditions for cotyledon development [20].

   – Use hormone-free basal medium supplemented with 6% sucrose; culture in the light for maturation [28].

   – For interior spruce (*P. glauca* and *P. engelmannii),* transfer callus to basal medium (AE) containing 1% activated charcoal, 3.4% sucrose and 0.54% Noble agar for 1 wk (16 h photoperiod, 25–354 μmol m^{-2} s^{-1}); transfer this pre-treated callus to basal medium without charcoal, but supplemented with 40–60 μM abscisic acid (ABA). Transfer biweekly until mature somatic embryos appear [34].

6. Germination of mature somatic embryos:

   – Transfer mature somatic embryos (Fig. 6) to 1/3 SH medium (obtainable from Sigma) for root development; roots (with root hairs) are visible within 10–14 days, and shoots within 3–4 weeks [28].

   – For interior spruce, partially dry embryos at high humidity (95%) before germination on 1/2 basal medium (AE) with 2–4% sucrose [33].

Notes

1. Stored cones (maximum of 2 months) give a higher response of 'embryogenic' explants [20].
2. Plantlets can also be generated using the suspension culture route [21]. AE medium formulation

(without agar) used for induction is used for suspension cultures; maturation is achieved by placing aliquots of suspension cultures on AE medium plus 15 mM NH_4NO_3, containing 90 mM sucrose and 7.6 μM ABA.

3. Cell suspensions have also been successfully cryopreserved for 1 year in liquid nitrogen. Calluses have been grown from these cultures and induced to form mature somatic embryos and subsequently, plantlets [26].

Media

MCM [7]	
	mg l^{-1}
CH_4N_2O	150
KNO_3	2000
$Ca(NO_3)_2 \cdot 4H_2O$	500
$(NH_4)_2SO_4$	400
KH_2PO_4	270
KCl	150
$MgSO_4 \cdot 7H_2O$	250
$NaFe \cdot EDTA \cdot 2H_2O$	37.5
H_3BO_3	1.5
$ZnSO_4 \cdot 7H_2O$	3.0
$MnSO_4 \cdot H_2O$	0.17
$Na_2MoO_4 \cdot 2H_2O$	0.25
KI	0.25
$CuSO_4 \cdot 5H_2O$	0.025
$CoCl_2 \cdot 6H_2O$	0.025
myo-inositol	90
glycine	2
thiamine-HCl	1.7
nicotinic acid	0.6
pantothenate	0.5
pyridoxine-HCl	1.2
folic acid	1.1
biotin	0.125
D-sucrose	30 000
0.7% w/v agar (Difco)	

AE [41]	
	mg l^{-1}
KH_2PO_4	340
KNO_3	1900
NH_4NO_3	1200
$MgSO_4 \cdot 7H_2O$	370
$CaCl_2 \cdot 2H_2O$	180
$MnSO_4 \cdot 4H_2O$	2.2
H_3BO_3	0.63
$Zn \cdot EDTA$	4.05
EDTA	19.0
$FeSO_4 \cdot 7H_2O$	14.0
KI	0.75
$Na_2MoO_4 \cdot 2H_2O$	0.025
$CuSO_4 \cdot 5H_2O$	0.0025
$CoCl_2 \cdot 6H_2O$	0.0025
L-glutamine	0.4
L-alanine	0.05
L-cysteine-HCl	0.02
L-arginine	0.01
L-leucine	0.01
L-phenylalanine	0.01
L-tyrosine	0.01
glycine	2.0
D-sucrose	34 200
D-glucose	180
D-xylose	150
L-arabinose	150
nicotinic acid	2.0
pyridoxine-HCl	1.0
thiamine-HCl	5.0
myo-inositol	100

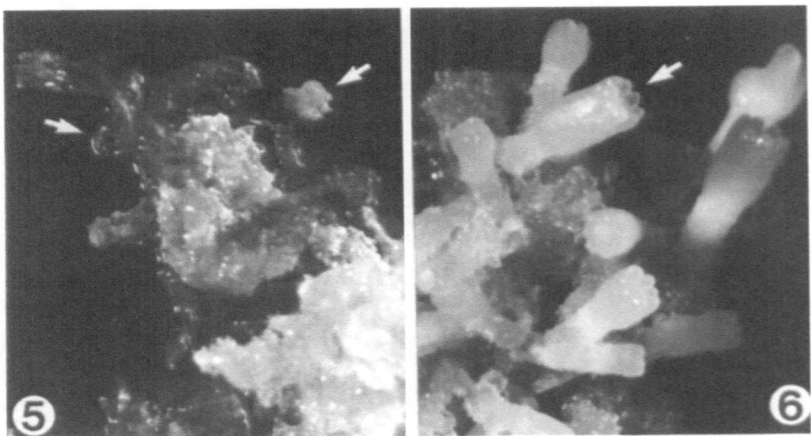

Fig. 5. Embryogenic callus of *Picea glauca* showing proembryos (arrows).

Fig. 6. Cotyledonary stage somatic embryos of *Picea glauca* (arrow).

References

1. Aitken-Christie J, Thorpe TA (1984) Clonal propagation: Gymnosperms. In: Vasil IK (Ed) Cell Culture and Somatic Cell Genetics of Plants, Vol 1 (pp 82–95) Academic Press, Orlando.
2. Aitken-Christie J, Singh AP, Davies H (1988) Multiplication of meristematic tissue: A new tissue culture system for radiata pine. In: Hanover JW, Keathley DE (Eds), Genetic Manipulation of Woody Plants (pp 413–432) Plenum Press, New York.
3. Becwar MR, Noland TL, Wyckoff JL (1989) Maturation, germination and conversion of Norway spruce (Picea abies L.) somatic embryos to plants. In Vitro Cell Dev Biol 25: 575–580.
4. Biondi S, Thorpe TA (1982) Clonal propagation of forest tree species. In: Rao AN (Ed) Tissue Culture of Economically Important Plants (pp 197–204) Costed & Asian Network of Biological Sciences, Singapore.
5. Bonga JM (1987) Clonal propagation of mature trees: Problems and possible solutions. In: Bonga JM, Durzan DJ (Eds) Cell and Tissue Culture in Forestry, Vol 1 (pp 249–271) Martinus Nijhoff Publ, Dordrecht.
6. Bonga JM, Durzan DJ (Eds) (1987) Cell and Tissue Culture in Forestry, Vol 1. Martinus Nijhoff, Dordrecht.
7. Bornman, CH (1983) Possibilities and constraints in the regeneration of trees from cotyledonary needles of Picea abies in vitro. Physiol Plant 57: 5–17.
8. Boulay M (1979) Multiplication et clonage rapide de Sequoia sempervirens par la culture in vitro. Ann AFOCEL, No 12: 49–56 (6/79).
9. Boulay M (1987) In vitro propagation of tree species. In: Green CE, Somers DA, Hackett WP, Biesboer DD (Eds) Plant Tissue and Cell Culture (pp 376–382) AR Liss, New York.
10. Carson MJ (1986) Advantages of clonal forestry for Pinus radiata – real or imagined? NZ J For Sci 16: 403–415.
11. Cohen DH (1990) Marketing strategies for the global market. In: Proceeding XIX IUFRO World Congress, Montreal Aug 1990, Vol 5 (pp 308–403), Can IUFRO World Congress Committee, Ottawa.
12. David A, David H, Faye M, Isemukali K (1979) Culture 'in vitro' et micropropagation du Pin maritime (Pinus pinaster Sol.). In: Micropropagation d'Arbres Forestiers No 12, 6/79 pp 33–40, AFOCEL, Nangis.
13. Dunstan DI (1988) Prospects and progress in conifer biotechnology. Can J For Res 18: 1497–1506.
14. Dunstan DI, Thorpe TA (1986) Regeneration in forest trees. In: Vasil IK (Ed) Cell Culture and Somatic Cell Genetics in Plants, Vol 3 (pp 223–241) Academic Press, New York.
15. Dunstan DI, Mohammed GH, Thorpe TA (1986) Shoot production and elongation on explants from vegetative buds excised from 17 to 20 year-old Douglas fir [Pseudotsuga menziesii (Mirb.) Franco] N Z J For Sci 16: 269–282.
16. Franclet A, Boulay M, Bekkaoui F, Fouret Y, Verschoore-Marouzet B, Walker N (1987) Rejuvenation In: Bonga JM, Durzan DJ (Eds) Cell and Tissue Culture in Forestry, Vol 1 (pp 232–248) Martinus Nijhoff, Dordrecht.
17. Gladfelter HJ, Phillips GC (1987) De novo shoot organogenesis of Pinus eldarica Medw. in vitro. I. Reproducible regeneration from long-term callus cultures. Plant Cell Rpt 6: 163–166.
18. Gupta PK, Durzan DJ (1987) Biotechnology of somatic polyembryogenesis and plantlet regeneration in loblolly pine. Bio/Technology 5: 147–151.
19. Haissig BE, Nelson ND, Kidd GH (1987) Trends in the use of tissue culture in forest improvement. Bio/Technology 5: 52–57.
20. Hakman I, Fowke LC (1987) Somatic embryogenesis in Picea glauca (white spruce) and Picea mariana (black spruce). Can J Bot 65: 656–659.
21. Hakman I, von Arnold S (1988) Somatic embryogenesis and plant regeneration from suspension cultures of Picea glauca (white spruce). Physiol Plant 72: 579–587.
22. Harry IS, Thompson MR, Lu C-Y, Thorpe TA (1987) In vitro plantlet formation from embryonic explants of eastern white cedar (Thuja occidentalis L.). Tree Physiol 3: 273–283.

23. Hasnain S, Cheliak W (1986) Tissue culture in forestry: Economic and genetic potential. For Chron 62: 219–225.

24. Haynes JD (1975) Botany: An Introductory Survey of the Plant Kingdom, (pp 365–421) John Wiley & Sons, New York.

25. Jarvis BC (1986) Endogenous control of adventitious rooting in non-woody cuttings. In: Jackson MB (Ed) New Root Formation in Plants and Cuttings (pp 191–222) Martinus Nijhoff, Boston.

26. Kartha KK, Fowke LC, Leung NL, Caswell KL, Hakman I (1988) Induction of somatic embryos and plantlets from cryopreserved cell cultures of white spruce (*Picea glauca*). J Plant Physiol 132: 529–539.

27. Kleinschmit J (1974) A programme for large-scale cutting propagation of Norway spruce. N Z J For Sci 4: 359–366.

28. Lu C-Y, Thorpe TA (1987) Somatic embryogenesis and plantlet regeneration in cultured immature embryos of *Picea glauca*. J Plant Physiol 128: 297–302.

29. Martinez Pulido C, Harry IS, Thorpe TA (1990) In vitro regeneration of plantlets of Canary Island pine (*Pinus canariensis*). Can J For Res 20: 1200–1211.

30. Mohammed GH, Vidaver WE (1988) Root production and plantlet development in tissue-cultured conifers. Plant Cell Tissue Organ Culture 14: 137–160.

31. Patel KR, Berlyn GP (1982) Genetic instability of multiple buds of *Pinus coulteri* regenerated from tissue culture. Can J For Res 12: 93–101.

32. Poissonier M, Franclet A, Dumant MJ, Gauntry JY (1980) Enracinement de tigelles in vitro de *Sequoia sempervirens*. Ann AFOCEL: 231–253.

33. Roberts DR, Sutton BCS, Flinn BS (1990) Synchronous and high frequency germination of interior spruce somatic embryos following partial drying at high relative humidity. Can J Bot 68: 1086–1090.

34. Roberts DR, Flinn BS, Webb DT, Webster FB, Sutton BCS (1990) Abscisic acid and indole-3-butyric acid regulation of maturation and accumulation of storage proteins in somatic embryos of interior spruce. Physiol Plant 78: 355–360.

35. Rumary C, Thorpe TA (1984) Plantlet formation in black and white spruce. I. In vitro techniques. Can J For Res 14: 10–16.

36. Schenk RV, Hildebrandt AC (1972) Medium and techniques for induction and growth of monocotyledonous and dicotyledonous plant cell cultures. Can J Bot 50: 199–204.

37. Thorpe TA, Biondi S (1984) Conifers. In: Sharp WR, Evans DA, Ammirato PV, Yamada Y (Eds) Handbook of Plant Cell Culture, Vol 22 (pp 435–470) Macmillan, New York.

38. Thorpe TA, Hasnain S (1988) Micropropagation of conifers: Methods opportunities and costs. In: Morgenstern EK, Boyle JB (Eds) Tree Improvement – Progressing Together, Proceedings 21st meeting of the Canadian Tree Improvement Association, Truro, NS, August 1987 (pp 68–84) Can For Ser., Ontario.

39. Thorpe TA, Patel KR (1984) Clonal propagation: Adventitious buds. In: Vasil IK (Ed) Cell Culture and Somatic Cell Genetics in Plants, Vol 1 (pp 49–60) Academic press, New York.

40. Thorpe TA, Harry IS, Kumar PP (1990) Application of micropropagation to forestry. In: Debergh P, Zimmerman RH (Eds) Micropropagation: Technology and Application (pp 311–336) Kluwer Academic Publishers, Dordrecht.

41. von Arnold S, Eriksson T (1980) In vitro studies of adventitious shoot formation in *Pinus contorta*. Can J Bot 59: 870–874.

Plant Tissue Culture Manual **C4**: 1–13, 1991.
© 1991 *Kluwer Academic Publishers*.

Cytological techniques

ANGELA KARP
*Department of Agricultural Sciences, University of Bristol, AFRC Institute of Arable Crops Research,
Long Ashton Research Station, Long Ashton, Bristol, BS18 9AF, UK*

Introduction

In an age of sophisticated recombinant DNA technology, it is easy to forget the more simple approaches of nuclear genome analysis that played such an important role in early genetics. Observation of mitotic chromosomes under the light microscope is a rapid and informative method of studying the genome in its entirety. Gross changes can be identified as changes in chromosome number and structure. If chromosomes are studied at meiosis, more detailed information can be obtained on structural rearrangements and on aspects of the genetic system such as pairing and recombination.

Now that techniques of protoplast fusion and transformation have been developed for direct genetic manipulation, such a simple method of looking at the genome should not be ignored. Most genetic manipulation approaches utilise tissue culture systems of plant regeneration from cultured cells. Although such systems are asexual, the regenerated plants may not be homogeneous due to the occurrence of somaclonal variation during the culture phase [1]. Genetically engineered plants may, therefore, contain uncontrolled genome modifications in addition to the changes engineered. The repercussions of genetic manipulation itself on the genetic system are also not yet known.

Basic cytological techniques, enabling accurate determination of chromosome number and structure, should be of standard use in plant tissue culture and genetic manipulation laboratories. The procedures outlined here include analysis of regenerated plants (mitotic and meiotic) and cell suspension and protoplast cultures. More advanced cytological procedures, that combine molecular biology with cytogenetics, will be described in a later chapter: *In situ* hybridisation to chromosomes.

Cytological techniques involve four basic stages – collection, fixation, staining and preparation of chromosome squashes. For analysis of mitotic chromosomes a fourth stage of pre-treatment is usually added, to arrest mitosis at metaphase, so that the chromosomes can be visualised in their most condensed form. Although the techniques are simple in outline, chromosomes vary enormously among plant species, making it impossible to present a single method which will account for all the problems that may be encountered. Some modifications of the basic procedures are, therefore, to be expected and general guidelines highlighting points of importance are given in the accompanying notes. Cytogeneticists themselves differ in the procedures they use and the ones given here represent a personal selection. More general methods are given in the references [2, 3].

Procedures

1. Root-tip squash preparations for mitotic chromosome analysis of regenerated plants

Steps in the procedure
1. Using a clean pair of forceps, collect healthy roots into a small vial containing distilled water.
2. Transfer the roots to a suitable pre-treatment as quickly as possible and incubate for the appropriate time and temperature (Table 1).

Table 1. Pre-treatments for arresting mitosis at metaphase

Chemical	Concentration	Time (hours)	Temp. (°C)	Examples of suitability
8-hydroxyquinoline (in distilled water)	0.29 g in a litre (dissolve at 60 °C)	$3\frac{1}{2}$–4 h	18–20 °C	potato, oilseed rape, sugar beet, tobacco
colchicine (in distilled water)	0.05 g in 100 ml	4–6 h	18°	*Vicia faba*, *Scilla*, Hyacinth
ice-cold water	–	24 h	4°	cereals
α-bromonaphthalene (in distilled water)	saturated	$3\frac{1}{2}$–4 h or	RT*	cereals
		18 h	4°	cereals
α-bromonaphthalene (in alcohol)	1 ml in 100 ml alcohol (stock) (Use 10 µl stock /10 ml water	$3\frac{1}{2}$–4 h or	RT*	cereals
		18 h	4°	cereals

* RT = room temperature.

3. Fix the roots by transferring them into 3 : 1 absolute alcohol: glacial acetic acid and incubate for at least 24 hours at 4 °C.
4. Hydrolyse the roots by incubation in 1M HCl at 60 °C for 4–9 minutes.
5. Wash the roots briefly in distilled water.
6. Transfer the roots to feulgen and leave to stain for 30 minutes.
7. Place a stained root onto a clean glass slide. Remove the translucent root cap at the extreme tip, cut below the stained meristematic zone and discard the unstained portion of the root.
8. Place the stained root-tip in a small drop of 45% acetic acid. Mix in a small drop of aceto-carmine (BDH).
9. Tap the root-tip thoroughly in the drop with a flat ended glass rod. Remove any remaining pieces with a needle.

10. Gently lower a coverslip over the drop. Place a piece of filter paper over the coverslip and press gently to remove any excess stain.
11. Keeping the filter paper in place with the fingers of one hand on either side of the preparation, squash vertically downwards using the thumb or forefinger of your other hand. Do not rock your thumb\finger during the squash as this will roll the cells.
12. Examine quickly under the microscope and squash again if necessary.
13. Ring the coverslip with rubber solution and examine once the solution had dried. Chromosome counts should be made from a minimum of five cells, where the chromosomes are well spread, and from more than one root.

Notes on collecting roots
Careful selection of healthy roots is essential for good chromosome preparations. Roots can be collected from small plantlets in culture medium, vermiculite or soil, or from germinated selfed seed. Collect roots into distilled water, but do not keep them in water for more than 15 minutes.

When collecting roots from regenerated plants in culture, ensure they are grown in a good depth of agar in the culture vessel. Collect roots 1–2 cm long as they grow down into the agar. Sub-culture at the same time. Do not use roots that have reached the bottom and are wrapped around the base.

When collecting roots from regenerated plants in soil, ensure that the plants are watered carefully in advance and that the roots are not too dry or rotting because the pot is standing in water. Place your hand over the pot, gently holding the plant between your fingers. Invert and tap the base of the pot to loosen and remove it. Collect healthy roots that are white in colour with translucent tips from the sides of the pot or just as they emerge from the base. Snip the root to 1–2 cm long from the tip and transfer to a clean tube of water. Collect a selection of roots of different sizes from each plant before replacing the pot. Collect only roots that are growing down and not pot-bound roots.

When using germinating seeds, surface sterilise the seed first in 10% bleach. Germinate the seed on filter paper that is moist to touch but has no excess surface water. Do not water the seeds immediately before collection. Collect the roots when they are 1–2 cm long.

Notes on pre-treatments
A range of pre-treatments is available and each should be tested for suitability (Table 1). It is important that the correct time and temperature are used as specified for each pre-treatment.

Notes on fixation
Roots in fixative should be kept at 4 °C where they will keep for months. A fixation of 2–3 days is optimal but preparations can be made after 24 hours if required.

Notes on staining
Although the roots are stained with feulgen, I also use a background stain of aceto-carmine. This colours the cytoplasm and allows you to determine whether the cell is intact. Background colour is also helpful in plants with small DNA contents, where faint cells may otherwise be difficult to locate. The density of the background can be varied by the mixture of acetic acid : aceto-carmine used on the slide. Aceto-carmine can be purchased as a prepared solution (eg from BDH) or can be made up as a 1% solution (1 g carmine (natural red4, e.g. Sigma) dissolved in 100 ml 45% acetic acid). After staining in feulgen, roots should be used within 12 hours. Stain only those roots you will analyse during the day, or at most leave the stained roots in the fridge at 4 °C overnight.

Notes on the chromosome preparation
For a good chromosome squash, ensure that the root-tip cells are completely dispersed before adding the coverslip. Work with a small drop and keep the tapper close to the drop to avoid excess splashing. Only if the roots are very tiny should the coverslip be added first and tapping done through the coverslip.

Plants differ in the amount of pressure required to spread the chromosomes flat. It is essential to achieve a good spread for accurate counting, particularly in species with small chromosomes. Similarly, a good photograph will only be obtained from a flat preparation. Although the preparation may be initially screened using low power objectives, chromosome counts and photographs should only be obtained using oil immersion high magnification objectives.

2. Meiotic chromosome preparations

Steps in the procedure

1. Choose a suitable inflorescence. Peel away the outer leaves or bracts and place the inflorescence into a tube containing Carnoys no. 2 fixative (6:3:1 alcohol : chloroform : acetic acid) to which are added a few drops of 10% w/v ferric chloride solution until a straw yellow colour is obtained). Fix for at least one month, replacing with fresh Carnoys after 2—3 weeks.
2. Transfer the inflorescence into a clean Petri dish containing 70% alcohol. Carefully dissect the anthers from one floret, noting from where they have been taken. Take one anther and place it on a clean slide.
3. Dab any excess alcohol with a piece of filter paper and add a drop of aceto-carmine.
4. Cut the anther into two or more pieces depending on the size. Tap them thoroughly in the drop of aceto-carmine. Remove any remaining pieces of anther wall.
5. Carry out steps 10—12 of procedure for root-tip squashes, but apply much less pressure and monitor carefully during the squash.
6. Identify the stage of meiosis and examine for abnormalities.

Notes on meiotic chromosome preparations

1. In cereals such as wheat and rye, meiosis occurs before complete emergence of the ear and it is relatively easy to select inflorescences at the correct stage. Feel how large the inflorescence is with your fingers and use the appearance of the flag leaf and the extent of emergence as indicators of the stage. In other plants, selection of the correct stage may be more difficult and trial and error may be the only way. Always check a few anthers by quickly squashing out in aceto-carmine and always collect a range of infloresences or buds younger and older than the correct stage.
2. If a more immediate analysis of meiosis is required (e.g., for screening prior to crossing), anthers can be fixed in 3 : 1 alcohol : acetic acid and the chromosomes stained in feulgen (see root-tip squash procedure).
3. Warming of the slide before squashing is advantageous in many species. The slide should be warmed gently over a small burner and then allowed to cool a little before squashing. Do not let the slide overheat or the stain will bubble and the meiotic cells burst. Check the temperature during heating by touching the back of your hand with the slide.
4. For help in interpreting meiosis see the general books [4, 5, 6].

3. Chromosome preparations from cell suspension cultures

Method 1

Steps in the procedure

1. Take a 50 ml aliquot of cell suspension 3–4 days after sub-culturing and transfer it to a clean conical flask. Add 50 ml of 0.02% colchicine and shake the mixture on an orbital shaker (100rpm) for 1–2 hours at 25 °C.
2. Transfer the contents to centrifuge tubes and centrifuge at 1100 rpm for 10 minutes. Discard the supernatant and resuspend the pellet in 20 ml of 3 : 1 ethanol : acetic acid fixative. Refrigerate overnight at 4 °C.
3. Centrifuge at 1100rpm for 10 minutes, remove the supernatant and resuspend the pellet in 0.1M sodium acetate (ph 4.5). Leave the mixture to settle for a few minutes. Centrifuge at 1100rpm for 10 minutes, discard the supernatant and transfer the pellet to a clean conical flask.
4. Resuspend the pellet in 20 ml of an enzyme mixture containing 0.25 g Onozuka cellulase, 0.25 g of Macerozyme R10 (Yakult Pharmaceutical Industry Co. Ltd., Japan) and 49.5 ml 0.1M sodium acetate buffer, pH 4.5. Incubate the mixture at 25 °C for 2 hours.
5. Wash the cells in 0.1M sodium acetate (pH 4.5) as described in step 3 but resuspend the pellet in 10 ml of 45% acetic acid. Leave overnight in the fridge at 4 °C.
6. Pipette 20 μl of the fixed supension onto a clean slide and allow the fixative to evaporate. Add a few drops of a 1% solution of Macerozyme R10 and gently mix the cells into the drop. After a few minutes, tap the cells in the drop and continue tapping until the cells are dispersed (this may take up to 10 minutes depending on the cell suspension).
7. Add a few drops of modified carbol fuchsin, mix the cells into the stain and leave for 4–5 minutes before adding a 22 × 50 mm coverslip.
8. Squash as for a root-tip preparation. If air bubbles appear, add a little more carbol fuchsin at the edge of the coverslip. Ring with rubber solution and examine.

Method 2

Steps in the procedure

1. Transfer a 10ml aliquot of cell suspension 3–4 days after sub-culturing into a glass vial.
2. Remove the culture fluid with a pasteur pipette.
3. Make a stock solution of 1ml alpha-bromonaphthalene in 100 ml absolute alcohol and shake vigorously. Add 10 μl of the stock solution to 10 mls of distilled water, mix and add to the cell suspension in the glass vial. Cap the vial, shake to mix and incubate at 4 °C overnight.
4. Remove the pre-treatment with a pasteur pipette.
5. Wash the cells with 3 : 1 absolute alcohol : acetic acid fixative by adding 5 ml to

the vial, shaking the mixture and then removing the fixative with a pasteur pipette. Repeat the wash step once.

6. Add a final 5 ml of fixative and leave to incubate for 24 hours at 4 °C.

7. Make chromosome preparations as described above. You will need to squash several times to cover the large area of the coverslip.

Notes

1. Two alternative methods are given here. The first method is based on Kao [7] and proved very successful with wheat [8]. Method 2 is quicker and particularly appropriate for experiments in which small samples need to be examined over a time course. However, the omission of an early enzyme treatment (step 4 of method 1) means that the cells can be more difficult to squash and for 'tough' cell suspensions method 1 may be necessary.

2. The most common difficulty encountered in examining chromosomes of cell suspensions is finding cells that are dividing. Try different time intervals after subculturing to establish when most divisions occur in your culture. Screen the slides systematically (the large coverslips are used to provide maximum material for screening) and have patience to cover all the slide thoroughly. Squash the preparation well so that when you do find chromosomes they can be counted accurately.

4. Chromosome preparations from protoplast cultures

Steps in the procedure

1. Transfer 5 ml of a protoplast culture 2—4 days after isolation into a small conical flask. Add 5 ml 0.02% colchicine solution and incubate for 6—10 hours at 25 °C on an orbital shaker (100 rpm).
2. Transfer to centrifuge tubes and centrifuge at 1000 rpm for 10 minutes.
3. Remove the colchicine supernatant and resuspend the pellet in 1 ml 3 : 1 ethanol: acetic acid fixative. Refrigerate at 4 °C overnight.
4. Centrifuge at 1100 rpm for 15 minutes, remove the supernatant and resuspend the pellet in a few drops of 45% acetic acid. Refrigerate overnight at 4 °C.
5. Pipette 20 µl onto a clean slide and allow the fixative to evaporate.
6. Add a few drops of modified carbol fuchsin and allow to stain for 1—2 minutes. Cover with a 22 × 50 mm coverslip and squash gently.
7. Seal with rubber solution and examine.

Notes

Unlike cell suspensions, protoplast cultures require virtually no squashing, however, finding cells in division is still difficult. Test samples fixed at time intervals after protoplast isolation or examine the protoplast cultures carefully to determine when most divisions occur in your material. Try different incubation times in colchicine to see which gives optimal results.

Solutions

— Feulgen:
- — 0.9 g basic fuchsin
- — 4.8 g sodium metabisulphite
- — 250 ml 0.15M HCl

Mix in a conical flask, cover with foil and shake for 24 hours. Add 5 g activated charcoal. Mix well. Filter (in fume hood). Repeat charcoal step until the filtrate is colourless. Store at 4 °C in the dark.

— Carbol fuchsin (9):
- — solution A: 3g basic fuchsin + 100ml 70% ethanol
- — solution B: 10 ml sol. A + 90ml 5% phenol
- — solution C: 45 ml sol. B + 6ml acetic acid + 6 ml 37% formaldehyde

Note: For modified carbol fuchsin
- — 2—10 ml carbol fuchsin (solution C above)
- — 90—98 ml 45% acetic acid
- — 1.8 g sorbitol

References

1. Karp A (1990) On the current understanding of somaclonal variation. In: Oxford Surveys of Plant Molecular and Cell Biology, Vol 7 (ed, Miflin BJ) Oxford University Press, Oxford pp. 1–58
2. Darlington CD, La Cour LF (1976) The Handling of Chromosomes. George Allen & Unwin Ltd., London.
3. Dyer AF (1979) Investigating Chromosomes. Edward Arnold, London.
4. Lewis KR, John B (1972) The Matter of Mendelian Heredity. Longman, London.
5. Rees H, Jones RN (1977) Chromosome Genetics. Edward Arnold, London.
6. Jones RN, Karp A (1986) Introducing Genetics. John Murray, London.
7. Kao (1975) A chromosomal staining method for cultured cells. In: Plant Tissue Culture Methods (eds, Gamborg OL, Wetter LR) Nat. Res. Council, Prairie Regional Laboratory, Saskatoon, Canada.
8. Karp A, Wu QS, Steele SH, Jones MGK (1987) Chromosome variation in dividing protoplasts and cell suspensions of wheat. Theor Appl Genet 74 : 140–146
9. Carr DH, Walker JE (1961) Carbol Fuchsin as a stain for human chromosomes. Stain Technol 36 : 233–236

Plant Tissue Culture Manual C5: 1–17, 1991.

Restriction fragment analysis of somaclones

ROBERT H. POTTER
The Agricultural University of Norway, Department of Biology & Nature Conservation, Box 14, N-1432 Ås-NLH, Norway

Introduction

Somaclonal variation is a feature common to all plant regeneration via a tissue culture or 'callus' phase [7, 10]. Whether the intention is to maximise variation (for production of variability in plant breeding) or minimise it (to reduce variability in mass propagation systems or transformation studies) it is very important to be able to identify variants as quickly as possible and to assess the amount of variation a particular treatment produces. Most identification of somaclones has been done on the basis of gross plant morphology, especially as the main interest in somaclonal variation was as a novel source of variability in plant breeding. Where this was not possible, then biochemical charac-terisation was performed – usually involving protein electrophoresis. Both of these techniques assess the phenotype of the plants and, as such, are affected by the environment in which the plants are grown, thus making it important to prepare the plants and grow them carefully before analysis.

Assessing the genotype of the plants is a more rapid and precise method of asessing somaclonal variation and much effort has been placed on chromo-some analysis in this respect (e.g. [8]). However, gross changes in chromosome number and morphology can not account for all of the observed variation at the whole plant level. The recently developed methods of restriction fragment analysis offer a much better method to analyse somaclones, both from the point of view of identifying subtle changes and also in the ability to analyse plants from different environments, even while still in culture.

Background

Restriction fragment length polymorphisms (RFLPs) as genetic markers have made a major impact in plant breeding and many excellent reviews of the theory and application of such techniques to breeding (e.g. [1, 15]) have been written. This article will focus on the specific application of the analysis of somaclonal material. The main difference is one of aim; instead of looking at the distribution of polymorphic alleles in a segregating population an essentially identical, clonal, population is being analysed for mutants. The analysis required is thus much simpler – the identification of altered band patterns instead of relating the presence or absence of alleles at different loci.

This does, however, lead to one or two differences in the methods used. The most important of these relates to the class of probe used to generate the restriction fragment patterns. Single copy probes are the ideal for RFLPs, while for analysis of somaclones probes covering larger areas of the genome will be more useful. This will increase the chances of finding variation, although it may be more difficult to identify variant patterns as the copy number – and so the complexity of the pattern – increases. As a result of this, early examples of somaclonal variation detected at the molecular level were found using repetitive clones as probes (e.g. [9, 16]) and these somaclones have been found to be the product of relatively large alterations in copy number of the target sequences. More subtle changes in DNA sequence are more difficult to find as the low-copy-number probes used to search for them only analyse small portions of the genome at one time. This is borne out in a study of a large number of somaclones of potato analysed with different classes of probe [13].

Another factor to consider in this work is the choice of restriction enzymes used to digest the DNA. Although in theory any enzyme will reveal differences in DNA sequence, in practice some enzymes have been found to be more useful, often depending on the class of enzyme used – whether a 4 or 6 base pair sequence is recognised for cleavage. 4-Base pair enzymes will cut DNA more often and give a smaller average fragment size, which will analyse more DNA sequence for single base alterations. The larger fragments produced by 6-base pair enzymes mean that more DNA per probe is analysed for insertions, deletions or copy number alterations. Although it is not known whether soma-clonal variants arise less from single base changes than from other DNA alterations, more polymophic alleles were found in rice using 6-base enzymes [12].

Using iso-schizomers (pairs of restriction enzymes that recognise the same base-pair sequence but are affected differently by methylation of bases within that sequence) the methylation state of the DNA can be established. This is an aspect that has aroused much interest in studies of somaclonal variation [3].

Apart from this the requirements for methodology are similar to RFLP analysis in that methods should be simple, reproducible, quick and as cheap as possible. The ability to extract and analyse DNA from very young material – even while still in culture – is one of the major advantages of this technique over the other methods for analysis of somaclonal variation. DNA sequence is not affected by the environment (at least not in the broad sense) and so direct comparisons can be made between unlike tissues. A further advantage is that the DNA can be stored for long periods to enable the direct comparison of starting material with treated material.

Procedures

Most of these methods are slight modifications of basic molecular biology techniques. The basic techniques can be found in most molecular biology laboratory manuals, such as [11].

DNA Extraction

This method is based on that of Dellaporta et al. [5] and has been used successfully on a wide range of plant species and tissues including in vitro potato shoots. Although not completely pure, the DNA is of sufficient quality for restriction analysis and has also been used for PCR amplification (R Kalla, pers. comm.).

Steps in the procedure

1. Freeze 1–2 g plant tissue (leaves or whole in vitro shoots) in liquid nitrogen. Transfer to a **cold** mortar and pestle along with more liquid nitrogen and grind well while still frozen. Well ground tissue gives a much better yield of DNA, but the tissue must be kept frozen until in the buffer to prevent degradation of the DNA.
2. Transfer to a 50 ml test tube (Oak Ridge tubes with caps are preferred) containing 15 ml of extraction buffer. Add 1 ml of 20% (w/v) sodium dodecyl sulphate (SDS), shake and incubate at 65 °C for 10 minutes.
3. Add 5 ml of 5M potassium acetate, shake and place in ice for 10 minutes before centrifugation at 10,000 rpm for 20 minutes at 4 °C. This precipitates the SDS and with it most of the protein and lipids which the SDS has solubilised.
4. Pour the supernatant through Miracloth (Calbiochem) into another tube containing 10 ml of isopropanol. Mix gently and incubate at −20 °C for 30 minutes before centrifugation at 10,000 rpm for 20 minutes at 4 °C. The low ratio of isopropanol (approx. 0.6 volumes) precipitates high molecular weight DNA but not carbohydrates or other impurities.
5. Pour off the supernatant and drain the tube for a few minutes before re-suspending the DNA pellet in 0.7 ml 50TE. Transfer this to an Eppendorf tube containing 7 μl of RNAse (10 mg/ml, boiled for 10 minutes to denature any DNAse activity) and incubate at 37 °C for one hour. Removal of RNA at this stage is not critical, but it aids later steps and also gives time for the DNA to be completely resuspended before insoluble impurities are pelleted in the next step.
6. Centrifuge the sample at high speed in a micro-centrifuge for 15 minutes and transfer the supernatant to a clean Eppendorf tube containing 75 μl of 3M sodium acetate pH 5.2. Mix and add 0.5 ml of isopropanol, mix again and keep at room temperature for 5 minutes. The DNA should be visible as a large clump at this stage and can be removed with a hooked pasteur pipette if preferred.
7. Pellet the DNA at high speed in a micro-centrifuge for 5 minutes and remove the supernatant. Wash the pellet with 75% ethanol, dry briefly and resuspend in 200 μl of TE overnight at 4 °C. Do not vortex the tube to resuspend or the DNA will shear.

Solutions

DNA extraction buffer: (autoclaved before addition of β-ME)
   100 mM Tris.Cl pH 8.0
   500 mM Sodium chloride
   50 mM Ethylene-diamine-tetra-acetic acid (EDTA) pH 8.0
   0.6% β-mercaptoethanol

50 TE: (autoclaved)
   50 mM Tris.Cl pH 8.0
   10 mM EDTA pH 8.0

TE: (autoclaved)
   10 mM Tris.Cl pH 8.0
   1 mM EDTA pH 8.0

Restriction endonuclease digestion of DNA

This is best carried out following the recommendations of the supplier of the enzymes, and most suppliers now provide stock buffer solutions optimised for each particular enzyme. However, when processing many samples, it saves a lot of time and effort to make up a stock of enzyme and buffer and then add an aliquot of this to each DNA sample. The following amounts are for 12 samples — as most micro-centrifuges have rotors for this number of tubes this is a convenient amount to work with.

Steps in the procedure:
1. Make up the buffer/enzyme stock without adding the enzyme and keep on ice.
2. Into separate labelled tubes, pipette 30 µl of the total DNA samples.
3. Add the enzyme to the stock, vortex briefly and add 70 µl to each of the tubes containing DNA. Touch the pipette to the wall of the tube above the level of the DNA solution while dispensing in order to use the same pipette tip but without contaminating the stock with DNA. In order not to run out of the buffer/enzyme mixture, only depress the pipette to the first stop to expel the contents and spin the tube containing the enzyme mixture for a few seconds to collect the liquid from the walls of the tube.
4. Briefly vortex the samples and incubate at 37 °C overnight. Use a cover or other insulation on top of the tubes to prevent excessive condensation on the lid of the tube.
5. After digestion, spin down the samples for a few seconds in a micro-centrifuge (to collect the drops of condensation etc.), vortex briefly and remove 5 µl into a clean tube.
6. Add 5 µl of gel loading buffer to this small aliquot and run the samples on a 0.8% agarose mini-gel (30 ml of gel in a Biorad 'mini-sub cell' or 80 ml of gel in a 'wide-mini-sub cell') in TAE buffer at 100 mA until the blue dye is 2/3 of the way down the gel. As markers use uncut extracted DNA and known amounts (0.1, 0.5, 1.0 µg) of sheared herring sperm DNA (or other accurately measured DNA) in order to estimate the concentration of the restricted DNA samples.
7. Soak the gel in a 0.5 µg/ml solution of ethidium bromide (EtBr — in distilled water) for 10–15 minutes and view under UV illumination to visualise the DNA. The restricted DNA will appear as a smear with perhaps some stronger bands — from organellar DNA — visible in the lighter portions. Compare the staining intensity with the known markers and so estimate the amount of DNA in each of the restriction reactions.
8. Precipitate the DNA in each of the restriction reactions by the addition of 50 µl of 3M sodium acetate, pH 5.2 and 300 µl of isopropanol. Keep at −20 °C for 5–10 minutes and then pellet the DNA in a microcentrifuge at high speed for 15 minutes.
9. Drain the pellet, wash with 75% ethanol and dry briefly. Resuspend the pellet in TE at a concentration of 0.5 µg/ml based on the estimated amount of DNA

from the gel. When completely resuspended, add a 1/2 volume of gel loading buffer, vortex, take 3 μl and run on another mini-gel in TAE buffer.

10. Stain the gel as before, visualise and check that the concentrations of the samples are roughly equal to 1 μg of sheared DNA run as a marker.

Solutions

Buffer/enzyme stock:
    120 μl 10× buffer solution
    24 μl 0.1 M spermidine trihydrochloride
    696 μl sterile distilled water (SDW)
    = 840 μl
    500 Units of restriction enzyme

Gel loading buffer:
    2 ml 0.5M EDTA, pH 8.0
    3 ml SDW
    5 ml Glycerol
    800 μl Bromo-phenol blue — saturated solution in SDW

50× TAE electrophoresis buffer: - 1 litre
    242 g Tris base
    57.1 ml glacial acetic acid
    100 ml 0.5M EDTA

Southern blotting

The preparation of filters for hybridisation is referred to as Southern blotting as a reference to Southern [14] who first developed the technique with nitrocellulose filters. Now many different varieties of nylon filters are used and the methods given by the manufacturers of any particular filter are likely to be the best for that particular type. The method that follows is a general one that will work for nearly all filters, including nitrocellulose, but can probably be improved upon by keeping strictly to the manufacturers protocols.

Steps in the procedure

1. Prepare a 0.75% agarose gel in TBE – 200 ml of gel in a 25 cm Biorad 'midi-sub cell' tray is a good size. If a wide-toothed comb is used (5.5 mm, 15 teeth) then 7–10 µg of restricted DNA should be loaded (as estimated from the gels run during the restriction procedure), but adequate results can be obtained using 3–5 µg of DNA per well when using a smaller toothed comb (2.5 mm, 30 teeth) and this allows many more samples to be analysed per gel, as well as reducing the amount of DNA required. Run the gel at around 50 mA for 14–16 hours, until the marker dye has migrated 20 cm. Stain and visualise as before, but increase the time in the EtBr to 20–30 minutes and destain in clean water for 15 minutes to get a clearer view.
2. Expose the gel to short wave UV (on the illuminator used for visualisation) for 20 minutes to 'nick' the DNA and so reduce the fragment size, which aids even transfer.
3. Denature the double stranded DNA by soaking the gel in 500 ml 0.5M sodium hydroxide, 1.5M sodium chloride for one hour, preferably with gentle shaking. This separates the DNA strands enabling hybridisation to labelled probe later on.
4. Neutralise the gel in 500 ml 3M sodium acetate, pH 5.2 for one hour. Nitrocellulose and many nylon filters are positively charged (and so able to bind DNA) only at low or neutral pH and so the sodium hydroxide must be neutralised before transfer. The DNA will not re-anneal in the gel or on the filter before it is bound and so will still be able to hybridise to the probe.
5. Wrap the gel tray in Whatman 3MM paper and lay this upside-down in a large dish. Place two more pieces of 3MM on top of this and soak the paper in 10 × SSC, filling the box but not covering the tray. Ensure the ends of the bottom layer of 3MM paper are in the 10 × SSC to act as a wick.
6. Carefully lay the gel on top of this, ensuring that no air bubbles are trapped underneath it, and then place a piece of pre-wet nitrocellulose or other nylon filter – cut to the exact size of the gel – on top of the gel. Place 3 layers of wet 3MM – cut slightly smaller than the gel/filter – on top of this and smooth out any air bubbles by rolling a clean pipette over the surface. Any air bubbles will prevent even flow of solution through the gel and so affect transfer.
7. Place strips of plastic around the gel on the lower layers of 3MM to prevent any contact between the upper and lower layers and so ensuring all solution goes

through the gel/filter. Place a 15 cm stack of absorbant paper on top of the sandwich, and place a weight on top of this.

8. Fill the dish with 10 × SSC to provide a reservoir and secure the stack against collapse (cling-film is useful as it also prevents evaporation as well). Let the transfer proceed for at least 16 hours, replacing the stack of paper if required. Although most small fragments will transfer in a few hours, larger fragments of DNA take longer and it is important to allow for this or the larger fragments will be under-represented on the filter.

9. After transfer, remove all the paper and place the filter — DNA side up — onto some absorbant paper and bake at 80 °C for 2 hours (nitrocellulose should be baked under vacuum to reduce the risk of fire). The dry filters can be stored in plastic bags at room temperature just about indefinitely.

Solutions

10× TBE electrophoresis buffer: — 1 litre
    108 g Tris base
    55 g Boric acid
    40 ml 0.5M EDTA

20× SSC: — 1 litre
    175.32 g Sodium chloride
    88.23 g tri-Sodium citrate

Probe isolation & labelling

The labelling of probe DNA used to require very pure preparations of DNA as a substrate, but now simpler protocols have been devloped which are not as sensitive to impurities and allow the use of relatively crude DNA samples. The protocols here are based simply on ease and speed of use, while at the same time repeatability is very good. The plasmid miniprep. is taken from Birnboim and Doly [2] and is described for normal *E. coli* strains such as JM101. The 'oligo-labelling' method of Feinberg and Vogelstein [6] is now standard and kits can be obtained containing all the ingredients for this method, however the basic method is given here as it is not too time-consuming to prepare all the stock solutions.

Steps in the procedure

1. From a single bacterial colony, make a broad streak on an L-plate containing suitable antibiotics for selection and grow overnight. The particular antibiotic depends on the vector used, but is usually 100 mg/l ampicillin when using plasmids of the pUC series which are most common.
2. Scrape bacteria off the plate using a sterile toothpick and resuspend in 100 μl of lysozyme solution in an Eppendorf tube. Incubate this on ice for 30 minutes, or at room temperature for 5 minutes, while the lysozyme lyses the cells. Do not allow this to proceed too long, so using ice is preferable if many samples are being processed.
3. Add 200 μl of 1% SDS, 0.2M sodium hydroxide and vortex gently, keep on ice for 5 minutes. This will complete lysis of the cells and the solution should clear slightly and become viscous due to the release of DNA etc.
4. Add 150 μl of 3M sodium acetate, pH 5.2, invert gently to mix and keep on ice for 60 minutes. This will precipitate the SDS and most of the proteins, lipids and high molecular weight RNA and DNA (the chromosomal DNA).
5. Spin in a microcentrifuge for 5 minutes at high speed and transfer 400 μl of the clear supernatant to a clean tube. Add 1 ml of ethanol and precipitate the plasmid DNA at −20 °C for 30 minutes.
6. Spin for 2 minutes, remove the supernatant with a pipette and resuspend the pellet in 100 μl of 0.1M sodium acetate, 50 mM Tris · Cl pH 8.0. Precipitate again with 200 μl of ethanol at −20 °C for 10 minutes.
7. Repeat step 6. Spin down again, and this time resuspend in 40 μl of SDW. Add 5 μl of RNAse, 5 μl of the appropriate 10 × restriction enzyme buffer and 25 units of the restriction enzyme used to clone the insert in the vector. Incubate for 2 hours at 37 °C. This will cut the plasmid and separate the vector sequence from the insert; only this insert will be used as the probe.
8. Pour a 1% low-gelling-temperature agarose mini-gel in TAE containing 1 μg/ml of EtBr (add after solubilisation of gel). Often referred to as 'sea-plaque' agarose, this is a purified form with a gelling temperature below 30 °C and is best left to set in a cold room. Take 15 μl of the restricted plasmid and add 5 μl of gel loading buffer and run this on the gel with restricted vector as a marker.

9. Run the gel at 50 mA for 30 minutes and then examine under UV — still on the gel tray if possible. If the insert can be distinguished from the vector (as compared to the cut vector used as a marker) then go to the next step. If not then return the gel to the tank and run longer. These gels are very fragile, especially when they have warmed up during electrophoresis, and so care must be taken when handling them.

10. When the insert can be clearly distinguished from the vector, carefully cut out the piece of gel containing the insert and place it in a pre-weighed Eppendorf tube. Re-weigh the tube to find the weight of the gel slice and add 3 ml of SDW per g of gel. Place the capped tube into a boiling water bath for 7 minutes and then store the sample at −20 °C. Estimate the amount of DNA by comparison to standards on the gel (use a known amount of a set of size markers).

11. Place 50 μCi of α^{32}PdATP into a siliconised Eppendorf tube and dry this down, preferably in an evacuated microcentrifuge.

12. Thaw the probe sample and remove an amount equivalent to 50 ng of DNA, add SDW to 60 μl and boil (in a water bath) for 3 minutes. Add 15 μl of OLB and 3 μl of nuclease-free Bovine serum albumin (BSA — molecular biology grade). Vortex this and keep on ice until the ^{32}P is dried down and then add the mixture to the tube containg the label.

13. Add 5 units of Klenow fragment (the large fragment of DNA Polymerase I) and pipette up and down gently to mix the reaction. Allow to proceed at 20–25 °C for at least 4 hours, or overnight if more convenient.

14. Prepare a gel-exclusion column with Ultrogel 34 or Sephadex G-50 (fine) in a disposable column or 1ml syringe (pack the bottom of the syringe with siliconised glass wool) and equilibrate with STE. Add 5 μl of 0.2M EDTA, 1% bromo-phenol blue to the reaction mixture (this stops the reaction and gives a marker for the unincorporated nucleotides coming off the column) and add to the top of the column.

15. Wash the reaction tube out with two 100 μl aliquots of STE (add this to the column) and then collect 100 μl fractions off the column by adding 100 μl of STE to the top and collecting the eluate. Continue until the collected fractions turn blue — from the bromo-phenol blue. Measure the radioactivity of the fractions (a GM tube is sufficient) and pool the early fractions which are highly radioactive — containing the labelled DNA. The activity in the fractions should go down and then up again as the blue dye is eluted, as this coincides with the non-incorporated nucleotides. Removal of the non-incorporated nucleotides reduces non-specific background hybridisation. The two fractions **before** the dye should also be discarded as these will contain only very short fragments of DNA which can also cause background.

Solutions

L-plates: — 1 litre (autoclave before addition of antibiotics)

    5 g Sodium chloride
    5 g yeast extract
    10 g Bactotryptone
    7 g Agar

Lysozyme solution: (autoclave before addition of lysozyme)
    50 mM Glucose
    25 mM Tris · Cl pH 8.0
    10 mM EDTA
    2 mg/ml Lysozyme — add just before use then keep on ice

Oligo-labelling buffer — OLB:
    Solution O: 1.25M Tris.Cl pH 8.0
                   0.125M Magnesium chloride
    Solution A: 940 μl solution O
                   5 μl 0.1M dCTP
                   5 μl 0.1M dGTP
                   5 μl 0.1M dTTP
                   18 μl β-mercaptoethanol
    Solution B: 2M HEPES buffer, pH 6.6
    Solution C: Hexa-deoxyribo-nucleotides, dissolved in TE at 90 OD units/ml

    mix in the ratio of: 50 A: 125 B: 75 C by volume

STE: (autoclaved)
    100 mM Sodium chloride
    10 mM Tris · Cl pH 8.0
    1 mM EDTA pH 8.0

Hybridisation & autoradiography

This is the final procedure where the filters are incubated with the labelled probe and specific bands are then visualised on X-ray sensitive film. Again, there are many different protocols, with each manufacturer recommending a slightly different hybridisation solution for their own filter and, for the most part, following the recommended procedures will give the best result. However, a very simple procedure developed by Church & Gilbert [4] and given here works very nicely — even with large numbers of filters. In order to reduce the volume of hybridisation solution (and so improve efficiency and speed), many people use plastic bags or rotating tubes for hybridisation. In this case, however, the traditional plastic sandwich box has been retained because it is much more convenient for the hybrididsation of many filters at one time, and in fact up to 24 filters have been hybridised in 100 ml of solution using this method.

Steps in the procedure
1. Wet the dry filters in SDW and place into a large volume (20 ml per filter) of pre-hybridisation solution (hybridisation solution without probe) at 65 °C for 10 minutes. Ensure that the filters are evenly covered and can move freely over one another.
2. Denature the probe by addition of 10M sodium hydroxide, to a final concentration of 0.5M. Heat to 65 °C for 10 minutes and then add to the hybridisation solution in a plastic sandwich box just large enough for the filters to lie flat in. A minimum of 20–30 ml solution is needed, but for many filters 5–10 ml per filter is adequate.
3. Add the filters to the box one at a time ensuring that the solution covers the surface of each filter before the next filter is added. Try to avoid air bubbles between the filters and do not press them down as a film of solution between each filter is required for even hybridisation.
4. Cover and seal the box to prevent evaporation and incubate at 65 °C for 24 hours, preferably in a shaking water bath or oven. The shaking helps to prevent filters sticking to each other, and keeps any air bubbles moving so as not to interfere with hybridisation.
5. Take the filters from the box — one at a time again — and place into a large volume (at least 50 ml per filter) of 2 × SSC, 0.2% SDS. After 10 minutes of gentle agitation, take the filters out and place into 1× SSC, 0.2% SDS at 60 °C for 30 minutes with shaking. A final 30 minute wash at 60 °C in 0.5 × SSC, 0.2% SDS should be sufficient, but check the corners (or other part where there should be no hybridisation) for any background — when they are below 2 counts-per-second the non-specific binding is taken to be washed off.
6. Wrap the filters — singly — in cling-film and place into an autoradiogram cassette with two intensifying screens. Load with film and expose in a −70 °C freezer overnight or longer (if no part of the filter is above 10 counts-per-second then start off with a 2 day exposure). The intensifying screens are needed when using ^{32}P as the emitted particle is of such high energy that it passes straight through

the film ionising very few particles. The intensifying screens stop the particle and emit many more slower particles back which make a stronger signal on the film.

7. Allow the cassette to warm up to room temperature before removal and developing of the film according to the manufacturers instructions. If good signals can be seen on the film then the filter can be stripped immediately, however a further exposure is usually required — either longer or shorter than the first depending on the amount of hybridisation. This is easy to perform by simply placing a new film into the cassette and re-exposing at $-70\ °C$.

8. When adequate autoradiograms have been produced (remember the half-life of ^{32}P is only 16 days) the probe is stripped from the filter by washing in TE at $90\ °C$ for 30 minutes and then in SDW at $90\ °C$ for 30 minutes. To make sure that the filters do not dry out after this, wrap in cling-film again and store at $-20\ °C$ until they are required for re-hybridisation. Nylon based filter membranes are usually good for up to 10 hybridisations, but nitro-cellulose is too fragile for this.

Solutions

Hybridisation solution:

    1% Bovine serum albumin, crystalline grade

    1 mM EDTA

    0.5M Sodium phosphate pH 7.2

    7% SDS

Notes

DNA extraction

1. Keep the mortar & pestle in the freezer at −20 °C overnight before use — this prevents cracking when liquid nitrogen is added. Add more liquid nitrogen while grinding to keep the tissue frozen.
3. 5M potassium acetate is 492 g in 1 litre of solution. The high pH is also important to precipitate SDS. The speed of centrifugation is not critical — 3500 rpm is adequate at this step.
4. The Miracloth can be replaced by 4–5 layers of muslin, but it is not ideal. If the speed of centrifugation is reduced here, then increase the time to compensate in order not to reduce yield of DNA.
6. Some impurities precipitated by the isopropanol are not soluble in the low ionic buffer used to resuspend the DNA. These are pelleted and removed here, before the addition of the sodium acetate and isopropanol for the second precipitation. Mix gently here — vortexing can shear the DNA.

Restriction endonuclease digestion

6. To prepare sheared marker DNA of known concentration, first dissolve the DNA in SDW then pass through a fine gauge hypodermic needle 5–10 times.
7. To check for complete restriction, look at the restricted samples to see if any DNA remains of the same size as the uncut sample. Although with some enzymes there is a heavier band at the very top of the smear (from highly repetitive DNA), this should be smaller (and so further down the gel) than the uncut sample. If there apears to be uncut DNA still present, add more enzyme to the sample and incubate further. If still not completely restricted then further purification of the restricted DNA may be required. Complete restriction is necessary to eliminate any artefactual differences in DNA banding patterns.

Southern blotting

1. Permanent records of gels are easily obtained using Polaroid cameras with type 660 b/w film and a red filter to block the UV. Molecular weight markers (restricted phage or plasmid DNA) are useful on these gels, but measure the migration distance in order to find their position again on the autoradiogram. At this stage the gel can be trimmed to a convenient size for blotting and/or hybridisation. Usually the top few centimetres do not contain DNA anyway and so can be discarded, but remember to mark the gel in such a way as to remember the orientation, as this will be difficult after the wells are removed.
2. This can also be performed by soaking in 0.2M hydrochloric acid for 20 minutes to de-purinate the DNA, but do not allow this to proceed too long or the DNA will be degraded too much.
3. The volume used here and for subsequent steps depends on the size of the gel and the dish used. Generally speaking, the gel needs to be completely submerged and able to move slightly within the dish to ensure even diffusion of solutions through the gel.
4. Many of the newer filters now bind DNA in alkaline solutions and so neutralisation is not required — check the instructions supplied by the filter manufacturer.
5. Any support can be used here, but the gel tray is a convenient size, and the support must be deep enough to allow for a large reservoir of 10× SSC to be placed in the dish without it coming over the top of the support.
9. Irradiation with UV light is supposed to bind the DNA to the filter more efficiently than baking, but the dose of UV needs to be controlled and so baking is simpler.

8. Adding EtBr to the gel saves time, especially when it is not certain how long to run the gel for. It can also be used for the other mini-gels, but is not recommended for the large gels used for blotting. Take care as EtBr is a powerful mutagen.

9. Completely restricted plasmid will appear as two bands — one the vector sequence and the other the insert (unless the insert has a restriction site within it). If the insert size is known then this can be checked using molecular size markers to be sure that the correct piece is taken for labelling. If the restriction is incomplete, a third band may be seen corresponding to the sum of the insert and vector, which will not be used for labelling. If more bands are seen then this may be due to uncut circular forms of the plasmid and will make it difficult to distinguish the separated insert. In this case the extraction should be repeated and more enzyme used or the DNA purified further before restriction.

10. Take the minimum amount of gel possible in order to reduce the volume and so keep the insert as concentrated as possible. It is best to photograph the gel and use the photo to estimate the amount of DNA as this reduces the time of exposure to the UV which can degrade DNA rapidly when bound to ethidium bromide

12. It may be necessary to boil the sample before removing the aliquot in order to fully disperse the gel again. The boiling after addition of SDW is to denature the DNA as single stranded DNA is required as substrate. For this reason keep the reaction on ice after boiling to prevent re-naturation before the buffer is added.

13. Some brands of Klenow fragment are more sensitive to the agarose than others. If labelling efficiency is low, try labelling whole plasmid (without restriction and gel separation) to determine if the gel is causing problems.

14. Many people use spun columns to remove unincorporated label which is much quicker, but this method is a good start and gives an impression of the labelling efficiency (by comparing the activity of the early and later fractions) without resorting to more complicated assaying techniques.

Hybridisation

2. If the probe volume is more than 1/20 of the hybridisation solution, then the probe should be neutralised before addition. Alternatively the probe can be **boiled** without alkali for 5 minutes to denature.

4. A good seal is essential here, as evaporation will quickly reduce the volume and if the filters dry out they are useless. If hybridisation volumes are high (and probe concentrations low) longer hybridisation times are called for (see [11] for a theoretical work up of this). In this case it is best to remove the filters and replace them in the box periodically (once a day) to prevent sticking and maintain even solution contact over all of the filters.

5. To remove background, the filters can be washed more stringently (lower salt concentrations and/or higher temperatures — see [11] for data on calculation of DNA melting temperatures) but it is better to wash as little as possible to retain as much specific signal as possible. It is also possible to wash the filters after autoradiography before exposing again if the background is too high.

7. Sometimes the film is frozen to the filter and if separated quickly generates small sparks of light which are enough to cause background exposures on the film.

8. To re-use filters it is important that they do not dry out. To rehybridise, simply follow the protocol above, without the first wetting in SDW as the filters are already wet.

References

1. Beckman, J.S. & Soller, M (1986) Restriction fragment length polymorphisms in plant genetic improvement. Oxford Surveys of Plant Molecular and Cell Biology, Vol.3 pp.196–250
2. Birnboim, H.C. & Doly, J. (1979) A rapid alkaline extraction procedure for screening recombinant plasmid DNA. Nucl. Acids Res. 7: 1513
3. Brown, P.T.H. & Lörz, H. (1985) Molecular changes and possible origins of somaclonal variation. In: Somaclonal variation and crop improvement. Ed. J Semal, Martinus Nijhoff. pp.148–159
4. Church, G.M. & Gilbert, W. (1984) Genomic sequencing. Proc. Nat. Acad. Sci. USA 81: 1991–1995
5. Dellaporta, S.L. Wood, J. Hicks, J.B. (1983) A plant DNA minipreparation, version II. Plant Mol. Biol. Rep. 1: 19–21
6. Feinberg, A.P. & Vogelstein, H. (1983) A technique for radiolabelling DNA restriction fragments to high specific activity. Anal. Biochem. 132: 6–13
7. Karp, A. & Bright, S.W.J. (1985) On the causes and origins of somaclonal variation. Oxford Surveys of Plant Molecular and Cell Biology. Vol 1 pp.199–234
8. Karp, A. & Maddock, S.E. (1984) Chromosome variation in wheat plants regenerated from cultured immature embryos. Theor. Appl. Genet. 67: 249–255
9. Landsmann, J. & Uhrig, H. (1985) Somaclonal variation in S. tuberosum detected at the molecular level. Theor. Appl. Genet. 71: 500–505
10. Larkin, P. & Scowcroft, W.R. (1981) Somaclonal variation – a novel source of variability from cell cultures for plant improvement. Theor. Appl. Genet. 60: 197–214
11. Maniatis, T. Fritsch, E.F. Sambrook, J. (1982) Molecular cloning: A laboratory handbook. Cold Spring Harbor Laboratory, New York.
12. McCough, S.R. Kochert, G.Y. Wang, Z.Y. Kush, G.S. Coffman, W.R. Tanksley, S.D. (1988) Molecular mapping of rice chromosomes. Theor. Appl. Genet. 76: 815–829
13. Potter, R.H. & Jones, M.G.K. (1991) Genetic stability of potato (Solanum tuberosum L.) after in vitro manipulation. Plant Science (in press).
14. Southern, E.M. (1975) Detection of specific sequences among DNA fragments separated by gel electrophoresis. J. Mol. Biol. 98: 503–517
15. Tanksley, S.D. Young, N.D. Paterson, A.H. Bonierbale, M.W. (1989) RFLP mapping in plant breeding – new tools for an old science. Biotechnology 7: 257–264
16. Zheng, K.L. Castiglione, S. Biasini, M.G. Biroli, A. Morandi, L. Sala, F. (1987) Nuclear DNA amplification in cultured cells of Oryza sativa. Theor. Appl. Genet. 74: 65–70

Plant Tissue Culture Manual C6: 1–12, 1992.
© 1992 *Kluwer Academic Publishers*.

Virus elimination and testing

MARY C. COLEMAN[1] & WAYNE POWELL
Scottish Crop Research Institute, Invergowrie, Dundee, Scotland DD2 5DA
[1]*Agricultural Genetics Company Ltd., Cambridge CB4 4BH, UK*

Introduction

In recent years there has been an increasing awareness of the value of germplasm conservation. As new varieties comprise an increasingly large proportion of cultivated crops, the genetic diversity of major crop species is being eroded. A number of initiatives have been undertaken by international organisations to collect, evaluate and preserve wild and primitive genotypes. For seed-propagated crops this is relatively easy but many crops have to be propagated asexually because the plants do not produce seeds or the crop depends on the performance of a selected genotype. Thus asexually propagated material needs to be maintained clonally by an appropriate means of vegetative propagation.

Over the past 50 years *in vitro* multiplication methods have been developed to facilitate the rapid introduction of new varieties, international exchange of germplasm and the storage of material without the risk of infections which can occur *in vivo*. A prerequisite for the use of *in vitro* multiplication for germplasm storage and exchange is the availability of 'elite' disease-free plants from which cultures can be initiated. In this chapter will be discussed methods of detecting and eliminating virus infection from stock plants before or during micro-propagation. It should also be borne in mind that such plants may be infected with bacteria and/or fungi and steps to eliminate them may also need to be considered.

Virus elimination

There are several methods available to produce virus-free plants, including heat treatment, meristem culture, heat treatment combined with meristem culture, adventitious regeneration in the presence of chemicals such as Virazole (1-D Ribofuranosyl-1,2,4 triazole-3-carboxamide) and culture of cells or protoplasts from non-infected cells.

Heat treatment is an effective way of inactivating isometric viruses [36], while meristem culture has gained importance due to the absence of virus in the apical dome in most species. The absence of virus has been attributed to competition in the meristem between production of virus particles and cell division, but it may also be due to a lack of vascular elements in the meristem, thus hindering the transport of the virus particles to the apical dome [36]. The absence of infected cells in mesophyll cell populations, and particularly in infections characterised by the presence of dark green areas within mosaic patterns, has offered the possibility of regenerating virus-free plants from such cells or areas [38, 15, 2, 34].

It is sometimes possible to suppress virus multiplication and eliminate viruses by the use of antiviral chemicals such as Virazole (syn Ribavirin) or vibarabine in the medium [43, 26, 29]. In a few cases such as lily [10] and apple [18] the use of Virazole has resulted in the production of virus-free plants. Incorporation of Virazole into potato explant and meristem culture media has been shown to give virus-free progeny from virus-infected explant and meristem donor plants [7]. Bittner *et al.* [4] showed that Ribavirin and 2,4-dioxohexa-hydro-1, 3, 5-triazine (DHT) significantly inhibited the replication of potato virus S. Meristem culture and adventitious regeneration in the presence of anti-viral chemicals is described below.

Methods of virus testing

Potential donor plants must be screened for the presence of infection, particularly if the virus is symptomless or if the plant is infected with more than one virus. The simplest methods involve the use of indicator plants which react promptly and characteristically to sap inoculation, usually with the formation of local lesions on the inoculated leaves [39]. The best 'general purpose' indicator plants are found in the Chenopodiaceae, particularly *Gomphrena globosa*. Virus particles can often be directly observed with an electron microscope, e.g. the quick leaf-dip method of Brandes [5]. The advantages of combining EM with serology have been elucidated by Gough & Shukla [17]. However, present methods limit the number of samples which can be examined. A widely used serological technique is enzyme linked immuno-sorbent assay (ELISA) [9]. It normally takes 1 to 2 days to complete and can handle very large numbers of samples.

More recent methods involving nucleic acid hybridisation to detect the viral genome are an alternative to ELISA. Nucleic acid hybridization has the advantage of greater sensitivity and a high degree of specificity.

Symptoms

Symptoms of viral infection vary from mosaics or ringspots to variegation, vein-banding or clearing. Internal symptoms may also be present, i.e. intracellular inclusions which are aggregates of virus particles large enough to be seen using an optical microscope. The two main types of inclusion are crystalline and amorphous, the latter being known as X bodies. Other structures may occur in infected cells, such as 'pinwheels', which are symptoms of infection with potato virus Y, although it is unlikely that they consist of virus

[19]. The symptoms associated with virus infections have been reviewed by Smith [39], Esau [14] and Matthews [30].

Test plants

Test plants are of considerable importance not only for detecting and identifying known viruses but also for revealing the presence of new ones. They are also of use in differentiating viruses and in separating certain virus complexes such as those that commonly occur in potato [20, 21, 11]. Viruses occurring in a complex can be separated by observing their reaction when challenged by a test plant. Test plants may be primary, secondary or tertiary separators, depending on whether in the course of a stepwise virus separation they are suitable for the simultaneous separation of two viruses (primary separator) or for the further separation of residual virus complexes (secondary and tertiary separators). A check-list of host plants for identifying and separating twelve potato viruses is given by Horvath [22].

Virus detection may be seasonally dependent, e.g. in *Pelargonium* it is most efficient in early spring, using mechanical inoculation of sap diluted in a phosphate buffer on *Chenopodium quinoa*. This is sensitive to all known *Pelargonium* viruses [44].

Electron microscopy

Viruses can be observed in crude sap using electron microscopy. The sample is mounted on a 3 mm copper grid covered with a thin film of plastic (Formvar). The grids contain a number of apertures (60 to 160 meshes/cm). The virus particles stick to the Formvar film, but as they are highly transparent to electrons shadow-casting or negative staining is essential [35]. Shadow casting may be performed with dried preparations, which can be prepared elsewhere and sent to a suitably equipped laboratory for examination.

Serology

The antigenic properties of viruses represent the single most useful criterion for their reliable detection and identification. Early serological detection techniques included the micro-precipitation test, the chloroplast agglutination test and the Ouchterlony agar double-diffusion test [35]. Enzyme-linked immunosorbent assay (ELISA) is a sensitive test which is widely used. Most workers use the direct double-antibody sandwich method which requires the preparation of a different antibody conjugate for each virus to be tested [8]. It is the method of choice when it is important to distinguish between closely related serotypes. It requires the preparation of one antiserum derived from a single animal species. The indirect ELISA method requires the use of two antisera prepared in different animal species (e.g. rabbit and chicken) and a single enzyme conjugate can be used for all virus systems. The indirect ELISA methods can detect a broader range of serologically related viruses and are usually more sensitive than the direct methods [40]. Multi-layered sandwich procedures are often the most sensitive, e.g. using an antibody from chicken, a viral antigen, an antibody from mouse, rabbit anti-mouse globulins and enzyme-labelled goat anti-rabbit globulins. The use of an avian antibody in one of the layers is advantageous [1] as the absence of cross-reaction between avian and mammalian globulins keeps background readings low [42].

Observing serological reaction with an electron microscope provides another means of virus identification. Derrick [12] applied the principle of a solid phase immunoassay to electron microscopy by trapping viruses on grids coated with specific antiserum. This technique is known as ISEM [37], and has been found to have a sensitivity similiar to that of ELISA. By pre-treating the grids with protein A (PALIEM), the antibody molecules can be anchored more efficiently and the sensitivity of virus detection may be slightly improved [28]. Serological procedures for detection of plant virus infections are described and reviewed by Van Regenmortel [41].

Presence of double-stranded RNA

The use of viral double-stranded (ds) RNA for disease detection arises from the fact that only plants infected with RNA viruses or virus-like agents contain homogeneous segments of high molecular weight ($70.1 + 10^6$ KDa) ds RNA [23, 31, 32]. Additional diagnostic information about the virus can be obtained from the pattern of genomic and subgenomic ds RNA species upon gel electrophoretic analysis of the isolated ds RNAs. For procedures see Morris & Dodds [31], Dodds & Bar-Joseph [13], Jordan *et al.* [25], Jordan & Dodds [24].

Nucleic acid probes

Nucleic acid hybridisation methods have been used to detect viral genomes with the cDNA probes capable of identifying viral target sequences in the 1 to 10 ng range. The method involves the hybridisation of ^{32}[P] labelled cDNA clones with crude sap spots which have been immobilised on a nitrocellulose membrane, and has been described by Baulcombe *et al.* [3] for potato virus X. The cDNA is prepared by reverse transcription of partially purified viral RNA, and the probes are radioactively labelled with ^{32}[P] by nick translation. Non-radioactively labelled probes such as biotin-labelled nucleotides and the fluorescein-conjugated avidin system overcome the safety problems associated with the use of radioactivity [27].

Conclusions

Detection and elimination methods should be rapid, reliable and sensitive and should at the same time give clear-cut results. Schemes for the production of virus-free planting stock and subsequent multiplication and distribution have been operating for some vegetatively-propagated horticultural crops in agriculturally advanced countries for many years. For example, virus-indexed strawberry plants have been available for over 50 years in Britain where the Nuclear Stock Association Ltd distributes virus tested material of flower bulbs, soft fruit and fruit trees [6]. The methods of detection depend on both the plant and the virus, and should use a combination of simple biological tests and more advanced tests such as ELISA and cDNA probes. Meristem culture and heat treatments should result in viral elimination.

The following sections outline some of the methods which can be used in virus elimination and testing.

Protocols

Meristem culture

1. Select starting material from seedlings, new buds or young shoots.
2. Excise apical and lateral buds.
3. Surface sterilise (optional) with 80% v/v alcohol for 30 seconds followed by 1% v/v sodium hypochlorite for 15 minutes. Rinse in sterile distilled water 3 times. All procedures after the alcohol sterilization are carried out in a laminar flow cabinet.
4. With the aid of a binocular microscope and using sterile forceps and scalpel remove the outer leaves and primordia until the glossy apical dome is visible.
5. Excise the dome and place on the culture medium.
6. Seal the tube/jar and place in a growth room with 16 h photoperiod, 160–180 μmol photons $m^{-2} s^{-1}$.

Notes

2. To eliminate viruses the size of the dome should not exceed 0.2–0.5 mm in length, depending on the plant species.
3. Surface sterilisation is optional and is not necessary with most material as the apical dome is normally very well protected and free from surface contaminants.
5. The composition of the nutrient medium is important. Each plant species and sometimes different cultivars within a species require special media. Murashige & Skoog [33] medium and Gamborg's B5 [16] are the most widely used.
   Some excised meristems (e.g. *Cattleya*) produce toxic substances, and in these cases a liquid medium should be used and the meristem placed on paper bridges.
   The percentage of isolated meristems developing into virus free plants is generally small.
   By culturing meristems in the presence of a suitable antiviral chemical it is possible to maintain inhibitory conditions for long enough to eliminate the virus.
6. The combination of heat treatment of the plant with meristem culture is more efficient if viruses are heat sensitive. Heat treatment varies; plants can be exposed to temperatures of 32–35 °C for 23 days, to 33–37 °C for 4 weeks or to 37–38 °C for 20 to 40 days depending on the plant and virus involved. The temperature and treatment time should be chosen to allow the plant to just survive while inactivating the virus. Heat treatment and meristem culture are commonly used with potato (*Solanum tuberosum*), carnation, strawberry and chrysanthemum.

Virus elimination using differentiating systems

1. Take internodal explants 2 cm in length from an actively growing donor plant.
2. Surface sterilise in 80% v/v alcohol for 30 sec followed by 1% v/v sodium hypochlorite for 15 minutes. Rinse three times in sterile water.
3. Cut off approximately 3 mm from each end of the explant and divide it longitudinally.
4. Place cut surface down on suitable regeneration medium containing an antiviral chemical e.g. 250 μm Ribavirin.

4. Ribavirin must be continuously present in the medium for virus elimination to occur.
   Plants take longer to differentiate and grow in the presence of Ribavirin.

ELISA [9]

1. Put 200 µl of antibody (α globulin) 1 µg/ml in coating buffer into each cell of a 96 well microtitre plate. Incubate for 2 hours at 37 °C.
2. Prepare samples by grinding in extraction buffer. Wash plate with phosphate buffered saline (PBS) plus Tween by flooding the plate and leave it for 3 minutes. Repeat the process three times.
3. Add 200 µl of sample in duplicate to the plate and incubate overnight at 4 °C.
4. Wash the plate 3 times.
5. Add 200 µl antibody-conjugate (enzyme labelled α globulin) and incubate at 37 °C for 3–6 hours.
6. Wash plate 3 times as in (2) above.
7. Add 200 µl freshly prepared substrate to each well. Incubate at room temperature (20–24 °C) for 1 hour.
8. Stop reaction with 50 µl of 3 M solution of NaOH.
9. Assess results (a) visual observation of colour reaction
   (b) measure absorbance at 405 nm.

Notes
1. Many commercial ELISA kits are available to detect viruses in agricultural and horticultural crops e.g. those by Boehringer (Mannheim).
   Preparation of antiserum is described by Noordan [35] and Clark & Adams [9].

Solutions
— Coating buffer:
  — NaHCO$_3$ 2.93 g
    Na$_2$CO$_3$ 1.59 g } in 1 litre, pH 9.6
— Phosphate buffered saline (PBS):
  — Na$_2$HPO$_4$.2H$_2$O (0.01 M)   8.9 g
  — NaH$_2$PO$_4$.2H$_2$O (0.01 M)   7.8 g } in 1 l, pH 7.0
  — NaCl   42.5 g,
— Extraction buffer:
  — PBS — Tween
  — 2% w/v polyvinylpyrrolidone (PVP)
— Washing buffer:
  — PBS — Tween (= PBS + 0.5 ml/l Tween)
— Conjugate buffer:
  — PBS — Tween
  — 2% w/v PVP
  — 0.2% w/v Ovalbumin
— Substrate buffer:
  — 0.1 M diethanolamine, pH 9.8

- Substrate:
  - 0.6 mg/ml 4-nitrophenylphosphate in substrate buffer

  Siliconise all glassware used for antiserum.

Protein A linked immuno-electron microscopy (PALIEM)

1. Coat copper electron microscope grids with 0.4% w/v Formvar in water-free ethylene dichloride.
2. Grip grid, coated side uppermost, with forceps and place a 5 μl drop of protein A (0.1 mg/l in 5 mM sodium phosphate buffer, pH 7.0) on it.
3. Incubate for 10 minutes at 20 °C.
4. Wash grid with approximately 20 drops of sodium phosphate buffer (5 mM, pH 7.0).
5. Place 5 μl of phosphate buffer on the grid and dip the cut edge of a leaf into it (incubate as above).
6. Wash grid with buffer (4).
7. Add 5 μl of dilute antiserum and incubate (3).
8. Wash the grid with buffer followed by distilled water (30 drops of each).
9. Stain by placing a drop of 2% w/v uranyl acetate in water on the slide.
10. Examine using an electron microscope for virus presence.

References

1. Al Moudallal Z, Altschuh D, Briand JP & Van Regenmortel MHV (1984) Comparative sensitivity of different ELISA procedures for detecting monoclonal antibodies. Journal of Immunological Methods 68: 35–43.
2. Atkinson PH & Matthews REF (1970) On the origin of dark green tissue in tobacco leaves infected with tobacco mosaic virus. Virology 90: 344–356.
3. Baulcombe D, Flavell RB, Boulton RE & Jellis GJ (1984). The sensitivity and specificity of a rapid nucleic acid hybridisation method for the detection of potato virus X in crude sap samples. Plant Pathology 33: 361–370.
4. Bittner H, Schenk G, Schuster G & Kluge S (1989) Elimination by chemotherapy of potato virus S from potato plants grown in vitro. Potato Research 32: 175–179.
5. Brandes J (1957) Eine elektronenmikroskopische Schnellmethode zum Nachweis faden und Stabchenformiger, Viren, insbesondere in Kartoffeldunkelkeimen. Nachr Bt Dr Pflschutzd 9: 157–152.
6. Brunt AA (1985) The production and distribution of virus-tested ornamental bulb crops in England: Principles practice and prognosis. Acta Horticulturae 164: 153–161.
7. Cassells AC & Long RD (1982) The elimination of potato viruses X, Y, S and M in meristem and explant cultures of potato in the presence of Virazole. Potato Research 25: 165–173.
8. Clark MF (1981) Immunosorbent assays in plant pathology. Annual Review of Phytopathology 19: 83–106.
9. Clark MF & Adams AN (1977) Characteristics of the microplate method of enzyme linked immunosorbent assay for the detection of plant viruses. Journal of General Virology 34: 475–483.
10. Cohen AJ (1986) Plant tissue cell culture abstracts. International Congress 6: 30.
11. De Bokx JA (1972) Viruses of potato and seed-potato production. Wageningen: Centre for Agricultural Publishing and Documentation.
12. Derrick KS (1973) Quantitative assay for plant viruses using serologically specific electron microscopy. Virology 56: 652–653.
13. Dodds JA & Bar-Joseph M (1983) Double-stranded RNA from plants infected with clostero viruses. Phytopathology 73: 419–423.
14. Esau K (1967) Anatomy of plant virus infections. Annual Review of Phytopathology 5: 45–76.
15. Fulton, R.W. (1951) Superinfection by strains of tobacco mosaic virus. Phytopathology 41, 579–592.
16. Gamborg DL, Miller RA & Ojima K (1968) Nutrient requirements of suspension cultures of soybean root cells. Experimental Cell Research 50: 151–158.
17. Gough KA & Shukla DD (1980) Further studies on the use of protein A in immuno electron microscopy for detecting virus particles. Journal of General Virology 51: 415–419.
18. Hansen AJ & Lane WD (1985) Elimination of apple chlorotic leafspot virus from apple shoot cultures. Plant Diseases 69: 134–135.
19. Hiebert E, Purcifull DE, Christie RG & Christie SR (1971) Partial purification of inclusions induced by tobacco etch virus and potato virus Y. Virology 43: 638–646.
20. Horvath J (1963) Neuere Beitrage zum Vorkommen von Kartoffelviren mit besonderer Rücksicht auf Komplexinfektionen. Acta Agronomica Academicae Scientiarum Hungarice 14: 67–81.
21. Horvath J (1967) Separation and determination of viruses pathogenic to potato virus Y. Acta Phytopathologica Academiae Scientiarum Hungaricae 2: 319–360.
22. Horvath J (1985) A check-list of new host plants for identification and separation of twelve potato viruses. Potato Research 28: 71–89.
23. Ikegami M & Fraenkel-Comrat H (1979) Characterisation of double stranded ribonucleic acid in tobacco leaves. Proceedings National Academy of Sciences USA 76: 3637–3640.
24. Jordan R & Dodds JA (1985) Double-stranded RNA in detection of diseases of known and unproven viral etiology. Acta Horticulture 164: 101–108.

25. Jordan RL, Heick JA, Dodds JA & Ohr H (1983) Rapid detection of sunblotch viroid RNA and virus-like double-stranded RNA in multiple avocado samples. Phytopathology 73: 791.(Abs.)

26. Kartha KK (1986) Production and indexing of disease-free plants. In: Plant tissue culture and its agricultural application. (LA Withers and PG Alderson, Eds). London, Butterworth Press.

27. Lange L (1986) The practical application of new developments in test procedures for the detection of viruses in seed. In: Developments in Applied Biology 1. Developments and Applications in Virus Testing. Eds RAC Jones and L Torrance. Association of Applied Biologists, Wellesbourne, UK.

28. Lesemann DE & Paul HL (1980) Conditions for the use of protein A in combination with the Derrick method of immuno electron microscopy. Acta Horticulturae 110: 119–128.

29. Long RD & Cassells AC (1986) Elimination of viruses from tissue culture in the presence of antiviral chemicals. In: Plant Tissue Culture and its Agricultural Application (LA Withers and PG Alderson, Eds), Butterworth Press.

30. Matthews REF (1970) Plant Virology, Academic Press, New York.

31. Morris TJ & Dodds JA (1979) Isolation and analysis of double-stranded RNA from virus infected plant and fungal tissue. Phytopathology 69: 854–858.

32. Morris TJ, Dodds JA, Hillman B, Jordan RL, Lommel SA & Tamalki SJ (1983) Viral specific ds RNA: diagnostic value for plant virus disease identification. Plant Molecular Biology Reporter 1: 27–30.

33. Murashige T & Skoog F (1962) A revised medium for rapid growth and bioassay with tobacco tissue cultures. Physiological Plantarum 15: 473–497.

34. Muraskishi HH & Carlson PS (1976) Regeneration of virus-free plants from dark green islands of tobacco mosaic virus-infected tobacco leaves. Phytopathology 66: 931–932.

35. Noordam D (1973) Identification of plant viruses. Methods and experiments. Centre for Agricultural Publishing Documentation, Wageningen.

36. Quak, F. (1977). Meristem culture and virus-free plants. In: Applied and fundamental aspects of plant cell, tissue and organ culture. J Reinert and YPS Bajaj (Eds). Springer-Verlag, Berlin.

37. Roberts IM, Milne RG, Van Regenmortel MHV (1982) Suggested terminology for virus-antibody interactions observed by electron microscopy. Intervirology 18: 147–149.

38. Shepard JF (1975) Regeneration of plants from protoplasts of potato virus X infected tobacco leaves. Virology 66: 492–501.

39. Smith KM (1974) Plant Viruses. Great Britain Northumberland Press.

40. Van Regenmortel MHV (1982) Serology and immunochemistry of plant viruses. Academic Press, New York.

41. Van Regenmortel MHV (1985) New serological procedures including the development and uses of monoclonal antibodies in virus detection and diagnosis. Acta Horticulturae 164: 187–194.

42. Van Regenmortel MHV (1986) The potential for using monoclonal antibodies in the detection of plant viruses. In: Developments in Applied Biology 1. Developments and Applications in Virus Testing. Eds RAC Jones and L Torrance. Association of Applied Biologists, Wellesbourne, UK.

43. Walkey DGA (1980) Production of virus-free plants by tissue culture. In: Tissue culture methods for plant pathologists (DS Ingram and JP Helgeson, Eds), 109–117. Blackwell Scientific Publications, Oxford.

44. Welvaert W & Samyn G (1985) Relative importance of *Pelargonium* viruses in cutting nurseries. Acta Horticulturae 164: 195–198.

Plant Tissue Culture Manual C7: 1–18, 1993.
© 1993 *Kluwer Academic Publishers. Printed in the Netherlands.*

Clonal Propagation of *Citrus*

T. S. RANGAN
Phytogen, 101 Waverly Dr., Pasadena, CA 91105, U.S.A.

Introduction

Fruit tree species are generally heterogenous and elite selections have been usually propagated vegetatively to maintain their genetic makeup and superior characteristics. With the exception of species such as *Betula, Populus* and *Prunus,* regeneration of woody taxa in cell and tissue cultures has been rather elusive despite several years of research. One possible reason is the absence of tissues with morphogenetic potential in mature woody plants. In *Citrus* and *Mangifera* and several other species (see [34, 55]) embryos arise adventitiously from cells of the nucellus or integument. It is believed that most of the viruses of *Citrus* are not transmitted through seedlings, whether of zygotic or nucellar origin. Research in the last 10–15 years has demonstrated the increasing utilization of cell and tissue culture in propagation, breeding and improvement of *Citrus.* This chapter deals with some of those aspects including shoot-tip grafting, embryogenesis, protoplast culture and genetic transformation.

1. Organogenesis

Tissue cultures can be established from several different tissues but the ability to regenerate varies with the source of the explant. The formation of adventitious shoot buds in cultures of *Citrus madurensis* [15] and *Poncirus trifoliata* [36] has been reported. Multiple shoots in nodal explants from both seedling and mature plants of *C. sinensis* × *P. trifoliata, C. limonia, C. reshnii, C. sinensis* and *P. trifoliata* have been induced and mature plants established [1, 7]. A gradient of bud forming ability was shown to exist in epicotyl segments of *C. sinensis.* This ability to form buds decreased as the distance from cotyledonary node increased, and the presence or absence of cotyledons also significantly affected bud formation [3].

Callus from *Citrus* albedo [42] and juice vesicles [32, 33, 42, 66] have failed to regenerate. However, Nito and Iwamasa [47] obtained somatic embryos and plantlets from juice vesicles of *C. unshiu.*

2. Shoot-tip grafting

A common method to obtain disease-free *Citrus* plants is through the selection of nucellar seedlings. However, nucellar seedlings have a long period of juvenility before becoming productive. Thermotherapy, yet another method, is ineffective in eliminating many diseases such as exocortis and xyloporosis. Shoot tip grafting is a technique now successfully employed in several laboratories to obtain disease-free plants.

Murashige et al. [43] were the first to obtain a few Citrus plants by grafting shoot tips in vitro. Subsequently Navarro [45] extensively used this technique to obtain disease-free Citrus. The technique involves the following steps: rootstock and scion preparation, grafting, culturing of grafted plants and finally, transfer of grafted plants to soil.

Root stocks are obtained by germinating the seeds of troyer citrange or any other root stock that is compatible with the scion. The shoot tips (apical meristem + 3 leaf primordia) are collected from vegetative flushes of field or greenhouse-grown plants. Shoot tips from dormant buds or buds growing in vitro can also be used.

The shoot tip is grafted on top of the decapitated rootstock and the grafted seedlings are cultured in vitro in liquid medium. After 4–6 weeks the successfully grafted seedlings are transferred to soil. Presently shoot-tip grafting is being used in several countries to establish disease-free Citrus clones for commercial propagation [44].

3. Nucellar embryogenesis

The phenomenon of nucellar polyembryony is of considerable interest to citriculturists. Although the polyembryonic citruses pose no problem in yielding nucellar embryos and seedlings, many of the commercially important monoembryonic varieties produce no nucellar embryos under normal conditions. It was, therefore, believed that monoembryonic Citrus could neither be freed from viruses nor rejuvenated through nucellar seedlings. In recent years the morphogenetic potential of nucellar cells of both poly- and monoembryonic citruses and their relatives has been convincingly demonstrated [52, 56, 62].

Except for a brief abstract [63], the first report on culture of nucellus has been that of Rangaswamy [54]. The nucellus of C. microcarpa, when cultured on a casein hydrolysate-supplemented medium, proliferated into a callus mass and differentiated into "pseudobulbils" which eventually developed into plants [35, 55]. Rangan et al. [52, 53] succeeded in inducing embryogenesis in monoembryonic Citrus – C. grandis, C. limon and C. reticulata × C. sinensis. Unlike in C. microcarpa or C. reticulata, the nucellar explants gave rise directly to organized embryos. Bitters et al. [2] extended these studies to other mono- and polyembryonic as well as seedless Citrus such as C. temple, C. reticulata, C. limon (Meyer lemon), C. maxima, C. sinensis (Robertson navel), C. latipes and C. latifolia (Bearss lime, a triploid seedless variety). Although embryos and plants could be obtained, the percentage of successful cultures and the age of the ovules at which the nucelli were most responsive varied from species to species. A diffusible, anti-embryogenic factor in the nucellus of monoembryonic citrus was reported by Esan (see [40, 64]) and the presence of this substance could be a major limiting factor in the initiation of embryogenesis *in vitro*.

Pollination and fertilization were generally considered as essential prerequisites for the initiation of nucellar embryogenesis. While in earlier investigations [52, 53, 55] nucelli were removed from fertilized ovules, Bitters *et al.* [2], Button and Bornmann [4], Kochba *et al.* [28], Kochba and Spiegel Roy [25] and Mitra and Chaturvedi [37] obtained embryogenic calluses from nucelli excised from unfertilized ovules. The feasibility of using unfertilized ovules as an explant source can not only be profitably exploited for propagating seedless varieties but also for importing seedless varieties without the risk of introducing new diseases [38]. Embryogenic calluses from nucelli and ovular explants of several *Citrus* cultivars have been established and plants regenerated [11, 29].

Substances promoting embryogenesis include malt extract [2, 4, 28, 37, 52, 53], adenine [4], adenine sulfate and kinetin [37], abscisic acid and ethephon [30]. Antibiotics such as chloramphenicol and methotrexate are also reported to induce embryogenesis in cultures of recalcitrant species such as "Key" lime [39]. In a series of papers Kochba and his co-workers reported the effects of the composition of culture medium [25], growth substances [26, 30], various sugars and the age of cultures [24, 31] and gamma irradiation [27, 60, 61] on embryogenesis in Shamouti nucellar callus.

The habituated growth habit of *Citrus* nucellar callus indicates the presence of adequate endogenous levels of auxin and possibly other growth substances. While addition of exogenous IAA, NAA or cytokinins such as kinetin, 6-benzyl-aminopurine or 2ip inhibits embryogenesis, auxin synthesis inhibitors such as 5-hydroxynitro-benzyl-bromide (HNB), 7-aza-indole (AZI), and inhibitors of gibberellin synthesis such as 2-chloroethyltrimethyl ammonium chloride (CCC), succinic acid 2,2-methyl-hydrazide (ALAR) stimulated embryogenesis [26].

By culturing unfertilized ovules of *C. sinensis* and *C. reticulata* × *C. paradisii* on a medium containing colchicine, Gmitter and Ling [12] obtained non-chimeric plants. The regenerants included both tetraploid and diploid individuals. This approach may provide a means to develop tetraploids from monoembryonic cultivars [14].

4. Endosperm

In conventional breeding, one of the methods to obtain triploid plants with nearly seedless fruits is by $2 \times -4 \times$ hybridization. The endosperm being a genetically unique tissue, regeneration of plants from the endosperm could be a novel way to establish triploid plants. Embryos and plants were obtained from endosperm cultures of *C. grandis* and *C. sinensis* [71]. Gmitter *et al.* [13] also reported embryos and triploid plantlets in cultures of *C. sinensis*, *C. paradisii* and *C. grandis*. As the shoots produced were weak they were micrografted *in vitro* onto citrange rootstock, and viable, verifiable triploids were obtained.

5. Suspension cultures

Many of the *Citrus* cultivars originated as bud mutations thus justifying the hope that the incidence of mutations can be increased by treatment with mutagenic agents. The critical requirement is the ability to select the mutant cell and avoid the loss of mutation due to chimera formation. Though nucellus and unorganized callus could be used as explants, the tedious procedure makes this approach impractical. Using callus cultures, cell lines tolerant to NaCl and 2,4-D have been isolated. However, suspension cultures coupled with suitable mutagenic treatments and selection procedures offer a superior approach. Suspension cultures could also serve as ideal material for isolation of protoplasts as well as genetic transformation experiments [5, 21, 23, 48, 49, 50, 59].

6. Protoplasts

With the unequivocal demonstration of the morphogenetic ability of the nucellar callus, it was only logical and desirable to develop a protoplast-to-plant technology for *Citrus*. In crosses where conventional breeding is problematic because of incompatibility, protoplast culture and somatic hybridization could play a significant role [67]. Vardi and her co-workers were the first to successfully isolate protoplasts and regenerate plants from cultures of *C. sinensis* [9, 67, 70]. Both embryogenic calluses and suspension cultures could be used as source material. The conditions during isolation, such as osmoticum and enzyme source strongly influence the yield and the ability of protoplasts to form colonies. For adequate cell division, a proper plating density ($8 \times 10^4 - 10^5$ protoplasts ml^{-1}) is essential. Grosser *et al.* [16, 17] obtained somatic hybrids of "key" lime (*C. aurantifolia*) and "Valencia" sweet orange (*C. sinensis*) through protoplast fusion. Such interspecific, allotetraploid plants have great potential in breeding. Similar interspecific hybrids between *C. sinensis* × *Poncirus trifoliata*, *C. sinensis* × *C. unshiu* [23], *C. sinensis* × Murcott tangor [22, 23], *C. sinensis* × *C. paradisii* [48], *C. sinensis* × *C. trifoliata* [50] *C. sinensis* and *C. limon* [65] have also been reported. Grosser *et al.* [18] obtained the first intergeneric hybrid between sexually incompatible *C. reticulata* and *Citropsis gilletiana*. Another intergeneric somatic hybrid between Troyer citrange and *P. trifoliata* [49] has also been obtained.

7. Genetic transformation

The progress in the manipulation of protoplasts has made possible not only the production of somatic hybrids but also the introduction of foreign DNA into protoplasts. Using the PEG-mediated DNA uptake method, Kobayashi and Uchimiya [21] introduced plasmid DNA containing the selectable marker gene for aminoglycoside phosphotransferase II [APH(3)II]. Colonies growing on a medium containing kanamycin showed APH(3) activity and also the presence of the intact gene. Vardi *et al.* [69] successfully introduced plasmid DNA containing two marker genes namely, chloramphenicol acetyltransferase (CAT) and neomycin phosphotransferase (NPT II) into protoplasts of *C. jambhiri*. The transgenic nature of the colonies selected on paromomycin was confirmed by the transient expression of CAT activity, stable expression of NPT II activity as well as by Southern hybridization. From two of the colonies transgenic plants were also regenerated. Hidaka *et al.* [20] transformed suspension culture cells of *C. sinensis* and *C. reticulata* using *Agrobacterium* carrying the marker genes for neomycin phosphotransferase and hygromycin phosphotransferase. The presence of the marker genes in callus and regenerated plantlets was confirmed by Southern hybridization.

Protocols

1. Nucellar embryogenesis in vitro

Steps in the procedure
1. Tag the flower buds on the day of anthesis to ensure the uniform age of the explant.
2. In some studies it may be essential to emasculate and pollinate with pollen of trifoliate orange (*Poncirus trifoliata*). The reason for controlled pollination is to provide a readily distinguishable marker so that any zygotic seedlings could be readily identified by their trifoliate leaves from nucellar seedlings.
3. Take samples of the developing ovules at weekly intervals to determine the suitable stage of culture, which will vary with different species.
4. Collect the developing fruits (after determining the optimal stage), wash in tap water, surface-sterilize in 10—15% calcium hypochlorite or commercial bleach for 10—15 minutes and rinse 3 times in sterile distilled water.
5. Cut open the fruits under aseptic conditions and remove the ovules. The isolation of nucellus is carried out under a binocular dissecting microscope equipped with 20—60× magnification. Hold the ovules at the chalazal end with a microforceps, and use a sharp scalpel to make an incision from the chalazal to the micropylar end through the integuments. Peel away the integuments, exposing the nucellus. The zygotic embryos and the endosperm are discarded. Sever the nucellus at the chalazal end where it is attached to the integuments, and place in culture.
6. The nutrient medium consists of Murashige and Skoog's [41] salts. The basal medium is supplemented with auxins, cytokinins and other growth substances such as casein hydrolysate or malt extract as required, according to the species. Some examples are listed in Table 1.
7. The cultures are maintained at $25 \pm 2\,°C$, 50—60% relative humidity and, under 16:8 hr light:dark regime in diffuse light (1000—1500 lux).
8. For culture of ovules (fertilized or unfertilized) the ovules are excised from sterilized young fruits (4—6 weeks after pollination depending on the cultivar).
9. About 20—25 ovules are cultured on Murashige and Tucker's [42] medium (for composition see Table 2).
10. When embryogenic callus develops, subculture it onto the same medium and maintain by subculture every 3—4 weeks.
11. For germination, transfer the somatic embryos (with cotyledons) to medium supplemented with gibberellic acid (1—5 mg/l) and grow for 5—7 weeks (see [29]).
12. To induce a good root system it may also be useful to transfer plantlets onto a filter paper bridge in liquid medium.
13. Transfer plantlets with well-developed root systems to pots with sterilized potting mixture, and cover with a plastic bag to maintain high humidity.
14. Upon further development, transfer the plants to the greenhouse and maintain under high humidity for 4—7 days, and gradually remove the plastic cover.

Table 1. Growth supplements used for initiation of nucellar embryogenesis in vitro*

Cultivar	Growth supplements (mg/l)	Reference
C. microcarpa	Casein hydrolysate (400)	[55]
C. reticulata X C. sinensis (Temple Orange) C. grandis (Pong yau Pummello) C. limon (Ponderosa lemon)	Adenine sulphate (25) Naphthaleneacetic acid (0.5) Casein hydrolysate (500) Malt extract (500)	(52)
C. sinensis (Washington Navel)	Ascorbic acid (40) Coconut milk (15% v/v) Adenine or Adenine sulphate (40) Malt extract (400) Casein hydrolysate (400)	[4]
C. sinensis (Valencia and Shamouti)	Malt extract (500) Kinetin (0.1–1.0) IAA (0.1–1.0) Coconut milk (5% v/v) Gibberellic acid (1)	
C. aurantifolia C. sinensis	Naphthaleneacetic acid (0.1) Adenine sulphate (25–50) Kinetin (0.5–2.0) Casein hydrolysate (200–600) Malt extract (100–300) Coconut milk (10–20% v/v)	[37]
C. sinensis (Hamlin, Pell Navel, Pineapple) C. paradisi (Marsh seedless) C. reticulata (Owari) C. paradisi X C. reticulata (Orlando) C. aurantifolia (Key) C. limon (Bearse lime)	Malt extract (500) 2,4-D (0.01) 6-Benzylaminopurine (0.1) Daminozide (0.1)	[11]

* Growth supplements are added individually or in combinations.

2. Isolation and culture of protoplasts

The protocol developed by Vardi and co-workers [67, 70] is very efficient for protoplast isolation and culture.

Table 2. Composition of Murashige and Tucker's medium

Constituent	Concentration
Mineral salts	Same as Murashige and Skoog [41]
2,4-D	3×10^{-6} M
or NAA	3×10^{-5} M
Kinetin	10^{-6} M
Gibberellic acid (optional)	3×10^{-8} to 3×10^{-6} M
Thiamine HCl	10 mg/l
Pyridoxine HCl	10 mg/l
Myo-inositol	100 mg/l
Glycine	2 mg/l
Sucrose	5% w/v
Bacto agar	1% w/v

Steps in the procedure
1. Nucellar embryogenic callus, subcultured at intervals of 2—3 weeks for at least 2 passages, is the ideal source for protoplasts.
2. Place about 0.5 g of the callus in 10 ml of the maceration medium for overnight incubation.
3. Isolate the protoplasts by filtering the mixture through nylon screens of 50 μm and 30 μm pore size.
4. Centrifuge the filtrate 3 times at 100 $\times$ g for 5 minutes.
5. Resuspend the protoplasts to a final density of 10^5 protoplasts/ml and plate on a culture medium solidified with 0.6% agar.
6. Incubate cultures in continuous light (about 2000 lux) at $26 \pm 1\,^\circ$C
7. After the colonies are formed, the same protocol as described for nucellar embryogenesis may be followed for both embryogenesis and regeneration.

Maceration medium

0.3% w/v pectinase (fungal, Koch-Light Laboratories) or
0.3% w/v Macerozyme R-10 (Kinki Yakult)
0.2% w/v cellulase Onozuka R-10 (Kinki Yakult)
0.1% w/v Driselase (Kyowa Hakko Kogyo Co.)
0.3M sucrose
0.4M mannitol
Half strength Murashige & Tucker's macrosalts (Table 2)

Filter sterilize prior to use

3. Genetic Transformation of Protoplasts

To-date there are only two reports of transformation of *Citrus* protoplasts [21, 69] and one of *Agrobacterium*-mediated transformation of suspension cultured cells [20]. The procedures followed for protoplast transformation are described below. Further refinements in the technique and experimental conditions will likely be forthcoming in the future.

Steps in the procedure

1. Using nucellar embryogenic callus [69] or suspension cultures [21], the protoplasts are isolated (see protocol — Isolation and culture of protoplasts).
2. Resuspend the protoplasts in nutrient medium at the appropriate density and osmoticum.
3. Mix about 0.9 ml of medium with protoplasts with 0.1 ml of plasmid DNA containing 10 μg of circular or linearized DNA with the marker gene (chloramphenicol acetyltransferase or neomycin phosphotransferase).
4. After 5 minutes, add 0.25 ml of Ma-Mg and 0.5 ml of polyethylene glycol (PEG) solution to the protoplast-DNA mixture.
5. After 10 minutes incubation at room temperature add 3 ml of 0.275 M $Ca(NO_3)_2$ and incubate the mixture again for another 10 minutes.
6. Add another 2 ml of $Ca(NO_3)_2$, centrifuge the protoplasts and resuspend in liquid medium.
7. Culture protoplasts (10^5/ml) in 3 ml of medium containing 0.35 M mannitol in a Petri dish sealed with parafilm and maintain at 26 °C under 16 hr/day illumination (500 lux).
8. After 2 weeks, dilute the medium with an equal volume of basal medium and grow the cultures for another 2 weeks.
9. Plate the protoplasts on solidified agar medium containing 25 μg/ml of either kanamycin sulfate or paromomycin sulfate. Plating on top of a feeder layer [68, 69, 70] may be beneficial.
10. After 6—8 weeks, recover the transformed colonies and maintain for molecular analysis and regeneration.
11. For transient expression determination (CAT), incubate the protoplasts for 72 hours, followed by assay performed according to the protocol of Herrera-Estrella *et al.* [19] or Seed and Shean [58].
12. Analysis for stable transformation (NPT II) is done following the protocol of Reiss *et al.* [57] or Firoozabady *et al.* [8].

Solutions

Ma-Mg
— 0.5 M mannitol
— 15 mM $MgCl_2$
— 0.2% w/v morpholinoethane sulfonic acid (MES)
— adjust pH to 5.6

PEG solution
- 0.4 M mannitol
- 0.1 M $Ca(NO_3)_2$
- 40% w/v polyethylene glycol (PEG) 6000
- adjust pH to 7.0

Conclusions

Tissue and cell culture has proved to be a very valuable tool in *Citrus* improvement programs. Although nucellar embryogenesis has been successful, the occurrence of phenotypic variation makes it less attractive as a means to obtain disease-free plants [46]. Shoot tip grafting offers a superior, dependable method to obtain true-to-type and virus-free plants. Embryogenic cell suspensions could be used to select cell lines with tolerance to NaCl, and 2,4-D, or other desirable traits. Such NaCl-tolerant plants would be of immense importance in areas with adverse soil conditions. Somatic hybridization through protoplast fusion of sexually incompatible *Citrus* spp. and its relatives has enabled the transfer of several desirable traits. A hybrid rootstock of commercial importance combining the tolerance to tristeza and exocortis, calcareous soils and cold hardiness (from *Citrus reticulata*), and resistance to foot rot and nematode (from *Citropsis gillettiana*) has been obtained. Such hybrids play a significant role in root stock improvement.

The rapid advances in rDNA technology open new avenues for *Citrus* improvement. The introduction of coat-protein gene of tobacco mosaic virus into tobacco and alfalfa mosaic virus into alfalfa has conferred significant resistance to diseases caused by these viruses. A similar approach to control viruses that affect *Citrus* might prove very rewarding.

The antifreeze protein (AFP) is an α-helical, alanine-rich protein produced by flounders during the winter season. The protein depresses the freezing point and thereby allows the flounder to survive the sub-zero ocean temperatures in its native habitat. It has been recently shown that AFP improved the cold hardiness of plants by lowering the freezing temperature of leaves [6]. A synthetic gene coding for AFP has been constructed, introduced and expressed in corn protoplasts [10]. The next logical step would be to develop *Citrus* plants expressing the AFP protein, which could be of great value in regions such as Florida where enormous losses are incurred due to freezing.

References

1. Barlass, M and Skene, KGM (1982) *In vitro* plantlet formation from *Citrus* species and hybrids. Sci Hort 17: 333–341.
2. Bitters, WP, Murashige, T, Rangan, TS and Nauer, E (1970) Investigations on established virus-free plants through tissue culture. Calif Citrus Nurserym Soc Yrbk 9: 27–30.
3. Burger, DW and Hackett, WP (1986) Gradients of adventitious bud formation on excised epicotyl and root sections of *Citrus*. Pl Sci 43: 229–232.
4. Button, J and Bornmann, CH (1971) Development of nucellar plants from unpollinated and unfertilized ovules of the Washington navel orange in vitro. J S Afr Bot 37: 127–134.
5. Button, J and Botha, CEJ (1975) Enzymic maceration of *Citrus* callus and the regeneration of plants from single cells. J exp Bot 26: 723–729.
6. Cutler, AJ, Saleem, M, Kendall, E, Gusta, LV, Georges, F and Fletcher, GL (1989) The winter flounder antifreeze protein improves the cold hardiness of plant tissues. J Pl Physiol 135: 351–354.
7. Duran Vila, N, Ortega, V and Navarro, L (1989) Morphogenesis and tissue cultures of three citrus species. Pl Cell Tis Org Cult 16: 123–133.
8. Firoozabady, E, DeBoer, Dl, Merlo, DJ, Halk, El, Amerson, LN, Rashka, KE and Murray, EE (1987) Transformation of cotton (*Gossypium hirsutum* L.) by *Agrobacterium tumefaciens* and regeneration of transgenic plants. Pl Mol Biol 10: 105–116.
9. Galun, E, Aviv, D, Raveh, D, Vardi, A and Zelcer, A (1977) Protoplasts in studies on cell genetics and morphogenesis. In: Reinhard, E, Alfermann, AW (Eds.) Proc in Life Sciences, pp 301–312, Springer-Verlag, Berlin.
10. Georges, F, Saleem, M and Cutler, AJ (1990) Design and cloning of a synthetic gene for the flounder antifreeze protein and its expression in plant cells. Gene 91: 159–165.
11. Gmitter, FG Jr and Moore, GA (1986) Plant regeneration from undeveloped ovules and embryogenic calli of Citrus: Embryo production, germination, and plant survival. Pl Cell Tis Org Cult 6: 139–147.
12. Gmitter, FG Jr, and Ling, X (1991) Embryogenesis *in vitro* and nonchimeric tetraploid plant recovery from undeveloped *Citrus* ovules treated with colchicine. J Am Soc Hort Sci 116: 317–321.
13. Gmitter, FG Jr, Ling, XB and Deng, XX (1990) Induction of triploid *Citrus* plants from endosperm calli *in vitro*. Theor Appl Genet 80: 785–790.
14. Gmitter, FG Jr, Ling, X, Cai, C and Grosser, JW (1991) Colchicine-induced polyploidy in *Citrus* embryogenic cultures, somatic embryos, and regenerated plants. Pl Sci 74: 135–141.
15. Grinblat, U (1972) Differentiation of *Citrus* stem *in vitro*. J Am Soc Hort Sci 97: 599–603.
16. Grosser, JW, Gmitter, FG Jr, and Chandler, JL (1988) Intergeneric somatic hybrid plants of *Citrus sinensis* cv Hamlin and *Poncirus trifoliata* cv Flying Dragon. Pl Cell Rep 7: 5–8.
17. Grosser, JW, Moore, GA and Gmitter, FG Jr (1989) Interspecific somatic hybrid plants from the fusion of "Key" lime (*Citrus aurantifolia*) with "Valencia" sweet orange (*Citrus sinensis*) protoplasts. Sci Hort 39: 23–29.
18. Grosser, JW, Gmitter, FG Jr, Tusa, N and Chandler, JL (1990) Somatic hybrid plants from sexually incompatible woody species: *Citrus reticulata* and *Citropsis gilletiana*. Pl Cell Rep 8: 656–659.
19. Herrera-Estrella, L, Depicker, A, Van Montagu, M and Schell, J (1983) Expression of chimeric genes transferred into plant cells using a Ti-plasmid-derived vector. Nature 303: 209–213.
20. Hidaka, T, Omura, M, Ugaki, M, Tomiyama, M, Kato, A, Ohshima, M and Motoyoshi, F (1990) *Agrobacterium*-mediated transformation and regeneration of *Citrus* spp. from suspension cells. Jap J Breed 40: 199–207.
21. Kobayashi, S and Uchimiya, H (1989) Expression and integration of a foreign gene in orange (*Citrus sinensis* Osb.) protoplasts by direct DNA transfer. Jap J Genet 64: 91–97.
22. Kobayashi, S, Ohgawara, T, Fujiwara, K and Oiyama, I (1991) Analysis of cytoplasmic

genomes in somatic hybrids between navel orange (*Citrus sinensis* Osb.) and "Murcott" tangor Theor Appl Genet 82: 6–10.

23. Kobayashi, S, Ohgawara, T, Ohgawara, E, Oiyama, I and Ishii, S (1988) A somatic hybrid plant obtained by protoplast fusion between navel orange (*Citrus sinensis*) and satsuma mandarin (*C. unshiu*). Pl Cell Tis Org Cult 14: 63–69.

24. Kochba, J and Button, J (1974) The stimulation of embryogenesis and embryoid development in habituated ovular callus from the "Shamouti" orange (*Citrus sinensis*) as affected by tissue age and sucrose concentration. Z Pflanzenphysiol 73: 415–421.

25. Kochba, J and Spiegel-Roy, P (1973) Effect of culture media on embryoid formation from ovular callus of "Shamouti" orange (*Citrus sinensis*). Z Pflanzenphysiol 69: 156–162.

26. Kochba, J and Spiegel-Roy, P (1977a) The effects of auxins, cytokinins and inhibitors on embryogenesis in habituated ovular callus of the "Shamouti" orange (*Citrus sinensis*). Z Pflanzenphysiol 81: 281–283.

27. Kochba, J and Spiegel-Roy, P (1977b) Embryogenesis in gamma-irradiated habituated ovular callus of the "Shamouti" orange as affected by auxin and by tissue age. Environ Exp Bot 17: 151–159.

28. Kochba, J, Spiegel-Roy, P and Safran, H (1972) Adventive plants from ovules and nucelli in *Citrus*. Planta 106: 237–245.

29. Kochba, J, Button, J, Spiegel-Roy, P, Bornmann, CH and Kochba, M (1974) Stimulation of rooting of *Citrus* embryoids by gibberellic acid and adenine sulphate. Ann Bot 38: 795–802.

30. Kochba, J, Spiegel-Roy, P, Neumann, H and Saad, S (1978) Stimulation of embryogenesis in citrus ovular callus by ABA, ethephon, CCC and Alar and its suppression by GA_3. Z Pflanzenphysiol 89: 427–432.

31. Kochba, J, Spiegel-Roy, P, Neumann, H and Saad, S (1982) Effect of carbohydrates on somatic embryogenesis in subcultured nucellar callus of *Citrus* cultivars. Z Pflanzenphysiol 105: 359–362.

32. Kordan, HA (1977) Mitosis and cell proliferation in lemon fruit explants incubated on attenuated nutrient solutions. New Phytol 79: 673–677.

33. Kulshrestha, VK, Chauhan, YS and Roberts, LW (1982) *In vitro* studies on xylogenesis in *Citrus* fruit vesicles I. Effects of auxin, cytokinin and citric acid on induction of xylem differentiation. Phytomorphology 32: 14–17.

34. Litz, RE (1987) Application of tissue culture to tropical trees. In: Green, CE, Somers, DA, Hackett, WP Biesboer, DD (Eds.) Plant Tissue and Cell Culture, pp 407–418, Alan R Liss, Inc New York.

35. Maheshwari, P and Rangaswamy, NS (1959) Polyembryony and *in vitro* cultures of embryos of *Citrus* and *Mangifera*. Indian J Hort 15: 275–282.

36. Matsumoto, K and Yamaguchi, H (1983) Induction of adventitious buds and globular embryoids on seedlings of trifoliate orange (*Poncirus trifoliata*). Jap J Breed 33: 123–129.

37. Mitra, GC and Chaturvedi, HC (1972) Embryoids and complete plants from unpollinated ovaries and from ovules of *in vivo*-grown emasculated flower buds of *Citrus* spp. Bull Torrey bot Cl 99: 184–189.

38. Moore, GA (1985) Factors affecting *in vitro* embryogenesis from undeveloped ovules of mature *Citrus* fruit. J Am Soc hort Sci 110: 66–70.

39. Moore, GA, Miller, MJ and Cline, K 1988. The effects of low levels of chloramphenicol and methotrexate on somatic embryogenesis in *Citrus*. *In vitro* Cell Dev Biol 24: 1205–1208.

40. Murashige, T (1974) Plant propagation through tissue culture. Annu Rev Pl Physiol 25: 135–165.

41. Murashige, T and Skoog, F (1962) A revised medium for rapid growth and bioassays with tobacco tissue cultures. Physiol Plant 15: 473–497.

42. Murashige, T and Tucker, DP (1969) Growth factor requirement of *Citrus* tissue culture. In: Chapman, HD (Ed.) Proc 1st Int *Citrus* Symp, pp 1155–1161, Univ. of California, Riverside.

43. Murashige, T, Bitters, WP, Rangan, TS, Nauer, EM Roistacher, CN and Holliday, BP (1972) A technique of shoot apex grafting and its utilization towards recovering virus-free *Citrus* clones. HortScience 7: 118–119.

44. Navarro, L (1987) Application of shoot-tip grafting *in vitro* to woody species. Acta Hort 227: 43–55.
45. Navarro, L 1981. *Citrus* shoot-tip grafting *in vitro* (STG) and its applications: A review. Proc Int Soc Citriculture 1: 452–456.
46. Navarro, L, Ortiz, JM and Juarez, J (1985) Aberrant citrus plants obtained by somatic embryogenesis of nucelli cultured in vitro. HortScience 20: 214–215.
47. Nito, N and Iwamasa, M (1990) *In vitro* plantlet formation from juice vesicle callus of satsuma (*Citrus unshiu* Marc.). Pl Cell Tis Org Cult 20: 137–140.
48. Ohgawara, T, Kobayashi, S, Ishii, I, Yoshinaga, K and Oiyama, I (1989) Somatic hybridization in *Citrus*: navel orange (*C. sinensis* Osb.) and grapefruit (*C. paradisi* Macf.). Theor Appl Genet 78: 609–612.
49. Ohgawara, T, Kobayashi, S, Ishii, I, Yoshinaga, K and Oiyama, I (1991) Fertile fruit trees obtained by somatic hybridization: navel orange (*Citrus sinensis*) and troyer citrange (*C. sinensis* × *Poncirus trifoliata*). Theor Appl Genet 81: 141–143.
50. Ohgawara, T, Kobayashi, S, Ohgawara, E, Uchimiya, H and Ishii, I (1985) Somatic hybrid plants obtained by protoplast fusion between *Citrus sinensis* and *Poncirus trifoliata*. Theor Appl Genet 71: 1–4.
51. Rangan, TS (1984) Clonal propagation: Somatic embryos of *Citrus*. In: Indra K Vasil (Ed.) Cell Culture and Somatic Cell Genetics of Plants, pp 68–73, Acad Press Inc, Orlando.
52. Rangan, TS, Murashige, T and Bitters, WP (1968) *In vitro* initiation of nucellar embryos in monoembryonic *Citrus*. HortScience 3: 226–227.
53. Rangan, TS, Murashige, T and Bitters, WP (1969) *In vitro* studies of zygotic and nucellar embryogenesis in *Citrus*. In: Chapman, HD (Ed.) Proc 1st Int Citrus Symposium, pp 225–229, University of California, Riverside.
54. Rangaswamy, NS (1958) Culture of nucellar tissue of *Citrus in vitro*. Experientia 14: 111–112.
55. Rangaswamy, NS (1961) Experimental studies on the female reproductive structures of *Citrus microcarpa* Bunge. Phytomorphology 11: 109–127.
56. Rangaswamy, NS (1981) Nucellus as an experimental system in basic and applied tissue culture research. In: Rao, AN (Ed.) Proc COSTED Symp on Tissue Culture of Economically Important Plants, pp 269–286, Singapore.
57. Reiss, B, Sprengel, R, Will, H and Schaller, H (1984) A new sensitive method for qualitative and quantitive assay of neomycin phosphotransferase in crude cell extracts. Gene 30: 211–218.
58. Seed, B and Shean, J-Y (1988) A simple phase-extraction assay for chloramphenicol acetyl-transferase activity. Gene 67: 271–277.
59. Sim, G-E, Loh, C-S and Goh, C-J (1988) Direct somatic embryogenesis from protoplasts of *Citrus mitis* Blanco. Pl Cell Rep 7: 418–420.
60. Spiegel-Roy, P and Kochba, J (1973) Stimulation of differentiation in orange (*Citrus sinensis*) ovular callus in relation to irradiation of the media. Rad Bot 13: 97–104.
61. Spiegel-Roy, P and Kochba, J (1980) Embryogenesis in *Citrus* tissue cultures. In: Fiechter, A (Ed.) Advances in Biochemical Engineering, pp 27–48. Springer-Verlag, Berlin.
62. Spiegel-Roy, P and Vardi, A (1984) *Citrus*. In: Ammirato, PV, Evans, DA, Sharp, WA, Yamada, Y (Eds.) Handbook of Plant Cell Culture, vol 3, pp 355–372, Macmillan Inc, New York.
63. Stevenson, FF (1956) The behaviour of citrus tissues and embryos *in vitro*. Diss Abstr 16: 2292–2293. (Biol Abstr 31: 3256, entry 35979, 1957.)
64. Tisserat, B and Murashige, T (1977) Effect of ethephon, ethylene, and 2,4-dichlorophenoxy-acetic acid on asexual embryogenesis *in vitro*. Pl Physiol 60: 437–439.
65. Tusa, N, Grosser, JW and Gmitter, FG Jr (1990) Plant regeneration of "Valencia" sweet orange, "Femminello" lemon, and the interspecific somatic hybrid following protoplast fusion. J Am Soc Hort Sci 115: 1043–1046.
66. Unger, JW and Feng, KA (1978) Growth and differentiation of juice vesicles of orange grown *in vitro*. Am J Bot 65; 511–515.

67. Vardi, A and Galun, E (1988) Recent advances in protoplast culture of horticultural crops: *Citrus*. Sci Hort 37: 217–230.
68. Vardi, A and Raveh, D (1976) Cross-feeder experiments between tobacco and orange protoplasts. Z Pflanzenphysiol 78: 350–359.
69. Vardi, A, Bleichman, S and Aviv, D (1990) Genetic transformation of *Citrus* protoplasts and regeneration of transgenic plants. Pl Sci 69: 199–206.
70. Vardi, A, Spiegel-Roy, P and Galun, E (1982) Plant regeneration from *Citrus* protoplasts: variability in methodological requirements among cultivars and species. Theor Appl Genet 62: 171–176.
71. Wang, D and Chang, CJ (1978) Triploid *Citrus* plantlet from endosperm culture (in Chinese). Sci Sinica 21: 822–827.

Plant Tissue Culture Manual C8, 1–24, 1995.

Clonal propagation of eucalypts

J.A. McCOMB

School of Biological and Environmental Sciences, Murdoch University, Perth, Western Australia, 6150

Introduction

The recent Flora of Australia [14] lists 513 species of eucalypts and some authorities consider they should be split into several separate genera. Consequently a review of micropropagation techniques for eucalypts must necessarily be more general than an account directed towards a single species. More than 50 species and many hybrids have been grown *in vitro*. Several reviews [24, 40, 50, 70] tabulate the media and the levels of success achieved with these species. Despite the diversity of the species it is possible to suggest media and techniques which should result in sufficient *in vitro* material with which to do meaningful experiments on optimising media and conditions.

One aspect of eucalypt propagation that is not given sufficient prominence in the literature is the great genotypic variation in response to tissue culture media. The genotypic variation within species for characteristics such as ability of cuttings to root [23, 84] is more frequently reported than the level of variation in response to tissue culture which is similar in extent. Data on *E. globulus* and *E. nitens* [83] illustrate the genotypic variation in *in vitro* performance particularly clearly (Table 1). Often, using a published medium for a particular species,

Table 1. The *in vitro* performance of explants derived from seedlings from different provenances of *E. globulus* and *E. nitens*. Data from [83].

Species	Provenance	Number Cultured	*In vitro* response (number of clones)			
			Necrotic	Poor	Acceptable	Excellent
E. globulus	Jeeralang	234	32	154	44	4
	Otways	234	1	157	62	14
	Moogara	234	20	149	59	6
E. nitens	Rubicon	182	44	100	35	3
	Erica	234	7	154	60	13
	Tallaganda	234	1	141	90	3

it is impossible to obtain good results for some genotypes of interest. It is usually more cost effective to survey of a large number of superior genotypes to identify those which respond well in tissue culture, than it is to do a sequence of experiments to try to improve the *in vitro* performance of intransigent genotypes.

The growth habits of eucalypts have implications for methods of obtaining suitable explants for tissue culture. The species are heterophyllous and leaf shape may change gradually or abruptly through the stages seedling, juvenile, intermediate and mature. In all but a few species, it is impossible to root cuttings unless they are taken from the lowest nodes of seedlings [63] and rooting inhibitors have been isolated from the shoots of mature trees [22, 64]. As the trees grow, they lay down axillary buds which proliferate under the bark as dormant bud strands (proventitious buds) as the tree increases in girth. At the base of the plant, the cotyledonary node together with 2–3 nodes above it, may swell to form a large lignotuber, and under the surface of this structure there are also many potential shoot buds. For most species, if a tree is defoliated by fire or insects, the epicormic shoots sprout in abundance from the trunk, and when the tree is felled or girdled, shoots arise in large numbers from the base of the trunk and the lignotuber. The epicormic or coppice shoots that arise from near the base of the plant show juvenile characteristics in leaf form and frequently at least some level of rooting ability.

When mature explants have been used to initiate *in vitro* shoot cultures they take many months to stabilise in culture and some genotypes never root [49]. Rooted plantlets from mature explants may show the mature leaf phenotype at ground level which indicates that the nexus between the juvenile appearance of the leaves and rooting ability can be broken [6]. Another unusual feature of eucalypts is that in a single leaf axil there may be a naked axillary bud, as well as concealed axillary buds. Thus in contrast to most plants in which once the visible axillary bud has sprouted and been excised no further growth can be expected from that leaf axil, in eucalypts further sprouting can occur without the buds being adventitious.

Preparation of explant

For a few purposes seeds are suitable starting material but it is usually desirable to obtain juvenile shoots for culture from superior trees 10 to 100 or more years old. Rejuvenation is achieved in a number of ways.

a) *Grafting.* Mature scions grafted to seedling rootstocks may change to the juvenile phenotype [26, 27]. Grafting is not always easy in eucalypts and it may be necessary to use a half sib seedling as the root stock to obtain compatibility. Further, a sequence of grafts onto seedling rootstocks may be needed to achieve full reversion to juvenility.

b) *Felling the tree.* This yields the greatest source of vigorous explants. The tree should be felled 10 cm or so from the ground, and initial experiments conducted with non-selected trees to determine the optimum time of the year for felling under local environmental conditions [52]. Not all species coppice, and amongst those that do coppice, not all trunks sprout. Sprouting has been enhanced in some species by the application of cytokinins [71].

c) *Partial girdling.* When a tree is too valuable to fell, some coppice sprouts may be obtained by partial girdling [52, 81] possibly accompanied by cytokinin treatment [48].

d) *Epicormic shoots from trunk sections.* Sections of trunk can be cut from a tree, sealed at the ends with paraffin wax, the surface sterilised to reduce fungal infection, and kept under conditions of high humidity in a glasshouse until epicormic shoots appear [34, 78]. This technique is particularly valuable if the trees of interest are in a remote locality or if there are problems with insect and fungal damage of young shoots sprouting on trees in the forest.

e) *Cuttings.* Juvenile material obtained using the above methods can sometimes be rooted as cuttings. If such a plant can be obtained and grown vigorously under glasshouse conditions it yields much better explants for tissue culture than do plants in the field. A mother plant in the glasshouse should be severely pruned so that it continually produces shoots from the basal 10–20 cm of the stem.

For all plants, establishment of sterile cultures is easiest if actively growing shoots are used. It is also helpful if the plant can be sprayed with a systemic fungicide a week or so before excision of explants and if care is taken not to splash the shoots while watering the plants.

Surface sterilisation and prevention of browning

Shoots are usually sterilised using a chlorine solution [50] or more rarely mercuric chloride [30]. Washing in running water for an hour before surface sterilising may improve results in some species, but in others, the bruising that occurs appears to increase contamination and can result in explant death. An initial dip in 70% alcohol for 30 sec is beneficial with those species that have a loose waxy surface, and it is usual to include a drop of detergent such as Tween 20 or Tween 80 in the surface sterilant. In common with most woody species, eucalypts can show problems with browning of explants. This can be minimised by soaking explants in water for several hours before insertion into media [20], keeping cultures in the dark for an initial period of 1–2 weeks [24], changing explants to fresh medium, a period of liquid culture [30] or incorporating an antioxidant such as PVP ($10 \text{ g } l^{-1}$) in the medium [16]. Methods for control of oxidation stress in plant tissue culture have been reveiwed recently [10].

The surface sterilised stem pieces are cut into apical segments and segments of 1–2 nodes. Unless it is known that the level of contamination is very low it is best to set up single explants in small vials. The apical section can be used in some species such as *E. citriodora* but is frequently too delicate to survive the rigours of surface sterilisation and should be discarded. Up to 3 or so pieces can be used from the same stem, usually as segments with 2–3 nodes each. Nodes lower than this usually show unacceptably high contamination rates, sprout less frequently, and if they do sprout, display less vigorous growth and more latent internal contamination than the higher nodes. The success of surface sterilisation can vary considerably within species. For example some jarrah trees resistant to insect damage (leaf miner) have a thicker cuticle and are far easier to surface sterilise than normal trees susceptible to leaf miners [49].

Shoot multiplication

A wide array of media have been described in the literature [50, 70]. Many reports give 'optimal' media unsubstantiated by data showing rigorous testing of the variables or statistical analysis. Some workers [31, 81] have used a hormone free pre-initiation medium in the sequence to screen for contaminants and reduce phenolic production. In our hands the delay of exposure to the growth regulators results in poorer sprouting of explants.

Once shoots have sprouted and a clump of shoots developed, they should not be bulked with others into larger containers until some tests for internal contaminants are carried out. The indicative basal white haze in the medium is easier to see using gelrite than agar. Internal contaminants can be encouraged to proliferate and the contaminated shoots identified by the cutting a slice from the base of the new shoot clump, mashing it and placing the mashed segment and the shoot clump side by side on fresh shoot multiplication medium supplemented with 1% tryptone. More complex media for screening for internal contaminants have been suggested [43, 57]. Despite early checks, internal contaminants may occur in later subcultures. Contaminated cultures can rarely be freed entirely of contaminants by use of antibiotics [25, 85] though treatment of eucalypt cultures with rifampicin ($100 \, mg \, l^{-1}$) may suppress evidence of the contaminant for several subcultures. Fungal contamination of initial explants can be suppressed by inclusion of $50 \, mg \, l^{-1}$ of Benomyl in the medium [16].

Basal mineral nutrient formulations used are MS [60], or MS with half strength macro nutrients. The McCown woody plant medium [44] or that of Quoiri and Lepoivre which has low levels of ammonium and chloride have occasionally been reported to be beneficial [77]. The cytokinins BAP, and kinetin or zeatin are most frequently used to enhance shoot multiplication. Although a particular cytokinin may appear excellent in early cultures its continued use may result in increasingly poorer growth until it is clearly toxic. This may be avoided by reducing the level of the cytokinin in the medium after the initial stimulation of growth [17, 41] or by giving one to several subcultures on a different cytokinin. For *E. globulus* it has been found that alternation of shoot cultures on BAP then kinetin is necessary for many genotypes, and this alternation has an impact on rooting ability [8, 9].

When subcultured, the shoots of many species respond badly to being cut into single stemmed pieces. It is better to trim any callus or discoloured parts from the base of shoot clumps and to divide them into clumps of 4–5 shoots for transfer. There is an optimum ratio between medium, plant material and gaseous head space in the culture vials and small culture vessels are best when only small amounts of material are available.

To obtain shoots of suitable height for rooting various techniques have been utilised. The addition of GA_3 to the medium [24] gives long shoots but as this substance has an inhibitory effect on rooting the subculture must be followed by one without GA_3, and possibly including the gibberellin inhibitor ancymidol. A passage on medium without hormones may be sufficient to cause shoots to elongate, or addition of charcoal to this medium may be necessary. Etiolation of shoots of *E. grandis* has been shown to enhance rooting of cuttings *in vivo* [15] and it improves rooting of *in vitro* shoots of species of another myrtaceous genus *Verticordia* [51]. However, such shoots are very delicate and difficult to handle without damage or desiccation during subculture.

The gelling agent used is mainly agar, gelrite or a mix of both (2.5% agar 2.5% gelrite). For *E. grandis* gelrite was found to be superior for both shoot multiplication and rooting [46]. Vitrification of shoots may be higher on gelrite media but when this is detected one or two passages on agar media can eliminate the problem.

Stabilisation of cultures can be a problem with woody plants [56] and in eucalypts cultures from mature trees this may involve a period of years [49]. Less time is required with coppice explants [81] and cultures established from seedlings show no, or a very short period of stabilisation. Before stabilisation, experiments done to optimise growth regulators in the medium give irreproducible results. Difficulty in reproducing some results from the literature giving 'suitable' media may be because experiments have been done before cultures have stabilised. Results are more reliable when authors report the length of time the material has been in culture before the experiments were conducted and when they show that they can repeat their results.

There is one report from *E. radiata* in which explants from branches of a mature tree showed a higher shoot multiplication rate than those from coppice. However it appears this was based on a single (not the same) genotype of each explant type and the coppice showed three times the rooting level of the mature explants [17]. It is more usual for cultures from adult tissues to multiply more slowly and root more poorly than those from basal coppice or seedling explants [49, 68].

Novel techniques reported for eucalypt shoot multiplication include use of a callus-like nodular mass from *E. grandis* hybrid hypocotyls [82] and induction of meristematic nodules on *E. camaldulensis* shoot cultures [13, 61]. The nodules can be regenerated to a clump of shoots. Single meristem explants can be grown to clumps of meristems using liquid media in rotated tubes [35]. Meristems are relatively easy to dissect in eucalypts, particularly using *in vitro* shoots, due to the decussate arrangement of the leaves at the apex.

Rooting

Individual shoots $1\frac{1}{2}$–2 cm are selected for rooting. Careful operators cut just below the node and trim off the lower leaves, but there are little hard data to substantiate that this is beneficial except that leaves touching the medium may root, which can cause problems at potting out. The mineral nutrients are frequently reduced for the rooting phase. The level of macronutrients in MS is reduced to $\frac{1}{2}$ or $\frac{1}{4}$ [8, 49, 50] with the reduction of NH_4 being particularly significant [33]. The beneficial effects of use of WPM [44] for the rooting phase [83] may be due to it having a more favourable NH_4^+ to NO_3^- ratio (1 : 3) than MS (1 : 2). For many species the inclusion of the vitamins and other organic addenda is unnecessary in rooting media [32].

The auxins most widely used for root induction are IBA and NAA, sometimes together. Cultures placed in the dark for the first week on auxin media show higher rooting than those in constant light conditions. If shoots are left in a medium containing auxin for 4 weeks a basal callus will grow while the roots are developing. This callus reduces transplant success. Treatment of shoots in an auxin medium for 2 days to 1 week then transfer of shoots to medium without hormones results in better root growth and reduced callus in species such as *E. regnans* [11] and *E. citriodora* [47]. The medium without hormones may contain charcoal [47]. These additional steps increase costs. Similarly methods that include a period of culture on filter paper bridges with liquid media [30, 68] are inappropriate for commercial production of clones. The range of genotypic variation in ability to root in *in vitro*, as *in vivo* may be from 0–100% and selection for genotypes with high rooting ability is rewarding. The ability of material to root as cuttings *in vivo* and as shoots *in vitro* is not always positively correlated. Sometimes material that cannot be rooted or shows low rooting *in vivo* will show higher levels of rooting *in vitro* [52]. The reverse is rare. The strategy of setting up new cultures from well rooted plantlets from tissue culture has in some cases raised the rooting percentage and quality [17].

Compounds in the shoot culture medium may have a carryover effect on rooting. The effect of gibberellin has been mentioned above. In *E. marginata* it was shown that shoots multiplied on media with BAP and NAA gave higher rooting than those multiplied on a medium with the cytokinin but no auxin [50]. For *E. globulus* shoots cultured on BAP gave low rooting compared with those on medium with kinetin [8, 9] while in *E. gunnii* BAP has also been shown to be inhibitory to subsequent rooting compared with use of kinetin or zeatin in shoot culture media [21]. The different cytokinins were found to induce different levels of flavonoids in the shoots, with a high level of quercitin glucosides being correlated with growth of shoots on kinetin or zeatin and subsequent high rooting success.

Roots can be induced on eucalypt shoots using *Agrobacterium rhizogenes* [1] but there is insufficient evidence as yet that this treatment can improve rooting of difficult-to-root lines, and whether the growth form and wood of a chimaeric transgenic tree is normal. Whilst isolated roots of eucalypts appear difficult, if not impossible to culture [2, 19] isolated transformed roots may grow prolifically [45].

Potting out and initial growth

Rooted plants removed from agar can be kept in the culture room in containers with sterile distilled water for several days if it is necessary to bulk up material before potting. It has been found that best survival is from shoots with 3 or more roots at 1–2 cm long and that for mass production, plantlets with 1 or 2 roots cannot be given the care that is necessary for survival of a worthwhile percentage. As for most plants grown in conventional tissue culture containers, the leaves of *in vitro* eucalypts are deficient in surface waxes and have superficial stomata, so that high humidity is required for the initial period after transplanting. Some species tolerate misting (*E. camaldulensis*) but others (*E. globulus, E. marginata*) only will grow well in a humid chamber where the spray is applied to the outside of a fabric tent over the plants.

Plants from mature jarrah (*E. marginata*) trees, initiated from the crown, have been shown to initially have the branched form of the crown of the tree and a shallower root system than seedlings. However, after 6 months the plants developed a leader shoot and grew into normal saplings [6]. Plants from explants from the crown of a tree may also show a degree of plaigiotropy and require initial staking.

Plants from seedling explants or rejuvenated shoots show vigorous orthotropic growth. An analysis of the root systems of *E. camaldulensis* clones and seedlings 9 months old showed that the clones each had a recognisable root architecture but there were no differences between clones and seedlings in distribution of sinker and minor roots or in total length. At 9 months old mean total root length of clones was 5.3 km, similar to that of seedlings [3].

Eucalypts grown from tissue culture do not develop a lignotuber except from explants of the cotyledon node and one or two nodes above it.

Callus cultures

Plants regenerated from callus cultures have a higher risk of somaclonal variation then those from multiple shoot cultures [66], so the callus system is unsuitable for micropropagation when clonal material is required. Callus from seedling pieces (particularly hypocotyl and cotyledon) or young leaves of stabilised shoot cultures can be induced to regenerate shoots [82]. Callus from floral organs (such as stamen filaments and styles) which are considered more 'juvenile' than vegetative organs on a mature tree, will also regenerate in *E. marginata* [5, 49]. For most species of eucalypts it is difficult to keep callus viable and regenerating for more than 6 months. Similarly there are few reports of long-term suspension cell cultures. Exceptions are from *E. gunnii* [75], *E. grandis* [65], *E. perrineana* [28] and *E. camaldulensis* [72].

Embryogenesis

Direct embryogenesis or embryo formation after callus formation is reported from hypocotyls and cotyledons and segments of leaves from *in vitro* shoot cultures [39, 58, 59, 62, 67]. It is possible that plants from embryoids may have less somaclonal variation than plants from callus, and if methods for bulk handling and the formation of artificial seeds were available they could result in much lower cost plants.

Protoplasts

There are a few reports of protoplasts being isolated from leaves of glasshouse-grown eucalypts. These gave at best a few cells with one or two divisions and did not regenerate [73, 80]. Research at present is concentrated on protoplasts from leaves of *in vitro* plants, callus and suspension cells [65, 73, 74], embryonic callus, and masses of shoot meristems formed after culture of meristems in rotating liquid culture [35]. Development to callus has been reported in *E. gunnii* [73] and whole plants have been regenerated in *E. saligna* [35,36]. Protoplasts have been used for studies of cold tolerance [42].

An unusual aspect of the work from Oji Paper Co Japan was the use of co-cultivation of kenaf and eucalypt protoplasts [35]. The kenaf (*Hibiscus cannabinus*) protoplasts grew to callus but did not regenerate plants. They acted as a nurse tissue for the eucalypt cells. Nurse cultures of eucalyptus suspension cultures have also been used successfully [73].

The level of somaclonal variation that is present amongst plants regenerated from protoplasts, callus, or via embryoids has not been determined in eucalypts.

Transformation

Apart from the transformation work using *Agrobacterium rhizogenes* transient expression of reporter genes was reported after treatment of eucalypt protoplasts or intact cells using electroporation [76]. Kawazu *et al.* (1991) were able to transform *E. saligna* protoplasts using electroporation and regenerated transgenic plants.

Physical conditions

The temperature for eucalypt culture growth is optimal between 25–28 °C. Constant light or a 16 hour day/night cycle is used. Some species have been reported to remain viable in dim light at 2 °C for up to 12 months [31]. There are no reports of an effect of light source and Gro lux or cool white fluorescent tubes are widely used. Culture aeration can be improved by use of holes and inserts in the lids [69], or replacing lids with Gladwrap® or Vitafilm®.

Automation and photoautotrophic growth

Reduction of the cost of micropropagated eucalypts for plantations might be achieved through photoautotrophic growth and automation of some steps that are labour intensive. Insertion of a robot into the sequence will be simplified if the medium is without sugar, and thus less liable to contamination. This requires a photoautotrophic system with high light intensity and enhanced CO_2 levels and will probably include an automated liquid feeding system. There is mention in the literature of eucalypts being propagated in this way and subcultured by robots but as yet details are unclear [38].

Applications

Clonal eucalypts from superior trees have been reported to yield higher biomass than seedlings in *E. tereticornis* and *E. torelliana* [37]. The resistance to jarrah leaf miner shown by some trees of *E. marginata* in the forest was displayed consistently by the tissue cultured ramets [7]. Tissue culture has also been used to propagate lines of *E. marginata* shown to be resistant to dieback caused by *Phytophthora cinnamomi* and resistant trees in field trials are growing well after inoculation with a number of strains of the pathogen [54, 55]. Eucalypts tolerant to salt and waterlogging are also valuable for reclamation of degraded land. Selection of salt-waterlogging tolerant genotypes of 10 species has been undertaken, and after cloning *in vitro* field trials of some 9 species have been established in Australia and other countries [53, 79]. At present the oldest field trials are at 5 years, and on sites where trees experience salt stress (soil EC [1 : 5] less than $25 \, mSm^{-1}$) there is evidence that the clonal plants are surviving and growing better than provenance-matched seedlings [4]. Other cloning work utilising eucalypts has had as its objective multiplication of trees with high oil yields, [47] and it was shown that the cloned trees had the same oil levels as the ortet. Some work in France on cold tolerance was unsuccessful because an unexpectedly severe frost killed the trees [12]. In countries with high labour costs, unless the cost of propagation of eucalypts *in vitro* can be reduced, it will not be possible to utilise tissue culture of eucalypts for mass propagation of trees for plantations. Until this is done tissue culture may be limited to providing the plants needed for seed orchards, or for mother plants for cuttings.

Procedures

Shoot sterilisation

Steps in the procedure
1. Select shoots 5–10 cm long, free from obvious insect fungal damage. Vigorously growing shoots with large naked, unopened axillary buds are best. Trim leaves to petioles $\frac{1}{2}$ cm long or cut sessile leaves appropriately.
2. Species with a rough waxy surface should be dipped in 70% ethanol for 30 s.
3. Surface sterilise by submerging shoots and swirling gently in freshly prepared – 2–5% sodium hypochlorite with a drop of Tween 80 for 10–20 min or in 1–2% benzalkonium chloride (zephiran) in 10% alcohol for 10–20 min.
4. Wash in 2–3 changes of sterile water. Allow shoots to remain in the last wash until they are cut up.
5. Cut 0.5 cm from the base of the stem, and trim cut ends of petioles. Cut off and discard apex.
6. Cut stem into pieces 1–2 cm long with 1–2 nodes and place in individual small tubes of shoot-induction medium.
7. Once shoots have sprouted, trim base of explant and transfer to fresh medium. Excise shoots from the initial explant only when a shoot or clump of shoots $1\frac{1}{2}$–2 cm long has formed, or when callus begins to form at the base of the shoot.
8. Test shoot clumps individually for internal contamination by including 1% tryptone in the medium before bulking shoots in larger containers.

Solutions
Shoot induction medium
— Murashige and Skoog medium [60]
— 2% sucrose
— 2.5 µM BAP
— 1.25 µM NAA
— adjust pH to 5.8 with 0.1 M KOH
— gell with 2.5 g/l agar and 2.5 g/l Gelrite

Shoot multiplication and rooting

Shoot multiplication is usually achieved with one of the cytokinins: BAP, kinetin or zeatin at 1–10 μM with NAA at 0.01 μM. A mixture of these cytokinins, or alternation of media with different cytokinins may be beneficial. Subculture time is usually 4 weeks. Some species, eg *E. camaldulensis* tolerate long periods without change of medium while others eg *E. globulus* and *E. marginata* deteriorate if not subcultured regularly at 3 and 4 weeks respectively. A passage in medium with 3 μM GA_3 or in media without growth regulators may be necessary to obtain long shoots for rooting.

Steps in the procedure
1. Once cultures have stabilised and long shoots are produced, excise individual shoots below a node and trim lower leaves so they do not touch the medium.
2. Place shoots for 1 week in the dark on a medium with $\frac{1}{2}$ MS major minerals full strength micro-elements, full strength iron, (but no vitamins or mesoinositol), and 1–15 μM IBA or a combination of IBA and NAA (2.5 μM), 2% sugar pH 5.8 and gelled as above. Up to 100 μM IBA has been used for some species [11].

Note
For some species this has to be adjusted. *E. marginata* roots best given $\frac{1}{4}$ strength macro elements. Other species respond to addition of the vitamins, while others require a reduction in NH_4^+.

3. After 1 week, transfer shoots to fresh medium of the same constitution as the rooting medium but without auxin, and grow in the light for 3 weeks.

Seed sterilisation

Steps in the procedure

1. Pick over seeds and place in 2−5% sodium hypochlorite, depending on the seed size and stir for 0.5 h or until the seed coat bleaches and is easy to remove. $HgCl_2$ (0.1% w/v) for 5−20 min may be used to sterilise seeds for which there is a particular problem with fungal contamination, but this mercuric compound is best avoided.

2. Strain the seeds and then wash in 2 changes of sterile water; remove the seed coat.

3. Germinate seeds in the dark or the light on $\frac{1}{2}$ strength MS major and minor minerals gelled with 0.8% agar (or agar/gelrite). Sugar 1−2% can be added if detection of contamination is a problem on the medium without sugar.

Callus cultures, regeneration and embryogenesis

1. Explants from *in vitro* shoot cultures are excellent starting material for callus and give less problems with phenolics than explants from *in vivo* plants. Explants of leaves and stems from juvenile or mature plants can be sterilised as described above for shoots. Whole unopened flower buds can be dipped in alcohol and flamed before being given 20 min in 2% sodium hypochlorite without damage to the anthers or styles which can be cut out and cultured.
2. Medium for callus induction is MS minerals with B5 vitamins and inositol levels [29], 3% sucrose, 10 μM kinetin, 10 μM NAA, 8 g·l^{-1} agar.
3. Media for regeneration include the same minerals, vitamins and inositol as for callus induction, but with 2% sucrose, 0.5–2.0 μM IAA and 5–10 μM zeatin or 2.5 μM BAP and 10% coconut milk.
4. Media for embryogenesis. It is impossible to recommend media for embryogenesis with any great confidence as any wide variety of media have been reported to induce embryogenesis, with none being consistently effective, which raises the suspicion that there is a large effect of genotype. Two methods which gave high numbers of embryoids are as follows:

Steps in the procedure
1. Induce callus and embryoid initiation on solid B$_5$ medium [29] with 15 μM NAA. Growth of an embryonic mass is sustained by culture on B$_5$ medium 25 μM NAA, casein hydrolysate (500 mg l^{-1}) and glutamine (500 mg l^{-1}). Further growth of isolated embryos is achieved on $\frac{1}{2}$ strength MS medium with no growth regulators [58, 59].
2. A suspension formed on a 2,4 D medium was transfered to liquid MS medium with 5 μM BAP, 5 μM kinetin, 5 μM NAA and 5 μM 2,4 D in Steward flasks. The fine granular structures obtained were transferred to MS 15 μM NAA and 5 μM BAP. Large numbers of embryoids were obtained but could not be germinated [39].

Note
The relatively newly available compounds 4 CPPU and thidiazuron appear to be very promising, and there is some suggestion that the polyamines might promote embryogenesis [67].

Potting out

1. Extract rooted plants from agar before the roots exceed 2 cm in length.
2. Plant into a steam sterilised, well drained mix such as *Sphagnum* peat : coarse river sand : perlite 1 : 1 : 1.
3. Put under high humidity for 2 weeks or so. Plants are ready for normal glasshouse conditions when roots appear through the bottom of the seedling tray, or when new leaves develop.
4. After 2 weeks, dilute fertiliser can be applied ($\frac{1}{4}$ strength of proprietary brands such as Thrive).
5. Each week plants should be sprayed with fungicide and the compounds used should be alternated (for example 0.5 g l^{-1} Benlate alternated with 0.5 g l^{-1} Ridomil or similar).

References

1. Adam S, Chrisqui D, David C (1987) Involvement of the physiological state of *Eucalyptus gunnii* in its ability to insert T-DNA of *Agrobacterium rhizogenes*. XIVth Bot Congress Berlin July 24–Aug 1 1987. Abstract 3–113–1, pp 461.
2. Bachelard EP, Stowe BB (1963) Growth *in vitro* of roots of *Acer rubrum L.* and *Eucalyptus camaldulensis* Dehn. Physiol Plant 16: 20–30.
3. Bell DT, van der Moezel P, Bennett IJ, McComb JA, Wilkins C, Marshall S, Morgan A (1993a) Comparisons of growth of *Eucalyptus camaldulensis* from seeds and tissue culture: root, shoot and leaf morphology of 9-month-old plants grown in deep sand and sand over clay. For Ecol Manag 57: 125–139.
4. Bell DT, McComb JA, Van der Moezel PG, Bennett IJ, Kabay ED (1993b) Comparisons of selected cloned plantlets against unselected seedlings for rehabilitation of waterlogged and saline discharge zones in Australian agricultural catchments. Aust For (in press).
5. Bennett IJ (1988) Use of Tissue Culture Techniques for the Improvement of *Eucalyptus marginata* Donn ex Sm. PhD thesis Murdoch University, Western Australia.
6. Bennett IJ, Tonkin CM, Wroth MM, Davison EM, McComb JA (1986) A comparison of the growth of seedling and micropropagated *Eucalyptus marginata* (jarrah). Early growth to 2 years. For Ecol Manag 14: 1–12.
7. Bennett IJ, McComb JA, Bradley JS (1992a) Testing the expression of resistance to insect attack: resistance of jarrah (*Eucalyptus marginata* Donn ex Sm) to jarrah leaf miner (*Perthida glyphopa* Common). For Ecol Manag 48: 99–105.
8. Bennett IJ, McComb JA, Tonkin CM, McDavid D (1992b) Effect of cytokinnins on multiplication and rooting of *Eucalyptus globulus* and other *Eucalyptus* species. In: AFOCEL Mass Production Technology for Genetically Improved Fast Growing Forest Tree Species. AFOCEL Bordeaux France Sept 14 1992 2: 195–201.
9. Bennett IJ, McComb JA, Tonkin CM, McDavid DAJ (1993) Alternating cytokinins in multiplication media stimulates *in vitro* shoot growth and rooting of *Eucalyptus globulus* Labill. Ann Bot (in press).
10. Benson EE (1990) Free radicals in stressed and aging plant tissue cultures. In Rodriquez R et al. (Eds.) Plant Aging: Basic and Applied Approaches, pp 269–275, Plenum Publ, New York.
11. Blomstead C, Cameron J, Whiteman P, Chandler S (1991). Micropropagation of juvenile *Eucalyptus regnans* mountain ash. Aust J Bot 39: 179–186.
12. Boulay M (1983) Micropropagation of frost resistant *Eucalyptus*. In: Proc Workshop on *Eucalyptus* in California. June 14–16, Sacramento California.
13. Boxus P, Terzi JM, Lievins C, Pylyser M, Ngaboyamahina P, Duhem K (1991) Improvement and perspectives of micropropagation techniques applied to some hot climate plants. Acta Hort 289: 55–64.
14. Bureau of Flora and Fauna (1987) Flora of Australia Vol 19 Myrtaceae – *Eucalyptus Angophora*. (Aust Publ Service Canberra).
15. Carter A, Slee MU (1991) Etiolation as an aid to rooting cuttings of *Eucalyptus grandis*. In: IUFRO Symposium on Intensive Forestry: The Role of Eucalypts (Durban South Africa Sept 1991) 1: 64–69.
16. Carvalho DD, Pinto JEBP, Pasqual M (1990) Use of fungicides and antioxidants for *in vitro* nodal segments culture of *Eucalyptus grandis* Hill ex Maiden. Cient Prat. 14: 97–106 (in Portuguese).
17. Chang SH, Donald DEM, Jacobs G (1992) Micropropagation of *Eucalyptus radiata* spp *radiata* using explants from mature and coppice material. S Afr For J 162: 43–47.
18. Cheng B, Peterson CM, Mitchell RJ (1992) The role of sucrose, auxin and explant source on the *in vitro* rooting of seedling explants of *Eucalyptus sideroxylon*. Plant Science 87: 207–214.
19. Cresswell R, de Fossard R (1974) Organ culture of *Eucalyptus grandis*. Aust For 37: 55–69.
20. Cresswell R, Nitch C (1975) Organ culture of *Eucalyptus grandis* L. Planta 125: 87–90.

21. Curir P, van Sumere CF, Termini A, Barthe P, Marchesini A, Doki M (1989) Flavonoid accumulation is correlated with adventitious roots formation. In: *Eucalyptus gunnii* Hook. micropropagated through axillary bud stimulation. Plant Physiol 92: 1148–1153.
22. Dhawan AK, Paton DM, Willing RR (1979) Occurrence and bioassay responses of G, a plant growth regulator in *Eucalyptus* and other Myrtaceae. Planta 146: 419–22.
23. Doran JC, Carter AS, Matheson AC (1992) Variation in root strike of Petford *Eucalyptus camaldulensis* clones. In: AFOCEL Mass Production Technology for Genetically Improved Fast Growing Forest Tree Species. AFOCEL Bordeaux France Sept 14 1992 1: 407–414.
24. Durand-Creswell R, Boulay M, Franclet A (1982) Vegetative propagation of *Eucalyptus*. In: Bonga JM, Durzan D (Eds.) Tissue Culture in Forestry, Martinus Nijhoff, The Hague.
25. Falkiner (1990) The criteria for choosing an antibiotic for control of bacteria in plant tissue culture. Intern Assoc Pl Tissue Culture Newsletter No 60 13–23.
26. Franclet A (1991) Biotechnology in 'rejuvenation': hope for the micropropagation of difficult woody plants. Acta Hort 289: 273–282.
27. Franclet A, Boulay M, Bekkaoui F, Fouret Y, Vershoore-Marouzet B, Walker N (1987) Rejuvenation. In: Bonga J, Durzan D (Eds.) Cell and Tissue Culture in Forestry 1: 232–248.
28. Furuya T, Orihara Y, Miyatake H (1989) Biotransformation of (−) methanol by *Eucalyptus perriniana* cultured cells. J Chem Soc Perkin Trans 1: 1171–1719.
29. Gamborg OL, Miller RA, Ojima K (1968) Nutrient requirements of suspension cultures of soybean root cells. Exp Cell Res 50: 151–158.
30. Gupta PK, Mascarenhas AF, Jagannathan V (1981) Tissue culture of forest trees – clonal propagation of mature trees of *Eucalyptus citriodora* Hook. by tissue culture. Pl Sci Lett 20: 195–201.
31. Hartney V (1982) Tissue culture of *Eucalyptus*. Comb Proc Intern Pl Prop Soc 32: 98–109.
32. Hartney V, Barker PK (1980) The vegetative propagation of eucalypts by tissue culture. In: Fast Growing Trees, IUFRO Congress Brazil.
33. Hyndman SE, Hasegana PM, Bressan RA (1982) The role of sucrose and nitrogen in adventitious root formation of cultured rose shoots. Pl Cell Tissue and Organ Cult 1: 229–238.
34. Ikemori YK (1987) Epicormic shoots from the branches of *Eucalyptus grandis* as an explant source for *in vitro* culture. Comm For Rev 66: 351–355.
35. Ito K, Doi K, Tatemichi Y, Shibata M (1990) Plant rejuvenation from protoplasts of *Eucalyptus*. Proc VIIth Intern Congress Plant Tissue and Cell Culture Amsterdam June 24–29 1990 Abstract A1–65.
36. Kawazu T, Doi K, Ohta T, Shinohara Y, Ito K, Shibata M (1990) Transformation of *Eucalyptus (Eucalyptus saligna)* using electroporation. Proc VIIth Intern Congress Plant Tissue and Cell Culture Amsterdam June 24–29 1990 Abstract A2–80.
37. Khuspe SS, Gupta PK, Kulkarni DK, Mehta V, Mascarenhas AF (1987) Increased biomass production by *Eucalyptus*. Can J For Res 17: 1361–1363.
38. Kozai T (1991) Micropropagation under photoautotrophic conditions. In: de Bergh P, Zimmerman R (Eds.) Micropropagation Technologies and Application, Kluwer Academic Publishers, Dordrecht.
39. Lakshmi Sita G (1986) Progress towards clonal propagation of *Eucalyptus grandis*. In: Withers LA, Alderson PG (Eds.) Plant Tissue Culture and its Agricultural Applications, pp 159–166, Butterworths.
40. Lakshmi Sita G (1992) Micropropagation of *Eucalyptus*. In: Ahija MR (Ed.) Micropropagation of Woody Plants, pp 263–80, Kluwer, Dordrecht, the Netherlands.
41. Lakshmi Sita G, Sukanya C (1987) Improvement of forest trees by tissue culture. In: Reddy GM (Ed) Plant Cell And Tissue Culture of Economically Important Plants. Proc National Symposium pp 195–198.
42. Leborgne N, Dupou-Cezanne L, Teulieres C, Canut H, Tocanne JF, Boudet AM (1992) Lateral and rotational mobilities of lipids in specific cellular membranes of *Eucalyptus gunnii* cultivars exhibiting different freezing tolerance. Pl Physiol 100: 246–254.

43. Leifert C, Waites WM (1990) Contaminants of plant tissue culture. International Assoc Pl Tissue Culture Newsletter 60: 2–13.
44. Lloyd G, McCown B (1981) Commercially feasible micropropagation of mountain laurel *Kalmia latifolia* by use of shoot tip culture. Comb Proc Int Pl Prop Soc 30: 421–427.
45. MacRae S (1991) *Agrobacterium* mediated transformation of eucalypts to improve rooting ability. In: IUFRO Symposium on Intensive Forestry: The Role of Eucalypts, Durban South Africa Sept 1: 117–125.
46. MacRae S, van Staden J (1990) *In vitro* culture of *Eucalyptus grandis* effects of gelling agents on propagation. J Plant Physiol 137: 249–251.
47. Mascarenhas AF, Itazara S, Potdar V, Kulkaini DK, Gupta PK (1982) Rapid clonal multiplication of mature forest trees through tissue culture. In: Fujiwara A (Ed) Plant Tissue Culture, pp 719–720, Japan.
48. Mazalewski RL, Hackett WP (1979) Cutting propagation of *Eucalyptus ficifola* using cytokinin induced basal trunk shoots. Comb Proc Inter Pl Prop Soc 29: 118–124.
49. McComb JA, Bennett IJ (1982) Vegetative propagation of *Eucalyptus* using tissue culture and its application to forest improvement in Western Australia. Proc 5th Int Cong Plant Tissue and Cell Culture. In: Fujiwara A (Ed.), Plant Tissue Culture, pp 721–2.
50. McComb JA, Bennett IJ (1986) Eucalyptus (*Eucalyptus* spp). In: Bajaj YPS (Ed) Biotechnology in Agriculture and Forestry 1: Trees pp 340–362.
51. McComb JA, Newell C, Arthur N (1986) Micropropagation of *Verticordia grandis* and other *Verticordia* species. The Plant Propagator 32: 12–14.
52. McComb JA, Wroth M (1986) Vegetative propagation of *Eucalyptus resinifera* and *E. maculata* using coppice cuttings and micropropagation. Aust For 16: 231–242.
53. McComb JA, Bennett IJ, van der Moezel PG, Bell DT (1989) Biotechnology enhances utilisation of Australian woody species for pulp fuel and land rehabilitation. Aust J Biotech 34: 297–301.
54. McComb JA, Bennett IJ, Stukely M, Crane C (1990) Selection and propagation of jarrah for dieback resistance. Comb Proc Intern Pl Prop Soc 40: 86–90.
55. McComb JA, Bennett IJ, Cahill (1992) Selection, propagation, field and laboratory testing of jarrah *(Eucalyptus marginata)* resistant to dieback *(Phytophthora cinnamomi)*. IUFRO/ AFOCEL Conference on Mass Production Technology for Genetically Improved Fast Growing Forest Tree Species Bordeaux France Sept 1992 2: 451–452.
56. McCown DD, McCown BH (1987) North American hardwoods. In: Bonga J, Durzan D (Eds.) Cell and Tissue Culture in Forestry, pp 247–260, Martinus Nijhoff, Dordrecht.
57. Mindali D, Vieira JD, Soares MM, Osti NM, Figueiredo DB, Ribeiro MC (1992) *In vitro Eucalyptus* species contaminating bacteria identification. In AFOCEL Mass Production Technology for Genetically Improved Fast Growing Forest Tree Species AFOCEL Bordeaux France Sept 14 1992 2: 245–253.
58. Muralidharan EM, Mascarenhas AF (1987) *In vitro* plantlet formation by organogenesis in *Eucalyptus camaldulensis* and by somatic embryogenesis in *Eucalyptus citriodora*. Plant Cell Rep 6: 256–259.
59. Muralidharan EM, Gupta PK, Mascarenhas AF (1989) Plantlet production through high frequency somatic embryogenesis in long term cultures of *Eucalyptus citriodora*. Plant Cell Reports 8: 41–43.
60. Murashige T, Skoog F (1962) A revised medium for rapid growth and bioassays with tobacco tissue culture. Physiol Plant 18: 473–497.
61. Ossor A, Boxus P (1992) Des aggolomérats de domes méristématiques: une nouvelle alternative dans la multiplication clonale d'*Eucalpytus camaldulensis*. In: AFOCEL Mass Production Technology for Genetically Improved Fast Growing Forest Tree Species, AFOCEL Bordeaux France Sept 14 2: 203–211.
62. Ouyang Q, Peng HZ, Li QQ (1981) Studies on the development of embryoids from *Eucalyptus* callus. Sci Silvae Sin 17: 1–7 (in Chinese).
63. Paton DM, Willings RR, Nicolls W, Pryor LD (1970) Rooting of stem cuttings of *Eucalyptus*; a rooting inhibitor in adult tissue. Aust J Bot 18: 175–183.

64. Paton DM, Willings RR, Pryor LD (1981) Root-shoot gradients in *Eucalyptus* ontogeny. Ann Bot 47: 835–8.

65. Penchel RM, Kirby EG (1990) Establishment of cell culture systems for selected superior hybrids of *Eucalyptus grandis* hybrids. VIIth Intern Congress Plant Tissue and Cell Culture Amsterdam 199 Abstract A1–114.

66. Potter R, Jones MKG (1991) An assessment of genetic stability of potato *in vitro* by molecular and phenotypic analysis. Pl Sci 76: 239–248.

67. Qin Chang-li, Kirby EG (1992) *In vitro* responses of cultured explants of *Eucalyptus* spp. In: AFOCEL Mass Production Technology for Genetically Improved Fast Growing Forest Tree Species AFOCEL Bordeaux France Sept 14 1992 2: 213–218.

68. Rao KS, Venkateswara R (1985) Tissue culture of forest trees clonal multiplication of *Eucalyptus grandis*. Plant Sci (Shannon) 40: 51–56.

69. Rossetto M, Dixon KW, Bunne E (1992) Aeration, a simple method to control vitrification and improve *in vitro* culture of rare Australian plants. *In vitro* Cell Div. Biol. Plant 28P:192–196.

70. Roux JJ le, Staden J van (1991) Micropropagation and tissue culture of *Eucalyptus* – a review. Tree Physiology 9: 435–477.

71. Salmon AD, (1990) Clonal propagation of the red flowering gum. Australian Horticulture 88: 63–68.

72. Sussex IM (1965) The origin and morphogenesis of *Eucalyptus* cell populations. In: White P, Groves A (Eds.) Proc Int Conf Pl Tissue Cult, pp 383–391, Berkley California.

73. Teulieres C, Boudet AM (1991) Isolation of protoplasts from different *Eucalyptus* species and preliminary studies on regeneration. Plant Cell Tissue and Organ Culture 25: 133–140.

74. Teulieres C, Alibert G, Marien JN, Boudet AM (1989a) Isolation and frost resistance screening of protoplasts from different clones of *Eucalyptus*. J Plant Physiol 134: 316–319.

75. Teulieres C, Feuillet C, Boudet AM (1989b) Differential characteristics of cell suspension cultures initiated from *Eucalyptus gunnii* clones differing by their frost tolerance. Pl Cell Rep 8: 407–410.

76. Teulieres C, Grima-Pettenati J, Curie C, Teissie J, Boudet AM (1991) Transient foreign gene expression in polyethylene/glucose treated or electropulsated *Eucalyptus gunnii* protoplasts. Pl Cell Tissue and Organ Culture 25: 125–132.

77. Texier F, Faucher M (1986) Culture *in vitro* d'apex d'eucalyptus age *Eucalyptus parvifolia* Camb Annales de Recherches Sylvicoles, AFOCEL France 1985 pp 7–23.

78. Trindade IT, Ferriera JG, Paris MS, Aloni R (1990) The role of cytokinin and auxin in rapid multiplication of shoots of *Eucalyptus globulus* grown *in vitro*. Australian Forestry 53: 221–223.

79. Van der Moezel PG, Bell DT, Bennett IJ, Strawbridge M, McComb JA (1990) The development of salt tolerant clonal trees in Australia. Comb Proc Intern Pl Prop Soc 40: 73–75.

80. Venketeswaran S, Gandhi V (1980) Protoplast isolation and culture from tree genera. In vitro 16: 232.

81. Warrag EI, Lesney MS, Rockwood DL (1990) Micropropagation of field tested superior *Eucalyptus grandis* hybrids. New For 4: 67–80.

82. Warrag EI, Lesney MS, Rockwood DL (1991) Nodule culture and regeneration of *Eucalyptus grandis* hybrids. Pl Cell Rep 9: 586–589.

83. Willyams D, Whiteman P, Cameron J, Chandler SF (1992) Inter- and intra-family variability of rooting capacity in micropropagated *Eucalyptus globulus* and *Eucalyptus nitens*. In: AFOCEL Mass Production Technology for Genetically Improved Fast Growing Forest Tree Species, AFOCEL Bordeaux France Sept 14 2: 177–181.

84. Wilson PJ (1992) The development of new clones of *Eucalyptus globulus* and *E. globulus* hybrids by stem cuttings propagation. In: AFOCEL Mass Production Technology for Genetically Improved Fast Growing Forest Tree Species, AFOCEL Bordeaux France Sept 14 1: 379–386.

85. Young PM, Hutchins AS, Canfield ML (1984) Use of antibiotics to control bacteria in shoot cultures of woody plants. Pl Sci Letts 34: 203–9.

Plant Tissue Culture Manual **C9**, 1–27, 1996.
© *1996 Kluwer Academic Publishers.*

Cryopreservation of plant tissue cultures: the example of embryogenic tissue cultures from conifers

P.J. CHAREST[1]*, J. BONGA[2] & K. KLIMASZEWSKA[1]
[1]*Molecular Genetics and Tissue Culture Group, Petawawa National Forestry Institute, Canadian Forest Service, NRCan, Chalk River, Ontario, Canada K0J 1J0, E-mail: pcharest@pfni.forestry.ca;*
[2]*Maritimes Region Canadian Forest Service, NRCan, Fredericton, New Brunswick, Canada E3B 5P7 (*Author for correspondence)*

Introduction

Several methods are available to maintain plant tissue cultures. Each method has advantages and disadvantages regarding the labour required, the frequency of necessary transfers, the availability of equipment, the potential for contamination, the risk of somaclonal variation, and the type of tissue to be preserved [1]. Subculturing is used in most laboratories because of its simplicity. However, it is risky because of the possibility of contamination and somaclonal variation and it is not practical for long-term preservation because it is labour intensive. Typically, subculture is required every 1 to 4 weeks. Another method is reduction of the growth rate (often called slow growth or minimal growth), which is the same as subculturing except that growth of tissue is inhibited by special environmental conditions [2, 3]. Artificial seed production is also used to maintain plant tissue cultures. It can be used with mature somatic embryos desiccated to a level that arrests metabolism [4, 5]. The limitation of the method is that it can only be used in species where somatic embryogenesis is available and desiccation can be applied successfully. Both slow growth maintenance and artificial seed production are medium-term approaches (2 months to a year). The best method for long term storage is the freezing of tissues at the temperature of liquid nitrogen (−196 °C) or its gaseous phase (−140 °C) which suspends all metabolic processes. This technique, called cryopreservation, has four types of protocols [6]. In all of the protocols, the critical point is to avoid intracellular ice crystal formation by ensuring that the cells and tissues are adequately dehydrated. The first type of protocol is referred to as conventional slow freezing. It includes exposure to chemical cryoprotectants, followed by ice inoculation and gradual slow freezing to −40 °C. The tissues are then placed in liquid nitrogen. The second type is called simple freezing and includes exposure to cryoprotectants at room temperature, followed by rapid freezing at −30 °C (temperature of a domestic freezer). The vials containing the tissues are then placed in liquid nitrogen. The third type is vitrification where the tissues are treated with high concentrations of cryoprotectants and then directly placed in liquid nitrogen. The resulting solution (cellular content and cryoprotectants) is supercooled, which transforms it into amorphous glass without ice crystallization. The fourth protocol includes the desiccation of the plant tissues to induce physiological drought followed by

Table 1. Types of plant tissues that have been cryopreserved

Protocols	Type of plant tissues
Conventional slow freezing	cultured cells [10, 11, 12], meristems [13] and somatic embryos [14]
Simple freezing	cultured cells [15, 16] and somatic embryos [17, 18]
Vitrification	winter hardy meristems and buds [19], meristems [20], orthodox pollen [21], somatic embryos [22], cultured cells [23]
Desiccation followed by vitrification	somatic embryos [24], lateral buds [25], apical meristems [26], shoot tips [27], excised zygotic embryos from recalcitrant seeds [28], recalcitrant pollen [29]

Orthodox = desiccation tolerant, Recalcitrant = desiccation sensitive, references are only examples of the protocols that were applied.

immersion in liquid nitrogen. Table 1 summarizes the four types of methods used and their potential application to different types of tissues. For more details on the theory and practice of cryopreservation applied to plant cells the reader should consult Kartha [7], Dresser et al. [8], and Withers [9].

The present chapter will describe the conventional slow freezing method that has been applied extensively in our laboratories to somatic embryogenic cultures of conifers [30, 31, 32]. As indicated in Table 2, it has been used with several conifer species and cell lines (genotypes) from which trees were regenerated after thawing. Cryopreservation of conifer embryogenic tissue does not appear to be genotype-dependent [32]. Conifer embryogenic cultures have been preserved for a period of 3 years at PNFI and they are still viable after thawing.

Table 2. Cryopreservation of conifer embryogenic lines at the Canadian Forest Service

Species	No. of lines in cryopreservation	Location
Larix decidua	9	PNFI
L. decidua (*haploid*)	4	PNFI
Larix laricina	12	PNFI
Larix laricina (*transgenic*)	2	PNFI
Larix leptolepis	3	PNFI
Larix leptolepis (*haploid*)	1	PNFI
Larix x eurolepis	11	PNFI
Larix x leptoeuropaea	26	PNFI
L. occidentalis	2	PNFI
Picea abies	5	PNFI
Picea glauca	54	PNFI
	450	CFS- Maritimes Region
Picea glauca (*transgenic*)	19	PNFI
P. glauca engelmannii complex	16	PNFI
Picea mariana	259	PNFI
P. mariana (*transgenic*)	11	PNFI
Picea rubens	15	PNFI
Pinus strobus	11	PNFI
	3	CFS– Quebec Region
Total number of species and lines	12 species and over 913 lines	

No. of lines cryopreserved successfully as of March 1995

Procedures

Cryopreservation of conifer somatic embryogenic tissues comprises five main steps: choice of tissues to be frozen, chemical pretreatment, freezing, thawing, and regrowth [30]. The choice of tissue is critical for the success of the process. The protocol described below is optimized for tissues that are typically embryogenic (Fig. 1). As a general rule, rapidly growing tissues with a population of small cells that are

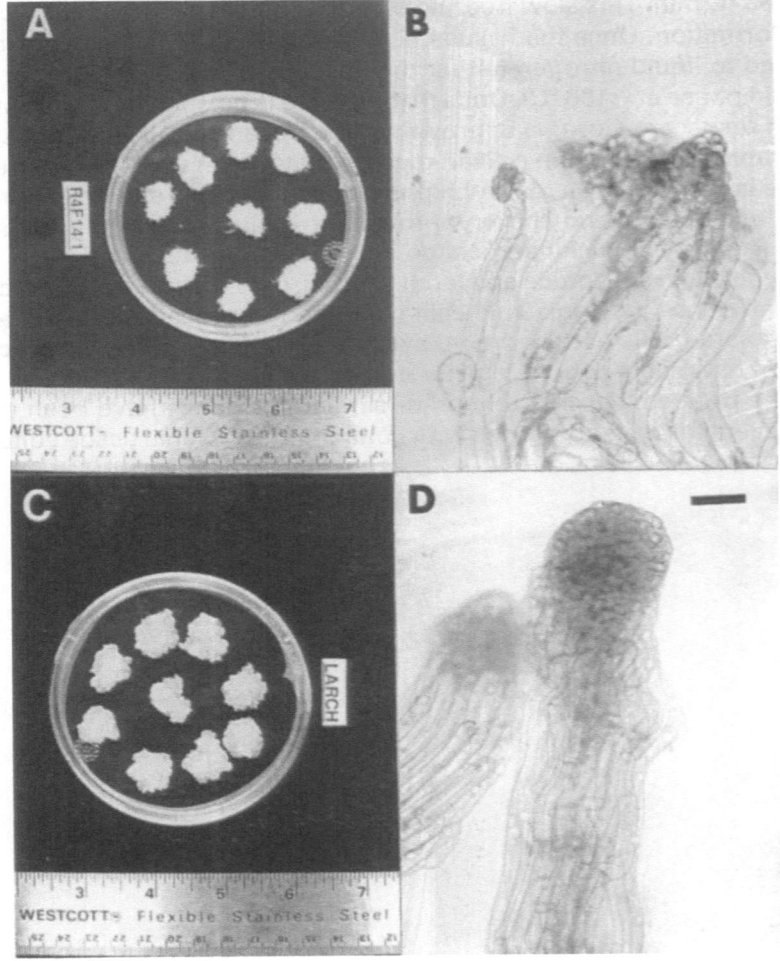

Fig. 1. Photograph of embryonal masses of A) *Picea mariana* and C) *Larix x eurolepis* at 2 weeks after subculture on maintenance media. Micrograph of embryonal masses showing isolated early embryos of B) *Picea mariana* and D) *Larix x eurolepis*. Two zones can be distinguished; the meristematic head (small dense cells) and the suspensor area (loose elongated cells). For B) and D), bar=32μm.

densely cytoplasmic will give the best results. One way to achieve rapid cell division is to use a cell suspension prepared from the embryonal masses.

Chemical pretreatment of the cells is required to dehydrate the tissues and to lower their freezing temperature. This is done by growing the tissues in media with high chemical osmoticum (e.g., polyethylene glycol or sorbitol) followed by a short treatment with a cryoprotectant such as dimethyl sulfoxide (DMSO). Following this treatment, the cells are gradually frozen from 0 °C to −40 °C at an average rate of −0.33 °C/min. This slow freezing is required to avoid internal ice crystal formation. Once the tissues have reached −40 °C, they are transferred to liquid nitrogen (either the gaseous phase at −160 °C or the liquid phase at −196 °C). Once frozen, there is no specific requirement with respect to storage. Regrowth of the frozen cells is accomplished by rapidly thawing the cells in a warm water bath (37 °C). The procedure is followed by a regrowth phase that includes a step to eliminate the osmoticum and the cryoprotectant. This reduces the osmotic shock and the toxicity associated with the cryoprotectant.

Typically, with spruce and larch species, only the meristematic cells (small densely cytoplasmic cells) survive the procedures. The large vacuolated cells of the embryo suspensor do not survive the process and a new suspensor is regenerated [31].

The protocol described here, or similar protocols, have been used in several other laboratories with conifer embryogenic tissue cultures [33–38].

Steps in the procedure

A) *Tissue culture protocols*

The tissue culture protocols for spruce and larch embryogenic cultures, including initiation of embryogenic cultures, maintenance, maturation of somatic embryos, germination, and transfer to soil have been described in Lelu et al. [39]. Furthermore, chapter C3 (pp. 1–16) of this manual by Thorpe and Harry [40] covers somatic embryogenesis of conifers. Table 2 gives a listing of species for which this protocol has been applied successfully. As examples, the recipes for *Picea mariana* and *Larix x eurolepis* for tissue culture maintenance and cryopreservation will be provided.

The embryonal masses of both species are maintained on solid medium (1/2 LM for *P. mariana* and MSG for *L. x eurolepis*) in Petri dishes (100 × 25mm; nine clumps per dish) and subcultured every 14 days. They can also be maintained in liquid medium (same media as above but without the gelling agent) in 125ml flasks containing 50ml of medium and subcultured every 7 days on a rotary shaker at 120 rpm.

B) Material to be frozen

The embryogenic tissue cultures should be in a vigorous phase of growth. Typically, 3-day-old embryonal masses from a cell suspension subcultured every 7 days or 4-day-old embryonal masses from a culture on solid medium subcultured every 7 or 14 days can be used.

Note
The embryogenic tissues should be monitored closely for their macroscopic and microscopic appearance. The manipulator must develop a knowledge of what constitutes a healthy tissue culture. Any changes could indicate that the line is not suitable for cryopreservation.

C) Pretreatment

1. If the cultures are maintained on gelled medium, suspend the embryonal masses in liquid tissue culture medium (1/2 LM for *P. mariana* and MSG for *L. x eurolepis*). Gently mix the suspended tissues and use a spatula if needed to break up large tissue pieces.
2. Collect the embryonal masses from the suspension by filtering through a nylon mesh (pore size 73 μm).
3. Transfer 1–2 g (fresh weight) of cells to 125 ml Erlenmeyer flasks each containing 7 ml of liquid medium (1/2 LM for *P. mariana* and MSG for *L. x eurolepis*) supplemented with 0.4 M sorbitol.
4. Place the flasks on a rotary shaker at 100–115 rpm for 20–24 hrs.
5. Place the flasks on ice.
6. Add stepwise, over a period of 30 min, 3 ml of a DMSO solution prepared in 0.4 M sorbitol medium so that the final concentration will reach 5–10% DMSO (volume/volume).
7. Mix gently and leave on ice for 30 min for the mixture temperature to equilibrate.

D) Freezing

8. Mix the pretreated suspension gently. Dispense 1.0 ml aliquots of suspended cells into 1.2 ml sterile cryogenic vials (Corning Lab. Science Co., USA).
9. Place vials in a programmable freezer. The freezer should be programmed to cool as follows: hold at 0 °C for 10 min and cool to −40 °C at a rate of 0.33 °C/min.
10. Remove the vials from the freezer and immerse in liquid nitrogen (−196 °C).

E) Thawing

11. Remove vials from storage and place them in a container with liquid nitrogen.
12. Prepare a warm water bath at 37–40 °C.
13. Remove vials from the liquid nitrogen, a few at a time, and thaw by swirling them in the warm water bath. The vials should be removed from the bath as soon as the ice pellet inside the vial has thawed, which should be within 2 min.
14. Allow the contents to equilibrate to room temperature (a few minutes) before opening the vial.

F) Regrowth

15. Pour contents of the vial onto two layers of sterile filter paper discs (Whatman #2, Whatman International Ltd., England) placed on the surface of the gelled medium in a Petri dish (1/2LM for *P. mariana* and MSG for *L. x eurolepis*).
16. Seal the Petri dish with parafilm and place in the dark at 25 °C.
17. After 24 hrs, transfer the upper filter paper disc with the cells to fresh medium to remove DMSO and excess liquid.
18. After 2 weeks, subculture as usual by transferring portions of the actively growing embryonal cell masses to a Petri dish containing gelled tissue culture medium (1/2 LM *for P. mariana* and MSG for *L. x eurolepis*).
19. Regenerate plantlets following the published protocols [39].

G) Vital staining of regrowing tissue cultures

The cell viability after thawing can be determined using fluorescein diacetate (FDA; [44]) by incubating the cells in 0.005% (weight/volume) of FDA for 30 minutes, followed by two washings in tissue culture medium. The cells are observed under a microscope with ultraviolet illumination. Living cells will produce fluorescence.

Notes

3. It is important that the embryonal cell mass density is such that after thawing there will be enough surviving cells with sustained growth. We found that increasing the fresh weight from 1 g to 2 g per 10 ml of the final medium volume will ensure a rapid regrowth in some lines.
10. Storage of the cryovials is preferable in the vapour phase of liquid nitrogen because this reduces the risks of the vials exploding. Ashwood-Smith and Friedmann [43] estimated that it would take up to 30 000 years before background radiation would have a significant effect on the genetic make up of the cultures stored at −196 °C.

Solutions

For 500 ml of half strength Litvay's medium (1/2LM; [41])

1/2 LM 5X frozen stock	100 ml
Casein hydrolysate (casamino acids)	0.5 g
Sucrose (1% final)	5.0 g
2,4-D (1 mg/ml stock solution)	1.1 ml
6-benzylaminopurine (0.5 mg/ml stock solution)	1.1 ml
Add distilled H_2O to 500 ml	
pH	5.7

Autoclave and add 10 ml of filter sterilized glutamine (25 mg/ml stock solution) to cooled medium.

If solid medium is required, add gelrite (2 g/500 ml) before adjusting pH and before autoclaving.

For 2 litres of Litvay's 5X stock (1/2 LM 5X frozen stock)

NH_4NO_3	8.25 g
KNO_3	9.5 g
$MgSO_4.7H_2O$	9.25 g
KH_2PO_4 (monobasic)	1.7 g
$CaCl_2.2H_2O$	0.11 g
LM micro nutrient stock (100X)	100 ml
LM vitamin stock (100X)	100 ml
Myo-inositol	1g
Fe diethylene triamine pentaacetate	0.28 g

Add distilled H_2O to 2000 ml and freeze in 100 ml sterile Twirlbags (Fisher Scientific Co, Canada) for further use.

For 1 litre of LM micronutrient stock 100X

KI	0.415 g
H_3BO_3	3.1 g
$MnSO_4.H_2O$	2.1 g
$ZnSO_4.7H_2O$	4.3 g
$Na_2MoO_4.2H_2O$	0.125 g
$CuSO_4.5H_2O$	0.05 g
$CoCl_2.6H_2O$	0.013 g

Add distilled H_2O to 1000 ml and freeze in 100 ml sterile Twirlbags for further use.

For 1 litre of LM vitamin stock 100X

Nicotinic acid	0.05 g
Pyridoxine HCl	0.01 g
Thiamine HCl	0.01 g

Add distilled H_2O to 1000 ml and freeze in 100 ml sterile Twirlbags for further use.

For 500 ml of MSG medium (MSG; [42])

MSG 5X frozen stock	100 ml
Sucrose (2%)	10.0 g
2,4-D (1mg/ml stock solution)	1.0 ml
6-benzylaminopurine	0.5 ml
Add distilled H_2O to 500 ml	5.8

Autoclave and add 29.2 ml of filter sterilized glutamine (25 mg/ml stock solution) to cooled medium.

If solid medium is required, add gelrite (2.0 g/500 ml) before autoclaving.

For 2 litres of MSG 5X stock

KNO_3	1.0 g
$CaCl_2.2H_2O$	4.4 g
$MgSO_4.7H_2O$	3.7 g
KH_2PO_4 (monobasic)	1.7 g
KCl	7.4 g
Fe diethylene triamine pentaacetate	0.4 g
MS micro nutrient stock (100X)	100 ml
MS vitamin stock (100X)	100 ml
KI (10 mg/ml stock solution)	0.83 ml

Add distilled H_2O to 2000 ml and freeze in 100 ml sterile Twirlbags for further use.

For 1 litre of MS micronutrient stock 100X

H_3BO_3	0.62 g
$MnSO_4.4H_2O$	2.23 g
ZnSO4.7H2O	0.86 g
$Na_2MoO_4.2H_2O$	0.025 g
$CuSO_4.5H_2O$	0.0025 g
$CoCl_2.6H_2O$	0.0025 g

Add distilled H_2O to 1000 ml and freeze in 100 ml sterile Twirlbags for further use.

For 1 litre of MS vitamin stock 100X

Myo-inositol	10.0 g
Nicotinic acid	0.05 g
Pyridoxine HCl	0.05 g
Thiamine HCl	0.01 g

Add distilled water to 1000 ml and freeze in 100 ml sterile Twirlbags for further use.

Concentrated dimethysulfoxide (DMSO) solution
For a solution of 10% DMSO in 10 ml, add 1.0 ml of tissue culture medium (1/2 LM for *P. mariana* and MSG for *L. x eurolepis*) with 0.4 M sorbitol to 1.0 ml double-strength tissue culture medium with 0.8 M sorbitol. Place on ice.

Gradually add 1.0 ml of filter sterilized DMSO and mix thoroughly. Make this solution just before adding to the pretreated cultures contained in 7.0 ml of tissue culture medium with 0.4 M sorbitol.

References

1. Wang BSP, Charest PJ, Downie B (1993) *Ex situ* storage of seeds, pollen and *in vitro* cultures of perennial woody plant species. FAO Forestry Paper #113. 83pp.
2. Banerjee N, de Langhe E (1985) A tissue culture technique for rapid clonal propagation and storage under minimal growth conditions of *Musa* (Banana and plantain). Plant Cell Rep. 4: 351–354.
3. Engelmann F (1990) Utilisation d'atmosphères à teneur en oxygène réduite pour la conservation de cultures d'embryons somatiques de palmier à huile (*Elaeis guineensis* Jacq.). C.R. Acad. Sci. Paris Série III 310: 679–684.
4. McKersie BD, Senaratna T, Bowley R, Brown DCW, Krochko JE, Bewley JD (1989) Application of artificial seed technology in the production of hybrid *alfalfa* (*Medicago sativa* L.). In Vitro Cell. Dev. Biol. 25: 1183–1188.
5. Redenbaugh K, Fujii JA, Slade D (1988) Encapsulated plant embryos. In: Biotechnology in Agriculture. A. Mizrahi ed. A.R. Liss New York, USA. pp. 225–248.
6. Sakai A (1993) Cryogenic strategies for survival of plant cultured cells and meristems cooled to −196 °C. JICA Group Ref. No. 6. pp. 5–26.
7. Kartha KK (1985) Cryopreservation of plant cells and organs. CRC Press Inc., Boca Raton, Florida. 276pp.
8. Dresser BL, Russell P, Pope CE, Pence V, Plair B, Long P (1992) Wildlife germplasm cryopreservation workshop manual. Center for Reproduction of Endangered Wildlife, Cincinnati Zoo & Botanical Garden, OH.
9. Withers LA (1980) Tissue culture storage for genetic conservation. International Board for Plant Genetic Resources Rep. Rome. 91pp.
10. Panis BJ, Withers LA, De Langhe EAL (1990) Cryopreservation of *Musa* suspension cultures and subsequent regeneration of plants. Cryo-Lett. 11: 337–350.
11. Gnanapragasam S, Vasil IK (1990) Plant regeneration from a cryopreserved embryogenic cell suspension of a commercial sugarcane hybrid (*Saccharum* sp.). Plant Cell Rep. 9: 419–423.
12. Wang ZY, Legris G, Nagel J, Potrykus I, Spangenber G (1994) Cryopreservation of embryogenic cell suspensions in *Festuca* and *Lolium* species. Plant Sci. 103: 93–106.
13. Reed BM (1990) Survival of in vitro-grown apical meristems of *Pyrus* following cryopreservation. Hort. Sci. 25: 111–113.
14. Bertrand-Desbrunais A, Fabre J, Engelmann F, Dereuddre J, Charrier A (1988) Reprise de l'embryogénèse adventive à partir d'embryons somatiques de caféier (*Coffea arabica* L.) après leur congélation dans l'azote liquide. C.R. Acad. Sci. Paris. Série III 307: 795–801.
15. Nishizawa S, Sakai A, Amano Y, Matsuzawa T (1992) Cryopreservation of asparagus (*Asparagus officinalis* L.) embryogenic suspension cells and subsequent plant regeneration by a simple freezing method. Cryo-Lett. 13: 379–388.
16. Sakai A, Kabayashi S, Oiyama I (1991) Cryopreservation of nucellar cells of navel orange (*Citrus sinensis* Osb.) by a simple freezing method. Plant Sci. 74: 243–248.
17. Tessereau H, Lecouteux C, Florin B, Schlienger C, Petiard V. (1991) Use of a simplified freezing process and dehydration for the storage of embryogenic cell lines and somatic embryos. Rev. Cytol. Biol. Végé. Bot. 14: 297–310.
18. Lecouteux C, Florin B, Tessereau H, Bollon H, Petiard V (1991) Cryopreservation of carrot somatic embryos using a simplified freezing process. Cryo-Lett. 12: 319–328.
19. Sakai A (1984) Cryopreservation of apical meristems. Hort. Rev. 6: 357–372.
20. Towill LE (1990) Cryopreservation of isolated mint shoot tips by vitrification. Plant Cell Rep. 9: 178–180.
21. Jorgensen J (1990) Conservation of valuable gene resources by cryopreservation in some forest tree species. J. Plant Physiol. 136: 373–376.
22. Engelmann F, Dereuddre J (1988) Cryopreservation of oil palm somatic embryos: importance of the freezing process. Cryo-Lett. 9: 220–235.

23. Sakai A, Kabayashi S, Oiyama I (1990) Cryopreservation of nucellar cells of navel orange (*Citrus sinensis Osb. var. brasiliensis* Tanada) by vitrification. Plant Cell Rep. 9: 30–33.

24. Dereuddre J, Tannoury M, Hassen N, Kaminski M, Vintejoux C (1992) Application de la technique d'enrobage à la congélation d'embryons somatiques de carotte (*Daucus carota* L.) dans l'azote liquide: étude cytologique. Bull. Soc. Bot. Fr. Lettres Bot. 139: 15–33.

25. Uragami A, Sakai A, Nagai M (1990) Cryopreservation of dried axillary buds from plantlets of *Asparagus officinalis* L. grown *in vitro*. Plant Cell Rep. 9: 328–331.

26. Poissonnier M, Monod V, Pagues M, Dereuddre J (1992) Cryopreservation dans l'azote liquide d'apex d'*Eucalyptus gunnii* (Hook. F.) cultivé *in vitro* après enrobage et déshydratation. Ann. Rech. Sylvicoles. Afocel. pp. 6–23.

27. Paulet F, Engelmann F, Glaszmann JC (1993) Cryopreservation of apices of *in vitro* plantlets of sugarcane (*Saccharum* sp. hybrids) using encapsulation/dehydration. Plant Cell Rep. 12: 525–529.

28. Pence VC (1990) Cryostorage of embryo axes of several large-seeded temperate tree species. Cryobiol. 27: 212–218.

29. Marchant R, Power JB, Davey JM, Chartier-Hollis JM, Lynch PT (1993) Cryopreservation of pollen from two rose cultivars. Euphytica 66: 235–241.

30. Charest PJ, Klimaszewska, K (1995) Cryopreservation of germplasm of *Larix* and *Picea* Species. In: Y.P.S. Bajaj (Ed.) Biotechnology in Agriculture and Forestry, vol. 32 Cryopreservation of plant germplasm – I, pp. 191–203. Springer-Verlag Berlin Heidelberg.

31. Klimaszewska K, Ward C, Cheliak M (1992) Cryopreservation and plant regeneration from embryogenic cultures of larch (*Larix x eurolepis*) and black spruce (*Picea mariana*). J. Exp. Bot. 43: 73–79.

32. Park YS, Pond SE, Bonga JM (1994) Somatic embryogenesis in white spruce (*Picea glauca*): genetic control in somatic embryos exposed to storage, maturation treatments, germination, and cryopreservation. Theor. Appl. Genet. 89: 742–750.

33. Gupta PK, Durzan DJ, Finkle BJ (1987) Somatic polyembryogenesis in embryogenic cell masses of *Picea abies* (Norway spruce) and *Pinus taeda* (loblolly pine) after thawing from liquid nitrogen. Can. J. For. Res. 17: 1130–1134

34. Laine E, Bade P, David A (1992) Recovery of plants from cryopreserved embryogenic cell suspensions of *Pinus caribaea*. Plant Cell Rep. 11: 295–298.

35. Kartha KK, Fowke LC, Leung NL, Caswell KL, Hakman I (1988) Induction of somatic embryos and plantlets from cryopreserved cell cultures of white spruce (*Picea glauca*). J. Plant Physiol. 132: 529–539.

36. Toivonen PMA, Kartha KK (1989) Cryopreservation of cotyledons of nongerminated white spruce [*Picea glauca* (Moench) Voss] embryos and subsequent plant regeneration. J. Plant Physiol. 134: 766–768.

37. Bercetche J, Galerne M, Dereuddre J (1990) Augmentation des capacités de régénération de cals embryogènes de *Picea abies* (L.) Karst après congélation dans l'azote liquide. C.R. Acad. Sci. Paris. Série III 310: 357–363.

38. Galerne M, Dereuddre J (1987) Survie de cals embryogènes d'épicea après congélation à −196 °C. Ann. Rech. Sylvicoles Afocel. pp. 7–32.

39. Lelu MA, Klimaszewska KK, Jones C, Ward C, von Aderkas P, Charest PJ (1993) A laboratory guide to somatic embryogenesis in spruce and larch. Petawawa National Forestry Institute Information Report PI-X-111. 57pp.

40. Thorpe TA, Harry IS (1991) Clonal propagation of conifers. Plant Tiss. Cult. Man. C3: 1–16.

41. Litvay JD, Johnson MA, Verma D, Einspahr D, Weyrauch K (1981) Conifer suspension culture medium development using analytical data from developing seeds. Inst. Pap. Chem. Tech. Pap. Serv. 115: 1–17.

42. Becwar MR, Nagmani R, Wann SR (1990) Initiation of embryogenic cultures and somatic embryo development in loblolly pine (*Pinus taeda*). Can. J. For. Res. 20: 810–817.

43. Ashwood-Smith MJ, Friedman JB (1979) Lethal and chromosomal effects of freezing, thawing, storage time and X-irradiation on mammalian cells preserved at -196 °C in dimethylsulfoxide. Cryobiol. 16: 132–140.
44. Wildholm SM (1972) The use of fluorescein diacetate and phenosafranine for determining viability of cultured plant cell. Stain Technol. 47: 189–193.

Section D:
Direct Gene Transfer & Protoplast Fusion

Direct Gene Transfer

Sect. D

Plant Tissue Culture
Manual

Plant Tissue Culture Manual **D1**: 1–12, 1991.

Gene transfer by particle bombardment

T. M. KLEIN[1], S. KNOWLTON[2], R. ARENTZEN[1]
[1]*Du Pont Agricultural Products, Wilmington, Delaware 19880-0402, USA*
[2]*Du Pont Medical Products, Box 122, Newark, Delaware 19714-6101, USA*

Introduction

Gene transfer into intact cells and tissues by particle bombardment is becoming an important tool in plant molecular biology [17]. The process has been used to stably transform major crop species including corn [7, 9], soybean [19], and cotton [6]. Genes have also been delivered into the chloroplast of *Chlamydomonas* [2] and the mitochodria of yeast [12]. Recent results indicate that genes can be introduced into the chloroplasts of higher plants [5] and that the foreign DNA can be stably maintained in the chloroplast and inherited by the progeny of the transgenic plants (Pal Maliga, personal communication). In addition to its application for the production of genetically transformed plants, particle bombardment has been used in transient expression experiments for the functional analysis of promoter elements that confer regulation to environmental [3] and tissue specific factors [16, 23 see also chapter by Twell et al., this volume]. These studies indicate that the technique can be utilized for gene expression studies in tissues as diverse as pollen, coleoptile, aleurone, and embryo. Regulatory genes that code for trans-acting factors have also been analyzed following their delivery into intact tissues [8, 18]. Gene transfer by particle bombardment provides a means of circumventing the use of protoplasts in transient assay systems. Therefore, the expression of reporter genes can be monitored directly in tissues which should exert proper regulation upon the introduced gene.

The efficiency of gene transfer by particle bombardment is highly dependent on various physical parameters that influence the particles' momentum and distribution upon the target tissue. The particles' momentum is related to its mass and velocity. The size and shape of the particle will also influence its potential for cell penetration. The number of particles per unit area that impact the target tissue is controlled to a large degree by the method of acceleration. At least five devices for accelerating particles into cells have been described in the literature [15, 19, 20, 21, 24]. Another important parameter controlling the efficiency of the process is the method used to adsorb DNA to the particles. Biological factors such as the size of the target cells, their turgor pressure, and stage of development are also of importance in particle bombardment.

Because of the large number of variables associated with the particle bombardment process, model systems that provide rapid feedback and are relatively simple to implement can provide valuable information. Results from a model system can be used as a benchmark to ensure that the technical aspects associated with particle bombardment, such as particle preparation and acceleration, are functional prior to experiments with a new cell type or tissue. This chapter describes several model systems that are useful in characterizing the particle bombardment process. These model systems should be applicable to the testing of various particle delivery systems. Specific instructions for the use of one particle acceleration system are given in this volume in the chapter on gene transfer to pollen by particle bombardment [22].

Procedures

Recipe 1. Yeast transformation

Auxotrophic strains of *Saccharomyces cerevisiae* can be bombarded with particles that are coated with plasmid DNA containing a gene that can complement the specific mutation. In a typical bombardment experiment, a yeast culture that has been grown to stationary phase in liquid medium is plated as a lawn on the surface of selective agar medium and bombarded with the appropriate gene. For example, auxotrophs that require uracil for growth can be bombarded with a gene that complements the specific mutation and permits prototrophic growth [1]. The number of transformants can then be readily scored. Visual inspection of the spatial distribution of the colonies that develop on the bombarded plate provides a means for determining the relative influence of various parameters on the impact distribution of the particles with the target cells.

In the following experiment *S. cerevisiae* strain 948 [1] which contains the non-reverting *ura3*-52 mutation is bombarded with plasmid YEp352 [10] which contains a wild type *URA3* gene.

Steps in the procedure
1. Prepare selective medium and YEPD liquid medium (see below).
2. Inoculate 5 ml of YEPD medium with a single yeast colony of strain 948. Grow yeast at 30 °C overnight with vigorous shaking.
3. Inoculate 50 ml of YEPD medium with 0.5 ml of the overnight culture. Grow at 30 °C for about 40 hours with vigorous shaking.
4. Determine the density of yeast cells with a hemacytometer. Adjust the density to 5×10^8 cells per ml.
5. Spread 0.2 ml of the yeast culture evenly over the surface of Petri dishes containing the selective medium. Allow the plates to dry in a laminar flow hood.
6. Bombard the plates with particles coated with plasmid YEp352 [10].
7. Incubate the plates for 2 to 3 days at 30 °C. Count the number of colonies. The bombardment parameters given by Twell and coworkers in this volume [22] should yield about 500–1000 uracil-independent colonies per bombardment.

Media
— Selective medium lacking uracil
  — Glucose stock solution
    — Add 50 g to 250 ml of water and autoclave.
  — Uracil dropout premix
    — Adenine, free base   0.4 g
    — Tryptophan   0.4 g
    — Histidine, free base   0.4 g
    — Arginine, free base   0.4 g
    — Methionine   0.4 g
    — Tyrosine, free base   0.6 g

— Leucine	1.2 g
— Lysine, free base	0.6 g
— Phenylalanine	1.0 g
— Threonine	4.0 g

Combine ingredients and mix thoroughly.
- — Preparation of selective medium

— Bacto agar (Difco)	15 g
— Yeast nitrogen base without amino acids (Difco)	6.7 g
— Uracil dropout premix (see above)	0.47 g
— Sorbitol	136 g
— Mannitol	136 g

Add to 900 ml of water, dissolve, and autoclave.
Add 100 ml of sterile glucose solution and mix.
Pour plates (100 mm in diameter) when cooled to 60 °C.
- — YEPD culture medium

— Yeast extract (Difco)	5 g
— Bacto peptone (Difco)	10 g
— Adenine	25 mg

Add 900 ml of water and autoclave. For solid medium, add 15 g of agar prior to autoclaving. Add 100 ml of sterile glucose solution (see above) and mix.

Recipe 2. Transient gene expression in cauliflower tissue

Most model systems used to study the physical parameters that influence the particle bombardment process rely on cell cultures or explants as the target tissue. However, each of these has drawbacks. When explants are used, many plants are usually needed to obtain enough target tissue for replicated experiments. Maintenance of these plants requires extensive growth chamber or greenhouse facilities to provide a continual source of explants. On the other hand, cell cultures do not require the growth and maintenance of whole plants. However, cell cultures often require frequent subculture and costly incubator space. In cases where cell division is slow, the tissue may be in short supply.

For these reasons, a plant model system for transient expression experiments has been developed with the following characteristics: 1) the system is simple to implement, 2) access to tissue does not require plant growth or culture facilities and does not involve extensive dissection or manipulation, 3) the tissue is biochemically compatible with the histochemical β-glucuronidase (GUS) assay [11] and does not contain significant amounts of chlorophyll so that the blue spots can be readily counted. Meeting these criteria, the outer layers of the main stalk of cauliflower heads were chosen as a model target tissue for DNA delivery by particle bombardment.

Steps in the procedure
1. Prepare agar (1.0%) Petri plates. Preparation of medium containing nutrients or hormones is not necessary.
2. Cauliflower can be obtained at the local grocery store. Prepare the cauliflower tissue by first separating the florets from the main stem. Sections (1 cm^2) of the main stem comprising the epidermal tissue and underlying tissue (to a depth of about 0.5 cm) are then made. Most of the pith tissue is discarded. The cauliflower sections are then placed on the agar with the epidermis side up. Nine such sections in a 3 by 3 array are placed in the center of a 100 mm Petri dish.
3. Bombard the tissue with particles coated with an appropriate GUS vector [11; vectors are also available from Clontech, Inc., Palo Alto, CA). Incubate the tissue for about 18 to 24 hours at 24 °C.
4. Treat the tissue with the GUS assay buffer (see below). Each piece of tissue can be placed in a separate well of an appropriate microtiter plate and the tissue submerged in the solution. A note can also be made of the original position of each piece of tissue on the bombarded Petri dish.
5. Seal the microtiter plates and incubate the tissue for 18 hours at 37 °C. Count the blue spots on each piece of tissue. A typical bombardment experiment can yield 500 blue cells on each 1 cm^2 section of tissue.

Solutions
- GUS assay buffer
    - 5 mM potassium ferricyanide
    - 5 mM potassium ferrocyanide
    - 0.05% (w/v) 5-bromo-4-chloro-3-indolyl-β-D-glucuronic acid
    - 0.06% (v/v) Triton X-100
    - 0.2 M sodium phosphate buffer, pH 7.0

Recipe 3. Transformation of tobacco by particle bombardment

Stable transformants of tobacco can be recovered from bombarded leaf tissue [14]. Whole leaves or leaf strips are bombarded with a gene that confers resistance to kanamycin. Kanamycin-resistant calluses develop on the leaf tissue after several weeks on the selective media. The following protocol provides a simple means for evaluating the effects of various bombardment parameters on the stable transformation of plant cells.

Steps in the procedure
1. Grow plants from surface sterilized seed on agar supplemented with MS salts (Sigma) and sucrose (30 g/l) in Magenta vials (eg from Sigma).
2. Leaves (3 to 5 cm in length) are excised from plants, placed abaxial side up on RMO medium (see below) in a Petri dish, and bombarded with microprojectiles carrying an appropriate kanamycin resistance gene [14].
3. Transfer the leaf tissue to RMO medium containing kanamycin (100 µg/ml) seven days after bombardment. Be sure that the leaf tissue remains in contact with the medium. Kanamycin resistant calluses are visible after 3 weeks. Subculture the calluses on kanamycin-containing medium.

Media
— RMO medium
    — MS salts (Sigma)
    — IAA   2.0 mg/l
    — BA   0.5 mg/l
    — thiamine   1.0 mg/l
    — inositol   100 mg/l
    — agarose   7 g/l
    — sucrose   30 g/l
    adjust pH to 5.8

Recipe 4. Analysis of gene transfer to maize cells using the anthocyanin markers

The enzymes necessary for anthocyanin production in maize are encoded by a set of genes that include *A1, A2, Bz1, Bz2,* and *C2.* The expression of the genes in this pathway are co-ordinately regulated by several regulatory loci, which condition the tissue-specific production of anthocyanin pigment [4]. In general, anthocyanin pigment formation requires the expression of at least two classes of regulatory genes. Both a functional *Pl* or *C1* gene plus a functional *R* or *B* gene are necessary for pigment formation. Bombardment of maize tissues with the appropriate regulatory gene or genes results in activation of the pathway and the production of pigmented cells [8, 18]. For example, embryogenic maize callus or suspension cultures derived from an appropriate genotype (i.e., crosses of inbred lines A188 and B73) can be bombarded with a plasmid that contains the cDNAs of *C1* and *B-I,* both under the control of the cauliflower mosaic virus 35S promoter and the nopaline synthase polyadenylation region [8]. Two days following bombardment red to purple cells can clearly be observed in the tissue.

The anthocyanin markers are valuable for observing the distribution of transiently expressing cells in the bombarded tissue. In addition, it should be possible to observe the populations of cells that can be penetrated by particles and to determine the number and size of particles that penetrate particular cell types. The anthocyanin genes have some advantages over GUS as a marker for transient gene expression. Unlike the product of the GUS reaction, the anthocyanin pigment remains within the cell expressing the gene; there is no bleeding of the pigment into surrounding cells. In addition, visualization of anthocyanin production does not require addition of substrates. Although the anthocyanin producing cells should be viable, it is not clear if they are capable of further division. Therefore, the usefulness of these genes as a visible marker for identifying stably transformed sectors developing in bombarded tissue cultures has yet to be determined.

Steps in the procedure
1. Embryogenic maize callus can be initiated from immature embryos as previously described [13].
2. Bombard the maize tissue with the *B-I* and *C1* genes [8].
3. Incubate the tissue in the dark for 24 to 48 hours. Count the purple cells. Between 5,000 and 10,000 purple cells can be observed in a field of cells covering a circle with a diameter of 5 cm.

References

1. Armaleo D, Ye GN, Klein TM, Shark KB, Sanford JC, Johnston SA (1990) Biolistic nuclear transformation of *Saccharomyces cerevisiae* and other fungi. Current Genetics 17: 97–103.
2. Boynton JE, Gillham NW, Harris EH, Hoster JP, Johnson AM, Jones AR, Randolf-Anderson BL, Robertson D, Klein TM, Shark KB, Sanford JC (1988) Chloroplast transformation in *Chlamydomonas* with high velocity microprojectiles. Science 240: 1534–1538.
3. Bruce WB, Christensen AH, Klein TM, Fromm ME, Quail PH (1989) Photoregulation of a phytochrome gene promoter from oat transferred into rice by particle bombardment. Proc Natl Acad Sci USA 86: 9692–9696.
4. Coe EH Jr, Neuffer MG, Hoisington DA (1988) The genetics of corn. In: Corn and Corn Improvement, Sprague GF, Dudley J, eds (American Society of Agronomy) pp. 81–258.
5. Daniell AD, Vivekananda J, Nielsen BL, Ye GN, Tewari KK, Sanford JC (1990) Transient foreign gene expression in chloroplasts of cultured tobacco cells after biolistic delivery of chloroplast vectors. Proc Natl Acad Sci USA 87: 88–92.
6. Finer JJ, McMullen MD (1990) Transformation of cotton (*Gossypium hisutum* L.) via particle bombardment. Plant Cell Reports 8: 586–589.
7. Fromm ME, Morrish F, Armstrong C, Williams R, Thomas J, Klein TM (1990) Inheritance and expression of chimeric genes in the progeny of transgenic maize. Bio/Technology 8: 833–839.
8. Goff SA, Klein TM, Roth BA, Fromm ME, Cone KC, Radicella JP, Chandler VL (1990) Transactivation of anthocyanin biosynthetic genes following transfer of *B* regulatory genes into maize tissues. EMBO J 9: 2517–2522.
9. Gordon-Kamm WJ, Spencer TM, Mangano ML, Adams TR, Daines RJ, Start WG, O'Brien JV, Chambers SA, Adams WR, Willetts NG, Rice TB, Mackey CJ, Krueger RW, Kausch AP, Lemaux PG (1990) Transformation of maize cells and regeneration of fertile transgenic plants. Plant Cell 2: 603–618.
10. Hill JE, Myers AM, Koerner TJ, Tzagoloff A (1986) *E. coli* shuttle vectors with multiple unique restriction sites. Yeast 2: 163–167.
11. Jefferson RA, Kavanagh TA, Bevan MW (1987) GUS fusions: beta-glucuronidase as a sensitive and versitile gene fusion marker in higher plants. EMBO J 6: 3901–3907.
12. Johnston SA, Anziano PQ, Shark KB, Sanford JC, Butow RA (1988) Mitochondria transformation in yeast by bombardment with microprojectiles. Science 240: 1538–1541.
13. Kamo KK, Hodges TK (1986) Establishment and characterization of long-term embryonic maize callus and cell suspension cultures. Plant Science 45: 111–117.
14. Klein TM, Harper EC, Svab Z, Sanford JC, Fromm ME (1988) Stable genetic transformation of intact *Nicotiana* cells by the particle bombardment process. Proc Natl Acad Sci USA 85: 8502–8505.
15. Klein TM, Gradziel T, Fromm ME, Sanford JC (1988) Factors influencing gene delivery into *Zea mays* cells by high-velocity microprojectiles. Bio/Technology 6: 559–563.
16. Klein TM, Roth BA, Fromm ME (1989) Regulation of anthocyanin biosynthetic genes introduced into intact maize tissues by microprojectiles. Proc Natl Acad Sci USA 86: 6681–6685.
17. Klein TM, Goff, SA, Roth BA, Fromm ME (1990) Applications of the particle gun in plant biology. In: Progress in Plant Cellular and Molecular Biology (eds. Nijkamp HJJ, van der Plas LHW, van Aartrijk J) Kluwer Academic Publishers. pp. 56–66.
18. Ludwig SR, Bowen B, Beach L, Wessler SR (1990) Regulatory gene as a novel visible marker for maize transformation. Science 247: 449–450.
19. McCabe DE, Swain WF, Marinell BJ, Christou P (1988) Stable transformation of soybean (*Glycine max*) by particle acceleration. Bio/Technology 6: 923–926.
20. Morikawa H, Iida A, Yamada Y (1989) Transient expression of foreign genes in plant cells and tissues obtained by a simple biolistic device (particle gun). Appl Microbiol Biotechnol 31: 320–322.

21. Oard JH, Paige DF, Simmonds JA, Gradziel TM (1990) Transient gene expression in maize rice and wheat cells using an airgun apparatus. Plant Physiol 92: 334–339.
22. Twell D, Klein TM, McCormick S (1991) Transformation of pollen by particle bombardment. Plant Tissue Culture Manual (this volume).
23. Twell D, Klein TM, Fromm ME, McMormick S (1989) Transient expression of chimeric genes delivered into pollen by microprojectile bombardment. Plant Physiol 91: 1270–1274.
24. Zumbrunn G, Schneider M and Rochaix J-D (1989) A simple particle gun for DNA-mediated cell transformation. Technique 1: 204–216.

cases led to the inhibition of pollen germination [28, 2]. In contrast Roeckel *et al.* [23] have developed incubation conditions for maize pollen under which DNA is not degraded and efficient fertilization is achieved. However no molecular evidence was obtained for stable transformation. In a more recent study Booy *et al.* [2] concluded,

> The problems concerning nuclease activity and the neccessity of DNA to be incorporated into the right sperm cell without a process of cell division make it doubtful that pollen-mediated transformation by simply incubating pollen with DNA will become an efficient transformation method in maize.

Alternative pollen-mediated transformation procedures that have been unsuccessful have recently been reviewed [22].

Transformation by particle bombardment

The development of particle bombardment as a means to deliver DNA into plant cells provides an alternative procedure to investigate pollen transformation. This procedure, developed by Klein *et al.* [10], involves the bombardment of plant cells with DNA coated, micron sized metal particles that have been accelerated to high velocity using a gunpowder charge. Using this method transient expression of introduced genes has been demonstrated in several plant species and in various cell types [10–13, 4]. It has been shown that the introduced DNA is capable of stable integration into the chromosomes of some cells, which under suitable selection conditions can be regenerated into stably transformed plants [11]. This transformation procedure has been used successfully to transform soybean [15] (via shoot meristem bombardment) and maize [6] (via bombardment of embryogenic suspension cultures), which have been difficult to transform using established procedures for direct gene transfer, or by *Agrobacterium* based methods.

DNA Delivery into pollen by particle bombardment

All transformation procedures require the use of suitable marker genes driven by functional regulatory sequences which allow the detection of cells containing the introduced DNA. Thus, regulatory sequences that function in pollen are required to investigate the efficacy of particle bombardment for the delivery of exogenous DNA into pollen. Recently, the promoter region of a gene from tomato (LAT52) that is expressed at high levels in maturing pollen of tomato [25] was shown to direct pollen-specific gene expression in transgenic plants [26]. Because the LAT52 promoter functions at high level in a pollen-specific manner in transgenic tomato, tobacco and *Arabidopsis* [27] it should be useful in developing pollen-mediated transformation procedures in different plant species. Similarly regulated genes have also been isolated from maize [7], *Oenothera* [3], and petunia [29], the promoters of which should provide

Plant Tissue Culture Manual **D2**: 1–14, 1991.
© 1991 *Kluwer Academic Publishers.*

Transformation of pollen by particle bombardment

DAVID TWELL[a], THEODORE M. KLEIN[b] and SHEILA McCORMICK
Plant Gene Expression Center, USDA-ARS/UC Berkeley, 800 Buchanan Street, Albany, CA 94710, USA
[a] *current address: Leicester Biocentre, University of Leicester, Leicester LE1 7RH*
[b] *current address: E. I. DuPont & Co. Medical Products Department, Glasgow Site, Box 122, Newark, DE 19714-6101, USA*

Introduction

The development of pollen as a vector for direct gene transfer would be a significant advance in our ability to introduce genes into plants. Such methodology should be of general utility for many plant species, and in particular for the major monocotyledonous crop plants such as maize, wheat and barley that are recalcitrant to protoplast regeneration and that are not amenable to *Agrobacterium* based transformation techniques. A further advantage would be the avoidance of tissue culture steps that are time consuming and known to result in undesirable somaclonal variation. The potential of pollen as a vector for direct gene transfer has long been realized. For more than 10 years numerous investigators have attempted pollen-mediated transformation, several of which have claimed success [5, 8, 19, 23]. However the ultimate proof that transformation has taken place, that is, the demonstration of integration of foreign DNA into the nuclear genome at the molecular level and the genetic transmission of this DNA, is still lacking. This chapter presents a summary of research that has been directed towards pollen-mediated gene transfer, a detailed protocol for the delivery of DNA into pollen using particle bombardment and a discussion of factors that may be important for the successful application of this technique to obtain stably transformed plants.

Direct DNA uptake studies in pollen

Most attempted pollen transformation procedures have involved the incubation in solution of mature or pregerminated pollen with plasmid or genomic DNA containing selectable or genetic markers. Subsequently this pollen has been used to fertilize acceptor plants via the normal pollination pathway, or by *in vitro* pollination. A major problem encountered with this approach has been the nuclease activity associated with germinating pollen, which has been shown to completely degrade exogenous DNA within 5 to 10 minutes. This phenomenon has been observed in tobacco, petunia and maize [14, 28, 17, 2,]. Procedures that have been used to inhibit nuclease activity of pollen have in most additional tools for the investigation of pollen transformation in other plant species.

Initial experiments to demonstrate DNA delivery into tobacco pollen by particle bombardment utilized the Cauliflower Mosaic Virus 35S (CaMV35S) promoter fused to the *E. coli* β-glucuronidase (GUS) gene. The GUS reporter gene was chosen as a sensitive reporter of transient expression (DNA delivery) because expression of GUS enzyme activity in single cells can be visualized using the indigogenic substrate X-Glu [9]. Following bombardment with the CaMV35S-GUS plasmid, pollen did not show blue staining after incubation with X-Glu. However, low levels of GUS activity were detected when extracts of the bombarded pollen were assayed with the sensitive fluorimetric GUS assay [9]. These results suggested that DNA could be delivered into pollen by particle bombardment. When tobacco pollen was bombarded with a similar gene fusion (LAT52-GUS) driven by the pollen-specific promoter LAT52, numerous blue staining pollen grains were detected with X-Glu, and high levels of GUS activity were detected in extracts of bombarded pollen [24]. These data confirm the utility of particle bombardment for the delivery of DNA directly into tobacco (cultivar Samsun NN) pollen. The same procedure has been used successfully to introduce and express the LAT52-GUS construct in tomato (cultivar VF36) pollen [24].

Considerations in developing pollen-mediated transformation

DNA delivery into pollen has been demonstrated using pollen-specific regulatory sequences linked to the reporter gene GUS [24]. Such constructs are useful tools to investigate key events that are neccessary for the successful application of particle bombardment to pollen transformation; these constructs maybe used to: 1) optimise DNA delivery into pollen, 2) monitor the *in vitro* function and 3) monitor the *in vivo* function of 'transformed' pollen containing the introduced DNA. The experimental application of this technology to investigate pollen transformation is discussed below:

1. DNA delivery into pollen
The efficiency of DNA delivery into tobacco pollen was assessed by X-Glu treatment of pollen bombarded with the LAT52-GUS gene fusion [24]. Approximately 0.1% of the bombarded pollen grains stained blue. This provided a minimum estimate of the number of pollen grains that received plasmid DNA. Given the relatively low frequency of DNA integration observed in direct gene transfer experiments [18, 11, 4, 15], it is desirable to increase the observed frequency of DNA delivery into pollen to increase the probability of obtaining stable integration events. The use of higher amounts of DNA per bombardment leads to a linear increase in transient gene expression between 2–15 μg DNA, with maximal expression being achieved at 20–25 μg DNA (Twell, Ushiba unpublished results). This may be due to an increase in the amount of DNA delivered per pollen grain and/or an increase in the number of pollen grains per

plate that receive DNA. Other parameters that may be changed to optimise DNA delivery into pollen include: the use of more pollen per bombardment, which also leads to increases in transient expression (Twell, Ushiba unpublished results), the use of particle preparations that vary in size or material (i.e. gold particles) or varying the scattering of particles after leaving the stopping plate; this may be achieved by using stopping plates drilled with holes of different diameter or varying the distance between the sample and the stopping plate (also see ref. 12).

For optimization of DNA delivery into pollen of other species it may be necessary to use regulatory sequences known to function in the species of interest. For example, when mature pollen grains of maize or lily were bombarded with the LAT52-GUS gene fusion, no X-Glu staining pollen or extractable GUS activity was detected (Twell, Klein unpublished results). Furthermore, such regulatory sequences must be active in mature hydrated pollen or in the early stages of pollen germination to detect DNA delivery. This has been demonstrated for the tomato LAT52 promoter, which is transiently expressed within 30 minutes of bombardment into tobacco pollen [24]. In this regard a pollen-specific gene with homology to the tomato LAT52 gene has been isolated from maize [7]. Other factors such as the viability of pollen under bombardment conditions will affect the ability to detect 'transformed' pollen grains using transient expression.

2. In vitro *function of 'transformed' pollen*

Once efficient DNA delivery into pollen has been established using an appropriate transiently expressed promoter-GUS fusion, it is valuable to assess the viability of the 'transformed' pollen grains. Following bombardment of tobacco pollen with the LAT52-GUS fusion and incubation overnight, 'transformed' pollen grains were identified by incubation with X-Glu. Under the conditions used (which were not optimal for pollen germination) many of the blue staining pollen grains had germinated and extended long pollen tubes [24]. This suggested that pollen grains that received DNA were not impaired in their ability to germinate *in vitro*. This procedure could be improved by recovery of pollen grains from filters after bombardment and incubation in optimised germination media to assess the viability of transformed pollen. Efficient *in vitro* pollen germination procedures are available for many species, which in conjunction with the GUS marker may be used as rapid viability tests of transformed pollen following bombardment.

3. In vivo *function of 'transformed' pollen*

Having established the viability of 'transformed' pollen using *in vitro* pollen germination, it is desirable to demonstrate pollen function *in vivo*. This may be achieved by the application of pollen bombarded with a promoter-GUS gene fusion to the stigma of untransformed plants. The ability of 'transformed' pollen grains to germinate and extend a pollen tube into the style could be rapidly assessed by preparing whole mounts or hand sections of the pollinated pistils, followed by treatment with X-Glu to identify blue staining pollen tubes of the 'transformed' pollen. This technique obviously depends upon the presence of GUS enzyme activity in the pollen tube, which has been demonstrated for the LAT52-GUS fusion expressed in tobacco pollen [24].

Since it has been demonstrated that bombarded pollen transiently expressing the LAT52-GUS construct is capable of germination [24], and that bombarded tobacco pollen recovered from plates is capable of fertilization and seed set (Twell, Klein, unpublished results) it is reasonable to expect stable integration events in a fraction of the bombarded pollen. Approximately 0.1% of 10^6 tobacco pollen grains received plasmid DNA after bombardment with the LAT52-GUS gene fusion [24]. Based upon rates of stable transformation in protoplasts [18] and in the particle gun transformation experiments [11], it is estimated that a maximum of 5% [11] of the transiently expressing pollen grains would become stably transformed. Thus, 1 in 20,000 bombarded pollen grains could potentially give rise to a stable integration event. Following fertilization with bombarded pollen these rare events could be recovered by simple selection at the seedling stage using an antibiotic resistance marker. Under these assumptions 50 transformants could potentially be recovered from a single bombardment of 10^6 pollen grains. Preliminary experiments in which tobacco pollen was bombarded with a gene construct conferring kanamycin resistance and used in fertilizations, indicated that the estimated frequency may actually be lower; no kanamycin resistant individuals were recovered from 3×10^4 seeds derived from the bombarded pollen (Twell, Klein, unpublished results). Thus, it is apparent that optimisation of several parameters is essential before pollen-mediated transformation can be acheived.

As discussed above, optimization of DNA delivery and careful evaluation of the effectiveness of 'transformed' pollen in fertilization can be investigated using an appropriate promoter-GUS gene fusion. A further consideration in pollen transformation studies is the requirement for DNA delivery and integration into the chromosomes of the (correct) sperm cell nucleus that leads to zygote formation. The demonstration of chloroplast transformation using particle bombardment [1] shows that DNA can be incorporated into different subcellular compartments. Thus, it should be possible to deliver DNA into the generative or sperm cells of mature pollen, which are completely enclosed within the cytoplasm of the vegetative cell. Unfortunately genes that function specifically in the generative or sperm cells have not yet been identified, so transient assays which demonstrate this point are not available. However, ultimate proof of DNA delivery into the correct sperm nucleus requires the genetic transmission of DNA into the progeny derived from pollen transformation.

Protocol for the introduction and assay of gene expression in pollen by particle bombardment

The following protocol was developed as a transient expression system to assay cis regulatory elements required for gene expression in tobacco (cultivar Samsun NN) pollen [24]. The same protocol has also been shown to work in tomato (cultivar VF36). The promoters used to drive high level expression of reporter genes in pollen were derived from the tomato genes LAT52 and LAT59, which are preferentially expressed in pollen [24−27, 30]. These promoters are activated after microspore mitosis and show a dramatic increase in activity until pollen maturation [26].

Steps in the procedure

Note: unless otherwise stated all steps are carried out at room temperature.

Pollen collection
1. Grow and maintain at least 25 mature tobacco plants under standard greenhouse conditions for pollen supplies. From 50 mature plants a daily supply of 0.25 g pollen can be harvested.
2. Pollen is most easily collected by the use of a small vacuum cleaner (ie. for an automobile) fitted with an extension tube that fits snugly into the flower over the dehiscent anthers. A nylon filter (Nitex) with a pore size of 20 μm is secured at the base of the vacuum tube upon which the pollen collects.
3. Following collection pollen is scraped from the filter, weighed and stored in microfuge tubes at −80 °C. Tobacco pollen may be stored for at least 3 months without significant loss of activity in transient expression assays or germination tests.

Preparation of DNA coated tungsten microprojectiles
4. Prepare a tungsten suspension by adding particles (M10: average diameter 1.0 μm) to absolute ethanol (50 mg per ml of ethanol). This suspension should be stored at −20 °C.
5. Wash particles by briefly spinning a 0.5 ml aliquot of the tungsten suspension in a microfuge. Draw off supernatant and resuspend particles in 0.5 ml sterile distilled water.
6. Repeat the washing step two times.
7. Distribute 25 μl aliquots of the particles into microfuge tubes.
8. Add 10 to 15 μg of plasmid DNA to each tube.
9. Add 25 μl of a 1.0 M solution of calcium chloride. The suspension should be vortexed while adding $CaCl_2$.
10. Add 10 μl of a 0.1 M solution of spermidine (free base). The suspension should be vortexed while adding spermidine.
11. Allow the suspension to stand for 10 minutes.

Preparation of pollen for bombardment

12. Prepare a 9 cm diameter Petri-plate containing sterile MS medium solidified with 0.8% bactoagar.
13. Place a circle of Whatman # 1 filter paper onto the agar surface and allow to wet evenly, avoiding trapped air bubbles.
14. Overlay the Whatman paper with a 4 cm^2 circle of nylon membrane and allow to wet evenly.
15. Prepare a suspension of tobacco pollen by adding dry pollen to liquid MS medium at a concentration of 40 mg per ml (ca. 4×10^6 pollen grains per ml).
16. Make sure the pollen is suspended evenly by several 5 second bursts of vortexing and inversions of the tube.
17. Immediately spread the pollen suspension onto the nylon membrane on the surface of a petri-plate prepared as described in steps 12–14. Make sure the pollen suspension is spread evenly by dispensing in a circular motion with a 1 ml micropipette.

Pollen bombardment

18. Remove 25 µl of supernatant from the DNA-tungsten suspension which has been allowed to settle for 10 minutes.
19. Place approximately 2 µl of the tungsten-DNA preparation on the front surface of the macroprojectile. Just before placing particles onto the macroprojectile, disperse the particles by briefly (1 sec) touching the outside of the microfuge tube to the probe (horn type; 1 cm diameter at the tip) of a sonicator.
20. Place the macroprojectile into the barrel of the particle gun, with the DNA side facing down. Push the macroprojectile down as far as it will go with the 'pushing tool'.
21. Place a blank charge (Remington # 1) into the barrel behind the macroprojectile.
22. Place the firing assembly over the barrel.
23. Place the stopping plate in position.
24. Place the sample into the firing chamber and close the door.
25. Turn on vacuum pump, fire when vacuum has reached 275 mm of mercury.
26. Release vacuum slowly, open door, remove sample.
27. Remove spent blank charge and force a small wad of cotton wool, moistened with 70% ethanol through the barrel to reduce the accumulation of deposits.

Transient expression assays

28. Following bombardment the plates are incubated at 26 °C under fluorescent light.
29. Harvest pollen for transient expression studies 6–12 h after bombardment by scraping pollen from the nylon filter and washing off with 0.5 ml cold extraction buffer (0.1 M potassium phosphate (pH 7.5), 1 mM dithiothreitol) into a precooled mortar.
30. Extract pollen by grinding with a pestle for 1 minute.
31. Spin in a microfuge at 4 °C for 5 minutes, remove the supernatant and assay immediately or store at −80 °C.

Visualization of GUS activity in pollen grains

To visualize transiently expressing pollen grains, transfer the nylon filter with the pollen in place to a petri plate containing a filter paper saturated with a solution containing 1 mM X-Glu, 0.1 M sodium phosphate (pH 7.0). Seal the plate with parafilm and incubate at room temperature. When tobacco pollen was bombarded with the LAT52-GUS fusion, incubated for 16 h, and subsequently treated with X-Glu, blue staining pollen grains were first observed within 15 minutes (Twell, unpublished results). It should be noted that pollen of some species such as tomato and tobacco contains low levels of endogenous GUS activity which leads to blue staining in the presence of X-Glu after extended incubation [21].

Fertilizations with bombarded pollen

Following bombardment tobacco pollen may be applied directly to the stigma of emasculated flowers using a small spatula or toothpick. For controlled pollinations, pollen may be recovered from the plate used for bombardment with a spatula, resuspended in liquid pollen germination medium (10% sucrose, 0.03% $CaCl_2$, 0.01% H_3BO_3, pH 6.4) counted using a haemocytometer and 5 µl of this solution (containing a known amount of pollen) applied to the stigma surface. Both of these procedures led to efficient seed set (Twell and Klein, unpublished results).

Notes

1. M10 tungsten microprojectiles are available from E. I. Dupont & Co., Delaware, USA. The particle delivery system used in these studies was a DuPont Biolistic™ PDS-1000.

8. Plasmid DNA should be purified by cesium chloride density gradient centrifugation and adjusted to a concentration of 1–2 mg per ml in TE. For higher amounts of DNA, the volumes of added solutions in the precipitation reaction should be scaled up proportionally. For quantitative transient expression assays a test plasmid (i.e. 7 μg promoter-GUS fusion) and a reference plasmid (i.e. 2 μg promoter-luciferase [20] fusion) are coprecipitated onto the microprojectiles and cobombarded into pollen.

10. It is important to use only the free base of spermidine, as the hydrochloride is not effective in this DNA precipitation.

11. DNA coated microprojectiles should be used for bombardment as soon as possible after preparation because of DNA degradation by tungsten. The efficacy of these preparations in transient expression is significantly diminished after overnight storage at $-20\,^\circ$C.

12. Agar solidified MS medium is prepared with Murashige and Skoog salts [16] containing 3% sucrose and B5 vitamins (pH 5.8) and sterilized by autoclaving. Agarose may improve pollen viability since agar is inhibitory to the growth of some cell types

14. The nylon membranes Genescreen (New England Nuclear) and Nytran (Schleicher and Scheull) allow easy recovery of pollen following bombardment.

17. Bombardment should be carried out as soon as possible (within one hour) after dispensing pollen onto the plates.

18. This step increases the effective tungsten-DNA concentration after resuspension by sonication.

19. The front surface of the macroprojectile is flat, while the back surface has an indentation. It is very important to pipette out the tungsten-DNA suspension immediately after sonication; any delay leads to clumping of the particles which subsequently cannot be pipetted easily.

29. The extraction buffer is suitable for extraction of both *E. coli* GUS [9] and firefly luciferase [20] enzyme activity.

References

1. Blowers AD, Bogorad L, Shark KB, Sanford JC (1989) Studies on *Chlamydomonas* chloroplast transformation: foreign DNA can be stably maintained in the chromosome. Plant Cell 1: 123–132.
2. Booy G, Krens FA, Huizing, HJ (1989) Attempted pollen-mediated transformation of maize. J. Plant Physiol. 135: 319–324
3. Brown SM, Crouch ML (1990) Characterization of a gene family abundantly expressed in *Oenothera organensis* pollen that shows sequence similarity to polygalacturonase. Plant Cell 2: 263–274.
4. Christou P, McCabe DE, Swain WF (1988) Stable transformation of soybean callus by DNA-coated gold particles. Plant Physiol. 87: 671–674.
5. De Wet JMJ, Bergquist JRR, Harlan JR, Brink DE, Cohen CE, Newell CA, De Wet AE (1985) Exogenous gene transfer in maize (*Zea mays*) using DNA-treated pollen. In: Chapman GP, Mantell SH, Daniels RW (Eds), Experimental manipulation of ovule tissues. Longman Inc., New York. pp. 197–209.
6. Fromm ME, Morrish F, Armstrong C, Williams R, Thomas J, Klein TM (1990) Inheritance and expression of chimeric genes in the progeny of transgenic maize plants. Bio/Technology (in press).
7. Hamilton DA, Bashe DM, Stinson JR, Mascarenhas JP (1989) Characterization of a pollen-specific genomic clone from maize. Sex Plant Reprod 2: 208-212.
8. Hess D (1987) Pollen based techniques in genetic manipulation. Int Rev Cytol. 107: 367–395.
9. Jefferson RA, Kavanagh TA, Bevan MW (1987) β-Glucuronidase as a sensitive and versatile gene fusion marker in higher plants. EMBO J 6: 3901–3907.
10. Klein TM, Wolf ED, Wu R, Sanford JC (1987) High velocity microprojectiles for delivering nucleic acids into living cells. Nature 327: 70–73.
11. Klein TM, Harper EC, Svab Z, Sanford JC, Fromm ME (1988) Stable genetic transformation of intact *Nicotiana* cells by the particle bombardment process. Proc Natl Acad Sci USA 85: 8502–8505.
12. Klein TM, Gradziel T, Fromm ME, Sanford JC (1988) Factors influencing gene delivery into *Zea mays* cells by high-velocity microprojectiles. Bio/technology 6: 559–566.
13. Klein TM, Roth BA, Fromm ME (1989) Regulation of anthocyanin biosynthetic genes introduced into intact maize tissues by microprojectiles. Proc Natl Acad Sci USA. 86: 6681–6685.
14. Matousek J, Tupy J (1983) The release of nucleases from tobacco pollen. Plant Sci Lett 30: 83–89.
15. McCabe DE, Swain WF, Marinell BJ, Christou P (1988) Stable transformation of soybean (*Glycine max*) by particle acceleration. Bio/Technology 6: 923–926.
16. Murashige T, Skoog F (1962) A revised medium for rapid growth and bioassays with tobacco tissue cultures. Physiol Plant 15: 473–497.
17. Negrutiu I, Heberle-Bors E, Potrykus I (1986) Attempts to transform for kanamycin resistance in mature pollen of tobacco. In: Mulcahy DL, Bergamini G, Ottaviano E (Eds). Biotechnology and ecology of pollen. Springer, New York. pp. 65–70.
18. Negrutiu I, Mouras A, Horth M, Jacobs M (1987) Direct gene transfer to plants: Present developments and some future perspectives. Plant Physiol Biochem 25: 493–503.
19. Ohta Y (1986) High efficiency transformation of maize by a mixture of pollen and exogenous DNA. Proc Natl Acad Sci USA 83: 715–719.
20. Ow DW, Wood KV, DeLuca M, DeWet JR, Helinski DR, Howell SH (1986) Transient and stable expression of the firefly luciferase gene in plant cells and transgenic plants. Science 234: 856–859.
21. Plegt L, Bino RJ (1989) β-Glucuronidase activity during development of the male gametophyte from transgenic and non-transgenic plants. Mol Gen Genet 216: 321–327.
22. Potrykus I (1990) Gene transfer to cereals: an assessment. Bio/Technology 8: 535–542.

23. Roeckel P, Heizman P, Dubois M, Dumas C (1988) Attempts to transform *Zea mays* via pollen grains. Effect of pollen and stigma nuclease activities. Sex Plant Reprod 1: 156–163.
24. Twell D, Klein TM, Fromm ME, McCormick S (1989) Transient expression of chimeric genes delivered into pollen by microprojectile bombardment. Plant Physiol 91: 1270–1274.
25. Twell D, Wing R, Yamaguchi J, McCormick S (1989) Isolation and expression of an anther-specific gene from tomato. Mol Gen Genet 217: 240–245.
26. Twell D, Yamaguchi J, McCormick S (1990) Pollen-specific expression in transgenic plants: coordinate regulation of two different tomato gene promoters during microsporogenesis. Development 109: 705–713.
27. Ursin VM, Yamaguchi J, McCormick S (1989) Gametophytic and sporophytic expression of anther-specific genes in developing tomato anthers. Plant Cell 1: 727–736.
28. Van der Westhuizen AJ, Gliemeroth AK, Wenzel W, Hess D (1987) Isolation and partial characterization of an extracellular nuclease from pollen of *Petunia hybrida*. J Plant Physiol 131: 373–384.
29. Van Tunen AJ, Mur LA, Brouns GS, Rienstra J-D, Koes RE, Mol JNM (1990) Pollen- and anther-specific *chi* promoters from petunia: tandem promoter regulation of the *chi*A gene. Plant Cell 2: 393–401.
30. Wing RA, Yamaguchi J, Larabell SK, Ursin VM, McCormick S (1989) Molecular and genetic characterization of two pollen-expressed genes that have sequence similarity to pectate lyases of the plant pathogen *Erwinia*. Plant Mol Biol 14: 17–28.

Plant Tissue Culture Manual **D3**: 1–11, 1991.
© 1991 *Kluwer Academic Publishers.*

Electrical fusion of protoplasts

MICHAEL G. K. JONES
*Plant Sciences, School of Biological & Environmental Sciences, Murdoch University, Murdoch, Perth,
W. Australia 6150*

Introduction

There are many potentially useful applications of protoplast fusion, in both
somatic cell genetics and in biotechnology [eg 1,2]. The two main approaches
to fusing plant protoplasts are based either on chemical or electrical procedures. Chemical fusion of protoplasts will be described elsewhere in this volume
[3]. Given the availability of suitable equipment, the advantage of electrofusion,
a physical approach, over chemical fusion is that it is rapid, simple, synchronous and more easily controlled, and the use of high concentrations of
potentially cytotoxic chemical fusogens are avoided [4]. In this chapter electrical approaches to fusion are described.

The first requirement in any experimental procedure that involves protoplasts is to be able to isolate protoplasts of good 'quality', that are viable and
free from contaminating debris. This is not always a trivial task, especially when
working with new species or genotypes not chosen for their response in culture.
It may require preliminary work to establish the correct physiological conditions for growing the source plants (or cultures), and the best combination of
wall degrading enzymes, culture media, osmotic conditions etc. In general,
starting with good quality protoplasts will lead to more consistent and reproducible results [5].

Background

When protoplasts that are in contact are subjected to a suitable external electric
field, which causes the formation of transient reversible pores in the plasma
membranes, they can fuse together [6, 7]. The plasma membrane normally acts
as an insulator, and has a high electrical resistance. When the potential difference across the membrane is increased, a point is reached when the membrane insulation breaks down (the membrane breakdown voltage), and a pore
is formed.

Bringing protoplasts together

Protoplasts normally have a negative surface charge, that helps repel neighbouring protoplasts. However, they must be brought into close contact before they can be fused. This can be achieved by both physical and chemical means. Physical means include mechanically pushing protoplasts together with micropipettes [6], gravity sedimentation to form layers of protoplasts on the base of a dish [8], or the more controllable process of dielectrophoresis [9, 10, 11, 12]. In dielectrophoresis, protoplasts in a medium of low conductivity (eg a mannitol solution of appropriate osmolarity) are placed between two electrodes, and a high frequency AC field (0.5–1.5 MHz) is applied (Fig. 1). The surface charges on the protoplasts become polarized, they act as dipoles and migrate along the electric field lines to regions of higher field intensity. As the protoplasts migrate, they contact other protoplasts and form chains parallel to the applied field lines (Fig. 2). There are a number of advantages in having linear arrays of contacting protoplasts; these include ease of observation and quantification of fusion, and the possibility of directing fusion between protoplasts with different electrofusion characteristics [12, 13]. Chemical aggregation of protoplasts can be achieved by allowing protoplasts to sediment out in the presence of a low concentration of PEG (2.5%, [14]) before application of the fusion pulse.

When dielectrophoresis is used to bring protoplasts into chains, a low conductivity medium is required. In contrast, protoplasts sedimented in layers can be kept in their normal culture medium. In practice, because of the short time taken for the whole procedure of dielectrophoresis and pulsing (< 60 sec), suspension of protoplasts for a short period in a low conductivity mannitol solution is not a problem, and after the procedure the protoplasts are simply diluted to the required density in culture medium.

Electrofusion pulse parameters

Both rectangular and capacitor discharge pulses can be used to induce pore formation and fusion. The transmembrane potential induced by an applied electrical field is given by the following equation:

$$V = 3/2rE \cos \theta$$

where r is the radius of the protoplast, E is the amplitude of the applied electric field, and θ is the angle with which a point on the membrane surface intersects the field [15]. As the applied field strength is increased, membrane breakdown occurs at the critical voltage V_{cr}. The value of V_{cr} is in the order of 0.5 to 1.5 V, and varies with membrane composition, cell type and origin. It appears that the mechanism of fusion does not involve complementary pores in each contacting protoplast, because pore formation, which occurs initially on the cell equators, is asymmetric. Pores first form opposite the anode [4]; in addition, poration can be induced before cells are brought together, and cells can still fuse.

The magnitude of the applied pulse required for fusion is related to protoplast size: lower voltage pulses are required as the radius increases (Fig. 3). The pulse duration is also important, and it is well established that shorter duration pulses require higher applied fields (Fig. 4). This may be because the critical break-down voltage varies with pulse duration, or that for shorter pulses there is not time for the protoplasts to charge up completely, so that a higher field strength must be applied to reach the critical voltage [16, 23]. For electrofusion, the viability and plating efficiency of protoplasts tends to be higher when shorter pulses at higher voltages are used, rather than longer pulses at lower voltages [11, 23].

Equipment

A number of commercially produced electrofusion machines are available, which provide controlled AC fields for dielectrophoretic cell alignment and rectangular fusion pulses of defined voltage and duration. These include the GCA Zimmermann Cell Fusion System and the Kruss GmbH Cell Fusion Apparatus. Alternatively various workshop built equipment has been used [11, 12, 17]. Similarly, simple capacitor discharge equipment can be made [e.g. 17]. In earlier work particularly with animal cells, wire electrodes stretched across microscope slides, with a separation of only 100 or 200 µm, were used. These are impractical for work with plant protoplasts. For analytical work, parallel wires 1 mm apart stretched across a slide are suitable [11], and for large scale fusions parallel stainless steel electrodes 1 mm apart are more commonly used [12, 13, 17]. A series of such parallel electrodes, 'lamellar electrodes' (Kruss GmbH) allows up to 1×10^6 protoplasts to be treated at once.

The methods described are mainly based on experience in fusing protoplasts of *Solanum* species, but are equally applicable to a range of other species (eg sugar beet [18], brassicas [11], and cereals [11]). Using the first two proce-dures described has led to the production of a series of somatic hybrid plants between potato (*Solanum tuberosum*) and the sexually incompatible South American wild species *Solanum brevidens* [19, 20, 21, 22].

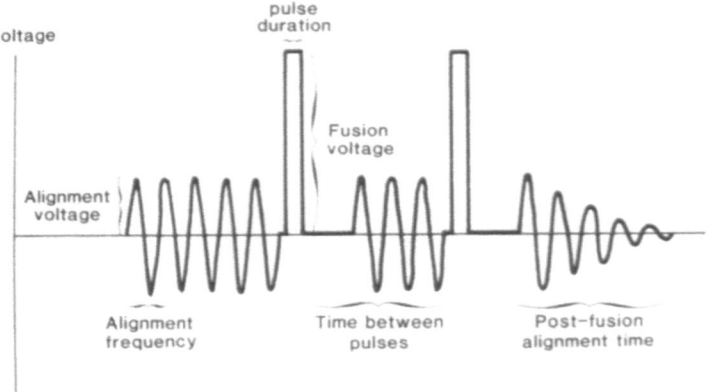

Fig. 1. Characteristics of the electric field applied to protoplasts for electrofusion with dielectro-phoretic alignment of protoplasts into chains. The high frequency alignment field is switched off when one or more fusion pulses are applied, then reapplied briefly to keep protoplasts in contact, but ramped down to zero during the post fusion period (modified from Zimmermann Cell Fusion Manual).

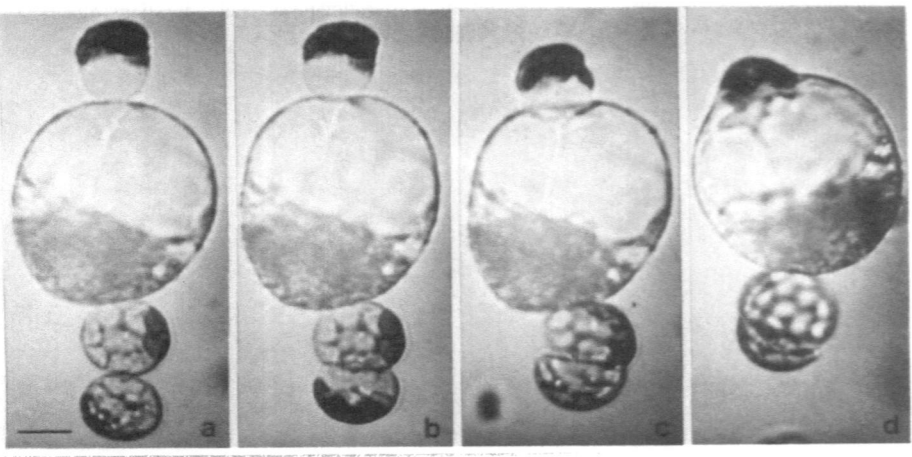

Fig. 2. Micrographs of protoplasts (a-c) dielectrophoretically aligned, 1-5min after fusion pulse, (d) 15min after pulse. The large protoplast is from *Brassica napus* hypocotyl; the others are *S. brevidens* mesophyll protoplasts, fusion in mannitol alone [11].

PTCM-D3/5

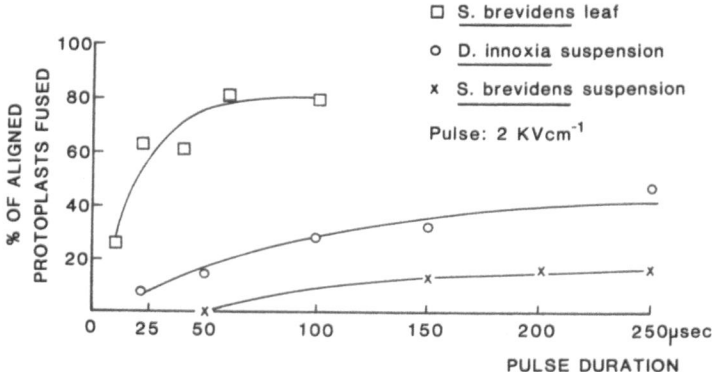

Fig. 3. Pulse duration-fusion response curves for *S. brevidens* mesophyll and suspension proto-plasts (23 μm diameter), and for *Datura innoxia* suspension protoplasts (44 μm diameter).

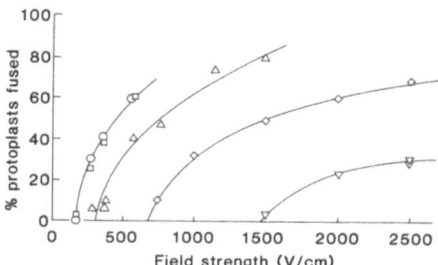

Fig. 4. The effect of pulse duration on electrofusion. This shows that the minimum pulse voltage that must be applied for fusion increases with shorter duration pulses. Pulse durations: 50 μs (▽), 100 μs (◇), 200 μs (△), 500 μs (○) and 1000 μs (□); [24].

Procedures

The procedure given below is for electrofusion following dielectophoretic alignment of protoplasts.

Pulse duration-fusion response curves (analytical electrofusion)

Optimum conditions for electrofusion can be established at an analytical level by constructing pulse duration-fusion response curves, in which fusion frequencies are measured for a number of different pulse voltages and durations.

Steps in the procedure
1. Prepare protoplasts and resuspend in fusion solution (mannitol at appropriate osmolarity, e.g. 0.5 M, with or without 1 mM $CaCl_2$).
2. Carefully pipette protoplasts at a suitable density between analytical electrodes 1 mm apart on a microscope slide placed on a microscope stage, so that the process can be monitored continuously. About 500 protoplasts should be present; gently cover with a coverslip.
3. Connect the electrodes to a Zimmermann or Kruss electrofusion apparatus.
4. Apply the dielectrophoretic AC field at 1 MHz and an amplitude of 15–30 V (if the protoplasts rotate alter the AC frequency until they do not). The protoplasts should align rapidly. Apply the fusion pulse when the average chain length is 5–10 protoplasts.
5. The fusion pulse voltage should be kept constant (eg 1500 V/cm), and its duration increased (eg starting with 5 µs, 10, 20, 25, 50 µs) in subsequent fusions.
6. To hold the chains in place, reduce the AC amplitude to 30% of the original value. Over the next few minutes the fusion characteristics can be recorded. Only protoplasts aligned in chains are scored. Count (1) total number of protoplasts in a chain, (2) number of unfused protoplasts, (3) number of binary (one to one) fusions, and fusions involving three, four, five or more protoplasts. From these numbers the following parameters can be worked out:
   – percentage of aligned protoplasts
   – percentage of aligned protoplasts fused
   – percentage of one-to-one fusions
   – percentage of total protoplasts involved in fusion events
7. For a given pulse voltage, plot percentage of total protoplasts involved in fusion events ('percentage fused') against pulse duration (microseconds). This can be repeated for different pulse voltages, giving curves as shown in Fig. 3.
8. The optimum fusion parameters can then be chosen; these are usually the values for the shortest pulses that give the highest fusion frequency.

Notes
4. If the protoplasts do not align rapidly, then the salt concentration may be too high. The protoplasts should be washed and resuspended in fresh electrofusion medium.

As a general rule, longer chains of protoplasts result in higher fusion frequencies and more multiple fusions.

8. Pulses of longer duration and higher voltage will give a higher overall fusion frequency, but there will be a greater proportion of multiple fusions. With excessive pulse voltages or durations, the membrane breakdown becomes irreversible, and the protoplasts die. The presence of 1mM $CaCl_2$ in the fusion medium increases the fusion frequency of suspension culture protoplasts, which are often less fusogenic than leaf protoplasts.

Large scale fusions

Using optimum conditions established above, fusions can be scaled up with only a slight reduction in fusion frequency (as a result of convection currents in the fusion chambers). The method involves parallel lamellar electrodes 1 mm apart (Kruss), or equivalent workshop made electrodes.

Steps in the procedure

1. Transfer 0.2–0.5 ml of protoplasts (density 5×10^5 to 2×10^6 per ml) in electrofusion solution to the fusion chamber, and insert the electrodes.
2. Apply the alignment and fusion pulse, using the optimum values obtained from small scale analytical fusions (e.g. pulse duration 10 μsec, voltage 1500 V/cm, optimum for dihaploid potato leaf protoplasts).
3. Dilute the protoplasts with culture medium to the correct density.

Electrofusion without dielectrophoretic collection

This fusion method does not require equipment to generate the high frequency collecting field. It is therefore simpler, but more difficult to quantify and control, and includes a low concentration of PEG to aid protoplast contact [14].

Steps in the procedure
1. Suspend protoplasts in culture medium with 2.5% w/v PEG. Transfer 0.2 ml to a Petri dish or pulsing chamber, and leave for 10 min to allow protoplasts to settle.
2. Bring electrodes into contact with the base of the dish with the protoplasts between the electrodes.
3. Apply the fusion pulse (rectangular pulse with parameters as above, or discharge from a suitable capacitor [17]), and leave for a few minutes at room temperature.
4. Adjust to required density with fresh culture medium. The fusion frequency and number of multiple fusion products is similarly influenced by protoplast density, pulse voltage and duration.

Culture and selection

The suspension of protoplasts obtained after electrofusion will consist of a mixture of unfused parental types, homokaryons (products of self fusion) and heterokaryons (products of fusion between the parental types). The fusion products will also be derived from two or more protoplasts. Heterokaryons may be selected at an early stage by a variety of approaches, including fluorescence activated cell sorting (FACS). Electrofusion is ideal for FACS because protoplasts do not stick to the dish, which often occurs with PEG fusions. However, because of the high fusion frequencies which can be obtained by electrofusion (eg up to 80% of the protoplasts involved in fusion events, and 50% one-to-one fusions at an overall fusion frequency of 60%, half of which will be heterokaryons), it may not be necessary to select heterokaryons after electrofusion, and after suitable culture, hybrid shoots can be identified visually, by isoenzyme analysis or by molecular techniques [19,21]. For example, 12.6% of shoots were hybrid following electrofusion and unselected culture of potato and *S. brevidens*, compared with 2.6% after PEG fusion.

References

1. Smith MA, Pay A, Dudits D (1989) Analysis of chloroplast and mitochondrial DNAs in asymmetric somatic hybrids between tobacco and carrot. Theor Appl Genet 77: 641–644.
2. Lindsey K, Jones MGK (1989) Plant Biotechnology in Agriculture, Open University Press, pp241.
3. Davey M (1991) Chemical fusion of protoplasts (this volume, in preparation).
4. Bates GW, Saunders JA, Sowers AE (1989) Electrofusion: principles and application. In Electroporation and Electrofusion in Cell Biology, eds Neumann E, Sowers AE and Jordan CA, Plenum, NY, pp387–395.
5. Jones MGK (1985) Developments in the culture of plant protoplasts and cells and their regeneration to plants. Symp Biotechnol and Crop Improvement and Protection, British Crop Protection Council Monograph no 34, pp 3–11.
6. Senda M, Takeda J, Abe S, Nakamura T (1979) Induction of cell fusion of plant protoplasts by electrical stimulation. Plant Cell Physiol 20: 1441–1443.
7. Zimmermann U, Scheurich (1981) High frequency fusion of plant protoplasts by electric fields. Planta 151: 26–32.
8. Morikawa H, Sugino K, Hayashi Y, Takeda J, Senda M, Hiai A, Yamada Y (1986) Interspecific plant hybridisation by electrofusion in *Nicotiana*.
9. Zimmermann U (1982) Electric field mediated fusion and related electrical phenomena. Biochem Biophys Acta 694: 227–277.
10. Zimmermann U (1986) Electrical breakdown, electropermeabilization and electrofusion. Rev Physiol Biochem Pharmacol 105: 175–256.
11. Tempelaar MJ, Jones MGK (1985a) Fusion characteristics of plant protoplasts in electric fields. Planta 165: 205–216.
12. Tempelaar MJ, Jones MGK (1985b) Directed electrofusion between protoplasts with different responses in a mass fusion system. Plant Cell Reports 4: 92–95.
13. Tempelaar MJ, Duyst A, de Vlas SY, Krol G, Symonds C, Jones M G K (1987) Modulation and direction of the electrofusion response in plant protoplasts. Plant Science 48: 99–105.
14. Montane M-H, Alibert G, Teissie (1987) Genetic investigations of somatic hybrids between tobacco and albino strains obtained by electrofusion. Studia Biophysica 119: 89–92.
15. Cole KS (1968) Membranes, ions and impulses. A chapter of classical biophysics. University of California Press, Berkeley.
16. Jones MGK (1988) Fusing plant protoplasts. Trends in Biotechnol 6: 153–158.
17. Watts JW and King JM (1984) A simple method for large scale electrofusion and culture of plant protoplasts. Biosci. Rep. 4: 335–342.
18. Eady CC (1989) Electrofusion and membrane labelling of sugar beet protoplasts. Ph D Thesis, University of Sheffield.
19. Fish N, Karp A, Jones MGK (1988) Production of somatic hybrid plants of *Solanum* by electrofusion. Theor. Appl. Genet. 76: 260–266.
20. Fish N, Steele S, Jones MGK (1988) Field characteristics of somatic hybrid plants of *S. tuberosum* and *S. brevidens*. Theor. Appl. Genet. 76: 113–117.
21. Pehu E, Karp A, Moore K, Jones MGK (1989) Molecular, cytological and morphological characterisation of somatic hybrids of dihaploid *Solanum tuberosum* and diploid *S. brevidens*. Theor. Appl. Genet. 78: 696–704.
22. Gibson RW, Jones MGK, Fish N (1988) Resistance to potato leaf roll virus and potato virus Y in somatic hybrids between dihaploid *S. tuberosum* and *S. brevidens*. Theor. Appl. Genet. 76: 113–117.
23. Mehrle W, Naton B, Hampp R (1990) Determination of physical membrane properties of plant cell protoplasts via the electrofusion technique: prediction of optimum fusion yields and protoplast viability. Plant Cell Rep. 8: 687–691.
24. Jones H, Tempelaar MJ, Jones MGK (1987) Recent advances in plant electroporation. In Oxford Surveys of Plant Molecular and Cell Biology, Ed Miflin BJ, OUP 4: 347–357.

Plant Tissue Culture Manual **D4**: 1–17, 1991.
© 1991 *Kluwer Academic Publishers.*

Cybrid production and selection

ESRA GALUN AND DVORA AVIV
Department of Plant Genetics, The Weizmann Institute of Science, Rehovot 76100, Israel

Overview

The term *cybrid cell* will be used below in the sense of a cell harboring a nuclear genome of a given species (the *recipient*) but containing an alien chloroplast genome (plastome) and/or a mitochondrial genome (chondriome) that is partially or totally alien (from the donor fusion-partner). A *cybrid plant* will mean a plant composed of cybrid cells; we shall use the term *cybridization* for the act of producing cybrid cells or cybrid plants. In angiosperm plants chloroplasts probably maintain their organelle-integrity; consequently when chloroplasts containing different plastomes reside in the same cell, they will very rarely exchange chloroplast DNA (ctDNA). On the other hand when mitochondria containing different chondriomes will be harbored in the very same cell they probably fuse and defuse allowing exchanges of their respective mitochondrial DNA (mtDNA). Therefore cybridization will usually have different consequences for the plastome and the chondriome compositions of the derived cybrid plants. Namely, the cybrids will contain only the donor's plastome, will be heteroplastomic (with the endogenous plastome as well as an alien plastome) or will retain only the recipient's plastome. Not so in respect to chondriome composition: the cybrid may contain only the donor's chondriome or only the recipient's chondriome but more commonly the cybrid will contain a novel chondriome derived from mtDNA exchanges between the donor's and the recipient's chondriome [1].

Background and introduction to procedures

Unlike mammal cells, the cells of almost all plant tissues have a rigid cell wall. Therefore, one of the biotechnological virtues of plant protoplasts is their ability to fuse. Protoplast fusion procedures have been through a continuous process of improvements. Presently either of two main methods are employed for protoplast fusion: electric-pulse induced fusion or fusion by first aggregating the protoplasts in the presence of a cationic polymer (e.g. polyethylene glycol-PEG) followed by dilution with a calcium-salt solution. The former method has certain advantages (e.g. providing a high level of heterofusion) but it does require quite elaborate instrumentation. We have tried both methods and routinely use the latter one. Therefore, we shall detail a procedure for PEG-mediated protoplast fusion [2, 3]. There are also two main approaches to utilize protoplast fusion in order to obtain cybrid cells and subsequently cybrid plants. The first approach is to use enucleated protoplasts as organelle *donors* [4] and fuse them with either normal protoplasts or miniprotoplasts. The other approach, termed by us 'donor-recipient protoplast-fusion', is based on the utilization of x- or γ-irradiated protoplasts as organelle donors [1, 5]. The irradiated *donor* protoplasts, in which nuclear division is arrested are then fused with *recipient* protoplasts that were previously treated with an antimetabolite. This latter treatment causes a complete metabolic arrest so that the recipient protoplasts lose their division capability – unless fused with the metabolically active donor protoplasts. Several additional procedures are required to select fusion derivatives, among the colonies, and to characterize the plastome and the chondriome compositions of the putative cybrids. These as well as some of the goals of cybridization were reviewed previously [6, 7]. We shall provide the protocols for cybridization and for the characterization of a specific fusion combination. Modifications required for other cybridizations, in *Nicotiana* and additional genera will be mentioned in the Notes.

Procedures

A. Sources of protoplasts

The protoplasts for our model shall be the following. *Nicotiana tabacum* cv. Xanthi shall be the source of *recipient* protoplasts and a *Nicotiana* Line 92 str^R7 (92 str^R7) shall be the source of organelle *donor* protoplasts. The latter line has an *N. tabacum* nuclear genome but its plastome and its chondriome are of *N. undulata*. This line has a plastome mutation conferring high resistance to streptomycin. These donor and recipient plants can be grown in a growth chamber or a greenhouse but preferably axenic shoot cultures should be used. Such cultures are established by the following procedure.

Steps in the procedure
1. Sterilize Magenta plastic boxes (eg obtainable from Sigma) and add to each box 50 ml of autoclaved Nitsch medium containing 1.0% agar and 2% sucrose; in parallel pipette 20 ml of the same Nitsch medium into each of two 9 cm Petri dishes.
2. Sterilize the seeds of the donor and of the recipient by putting 100—300 seeds into a 3 ml syringe; add 2 ml of a hypochlorite solution (1.5%) and after 15 min expel the hypochlorite; continue the work in a laminar hood and suck sterile water into the syringe. Repeat the water-wash at least three times. Subsequent work should be done under sterile conditions.
3. Plate about 100 seeds in each Petri dish containing Nitsch medium and germinate $(24 \pm 2 \, °C)$ under light (light fluence is not critical, ca 10—20 $\mu mol \cdot m^{-2} \, sec^{-1}$ is satisfactory).
4. When the seedlings reach the stage of the first true leaf cut above the roots and transfer 2 seedlings to each of the Nitsch medium containing Magenta boxes and maintain the boxes at $24 \pm 2 \, °C$ in the light (25 $\mu mol \cdot m^{-2} \, sec^{-1}$).
5. When the shoot cultures develop 6 to 8 unfolded leaves (about 4 weeks after transfer of seedlings), leaves may be harvested for protoplast isolation. Alternatively the shoots may be propagated by dissection and replanting in Magenta boxes containing the same medium.

Notes
1. Other containers can be used: (1) other plastic or glass containers that can be autoclaved; (2) sterile, disposable plastic containers.
2. When a wild *Nicotiana* species is used, as either donor or recipient the respective seeds should be incubated (before hypochlorite sterilization) for 10 min in a germination enhancement solution. This treatment may also facilitate the germination of seeds from other genera.

Media and solutions
— Nitsch medium: can be purchased from 'Sigma'.
— Germination enhancement solution;
  — 50 mg/l gibberellic acid
  — 20 g /l KNO_3

B. Isolation of protoplasts

The procedure described below is efficiently applicable to our model fusion-combination as well as to other cybridizations in *Nicotiana*. For protoplast isolation from other genera textbooks such as [8, 9, 10, 11] and other sources of information should be consulted. The isolation of protoplasts from cultured cell-suspensions, rather than from leaves requires different maceration fluids[a].

Steps in the procedure
1. Filter sterilize (through 0.45 μm filters) the maceration fluid and pipette 10 ml of it into each 9 cm Petri dish.
2. Harvest the shoot-culture leaves (avoid the very small and the senescing leaves) and layer them over a pad of wetted (and sterilized) Miracloth ('Calbiochem') tissue.
3. Perforate the leaves by pricking (with a multi-needle device or any other mean) and transfer the pricked leaves into the Petri dishes that contain the maceration fluid.
4. Incubate overnight (25 ± 2 °C, dark).
5. Use a wide-tip 10 ml pipette to collect the suspension of protoplasts and debris; filter this suspension through a funnel with a 100 μm nylon (or stainless steel) sieve and collect in 50 ml tubes.
6. Centrifuge the tubes at low speed (ca $200 \times g$, 5 min); due to the sucrose in the maceration fluid the protoplasts will form an upper layer in the tubes.
7. Use a Pasteur pipette connected to a vacuum source with a rubber tube; bend the tube to avoid the vacuum and insert the pipette to the bottom of the tube; then gently and slowly suck the suspension retaining the upper protoplast layer.
8. Resuspend the protoplasts with 20 ml CPW-wash solution, centrifuge again at even lower speed [ca $50 \times g$, 5 min]; the protoplasts should now produce a loose pellet.
9. Remove the supernatant and resuspend the protoplast pellet in CPW-wash solution to attain a density of about 10^5/ml.

Notes
a. The maceration fluid for most *Nicotiana* cell-suspension cultures contains 1% 'Driselase' ('Fluka') as the only cell-wall degrading enzyme preparation.
5. From this stage up to after fusion, the protoplasts should be handled with care: avoid pipetting with narrow-opening pipettes and high-speed centrifugations.
9. Counting of protoplast at this step and in subsequent steps can be performed in a hemacytometer having a depth of 0.2 mm (rather than the standard 0.1 mm).

Solutions
CPW for maceration fluid
— 27 mg KH_2PO_4
— 1130 mg $CaCl_2$
— 246 mg $MgSO_4 \cdot 7H_2O$

- 101 mg KNO_3
- dissolve in 1 l of 0.45M sucrose

CPW-wash solution
- as CPW for maceration fluid but with 0.55M mannitol (rather than sucrose)

Maceration fluid
- 0.03% Macerozyme (Yakult Honsha Co., Japan)
- 0.16% Cellulase R10 (Yakult Honsha Co., Japan)
- 0.08% Driselase (Fluka Chemie, Switzerland)
- suspend and mix in CPW for maceration fluid.

C. Prefusion treatment of protoplasts

Steps in procedure
1. Before treatment of the donor and the recipient protoplasts remove small aliquots from each of these two suspensions and keep for plating of 'control untreated protoplasts' (see below).
2. Pipette 3 ml aliquots of donor protoplast suspension into each (of one or more) 5.5 cm plastic Petri dishes and expose the dishes to 10 krad (100 Gy) gamma irradiation.
3. Wash the irradiated protoplasts: transfer the suspension into tubes and centrifuge (50 × g, 5 min), remove supernatant and resuspend the protoplasts with CPW-wash solution up to a density of 10^6 protoplast/ml.
4. Remove a small aliquot from the washed irradiated protoplast suspension, repeat the CPW washing and resuspension and keep this suspension for plating of 'irradiated donor protoplasts' (see below) as well as for 'feeder protoplasts' (see note 5 after 'Fusion and culture of fused protoplast')
5. Centrifuge the recipient protoplast suspension (50 × g, 5 min), remove the supernatant and resuspend in 5 ml of 0.25 mM iodoacetate (recrystallized) in CPW wash solution. After 25 min centrifuge (50 × g, 5 min) and resuspend the protoplasts in CPW wash solution (up to 10^6 protoplast/ml).
6. Remove a small aliquot from the suspension of iodoacetate-treated recipient protoplasts and keep for plating 'iodoacetate treated recipient protoplasts' (see below).

Notes
2. An X-ray source can be used instead of a γ-ray source; a rate of above 1000–2000 rad per minute may harm the protoplasts; mesophyll protoplasts of other genera may require a different radiation dose in order to cause cell-division arrest; protoplasts from cell-suspension cultures require much higher radiation doses (20–50 krad) to arrest cell division.
5. The appropriate iodoacetate concentration and the duration of exposure to this antimetabolite varies for different protoplasts; the lowest concentration (and duration of exposure) that causes death of plated protoplasts should be employed.
6. Rhodamine-6G is an alternative to iodoacetate for causing metabolic arrest. It is possible to treat protoplasts of one species with iodoacetate, treat protoplasts of another species with rhodamine 6-G and then fuse the protoplasts, to obtain somatic hybrids [12].

D. Fusion and culture of fused protoplasts

Steps in the procedure
1. Mix an aliquot (1 ml or more) of the irradiated donor protoplast suspension with an equal aliquot of the iodoacetate-treated recipient-protoplast suspension. Remove 0.1 ml of this mix and keep for plating of 'mixed protoplast' (see below).
2. Pipette, into each of several 5.5 cm plastic Petri dishes, 2 drops (of 0.15 ml each) of PEG solution; place these drops about 1 cm apart, in the center of the plate. Then pipette between these drops a 'drop' of 0.3 ml of the mixed donor and recipient protoplast suspension (note that the density of the mix should be about 10^6 protoplasts/ml).
3. Connect the three drops (in each Petri dish) gradually and leave for 10 min.
4. Add to the combined 'drop', gradually, 3 ml of 0.275M $Ca(NO_3)_2$, leave for 10 min, then add another 3 ml of 0.275M $Ca(NO_3)_2$ and transfer the contents of each Petri dish into a centrifuge tube and centrifuge ($\sim 40 \times g$, 5 min)
5. Remove the supernatant and resuspend the fused protoplasts in VKM culture medium containing 0.5M glucose, up to a density of $5 \cdot 10^4$ protoplasts/ml. Then plate 3 ml of this suspension in each (tissue-culture quality) 5.5 cm Petri dish. Place the dishes for 2–3 h in 2–4 °C (dark).
6. Maintain the Petri dishes in a culture chamber (25 $\pm$ 1 °C). Cover the dishes, for the first 2 days, to avoid excessive light then expose to light (25 μmol $\cdot$ m^{-2} sec^{-1}).
7. Plate the following five control protoplast suspensions:
   – donor protoplasts;
   – recipient protoplasts;
   – irradiated donor protoplasts;
   – iodoacetate-treated recipient protoplasts;
   – mixed protoplasts
in VKM medium, containing 0.5M glucose as described above (D/5 and D/7).

Solutions and culture medium
   – 0.275M $Ca(NO_3)_2$
     – 4.5 g $Ca(NO_3)_2$ in 100 ml deionized water
     – adjust pH to 7.0 and sterilize by filtration (0.45 μM)
   – PEG solution
     – 2 g glucose
     – 1.5 g $Ca(NO_3)_2$
     – 40 g polyethylene glycol ('3500', 'Sigma')
     – dissolve in 100 ml deionized water, adjust to pH 7.0 and filter-sterilize (0.45 μM).

VKM Culture Medium

Macroelements	mg/l
KNO_3	1480
NH_4NO_3	1440
$MgSO_4 \cdot 7H_2O$	984
$CaCl_2 \cdot 2H_2O$	735
KH_2PO_4	68
$[CH_2N(CH_2COO)_2]_2FeNa$	35

Microelements	
H_3BO_3	3
$MnSO_4 \cdot H_2O$	10
$ZnSO_4 \cdot 7H_2O$	2.0
$Na_2MoO_4\text{-}2H_2O$	0.25
$CuSO_4 \cdot 5H_2O$	0.025
$CoCl_2 \cdot 6H_2O$	0.025
KI	0.75

Vitamins	
d-Calcium panthothenate	1.0
Cholin chloride	1.0
p-Aminobenzoic acid	0.02
Nicotinic acid	1.0
Pyridoxin hydrochloride	1.0
Thiamin hydrochloride	10.0
Folic acid	0.4
Biotin	0.01

Sugars	
Mannitol	250
Sorbitol	250
Sucrose	250
Fructose	250
Ribose	250
Xylose	250
Mannose	250
Rhamnose	250
Cellobiose	250

Organic acids	
Pyruvic acid	20
Fumaric acid	40
Citric acid	40
Malic acid	40

Other organic supplements
Casein hydrolysate   250
Myo-inositol   100
Coconut water   20 ml/l

Hormones
BAP   0.5
NAA   2

Glucose   100 g/l

Notes

2,3,4. PEG mediated fusion causes severe morphological changes in the protoplasts: first the proto-plasts aggregate attaining a rather deformed shape; then during dilution with $Ca(NO_3)_2$, the aggregrates separate and the protoplasts regain their round shape but are rather fragile. During cold storage the protoplasts gradually return to normal robustness.

5.   The VKM medium provides optimal cell division of *Nicotiana* protoplasts; but other, less complicated culture media can also be used for *Nicotiana*. For culture of fused protoplasts, of other genera, use media recommended for the culture of the respective protoplasts.

To assure optimal division of fused protoplast (especially if the PEG treatment causes damage) there is an option to add irradiated donor protoplasts $(1 \cdot 10^4/ml)$ to the Petri dishes that contain the fused protoplasts.

E. Selection and regeneration of cybrids

In our fusion model the donor has streptomycin-resistant chloroplasts; thus cybrids with donor plastomes can readily be selected as described below. Other plastome-coded resistances can be used similarly (e.g. resistance to spectinomycin, to lincomycin, to tentoxin). We also used the chondriome-coded oligomycin resistance. Another means of selection is to use a recipient with a plastome-coded pigmentation deficiency (albino). Green cybrids should then contain the donor's plastome.

Steps in the procedure

1. Seven days after plating of fused protoplasts, start to reduce the osmolarity of the cultures by the addition, every other day, of about 0.5ml culture medium (VKM) containing only 0.2M glucose. These medium additions should reduce the osmolarity to half of its initial value, during the next two weeks.
2. Prepare MS medium containing 1.6% agar, 3 µg/ml NAA and 1 µg/ml BAP, autoclave and after partial cooling add streptomycin up to 1 mg/ml; transfer to vials and keep the vials in a 42°C water bath.
3. Mix equal volumes of cultured (fused) protoplasts with the streptomycin containing MS medium by adding the MS to the suspension (not *vice versa*) and pipette 4 ml of this mixture into each 5.5 cm plastic Petri dish.
4. Return the dishes to the growth chamber (same conditions as in D/6) for further growth.
5. Pipette 20 ml of MS medium containing 2 µg/ml kinetin, 0.8 µg ml/IAA, 500 µg/ml streptomycin and 10 mg/ml agar into each of several 9 cm plastic Petri dishes. This is the regeneration medium.
6. When some of the cultured colonies green and attain a size of ca. 1 mm, pick these colonies individually and transfer them onto the regeneration medium. Mark and list each colony because several rooted shoots usually regenerate from any one colony-derived callus. For rooting transfer the shoots to Nitsch medium (without growth regulators).
7. When the shoots are rooted and contain several leaves, plant them individually into Jiffy (No. 7) turf pots and maintain the pots for several days on trays in the laboratory. Then transfer to the greenhouse and grow in regular pots (provide at least a 10/14 h light/dark regime).

Notes

2. MS medium = Murashige and Skoog medium, obtainable e.g. from Sigma or Flow Laboratories.
6. While culture in streptomycin containing medium gives a selective advantage to cybrids with the donor's plastome there may be escapes. Therefore it is advisable to verify the plastome composition of the cybrids by procedures noted below.

F. Analyses of the nuclear genome, the plastome and the chondriome composition of cybrids

The procedures described above should result in cybrids having the nuclear genome of the recipient protoplasts but their plasmone (plastome and chondriome) may be identical to the plasmone of the donor, the plasmone of the recipient, or (in the majority of the cybrids) the plasmone will not be identical to either of the fusion partners. In practice, nuclear components of the donor may integrate into the cybrids nuclear genome. Moreover, in spite of treatments and selection there may be escapes of either the donor or the recipient. Therefore, for reliable molecular characterization of cybrids, all the three genomes should be analyzed. All the analyses except for the determination of streptomycin resistance are based on established molecular protocols [13, 14] and protocols for plastome and chondriome analyses were detailed by us previously [1].

Principles of nuclear and organelle analyses
1. *Nuclear genome analysis.* With the development of restriction fragment-length polymorphism (RFLP) methods (see chapter by R. Potter, this volume) the RFLP analysis became the most reliable means to evaluate the similarity of the nuclear genome of cybrid plants to the respective genome of the recipient fusion-partner. This analysis requires the isolation of total DNA from the cybrid plants, digestion of this DNA with endonucleases and its electrophoretic fractionation followed by Southern blotting and hybridization with appropriate labelled RFLP clones. Obviously, when the nuclear-coded morphological features of the recipient differ substantially from the respective features of the donor, these features will reflect the nuclear composition of the cybrid plants.
2. *Plastome analysis.* In the model cybridization described above the simplest means to identify the plastome of a cybrid plant is to wait for seed maturation[a]. The seeds are then tested as follows:
— Sterilize the seeds as in A/2.
— Add 20 ml of Nitsch medium containing 0.8% agar and 1 mg/ml streptomycin (selective medium) into 9 cm Petri dishes.
— Plant about 20 seeds of each cybrid in a separate Petri dish (that contains the selective medium).
— For controls, plant 10 seeds of *N. tabacum* cv Xanthi in half of a Petri dish and 10 seeds of Line 92 (*str*[R]7) in the other half of the same Petri dish.
— Germinate under light (25 μmol $\cdot$ m^{-2} sec^{-1}, 25 $\pm$ 2 °C.)
  Plants with the donor's plastome should yield seeds that will germinate green seedlings in this selective medium; plants with the recipient's plastome should yield seeds that germinate chlorotic seedlings.
For molecular analysis of the plastome, either of two procedures should be undertaken[b]:
  (1) Extract chloroplast DNA, (from 10—20 g tissue) digest with an endonuclease, fractionate by gel electrophoresis and stain with ethidium bromide to obtain the

ctDNA restriction profile. Always load on the same gel the digested ctDNAs of the two fusion partners.

(2) Extract the total DNA (from 0.2–1.0 g tissue) digest and separate by gel-electrophoresis as above, then blot (Southern), hybridize with labelled ctDNA clones and obtain autoradiographs.

3. *Chondriome analysis.* This analysis requires more plant tissue than either nuclear genome or plastome analysis. Therefore either the cybrid plants themselves can be used for chondriome analysis (but this will require harvesting many leaves from each cybrid) or the seeds of the cybrids can be planted and the DNA extracted from the sexual progeny of each cybrid. As with the molecular analysis of the plastome, either of two procedures can be followed but note that for mtDNA samples, that are sufficient for several restriction profiles, up to 40 g tissue are required. Much less tissue is required for Southern blot analysis of mtDNA.

For obtaining mtDNA restriction profiles the mtDNA is isolated [1] from 20–40 g of leaf-tissue, digested by restriction endonucleases, fractionated by gel electro-phoresis and stained by ethidium bromide. In each gel cybrid mtDNA as well as mtDNA from both fusion partners should be included for reference.

For Southern blotting of mtDNA, the total DNA from 0.5 to 1.0 g leaf tissue is extracted [15], fractionated and run on gels as above and then blotted (Southern), hybridized with labelled mtDNA clones and autoradiographs are obtained.

Notes

a. *N. tabacum* is normally self-pollinated, unless cross-pollinated manually. Therefore the seeds harvested from pods of cybrid plants will usually represent the selfed sexual progeny of such plants. When the recipient is alloplasmic male-sterile, hand pollination with a pollinator is required. A cybrid plant may be heteroplastomic (sorting-out of chloroplasts was not completed). Thus, it is advisable to harvest seed samples from several pods and to test these samples individually.

b. For details of this analysis and the molecular analysis of plastomes and chondriomes consult the respective protocols in the chapter by Pehu *et al.*, this volume, and [1, 13, 14, 15].

References

1. Galun E, Aviv D (1986) Organelle transfer. Methods in Enzymology 118: 595–611.
2. Kao KN, Michayluk MR (1974) A method for high-frequency intergeneric fusion of plant protoplasts. Planta 115: 335–367.
3. Wallin A, Glimelius K, Eriksson T (1974) The induction of aggregation and fusion of *Daucus carota* protoplasts by polyethylene glycol. Z. Pflanzenphysiol. 74: 68–80.
4. Lörz H, Paszkowsky I, Dierks-Ventling C, Potrykus I (1981) Isolation and characterization of cytoplasts and miniprotoplasts derived from protoplasts of cultured cells. Physiol. Plant 53: 385–391.
5. Zelcer A, Aviv D, Galun E (1978) Interspecific transfer of cytoplasmic male sterility by fusion between protoplasts of normal *Nicotiana sylvestris* and X-ray irradiated protoplasts of male sterile N. *tabacum.* Z. Pflanzenphysiol. 90: 397–407.
6. Galun E, Aviv D, Breiman A, Fromm H, Perl A, Vardi A (1987) Cybrids in *Nicotiana, Solanum* and *Citrus*: isolation and characterization of plastome mutants, pre-fusion treatment, selection and analysis of cybrids. In: von Wettstein D, Chua N-H (eds) Plant Molecular Biology (NATO Adv. Study Inst.) Ser. A, Life Sciences 140: 199–207
7. Galun E, Perl A, Aviv D (1988) Protoplast fusion mediated transfer of male sterility and other plasmone controlled traits. In: Applications of Plant Cell and Tissue Culture. (Ciba Foundation Symp), Wiley, Chichester, pp: 97–112.
8. Wetter LR, Constabel F (1982) Plant Cell Culture Methods. National Res Council Canada, Saskatoon, 146 p.
9. Evans DA, Sharp WR, Ammirato PV, Yamada Y (1983) Handbook of Plant Cell Culture Vol. I, Macmillan, New York 979 p (see also additional volumes of this handbook).
10. Vasil IK (1984) Cell Culture and Somatic Genetics in Plants. Volume 1, Academic Press, Orlando, 825 p (see additional volumes in this treatise).
11. Bajaj YSP (1989) Plant Protoplasts and Genetic Engineering, I, Springer Verlag, Berlin, Heidelberg, 444 p.
12. Bottcher UF, Aviv D, Galun E (1989) Complementation between protoplasts treated with either of two metabolic inhibitors results in somatic hybrid plants. Plant Science 63: 289–292.
13. Gelvin SG, Schilperoort RA (1988) Plant Molecular Biology Manual. Kluwer Acad. Pub. Dordrecht.
14. Ausubel FA, Brent R, Kingston RE, Moore DD, Seidman JG, Smith JA, Struhl K (1990) Current Protocols in Molecular Biology. Greene Pub. and Wiley-Interscience, New York.
15. Mettler IJ (1988) A simple and rapid method for minipreparation of DNA from tissue cultured plant cells. Plant Mol. Biol. Reporter 5: 346–349.

Plant Tissue Culture Manual **D5**: 1–19, 1991.
© 1991 *Kluwer Academic Publishers.*

Fluorescence-activated analysis and sorting of protoplasts and somatic hybrids

DAVID W. GALBRAITH
University of Arizona, Department of Plant Sciences, Tucson, Arizona 85721 USA

Abbreviations

MES – 2-[N-morpholino]ethanesulfonic acid
FDA – fluorescein diacetate
FITC – fluorescein isothiocyanate

Brief Overview

The first accounts of the application of flow cytometry and cell sorting to higher plant systems were published in 1982 (for reviews, see [3–5]). Since that time, a large variety of further reports have appeared covering such diverse subject areas as somatic hybridization, characterization of cell type, analysis of gene expression, cell cycle measurements, chromosome sorting, and secondary product accumulation. This diversity derives from a recognition of the unique analytical attributes of flow cytometry and cell sorting, which enable this technology to provide insights otherwise impossible using other techniques.

This article outlines a series of procedures developed for the analysis and sorting of plant protoplasts. These were originally devised for the purification of heterokaryons produced through induced fusion of protoplasts [1]. They have also led to methods for the isolation and characterization of the different types of protoplasts produced from leaves [8], and we envisage that these techniques will have important applications in the future including the isolation and characterization of mutants, and the identification of genes whose expression is regulated in a cell-specific manner [11].

Background and theory

Flow cytometry is a technique that involves the analysis of the optical properties of populations of single cells [12, 13, 16]. The technique in its most popular configuration centres around a jeweled orifice through which the cell suspension is expelled as a jet under pressure. This fluid stream intersects a region of focused light, typically provided by a tunable laser. As each rapidly-moving cell passes through this illumination zone it absorbs and/or scatters light. Should fluorochromes be present in the cell, part of the absorbed light is subsequently emitted in the form of fluorescence. The positioning of specific collection lenses, mirrors and wavelength-specific light filters permits the selective detection of these optical signals, including both light-scatter and fluorescence emission signals. These are accurately measured using photomultipliers, and are digitized and stored in the memory of a microprocessor. The processed signals form the basis for the accumulation of uni- or multiparametric frequency distributions, in which the intensities of the various signals, each derived from a single cell, are displayed in the form of population histograms. The type of information provided by flow cytometry therefore is uniquely different from that provided by conventional biochemical measurements, in which the data represents population averages. Flow cytometric measurements are also uniquely accurate and, since the flow rates are typically high, are remarkably rapid. Since fluorescent tags that report a wide variety of different cellular characteristics have been devised, flow cytometry has a unique niche in the experimental armamentum of the modern biologist [13].

Fluorescence-activated cell sorting complements the analytical capabilities of flow cytometry. The behavior of jets-in-air has long been known to involve the eventual production of droplets, due to surface tension effects [12, 16]. Synchrony of the point of formation of these droplets can readily be achieved through the imposition of a slight periodic disturbance on the jet orifice using a piezo-electric crystal. Application of a charge to the flow stream at the point at which a desired cell is about to enter the forming droplet results in the placing of a residual charge on the resultant droplet, which then can be deflected electrostatically. This ability to rapidly analyze, charge and select droplets containing desired cells using simple (at least by today's standards) electronic and computer technologies has led to the widespread development and use of fluorescence-activated cell sorters [13].

It should be emphasized that flow cytometry is not restricted in theory by cell type; it simply requires that the cells of interest possess some distinguishing and detectable optical property, and that they exist in the form of a single cell suspension. Cell sorting is similarly not restricted by cell type or size, although recent studies have illustrated the types of problems that are provided by large particles, of which plant protoplasts are the a prototypical example, and how these problems can be empirically resolved [8, 10].

Procedures

Flow cytometric analysis and sorting of protoplasts

Protoplast isolation and purification
The methods described here are those that we currently employ for the isolation of tobacco leaf and cell suspension cultures and for the isolation of *Arabidopsis* leaf protoplasts. For other plant systems, detailed procedures can be found elsewhere in this volume.

Tobacco leaf protoplasts

Steps in the procedure
1. Maintain tobacco plants under sterile conditions in Magenta boxes (obtainable e.g., from Sigma) at 20–25 °C under continuous illumination on basal MS medium (e.g., from Sigma) [14] containing 3% sucrose and 0.8% agar. Subculture the plantlets by excision of the apical portion of the plant (approximately 2.5 cm, containing the meristem and two expanded leaves). Leaves suitable for protoplast preparation are about 3–4 cm in length and 1.5–3 cm in width, and can usually be obtained one to two weeks after subculture.
2. Excise the leaves under sterile conditions in a laminar flow cabinet. Slice them into 1 × 10 mm pieces, and transfer them into sterile 85 mm diameter plastic Petri dishes containing 10 ml of LP medium containing 0.1% (w/v) each of cellulysin, macerase and driselase. Continue incubation at 22 °C for 16–18 h.
3. Swirl the digest gently; filter the released protoplasts through two layers of sterilized cheesecloth into a sterile 50 ml plastic centrifuge tube. Add two volumes of medium W5, and collect the protoplasts by centrifugation at 50 × g for 8 min.
4. Gently resuspend the protoplasts in 5–10 m of PF medium using a Pasteur pipette. Overlay this suspension with 5 ml of W5 medium, and centrifuge it at 50 × g for 10 min. The intact protoplasts accumulate at the interface, whereas damaged protoplasts and cell debris are found in the pellet.
5. Remove the protoplasts at the interface using a Pasteur pipette, dilute with an equal volume of W5 medium, and recover them by centrifugation at 50 × g for 5 min. Resuspend the purified protoplasts to a concentration of 10^5/ml using a hemocytometer to estimate this value. The protoplasts are derived from both the dermal and ground tissues of the leaf [11].
6. Determine protoplast viability by addition of a 1/1000 dilution of FDA (0.5 mg/ml in acetone), scoring under the fluorescence microscope in terms of the proportions of protoplasts exhibiting intense cytoplasmic fluorescence after 10 min of incubation at 20 °C. Preparations should be 80–90% viable.

Protoplasts from tobacco cell suspension cultures

Maintain the cell suspension culture (*N. tabacum* cv Xanthi) in darkness under sterile conditions at 25 °C as 100 ml aliquots in 500 ml Erlenmeyer flasks with constant orbital agitation (100 rpm), in basal MS medium at pH 5.7 [14] with 3% (w/v) sucrose and 1 mg/l 2,4-dichlorophenoxyacetic acid. Subculture the cells at five-day intervals, through transfer of approximately 50 ml of cells into 100 ml fresh medium.

Steps in the procedure
1. Divide the contents of one flask between three 50 ml sterile plastic culture tubes, and allow the cells to settle for 40 min.
2. Aspirate the supernatant, and replace it with four volumes of SP digestion medium. Transfer the suspension into sterile 90 mm diameter plastic Petri dishes, and continue the incubation at 22 °C for 16—18 h.
3. Filter the protoplasts through two layers of sterile cheesecloth into a 50 ml plastic culture tube. Add an equal volume of 0.2 M KCl, mix gently, and recover the protoplasts by centrifugation at 50 × g for 4 min.
4. Resuspend the protoplasts in 6 ml of a solution comprising 0.4 M sucrose dissolved in 10 mM $CaCl_2$, and 10 mM MES, pH 5.7 (sucrose-solution). Overlay this suspension successively with 2 ml of sucrose-solution diluted 1 : 1 with 0.4 M KCl dissolved in 10 mM $CaCl_2$, and 10 mM MES, pH 5.7 (KCl-solution), with 2 ml of sucrose-solution diluted 1 : 3 with KCl-solution, and finally with 2 ml of 0.4 M KCl-solution. Centrifuge this step gradient at 50 × g for 8 min.
5. Separately remove the interfaces and score them for protoplast integrity and purity and protoplast yield using a light microscope. Score them also for viability using FDA, as described above. Recover the protoplasts by centrifugation at 50xg for 5 min after dilution with 2—3 vols of KCl-solution. Protoplast viabilities typically should be 80—90%

Solutions
— LP medium
    — macro and microsalts from T0 medium [2]
    — 0.5 M mannitol
    — 10 mM $CaCl_2$
    — 10 mM MES, pH 5.7
    — 0.1% Cellulysin (e.g. from Calbiochem)
    — 0.1% Macerase (e.g. from Calbiochem)
    — 0.1% Driselase (e.g. from Sigma)

— SP medium
    — macro and microsalts from T0 medium [2]
    — 0.35 M KCl
    — 10 mM $CaCl_2$
    — 10 mM MES, pH 5.7
    — 0.2% Cellulysin (e.g. from Calbiochem)

- 0.2% Driselase (e.g. from Sigma)
- 0.1% Rhozyme (e.g. from Rhom and Haas)
- 0.0125% Pectolyase Y-23 (e.g. from Seichin Pharmacentical Co., TOKYO).

— PF medium
  - 25% (w/v) sucrose dissolved in medium TO [2]

— W5 medium [15]
  - 154 mM NaCl
  - 125 mM $CaCl_2 \cdot 2H_2O$
  - 5 mM KCl
  - 5 mM glucose
  - adjust pH to 5.7; autoclave for 20 minutes.

Standards for flow cytometer alignment and calibration

Alignment and calibration procedures require the availability of fluorescent standard particles that approximate the size of the protoplasts that will be analyzed and sorted. This is due to the fact that interactions between the particles and the flow tips, as the size of the particles approaches the diameter of the flow tip, alter the optimal conditions for analysis and sorting [9].

Steps in the procedure

1. Prepare fluorescent standards by fixation of protoplasts. Prepare the fixative by the addition of 0.8 g paraformaldehyde to 20 ml H_2O; heat the slurry with stirring to 60 °C; add 1 M NaOH until it clears; cool and add an equal volume of (2 × concentration) medium NTTO [9]. Resuspend freshly-isolated leaf protoplasts in a large excess of fixative. Continue fixation for 60 min. at room temperature. Recover the protoplasts by centrifugation at 50 × g, and store at 4 °C in phosphate-buffered saline (obtainable from Sigma) containing 0.02% sodium azide.

2. Alternatively, pollen can be employed as a fluorescent standard. Resuspend the pollen (2.5 mg/ml) in 0.1% (w/v) aniline blue (obtainable from Sigma) prepared in phosphate-buffered saline, pH 9.0. The pollen can be stored indefinitely at 4 °C.

Flow cytometer system set-up.

For all the listed applications, our work has employed Coulter EPICS V and 753 flow cytometers (Coulter Electronics, Hialeah, FL). For analysis of plant protoplasts and pollen, the flow cytometer is routinely operated using a 200 μm (nominal internal diameter) flow tip at a sheath fluid pressure of 6 psi, using GMSF medium as the sheath fluid. Protoplasts are analyzed at a flow rate of no greater than 300/sec. This rate is adjusted by alteration of the sample differential pressure; this value is typically 5 psi for protoplasts at a density of 10^5/ml. One parameter histograms are accumulated to a total count of 20,000.

Solutions

- GMSF medium
    - 50.5 g mannitol
    - 68.4 g glucose
    - 3 mM MES
    to one liter and pH 5.7.

NTTO Medium

	mg/l
NH_4NO_3	825
KNO_3	950
$CaCl_2 \cdot 2H_2O$	220
$MgSO_4 \cdot 7H_2O$	185

KH_2PO_4	85
$FeSO_4 \cdot 7H_2O$	27.85
Na_2EDTA	37.25
$ZnSO_4$	1
H_2BO_3	1
$MnSO_4 \cdot 4H_2O$	0.1
$CuSO_4 \cdot 5H_2O$	0.03
$AlCl_3$	0.03
$NiCl_2 \cdot 6H_2O$	0.03
KI	0.01
Inositol	100
Ca pantothenate	1
Biotin	1
pyridoxine-HCl	1
Thiamine-HCl	1
Naphthalene acetic acid	3
Benzyladenine	1
Mannitol	130,000

pH 5.5, adjusted before autoclaving

Flow analyses of protoplasts and pollen: fluorescence emission intensities

Steps in the procedure

1. For stimulation of chlorophyll autofluorescence, adjust the laser excitation to 100 mW at 457 nm. Interpose absorbance filters with half-maximal transmittance at 510 and 515 nm (LP510 and LP515) in order to exclude scattered light. Place a further absorbance filter (LP610) to allow only fluorescence derived from chlorophyll emission to reach the red channel photomultiplier. Accumulate integral or log integral red fluorescence, one-parameter frequency distributions.

2. For selective flow analysis of viable protoplasts, examine protoplasts for emission of fluorescein fluorescence following FDA staining; it is best that this analysis occur within 60 min of staining. Operate the flow cytometer as described above, except detecting fluorescence emission from fluorescein using the green channel photomultiplier. Eliminate scattered laser light using absorbance filters LP510 and LP515. Direct fluorescein fluorescence into the green detector using either a fully silvered mirror, or if simultaneous analysis of chlorophyll fluorescence is required, a dichroic mirror (DC590) that splits at a wavelength of 590 nm. Eliminate passage of chlorophyll autofluorescence into the green channel photomultiplier using blue glass filter BG38. Accumulate one or two-parameter histograms.

3. Protoplasts from cell suspension cultures can be also be analyzed following staining with FDA (or with FITC [7]). In this case, the BG38 filters become redundant.

Flow cytometric analysis of pollen.

For stimulation of aniline-blue fluorescence from stained pollen, adjust laser excitation to 200 mW at 514 nm. Screen the green channel photomultiplier from scattered laser light using barrier filters LP530 and LP540. Accumulate one parameter histograms of integral or log integral green fluorescence.

Flow analyses of protoplasts and pollen: time-of-flight.

Analysis of the time taken for the protoplasts to pass through the laser beam, measured as the amount of time taken for the rise and fall of the pulse of fluorescence generated by passage of the protoplasts through the beam (the pulse-width "time-of-flight", TOF), can be used for a measurement of protoplast size [8]. For achlorophyllous protoplasts, time-of-flight (TOF) measurements require the analysis of green fluorescence (peak signal) derived from FDA fluorochromasia; for protoplasts containing mature chloroplasts, analysis of chlorophyll autofluorescence provides an accurate measure of protoplast size, whereas analysis of FDA fluorochromasia by the same cells does not [8]. This is probably due to quenching of FDA fluorescence by the chloroplast pigments.

1. Adjust the TOF module to analyze the time that the pulses of fluorescence remain above specific thresholds; these thresholds can be set to values between 5 and 50% of the peak fluorescence value. Accumulate one-parameter frequency distributions.

2. Since the TOF parameter is a function of the velocity of the fluid stream, it is

therefore also a function of sheath pressure. In order to perform an internal size calibration, accumulate TOF histograms using stained pollen from different plant species as size standards, obtaining their absolute diameters using microscopy and a stage micrometer.

Instrument Sterilization

Steps in the procedure
1. Autoclave the sheath fluid filter (Pall, Ultipor Type DFA4001ARP), the sample container, the sample pick-up tube, and the sample introduction line.
2. Sterilize the sheath tanks and sheath lines by soaking in a 10% commercial bleach solution for 20 min, followed by rinsing with 70% ethanol and sterile H_2O.
3. Presterilize the sheath fluid either by autoclaving or by passage through Millipore 0.22 µm filters.

Sort alignment

A range of combinations of sheath pressures and bimorphic crystal drive frequencies are suitable for sorting of protoplasts using various flow tip sizes [10]. For the 200 µm flow tip, at 6 PSI sheath pressure, we employ a drive frequency of 8 kHz.

Steps in the procedure
1. Adjust the frequency and amplitude of the bimorphic crystal drive to provide a uniform and stable sorted stream, visualized using the sort test mode.
2. Count the number of undulations prior to the position of droplet formation in order to approximately set the charge delay setting.
3. Initiate analysis using standard particles that approximate the size of the proto-plasts that are to be sorted (either fixed protoplasts or pollen). Position sort windows on the histograms in order to span the relevant peaks.
4. Perform a sort matrix analysis; this involves systematically adjusting the charge delay setting, followed by sorting of a defined number of particles (50—100). Count the numbers of particles actually sorted, using microscopy, in order to determine the adjustment of the charge delay setting at which a sort efficiency of 100% is obtained.
5. Start flow analysis of the experimental sample, in order to accumulate a one or two-dimensional flow histogram on which to position the appropriate sort windows. Position the sort windows and initiate fluorescence-activated sorting, using the sort conditions previously defined with the standard particles.

We routinely perform sorting into the wells of 96-well tissue culture plates prefilled with 50—100 µl of sterile growth medium, using the Coulter Autoclone. Protoplasts are sorted to a density appropriate for further culture; feeder cells can also be included within the wells. Further details on culturing of sorted protoplasts can be found elsewhere [1, 3—7, 9].

Flow analysis and sorting of heterokaryons formed by protoplast fusion

The following protocol is outlined for fusion of tobacco leaf mesophyll (green) and albino (white) leaf protoplasts.

Protoplast isolation

Steps in the procedure
1. Propagate *Nicotiana tabacum* cv Xanthi apical cuttings as plantlets on MS [2] medium containing 4 mg/l Norflurazon [4-chloro-5-methylamino-2-(α, α, α-tri-fluoro-m-tolyl)-3-(2H)-pyridazinone; Sandoz]. Culture these and grown under continuous white light (ca. 2000 lux) at 20–25 °C. Excise newly emergent white shoots and leaves.
2. Release and purify protoplasts from the white leaves using conditions identical to those described for green leaves, except including FITC (10 µl of a 3 mg/ml stock in ethanol) per 10 ml of the protoplast digestion mixture.
3. Isolate green mesophyll protoplasts from wild-type tobacco plantlets as previously described.

Protoplast fusion protocol [17]

Steps in the procedure
1. Determine the numbers of both sets of purified protoplasts using hemo-cytometry; adjust these numbers to 2×10^6/ml in solution W5. Set aside a small portion (about 10^5) of each protoplast type prior to mixing; these will be used for instrument set-up prior to heterokaryon sorting. Mix the two protoplast populations in a 1 : 1 ratio to a final concentration of 2×10^6/ml in solution W5.
2. The fusion solution is made fresh prior to each use. Dissolve 1.5 g polyethylene glycol (PEG MW 8,000), 88 mg $CaCl_2 \cdot 2H_2O$, and 180 mg mannitol in 8 ml of H_2O, and filter-sterilize (Solution A). Dissolve 468 mg glycine in 25 ml of H_2O, adjust the pH to 10 using 10M NaOH, and filter-sterilize (Solution B). Combine 0.8 ml of Solution A with 0.1 ml Solution B and 0.1 ml of dimethyl sulphoxide (DMSO) (Fusion Solution).
3. In a sterile 60 × 15 mm Petri dish, place eight equal droplets (approximately 30 µl) of fusion solution in pairs across the dish, each pair being separated by approximately 15 mm. Place one droplet of protoplast mixture (about 30 µl) between each pair, resulting in coalescence of the droplets. Add a further two droplets of fusion solution to the periphery of the coalesced drops.
4. Incubate the Petri dishes for 20 minutes at room temperature. Dropwise, over a period of five minutes, add a total of 2 ml of a solution comprising medium W5 buffered to pH 5.5 with 50 mM MES.
5. Continue incubation at room temperature for a period of two hours.

Instrument settings

Steps in the procedure

1. Tune the laser to 457 nm, and a power output of 200 mW. The filter combinations include a LP457-502 band pass filter to eliminate scattered laser light (this filter has 0% transmittance between 457 and 502 nm, and about 80−90% transmittance above 502 nm). Red/green beam splitting is achieved using dichroic mirror DC590. The green channel photomultiplier is screened by blue glass filter BG38 to eliminate chlorophyll autofluorescence. The red channel photomultiplier is screened by red glass filter LP610, to eliminate fluorescein fluorescence. A 200 μm flow cell tip is utilized under sterile conditions, with a sheath pressure of 6 psi and a crystal drive frequency of 8.0 KHz. Sort alignment is achieved as described previously (above and [10]).

2. Adjust the subtraction module (Red/Green channel compensation) using the small aliquots of the mesophyll and FITC-labeled protoplasts in order eliminate cross-contributions of the two fluorochromes to the two photomultiplier channels. This process requires separate analysis of the two red and green fluorescent samples, accumulating one parameter histograms. Adjust the high-voltages and amplifications of the two photomultipliers so that the mode of the distributions is mid scale. Next, accumulate a two-parameter histogram (red versus green) using the red fluorescence sample. Adjust the red-minus-green (R-G) percentage setting on the subtraction module until two-parameter distributions are obtained in which a red signal is registered, but with little or no green signal. Under this circumstance, the distribution should fall on or near the appropriate (red) axis; this can be facilitated by examination of the red and the green traces in real time on the oscilloscope display, adjusting the subtraction setting until the amplitude of the green signal is at its lowest. Note: if **too much** subtraction is applied, an artifactual signal can be generated in the green channel by overshoot. The trace of the green signal can be observed to be obviously irregular. Repeat the above process using the green sample, adjusting the green-minus-red (G-R) percentage setting on the subtraction module until two-parameter distributions are obtained in which a green signal is registered, but with little or no red signal. Correspondingly, the two-dimensional distribution should fall on or near the green axis.

For optimal adjustment of the subtraction module, it is helpful if the emission intensities of the two fluorochromes be roughly equal in intensities as perceived by the photomultipliers. This can be achieved by suitable adjustment of the wavelengths of laser excitation and of the types of filters employed, or ideally by the use of twin laser excitation, if available.

Fluorescence-activated sorting of fused protoplasts

Steps in the procedure

1. Once instrument alignment is completed, mix the protoplasts remaining in the pre-fusion aliquots and accumulate a red versus green two-parameter histogram to a total of 20,000 counts. Place sort windows so that they exclude the twin

red and green protoplast populations; no (or very few) sort positive signals should register within the region boxed by the windows.

2. Transfer the diluted fusion mixture to a sterile sample tube and accumulate a two-parameter histogram to a total of 20,000 counts. Integrate the region encompassed by the sort windows to provide information on the fusion efficiency.

3. Add 0.15 ml growth medium (TO [2] medium supplemented with 0.38 M glucose, 0.154 M mannitol, and 75 mg/l ampicillin) to each well of a 96-well Costar microtiter plate. Sort fused protoplasts to a concentration of 1,000/well, at a flow rate of approximately 200/sec.

4. Culture the isolated protoplasts in darkness at 20–25 °C. Further culture details can be found elsewhere [1,3–7,9].

Troubleshooting

General concerns. In view of the demographics of flow cytometric instrumentation, in most research situations, the plant biologist wishing to perform flow analysis and/or sorting or plant protoplasts will be working within the context of an established flow laboratory, most probably in the form of a multiple-use facility, and frequently located in medical school or hospitals. Under these circumstances the facility personnel usually have extensive experience in the analysis and (albeit less frequently) sorting of animal cells, particularly those of lymphoid origin. Infrequently will they have encountered protoplasts, or even animal cells that deviate significantly from 10–20 µm in diameter. As a consequence, we anticipate that most problems will relate to unfamiliarity with the biological material under study rather than in the principles underlying flow cytometric analysis and sorting (some exceptions are noted below). Evidently, success in the analysis and sorting of protoplasts requires an understanding of the procedures of protoplast preparation and culture, particularly since protoplasts are fragile and are large cells [8]. Critical is the ability to recognize, using light microscopy, the visual characteristics indicative of protoplast viability, since attempts to apply flow cytometric techniques to partially viable or non-viable protoplasts are futile. We and others emphasize the important role played by the physiological status of the donor plant or cell type on protoplast yield and viability. Controlling for variation in light intensity, day length, plant stage and leaf number, particularly through in vitro plant propagation is strongly recommended. It is also important to employ those methods for protoplast preparation and culture that have previously been shown to be optimal for the tissue type under study.

Sample preparation. Flow cytometric analysis typically involves an examination of all particles present within cell suspensions, whether or not these represent intact protoplasts, cellular debris or other components such as subcellular organelles. This is because the flow analysis of fluorescent pulses is usually triggered on light-scatter; since plant cells contain significant numbers of intracellular organelles (for example, an average of 72 chloroplasts per mesophyll tobacco protoplast [9]), it can prove difficult at first to identify the meaningful signals within those derived from the starting materials, particularly since broken protoplasts can contribute large amounts of subcellular debris. For this reason, we routinely purify the protoplasts using sucrose gradient centrifugation. The flow cytometer can also be adjusted to trigger on fluorescence, cell size (TOF) or other parameters that unambiguously define the population of interest. Examination of the protoplast population using a fluorescence microscope and excitation/emission filters comparable to those within the flow cytometer is also helpful in order to determine what the optical properties of the sample actually comprise. For example, higher plant cells frequently contain the variety of intracellular pigments associated with photosynthesis. These can affect flow analyses either by absorbing the incident light, or by re-emitting it in the form of fluorescence. The researcher should be prepared to deal with these potential problems through the correct selection of appropriate fluorochromes, fixation procedures, and wavelengths of excitation and emission.

References

1. Afonso CL, Harkins KR, Thomas-Compton M, Krejci A, Galbraith DW (1985) Production of somatic hybrid plants through fluorescence-activated sorting of protoplasts. Bio/Technology 3: 811–816.
2. Chupeau Y, Missonier C, Hommel M-C, Goujard J (1978) Somatic hybrids of plants by fusion of protoplasts. Molec Gen Genet 165: 239–245.
3. Fox MH, Galbraith DW (1990) The application of flow cytometry and sorting to higher plant systems. In: Melamed MR, Lindmo T, Mendelsohn ML (eds) Flow Cytometry and Cell Sorting (second edition), pp. 633–650. New York: John Wiley.
4. Galbraith DW (1989) Analysis of higher plants by flow cytometry and cell sorting. Intl Rev Cytol 116: 165–228.
5. Galbraith DW (1990) Flow cytometric analysis and sorting of somatic hybrid and transformed protoplasts. In: Bajaj YPS (ed) Biotechnology in Agriculture and Forestry Vol. 9, Plant Protoplasts and Genetic Engineering, pp. 304–327. New York: Springer-Verlag.
6. Galbraith DW (1990) Isolation and flow cytometric characterization of plant protoplasts. Methods Cell Biol 33: 527–547.
7. Galbraith DW, Afonso CL, Harkins KR (1984) Flow sorting and culture of protoplasts: Conditions for high-frequency recovery and growth of sorted protoplasts of suspension cultures of *Nicotiana*. Plant Cell Rep 3: 151–155.
8. Galbraith DW, Harkins KR, Jefferson RA (1988) Flow cytometric characterization of the chlorophyll contents and size distributions of plant protoplasts. Cytometry 9: 75–83.
9. Harkins KR, Galbraith DW (1984) Flow sorting and culture of plant protoplasts. Physiol Plant 60: 43–52.
10. Harkins KR, Galbraith DW (1987). Factors governing the flow cytometric analysis and sorting of large biological particles. Cytometry 8: 60–71.
11. Harkins KR, Jefferson RA, Kavanagh TA, Bevan MW, Galbraith DW (1990) Expression of photosynthesis-related gene fusions is restricted by cell-type in transgenic plants and in transfected protoplasts. Proc Natl Acad Sci USA 87: 816–820.
12. Kachel V, Fellner-Feldegg H, Menke E (1990). Hydrodynamic properties of flow cytometry instruments. In: Melamed MR, Lindmo T, Mendelsohn ML (eds) Flow Cytometry and Cell Sorting (second edition), pp. 27–44. New York: John Wiley.
13. Melamed MR, Mullaney PF, Shapiro HM (1990). An historical review of the development of flow cytometers and sorters. In: Melamed MR, Lindmo T, Mendelsohn ML (eds) Flow Cytometry and Cell Sorting (second edition), pp. 1–9. New York: John Wiley.
14. Murashige T, Skoog F. (1962). A revised medium for rapid growth and bioassays with tobacco tissue cultures. Physiol Plant 15: 473–497.
15. Negrutiu I, Shillito R, Potrykus I, Biasini G, Sala F (1987). Hybrid genes in the analysis of transformation conditions. Plant Mol Biol 8: 363–373.
16. Steen HB. Characteristics of flow cytometers. In: Melamed MR, Lindmo T, Mendelsohn ML (eds) Flow Cytometry and Cell Sorting (second edition), pp. 11–25. New York: John Wiley.
17. Thomzik JE, Hain R (1988). Transfer and segregation of triazine tolerant chloroplasts in *Brassica napus*. Theor Appl Genet 76: 165–171.

Plant Tissue Culture Manual **D6**: 1–8, 1991.
© 1991 *Kluwer Academic Publishers.*

RFLP analysis of organellar genomes in somatic hybrids.

E. PEHU

University of Helsinki, Dept. of Crop Husbandry, SF-00710 Helsinki, Finland

Introduction

Somatic hybridization can be used to produce various combinations of organellar and nuclear genomes [20]. Furthermore, the fact that organelle populations from both fusion partners are present, at least initially in the same heterokaryon, creates an opportunity for interspecific recombination in the organelle genomes. This has made protoplast fusion a useful approach to produce hybrids for both compatible [23, 25, 26] and incompatible species [1, 7].

It has been shown in various studies of organelle composition of somatic hybrids that in the majority of cases chloroplasts of one fusion parent sort out, as each hybrid has been found to contain only one parental type of chloroplast, although both types may be present in the different plants of the hybrid population [2, 8, 9, 20, 26]. Only few cases of interspecific chloroplast DNA recombination have been reported [16, 17]. In contrast, examination of the mitochondrial DNAs in somatic hybrids has shown that although both parental types may be present, novel types as a result of recombination can also occur quite frequently [3, 12, 15, 18].

The donor-recipient protoplast fusion system has been developed to effect the transfer of useful cytoplasmic traits such as male sterility and herbicide resistance or transfer of only a few chromosomes of the donor to the recipient [5, 22, 24, 27, 28]. These asymmetric hybrids have provided a unique opportunity to study interspecific and intergeneric nuclear-organelle interactions.

In somatic hybridization experiments the population of hybrids to be characterized should be fairly large, to increase the likelihood of including all possible nuclear and organelle combinations, and to uncover the incidental products of recombinations. Over the past ten years several procedures have been published on the isolation of organellar DNA or its study by DNA hybridization.

The sequence of extraction of both chloroplast and mitochondrial DNA involves 1) tissue disruption; 2) density gradient centrifugation to isolate the organelle population; 3) lysis of the organelles and 4) deproteinization and precipitation of the DNA. In most cases the purity of the DNA after the above listed steps is sufficient for obtaining clear restriction fragment patterns. If needed the preparation can be further purified by CsCl centrifugation because the buoyant densities of the nuclear and organellar DNAs (especially mtDNA) are different.

In some species, especially those rich in phenolic compounds, an extraction buffer of high ionic strength has increased the yield of covalently closed circular forms of mtDNA [21]. For simultaneous isolation of both chloroplasts and mitochondria Pay and Smith [19] have developed a method that includes a self-generating percoll gradient. In some species the yield of the organellar DNA can be very low as has been found to be the case for several grasses. In these cases one could label the ends of the fragments after restriction digest as reported by Lehväslaiho *et al.* [13] and visualize the RFLPs by autoradiography.

In addition to RFLPs of restricted organellar DNAs separated on agarose gels and visualized by ethidium bromide, one can also use organellar DNA-specific probes for Southern analysis against restricted total cellular DNA of the somatic hybrids. This is more important in analysing mtDNA as it is often more difficult to get a good yield of restrictable mtDNA, and the amount of tissue required is often tens of grams. When using this approach one should make sure to load at least 5 µg of cellular DNA/lane to get a strong and reproducible signal in the hybridization. Obviously the portion of the organelle genome covered by using specific probes is more limited when compared to the restriction fragment pattern of the organellar DNA visualized in the gel. Furthermore, different sequences used as probes can vary in their power to uncover sequence rearrangements, i.e. some sequences represent sections of the mtDNA more likely to vary than others [5, 6, 12]. It is therefore useful to carry out the hybridization experiments with several different probes.

The following cpDNA extraction and restriction analysis is based on the procedure developed by Hosaka [10] as modified by Pehu *et al.* [20] to analyse somatic hybrids of *Solanum tuberosum* and *S. brevidens*. The mitochondrial DNA analysis method is based on the work of Bland *et al.* [4] as adapted to study somatic hybrids in the genus *Brassica* by Håkkanson *et al.* [11].

Procedures

Chloroplast DNA extraction

Steps in the procedure
1. Weigh 4.0 g (if plant material is limited 2.5 g) of young leaves (placed in the dark for 48h prior to extraction).
2. Add 30ml (or 15 ml) of buffer A.
3. Disrupt the tissue by three 1 sec bursts (high speed) in a Wareing blender (110ml cup for 4 g and 37 ml cup for 2.5 g of leaf tissue).
4. Pass the homogenate through 4 layers of cheesecloth and 2 layers of miracloth (Calbiochem).
5. Centrifuge at 1000 × g for 10 minutes at 4 °C.
6. Resuspend the pellet in 20 ml of buffer B gently with the help of a soft paintbrush.
7. Centrifuge at 2500 × g for 20 min.
8. Remove the supernatant.
9. Dilute the supernatant gently in exactly 3 volumes (60 ml) of buffer C (pour the supernatant into a conical flask on ice and place the ice bucket on a stirring plate; add buffer C gradually over a period of 3–4 min); place in two 50 ml tubes.
10. Centrifuge at 2000 × g for 10 min at 4 °C.
11. Resuspend the pellet into 1 ml (0.5 ml/tube) of buffer C.
12. Lyse the organelles by addition of 200 µl of 10% sarkosine (5–10 min); transfer the lysate into two Eppendorf tubes.
13. Add 0.5 ml of phenol/tube, invert the tube several times, centrifuge for a few minutes at 10,000 × g and collect the aqueous (top) layer.
14. Add 0.5 ml of chloroform/tube, invert the tube several times, centrifuge for a few minutes at 10,000 × g and collect the top layer.
15. Precipitate the DNA by addition of 75 µl of 3.5M sodium acetate (pH 5.3) and 0.7 ml cold isopropanol.
16. Wash the DNA pellet with 70% ethanol.
17. Resuspend the pellet in 40 µl of TE.
    A yield of 20 µg of chloroplast DNA/preparation can be obtained routinely by this protocol.
18. Restrict the chloroplast DNA with an enzyme of your choice according to suppliers instructions using 10 µl of the chloroplast DNA solution.
19. Separate the DNA fragments by agarose gel (0.8–1.0%) electrophoresis in TBE or TAE buffer [14]. Run the gel at room temperature at 2V/cm. Stain the gel with 0.5 µg/ml ethidium bromide for 45 min and photograph it using standard procedures.

Buffers: 500 ml
  A (Saltz and Beckman) buffer:
    2M Tris-Cl (pH 8.0)   12.5 ml

Sucrose	59.9 g
0.5M EDTA (pH 8.0)	7.0 ml
add water to 500 ml	
after autoclaving, add	
1M mercaptoethanol	2.5 ml
Bovine Serum Albumin	0.5 g

B buffer:

2M Tris-Cl (pH 8.0)	12.5 ml
0.5M EDTA (pH 8.0)	20.0 ml
Sucrose	250.0 g
add water to 500 ml, autoclave	

C Buffer:

2M Tris-Cl (pH 8.0)	12.5 ml
0.5M EDTA (pH 8.0)	20.0 ml
add water to 500 ml, autoclave	

Isolation of mitochondrial DNA

DAY 1.

1. Harvest 50–100 g young leaves (give a 24–48 h dark treatment to the source plants before harvesting).
2. Submerge the leaves in 5% chlorox for 5 min followed by 2–3 rinses in distilled water.
3. Remove large middle veins.
4. Weigh the leaves, yield should be 40–90 g. Divide into 4–5 portions.
5. Put 1 portion of the leaves into a precooled mortar and tear the leaves into small pieces. Add 100 ml of grinding buffer and grind the leaf tissue. Filter the homogenized tissue through 6 layers of cheese cloth. Rinse the mortar with 50 ml of grinding buffer. Return the tissue to the mortar and repeat the grinding 2 times with each portion. Filter the filtrate through 2 layers of Miracloth (Calbiochem).
6. Pour the filtrate into 250 ml bottles and centrifuge at $1000 \times g$ for 10 min.
7. Decant the supernatant into fresh bottles and centrifuge at $2000 \times g$ for 10 min.
8. Decant the supernatant again into fresh bottles and spin down the organelle pellet at $18,000 \times g$ for 20 min.
9. Discard the supernatant and add 3–4 ml of buffer B and resuspend the pellet gently with the help of a soft paintbrush.
10. Transfer the suspension from all bottles into a small beaker. Increase the final volume of the suspension into 40ml by rinsing the buckets with buffer B.
11. Add 1 ml of DNAse I (stock 2000 u/ml) and incubate in room temperature for 1 h.
12. Divide the suspension into 4×50ml tubes and fill them with buffer C. Centrifuge at $17,400 \times g$ for 12 min.
13. Resuspend the pellets in buffer C and combine into 2×50ml tubes. Fill the tubes with buffer C and recentrifuge.
14. Resuspend the pellets and combine into one 50 ml tube, fill up with buffer C and repeat the centrifugation.
15. Resuspend the pellet in 3 ml of buffer C and freeze at $-20\ ^{\circ}$C. The solution can be stored for several weeks.

DAY 2

16. Thaw the mitochondria suspension and add 360 µl of Proteinase K (stock 2 mg/ml in lysis buffer) and 300 µl of 20% Sarkosyl (Na-lauryl sarcosine dissolved in lysis buffer). Bring the total volume up to 6.8 ml with lysis buffer.
17. Place the tube in 60 °C water bath for 1 h.
18. Pour the lysate into a vial containing 7.5 g CsCl. Mix gently until the CsCl is dissolved.
19. Add 1.2 ml EtBr (stock 700 µg/ml lysis buffer). Mix gently.
20. Divide the sample into 2 Quick seal tubes. Weigh 1 ml of the solution to assure that the density of the CsCl solution is between 1.59–1.60g/ml (for *Brassica; modify for other species*). Fill the tubes with mineral oil and seal the tubes.

21. Spin the tubes at 40,000 rpm in a RPV 65 T rotor at 20 °C, 20 h (LKB ultra centrifuge).
22. Take out the DNA band under UV-light using a 2 ml syringe and a 0.9 × 40 mm needle to remove the band.
23. Extract EtBr with isoamyl alcohol (saturated with 0.01M EDTA) 3—4 times until the aqueous phase is colourless.
24. Transfer the aqueous phase into a 30 ml tube. Add 2 volumes of TE buffer and 6 volumes of 95% ethanol. Precipitate the DNA overnight in −20 °C.
25. Pellet the DNA at 4 °C, 12,000 × g, decant the supernatant and dry the pellet.
26. Dissolve the DNA with 400 μl of TE buffer (takes several hours).
27. Reprecipitate the DNA with 2.5 volumes of ethanol in an Eppendorf tube in −20 °C overnight.
28. Pellet the DNA at 4 °C, 10,000 × g for 10 min.
29. Wash the pellet with 70% ethanol, decant the supernatant and dry the pellet
30. Dissolve the DNA in 50 μl of TE buffer. Usual yield of mitochondrial DNA ranges between 10—40 μg.

Carry out the restriction and electrophoresis as described for chloroplast DNA.

Buffers

Lysis buffer:	250 ml
0.1M Tris	3.03 g
0.1M Na EDTA	9.31 g
adjust to pH 8	

Grinding buffer:	2.0 l
0.3M mannitol	110 g
0.05M Tris (2 M stock, pH 7.5)	50 ml
0.003M Na EDTA (0.1M stock, pH 7.5)	60 ml
Add just before use:	
0.01M mercaptoethanol	1.6 ml
0.1% BSA	2.0 g
15mM Na S O	1.44 g
10 mg/ml PVP	20 g
stir gently to get into solution	

Buffer B:	1 l
0.3M mannitol	54.7 g
0.05M Tris	25 ml
0.01M MgCl (0.1M stock)	100 ml
0.1% BSA	1 g

Buffer C:	1 l
0.3M sucrose	102.7 g
0.1M Na EDTA	37.2 g
adjust the pH to 8	

References

1. Austin S, Baer MA, Helgeson JP (1985) Transfer of resistance to potato leaf roll virus from *Solanum brevidens* into *Solanum tuberosum* by somatic fusion. Plant Sci 39: 75–82.
2. Barsby TL, Shepard JF, Kemble RJ, Wong R (1984) Somatic hybridization in the genus *Solanum*: *S. tuberosum* and *S. brevidens*. Plant Cell Rep 3: 165–167.
3. Belliard G, Vedel F, Pelletier G (1979) Mitochondrial recombination in cytoplasmic hybrids of *Nicotiana tabacum* by protoplast fusion. Nature 281: 401–403.
4. Bland MM, Matzinger DF, Levings CS (1985) Comparison of the mitochondrial genome of *Nicotiana tabacum* with its progenitor species. Theor Appl Genet 69: 535–541.
5. Bonnett HT, Glimelius K (1990) Cybrids of *Nicotiana tabacum* and *Petunia hybrida* have an intergeneric mixture of chloroplasts from *P. hybrida* and mitochondria identical or similar to *N. tabacum*. Theor Appl Genet 79: 550–555.
6. Chowdhury MKU, Schaeffer GW, Smith RL, DeBonte LR, Matthews BF (1990) Mitochondrial DNA variation in long-term tissue cultured rice lines. Theor Appl Genet 80: 81–87.
7. Fish N, Karp A, Jones MGK (1987) Improved isolation of dihaploid *S. tuberosum* protoplasts and the production of somatic hybrids between dihaploid *S. tuberosum* and *S. brevidens*. In Vitro 23: 575–580.
8. Gleddie S, Keller WA, Setterfield G (1986) Production and characterization of somatic hybrids between *Solanum melongena* L. and *S. sisymbriifolium* Lam. Theor Appl Genet 71: 613–621.
9. Gressel J, Cohen N, Binding H (1984) Somatic hybridization of an atrazine-resistant biotype of *Solanum nigrum* with *Solanum tuberosum*. 2. Segregation of plastomes. Theor Appl Genet 67: 131–134.
10. Hosaka K (1986) Who is the mother of potato?-restriction endonuclease analysis of chloroplast DNA of cultivated potatoes. Theor Appl Genet 72: 606–618.
11. Håkansson, G, van der Mark, F, Bonnett, HT, Glimelius, K (1988) Variant mitochondrial protein and DNA patterns associated with cytoplasmic male-sterile lines of *Nicotiana*. Theor Appl Genet 76: 431–437.
12. Kemble RJ, Barsby TL, Wong RSC, Shepard JF (1986) Mitochondrial DNA rearrangements in somatic hybrids of *Solanum tuberosum* and *Solanum brevidens*. Theor Appl Genet 72: 787–793.
13. Lehväslaiho H, Saura A, Lokki J (1987) Chloroplast DNA variation in the grass tribe Festuceae. Theor Appl Genet 74: 298–302.
14. Maniatis T, Fritsch EF, Sambrock J (1982) Molecular cloning: a laboratory manual. Cold Spring Harbor Laboratory Press, Cold Spring Harbor/NY.
15. Matthews BF, Widholm JM (1985) Organelle DNA compositions and isoenzyme expression in an interspecific hybrid of *Daucus*. Mol Gen Genet 198: 371–376.
16. Medgyesy P, Fejes E, Maliga P (1985) Interspecific chloroplast recombination in a *Nicotiana* somatic hybrid. Proc Natl Acad Sci USA 82: 6960–6964.
17. Medgyesy P, Thanh ND, Horvath V, Buzas, B (1990) Chloroplast or mitochondrial recombination after protoplast fusion forms an accommodated new organelle genome in incompatible nucleus-organelle combinations. VIIth International Congress on Plant Tissue and Cell Culture. Abstracts, p. 204.
18. Nagy F, Torok I, Maliga P (1981) Extensive rearrangements in the mitochondrial DNA in somatic hybrids of *Nicotiana tabacum* and *Nicotiana knightiana*. Mol Gen Genet 183: 437–439.
19. Pay A, Smith MA (1988) A rapid method for purification of organelles for DNA isolation: self-generating percoll gradients. Plant Cell Rep 7: 96–99.
20. Pehu E, Karp A, Moore K, Steele S, Dunckley R, Jones MGK (1989) Molecular, cytogenetic and morphological characterization of somatic hybrids of dihaploid *Solanum tuberosum* and diploid *S. brevidens*. Theor Appl Genet 78: 696–704.
21. Perez C, Bonavent J-F, Berville A (1990) Preparation of mitochondrial DNAs from sunflowers (*Helianthus annuus* L.) and from beets (*Beta vulgaris*) using a medium with a high ionic strength. Plant Mol Biol Rep 8: 105–113.

22. Perl A, Aviv D, Galun E (1990) Protoplast-fusion-derived *Solanum* cybrids: application and phylogenetic limitations. Theor Appl Genet 79: 632–640.
23. Rosen B, Hallden C, Heneen WK (1988) Diploid *Brassica napus* somatic hybrids: characterization of nuclear and organellar DNA. Theor Appl Genet 76: 197–203.
24. Smith MA, Pay A, Dudits D (1989) Analysis of chloroplast and mitochondrial DNAs in asymmetric somatic hybrids between tobacco and carrot. Theor Appl Genet 77: 641–644.
25. Sundberg E, Glimelius K (1986) A method for production of interspecific hybrids within Brassicaceae via somatic hybridization using resynthesis of *Brassica napus* as a model. Plant Sci 43: 155–162.
26. Sundberg E, Landgren M, Glimelius K (1987) Fertility and chromosome stability in *Brassica napus* resynthesised by protoplast fusion. Theor Appl Genet 75: 96–104.
27. Thanh ND, Pay A, Smith MA, Medgyesy P, Marton L (1988) Intertribal chloroplast transfer by protoplast fusion between *Nicotiana tabacum* and *Salpiglossis sinuata*. Mol Gen Genet 213: 186–190.
28. Thanh ND, Medgyesy P (1989) Limited chloroplast gene transfer via recombination overcomes plastome-genome incompatibility between *Nicotiana tabacum* and *Solanum tuberosum*. Plant Mol Biol 12: 87–93.

Plant Tissue Culture Manual **D7**: 1–20, 1992.

Isolation and uptake of plant nuclei

PRAVEEN K. SAXENA[1] & JOHN KING[2]

[1] *Department of Horticultural Science, University of Guelph, Guelph, Ontario, Canada, N1G 2W1;*
[2] *Department of Biology, University of Saskatchewan, Saskatoon, Saskatchewan, Canada, S7N 0W0*

Introduction

Genetic transformation of animal cells by transplantation of isolated organelles, nuclei, and chromosomes is well documented and has played a
significant role in investigating chromosome mapping and the regulation of gene
expression [11, 13]. In plants, direct DNA transfer using isolated organelles
was attempted following the discovery of techniques to eliminate the plant-
specific barrier, the cell wall, which previously hampered the introduction of
organelles into cells. Predictably, as soon as plant cells succumbed to procedures capable of enzymic degradation of cell walls, a whole new concept of
handling naked plant cells (protoplasts) emerged as well as a genuine hope of
being able to transform plant cells by foreign DNA introduction [15]. Thus,
many workers studied the uptake of a variety of macromolecules and
demonstrated the ability of plant protoplasts to accept foreign particles such
as ferritin, bacteria, and isolated organelles like nuclei and chloroplasts (see
references 7, 14, 21 for extensive reviews).

The selective transfer of large amounts of DNA into the protoplasts via a
whole nucleus, chromosome, or a part thereof as opposed to the fusion of two
entire protoplasts holds great potential in developing ways to improve commercial crops. For example, many traits of economic significance such as yield and
tolerance to salinity, drought, and extreme temperatures, are encoded by more
than one gene. Although impressive progress has been made recently in
transferring foreign genes into plant cells, a number of problems associated with
the application of this technology for crop improvement still remain to be
resolved, particularly the number of genes which can be transferred.

The current gene transfer strategies only allow the transfer of one or two
genes using either indirect (*Agrobacterium*-mediated) or direct (plasmid DNA)
methods of transformation. Construction of plasmids harbouring multiple
genes at present is difficult and it does not seem likely that such large plasmids
could be transferred effectively. Until such time that the genes regulating
characters of economic importance are identified, cloned, and vehicle-bound
(in plasmids), the transfer of isolated organelles with mapped and identified
genes of interest would seem to be an attractive approach in incorporating
polygenic traits. Furthermore, nuclear transplantation allows selective transfer
of nuclei in an unmodified cytoplasm, that of the recipient, as opposed to

protoplast fusion which results in mixing of cytoplasms as well. Selective transfer of nuclei may help avoid the problems of cytoplasmic incompatibility and the transfer of undesirable traits of cytoplasmic origin. Another important application of organelle transfer could be the induction of genetic variability into highly inbred germplasms particularly in self-pollinated crops.

In this chapter, the techniques for the isolation of nuclei and their transfer into plant protoplasts will be discussed.

Isolation of nuclei

For an efficient nuclear transplantation experiment, the nuclei should be available in large numbers. It is also important that nuclear preparations be free of cytoplasmic contamination and contain morphologically intact and biologically active nuclei. These fundamental requirements have been the issue of several investigations over the years [3, 22, 23, 28, 29]. There seems to be a general agreement that source material should be soft and having a minimum of starch, phenols, and tannins. Various favourable sources recognized include meristematic, etiolated, or embryonic cells. In this regard, cell suspension cultures seem particularly suitable as they provide large populations of actively growing cells which are an excellent source of protoplasts. Protoplasts have been found to be the most suitable source for nuclei isolation because of the absence of cell walls which makes it easier to release the nucleus without contamination by cell wall fragments.

Protocol 2 (see later) describes the method developed in our laboratory for isolating nuclei from protoplasts of cell suspensions. To achieve good nuclear preparations using protoplasts, a controlled denaturing of cell membrane is essential. In this connection, Triton X-100 is the common detergent used to solubilize cell membranes and to remove other cytoplasmic debris. However, the concentrations found to be effective (0.1–1%) in rupturing the protoplasts were associated with deleterious effects on the nuclear envelope [9, 10]. A solution to this problem has been presented [22, 23] by employing a two-step destabilization of protoplast membranes. Firstly, the protoplasts are deplasmolysed in a hypotonic nuclei isolation buffer containing a low Triton X-100 concentration (0.01–0.02%). In the second step, these deplasmolysed protoplasts are homogenized in a glass homogenizer by applying 10–15 gentle strokes or by passing through needles of 18–26 gauge. With this modification in the procedure, Saxena et al. [22] were able to obtain clean nuclei (Fig. 1A) which retained a high degree of nuclear envelope integrity. The most crucial factor which allowed the lysis of protoplasts using lower concentrations of Triton X-100 was the pH of the isolation buffer. A narrow range of pH, between 5.2 and 5.5 only, was found to be suitable. At pH values lower or higher than 5.2–5.5, the nuclear yields as well as the quality of the nuclei were unsatisfactory.

Once isolated, nuclei need to be stabilized osmotically in order to insure against lysis and membrane damage. Various stabilizers that have been used include sucrose, mannitol, dextran, and ficoll. Divalent cations like Mg^{2+} and Ca^{2+} have also been used as protectants but they cause clumping of nuclei, but this can be avoided by replacing these with Na^+ and K^+ [22, 29]. Other compounds which have been used for the overall good nuclear preparation are ethylene diamine tetracetic acid (EDTA) to block the action of DNAse in the preparation [1], and to facilitate protoplast lysis [12], and polyvinylpoly-pyrrolidone, diethyldithiocarbamate, and mercaptobenzthiazol when added to the isolation buffer help remove phenolics and inhibit phenol oxidases.

Some additional thoughts on nuclei isolation

1. While preparing protoplasts for subsequent nuclei isolation, optimization of the type and concentration of enzymes used for cell wall digestion is important as the requirements vary from one system to another. It is also worth looking into the relationship between conditions of protoplast isolation and rupture. In the first step, the emphasis is given on creating conditions for maximal yield and stability of protoplasts but the opposite is the aim in the next step where the nuclei are released by inducing rupture of protoplasts. It may be useful to isolate protoplasts in isolation solutions which favour high yields but relatively low stability of protoplasts. The stability of protoplasts refers to their ability to remain intact and to survive repeated centrifugations necessary for their purification which may result in their lysis. In certain cultures (e.g. *Vicia hajastana*), protoplasts isolated in the presence of mannitol as an osmoticum are relatively more stable than those isolated with sucrose or a mixture of NaCl and KCl. In our recent experiments (unpublished), when mannitol was replaced by sucrose or salts as the osmoticum during isolation, subsequent rupture of protoplasts produced very little debris providing a much cleaner nuclear preparation. A variation in the type of cell wall digesting enzymes, the pH of the isolation mixture, and the incubation temperature may be rewarding as these factors greatly influence protoplast stability.
2. In order to obtain nuclei with maximum nuclear envelope integrity, the detergent concentration should be optimized. Optimization in this case refers to finding the minimal effective concentration of the detergent and the optimum pH which allows the effective rupture of the protoplasts. An increase in the period of deplasmolysis from 5 to 15 min in the NIB (Protocol 2) may prove beneficial in situations where protoplasts are tough to break at lower detergent concentrations.

Uptake of Nuclei

Experiments to induce nuclear uptake as a method of gene transfer were first conducted by Potrykus and Hoffman [20] who were able to introduce foreign nuclei in about 0.5% of the protoplasts by using membrane modifiers (lysozyme, sodium nitrate) together with centrifugation. The percentage of uptake of nuclei by protoplasts was considerably improved (to about 5%) when the uptake was induced using PEG, high pH and Ca^{2+} [16]. The determination of uptake frequencies in the above studies was based on microscopic estimations of host protoplasts containing the transferred nuclei stained with a fluorescent dye. For an accurate estimation of uptake frequencies, it is essential to be able to differentiate the nuclei actually taken-up by protoplasts from those merely attached to the surface of the protoplasts. In previous studies, the location of the nuclei, i.e. outside or inside the protoplasts, was determined by rolling a sample of the protoplast population after uptake between the slide and the coverslip, a process which may not always be accurate. Saxena *et al.* [25] devised a simple procedure based on differential staining of nuclei to follow the uptake of nuclei (Fig. 1B). In this method, nuclear uptake is carried out using the nuclei stained with a fluorescent dye (Hoechst 33258). The entire population of protoplasts is then stained with Evan's Blue. The nuclei attached to the surface of the protoplasts appear blue when examined under a light microscope in bright field and the nuclei taken-up by protoplasts show fluorescence under UV light (Fig. 2). Uptake frequencies ranging from 4 to 6% were obtained with this procedure [25].

The mechanism of nuclear uptake by protoplasts is not well understood, but various possibilities can be discussed based upon indirect evidence available from other studies. Ferritin, representing the smaller class of particles (0.1 μm) has been shown to gain access to protoplasts through endocytosis via coated vesicles. Particles in the middle range (0.1–3 μm), such as polystyrene latex spheres and bacteria, were accepted by membrane invagination and endocytosis. Larger particles (3–16 μm), such as protoplasts, have been shown to be taken up by fusion (for a review see 6). Therefore, nuclear uptake seems likely to occur either by fusion or through plasma membrane invagination by endocytosis (Fig. 3). The size of the nuclei to be introduced relative to the size of recipient protoplast may determine the mode of uptake which may involve invagination of plasma membrane in the case of smaller nuclei and fusion where the donor and recipient are not very different in size. The events following the uptake of nuclei into protoplasts are also unknown. Figure 3 shows various possible ways in which transfer and integration of nuclear DNA may occur.

Integration of nuclear DNA

In the first investigation to follow the integration of the foreign nuclei introduced into protoplasts, Lörz and Potrykus [16] utilized two chlorophyll-deficient, light sensitive tobacco mutants referred to as *sublethal* and *virescent*, as donors of nuclei and recipient protoplasts. These recessive mutants were previously shown to complement successfully by protoplast fusion [17]. The hybrids recovered following transplantation of *virescent* nuclei into *sublethal* protoplasts were identifiable from the parental cell colonies under high light, but the plants regenerated from such cell colonies were all of a *sublethal* type, hence no biological proof could be gathered regarding the integration of the transferred nuclear genes [16]. Success in this area has since been reported by Saxena *et al.* [24] who induced nuclear uptake into protoplasts using PEG treatment and gathered convincing evidence to show that following nuclear uptake, genetic integration did occur in the recipient cells. They [24] utilized, as the recipient, a *Datura innoxia* auxotrophic mutant (*Pn-1*) which had an absolute requirement for pantothenate for growth, and *Vicia hajastana* as the donor of nuclei. Following uptake of *Vicia* nuclei, prototrophic clones were selected on a medium lacking pantothenate. When subjected to dot blot hybridization, the genomic DNA of the putative transformed prototrophic clones showed the presence of donor (*Vicia*) DNA providing evidence that *Vicia* DNA did integrate into the host genome following the uptake. Further, the restoration of morphogenic ability in a non-morphogenic cell line also strengthened the confidence that genetic transfer and integration only could account for this observation. It should, however, be noted that in this study, the correction of auxotrophy was used as the selective marker, a particularly stringent selection pressure. It is noteworthy that in all previous successful attempts on nuclear transplantation, in yeast and mammalian cells, auxotrophy was the marker used in selection [2, 4, 13].

Perhaps the key to successful nuclear transplantation is the choice of an appropriate marker for selection. Unfortunately, a transformation system describing the transfer of specific gene(s) via nuclei as carriers has not been developed yet. The use of other selective markers commonly employed in model plant gene transfer systems, such as the resistance to an antibiotic (e.g. kanamycin), may not be very effective in nuclear transplantation experiments because introduced nuclei would release large amounts of undefined, but very little of specific, DNA. For example, the nuclei isolated from cells of transgenic kanamycin-resistant plants are likely to contain, at best, a few copies of the gene which confers the resistance. Such a low copy number, coupled with generally low frequency of nuclear uptake (compared to up to 90% with other methods), and a reduced viability of the host protoplasts after uptake treatment, pose serious problems in recovering nuclear hybrids. In this context, it may be rewarding to use markers such as kanamycin, chlorsulfuron, and methotrexate, if nuclei are isolated from cells containing multiple copies of the marker gene [30].

In addition to the choice of marker, the relative sizes of the recipient protoplasts and transferred nuclei may also be critical in the recovery of transformed cells. Larger protoplasts are more likely to survive repeated centrifugations required during purification steps prior to culturing when they contain one or more introduced nuclei. The presence of starch grains in the cytoplasm, the density of the cytoplasm, and the number of other organelles in the protoplast (particularly the chloroplasts) will also affect the survival of recipients through various cultural manipulations requiring centrifugation. A high frequency of survival and further development of recipient protoplasts is essential for successful isolation of transformed clones considering the low frequency of nuclear uptake. Thus, the protoplasts from cell suspension cultures which have less dense cytoplasm and no or fewer chloroplasts are ideal recipients.

Another facet of nuclear-based, limited but discrete, gene transfer is the formation of micronuclei, which are essentially spheres containing only a few to single chromosomes surrounded by a thin layer of cytoplasm and plasma membrane. In experiments with animal cells, the fusion of these structures with the desired recipient results in hybrid clones containing only one or a few introduced chromosomes [5]. The induction of micronuclei and their isolation from plant cells has been reported [26]. However, the transfer and integration of foreign DNA by transplanting micronuclei into host cells or protoplasts has not been achieved. On the other hand, the efforts to isolate and transplant chromosomes into plant cells have met with some success. Many workers have reported the successful isolation of chromosomes from a variety of cell cultures e.g. *Triticum, Papaver, Vicia, Petunia*, [9, 18]. The uptake of wheat and parsley chromosomes by wheat, parsley, and maize protoplasts has been described with convincing cytological evidence of their incorporation into the recipient protoplasts [27]. Recently, Griesbach [8] reported the transfer of isolated *Petunia alpicola* chromosomes to *P. hybrida* protoplasts by microinjection. The transformants were shown to have the donor-specific marker proteins and the enzymes of flavonoid pathway as evidence of gene transfer. The characteristic flavonoid enzymes incorporated in the recipient species were transmitted in Mendelian fashion [8].

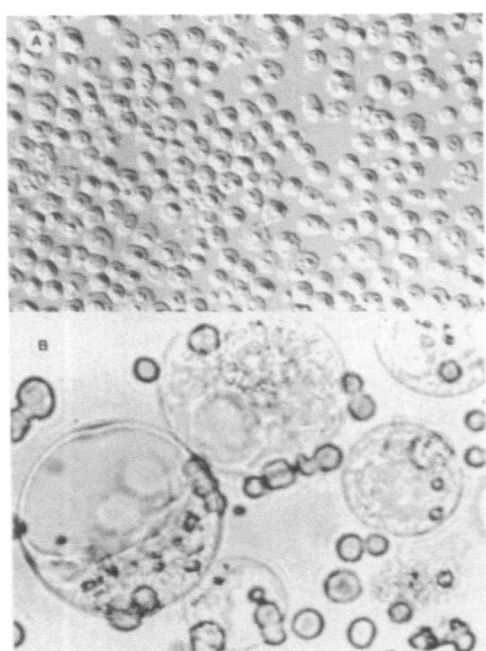

Fig. 1. Isolation and uptake of nuclei. A. Nuclei isolated by the procedure described in protocol 2. B. Induction of nuclear uptake by protoplasts in the presence of uptake inducing solution.

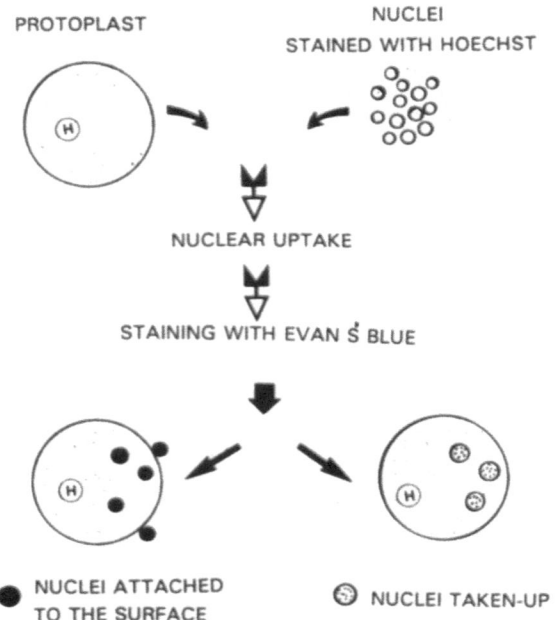

PROTOPLAST

NUCLEI
STAINED WITH HOECHST

NUCLEAR UPTAKE

STAINING WITH EVAN Ś BLUE

● NUCLEI ATTACHED
TO THE SURFACE

⊚ NUCLEI TAKEN-UP

Fig. 2. A diagram showing the technique of differential staining to determine the frequency of nuclear uptake. H denotes the host nucleus.

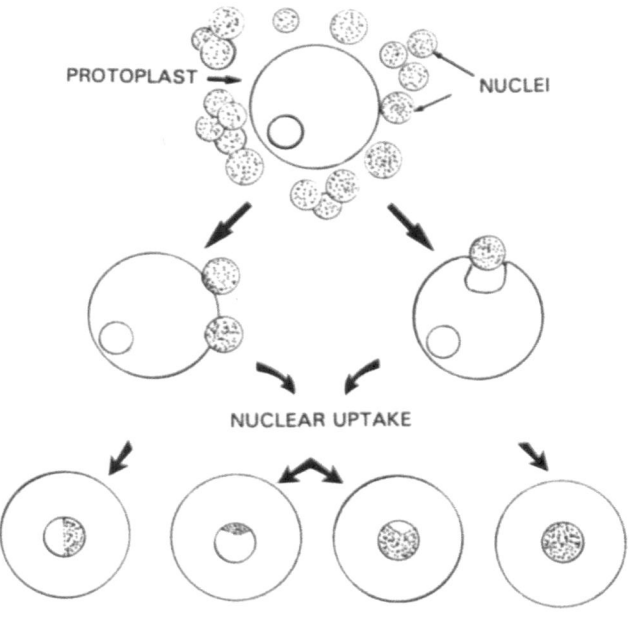

PROTOPLAST →   ← NUCLEI

NUCLEAR UPTAKE

NUCLEAR INTEGRATION

Fig. 3. A diagram showing the possible mechanisms of nuclear uptake and integration.

Procedures

The protocols described here are based on our studies with isolation of nuclei from protoplasts of *Vicia hajastana* and *Brassica nigra*, and their transplantation into the protoplasts of a pantothenate-requiring auxotroph of *Datura innoxia (Pn-1)* [22, 24].

Protocol 1. Isolation of protoplasts

Steps in the procedure

1. Arrange exponential phase (2 to 3-day-old) cell suspension cultures.
2. Using vacuum filtration, collect the cells over Miracloth and transfer 2—2.5 g of *Pn-1* cells to 10 ml of enzyme solution A dispensed into a 10 cm diameter Petri dish.
3. Use 10 ml enzyme solution B to digest 2—2.5 g *B. nigra* or *V. hajastana* cells. Prepare five Petri dishes.
4. Incubate the dishes for 2—3 h on a horizontal shaker (50 rpm) at 25 °C in the dark. Periodically examine the dishes under the microscope to ascertain complete cell wall digestion. When complete digestion had occurred, filter the suspensions through 85 µm nylon mesh and centrifuge at 120 × g for 5 min. Discard the supernatant. Suspend the pellet in 5 ml of 0.6 M sucrose, and layer 1 ml of mannitol solution on top of sucrose. Mannitol concentration for protoplasts to be used later for nuclei isolation should be 0.6 M and for recipient protoplasts it should be 0.5 M. Centrifuge for 5 min at 120 × g.
5. Protoplasts should gather at the interface of sucrose and mannitol. Remove the protoplasts with a Pasteur pipette and dilute the protoplast suspension with 8—10 ml of 0.5 or 0.6 M mannitol and centrifuge for 5 min at 120 × g.
6. Suspend the resulting pellet in appropriate mannitol solution (0.5 or 0.6 M).

Cell lines

— *Datura innoxia* P. Mill *(Pn-1)* cell suspension — a pantothenate-requiring auxotroph
— *Vicia hajastana* and *Brassica nigra* cell suspension cultures

Solutions
— Enzyme solution A: (in a 0.5 M mannitol solution)
   — 1% Cellulase Onozuka "R-10"
   — 1% Cellulase "RS"
   — 0.5% Macerozyme
   — 0.5% Rhozyme "HP150"
   — 5 mM Calcium chloride

- Enzyme solution B: (in a 0.6 M mannitol solution)
    - 1% Cellulase Onozuka "R-10"
    - 0.5% Cellulase "RS"
    - 0.5% Driselase
    - 0.5% Macerozyme
    - 0.5% Rhozyme "HP-150"
    - 5 mM Calcium chloride

Centrifuge the enzyme mixtures at 2500 × g for 15 min at 4 °C, collect the supernatant, adjust the pH to 5.7 and filter sterilize the solutions.
- A solution of 0.6 M sucrose with 5 mM calcium chloride.
- A solution of 0.6 M mannitol.
- A solution of 0.5 M mannitol.

Protocol 2. Isolation of nuclei

Steps in the procedure

1. Centrifuge the tubes containing protoplasts for nuclei isolation for 5 min at 120 × g. Transfer 15 ml of ice-cold NIB to centrifuge tubes each containing 0.2 ml of pelleted protoplasts. Gently mix the suspension by tilting the tubes and leave for 5–7 min on ice.
2. Homogenize the suspension in a glass homogenizer by applying 10–15 gentle strokes and filter the homogenate through two layers of Miracloth, pre-soaked in NIB. This step will remove large debris and partially rupture protoplasts.
3. Filter the partially clean filtrate through a polycarbonate filter with a pore size of 12 μm (filter pore size should be chosen according to the size of nuclei which would vary from system to system).
4. Filter the suspension again, this time using a smaller mesh size (in this case 10 μm as the nuclei are about 8 μm in diameter, hence the decision to employ 12 and 10 μm mesh). Centrifuge at 120–150 × g. The selection of g force for centrifugation will depend upon the nuclear size. Transfer the supernatant to fresh tubes and centrifuge again. This two step centrifugation would ensure a higher yield of nuclei.
5. At this point, pool nuclear pellets from different tubes and add 5 ml of NIB-1. Centrifuge, discard the supernatant, and add 5 ml of NIB-1 again. Centrifuge to obtain the pellet composed of relatively pure assembly of nuclei.

Notes

5. If the final pellet is off-white, it will indicate that nuclei are more or less free of starch grains, but a white pellet indicates the presence of higher numbers of starch grains.

   If the nuclei are prepared for uptake experiments, the presence of starch grains can be ignored, but if desired, removal of starch grains can be achieved by 2 to 4 washes with NIB-1 employing centrifugation.

   Isolated nuclei can be stored in the refrigerator at 4 °C for later use or can be kept on ice for immediate experimentation.

   This procedure yields nuclei which are intact, biologically active, and free of cytoplasmic contamination (Fig. 1A)

Solutions
— Nuclei Isolation Buffer (NIB):
    — 10 mM MES
    — 0.2 M sucrose
    — 0.02% Triton X-100
    — 2.5 mM EDTA
    — 2.5 mM dithiothreitol
    — 0.1 mM spermine
    — 10 mM Sodium chloride
    — 10 mM Potassium chloride

Adjust the pH to 5.3, filter sterilize and store at 4 °C.

At this temperature this buffer can be stored for a week.
— NIB-1: Same buffer without the Triton X-100.

Protocol 3. Uptake of nuclei

Steps in the procedure
1. Obtain a density of about 2×10^6/ml of the recipient protoplasts (in this example, *Pn-1*) and a density of 10^8/ml of the nuclei (*B. nigra* or *V. hajastana*, the donor in this example), and mix the two suspensions to achieve a ratio of 1:50 (protoplast:nuclei) to have a better uptake, but the system can accommodate a ratio as low as 1:20 (protoplast:nuclei) without seriously jeopardizing the experiment.
2. Transfer a drop (500 µl) of protoplast/nuclei mixture to a 35 mm diameter Petri dish and wait for 5 min and then add 100 µl of UIS at the periphery of the mixture drop. Add four more drops (100 µl each) of UIS in a similar fashion at 30 sec. intervals, and incubate for 15–20 min.
3. Following the incubation, gradually dilute the UIS with 10 ml of 0.35 M mannitol, using 2 ml aliquots every 5 min.
4. Centrifuge the suspension at 120 × g for 5 min.
5. Purify the protoplasts over 0.6 M sucrose, as described earlier, and suspend in 0.5 M mannitol.

Solutions
- Uptake Inducing Solution (UIS):
  - 30% Polyethylene glycol
  - 0.1 M Calcium chloride
  - 5% dimethylsulphoxide

Adjust the pH of this solution to 6.8.
- 0.35 M Mannitol with 5 mM calcium chloride.

Protocol 4. Determination of nuclear uptake frequency

Steps in the protocol
1. Stain isolated nuclei with 0.02% Hoechst"33258" for 5—10 min at 0 °C and remove excess stain by washing 3—4 times with NIB-1.
2. Perform the uptake procedure as described above.
3. To a small volume (0.5—1 ml) of protoplast suspension from step 5 above, add an equal volume of 0.1% Evan's Blue solution. After 5 min, wash the suspension with 0.5 M mannitol to remove excess stain.
4. View a drop of suspension (placed on a glass slide with a cover-slip on) under the UV and BF of a fluorescence microscope. Use excitation filters UG I and BG 38 in combination with a barrier filter K 575.
5. The nuclei appearing blue and fluorescing will be the ones outside the protoplasts, whereas those taken up by protoplasts will show only the fluorescence and no blue colour. Score at least 500 protoplasts to calculate the frequency of nuclear uptake.

Solutions
- 0.02% Hoechst "33258" to be stored at 4 °C in the dark.
- 0.1% Evan's Blue in 0.5 M mannitol.

Protocol 5. Selection of prototrophic clones

Steps in the Protocol

1. Culture the protoplasts after nuclear uptake at a density of 10^5 ml^{-1} in 2.5 ml of MS-1 medium and after a week add 2 ml of MS-2 medium.
2. Transfer the cultures containing growing cell colonies to centrifuge tubes, allow to stand for 15 min, and remove the supernatant. Adjust the density of cell colonies to approx. 1000/ml of MS-2 medium.
3. Transfer 1—2 ml of the suspension obtained in step 2 to selection plates. To prepare selection plates, collect cells from a 2 to 3-day-old fast growing wild-type cell suspension culture (*Datura innoxia* in this example), suspend them in MS-2 at a density of 200 mg/ml and transfer 1 ml aliquots to 10 cm diameter Petri dishes containing 25 ml of MS-3. Swirl the Petri dishes gently to distribute the suspension evenly and after 5 h place a snugly-fitting filter paper (10 cm) on top of the cells. Place another filter paper (7 cm) on top of the first one and plate protoplast-derived colonies on this filter paper. Plating on filter paper facilitates the transfer of colonies from one plate to another. Black filter paper may be used to discern the surviving colonies easily. Isolate surviving prototrophic colonies after 2—3 weeks and subculture on MS-3 medium. These are putative trans-formed clones.

Solutions
— MS-1, MS [19] medium containing 0.2 M mannitol, 3% sucrose, 1 mg/l 2,4-D, 0.5 mg/l benzyladenine, and 0.5 mg/l Ca-D-pantothenate.
— MS-2, same as MS-1 but lacking Ca-D-pantothenate.
— MS-3, same as MS-2 but containing 0.8% Difco agar.

References

1. Anderson JW (1986) Extraction of enzymes and subcellular organelles from plant tissues. Phytochemistry 7: 1973–1988.
2. Becher D, Conrad B, Bottcher F (1982) Genetic transfer mediated by isolated nuclei in *Saccharomyces cerevisiae*. Current Genetics 6: 163–165.
3. Dunham VL, Bryant JA (1983) Nuclei. In: Hall JZ, Moore AL (eds) Isolation of membranes and organelles from plant cells, pp 237–275. New York: Academic Press.
4. Ferenczy L, Pesti M (1982) Transfer of isolated nuclei into protoplasts of *Saccharomyces cerevisiae*. Curr Microbiol 7: 157–160.
5. Fournier REK (1982) Microcell-mediated chromosone transfer. In: Shay JW (ed) Techniques in somatic cell genetics, pp 309–327. New York London: Plenum Press.
6. Fowke LC (1986) Ultrastructural cytology of cultured plant tissues, cells, and protoplasts. In: Vasil IK (ed) Cell culture and somatic cell genetics of plants, pp 323–342. New York: Academic Press.
7. Fowke LC, Gamborg OL (1980) Applications of protoplasts to the study of plant cells. Int Rev Cytol 68: 9–51.
8. Griesbach RJ (1987) Chromosome-mediated transformation via microinjection. Plant Sci 50: 69–77.
9. Hadlaczky G, Bisztray G, Praznovszky T, Dudits D (1983) Mass isolation of plant chromosomes and nuclei. Planta 157: 278–285.
10. Hughes BG, Hess WM, Smith MA (1977) Ultrastructure of nuclei isolated from plant protoplasts. Protoplasma 93: 267–274.
11. Kao FT (1983) Somatic cell genetics and gene mapping. Int Rev Cytol 85: 109–146.
12. Kobza J, Edwards GE (1984) Isolation of organelles: Chloroplasts. In: Vasil IK (ed) Cell culture and somatic cell genetics of plants, Vol 1, pp 471–482. New York: Academic Press.
13. Kondorosi E, Duda E (1980) Introduction of foreign genetic material into cultured mammalian cells by liposomes loaded with isolated nuclei. FEBS Lett 120: 37–40.
14. Lörz H (1985) Isolated cell organelles and subprotoplasts – their role in somatic cell genetics. In: Dodds JH (ed) Plant genetic engineering, pp 27–59. London: Cambridge University Press.
15. Lörz H, Potrykus I (1976) Uptake of nuclei into higher plant protoplasts. In: Dudits D, Farkas GL, Maliga P (eds) Uptake of nuclei into higher plant protoplasts, pp 239–244. Budapest: Akadémiai Kiadó.
16. Lörz H, Potrykus I (1978) Investigations on the transfer of isolated nuclei into plant protoplasts. Theor Appl Genet 53: 251–256.
17. Melchers G, Labib G (1974) Somatic hybridization of plants by fusion of protoplasts. I. Selection of light resistant hybrids of "haploid" light sensitive varieties of tobacco. Mol Gen Genet 135: 277–294.
18. Mii M, Saxena PK, Fowke LC, King J (1987) Isolation of chromosomes from cell suspension cultures of *Vicia hajastana* Grossh. Cytologia 52: 523–528.
19. Murashige T, Skoog F (1962) A revised medium for rapid growth and bio assays with tobacco tissue cultures. Physiologia Plantarum 15: 473–497.
20. Potrykus I, Hoffmann F (1973) Transplantation of nuclei into protoplasts of higher plants. Z Pflanzenphysiol 69: 287–289.
21. Saxena PK, King J (1989) Isolation of nuclei and their transplantation into plant protoplasts. In: Bajaj YPS (ed) Biotechnology in Agriculture and Forestry, Vol 9, pp 328–342. Berlin Heidelberg: Springer-Verlag.
22. Saxena PK, Fowke LC, King J (1985a) An efficient procedure for isolation of nuclei from plant protoplasts. Protoplasma 128: 184–189.
23. Saxena PK, Liu Y, Mii M, Fowke LC, King J (1985b) High nuclear yields from protoplasts of several plants. J. Plant Physiol 121: 193–197.
24. Saxena PK, Mii M, Crosby WR, Fowke LC, King J (1986) Transplantation of isolated nuclei into plant protoplasts – A novel technique for introducing foreign DNA into plant cells. Planta 168: 29–35.

25. Saxena PK, Liu Y, King J (1987) Nuclear transplantation into protoplasts: Optimal conditions for induction and determination of nuclear uptake. J Plant Physiol 128: 451–460.
26. Sree Ramulu K, Verhoeven HA, Dijkhuis P (1988) Mitotic dynamics of micronuclei induced by amiprophos-methyl and prospects from chromosome-mediated gene transfer in plants. Theor Appl Genet 75: 575–584.
27. Szabados L, Hadlaczky G, Dudits D (1981) Uptake of isolated plant chromosomes by plant protoplasts. Planta 151: 141–145.
28. Tallman G, Reeck GR (1980) Isolation of nuclei from plant protoplasts without the use of a detergent. Plant Sci Lett 18: 271–275.
29. Willmitzer L, Wagner KG (1981) The isolation of nuclei from tissue-cultured plant cells. Exp Cell Res 135: 69–77.
30. Xiao W, Saxena PK, King J, Rank GH (1987) A transient duplication of the acetolactate synthase gene in a cell culture of *Datura innoxia*. Theor Appl Genet 74: 417–422.

Plant Tissue Culture Manual **D8**: 1–15, 1993.
© 1993 *Kluwer Academic Publishers.*

In situ hybridization to plant metaphase chromosomes: radioactive and non-radioactive detection of repetitive and low copy number genes

J. VEUSKENS[1], S. HINNISDAELS[1], A. MOURAS[2] *Free University of Brussels,*
Institute for Molecular Biology, Paardenstraat, 65, B-1640 St-Genesius-Rode, Belgium
[2] *Université de Bordeaux II, Laboratoire de Biologie Cellulaire, Av. des Facultés, F-33405 Talence-Cedex, France*

Introduction

In situ hybridization is the direct hybridization of a specific nucleic acid probe to cytological preparations such as interphase nuclei, metaphase chromosomes or sectioned tissues. Since it was first established by Gall and Pardue [4], the technique has been improved continuously and considerable progress in the detection procedure has made chromosomal localization of specific sequences on human chromosomes possible [2, 19]. Moreover, *in situ* hybridization has also been used to identify chromosomes, to detect chromosomal abnormalities, to study the spatial organization of genes in interphase nuclei [9] and to investigate the mechanisms of gene amplification [18]. Unfortunately, such progress and practical applications have been restricted essentially to cytogenetic studies of human and other mammalian systems. The development of *in situ* hybridization strategies in plants still lags behind. The major attributes to the slow development of this technique in plant cytogenetics are:
- the presence of a cell wall which prevents proper accessibility of probe DNA to target DNA.
- the difficulty to produce high quality chromosome spreads.
- the lack of marker chromosomes and reproducible banding procedures makes karyotyping of plant chromosomes extremely difficult.

Major improvements in methods for plant chromosome preparations [11, 15] and numerous refinements of the detection procedure have made the *in situ* localization of repetitive as well as single copy DNA sequences onto plant chromosomes possible [1, 6, 7, 8, 12].

In situ hybridization, the use of total genomic DNA or species specific repetitive DNA sequences as probe, has become an important tool for the analysis of alien chromosomes or chromosome fragments in sexual and somatic hybrids [5, 10]. Moreover, *in situ* hybridization has been succesfully used for the detection of transferred genes in genetically modified plants [1, 12, 14].

In this chapter we describe protocols for high quality plant metaphase chromosome preparations and for the *in situ* hybridization/detection of repetitive as well as low copy number genes.

Materials

1. Preparation of metaphase spreads

- Petri dishes (10 cm).
- Vacuum pump.
- Gyratory shaker, 50 rpm, 25–28 °C.
- Stainless sieves 50 μm.
- Conical centrifuge tubes (15 ml).
- Table centrifuge.
- Pasteur pipettes.
- Antimitotic agents (oryzalin, α-bromonaphtalene, colchicine).
- Enzyme mix (0.15 M sorbitol, 0.05 M Na$_3$ citrate, 2% cellulase Onozuka R10, 0.5–1% driselase, 0.3% pectolyase Y23, pH 4.8–5.0).
- W5 salt solution (154 mM NaCl, 125 mM CaCl$_2 \cdot$ 2H$_2$O, 5 mM KCl, 5 mM glucose).
- Fixative (ethanol/glacial acetic acid 3:1 v/v).
- 5N HCl.
- Giemsa stain.
- Sørensen's buffer, pH 6.9.

Sørensen's Buffer
Stock A: 0.2M NaH$_2$PO$_4 \cdot$ 2H$_2$O

Stock B: 0.2M Na$_2$HPO$_4 \cdot$ 7H$_2$O
  or 0.2M Na$_2$HPO$_4 \cdot$ 12H$_2$O
  or 0.2M Na$_2$HPO$_4 \cdot$ 2H$_2$O
Mix 39 ml Stock A, 61 ml Stock B, and 100 ml H$_2$O (total 200 ml).

- Clean microscopic slides (washed in ethanol/diethylether 1:1 v/v).

2. Probe preparation

Radioactive labelling
- ^{3}H-dXTP with highest specific activity (dATP, dCTP, dTTP).
- non radioactive dGTP.
- 0.2M EDTA, pH 8.0.
- NTB: 50 mM Tris-Hcl, pH 7.8, 5 mM MgCl$_2$, 10 mM beta-mercaptoethanol.
- DNAse I from bovine pancreas and *E. coli* DNA polymerase I.
- Sephadex G-50 or G-100 gel in 1 ml plastic syringe. The gel is previously equilibrated with buffer (1 × TE).
- 1 × TE, pH 8.0 (10 mM Tris-HCl, 1 mM EDTA).

Non radioactive labelling
- digoxigenin-11-dUTP labelling Kit (Boeringher Mannheim).
- 0.2 M EDTA, pH 8.0.
- 4 M LiCl.
- prechilled absolute ethanol.
- 1 × TE, pH 8.0.

3. Hybridization

- coating of microscopic slides:
  immerse slides for 5 min in 1N HCl at 95 °C., neutralize for 1 min in 1 M Tris-HCl pH 7.0, wash briefly in distilled water, dehydrate in 95% ethanol, air dry and incubate in 3 × SSC with 1 × Denhardt's solution for 4 hours at 65 °C. Rinse the slides briefly in water, fix in ethanol/acetic acid 3:1 v/v for 20 min and then air dry. The slides can be stored at 4 °C in dust free slide containers for several months.
- coverslip preparation (24 × 40 mm):
  wash in 1 N HCl, 95 °C for 5 min, rinse in Tris-HCl 1 M, pH 7.0 and distilled water, store in 95% ethanol at 4 °C.
- RNAse A (100 μg/ml 2 × SSC):
  boil for 10 min and store in 1 ml aliquots at − 20 °C.
- RNAse solution:
  mixture of RNAse A (100 μg/ml) and RNAse T1 (500 units/ml) in 2 × SSC, 2 × Denhardt's solution.
- siliconized coverslips (22 × 32 mm):
  siliconized coverslips are obtained by dipping them into Sigmacote solution (sigmacote SL-2, Sigma) followed by 3 washes in distilled water and baking at 180 °C for 120 min.
- 50 × Denhardt's (1% ficoll 400, 1% polyvinylpyrrolidone 360, 1% bovine serum albumin).
- 20 × SSC (3 M NaCl, 0.3 M sodium citrate, pH 7.0).
- 70% ethanol, 95% ethanol.
- 10% (w/v) dextransulphate solution.
- 20% SDS in distilled water.
- deionised formamide.
- rubber cement.
- sonicated or sheared carrier DNA (salmon sperm DNA or prokaryotic DNA chosen as function of target DNA).
- moist chamber.
- ovens set at 25 °C and 40 °C.

4. Detection and visualization

Radioactive probes
- 20 × SSC.
- 0.3 M ammonium acetate in 70% ethanol and in 95% ethanol.
- Kodak NTB-2 nuclear track emulsion.
- Kodak D-19 developer, diluted 1:1 with distilled water.
- Kodak rapid fixer.
- slide boxes (light-tight).
- water bath.
- dark room with safelight.

Non radioactive probes (Boeringher Mannheim)
- 20 × SSC.
- 70% and 95% ethanol.
- buffer 1 (100 mM Tris-HCl, 150 mM NaCl, pH 7.5).
- buffer 2 (0.5% w/v blocking reagent in buffer 1)
- buffer 3 (100 mM Tris-HCl, 100 mM NaCl, 50 mM $MgCl_2$, pH 9.5).
- buffer 4 (10mM Tris-HCl, 1 mM EDTA, pH 8.0).
- antibody-conjugate and NBT/X-phosphate solutions (Boeringher Mannheim).
- water bath.

Methods

1. Preparation of metaphase spreads

The preparation of high quality chromosome spreads free of cell wall and cellular debris is a prerequisite for successful *in situ* hybridization. The best source of material can be obtained by protoplasting plant root meristems. If only a little amount of material is available, the fixed meristems can be transformed into protoplasts by the method developed by Pijnacker and Ferwerda [15] or as described by Ambros *et al.* [1]. In our hands, both techniques have led to suitable chromosome spreads but the results may vary between and within plant species. Besides, a lot of experience is needed to manipulate meristematic cells. We therefore recommend a more convenient technique [11]. The use of *Agrobacterium rhizogenes* transformed root cultures provide large amounts of dividing cells to prepare protoplasts. The advantages of the method are that a lot of material is available and that the root cultures can be maintained on solid medium for years without any effect on chromosome number and chromosome morphology.

Steps in the procedure (adapted and modified from Mouras *et al.* [11])

1. Arrest young growing roots in metaphase by adding, according to the plant species, 30 μM oryzalin (Dow Elenco) or 0.05% alpha-chloro- or alpha-bromo-naphthalene in distilled water. Place on a gyratory shaker at 25 °C for 60 min. The correct time of prefixation has to be determined experimentally.
2. Collect roots, wash in distilled water and transfer to 10 ml enzyme mix. Enhance penetration of enzymes by slight vacuum infiltration for 3−5 min.
3. Incubate on a gyratory shaker (50 rpm, 25 °C). Protoplasts can be harvested when most of the root tips are detached from the rest of the roots (maximum 150 min).
4. Pour gently the protoplast suspension through a stainless sieve of 50 or 63 μm into a beaker. Rinse the remaining roots with 5 ml of 25% W5 solution to collect most of the protoplasts.
5. Transfer the filtered protoplast suspension into 15 ml conical centrifuge tubes and spin at 700 rpm (95 × g) for 5 min.
6. Remove the supernatant and submit cells to hypotonic shock by resuspending the pellet in 10 ml 12.5% W5 solution. Leave at room temperature for 10 min and spin as above.
7. Discard the supernatant and gently resuspend the pellet in 0.5 ml of the hypotonic solution. Add slowly 10 ml of freshly prepared cold fixative (3 : 1 v/v ethanol or methanol/glacial acetic acid) and mix gently. Spin at 700 rpm (95 × g) for 5 min.
8. Remove fixative, and resuspend the protoplasts in 10 ml of cold fresh fixative. Allow fixation to proceed for at least 30 min at −20 °C.
9. Leave 5−10 min at room temperature and spin at 700 rpm (95 × g) for 10 min. Remove excess fixative and adjust the volume to a final protoplast density of 10^6 ppl/ml.

10. With a Pasteur pipette let 1 to 2 drops of protoplast suspension in fixative fall onto a cleaned microscopic slide and allow to air dry. The rest of the protoplast suspension can be stored in fixative at −20 °C. for months.
11. Hydrolyse the preparation by immersion in 5N HCl for 20 min at room temperature and rinse thoroughly with tap water.
12. Stain the slide in 2% Giemsa solution in Sørensen's buffer (pH 6.9) and examine under the microscope. Unstained preparations can be examined by phase contrast microscopy.

2. Probe preparation

Choice of probe and probe labelling

Recombinant DNA technology provides the opportunity to obtain DNA or RNA probes from genomic or cDNA libraries of virtually any desired sequence. Such cloned DNA and RNA probes have been used successfully in *in situ* hybridization experiments. At present both radioactive and non radioactive methods are available for the labelling of DNA and RNA probes. The isotopes ^{3}H and ^{35}S have been the most commonly used for *in situ* hybridization [13]. The best isotope to use for accurate localization of sequences is ^{3}H due to its relatively low energy. However, long autoradiographic exposure times are needed for detection of single copy DNA sequences. Recently, non-radioactive labels such as biotin and digoxigenin have been used successfully in *in situ* hybridization experiments [6, 16]. Probes can be stored for longer times, the detection is fast, accurate and does not require the need to handle radioactive material. For the efficient incorporation of radioactive or non-radioactive nucleotides, both the modified nick translation [17] as well as the random priming labelling method [3] can be used.

Steps in the procedure

Radioactive labelling (Nick translation)
1. Dry down aliquots of 10 μM of each of the three labelled nucleotides (^{3}H dATP, ^{3}H dCTP, ^{3}H dTTP) with the highest specific activity available in microfuge tubes with a speedvac under vacuum in the cold room.
2. Place on ice and resuspend in 40 μl H$_2$O.
3. Add 5 μl 10 × NTB.
4. Add 60 μM cold dGTP.
5. Add 300–500 ng probe DNA.
6. Add 4 ng/ml DNAse I.
7. Add 5 units DNA polymerase I.
8. Incubate for 90 min at 15 °C.
9. Stop the reaction by adding 5 μl of 0.2 M EDTA, pH 8.0.
10. Add 30 μl 0.1% Orange G stain solution to monitor the reaction and separate unincorporated deoxyribonucleotide triphosphates from labelled DNA by chromatography on a 1 ml Sephadex G50 or G100 column. Collect fractions of 2

drops in Eppendorf tubes and measure the radioactivity of a sample (2 μl) of each fraction by liquid scintillation counting.

11. Select the tubes that contain labelled probe and pool them together.

Notes
1. Alternatively, DNA can be labelled using the random primer labelling method. Higher specific activities are obtained but due to the smaller generated fragments, a higher background hybridization can be observed.
2. For repetitive DNA probes a specific activity of 3×10^7 cpm/μg is suitable; single copy sequences require specific activities higher than 10^8 cpm/μg.

Non radioactive labelling (Random priming)
Protocol for labelling probe DNA with digoxigenin-11-dUTP by random priming according to Boeringher Mannheim (Cat. No. 1093657).

Steps in the procedure
1. Add the following compounds to a microfuge tube on ice:
   - 300—500 ng linearized DNA.
   - denature for 10 min at 95 °C and chill immediately on ice.
   - 2 μl 10 × hexanucleotide mixture.
   - 2 μl 10 × dNTP labelling mixture.
   - sterile water up to 19 μl.
   - 1 μl Klenow enzyme (2 units).
2. Centrifuge briefly and incubate overnight at room temperature.
3. Next morning add:
   - 2 μl 0.2M EDTA, pH 8.0.
   - 2.5 μl 4M LiCl.
   - 75 μl prechilled ethanol.
4. Mix well and leave for 60 min at -70 °C.
5. Centrifuge for 10 min at 12000 × g.
6. Wash pellet with cold ethanol (70% v/v), and centrifuge again.
7. Air dry and dissolve at 37 °C in 50 μl TE, pH 8.0.

Note
Labelled DNA can be stored at -20 °C for several months.

3. Preparation of hybridization mix

Calculate the volume of hybridization mix according to the number of slides which will be used. The volume of hybridization mix required per slide depends on the size of the coverslip. For a coverslip of 22 × 32 mm, 35 μl of hybridization mix is recommended. The hybridization mix contains:
- labelled DNA probe, 0.2—1 ng/μl hybridization mix.
- carrier DNA (calf thymus DNA, herring sperm DNA or lambda DNA), 500—1000× in excess to probe DNA.
- 50% (v/v) formamide (BRL 5515UA).

- 4–5 × SSC.
- 1× Denhardt's solution.
- 0.25% SDS.
- 10% dextran sulphate.
- adjust final volume with distilled water.

4. Hybridization of labelled DNA with nuclei and chromosomes

Steps in the procedure

1. Adjust the protoplast suspension to about 5×10^5 to 10^6 cells/ml; with a Pasteur pipette let 1 to 2 drops of protoplast suspension in fixative fall onto the surface of precoated microscopic slides and let air dry.
2. Place 80 μl of RNAse solution on slides and cover with pretreated coverslips (24 × 40 mm). Incubate for 60 min at 37 °C in a moist chamber containing 2 × SSC.
3. Wash the slides in 2 × SSC, 3 times, 5 min each.
4. Dehydrate the preparations successively in 70% and 95% ethanol for 10 min and let air dry.
5. Meanwhile, prepare hybridization mix containing probe DNA; denature by heating at 70 °C for 15 min and chill quickly on ice.
6. Place 35 μl of hybridization mix on each slide and cover with a siliconized coverslip (22 × 32 mm).
7. Seal with rubber cement and leave to dry completely.
8. Place slides in plastic bags and denature the target and probe DNA on the slides by incubation in a water bath at 70 °C or 80 °C, according to the plant species; temperature and time of denaturation have to be determined experimentally.
9. Hybridization is carried out overnight up to 48 hours in a moist chamber at 27 or 37 °C depending on the plant species.

5. Detection and visualization

Steps in the procedure

1. Remove carefully the plastic bag and peel off the rubber cement.
2. Dip slides in prewarmed (40 °C) 2 × SSC to remove the coverslips.
3. Remove excess and nonspecifically-bound probe by washes of different stringency as a function of DNA probe used during the hybridization. As a guide to start, the following washes can be recommended:
   - three times 2 × SSC at 37 °C, 10 min each
   - once 1 × SSC at 65 °C for 15 min.
   - three times 2 × SSC at 37 °C, 10 min each.
   - three times 2 × SSC at room temperature, 10 min each.
4. Dip the slides in distilled water and subsequently dehydrate the preparations for 10 min in 70% and 95% ethanol (containing 0.3 M ammonium acetate if slides have to be coated with NTB-2 emulsion).
5. Air dry slides and proceed with the detection.

Detection of hybridization signals

Radioactive probes
All manipulations should be done in a light-tight room!

Steps in the procedure
1. Kodak nuclear track emulsion NTB-2 is melted at 43 °C–45 °C and diluted 1:1 with distilled water containing 0.6 M ammonium acetate. The emulsion is aliquoted into 10 ml samples and stored in plastic light-proof containers at 4 °C until use. The emulsion can be kept for a maximum of 1 year.
2. Place the dipping chamber and an aliquot of emulsion in a water bath set at 45 °C. Pour the melted emulsion slowly into the dipping chamber; avoid air bubbles.
3. Coat slides by dipping them slowly two to three times into emulsion, clean the back of the slides with a paper towel and place them on a precooled glass plate (4 °C). Allow the emulsion to solidify and dry for 1 hour on a hot plate set at 37 °C.
4. Put the slides in light-tight boxes containing silicagel and store at 4 °C for exposure. Autoradiographic exposure times may vary from a few days (repetitive sequences) to several weeks (low copy number genes).
5. Develop the coated slides in developer solution for 2 to 3 min at 18 °C.
6. Wash briefly in distilled water.
7. Fix in rapid fixer for 5 min and rinse with tap water.
8. Air dry slides and stain with freshly prepared 5% Giemsa solution. The chromosomes may not be overstained so that silver grains remain clearly visible. Therefore, staining should be monitored continuously under an inverted microscope.
9. Rinse the slides in running tap water and let dry.
10. Examine the slides under the microscope (e.g. Zeiss Axiophot) and take pictures (black and white: Agfapan APX 25, color slides: Agfachrome 50 RS).

Non radioactive probes (according to Boeringher Mannheim)

Steps in the procedure
1. Wash the slides briefly in buffer 1.
2. Incubate for 30 min in buffer 2.
3. Wash the slides briefly in buffer 1.
4. Dilute antibody-conjugate in buffer 1 (150 mU/ml).
5. Incubate slides for 60 min with diluted antibody-conjugate solution.
6. Remove unbound antibody conjugate by washing 2–3 times 10 min each with buffer 1.
7. Incubate the slides in buffer 3 for 2 min.
8. Meanwhile, prepare color solution: 4.5 μl NBT solution and 3.5 μl X-phosphate solution in 1 ml buffer 3.
9. Incubate the slides with freshly prepared color solution for at least four hours

at room temperature in the dark. Do not shake or mix when color is developing.

10. Stop the color reaction by washing the slides in buffer 4.
11. Stain and examine the slides as described for radioactive detection.

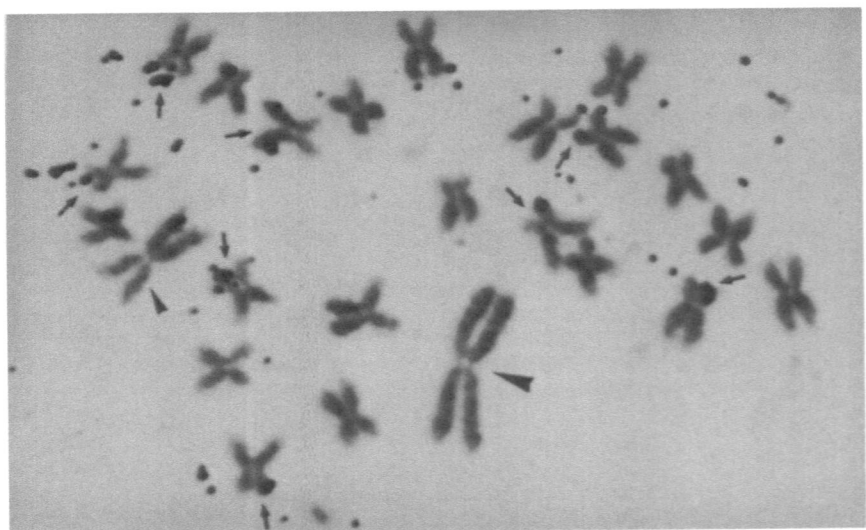

Fig. 1. In situ localization of rDNA genes on metaphase chromosomes of a *Melandrium album* male line (24, XY) by means of a ^{3}H labelled probe. Hybridization signals are seen on the telomeric regions of four chromosome pairs (arrows). The X and Y sex chromosomes are shown by small and large arrowheads, respectively.

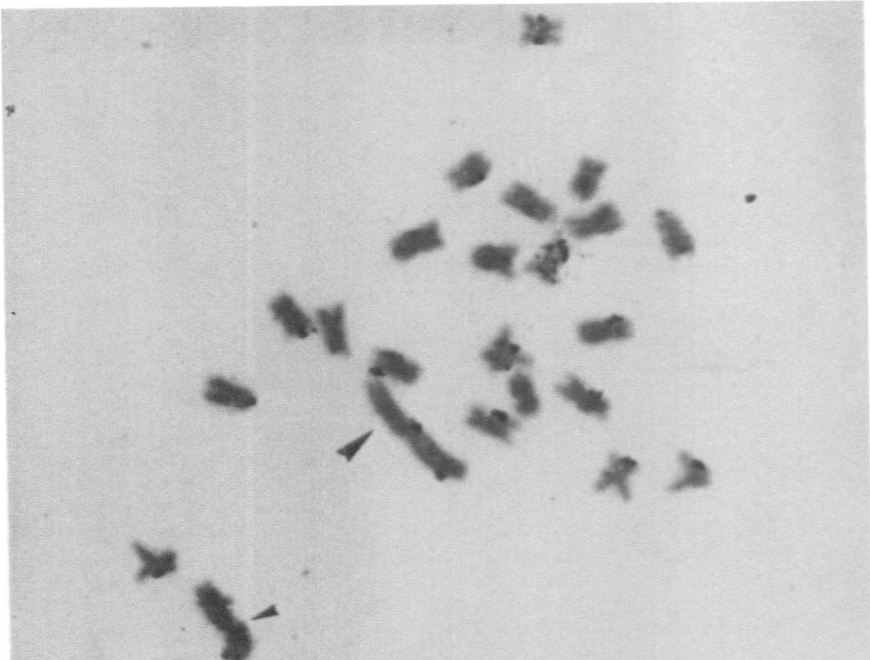

Fig. 2. Partial metaphase plate of *Melandrium album* after *in situ* hybridization with digoxigenin labelled β-tubulin genes. The X and Y sex chromosomes are shown by small and large arrowheads, respectively.

Trouble-shooting guide

Problems	Possible reasons and solutions
Poor chromosome spreading.	– *Improper hypotonic shock.* Adjust osmolarity of the hypotonic solution. Observe swelling (not bursting!) of protoplasts continuously under inverted microscope during hypotonic shock. – *Improperly fixed cells.* Change fixative until pellet becomes white. – *Improper evaporation of fixative.* Blowing short air puffs across the surface of the slide tends to improve the chromosome spreading. Inclination of the slide when adding protoplast suspension is another good alternative.
Improper probe labelling.	– *Hydrolysis of ^3H dNTP.* Dry labelled nucleotides under vacuum in a speedvac at 4 °C. – *High plasmid impurity and presence of enzyme inhibitors.* Further purificate plasmid by $CsCl_2$ gradient centrifugation. Titrate DNAse I for optimal conditions.
No or very weak hybridization signals.	– *Improper (insufficient or excessive) chromosome denaturation.* A balance must be found between hybridization efficiency and chromosome structure preservation. – *Too high stringency washes.* Decrease stringency conditions. – *Insufficient network formation.* Use recombinant molecules as probe (insert + vector) since the presence of vector sequences can contribute to the formation of networks.
Excessive labelling of chromosomes and high background formation.	– *Too low stringency washes.* Increase stringency conditions to remove aspecific hybridization signals. – *Insufficient amount of carrier DNA.* Use excess of carrier DNA (500–1000 ×) to probe DNA. – *High background signals.* Always use siliconized slides.

Light leaks in the darkroom or emulsion was outdated or exposed to radiation. Always check the quality of the emulsion prior to use.
Enhance blocking efficiency to avoid aspecific binding of antibody-conjugate during non radioactive detection.

Acknowledgments

J. Veuskens is a PhD student financed by I.W.O.N.L (880249). Dr. S. Hinnisdaels is a grantee of the National Foundation of Scientific Research, Belgium (Aangesteld Navorser N.F.W.O.).

References

1. Ambros, PF, Matzke, MA, Matzke, AJM (1986) Detection of a 17 kb unique sequence (T-DNA) in plant chromosomes by *in situ* hybridization. Chromosoma 94: 11–18.
2. Fan, YS, Davis, LM, Shows, TB (1990) Mapping small DNA sequences by fluorescence *in situ* hybridization directly on banded metaphase chromosomes. Proc Natl Acad Sci USA 87: 6223–6227.
3. Feinberg, AP, Vogelstein, B (1983) A technique for radiolabeling DNA restriction endonuclease fragments to high specific activity. Anal Biochem 132: 6–13.
4. Gall, JG, Pardue, ML (1969) Formation and detection of RNA-DNA hybrid molecules in cytological preparations. Proc Natl Acad Sci USA 63: 378–383.
5. Guidet, F, Rogowsky, P, Taylor, C, Song, W, Langridge, P (1991) Cloning and characterization of a new rye-specific repeated sequence. Genome 34: 81–87.
6. Gustafson, JP, Butler, E, McIntyre, CL (1990) Physical mapping of a low-copy DNA sequence in rye (Secale cereale L.). Proc Natl Acad Sci USA 87: 4971–4975.
7. Huang, PL, Hahlbrock, K, Somssich, IE (1988) Detection of a single-copy gene on plant chromosomes by *in situ* hybridization. Mol Gen Genet 211: 143–147.
8. Hutchinson, J, Lonsdale, LM (1982) The chromosomal distribution of cloned highly repetitive sequences from hexaploid wheat. Heredity 48: 371–376.
9. Lawrence, JB, Villnave, CA, Singer, RH (1988) Sensitive, high resolution chromatin and chromosome mapping *in situ*: presence and orientation of two closely integrated copies of EBV in a lymphoma line. Cell 52: 51–61.
10. Le, HT, Armstrong, KC, Miki, B (1989) Detection of rye DNA in wheat-rye hybrids and wheat translocation stocks using total genomic DNA as a probe. Plant Mol Biol Reporter 7(2): 150–158.
11. Mouras, A, Salesses, G, Lutz, A (1978) Sur l'utilisation des protoplastes en cytologie: amélioration d'une méthode récente en vue de l'identification des chromosomes mitotiques des genres *Nicotiana* et *Prunus*. Caryologia 31: 117–127.
12. Mouras, A, Saul, MW, Essad, S, Potrykus, I (1987) Localization by *in situ* hybridization of a low copy chimaeric resistance gene introduced into plants by direct gene transfer. Mol Gen Genet 207: 204–209.
13. Mouras, A, Negrutiu, I, Horth, M, Jacobs, M (1989) From repetitive DNA sequences to single copy gene mapping in plant chromosomes by *in situ* hybridization. Plant Physiol Biochem 27(2): 161–168.
14. Mouras, A, Negrutiu, I (1989) Localization of the T-DNA on marker chromosomes in transformed tobacco cells by *in situ* hybridization. Theor Appl Genet 78: 715–720.
15. Pijnacker, LP, Ferwerda, MA (1984) Giemsa C-banding of potato chromosomes. Can J Genet Cytol 26: 415–419.
16. Rayburn, AL, Gill, BS (1985) Use of biotin-labelled probes to map specific DNA sequences on wheat chromosomes. The Journal of Heredity 76: 78–81.
17. Rigby, PWJ, Dieckmann, M, Rhodes, C, Berg, P (1977) Labeling deoxyribonucleic acid to a high specific activity *in vitro* by nick translation with DNA polymerase I. J Mol Biol 113: 237–251.
18. Trask, BJ, Hamlin, JL (1989) Early dihydrofolate reductase gene amplification events in CHO cells usually occur on the same chromosome arm as the original locus. Genes & Development 3: 1913–1925.
19. Viegas-Pequignot, E, Dutrillaux, B, Magdelenat, H, Coppey-Moisan, M (1989) Mapping of single-copy DNA sequences on human chromosomes by *in situ* hybridization with biotinylated probes: Enhancement of detection sensitivity by intensified-fluorescence digital-imaging microscopy. Proc Natl Acad Sci USA 86: 582–586.

Plant Tissue Culture Manual **D9**, 1–15, 1995.

Chemical fusion of protoplasts

P. ANTHONY, R. MARCHANT, N.W. BLACKHALL, J.B. POWER &
M.R. DAVEY
Plant Genetic Manipulation Group, Department of Life Science, University of Nottingham, University Park, Nottingham, NG7 2RD, U.K.

Introduction

Somatic hybridisation of plants offers a novel method of circumventing sexual
barriers in plant breeding. It not only permits the production of new hybrids between sexually incompatible genotypes [14], but also serves as a method for the
genetic modification of vegetatively propagated crops, sterile or subfertile species and plants with relatively long life cycles [5], such as tree species. Somatic
hybridisation involves four distinct stages; protoplast isolation, protoplast fusion,
selection and plant regeneration followed by cytological and molecular analysis
of regenerated plants.

The successful isolation of protoplasts is dependent on several factors such as
tissue source (e.g. leaves, cell suspensions, cotyledons, roots, pollen tetrads), the
plant species and cultivar, together with the enzyme mixture and the physiological status of the source material, including the nature of the cell wall. The over-
riding consideration is protoplast viability, not only after isolation, but also after
subsequent fusion.

Fusion can be mediated by either chemical or electrical techniques [21; see
also chapter D3, this volume]. In the case of chemical fusion, relatively high
concentrations of fusogens such as sodium nitrate [17], calcium nitrate [22],
polyvinyl alcohol [12], dextran sulphate [8] or polyethylene glycol (PEG) [9]
have been employed, sometimes in combination with high pH/Ca^{2+} [1]. When
in close contact, the protoplast plasma membranes are temporarily destabilized
by electrical stimulation or the action of the fusogen, which results in pore formation and permits cytoplasmic linkage between adjacent protoplasts. These
linkages are thought to inhibit pore closure and permit randomly orientated lipid
molecules on the periphery of the pores to align and form membrane bridges
between adjacent protoplasts [2]. Subsequent mixing of cytoplasms occurs, followed by nuclear fusion, or, more frequently, by common spindle formation
during the first mitosis. There is evidence to suggest that the pretreatment of
protoplasts with non-ionic surfactants, or their inclusion in the fusogen solution,
may promote protoplast fusion.

Protoplast fusion results in the production of a mixed population of heterokaryons, homokaryons and unfused parental protoplasts. Under optimum culture conditions, heterokaryons undergo cell wall formation and mitosis, often
resulting in somatic hybrid cells by days 2–4 of culture. It is frequently necessary to utilise a selection strategy, involving either a means of isolating the het-

erokaryons, or a genetic complementation system designed to preferentially select for hybrid cells. Methods for the selection of heterokaryons have employed techniques such as manual micromanipulation [10] or automated flow cytometry [6]. Approaches for the selection of hybrid cells include those based on resistance to antibiotics, herbicides and amino acid analogues, and complementation involving the use of iodoacetamide treatment [13], x-ray or γ irradiation [20], or albino mutants [19]. Hybrid cells must be totipotent and capable of development, via embryogenesis or organogenesis, into somatic hybrid plants. In practical terms, shoot regeneration capacity need only be established for one of the parental protoplast systems.

Somatic hybrid plants can be characterised at the morphological, cytological, biochemical and molecular levels. Routine biochemical characterisation includes analyses of isozymes and fraction 1 protein, and the resistance of plants to viral infection and fungal toxins, together with their sensitivity to herbicides and antibiotics. Genetic analysis can be performed, provided hybrid plants are fertile [18]. More recently, the molecular techniques of RFLP and RAPD analyses have been adopted [4, 15, 16], while flow cytometry provides a rapid method for analysing the nuclear DNA content for ploidy determinations [7].

Techniques for large-scale chemically induced protoplast fusion

The following procedures are routine, well proven methods for the isolation and chemically-induced fusion of leaf mesophyll protoplasts of *Petunia parodii* with cell suspension-derived protoplasts of *P. hybrida* [18]. They can be adapted readily to any combination of protoplast systems. In this model system, heterokaryons are easily identified as they contain chloroplasts from the leaf mesophyll-derived protoplast partner, in the highly cytoplasmic background which originates from the cell suspension-derived protoplasts. Such a visual marker system permits optimisation of fusion conditions, including the ratio of the parental protoplasts in the fusion mixture and the temperature and duration of exposure to the fusogen. Despite the cytotoxic effects of PEG to protoplasts of some species, a method using purified PEG, with a reduced carbonyl content, ensures both higher frequencies of heterokaryon formation and viability than can be obtained using unpurified PEG preparations [3]. Autoclaving of PEG causes an increase in the carbonyl content. Consequently, it is advisable to filter sterilise PEG solutions. Storage should be in the dark at 4 °C for as short a time as possible, ideally less than a week. Some manufacturers now supply purified PEG as a sterile solution protected from the light in a suitable container and under nitrogen. This is a particularly convenient way of handling PEG.

Procedures

Isolation of leaf mesophyll protoplasts of Petunia parodii

Steps in the procedure
1. Detach several fully expanded basal leaves, including petioles, from plants (approx. 40 days old from seed germination) and place, together with two disposable medical gloves, in a sterile casserole dish. Surface sterilise leaves and the gloves by immersion in 7.5% (v/v) "Domestos" bleach solution (or the equivalent hypochlorite solution with wetting agent) for 30 min.
2. Using one sterilised glove, transfer the leaves to a second casserole dish and rinse both the leaves and the gloves three times in sterile tap water. Decant the water between each rinse.
3. Place the leaves on a rigid sterile surface (e.g. a sterile white tile). Remove the lower epidermis of the leaves by peeling with fine forceps, whilst holding each leaf flat with a gloved hand; discard the epidermis. Place the peeled leaves, adaxial side downwards, on the surface of approx. 30 ml of CPW13M solution (see Solutions) contained in a 14 cm Petri dish; completely cover the liquid surface with peeled leaf explants, but avoid overlapping the peeled explants. When yield per g. fresh weight is required, the leaf pieces can be weighed by placing them in a pre-weighed Petri dish containing CPW13M solution.
4. Leave the peeled explants on CPW13M solution for 30 min to 1 h; pipette off the CPW13M solution and replace immediately with 25 ml of the enzyme solution (see Solutions).
5. Incubate the leaves stationary overnight at 28 °C, in the dark.
6. Following incubation, remove the enzyme solution with a pipette, taking care not to disturb the digested leaf pieces. Replace the enzyme mixture with 20 ml of CPW21S solution (see Solutions). Release the protoplasts into the sucrose solution by gently squeezing the leaf material with a Pasteur pipette against the side of the Petri dish.
7. Transfer, by Pasteur pipette, the protoplast enzyme-mixture to 16 ml screw-cap centrifuge tubes. Adjust the volume of the liquid in the tubes, as necessary, with CPW21S solution, and centrifuge at $100 \times g$ for 10 min. Pipette the floating protoplasts (now free of debris) from the surface of the liquid and transfer to a clean centrifuge tube.
8. Wash the protoplast suspension once with CPW13M solution by resuspension and centrifugation ($80 \times g$ for 10 min). Discard the supernatant and resuspend the protoplast pellet in 10 ml (or a known volume) of CPW13M solution for the estimation of total pro-

toplast yield or yield per g. fresh weight, if the latter is required. Count the protoplasts using a haemocytometer.

9. Pellet the protoplasts by centrifugation ($80 \times g$ for 10 min), remove the supernatant and resuspend the protoplasts at 2.0×10^5 ml^{-1} in MSP19M liquid culture medium.

Isolation of cell suspension protoplasts of albino Petunia hybrida

Steps in the procedure
1. Use a cell suspension 3–5 d after subculture, at which time the cell walls are relatively thin.
2. Allow the cells to settle to the bottom of a 250 ml Erlenmeyer flask, decant the medium and remove any residual liquid with a Pasteur pipette. Alternatively, cells can be harvested on a 64 μm nylon sieve and transferred into a 250 ml flask. Add 30 ml of enzyme solution to 5–10 g fresh weight of cells (see Solutions) and incubate overnight (25 °C) at 40 rpm (horizontal shaker) in the dark.
3. Transfer, by pouring, the digested cells to 16 ml screw-capped tubes (top up with CPW13M solution as required) and centrifuge (80 × g for 10 min). Discard the supernatants and resuspend the pellets, containing protoplasts, in 5 ml per tube of CPW21S solution.
4. Pour the contents of 2–3 centrifuge tubes through a sterile 64 μm nylon sieve placed in the base of a 9 cm diameter Petri dish. Wash the sieve with 1–2 ml aliquots of CPW21S solution (several changes) until all the released protoplasts have passed through the sieve.
5. Transfer the filtrate (with protoplasts) to 16 ml screw-capped tubes (top up with CPW21S solution) and centrifuge (100 × g for 10 min).
6. Collect the floating protoplasts from the surface of the liquid in each tube and pool the samples in one clean centrifuge tube.
7. Wash the protoplasts once with CPW13M solution by resuspension and centrifugation (80 × g for 10 min). Discard the supernatant and resuspend the pellet in 10 ml of CPW13M solution; estimate the protoplast yield using a haemocytometer.
8. Pellet the protoplasts by centrifugation (80 × g for 10 min), remove the supernatant and resuspend the protoplasts at 2.0×10^5 ml^{-1} in MSP19M liquid culture medium.

General procedure and experimental design for the chemical fusion of protoplasts (volumes/replicates will vary depending on yield when applied to other species combinations)

Steps in the procedure

1. Suspend 7.2×10^6 protoplasts of *P. parodii* and *P. hybrida*, each at a density of 2.0×10^5 ml^{-1}, in 36 ml of MSP19M liquid culture medium (see Solutions and reference 18).

2. Dispense 4 ml of each protoplast suspension into two separate 16 ml centrifuge tubes to act as viability controls. A further 8 ml of each preparation should be dispensed into two separate centrifuge tubes to act as self-fusion controls to monitor viability and potential cross-feeding following culture medium-based selection for hybrids [18]. Add 4 ml of each of the two protoplast suspensions to three additional centrifuge tubes for the fusion treatments. More fusion tubes can be set up if protoplast yield permits.

3. Centrifuge all the tubes (except the viability controls) at $100 \times g$ for 5 min and discard the supernatants. Gently resuspend the pelleted protoplasts in 0.5 ml aliquots of MSP19M liquid medium.

4. Add the fusogen as detailed in the individual fusion protocols.

5. Following fusion (see separate protocols) centrifuge the tubes ($80 \times g$ for 10 min) and discard the supernatants. Resuspend the pelleted protoplasts in 16 ml per tube of MSP19M liquid culture medium (protoplasts are now at a density of 1.0×10^5 ml^{-1}).

6. Prepare several 9 cm Petri dishes (at least 12), each with 8 ml of agar-solidified MSP19M medium. From the fusion tubes (mixed populations of protoplasts) dispense 8 ml of suspended protoplasts per dish onto the surface of the medium. This gives a final, overall plating density of 5×10^{-4} protoplasts ml^{-1} of medium. For other species combinations, this may not be optimum plating density and the final protoplast plating density may have to be adjusted accordingly. Additionally, other culture approaches can be adopted, such as the use of liquid medium alone, and the embedding of protoplasts in agar or agarose droplets or layers.

7. Dilute each of the unfused viability controls with a further 4 ml of MSP19M medium (density now 1.0×10^5 ml^{-1}; 8 ml) and plate over 8 ml of agar-solidified medium as described earlier. Strictly, these controls should be subjected to identical washing and centrifugation procedures as for the fusion treated samples.

8. For the self-fusion controls, following fusion treatment and resuspension in 16 ml per tube of MSP19M medium, one 8 ml aliquot should be dispensed directly onto the agar-solidified medium counterpart (one dish per species-species fusion viability controls), whilst the remaining 8 ml aliquots of each self-fused species (at a density of 1×10^5 ml^{-1}) are mixed (1 : 1) to give two, 8 ml aliquots

for plating (8 ml per dish) over the agar-solidified medium. These crucial controls monitor not only post-fusion viability but, in the context of selection, possible crossfeeding and/or reversion in the albino *P. hybrida* parent. In other selection strategies, which rely on inhibition of growth of one or both parental protoplasts and their respective homokaryons, this type of control is equally essential.

9. Seal all Petri dishes with Nescofilm and culture at 25°C under 50 μmol m^{-2} s^{-1} of continuous "Daylight" fluorescent illumination.

Fusion protocols

I. Large-scale chemical fusion of protoplasts using PEG

Steps in the procedure
1. Prepare protoplasts for fusion according to steps 1–3 of the general procedure above.
2. Centrifuge all the tubes (80 × *g* for 10 min), except the viability controls, and resuspend the protoplasts in 0.5 ml of CPW11M solution.
3. Add 1.0 ml of 30% (w/v) PEG 1500 (see Solutions) and incubate the tubes for 20 min at 4 °C.
4. Add 1.5 ml of hypotonic solution dropwise with gentle agitation, to ensure rapid mixing. Repeat this addition twice at 2 min intervals.
5. Add 4 ml of MSP19M solution and incubate at 4 °C for 2 h in the fridge.
6. Follow steps 5–9 of the general procedure.

II. High pH/Ca^{2+} fusion

Steps in the procedure
1. Follow steps 1–3 of the general procedure.
2. Add 4 ml of high pH/Ca^{2+} fusion solution (see Solutions) to each of the tubes, except the viability controls.
3. Incubate the tubes at 30 °C for 10 min.
4. Centrifuge the tubes (50 × *g* for 4 min) and wash the protoplast pellets with CPW9M solution by resuspension and centrifugation (100 × g for 10 min).
5. Follow steps 5–9 of the general procedure.

III. PEG/high pH fusion

Steps in the procedure
1. Follow steps 1–3 of the general procedure.
2. Add 2 ml of PEG solution to each of the tubes, except the viability controls.
3. After 10 min incubation (23 °C), add 8 ml of high pH/Ca^{2+} solution to each tube and mix gently.
4. Following incubation (under identical conditions) for a further 10 min, centrifuge the tubes (50 × *g* for 4 min) and discard the supernatants. Wash the protoplast pellets once with CPW11M solution.
5. Follow steps 5–9 of the general procedure.

IV. Small-scale chemical fusion of protoplasts using PEG (for low yielding protoplast systems)

Steps in the procedure

1. Adjust the two parental protoplast suspensions to a density of 2.5×10^5 ml^{-1} in CPW13M solution. Mix equal volumes of the two protoplast suspensions in a centrifuge tube, using a minimum of 2 ml of each preparation.
2. Place a small drop (200 μl) of the protoplast mixture on a sterile coverslip contained in the bottom of a 5 cm Petri dish. Replace the lid and allow the protoplasts to settle for 5–10 min.
3. Dispense two 200 μl drops of 22.2% (w/v) PEG solution (see Solutions) on to the coverslip on each side of the protoplast mixture; allow the drops to coalesce with the CPW13M solution covering the settled protoplasts.
4. Leave the protoplasts undisturbed for 20–25 min, after which 200 μl of fusion solution is withdrawn and replaced with the same volume of Wash solution (see Solutions). Repeat this process at 5 min intervals over a 20 min period with a minimum of disturbance to the settled protoplasts.
5. Following elution of the fusion solution, replace the Wash solution with an equal volume of liquid MSP19M liquid medium (or alternative culture medium for other protoplast systems). Wash the protoplasts a minimum of three times, as in step (4).
6. Add a further 200 μl of liquid MSP19M liquid medium to give a final volume of 800 μl. Dispense the fusion-treated protoplasts in 100–200 μl drops in the base of the Petri dish. Seal the Petri dish with Nescofilm and culture at 25 °C under 50 μmol m^{-2} s^{-1} of continuous "Daylight" fluorescent illumination (or an appropriate light régime).

Notes

The selection scheme for this combination is based on differential parental protoplast growth responses to MSP19M medium. In this medium, *P. hybrida* protoplasts fail to divide, whereas *P. parodii* protoplasts form slow growing micro-calli. Hybrid cells form faster growing colonies due to heterosis; these colonies are easily identified.

Plant regeneration from somatic hybrid tissues, following chemical fusion and culture, can be achieved by the transfer (after 2–3 months) of protoplast-derived colonies onto agar-solidified MSZ medium (see Solutions). The resulting shoots are excised from the callus and rooted on agar-solidified MS-based medium lacking growth regulators. Rooted shoots are transferred to the glasshouse for *ex vitro* acclimation.

Solutions and culture media

- *CPW salts solution*
  1.48 g l^{-1} $CaCl_2 \cdot 2H_2O$
  27.20 mg l^{-1} KH_2PO_4
  101.0 mg l^{-1} KNO_3
  246.0 mg l^{-1} $MgSO_4 \cdot 7H_2O$
  0.16 mg l^{-1} KI
  0.025 mg l^{-1} $CuSO_4 \cdot 5H_2O$
  CPW9M, CPW11M, CPW13M and CPW21S are CPW salt solutions, with 9%, 11%, 13% (w/v) mannitol or 21% (w/v) sucrose respectively. Adjust the pH to 5.8.

- *Enzyme solution for protoplast isolation from leaves of* P. parodii
  1.5% (w/v) Meicelase.
  0.05% (w/v) Macerozyme R10.
  400 mg l^{-1} Ampicillin
  10 mg l^{-1} Gentamycin
  10 mg l^{-1} Tetracycline
  CPW13M solution, pH 5.8

- *Enzyme solution for the isolation of protoplasts from albino cell suspensions of* P. hybrida
  2.0% (w/v) Rhozyme HP 150
  2.0% (w/v) Meicelase
  0.03% (w/v) Macerozyme R10
  CPW13M solution, pH 5.8

- *PEG solution for large-scale protoplast fusion*
  30% (w/v) low carbonyl content PEG (Mol. Wt. 1500) in pH 8.0 buffer (19.5 g l^{-1} HEPES sodium salt)

- *Hypotonic solution*
  90 g l^{-1} mannitol
  2.0 g l^{-1} bovine serum albumin
  Adjust to pH 5.8 and filter sterilise

- *High pH/Ca^{2+} solution*
  0.05 M glycine-NaOH buffer
  90 g l^{-1} mannitol
  11 g l^{-1} $CaCl_2.2H_2O$
  pH 10.4
  Sterilise by filtration. Use only freshly made preparations.

– *PEG solution for small-scale protoplast fusion*
22.2% (w/v) PEG (Mol. Wt. 1500)
18.0 g l^{-1} sucrose
1.54 g l^{-1} $CaCl_2.2H_2O$
95.2 mg l^{-1} KH_2PO_4
pH 5.8

– *Wash solution*
110 g l^{-1} sucrose
7.4 g l^{-1} $CaCl_2.2H_2O$
3.75 g l^{-1} glycine
pH 5.8

– *UM liquid culture medium for cell suspension cultures of*
P. hybrida
MS salts [9]
30 g l^{-1} sucrose
2.0 mg l^{-1} 2,4-dichlorophenoxyacetic acid
0.25 mg l^{-1} kinetin
9.9 mg l^{-1} thiamine HCl
9.5 mg l^{-1} pyridoxine HCl
4.5 mg l^{-1} nicotinic acid
2.0 g l^{-1} casein enzymatic hydrolysate
pH 5.8

– *MSP19M protoplast culture medium*
MS salts [9]
30 g l^{-1} sucrose
90 g l^{-1} mannitol
2.0 mg l^{-1} α-naphthaleneacetic acid
0.5 mg l^{-1} 6-benzylaminopurine
pH 5.8
MSPI9M agar-solidified medium is as above, but with 0.8% (w/v) agar.

– *MSZ plant regeneration medium*
MS salts [9]
30 g l^{-1} sucrose
1.0 mg l^{-1} zeatin
8 g l^{-1} agar
pH 5.8

Materials and suppliers

1. 'Domestos' bleach: Lever Bros. Ltd., Kingston-upon-Thames, UK.
2. Disposable gloves: Becton Dickinson, Ltd., Wembley, Middlesex, UK.
3. Petri dishes: Sterilin Ltd., Teddington, Middlesex, UK.
4. Centrifuge tubes: Corning Glass Works, Corning, New York, USA.
5. Nylon sieves: Wilson Sieves, 2 Long Acre, Common Lane, Hucknall, Nottingham, UK.
6. Nescofilm: Bando Chemical Ind. Ltd., Kobe, Japan.
7. Rhozyme HP 150: Röhm and Haas, Philadelphia, USA.
8. Meicelase: Meiji Seika Kaisha, Tokyo, Japan.
9. Macerozyme R10: Kinki Yakult, Nishinomiya, Japan.
10. PEG Mol. Wt. 1500: Boehringer Mannheim UK Ltd., Lewes, East Sussex, UK.

References

1. Ahuja PS, Laiq-ur-Rahman, Bhargava SC, Banerjee S (1993) Regeneration of somatic hybrid plants between *Atropa belladonna* L. and *Hyoscyamus muticus* L. Plant Sci 92: 91–98.
2. Boni LT, Hui SW (1987) The mechanism of polyethylene glycol-induced fusion in model membranes. In: Sowers AE (Ed.) Cell Fusion, pp 119–124, Plenum Press, New York.
3. Chand PK, Davey MR, Power JB, Cocking EC (1988) An improved procedure for protoplast fusion using polyethylene glycol. J Plant Physiol 133: 480–485.
4. Craig AL, Morrison I, Baird E, Waugh R, Coleman M, Davie P, Powell W (1994) Expression of reducing sugar accumulation in interspecific somatic hybrids of potato. Plant Cell Rep 13: 401–405.
5. Gleba YY, Sytnik KM (1984) Protoplast fusion and parasexual hybridisation of higher plants. In: Shoeman R (Ed.) Protoplast Fusion, pp 36–62, Springer-Verlag, Berlin.
6. Hammatt N, Lister A, Blackhall NW, Gartland J, Ghose TK, Gilmour DM, Power JB, Davey MR, Cocking EC (1990) Selection of plant heterokaryons from diverse origins by flow cytometry. Protoplasma 154: 34–44.
7. Hammatt N, Blackhall NW, Davey MR (1991) Variation in the DNA content of *Glycine* species. J Exp Bot 42: 659–665.
8. Kishinami I, Widholm F (1987) Auxotrophic complementation in intergeneric hybrid cells obtained by electrical and dextran-induced protoplast fusion. Plant Cell Physiol 28: 211–218.
9. Louzada ES, Grosser JW, Gmitter FG (1993) Intergeneric somatic hybridisation of sexually incompatible parents: *Citrus sinensis* and *Atalantia ceylanica*. Plant Cell Rep 12: 687–690.
10. Mendis MH, Power JB, Davey MR (1991) Somatic hybrids of the forage legumes *Medicago sativa* L. and *M. falcata* L. J Exp Bot 42: 1565–1573.
11. Murashige T, Skoog F (1962) A revised medium for rapid growth and bioassays with tobacco tissue cultures. Physiol Plant 15: 473–497.
12. Nagata T (1978) A novel cell fusion method of protoplasts by polyvinyl alcohol. Naturwissenschaften 65: 263–264.
13. Nakano M, Mii M (1993) Interspecific somatic hybridisation in *Dianthus*: selection of hybrids by the use of iodoacetamide inactivation and regeneration ability. Plant Sci 88: 203–208.
14. Patil RS, Latif M, Vaz FBD, Davey MR, Power JB (1993) Hybridisation, through culture of embryos and immature seeds, of a range of tomato cultivars with a tomato somatic hybrid (*Lycopersicon esculentum* (+) *L. peruvianum*): emergence of a possible new marker gene for tomato breeding. Plant Breed 111: 273–282.
15. Pehu E, Thomas M, Poutala T, Karp A, Jones MGK (1990) Species-specific sequences in the genus *Solanum* – identification, characterisation and application to study somatic hybrids of *Solanum brevidens* and *Solanum tuberosum*. Theor App Genet 80: 693–698.
16. Polgar Z, Preiszner J, Dudits D, Feher A (1993) Vigorous growth of fusion products allows highly efficient selection of interspecific potato somatic hybrids – molecular proofs. Plant Cell Rep 12: 399–402.
17. Power JB, Cummins SE, Cocking EC (1970) Fusion of isolated plant protoplasts. Nature 225: 1016–1018.
18. Power JB, Frearson EM, Haywood C, George D, Evans PK, Berry SF, Cocking EC (1976) Somatic hybridisation of *Petunia hybrida* and *P. parodii*. Nature 263: 500–502.
19. Schoenmakers HCH, Wolters AMA, Nobel EM, de Klein CMJ, Koornneef M (1993) Allotriploid somatic hybrids of diploid tomato (*Lycoperiscon esculentum* Mill.) and monoploid potato (*Solanum tuberosum* L.). Theor Appl Genet 87: 328–336.
20. Schoenmakers HCH, Wolters AMA, de Haan A, Saiedi AK, Koornneef M (1994) Asymmetric somatic hybridisation between tomato (*Lycopersicon esculentum* Mill) and gamma-irradiated potato (*Solanum tuberosum* L.): a quantitative analysis. Theor Appl Genet 87: 713–720.
21. Stattmann M, Gerick E, Wenzel G (1994) Interspecific somatic hybrids between *Solanum khasianum* and *S. aculeatissimum* produced by electrofusion. Plant Cell Rep 13: 193–196.
22. Wang GR, Binding H (1993) Somatic hybridisation between *Senecio fuchsii* and *S. jacobaea*. Acta Hort 36: 315–320.

Springer Science+Business Media, LLC

Dear Reader

We would very much appreciate receiving your suggestions and criticisms on the *Plant Tissue Culture Manual*. They will be most helpful during our preparations for future supplements.

Would you please answer the questions listed below, and send your comments with any further suggestions you may have, to *Ir. A. Plaizier* at the above-mentioned address.

Thank you for your assistance!

A. Plaizier
Publisher

— —

PLANT TISSUE CULTURE MANUAL

1. What errors have you found? (list page numbers and describe mistakes)
2. What protocols do you find to be confusing or lacking in detail? (list chapter numbers and page numbers and describe problems)
3. What protocols do you feel should be replaced in future supplements with newer (better) methods?
4. What new topics or other material would you like to see included in future supplements?

Please print or type your answers in the space below, and continue overleaf.

Name:

Date:

Address:

PLANT TISSUE CULTURE MANUAL
Supplement 7

Edited by:

K. LINDSEY
Department of Biological Sciences, University of Durham, U.K.

Springer Science+Business Media, LLC

Library of Congress Cataloging-in-Publication Data

Plant tissue culture manual: fundamentals and applications / edited by K. Lindsey
    p.   cm.
    Includes bibliographical references and index.

    1. Plant tissue culture—Laboratory manuals.   I. Lindsey, K.
QK725.P587   1991   90–26765
581'.0724—dc20

Manual
ISBN 978-94-011-7658-3

ISBN 978-94-011-7658-3   ISBN 978-94-009-0103-2 (eBook)
DOI 10.1007/978-94-009-0103-2

Printed on acid-free paper

Contents

Preface

SECTION A: BASIC TECHNIQUES – CELL & TISSUE CULTURE OF MODEL SPECIES

1. Media preparation
   O.L. Gamborg, Fort Collins, USA

2. The initiation and maintenance of callus cultures of carrot & tobacco
   R.D. Hall, Wageningen, The Netherlands

3. The initiation and maintenance of plant cell suspension cultures
   R.D. Hall, Wageningen, The Netherlands

4. Shoot cultures and root cultures of tobacco
   J.F. Topping and K. Lindsey, Leicester, UK

5. Somatic embryogenesis in orchardgrass
   M.E. Horn, Davis, USA

6. *Arabidopsis* regeneration and transformation (leaf & cotyledon explant system)
   R. Schmidt and L. Willmitzer, Berlin, Germany

7. *Arabidopsis* protoplast transformation and regeneration
   B. Damm and L. Willmitzer, Berlin, Germany

8. *Arabidopsis* regeneration and transformation (root explant system)
   D. Valvekens, M. van Lijsebettens and M. van Montagu, Gent, Belgium

9. Somatic embryogenesis in carrot
   A.D. Krikorian and D.L. Smith, New York, USA

10. Low density cultures: microdroplets and single-cell nurse cultures
    G. Spangenberg, Ladenburg, Germany, and HU Koop, Munich, Germany

11. Tobacco protoplast isolation, culture and regeneration
    I. Negrutiu, Gembloux, Belgium

SECTION B: TISSUE CULTURE & TRANSFORMATION OF CROP SPECIES

1. Embryogenic callus, cell suspension and protoplast cultures of cereals
   I.K. Vasil and V. Vasil, Gainesville, USA

2. Transformation and regeneration of rice protoplasts
   J. Kyozuka and K. Shimamoto, Yokohama, Japan

3. Transformation and regeneration of orchardgrass protoplasts
   M.E. Horn, Davis, USA

4. Transformation and regeneration of oilseed rape protoplasts
   D. Rouan and P. Guerche, Versailles, France

5. Regeneration and transformation of potato by *Agrobacterium tumefaciens*
   R.G.F. Visser, Wageningen, Netherlands

6. Transformation of tomato with *Agrobacterium tumefaciens*
   S. McCormick, Albany, USA

7. Regeneration and transformation of sugarbeet by *Agrobacterium tumefaciens*
   K. Lindsey, P. Gallois and C. Eady, Leicester, UK

8. Regeneration and transformation of apple (*Malus pumila* Mill.)
   D.J. James and A.M. Dandekar, Davis, USA

9. Transformation and regeneration of maize protoplasts
   C.A. Rhodes and D.W. Gray, Palo Alto, USA

10. Regeneration and transformation of barley protoplasts
    P. Lazzeri, A. Jähne and H. Lörz, Hamburg, Germany

11. *Agrobacterium*-mediated transformation of potato stem and tuber tissue, regeneration and PCR screening for transformation
    J.P. Spychalla and M.W. Bevan, Norwich, UK

12. Production of fertile transgenic wheat by microprojectile bombardment
    Dirk Becker and Horst Lörz, Hamburg, Germany

13. Transient gene expression and stable genetic transformation into conifer tissues by microprojectile bombardment
    Armand Séguin, Denis Lachance and Pierre J. Charest, Ontario, Canada

SECTION C: PROPAGATION & CONSERVATION OF GERMPLASM

1. Clonal propagation of orchids
   Y. Sagawa, Honolulu, USA

2. Clonal propagation of palms
   B. Tisserat, Pasadena, USA

3. Clonal propagation of conifers
   T.A. Thorpe and I.S. Harry, Alberta, Canada

4. Cytological techniques
   A. Karp, Long Ashton, UK

5. Restriction fragment analysis of somaclones
   R.H. Potter, As, Norway

6. Virus elimination and testing
   M.C. Coleman and W. Powell, Dundee, UK

7. Clonal propagation of *Citrus*
   T.S. Rangan, Pasadena, USA

8. Clonal propagation of eucalypts
   J.A. McComb, Perth, Australia

9. Cryopreservation of plant tissue cultures: the example of embryogenic
   tissue cultures from conifers
   P.J. Charest, J. Bongs and K. Klimaszewska, Ontario and New
   Brunswick, Canada

SECTION D: DIRECT GENE TRANSFER & PROTOPLAST FUSION

1. Gene transfer by particle bombardment
   T.M. Klein, S. Knowlton and R. Arentzen, Wilmington, USA

2. Transformation of pollen by particle bombardment
   D. Twell, T.M. Klein and S. McCormick, Albany, USA

3. Electrical fusion of protoplasts
   M.G.K. Jones, Murdoch, Australia

4. Cybrid production and selection
   E. Galun and D. Aviv, Rehovot, Israel

5. Fluorescence-activated analysis and sorting of protoplasts and somatic
   hybrids
   D.W. Galbraith, Lincoln, USA

6. RFLP analysis of organellar genomes in somatic hybrids
   E. Pehu, Helsinki, Finland

7. Isolation and uptake of plant nuclei
   P.K. Saxena and J. King, Saskatoon, Canada

8. *In situ* hybridization to plant metaphase chromosomes: Radioactive and
   non-radioactive detection of repetitive and low copy number genes
   J. Veuskens, S. Hinnisdaels and A. Mouras, Talence, France

9. Chemical fusion of protoplasts
   P. Anthony, R. Marchant, N.W. Blackhall, J.B. Power, M.R. Davey,
   Nottingham, UK

SECTION E: REPRODUCTIVE TISSUES

1. *In vitro* fertilisation of maize
   E. Kranz, Hamburg, Germany

2. Endosperm culture
   S. Stirn and H.-J. Jacobsen, Bonn, Germany

3. Endosperm culture
   B.M. Johri and P.S. Srivastava, Delhi, India

4. Hybrid embryo rescue
   A. Agnihotri, New Delhi, India

5. *In vitro* culture of *Brassica juncea* zygotic proembryos
   Chun-ming Liu, Zhi-hong Xu and Nam-Hai Chua,
   Singapore and New York, USA

*6. Production of haploids in *Brassica* spp. via microspore culture
   A.M.R. Ferrie and W.A. Keller, Saskatoon, Canada

SECTION F: MUTANT SELECTION

1. Use of chemical and physical mutagens *in vitro*
   P.J. Dix, Maynooth, Ireland

2. *In vitro* culture, mutant selection, genetic analysis and transformation of *Physcomitrella patens*
   David Cove, Leeds, UK

SECTION G: SECONDARY METABOLITES

1. Tropane alkaloid biosynthesis *in vitro*
   R. Robins, Colney, Norwich, UK

2. Anthocyanin biosynthesis *in vitro*
   A. Komamine and K. Kakegawa, Tsukuba City, Ibaraki, Japan

3. Biosynthesis of monoterpene indole alkaloids *in vitro*
   W.G.W. Kurz, K. Constabel, R. Tyler, Saskatchewan, Canada

SECTION H: TISSUE CULTURE TECHNIQUES FOR FUNDAMENTAL STUDIES

1. Establishment of photoautotrophic cell cultures
   W. Hüsemann, Münster, Germany

2. *Zinnia* mesophyll culture system to study xylogenesis
   M. Sugiyama and H. Fukuda, Sendai, Japan

---

* Included in Supplement 7.

3. Cell cycle studies: induction of synchrony in suspension cultures of *Catharanthus roseus* cells
   H. Kodama and A. Komamine

4. Thin Cell Layer (TCL) method to programme morphogenetic patterns
   K. Tran Thanh Van and C. Gendy, Paris, France

5. *In vitro* infection of *Arabidopsis* with nematodes
   Joke C. Klap and Peter C. Sijmons, Leiden and Wageningen,
   The Netherlands

*6. *Asparagus* cell cultures as a source of wound-inducible genes
   R.M. Darby and J. Draper, Leicester, U.K.

*7. Use of video cell tracking to identify embryogenic cultured cells
   Marcel A.J. Toonen and Sacco C. de Vries, Wageningen,
   The Netherlands

* Index

---

* Included in Supplement 7.

Preface

Plant tissue culture has a long history, dating back to the work of Gottlieb Haberlandt and others at the end of the 19th century, but the associated concepts and techniques have reached a level of usefulness and application which has never been greater. The technical innovations have given new insights into fundamental aspects of plant differentiation and development, and have paved the way to the identification of strategies for the genetic manipulation of plants. It is the aim of this manual to deliver a broad range of these techniques in a form which is accessible to students and research scientists of diverse backgrounds, including those with little or no previous experience. The themes of the manual aim to reflect those research areas which have been advanced by tissue culture technology.

As was the case for the sister volume *Plant Molecular Biology Manual*, the objective has been from the start to produce a manual which is at home on the laboratory bench. The plastic-covered, ring-bound format has proved to be most popular and is retained here. Equally, the emphasis has been on producing a collection of detailed step-by-step protocols, each supplemented with an introductory text and practical footnotes, to provide the next best thing to a supervisor at one's shoulder. Each author was chosen as one actively using the respective technique in his or her own laboratory, in order to give an authoritative account, with the most common difficulties or pitfalls highlighted for the benefit of the newcomer to the field.

The manual is published initially as a core text of basic techniques, to which will be added supplementary chapters and sections at regular intervals. One characteristic of plant tissue culture methodology is the fact that what works for one species quite possibly will not work for another (or even for a different genotype of the same species): each requires a carefully optimized protocol, whether it be for callus culture initiation, protoplast isolation and culture or plant regeneration following *Agrobacterium*-mediated transformation. With this in mind, we have decided on the one hand to generalize experimental approaches for some basic techniques, by reference to the use of model species. This should at least give the researcher a feel for the techniques, and indeed provide a starting point for species-specific optimization of protocols. On the other hand, we have provided detailed protocols for the manipulation of a number of specific crop species, which have been difficult to optimize and for which there may be particular interest. A detailed bibliography is provided in each chapter, so that access to the relevant literature is possible.

I would like to thank the authors both for their enthusiasm to contribute and for the rapidity with which they submitted their manuscripts. Thanks go in particular to Dr Mike Brewis of Kluwer for his valuable discussions during the embryonic stages of the manual and for his enormous help in taking on the administrative tasks.

Leicester, November 1990   K. LINDSEY

Section E:
Reproductive Tissues

Reproductive Tissues

Sect. E

Plant Molecular Biology
Manual

Plant Tissue Culture Manual **E1**: 1–12, 1992.
© 1992 *Kluwer Academic Publishers.*

In vitro fertilization of maize mediated by electrofusion of single gametes

E. KRANZ
Universität Hamburg, Institut für Allgemeine Botanik, AMP II, Ohnhorststr. 18, D-2000 Hamburg 52, Germany.

Introduction

In vitro fertilization with single gametes has been performed using isolated egg cells from embryo sacs and sperm cells from mature pollen grains [12, 13]. Pairs of gametes have been fused individually by an electrical pulse after dielectrophoretical alignment under microscopic observation (Fig. 1, Fig. 2). In addition, fusion products have been created by fusion of single sperm cells with single synergids or central cells.

The technology involving the manipulation of isolated gametes and non-gametic cells of the embryo sac permits studies of fertilization processes at the single cell level.

In vitro-created zygotes have a 'natural' competence for division and might therefore be useful for studies on cell cycle, embryogenesis and differentiation. The growth of the artificially-produced zygotes as well as non-gametic cells of the embryo sac will promote studies of early events of zygote development.

The detection of development-specific genes in only one or a few reproductive cells or zygotes might become feasible by using the polymerase chain reaction (PCR) method for amplification of mRNAs from single cells allowing the study of gene expression [1, 4, 15]. The transfer of foreign genes into gametes (via electroporation) as well as into zygotes (via microinjection) before the first division takes place, seems to be promising, as synchronized single cells can be used. High survival and division rates of artificially-produced zygotes after injection with plasmid DNA have been observed (E. Kranz, unpublished).

The transmission of cytoplasm through the *in vitro* fertilization process will facilitate studies on cytoplasmic inheritance. The fate of organelles (plastids and mitochondria) transmitted by sperm cells or additionally by cytoplasts from somatic cells can be studied with this system [14].

In this chapter are described detailed procedures for the isolation of maize sperm cells, egg cells and non-gametic cells of the embryo sac, the controlled electrofusion of defined pairs of gametes as well as culture procedures for the zygotes.

Isolation of egg and sperm cells of maize

Enzymes (pectinase, pectolyase, hemicellulase, cellulase) generally used for protoplast isolation of somatic cells have also been applied to the isolation of the egg cells of maize [12, 13, 14, 21, 22]. Cytohelicase has also been used for the isolation of embryo sacs of maize [21, 22]. Driselase [5], macerozyme R10 [17], snailase [5, 23, 24, 25] and β-glucuronidase [6] have also been used in enzyme mixtures for embryo sac and egg cell isolation of various plant species.

Ovules are selected from the middle part of the ears under a dissecting microscope. The ears are bagged before silk emergence and the outer leaves are surface sterilized with ethanol (70%). Twenty to thirty pieces of ovule tissue (containing the embryo sac) are collected in a mannitol solution (570 mosmol/kg H_2O). After enzyme treatment the egg cells of maize are mechanically isolated.

Routinely about 5 egg cells per 20 ovule pieces can be isolated using the combined method of enzymatic digestion and manual dissection. This can be achieved with ovular tissue from various developmental stages of the ears (indicated by silk emergence lengths of 3.5 to 17.0 cm, line A 188). The yield is ultimately determined by the manual isolation step.

The sperm cells are released from the pollen grains after rupture by osmotic shock [2, 13, 16, 19]. Pollen is collected from freshly dehisced anthers and used immediately or stored for some hours in plastic dishes sealed with Parafilm. The air is moistened by a piece of wet filter paper fixed to the lid of the dish [11]. The pollen grains are incubated in the fusion medium (0.55 M mannitol) for bursting.

Fusion of isolated gametes

Isolated sperm cells, egg cells and non-gametic cells of the embryo sac are individually selected under microscopic observation and transferred by micro-capillaries, using a hydraulic system. The capillaries are connected by means of Teflon tubing filled with mineral oil to a computer-controlled dispenser/dilutor [8]. The uptake and release of the cells from the capillary is performed by this pump. Steps for suction or delivery, the speed and the volume can be programmed. The volume to be used depends on the diameter of the capillary. A sliding stage containing openings both for a 3.5 cm dish, which contains the ovular tissue pieces or the pollen grains, respectively, as well as a coverslip (24 × 40 mm) containing the fusion droplets, is used. For the isolation and selection procedure, cell transfer and fusion, the stage is moved manually.

Electrofusion of selected single gametes is performed using the individual cell fusion method [9, 10, 20]. Selected single gametes are transferred to fusion droplets of 2 μl each (0.55M mannitol) which are overlayered by mineral oil on a coverslip with siliconized edges. Electrofusion is performed with single pairs of gametes under microscopic observation. A pair of electrodes are fixed to an electrode support mounted under the condensor of the microscope. The electrodes are moved along the z-axis by a positioning-system-controlled motor. Fusion is induced by single or multiple negative DC pulses after dielectrophoretic alignment on one of the electrodes for a few seconds using an electrofusion apparatus. The conditions generally used for the electrical fusion of protoplasts of somatic cells are also applied to the protoplasts of the gametic cells of maize. The mean fusion frequency is 79%.

Culture experiments

The fusion products are cultured on the semipermeable membrane of 'Millicell-CM' dishes, which are inserted in plastic dishes, previously filled with 1.5 ml of a maize feeder suspension. Fusion products as well as feeder cells are cultivated in a modified MS liquid medium [18], containing 1.0 mg/l 2,4-D and 0.02 mg/l kinetin. The fusion products start to divide within 2.5–3 days (mean frequency of 83%). High frequencies of formation of multicellular structures can be found after fusion combinations using sperm cells and egg cells from different lines [13]. Despite the relatively low yield of egg cells, which can be isolated in 3 hours/exprimenter (about 15–20 egg cells, depending on the quality of the plant material), the high frequencies of fusion as well as cell division rates allow this method to be used routinely.

In vitro fertilization
Isolation

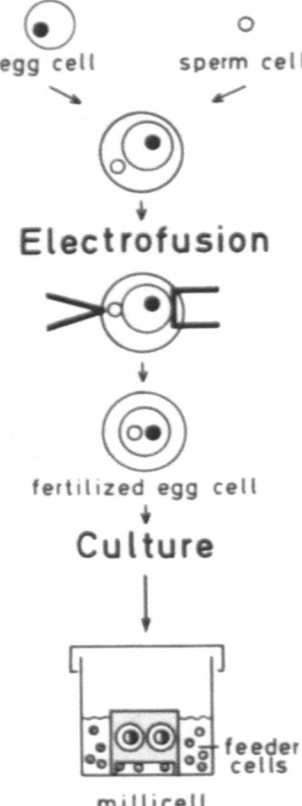

egg cell   sperm cell

Electrofusion

fertilized egg cell

Culture

feeder cells

millicell

Fig. 1. Electrofusion-mediated *in vitro* fertilization method. After isolation single sperm and egg cells are transferred into the fusion droplet. Pairs of individually-selected gametes are fused electrically after dielectrophoretical alignment on one of the electrodes. For culture, fusion products are transferred into 'Millicell' inserts surrounded by a feeder cell suspension. Taken from [13]. Used with permission.

Fig. 2. Components of the set-up according to Koop *et al.* [8] and Schweiger *et al.* [20] and additional modifications.
(1) Inverted microscope; (2) computer-controlled pump; (3) computer and monitor; (4) electronics for positioning-system; (5) power supply for microscope lamp; (6) electrofusion apparatus; (7) external trigger box for electrofusion; (8) joy-stick; (9) *z*-axis-step motor.

Procedures

Steps in the procedure

1. *Preparation of coverslips*

   Siliconize the edges of the coverslip with repel silane. Sterilize coverslips under UV-light (15 min). Overlay the coverslip with a layer of mineral oil. Inject 30 fusion droplets of 2 μl each, under the mineral oil layer each consisting of 0.55M mannitol in 3 rows by a microcapillary.

2. *Pollen and ear collection*

   Collect pollen from freshly dehisced anthers (between 9 and 12 a.m.) and use it immediately. Pollen can be stored for some hours at room temperature in plastic dishes sealed with parafilm. Moisten the air by a piece of wet filter paper fixed to the lid of the dish.

   Bag ears before silk emergence. Collect ears after silk emergence and sterilize the outer leaves of the ears with ethanol (70%). Select ovules from the middle part of the ears.

3. *Isolation and selection of egg cells*

   Dissect 20–30 pieces of tissue (containing the embryo sac) from the ovules under a dissecting microscope and collect them in mannitol solution (0.75 ml, 570 mosmol/kg H_2O) in 3.5-cm diameter plastic dishes. After collecting the tissue pieces in mannitol solution add 0.75 ml of enzyme solution. Incubate the ovular tissue-enzyme mixture at $24° \pm 0.5 °C$ for about 30 min (depending on the 'quality' of the plant material) without shaking. Isolate egg cells and non-gametic cells manually after incubation, with a tiny glass-needle or with the microcapillary, directly in the incubation dish under an inverted microscope. Select isolated egg cells and transfer them individually, using a microcapillary with a hand-drawn tip opening of about 200 μm, into the fusion droplets. Transfer the selected cell from one droplet into another for washing.

4. *Isolation and selection of sperm cells*

   Incubate about 1000 pollen grains in 1.5 ml of 0.55M mannitol in 3.5 cm-diameter plastic dishes. Select individual sperm cells under microscopic obser-vation by a microcapillary (diameter of the tip opening about 20 μm) after pollen grains rupture by osmotic shock. Transfer sperm cells into the fusion droplets.

   Sperm and egg cells can be selected by use of a hydraulic system, using a dispenser/diluter.

5. *Individual electrical fusion*

   Fix a pair of electrodes (platinum wire, diameter 50 μm) to an electrode support mounted under the condenser of the inverted microscope. Move the electrodes to the crosshairs position. Lower the electrodes along the z-axis onto the coverslip and into the fusion droplet. Fix the two gametes at one electrode. Move the egg cell towards the electrode (by moving the microscope stage). The final contact between the egg cell and the electrode is accomplished by dielectrophoresis (1 MHz, 71 V × cm^{-1}). Repeat this procedure for the sperm cell. The egg cell must be in contact with the electrode, with the sperm cell being placed onto the egg cell. After alignment of the gametes the final distance between the two electrodes has to be adjusted by moving the electrodes along the z-axis. The distance of the electrodes should be 2× the diameter of the two cells. Induce fusion by applying single or multiple (2–3) negative DC-pulses (50 µs; 0.9–1.0 kV × cm^{-1}).

6. *Culture procedures*

Fill 3.5 cm plastic dishes with 1.5 ml of a maize feeder suspension and insert a 'Millicell-CM' dish (diameter 12 mm, Millipore). Fill the 'Millicell-CM' insert with 200 µl mannitol (0.55M) and transfer the fusion products into the insert by a microcapillary. Culture conditions: 24° ± 0.5 °C in the dark or a light/dark cycle of 16/8 h and a light intensity of approximately 50 µEm^{-2}s^{-1}. Shaking on a rotary shaker at 70 rpm may be advantageous.

Note

Minimizing the exposure of the ovular pieces to the enzyme solution prevents uncontrolled fusion of the cells of the embryo sac. The enzyme concentration sometimes has to be reduced to a half, especially when the maize plants are grown under suboptimal growth conditions (e.g. low light during winter season) resulting in only small ovules. Depending of the 'quality' of the plant material, the enzyme concentration and the incubation time must be determined exactly. After enzyme treatment the partially-macerated ovular pieces can be stored in the refrigerator at 5 °C for some hours.

Solutions

Enzyme solution

1.5% (w/v) pectinase (Serva, Heidelberg, Germany)
0.5% (w/v) pectolyase Y-23 (Seishin, Tokyo, Japan)
1.0% (w/v) hemicellulase (Sigma, Deisenhofen, Germany)
1.0% (w/v) cellulase 'Onozuka' RS (Yakult Honsha, Tokyo, Japan).

Adjust pH to 5.0, and adjust osmolarity to 570 mosmol/kg H_2O with mannitol.

Maize feeder suspension culture medium

Murashige and Skoog [18] basal medium (eg Sigma)
Supplemented with:

1.0 mg/l 2,4-dichlorophenoxyacetic acid (2,4-D)
0.02 mg/l kinetin
Adjust pH 5.7, and adjust osmolarity to 600 mosmol/kg H_2O with glucose.

Materials and Suppliers

1. Inverted microscope: Axiovert 35 M, Carl Zeiss, W-7082 Oberkochen, Germany
2. Positionierungssteuerung: MCL or MCC 13 JS, Lang Electronic, W-8338 Hüttenberg, Germany
3. Microlab M: dispenser/dilutor, Hamilton Deutschland GmbH, W-6100 Darmstadt, Germany
4. Electrofusion apparatus: CFA 500, Krüss GmbH, W-2000 Hamburg 61, Borsteler Chaussee 85-99a, Germany
5. Microelectrode puller: Leitz, W-6330 Wetzlar, Germany or Narashige Scientific Instrumental Lab., 27-9 Minami-Karasuyama 4-Chome, Setagaya-Ku, Tokyo, Japan
6. Einmal-Mikropipetten: Collor-Code green, 50 µl, Cat. No. 708733 Brand, Germany
7. Mineral il: Paraffin flüssig für Spektroskopie, Merck, Cat. No. 7161, W-6100 Darmstadt, Germany
8. Repel Silane: (dimethyldichlorosilane solution, 2% in 1,1,1-trichloro-ethane), LKB, Cat. No. 1850–252
9. Culture plate inserts: Millicell™-CM, 0.4 µm, 12mm diameter, Cat. No. PICM 012 50, Millipore Products Division, Bedford, MA 01730, USA

References

1. Belyavsky B, Vinogradova T, Rajewsky K (1989) PCR-based cDNA library construction: general cDNA libraries at the level of a few cells. Nucl Acids Res 17: 2919–2932.
2. Cass DD, Fabi GC (1988) Structure and properties of sperm cells isolated from the pollen of *Zea mays*. Can J Bot 66: 819–825.
3. Dupuis I, Roeckel P, Matthys-Rochon E, Dumas C (1987) Procedure to isolate viable sperm cells from corn (*Zea mays* L.) pollen grains. Plant Physiol 85: 876–878.
4. Erlich HA (1989) PCR Technology. Stockton Press, New York.
5. Hu SY, Li LG, Zhou C (1985) Isolation of viable embryo sacs and their protoplasts of *Nicotiana tabacum*. Acta Bot Sin 27: 337–344.
6. Huang BQ, Russel SD (1989) Isolation of fixed and viable eggs, central cells and embryo sacs from ovules of *Plumbago zeylanica*. Plant Physiol 90: 9–12.
7. Keijzer CJ, Reinders MC, Leferink-ten Klooster HB (1988) A micromanipulation method for artificial fertilization in *Torenia*. In: Cresti M, Gori P, Pacini E (eds) Sexual Reproduction in higher plants, pp 119–124. Berlin Heidelberg New York: Springer.
8. Koop HU, Schweiger HG (1985a) Regeneration of plants from individually cultivated protoplasts using an improved microculture system. J Plant Physiol 121: 245–257.
9. Koop HU, Schweiger HG (1985b) Regeneration of plants after electrofusion of selected pairs of protoplasts. Eur J Cell Biol 39: 46–49.
10. Koop HU, Dirk, J, Wolff D, Schweiger HG (1983) Somatic hybridization of two selected single cells. Cell Biol Int Rep 7: 1123–1128.
11. Kranz E, Lörz H (1990) Micromanipulation and *in vitro* fertilization with single pollen grains of maize. Sex Plant Reprod 3: 160–169.
12. Kranz E, Bautor J, Lörz H (1990) *In vitro* fertilization of single, isolated gametes, transmission of cytoplasmic organelles and cell reconstitution of maize (*Zea mays* L.). In: Nijkamp HJJ, Van der Plas LHW, Van Aartrijk J (eds) Progress in plant cellular and molecular biology. Proceedings of the VIIth International Congress on Plant Tissue and Cell Culture, Amsterdam, The Netherlands, 24–29 June 1990, pp 252–257. Dordrecht Boston London: Kluwer Academic Publishers.
13. Kranz E, Bautor J, Lörz H (1991a) *In vitro* fertilization of single, isolated gametes of maize mediated by electrofusion. Sex Plant Reprod 4: 12–16.
14. Kranz E, Bautor J, Lörz H (1991b) Electrofusion-mediated transmission of cytoplasmic organelles through the *in vitro* fertilization process, fusion of sperm cells with synergids and central cells, and cell reconstitution in maize. Sex Plant Reprod 4: 17–21.
15. Li H, Gyllensten UB, Cui X, Saiki RK, Erlich HA, Arnheim N (1988) Amplification and analysis of DNA sequences in single human sperm and diploid cells. Nature 335: 414–417.
16. Matthys-Rochon E, Vergne P, Detchepare S, Dumas C (1987) Male germ unit isolation from three tricellular pollen species: *Brassica oleracea*, *Zea mays*, and *Triticum aestivum*. Plant Physiol 83: 464–466.
17. Mol R (1986) Isolation of protoplasts from female gametophytes of *Torenia fournieri*. Plant Cell Rep 3: 202–206.
18. Murashige T, Skoog F (1962) A revised medium for rapid growth and bioassays with tobacco tissue cultures. Physiol Plant 15: 473–497.
19. Russel SD (1986) Isolation of sperm cells from the pollen of *Plumbago zeylanica*. Plant Physiol 81: 317–319.
20. Schweiger HG, Dirk J, Koop HU, Kranz E, Neuhaus G, Spangenberg G, Wolff D (1987) Individual selection, culture and manipulation of higher plant cells. Theor Appl Genet 73: 769–783.
21. Wagner VT, Song Y, Matthys-Rochon E, Dumas C (1988) The isolated embryo sac of *Zea mays*: structural and ultrastructural observations. In: Cresti M, Gori P, Pacini E (eds) Sexual reproduction in higher plants, pp 125–130. Berlin Heidelberg New York: Springer.
22. Wagner VT, Song YC, Matthys-Rochon E, Dumas C (1989) Observations on the isolated embryo sac of *Zea mays* L. Plant Sci 59: 127–132.
23. Zhou C (1987) A study of fertilization events in living embryo sacs isolated from sunflower ovules. Plant Sci 52: 147–151.

24. Zhou C, Yang HY (1985) Observations on enzymatically isolated, living and fixed embryo sacs in several angiosperm species. Planta 165: 225–231.
25. Zhou C, Yang HY (1986) Isolation of embryo sacs by enzymatic maceration and its potential in haploid study. In: Hu H, Yang HY (eds) Haploids of higher plants *in vitro*, pp 192–203. Berlin: Springer.

Plant Tissue Culture Manual **E2**: 1–10, 1992.

Monoclonal antibodies against marker proteins for somatic embryogenesis

SUSANNE STIRN[1] & HANS-JÖRG JACOBSEN[2]

[1]*Institut für Allgemeine Botanik, Angewandte Molekularbiologie der Pflanzen II, Ohnhorststr. 18, 2000 Hamburg 52, Germany.*
[2]*Institut für Molekulaire Genetik, Herrenhäuserstr. 2, 3000 Hannover, Germany.*

Introduction

The unique ability of somatic plant cells to regenerate intact plants via somatic embryogenesis is an elegant tool to obtain plants with selected characters (e.g. resistance to fungi, improved quality). Besides this, somatic embryos will provide easily accessible material to study developmental processes in plants.

As a limiting factor, however, somatic embryogenesis, especially in important crop plants, seems to be genotype-dependent. This problem can only be overcome at present by laborious, time- and money-consuming optimization protocols for recalcitrant genotypes and cultivars.

As an alternative approach, the better understanding of fundamental processes which trigger and control somatic embryogenesis will lead to more rational regeneration protocols.

Despite taxonomic differences, plant regeneration via somatic embryogenesis can be divided into two distinguishable phases according to Komamine *et al.* [8]:

– induction (single cell → globular cell clusters or proembryogenic masses)
– embryo development (globular → bipolar embryos)

In most cases, the induction of somatic embryos requires the application of 'strong' auxins, triggering embryo-competent cells to form globular structures (for a review see Kohlenbach [5]), while in later stages auxin is inhibitory.

The globular stage of somatic embryos of carrot seems to be a critical one as can be suggested from the peculiar heat shock response of mid-globular stage embryos [18] indicating that the subsequent development beyond the globular stage requires additional or new signals.

The development of cells to embryogenic cell clusters and afterwards to somatic embryos is accompanied by specific changes in protein pattern: new proteins are synthesized, others decrease and finally disappear.

The characterization and functional analysis of proteins which are expressed in close correlation with developmental events ('marker proteins') is one of the possibilities to gain further information on the molecular basis of induction and differentiation of plant cells. Furthermore, analysis of protein patterns will give distinct information about the actual status of plant cells in culture.

While some proteins seem to be involved in the process of somatic embryogenesis, some others – like the storage proteins – are markers reflecting several stages of embryo development. So we have to distinguish between causally related markers and stage-specific ones. In order to define certain stages of embryo development, the latter group of proteins will provide some information, while understanding of somatic embryogenesis requires knowledge of proteins associated with the control of differentiation. It is obvious that these proteins are likely to be of low abundance and thus might be difficult to detect.

There are a number of papers reporting changes in polypeptide pattern during somatic embryogenesis in carrot [15], clover [10], orchard grass [4], pea [14] and rice [3]. It is hard to compare the markers found because of only vague homologies regarding their respective molecular weights and isoelectric points.

Therefore, we decided to select antibodies against marker proteins by differential hybridoma screening. This technique has two major advantages: first, the protein analysis with antibodies is much more sensitive than, for instance, 2-D gel electrophoresis, so that there are good changes to detect low-abundant proteins and second, it produces antibodies which may be used to monitor marker protein expression in different tissues and species.

Procedures

Differential hybridoma screening is comparable to differential cDNA screening concerning the methodological background: in both cases one needs a sample containing the protein/cDNA of interest and a control sample to remove all components not specific to the first sample but common to other tissues.

For the detection of embryo-specific proteins we applied the somatic embryogenesis system for pea, as reported by Kysely [6, 7]. In brief:

Immature zygotic embryos as well as shoot apices of 5-day old pea seedlings were used as explants for the induction of somatic embryos. On MS-medium with high auxin (4 µM 2,4-D), embryogenic calluses developing 1−5 somatic embryos are formed within 4−5 weeks. After subculturing on a medium with 4.4 µM kinetin, the calluses continue to form somatic embryos as well as meristematic centres beneath the surface of the calluses (embryogenic callus). As we were interested in proteins expressed during early stages of somatic embryogenesis, the embryogenic callus was used to extract proteins from and to raise monoclonal antibodies against.

As control, epicotyl segments of 5-day-old pea seedlings were grown in the presence of 4.4 µM 2,4-D and 11.1 µM kinetin. A non-embryogenic callus without any meristematic centres was formed. Ideally, the negative control should only differ in one aspect (formation of embryos versus no formation of embryos) while all other parameters should remain stable (explant source, type of phytohormone, time in culture). In meristematic tissues of pea, the simultaneous addition of auxin and cytokinin does not prevent the *induction* of somatic embryos, only their further development. We therefore chose epicotyls as explants, in which we never observed any sign of somatic embryogenesis, whatever conditions were applied.

Tips, leaves, epicotyls and roots of 10-day-old pea seedlings were harvested and used for control experiments. From these tissues soluble proteins were extracted.

Steps in the procedure

1. Homogenize somatic embryos and tissues of young pea seedlings in an agate mortar with extraction buffer [17]. The tissue to buffer ratio should be around 1 mg to 2−4 µl in the case of embryos and 1 mg to 0.5−1 µl in the case of tissues of young pea seedlings.

   Callus material can be extracted in a Waring blender with the same extraction buffer (1 g callus: 0.5 ml buffer).

2. Centrifuge all extracts at 148000 × g for 2 hours at 4 °C to obtain soluble protein. In the case of callus, desalt supernatants and concentrate by ultrafiltration (Amicon, exclusion size of the membrane 10 kD).

3. Determine the protein content in the supernatants according to Bradford [2].

Notes

It is advisable to add as little buffer as possible to tissues where only limited amounts are available. This will avoid having to concentrate the probes prior to immunization and screening experiments. With each concentration step you will lose protein.

Protein determination in small volumes is possible with a 'micro–Bradford' (1−20 µg protein). Here, you add 0.2 ml of dye reagent to 0.8 ml of diluted sample (2−3 µl) supernatant in sample buffer) [1]. As usual, prepare a standard curve each time the assay is performed.

Immunization

1. Sixty-day-old female balb-c mice are immunized with 100 µg protein (from embryos and embryogenic callus, respectively) in *Complete* Freund's Adjuvant. Mix the antigen with an equal volume of *Complete* Freund's Adjuvant. The best antigen concentration is around 1 mg/ml (dilution, if necessary, is done with PBS) so that the total volume does not exceed 200 µl. Inject the antigen subcutaneously, in four spots in the abdominal region ('prime').
2. Repeat the immunization after 4 weeks by injecting 100 µg protein in *Incomplete* Freund's Adjuvant. The method is exactly the same as for the first immunization except that the protein is mixed with an equal of *Incomplete* Freund's Adjuvant. Inject intraperitoneally. Repeat the immunization in the same manner after 4 weeks for at least three times.
3. On days -7, -5 and -3 before fusion of mice spleen cells with myeloma cells (SP2/O-Ag14 [13]) (day 0), boost the mice with 100 µg pure protein (volume: around 100 µl), intraperitoneally.

Notes

Double-ended needles are the most convenient method to mix antigen and adjuvant until they are emulsified.

Mice should be kept according to the laws for protection of animals. Only healthy mice will give a good immune response.

Fusion

Perform the fusion according to Pratt [12] with polyethylene glycol. In brief:

1. Kill a mouse by cervical dislocation.
2. Immediately after having sterilized the body, take blood from heart to determine the serum titer and to have a positive control for screening experiments.
3. Collect and count myelomas and resuspend at a concentration of 10^7 in 10 ml IMDM (Gibco 074-2200).
4. After having removed the outer skin of the mouse, wash the peritoneal cavity 3–4 times with 1 ml of ice-cold sucrose solution (10% in PBS [Gibco 041-4190], sterile). Collect macrophages by centrifugation (300 × g, 5 min) and resuspended in 90 ml medium (IMDM with 20% FCS [Gibco 011-6290]). Plate with 1 drop/well in 96-well microtiter plates (12–13 plates).
5. In parallel, remove the spleen aseptically and place it in 8 ml PBS. With the help of forceps, macerate the spleen gently. Then transfer spleen cells to a centrifugation tube where larger pieces are sheared by taking the suspension up and down in a Pasteur pipette. Collect spleen cells by centrifugation (300 × g, 5 min) and discard the supernatant.
6. Add the prepared myeloma cells (10 ml) to the spleen cell pellet and mix. After another centrifugation step (see above), pour off the supernatant completely, then loosen the pellet by tapping the centrifugation tube on the bench.
7. While swirling in a water bath, add 0.7 ml PEG (Merck PEG 4000 GC) over 1 minute (1 drop per 2 seconds). After allowing to stand for 90 seconds, add 15 ml

IMDM over 3 minutes, starting very slowly and going more rapidly towards the end.

8. Centrifuge the fusion products (300 × g, 5 min) with slow manual acceleration.

9. Remove the supernatant completely by taking off the last droplets with a Pasteur pipette. Resuspend the hybridomas in 90 ml IMDM with 20% FCS and 4% HAT (Gibco 043-1060) and plate on the macrophages (1 drop/well).

Notes

Thaw myeloma cells (SP2/O-Ag14) from frozen stock at least one week prior to fusion. The cells should be in the exponential growth phase. Therefore subculture the cells daily during the last 3–4 days.

You can use the same mouse for preparation of macrophages and for the fusion experiment. It is therefore better to have two people working, one plating out macrophages while the other is continuing the fusion.

All solutions and media needed in the course of the fusion should be warmed to 37 °C. All centrifugation steps should be performed at room temperature.

To the outer rows of a microtiter plate we add two drops of hybridomas. This is to prevent drying of the wells caused by evaporation and to test another plating density.

Antibody screening

The screening protocol consists of two subsequent ELISA tests and one immu-nostain:

1. In the *first ELISA*, hybridomas producing antibodies against plant proteins will be selected. Half of the wells of a microtiter plate (rows A,C,E,G) are incubated overnight (4 °C) with proteins used for immunization (1 µg protein/50 µl borate buffer), the other half with 50 µl borate buffer as negative control.

2. The next morning, wash the microtiter plate with ELISA-wash and block unspecific binding by filling the wells with ELISA-block (30 min, RT).

3. As the next step, incubate wells with 50 µl of hybridoma supernatants (protein content around 1 µg/ml) for 2 h at room temperature (RT).

4. This is followed by incubating the wells with 50 µl of rabbit-anti-mouse anti-bodies coupled with alkaline phosphatase (Sigma A-1902) (2 h, RT). The working dilution for the antibody usually is 1 : 500 in ELISA-diluent.

5. Start enzyme reaction by adding 50 µl of substrate (Sigma N 9389, 1 tablet in 8.66 ml substrate buffer) to the wells and stop after 30 minutes by adding 50 µl 3 N NaOH.

6. For quantification, determine the extinction at two different wavelengths (405/492 nm).

Only those hybridoma supernatants will be subcultured which give positive reactions.

7. In the *second ELISA*, you select antibodies which react preferentially with proteins from embryogenic callus. Therefore, hybridoma supernatants will be tested simultaneously for a reaction with proteins from embryogenic callus and proteins from nonembryogenic callus.

8. In parallel to the first ELISA, coat rows A,D with proteins from embryogenic

alluses and rows B,E with borate buffer. Additionally, incubate rows C,F with proteins from nonembryogenic calluses.

9. The washing and enhancer steps are the same as in the first ELISA (repeating steps 3—9).

10. Only those hybridoma cell lines will be cultured further, which gave at least a fourfold higher extinction value with proteins from embryogenic calluses in comparison to proteins from nonembryogenic calluses (see Table 1).

Notes

For both ELISAs, pretreat microtiterplates (Flow 76-317-05) with 70% ethanol for 2 h at room temperature. This is to standardize binding of proteins to the polystyrene surface.

Table 1. Comparison of extinction values obtained with the second ELISA

Example		Extinction (405/492 nm)		Percentage of hybridoma cell lines
	Proteins from embryog. callus	Buffer	Proteins from non-embr. callus	
Ia	0.752	0.054	0.135	
Ib	0.495	0.042	0.046	16.3%
II	0.437	0.199	0.263	83.7%
Serum	0.449	0.064		
NIM	0.068	0.047		

NIM = non-immune mouse serum.

Immunostain

1. Perform as *third selection step* an immunostain: Separate proteins from embryos, embryogenic and non-embryogenic calluses by SDS-PAGE [9, 11].

2. Transfer separated proteins from SDS-gels to nitrocellulose by semi-dry blot [16].

3. Block unspecific binding in nitrocellulose blocking solution.

4. Incubate the nitrocellulose with hybridoma supernatants (recommended concentration depends on the antibody: between 1 : 100 to 1 : 1000) for 2 h at RT.

5. Enhance signals with rabbit-anti-mouse antibodies (Sigma M-9637, 1.1 µl/ml dot blot diluent) and goat-anti-rabbit antibodies coupled with alkaline phosphatase (Sigma A-7778, 1 : 1000 in dot blot diluent).

6. Visualize reaction by adding substrate (10% Nitrostock and 1% BCIP in nitrocellulose substrate buffer) and stop with water.

7. Using this scheme it is possible to detect antibodies recognizing proteins only specific of embryogenic tissue, e.g. the marker proteins for somatic embryogenesis.

In our system we analysed 1830 hybridoma clones resulting from seven fusions, and found two embryo-specific clones recognizing polypeptides of molecular weights of 50 000 and 20 000.

To save antibody, seal nitrocellulose between plastic foil and incubate on a rotary shaker. With this procedure only 3–5 ml of antibody solution is needed for 10 × 10 cm nitrocellulose.

Solutions

Extraction buffer

0.2 M	boric acid
2 mM	borax
0.25 M	KCl
20 mM	Tris-HCl, pH 6.8
4 mM	diethyldithiocarbamate and
2.8%	(w/v) Polyclar AT are added prior to extraction

Solutions for ELISA

Borate buffer

100 mM	boric acid
50 mM	borax
75 mM	sodium chloride
	pH 8.5

ELISA wash

10 mM	Tris-HCl, pH 8.0
0.02%	sodium azide
0.05%	Tween 20

ELISA block

5 mM	phosphate buffer, pH 7.4
140 mM	sodium chloride
0.02%	sodium azide
1%	BSA
0.05%	Tween 20

ELISA diluent

5 mM	phosphate buffer, pH 7.4
140 mM	sodium chloride
0.02%	sodium azide
1%	BSA

ELISA substrate buffer

1 M	diethanolamine
0.5 mM	magnesium chloride
	pH 9.8

Solutions for immunoblot

nitrocellulose block

5 mM	phosphate buffer pH 7,4
140 mM	sodium chloride
1%	BSA
0,02%	sodium azide
1%	lamb serum (Gibco 035-6070)

Dot blot wash

10 mM	Tris/HCl, pH 8
0,01%	BSA
0,02%	sodium azide
1%	lamb serum
0,05%	Tween 20

Dot blot diluent

5 mM	phosphate buffer pH 7,4
140 mM	sodium chloride
0.05%	BSA
0.01%	sodium azide
5%	lamb serum
0.025%	Tween 20

nitrocellulose substrate buffer

0.1 M	sodium hydrogen-carbonate
0.05 M	sodium carbonate
4 mM	magnesium chloride

nitro stock

1 mg	4-nitro blue tetrazolium in
1 ml	nitrocellulose substrate buffer

BCIP

5 mg	4-bromo-4-chloro-3-indolylphosphate in
1 ml	dimethyl sufoxide

References

1. Bio-Rad Protein Assay Instruction Manual (1986) Microassay Procedure, Bio-Rad, Munich, 9.
2. Bradford M (1976) A rapid and sensitive method for the quantification of microgram quantities of protein utilizing the principle of protein-dye binding. Anal Biochem 77: 248–254.
3. Chen LJ, Luthe DS (1987) Analysis of proteins from embryogenic and non-embryogenic rice (*Oryza sativa Ll*) calli. Plant Science 48: 181–188.
4. Hahne G, Mayer JE, Lörz H (1988) Embryogenic and callus-specific proteins in somatic embryogenesis of the grass, *Dactylis glomerata* L. Plant Science 55: 267–279.
5. Kohlenbach HW (1985) Fundamental and applied aspects of *in vitro* plant regeneration by somatic embryogenesis. In: *In vitro* techniques, propagation and long term storage, A. Schäfer-Menuhr (ed.), Martinus Nijhoff, W. Junk Pub., Dordrecht, The Netherlands, pp 101–109.
6. Kysely W, Myers J, Lazzeri PA, Collins GB, Jacobsen HJ (1987) Plant regeneration via somatic embryogenesis in pea. Plant Cell Rep 6: 305–308.
7. Kysely W, Jacobsen HJ (1990) Somatic embryogenesis from pea embryos and shoot apices. Plant Cell Tiss Org Cult 20: 7–14.
8. Komamine A, Matsumoto M, Tsukahara M, Fujiwara A, Kawahara R, Ito M, Smith J, Nomura K, Fujimura T (1990) Mechanisms of somatic embryogenesis in cell cultures – physiology, biochemistry and molecular biology. Progress in Plant Cellular and Molecular Biology, Amsterdam, pp 307–313.
9. Laemmli UK (1970) Cleavage of structural proteins during the assembly of the head of bacteriophage T4. Nature 227: 680–685.
10. McGee JD, Williams EG, Collins GB, Hildebrand DF (1989) Somatic embryogenesis in *Trifolium*: protein profiles associated with high- and low-frequency regeneration. J Plant Physiol 135: 306–312.
11. Murray MG, Key JL (1978) 2,4-dichlorophenoxyacetic acid enhanced phosphorylation of soybean nuclear proteins. Plant Physiol 61: 190–198.
12. Pratt LH (1984) Phytochrome Immunocytochemistry. In: Techniques in Photomorphogenesis, Academic Press, London, pp 201–226.
13. Shulman M, Wilde CD, Köhler G (1978) A better cell line for making hybridomas secreting specific antibodies. Nature 276: 269–270.
14. Stirn S, Jacobsen HJ (1987) Marker proteins for embryogenic differentiation patterns in pea callus. Plant Cell Rep 6: 50–54.
15. Sung ZR, Okimoto R (1981) Embryogenic proteins in somatic embryos of carrot. Proc Natl Acad Sci USA 78: 3683–3687.
16. Towbin H, Staehelin T, Gordon J (1983) Electrophoretic transfer of proteins from polyacrylamide gels to nitrocellulose sheets: Procedure and applications. Proc Natl Acad Sci USA 76: 4350–4354.
17. van der Linde PCG, Bouman H, Mennes AM, Libbenga KR (1984) A soluble auxin-binding protein from tobacco tissues stimulates RNA synthesis *in vitro*. Planta 160: 102–108.
18. Zimmerman JL, Apuya N, Darwish K, O'Carrol C (1989) Novel regulation of heat shock genes during carrot somatic embryo development. The Plant Cell 1: 1137–1146.

Plant Tissue Culture Manual E3: 1–21, 1992.
© 1992 *Kluwer Academic Publishers*.

Endosperm culture

P.S. SRIVASTAVA[1] & B.M. JOHRI[2]
[1] *Department of Botany, Hamdard University, Hamdard Nagar, New Delhi 110062, India.*
[2] *Department of Botany, University of Delhi, Delhi 110007, India.*

Introduction

In over 81% families of flowering plants, the developing seeds have a nutritive triploid endosperm tissue. This results from the fusion of usually two polar haploid nuclei contributed by the female parent and one haploid male gamete. The endosperm provides nourishment to the developing embryo and influences its differentiation [1]. The failure of endosperm development causes abortion of the embryo and the resulting seeds are sterile [2]. A mature seed is nonendospermous (exalbuminous) when the embryo consumes the entire endosperm during development, but endospermous (albuminous) when it persists as a reservoir of food reserves (starch, fat, and proteins) and is utilized during seed germination.

Irrespective of the number of polar nuclei (and their genetic constitution) taking part in fusion, the primary endosperm nucleus has the same nutritive function (see Ref. 3). It is diploid (*Butomopsis*), pentaploid (*Fritillaria*) or polyploid (*Acalypha indica*, *Peperomia*). In apomicts, endosperm may develop autonomously (without fertilization of the secondary nucleus) as in *Taraxacum* and *Erigeron* [4].

The unique genetic constitution of the endosperm perhaps renders it incapable of *in vivo* organogenesis/embryogenesis, while other ovular tissues (integument/s, nucellus, synergids) often produce supernumerary 'embryos'. For a long time the cellular mature endosperm has been considered as a 'dead' tissue. Johri and Bhojwani, for the first time, demonstrated the *in vitro* differentiation of vascularised shoot buds (Fig. 1A) directly from the mature endosperm of *Exocarpus cupressiformis* [5].

The endosperm tissue (with or without the embryo intact) cultured *in vitro* may proliferate and form non-vascularized or vascularized callus. Embryoids, shoot buds, roots and plantlets differentiate from the callus (see Table 1).

The differentiation of plantlets from the endosperm of monocotyledons is less commonly observed than from the endosperm of dicotyledons. The mature endosperm of some taxa, such as *Petroselinum hortens* [6] and *Oryza sativa* [7], are reported to regenerate plantlets (see Table 1). Until now, only the plantlets from *Santalum album* [8] and *Juglans regia* [9] have been successfully transplanted to soil (Fig. 2). Since the earliest attempt to culture the endosperm of maize *in vitro* in 1933 [56], it has now been successfully cultured in about 40 species of angiosperms: from immature endosperm in 11 taxa (plantlets in five species; only *Asparagus* shows embryogenesis), and from mature endosperm

Table 1. *In vitro* growth responses of endosperm

Taxa	Explant	Medium	Response	Reference
Achras sapota	MW	BM + 2,4-D (2 mg/l) + KN (5 mg/l) + CM (15%) + CH (400 mg/l)	Profuse callusing	[34]
Actinidia chinensis	MW	MS + ZN (1 mg/l) + CH (400 mg/l)	Plantlets	[38]
Actinidia chinensis × A. melanadra A. arguta × A. deliciosa	MW	BM + IAA (0.3 mg/l) + ZN (5 mg/l) + CH (400 mg/l)	Plantlets*	[39]
Annona squamosa	ME	BM + IAA (0.5 mg/l) + BAP (20 mg/l)	Plantlets	[29]
Asimina triloba	IW	BM	Profuse callusing	[15]
Asparagus officinalis	IW	MS + NAA (0.5 mg/l) + BAP (1.0 mg/l)	Plantlets*	[26]
Citrus grandis	IE	BM (high mineral concentration) + GA (1 mg/l)	Plantlets*	[24]
Coffea arabica	IW	MS + ME (500 mg/l) + CH (200 mg/l)	Callusing	[19, 61]
Cocos nucifera	IW	BM + 2,4-D (50 mg/l) + KN (2 mg/l)	Profuse callusing	[41]
Codiaeum variegatum	ME	BM + 2,4-D (1 mg/l) + CH (500 mg/l) + CM (10%)	Plantlets	[62]
Croton bonplandianum	ME	BM + CH (500 mg/l)	Roots	[36]
Cucumis sativus	IW	BM + IAA (2 mg/l) + diphenylurea (5 mg/l) + CH (500 mg/l)	Profuse callusing	[43]
Dendrophthoe falcata	ME	BM + IAA (2.5 mg/l) + KN (5 mg/l)	Shoots	[17]
Emblica officinalis	MW	MS + BAP (0.2 mg/l) + NAA (0.001 mg/l)	Plantlets*	[30]
Euphorbia geniculata	MW	MS + NAA/IAA (0.1 mg/l)	Roots	[27]
Exocarpus cupressiformis	ME	BM + IAA (1 mg/l) + KN (1 mg/l) + CH (400 mg/l)	Shoot buds	[5]
Hordeum vulgare	IW	BM + IAA (1 mg/l) + CH (0.25%)	Profuse callusing, plantlets	[25, 44]
Jatropha panduraefolia	ME	BM + IAA (0.5 mg/l) + KN (1 mg/l) + CH (1000 mg/l)	Root, shoots	[18]
Juglans regia	MW	BM + IBA (0.01 mg/l) + KN (0.2 mg/l) + BAP (0.1 mg/l) + glutamine	Callus, plantlets*	[9]
Leptomeria acida	ME	BM + IAA (2 mg/l) + KN (5 mg/l)	Shoot buds	[17]
Lolium perenne	IW	BM + YE (0.5%) + IAA (5.7 µM)	Profuse callusing	[20]
L. multiflorum	IW	BM + YE (0.5%) + IAA (5.7 µM)	Callus (in suspension culture)	[42]
Lycium chinensis	MW	BM + NAA (0.1 mg/l) + BAP (0.5 mg/l)	Plantlets	[31]

Table 1. (continued)

Taxa	Explant	Medium	Response	Reference
Lycopersicon esculentum	MW	MS + 2,4-D (1 mg/l) + BAP (0.1 mg/l) + GA (10 mg/l)	Profuse callusing	[63]
Malus pumila	IW	BM + 2,4-D (0.1–0.5 mg/l) MS + BAP (0.1–1 mg/l) + NAA (0.01– ' 0.05 mg/l) + CH (500 mg/l)	Plantlets	[23, 45]
Nigella damascena	ME	BM + 2,4-D (5 × 10^{-6} M)	Callusing	[28]
Nuytsia floribunda	ME	BM + IBA (10 mg/l) + KN (5 mg/l) + CH (400 mg/l)	Callusing	[17]
Oryza sativa	MW IW	BM + 2,4-D (10^{-5} M) + YE (0.4%), MS + IAA (4 mg/l) + KN (2 mg/l)	Plantlets	[7, 32]
Petroselinum hortens	ME	MS	Plantlets*	[6]
Putranjiva roxburghii	ME	BM + IAA (2 mg/l) + KN (5 mg/l) + CH (1000 mg/l)	Callus, plantlets	[40]
Prunus persica	MW	MS + BAP (1 mg/l) + CH (500 mg/l)	Callus, embryoids	[33]
Ricinus communis	ME	BM + 2,4-D (2 mg/l) + KN (5 mg/l) + YE (2500 mg/l)	Callus	[11, 16]
Santalum album	ME	BM + 2,4-D (2 mg/l) + KN (5 mg/l) + YE (2500 mg/l) MS + NAA (0.1 mg/l) + GA (1 mg/l)	Callus plantlets*	[8, 37]
Scurrula pulverulenta	ME	BM + IAA (1 mg/l) + KN (1 mg/l) + CH (400 mg/l)	Shoot buds, haustoria	[22]
Taxillus cuneatus *T. vestitus*	MW	BM + KN (5 mg/l)	Shoot buds	[21]
Triticum aestivum	IW	BM + 2,4-D (1 mg/l) + KN (0.5 mg/l) + CM (10%)	Callusing	[44, 55]
Vitis vinifera	ME	BM + 2,4-D (2 mg/l) + KN (1 mg/l) + CM (10%)	Profuse callusing	[35]
Zea mays	IW	BM + Potato extract BM + YE (1.0%) BM + Tomato juice (2%) BM + YE (0.5%)	Callus, plantlets	[10, 12, 49–54,56, 64–68]

IW = immature endosperm without embryo   IE = immature endosperm with embryo
MW = mature endosperm without embryo   ME = mature endosperm with embryo
*   = through embryogenesis   BM = basal medium, comprising inorganic mineral salts, amino acids and vitamins, and an iron source

in 29 taxa, including semi-parasitic angiosperms (plantlets differentiate in 16 taxa but only in three taxa through embryogenesis).

The inability of earlier investigators to induce organogenesis and embryogenesis can be attributed to the lack of selection of endosperm of appropriate age and the appropriate nutrient media. The commonly used White basal medium (WB) contains only a limited amount of reduced nitrogen and, therefore, complex chemicals such as coconut milk, casein hydrolysate or yeast extract have to be added. With the increased use of Murashige and Skoog medium (MS), many research workers have succeeded in promoting the growth of endosperm in cultures. Consequently, the assumption that the endosperm lacks morphogenic potential has been disproved.

Endosperm callus culture

For initiating endosperm cultures, a detailed knowledge about the development and structure of the experimental material, the age of the explant, and the selection of appropriate media are absolutely essential. In some plants, only the immature endosperm can be cultured, but mature endosperm in Euphorbiaceae, Loranthaceae and Viscaceae.

The commonly used nutrients are the modified White, and Murashige and Skoog media. The addition of growth hormones and other supplements to the media varies from explant to explant and depends on the response for callus formation and differentiation.

Note

During germination, gibberellin-like substances are known to be released from the embryos of many species, such as barley [47] and castor [48]. The mature endosperm culture alongwith the embryo proliferates (the embryo is removed as soon as germination ensues) and produces a subculturable callus in several cases. The callus subsequently differentiates shoots on a new medium. The role of embryo for initial proliferation and differentiation can be replaced by GA_3 (see Ref. 36).

In some taxa, initially the endosperm shows a positive response only if the embryo remains intact. However, if the embryo is to be removed at the initial explantation stage itself, the endosperm has to be split open into two halves.

The immature endosperm requires for culture a careful selection of seeds of appropriate age (days after pollination) as the mature seeds may be nonendospermous.

The free-nuclear stage does not show any response; only the cellular endosperm is amenable to culture *in vitro*.

The media and response

Immature Endosperm

Plantlets from the immature endosperm are reported only in *Oryza sativa* [7], *Malus pumila* [23], *Citrus grandis* [24], *Hordeum vulgare* [25] and *Asparagus officinalis* [26].

The immature endosperm tissue generally requires, in addition to the mineral salts and vitamins, an additional source of nitrogen, yeast extract (YE), casein hydrolysate (CH), coconut milk (CM) or tomato juice (TJ).

The callus from immature endosperm undergoes organogenesis only if the medium contains an auxin, or cytokinin, or both. The endosperm of *Asimina* [15] calluses profusely on WB alone. The addition of an auxin induces callus in the endosperm of *Euphorbia* [27] and *Nigella* [28]. *Annona* [29], *Emblica* [30], *Juglans* [9] and *Lycium* [31] require both an auxin and cytokinin, and for *Prunus*, [33] benzylaminopurine (BAP) + CH. The immature endosperm of *Cocos* [41] calluses on WB + 2,4-dichlorophenoxyacetic acid (2,4-D, 50 mg/l) + Kinetin (KN, 2 mg/l), *Coffea* [19] on MS + malt extract (50 mg/l) + CH (200 mg/l), *Lolium* [20, 42] on WB + indoleacetic acid (IAA, 5.7 μm) + YE (0.5% w/v), *Cucumis* [43] on WB + IAA (2 mg/l) + diphenylurea (5 mg/l) + CH (500 mg/l), and of *Triticum* [44] on WB + 2,4-D (1 mg/l) + KN (0.5 mg/l) + CM (10% v/v).

For plantlet differentiation from the immature endosperm of *Asparagus* [26], the addition of naphthaleneacetic acid (NAA, 0.5 mg/l) + BAP (1.0 mg/l) to MS is essential, and for *Citrus* WB + gibberellic acid (GA$_3$, 1 mg/l). Regeneration of plantlets in *Hordeum* [25] occurs on MS + NAA (1 mg/l) + CH or YE (0.25% w/v), in *Malus* [23, 45] on MS + NAA (0.01–0.05 mg/l) + BAP (0.1–1 mg/l) + CH (500 mg/l), and in *Oryza* [7] on MS + IAA (4 mg/l) + KN (2 mg/l) + YE (0.4% w/v). The immature endosperm (excised 4–7 days after pollination, DAP) of *Oryza* calluses on WB fortified either only with YE (0.4% w/v), or 2,4-D (0.2 mg/l) and YE (0.4% w/v), or 2,4-D (2 mg/l) with CM (15% v/v). Plantlet regeneration occurs on WB + YE (0.4% w/v) + IAA (10^{-5} M), or on WB + YE (0.4% w/v) + KN. Only the immature endosperm of *Petroselinum* shows embryogenesis on MS without any growth hormones [6]. Thus, for the majority of taxa, an auxin (preferably 2,4-D (1–2 mg/l) + KN (2–5 mg/l) + CH or YE (400–2500 mg/l) are essential for the initiation of callus. For regeneration, these calluses have to be transferred to a medium lacking 2,4-D.

Mature endosperm

The mature endosperm of more than 22 species of angiosperms have been cultured *in vitro*. Of these, 16 show organogenesis (including plantlet formation), the remaining seven produce only callus (Table 1). Callusing usually precedes organogenesis. The endosperm of autotrophic taxa require an auxin, a cytokinin, and CH or YE. On White's medium (WB) with 2,4-D (2 mg/l) + KN (5 mg/l) + YE (2500 mg/l), a continuously growing friable creamish callus develops from the mature endosperm of *Croton, Jatropha* and *Ricinus* [13, 14]. Upon repeated subcultures, the callus becomes compact and somewhat hard. In *Putranjiva*, callusing as well as organogenesis occurs on WB + IAA (2 mg/l) + KN (5 mg/l) + CH (1000 mg/l).

The optimal concentration of CH is 1500 mg/l for the induction of shoot buds in 87% cultures of endosperm of *Putranjiva*. The growth of callus, however, is much better with CH (2000 to 2500 mg/l). The promotive effect of CH can be replaced by the use of amino acids, leucine and valine, either alone or together. Valine (5×10^{-5} M) alone induces shoot differentiation in 70% cultures (Fig. 3B a–d).

Of the different substituted aminopurines, 6-(γ, γ-dimethylallylamino) purine, and zeatin (ZN, 3×10^{-5} M) promote shoot differentiation in 87% cultures (Fig. 3C a–g), and also bring about maximal increase in the fresh weight of callus. The promotory effect of auxins is shown in (Fig. 3D a–d).

Note

The components of WB also influence the response. If KNO_3, $Ca(NO_3)_2$, and KH_2PO_4 are removed from WB, the growth of callus and differentiation of shoots is adversely affected. But, without $MgSO_4$, there is excellent growth of callus and 49% cultures show shoot buds (Fig. 3A a–g).

Endosperm of semi-parasitic angiosperms

In six semi-parasitic species, organ differentiation has so far been demonstrated: *Exocarpus* (Fig. 1A), *Taxillus* (Fig. 1B), *Leptomeria* (Fig. 1C), *Scurrula* (Fig. 1D, E), *Dendrophthoe* (Fig. 1F) and *Santalum* (Fig. 2A–C) [5, 8, 17, 21, 22]. Except in *Santalum*, in other semi-parasitic members the formation of shoot buds from mature endosperm occurs after negligible callusing or without callusing.

The endosperm shows optimal growth on WB containing either only a cytokinin, or a cytokinin alongwith an auxin, e.g. direct differentiation occurs in *Taxillus* on WB + KN (10 mg/l); in *Exocarpus* on WB + IAA (1 mg/l) + KN (1 mg/l). In fact, 5-week-old cultures of *Exocarpus cupressiformis* [5] exhibit shoot buds on WB + IAA (1 mg/l) + CH (400 mg/l). The embryo develops into a normal seedling.

The omission of CH from the medium results in an increased number of shoot buds. Shoot buds excised and planted on the same medium produce callus from the cut end after 11 weeks, and shoot buds and haustoria redifferentiate. The *in vitro*-produced leaves, however, lack the characteristic natural trichomes. The endosperm of *Scurrula pulverulenta* exhibits chlorophyllous buds when grown on WB supplemented only with ZN (10^{-5} M). When subcultured, haustoria differentiate from these buds.

The presence of an aminopurine in WB induces shoot buds and haustoria in *Taxillus vestitus* and *T. cuneatus* [21]. In *T. vestitus*, the orientation of the endosperm explant on the medium influences the formation of shoot buds. The half-split endosperm (without the embryo) implanted with the cut surface in contact with WB + 6-(γ, γ-dimethylallylamino) purine (2×10^{-5} M) shows shoot buds in almost 100% cultures after 8 weeks. In fact, this response also depends upon the specific cytokinin added to the medium. On WB + KN (5 mg/l), shoot buds appear after 9 weeks, with 10 mg/l in 6 weeks.

The presence of IAA or IBA + KN reduces the potentiality for bud formation. The suppressive effect of auxin can be overcome by enhancing the proportion of cytokinin. Thus, while only 30% cultures produce buds [6] on WB + IAA (2 mg/l) + KN (2 mg/l), KN (10 mg/l) induces buds in 80% cultures. Indolebutyric acid (IBA) at 5 mg/l totally suppresses callusing and bud induction.

On WB + IAA (5 mg/l) + KN (5 mg/l) + CH (2000 mg/l), the endosperm of *T. vestitus* forms shoot buds in 85% cultures after 7 weeks; another week later 45% cultures show the differentiation of haustoria. With reduced IAA (2.5 mg/l), *T. cuneatus* shows buds in 85% and haustoria in 40% cultures, after 10 weeks. IBA (5 mg/l) affects bud formation in 55% and haustoria in 60% cultures [17, 21]. Thus, IBA promotes haustoria formation in *T. cuneatus*, but is ineffective for *T. vestitus*.

Note

A cytokinin is indispensible for bud induction. The endosperm pieces if soaked only in KN (0.025 mg/l) for 4 h (and cultured on WB alone) form buds all over the surface. In *T. cuneatus*, callusing invariably precedes bud differentiation; the buds are comparatively longer (12 mm) than those of *T. vestitus* (5 mm).

The endosperm of *Dendrophthoe falcata* calluses and differentiates shoot buds on WB + IAA (2.5 mg/l) + KN (5 mg/l); the buds grow into well developed shoots. WB + IAA (5 mg/l) + KN (10 mg/l) + CH (2000 mg/l) induces buds in 46% cultures after 14 weeks. The substitution of KN with 6-(γ, γ-dimethylallylamino) purine (5 mg/l) promotes bud formation in 56% cultures. IBA (5 mg/l) with KN (5 mg/l) + CH (2000 mg/l) induce profuse growth of callus [17].

Maximal bud induction (92%) occurs on WB + IAA (2 mg/l) + KN (10 mg/l) after 12 weeks, in *Leptomeria acida*, rapid growth of endosperm callus occurs on WB + IBA (5 mg/l) + KN (5 mg/l) + CH (2000 mg/l).

The endosperm callus of *Nuytsia floribunda* – obtained on WB + IBA (5 mg/l) + KN (5 mg/l) or adenine (40 mg/l) + CH (2000 mg/l) – can be maintained satisfactorily for about 40 months, subcultured every 12 weeks.

The endosperm of *Santalum album* produces callus on a variety of media, but MS medium supplemented with BAP and NAA elicit better results than with 2,4-D. Embryoid formation occurs on MS + GA$_3$ (1–2 mg/l), MS + BAP (0.3–2.0 mg/l) + IAA (1 mg/l), and MS + KN (0.3–2 mg/l) + GA (1 mg/l). Extensive embryogenesis has been observed on MS + BAP (0.3 mg/l) + IAA (1 mg/l), and on MS + GA (1 mg/l) + KN (0.3 mg/l). The embryoids develop further upon transfer to WB [8].

The regeneration of plantlets through organogenesis in endosperm cultures became well established as early as 1973 [see 14]; but embryogenesis remained elusive. In fact, until 1980, the only convincing report of endosperm callus which showed embryogenesis is that of a tree species, *Santalum album* [8]. The differentiated embryoids developed into plantlets, and were successfully transplanted first to garden soil and then to a forest nursery.

Plantlets differentiate, through somatic embryogenesis, from the mature endosperm tissue of open-pollinated seeds of *Juglans regia* [9] on the same medium [46] that supports the regeneration of plantlets from the cotyledons. The medium contains BAP (0.1 mg/l) + KN (0.2 mg/l) + IBA (1 mg/l) + glutamine (125 mg/l). It has been possible to isolate a repetitive embryogenic line by continuous subculture on the basal medium [9, 46] with 6% sucrose. The roots and hypocotyl of the *in vitro* seedlings regenerate additional somatic embryos. The plantlets could be grown to the young sapling stage in soil in the field. These are now several years old and appear healthy (as also in *Santalum*).

Callus formation with a high embryogenic potential can be induced from the endosperm of parsley on MS without any hormones. The results do suggest that once the endosperm undergoes a shift related to the dependency upon the growth substances, it calluses and ultimately undergoes embryogenesis [8].

The regeneration of plantlets through organogenesis and embryogenesis in endosperm cultures of F_1 and F_2 seeds of *Actinidia chinensis* × *A. melanandra*, *A. arguta* var. *arguta* × *A. melanandra* var. *melanandra* (organogenesis only), and *A. arguta* var. *arguta* × *A. deliciosa* var. *deliciosa* is variable and genotype dependent.

Embryoid formation in immature endosperm cultures is possible in *Citrus grandis* Osbeck [24]. Callus can be induced on WB + 2,4-D (2 mg/l) + benzyladenine (BA, 0.25 mg/l) + CH (1000 mg/l). The addition of GA (4–15 mg/l) to the medium brings about the differentiation of embryoids which germinate and form complete plantlets. Higher concentrations (10–15 mg/l) of GA are necessary to promote further growth of embryoids from heart-shaped to cotyledonary stage.

Applications

The culture of endosperm is now being used increasingly for physiological, biochemical, and genetical studies related to the production of enzymes in a cell-free system, detection of inhibitors present in the endosperm, use of anthocyanin and proteins as biochemical markers, movement of compounds from maternal tissue to the developing seeds, and identification of genes encoding for specific proteins in the endosperm [49].

It has been possible to induce callusing of different mutants of maize with a varying content of IAA [50]. Such studies would finally unveil the correlation between the mutants and the deficient biochemical pathways of certain growth regulators. It also serves as a model system to study *in vitro* synthesis of starch

[51, 52]. The transcription of the genes for storage protein synthesis is strictly tissue-specific and developmentally regulated in maize [53]. When 10-day-old endosperm is cultured, zein accumulation begins and proceeds linearly for several days [53]. During this period of *in vitro* growth, both zein synthesis and its regulation appear to occur in the cultured callus as in the *in vivo* kernels. Thus, long term endosperm cultures appear to represent a good experimental system for the study of starch synthesis.

Cultures of maize endosperm from strains homozygous for all genes required for anthocyanin synthesis developed an intense pigmentation. Colourless areas have also been frequently observed that show a higher growth rate. The difference in the rate of growth associated with the ability of cells to accumulate pigment suggests different physiological states of the culture. The unpigmented sectors show a more active metabolism represented by a faster growth rate and a higher salt soluble protein content [54]. The presence of zein, both in pigmented and nonpigmented callus, indicates that the endosperm cells in culture are able to synthesize a storage protein while continuing active proliferation. In contrast, under *in vivo* conditions, zein and anthocyanin synthesis starts soon after cell division ceases. Zein synthesis, in endosperm cultures, seems to be produced in a constitutive manner [54].

The *in vitro* culture of endosperm is, therefore, very significant. The technique can also be fruitfully exploited for a better understanding of the embryo-endosperm relationship. The present evidences indicate the uptake from nutrient media into cob and translocation of various ^{14}C compounds into developing maize seeds in caryopsis cultures. Similarly, the presence of an inhibitor in wheat endosperm has been revealed by embryo transplantation in irradiated seeds [55].

Another interesting area of study would be of somatic hybrids resulting from the fusion of diploid–diploid, diploid–triploid and triploid–triploid protoplasts. Preliminary trials do indicate that the protoplasts from the cells of cultured endosperm of, for example, *Lolium multiflorum* [59] and *Santalum album* [60] can be isolated and grown on nutrient media, but, as yet, fail to regenerate plants.

The tissue cultures of endosperm have also been used to evaluate the production of caffeine and its release into the medium. Feeding the endosperm with $NaH^{14}CO_3$ leads to the synthesis of ring-labelled caffeine [61].

The triploid endosperm is therefore an excellent tissue for *in vitro* morphogenic studies and for raising triploid plants. This procedure opens up a new vista for plant breeders, especially when conventional breeding methods are unsuccessful, or are too cumbersome. The edible fruit crops can be exploited by *in vitro* endosperm culture techniques to raise highly prized seedless varieties.

The potentials of the *in vitro* cultured endosperm tissue, cells and protoplasts, and the synthesis of biochemicals have not yet been fully exploited.

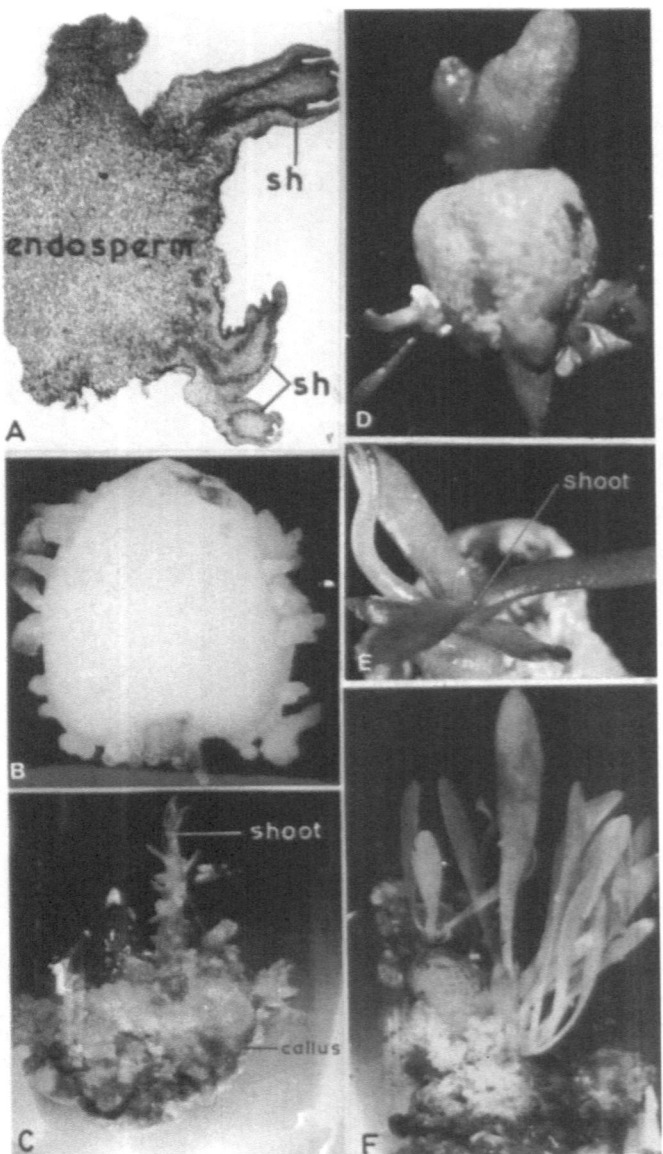

Fig. 1. *A−F.* Differentiation of shoot buds from mature endosperm of semi-parasitic angiosperms. (A) *Exocarpus cupressiformis*, longisection with three vascularized shoot buds, on WB + IAA (1 mg/l) + KN (1 mg/l) + CH (400 mg/l), 11-wk-old culture, *sh*, shoot bud [5]. (B) *Taxillus vestitus*, 10-wk-old culture of endosperm half with cut surface in contact with the medium, WB + KN (10 mg/l), numerous shoot buds develop all along the periphery [21]. (C) *Leptomeria acida*, differentiation of shoot in 12-wk-old subcultured (10 passages of 12 weeks each) endosperm callus transferred from (WB + IBA (5 mg/l) + KN (5 mg/l) + CH (2000 mg/l) to WB + IAA (2.5 mg/l) + KN (5 mg/l) + CH (2000 mg/l) [17]. (D,E) *Scurrula pulverulenta*. (D) 16-wk-old culture, on WB + ZN (10^{-5} M), with shoot buds from endosperm. (E) Callusing of *in vitro-*

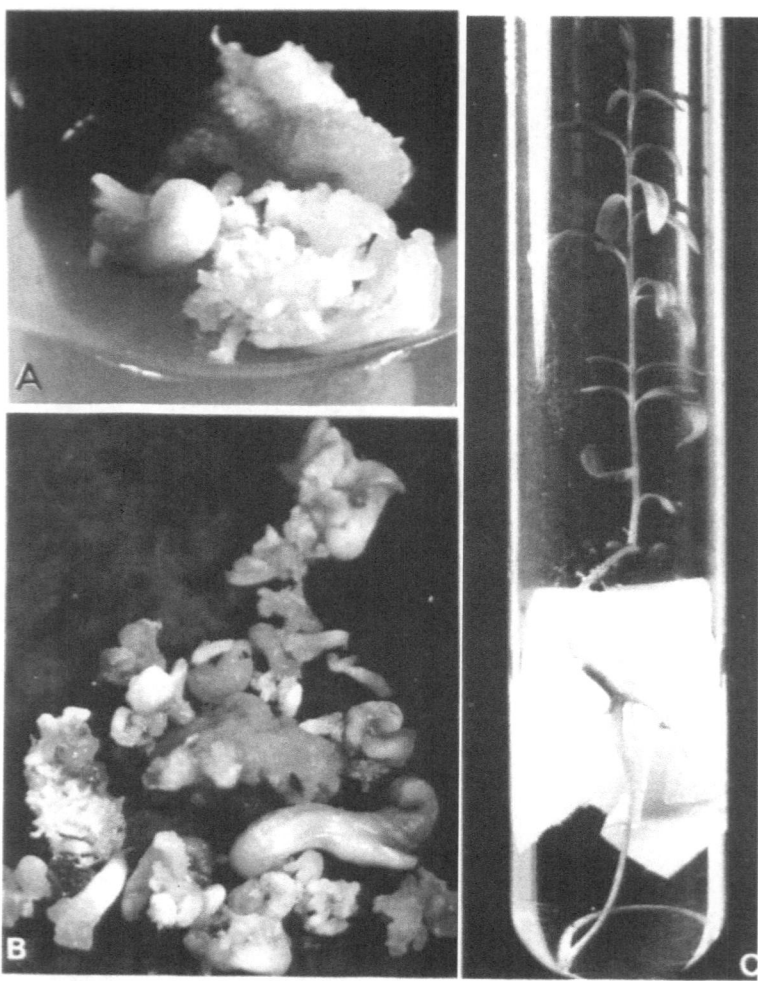

Fig. 2. A–C. Santalum album, embryoid differentiation and plantlet formation from endosperm callus on MS + GA (1 mg/l) + KN (0.3 mg/l). (A) Embryoids. (B) Later stage, mostly dicotyledonous embryoids. (C) Germination and rooting of embryoids on filter paper bridge; the plantlets could be transplanted to soil [8].

produced buds and differentiation of new shoots after 11 weeks, on WB + IAA (1 mg/l) + KN (1 mg/l) + CH (400 mg/l) [22]. (F) *Dendrophthoe falcata*, 15-wk-old culture with shoot and many leaves, on WB + IAA (5 mg/l) + KN (10 mg/l) + CH (2000 mg/l) [17].

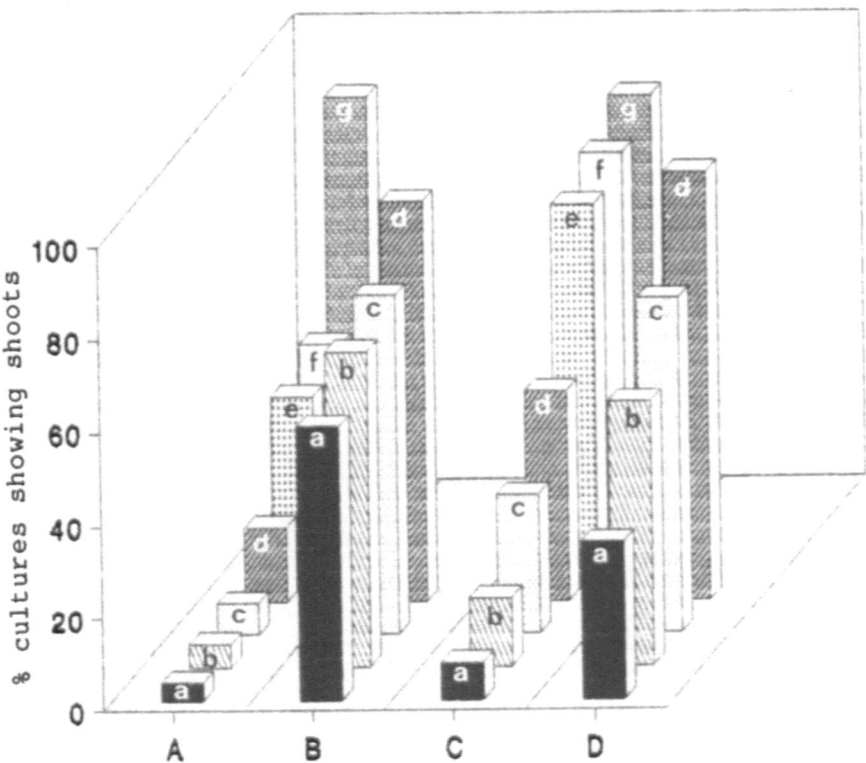

Fig. 3. A–D. Putranjiva roxburghii, differentiation of shoots (from mature endosperm) in 12-wk-old cultures. (A) (a–g). WB + IAA (2 mg/l) + KN (5 mg/l) + CH (1000 mg/l) (g). WB without $Ca(NO_3)_2$ (a), without KNO_3 (b), without KH_2PO_4 (c), without KCl (d), without Na_2SO_4 (e), without $MgSO_4$ (f), (B) (a–d). Response of endosperm callus on WB + IAA (2 mg/l) + KN (5 mg/l) + CH (1000 mg/l) (d), CH replaced by leucine (a), by leucine and valine (b), by valine (c), (C) (a–g). Effect of some substituted aminopurines, and adenine. Medium: WB + IAA (1.2×10^{-5} M) + aminopurine (3×10^{-5} M) + CH (1000 mg/l); triacanthine (a), adenine (b), SD 8339 (c), benzyladenine (d), KN (e), ZN (f), 6-(γ, γ-dimethylallylamino) purine (g). (D) (a–d). Effect of different auxins. Medium: WB + Auxin (1.2×10^{-5} M) + KN (3×10^{-5} M) + CH (1000 mg/l). NAA (a), IPA (b), IBA (c), IAA (d) [40].

Procedures

Culture of immature endosperm

Steps in the procedure
1. Collect young fruits of appropriate stage (in relation to days after pollination, DAP). Only the cellular endosperm is amenable to culture. Nuclear endosperm does not respond.
2. Wash the fruits thoroughly in running tap water followed by 5—10 min in 70% ethanol.
3. Remove the fruit wall and sterilize the seed with sodium hypochlorite (1—2% v/v) solution for 5—10 min followed by 70% ethanol for 2—5 min.
4. After removing the seed coat carefully excise the endosperm aseptically under a stereoscopic microscope; from larger seeds, the endosperm can be easily removed even without the use of a microscope.
5. Place the entire endosperm, or pieces thereof on a suitable gelled nutrient medium.
6. Maintain cultures at 25 °C under 16/8 h diffuse day light/dark regime.
7. Callusing or any morphogenic response can be observed after the first week.

Notes

1. Soon after fertilization, the developing fruits are bagged and developmental stage of endosperm ascertained through dissections; microtome sections are necessary for small seeds.
5. The immature endosperm calluses and undergoes organogenesis without any pretreatment with GA. Results with *Lycopersicon*, *Malus* and *Santalum* indicate that the contact with the embryo during the initial stages is not essential.

Culture of mature endosperm

Steps in the procedure
1. Use, preferably, fresh mature fruits of endospermous seeds.
2. Wash the fruits in running tap water followed by immersion in 70% ethanol for 5 min.
3. Remove the fruit wall and sterilize the seed with 70% ethanol for 5 min.
4. Remove the seed coat; use a nutcracker for hard seeds, such as in Euphorbiaceae.
5. Sterilize the entire mature endosperm with 1—2% (v/v) sodium hypochlorite solution for 5—10 min followed by a quick dip in 70% ethanol.
6. Wash the endosperm repeatedly with sterile distilled water and implant aseptically either the entire endosperm with embryo, or small segments of endosperm (without embryo) on a suitable nutrient medium.
7. When cultured along with the intact embryo, remove the embryo immediately after initiation of proliferation of the endosperm.
8. Maintain the cultures at 25 °C under 16/8 h diffuse day light/dark regime or under continuous diffuse day light.
9. Morphogenic response can be observed after 7—10 days.

. In *Putranjiva* and *Ricinus* the seed coat is quite hard and the endosperm is solid and massive.

. In *Santalum* and *Citrus*, the addition of GA to the medium stimulates callusing, organogenesis and also embryogenesis. In *Juglans* and *Petroselinum* the endosperm is excised from germinating seeds to get positive results. The promotory role of GA is implicated and, consequently, the stimulatory effect provided by the presence of the embryo.

ulture of endosperm of semi-parasitic plants

teps in the procedure

. Procure fresh mature fruits of the desired taxa.

. The seeds of semi-parasites are naked and lack a coat; the massive endosperm is directly surrounded by the fruit wall.

. Wash the fruits in running tap water, rinse thoroughly in dilute solution of a detergent, followed by immersion in 70% ethanol for 5–10 min.

. While handling the endosperm of Loranthaceae, wash the hands and dissection instruments in 90% ethanol frequently to remove the sticky viscin which hampers the dissection.

. Remove the fruit wall under aseptic conditions, and plant the endosperm (with or without the embryo intact) on a suitable nutrient medium.

. Maintain cultures at 25 °C, preferably under 12/12 h diffuse day light/dark regime.

. The endosperm swells after 5–7 days, and shoot buds differentiate after 2–3 weeks.

ote

Where the mature endosperm is cultured with the embryo intact, the diploid and triploid shoot buds and leaves can be easily identified by their position and confirmed on the basis of chromosomes in squash mounts.

ulture of endosperm of cereals

teps in the procedure

Use kernels of specific age for successful response.

Depending on the species, take inflorescences with young caryopses 4–12 DAP; remove sterile regions with forceps.

Sterilize the remaining inflorescence with 70% ethanol by dipping for 1 min followed by 10–15 min in 10–15% (v/v) sodium hypochlorite solution. One or two drops of wetting agent, such as Tween 20, can also be added.

. Wash repeatedly with sterile distilled water in a laminar flow chamber.

. From the fertile caryopsis, remove the wall aseptically then dissect the embryo.

. Implant the entire endosperm, or a portion thereof, on a suitable nutrient medium.

. Maintain cultures at 25 °C in dark.

Proliferation can be observed after 7 days.

ote

The responsive age for endosperm of *Oryza* is 4–7 DAP, for *Triticum* and *Hordeum* 8 DAP, and for *Zea* 8–11 DAP.

White's Basal Salts (obtainable eg from Sigma)

	mg/l
H_3BO_3	1.5
$Ca(NO_3)_2 \cdot 4H_2O$	208.4
$CuSO_4 \cdot 5H_2O$	0.01
$Fe_2(SO_4)_3$	2.5
$MgSO_4$	366.2
$MnSO_4 \cdot H_2O$	3.788
MoO_3	0.001
KCl	65.0
KI	0.75
KNO_3	80.0
NaH_2PO_4	16.5
Na_2SO_4	200.0
$ZnSO_4 \cdot 7H_2O$	3.0

References

1. Williams E, de Latour G (1980) The use of embryo culture with transplanted nurse endosperm for the production of interspecific hybrids in pasture legumes. Bot Gaz 141: 252–257.
2. Johnston SA, den Nijs TPM, Peloquin SJ, Hanneman Jr RE (1980) The significance of genic balance to endosperm development in interspecific crosses. Theoret Appl Genet 57: 5–9.
3. Johri BM (1990) The role and development of endosperm in angiosperms. In: Purkayastha RP (ed) Economic plants and microbes. Today & Tomorrow, New Delhi, pp 251–254.
4. Battaglia E (1963) Apomixis. In: Maheshwari P (ed) Recent advances in the embryology of angiosperms. Intl Soc Pl Morphologists, Univ Delhi, pp 221–264.
5. Johri BM, Bhojwani SS (1965) Growth response of mature endosperm in cultures. Nature 298: 1345–1347.
6. Masuda K, Koda Y, Okazawa Y (1977) Callus formation and embryogenesis of endosperm tissue of parsley seed cultured on hormone-free medium. Physiologia Plant 41: 135–138.
7. Nakano H, Tashiro T, Maeda E (1975) Plant differentiation in callus tissue induced from immature endosperm of *Oryza sativa* L. Z Pflanzenphysiol 76: 444–449.
8. Lakshmi Sita G, Raghava Rani NV, Vaidyanathan CS (1980) Triploid plants from endosperm culture of sandalwood by experimental embryogenesis. Pl Sci Lett 20: 63–69.
9. Tulecke W, McGranhan G, Ahmadi H (1988) Regeneration by somatic embryogenesis of triploid plants from endosperm of walnut, *Juglans regia* L. cv. Manregian. Pl Cell Rep 7: 301–304.
10. LaRue CD (1949) Culture of the endosperm of maize. Amer J Bot 36: 798.
11. Satsangi A, Mohan Ram HY (1965) A continuously-growing tissue culture from the mature endosperm of *Ricinus communis* L. Phytomorphology 15: 26–30.
12. Straus J, LaRue CD (1954) Maize endosperm tissue grown *in vitro* 1: Culture requirements. Amer J Bot 41: 687–694.
13. Johri BM, Srivastava PS (1973) Morphogenesis in endosperm cultures. Z Pflanzenphysiol 70: 285–304.
14. Srivastava PS (1982) Endosperm culture. In: Johri BM (ed) Experimental Embryology of Vascular Plants. Springer, Heidelberg, pp 175–193.
15. Lampton RK (1952) Developmental and experimental morphology of the ovule and seed of *Asimina triloba*. PhD Thesis, Univ Michigan, Ann Arbor.
16. Johri BM, Srivastava PS (1972) *In vitro* growth response of mature endosperm of *Ricinus communis* L. In: Murty YS, Johri BM, Mohan Ram HY, Verghese TM (eds) Advances in Plant Morphology (Prof V Puri Comm Vol). Sarita Prakashan, Meerut (India), pp 339–358.
17. Nag KK, Johri BM (1971) Morphogenic studies on endosperm of some parasitic angiosperms. Phytomorphology 21: 202–218.
18. Srivastava PS, Johri BM (1974) Morphogenesis in mature endosperm cultures of *Jatropha panduraefolia*. Beitr Biol Pflanzen 50: 225–268.
19. Monaco LC, Sandahl MR, Carvalho A, Crocomo OJ, Sharp WR (1977) Applications of tissue culture in the improvement of coffee. In: Reinert J, Bajaj YPS (eds) Applied and Fundamental Aspects of Plant Cell, Tissue, and Organ Culture. Springer, Heidelberg, pp 109–126.
20. Norstog K (1956) Growth of rye grass endosperm *in vitro*. Bot Gaz 117: 253–259.
21. Johri BM, Nag KK (1970) Endosperm of *Taxillus vestitus*: A system to study the effect of cytokinins *in vitro* in shoot-bud formation. Curr Sci 39: 177–179.
22. Bhojwani SS, Johri BM (1970) Cytokinin-induced shoot bud differentiation in mature endosperm of *Scurrula pulverulenta*. Z Pflanzenphysiol 63: 269–275.
23. Mu S-K, Liu S-q, Zhou Y-K, Qian N-f, Zhang P, Xie H-X, Zhang F-S, Yan Z-L (1977) Induction of callus from apple endosperm and differentiation of the endosperm plantlet. Scientia Sin 20: 370–376.
24. Wang T-Y, Chang C-I (1978) Triploid citrus plantlet from endosperm culture. Scientia Sin 21: 823–827.
25. Sun C-S, Chu C-E (1981) The induction of endosperm plantlets and their ploidy of barley *in vitro*. Acta Bot Sin 23: 265.

26. Liu S-Q, Gui Y-I, Gu S-R, Xu T-Y (1987) Induction of endosperm calluses and regeneration of endosperm plantlets of *Asparagus officinalis*. Acta Bot Sin 29: 373–376.
27. Sehgal CB, Narang KH, Sunila A (1981) Growth responses of mature endosperm of *Euphorbia geniculata* Orteg. in cultures. Beitr Biol Pflanzen 55: 385–392.
28. Sethi Meenakshi, Rangaswamy NS (1976) Endosperm embryoids in culture of *Nigella damascena*. Curr Sci 43: 109–111.
29. Nair S, Shirgurkar MV, Mascarenhas AF (1986) Studies on endosperm culture of *Annona squamosa* Linn. Pl Cell Rep 5: 132–135.
30. Sehgal CB, Sunila K (1985) Morphogenesis and plant regeneration from cultured endosperm of *Emblica officinalis* Gaertn. Pl Cell Rep 4: 263–266.
31. Gu S-R, Gui Y-I, Xu T-Y (1985) Induction of endosperm plantlets in *Lycium*. Acta Bot Sin 27: 106–109.
32. Bajaj YPS, Saini SS, Bidani M (1980) Production of triploid plants from the immature and mature endosperm cultures of rice. Theoret Appl Genet 58: 17–18.
33. Liu S-Q, Liu J-Q (1980) Callus induction and embryoid formation in endosperm culture of *Prunus persica*. Acta Bot Sin 22: 198–199.
34. Bapat VA, Narayanaswami S (1977) Mesocarp and endosperm culture of *Achras sapota* Linn. *in vitro*. Indian J Exptl Biol 15: 294–296.
35. Mu S-K, Kwei Y-L, Liu S-K, Chang F-C, Lo F-M, Yang M-Y and Wang F-R (1977) Induction of callus in *Vitis* endosperm cultured *in vitro*. Scientia Sin 19: 18–19.
36. Bhojwani SS, Johri BM (1971) Morphogenetic studies on cultured mature endosperm of *Croton bonplandianum*. New Phytol 70: 761–766.
37. Rangaswamy NS, Rao PS (1963) Experimental studies on *Santalum album* L: Establishment of tissue cultures of endosperm. Phytomorphology 13: 450–454.
38. Gui Y-I, Mu X-J, Xu T-Y (1982) Studies on morphological differentiation of endosperm plantlets of Chinese gooseberry *in vitro*. Acta Bot Sin 24: 216–221.
39. Kin MS, Fraser LG, Harvey CF (1990) Initiation of callus and regeneration of plantlets from endosperm of *Actinidia* interspecific hybrids. Scientia Hortic 44: 107–117.
40. Srivastava PS, Johri BM (1978) Triploid plants of *Putranjiva roxburghii* from endosperm. Beitr Biol Pflanzen 54: 381–397.
41. Kumar PP, Raju CR, Chandra Mohan M, Iyer RD (1985) Induction and maintenance of friable callus from cellular endosperm of *Cocos nucifera* L. Pl Sci 40: 203–207.
42. Smith MM, Stone BA (1973) Studies on *Lolium multiflorum* endosperm in tissue culture. Aust J Biol Sci 26: 123–133.
43. Nakajima T (1962) Physiological studies of seed development. Bull Univ Osaka Prefect Ser B 13: 13–48.
44. Sehgal CB (1974) Growth of barley and wheat endosperm in cultures. *Curr Sci* 43: 38–40.
45. Mu S-K, Liu S-C (1978) Cytological observations on calluses derived from apple endosperm cultured *in vitro*. In: Proc Symp Plant Tissue Culture. Sci Press, Peking, pp 507–510.
46. Tulecke W, McGranhan G (1985) Somatic embryogenesis and plant regeneration from cotyledons of walnut. Pl Sci 49: 57–63.
47. Paleg LG (1960) Physiological effects of gibberellic acid 2: On starch hydrolyzing enzymes of barley endosperm. Pl Physiol 35: 902–906.
48. Brown DJ, Canvin DT, Zilkey BF (1970) Growth and metabolism of *Ricinus communis* endosperm in tissue culture. Canad J Bot 48: 2323–2331.
49. Graebe JE, Novelli GD (1966) Amino acid incorporation in cell-free system from submerged tissue cultures of *Zea mays* L. Exptl Cell Res 41: 521–534.
50. Castelli S, Manzocchi LA, Torri G (1988) Callus induction in endosperm from maize mutant with different IAA content. In: Pais MSS, Mavituna F, Novais JM (eds) Plant Cell Biotechnology. Springer, Berlin, pp 63–67.
51. Chu LC, Shannon JC (1975) *In vitro* culture of maize endosperm: A model system for studying *in vitro* starch biosynthesis. Crop Sci 15: 814–819.
52. Sarawitz CH, Boyer CD (1987) Starch characteristics in culture of normal and mutant maize endosperm. Scientia Sin 12: 489–495.
53. Manzocchi LA, Bianchi MW, Viotti A (1989) Expression of zein in long-term cultures of wild type and opaque-2 maize endosperms. Pl Cell Rep 7: 639–643.

54. Racchi ML, Manzocchi LA (1988) Anthocyanin and proteins as biochemical markers in maize endosperm cultures. Pl Cell Rep 7: 78–81.
55. Meletti P, Floris C, D'Amato F (1964) Occurrence of an inhibitor in wheat endosperm as revealed by embryo transplantation in irradiated seeds. Rad Bot 4: 497–502.
56. Lampe L, Mills CO (1933) Growth and development of isolated endosperm and embryo of maize. Abstr Papers, Bot Soc Boston.
57. Lakshmi Sita G (1987) Triploids. In: Bonga JM, Durzan DJ (eds) Cell and Tissue Culture in Forestry, Vol 2. Specific Principles and Methods: Growth and Development. Martinus Nijhoff, Dordrecht (The Netherlands) pp 269–284.
58. Johri BM, Nag KK (1974) Cytology and morphogenesis of embryo and endosperm tissue of *Dendrophthoe* and *Taxillus*. Cytologia 39: 801–813.
59. Keller F, Stone BA (1978) Preparation of *Lolium* protoplasts and their purification using an anti-galactansepharose conjugate. Z Pflanzenphysiol 87: 167–172.
60. Lakshmi Sita G, Rani HS (1983) Preliminary studies on isolation and culture of protoplasts from sandalwood. Experientia 45: 4–5.
61. Keller H, Wanner H, Baumann TW (1972) Kaffeinsynthese in Fruchten und Gewebekulturen von *Coffea arabica*. Planta 108: 339–350.
62. Gayatri MC (1978) *In vitro* studies on *Codiaeum variegatum*: Growth and organogenesis in endosperm tissue. Phytomorphology 28: 295–400.
63. Zur VK, Mills D, Mizrahi Y (1990) Callus formation from tomato endosperm. Acta Hortic 280: 139–141.
64. LaRue CD (1947) Growth and regeneration of the endosperm of maize in culture. Amer J Bot 34: 585–586.
65. Straus J (1954) Maize endosperm tissue grown *in vitro* 2. Morphology and cytology. Amer J Bot 41: 833–839.
66. Straus J (1960) Maize endosperm tissue grown *in vitro* 3. Development of a synthetic medium. Amer J Bot 47: 641–647.
67. Tamaoki T, Ullstrup AJ (1958) Cultivation *in vitro* of excised endosperm and meristem tissue of corn. Bull Torrey Bot Club 85: 260–272.
68. Zhu Q, Chen X, Li W, Chen Y (1988) *In vitro* regeneration of plantlets from immature endosperm of maize (*Zea mays*). In: Genetic Manipulation in Crops. Cassell, Tycooly (UK) pp 370–371.

Plant Tissue Culture Manual **E4**: 1–8, 1993.
© 1993 *Kluwer Academic Publishers.*

Hybrid embryo rescue

ABHA AGNIHOTRI
Biotechnology Division, Tata Energy Research Institute, New Delhi 110 003, India

Introduction

Hybridization between widely diverse plant species is of value for introgression of useful genes into cultivars and for the development of new and useful species with better yield and resistance to biotic and abiotic stresses. Many agriculturally useful traits have been transferred from wild species to the cultivated species belonging to many families [12].

Most of the wide crosses are difficult to obtain by conventional methods. Non-realization of hybrids may be due to pre- or post-fertilization barriers [7, 31]. In most of the crosses where fertilization does take place, the hybrid embryo aborts before maturation [10, 31]. Even when seeds are realized, they either fail to germinate or produce only weak seedlings which do not survive. This is mainly due to disharmony between parental genomes resulting in embryo mortality, endosperm breakdown, seed inviability and hybrid sterility [34].

Lack of endosperm development or its early degeneration results in insufficient nourishment of the embryo and is the major cause for hybrid embryo abortion [11, 20, 23]. This can be overcome by (a) hybrid embryo implantation on normal endosperm, (b) culturing of hybrid embryos, ovules and ovaries and (c) organogenesis or somatic embryogenesis from callus derived from hybrid embryos [23, 31]. Laibach [18] successfully raised seedlings from the interspecific cross, *Linum perenne* × *L. austriacum*, through embryo rescue. Since then a large number of hybrids have been produced through embryo culture and the procedures have been discussed comprehensively by several workers [8, 9, 30, 33].

Brassica is an important oilseed crop. Losses in yield due to drought, pests and diseases are high. There is an imperative need to develop new varieties for higher yield and resistance to biotic and abiotic stresses. A large number of interspecific and intergeneric hybrids have been produced [14, 28]. In brassicas embryo culture was first used by Nishi *et al.* [27] for obtaining interspecific hybrids of *B. oleracea* × *B. campestris*. Following this many interspecific and intergeneric hybrids have been raised through embryo culture [4, 13, 14, 21, 29, 32].

In many crosses embryo abortion occurs at a very young stage. Owing to the difficulties associated with excision and culture of very young embryos, embryo rescue attempts have been made through the culture of ovule or the entire ovary

[5, 15, 16, 17, 19, 35]. Retention of more of the paternal tissue in the cultured explant favours the growth of the young embryo [20]. It is difficult to excise and grow young embryos requiring a balanced combination of growth hormones [22, 31]. It is comparatively easier to excise embryos after two to three weeks of growth *in vitro* of ovary/ovule and then grow them on a nutrient medium containing essential minerals, sucrose, amino acids/casein hydrolysate which is far less complex than the media required for young embryos.

In many interspecific and intergeneric hybrids in which embryo abortion occurs early, successive culture of ovary, ovule and seed/embryo has been reported to be more effective than culture only of ovary or ovule; examples include *B. fruticulosa* × *B. campestris* [26], *Erucastrum gallicum* × *B. juncea/ B. napus* [6], *B. juncea/B. napus* × *B. gravinae* [25], *E. sativa* × *B. campestris* [1], *B. spinescens* × *B. campestris* [2] and *B. napus* × *Raphanobrassica* [3]. The details of embryo rescue by sequential culture, essentially based on our work on wide hybrids in brassicas [1, 2, 3], are presented here.

Procedures

Raising of plants

Steps in the procedure
1. Sow seeds in pots or in the field. Fill up pots (12" diameter) with soil, peat moss and farm yard manure (2:1:1) mixture. Water the plants on alternate days. The field preparation should be done according to standard practices with farm yard manure and NPK at 60 Kg N, 40 Kg P_2O_5 and 40 Kg K_2O per hectare. Sow the seeds in rows 60 cm apart. Carry out thinning to leave plant-to-plant distance of 10–15 cm. Irrigate the fields thrice: 15 days after sowing, at the times of initiation of flowering, and during pod filling. The seeds are generally sown in October when average temperature is 20–25 °C. Pollinations should be carried out during the peak of flowering period.

Emasculation and pollination

Steps in the procedure
1. Emasculate the flower buds manually with the help of forceps (on selected inflorescences of the female parent) which would open the next day.
2. Remove the remaining flowers, fruits and younger buds and bag the inflorescence with paper bags. Bags should be clipped to avoid contamination through wind or insects.
3. Collect the freshly opened flowers of the male parent in the morning (around 9 a.m.) and keep in the sun in a covered glass Petri plate until anthers dehisce (around 11 a.m.).
4. Use the dehisced anthers for pollinating the pistils (on the day of anthesis) of emasculated and bagged flowers. The flowers should be rebagged after pollination.

Surface sterilization and explant preparation

Steps in the procedure

All *in vitro* aseptic operations are carried out inside a horizontal Laminar Air Flow Bench. Cultures are maintained at 25 ± 2 °C at 16 h light (2000 lux) – 8 h dark period.
1. Rinse the pollinated ovaries with water and treat with 0.05% mercuric chloride solution for 5 min.
2. Wash them 3 or 4 times with sterile distilled water to remove all traces of mercuric chloride.

Ovary, ovule and embryo cultures (sequential culture)

Steps in the procedure

1. Culture two or three surface-sterilized ovaries in tubes (150 × 25 mm) containing Medium E1.
2. After two weeks of culture dissect the ovaries under a hand lens. Excise the enlarged ovules and culture on fresh medium of the same composition (Medium E1).
3. After about 2 weeks excise the embryos from cultured ovules and reculture on fresh medium of the same composition. Allow them to grow for two to three weeks and then transfer the enlarged embryos to MS medium containing one tenth of the original concentrations of the hormones (Medium E2).
4. After further growth of the embryos for about two more weeks reduce the hormone concentration (Medium E3).
5. Transfer the embryos to Medium B1 and allow to grow into plantlets.

Callusing and embryogenesis

Steps in the procedure

1. Use leaf discs of the hybrid plantlets for induction of callusing and embryogenesis on Medium E1. Also, hypocotyl segments of some of the embryos which fail to develop any further can be used for the induction of callus on Medium E5.
2. When the callus is transferred to Medium E1, a large number of somatic embryos are produced within two weeks.
3. The callus can be maintained for up to 1 year by subculturing every 3 to 4 weeks and several embryos are produced during each subculture.

Micropropagation

Steps in the procedure

1. Use surface sterilized single node explants for micropropagation of hybrids. Culture only one explant in each tube containing Medium E4. It is advisable to grow the plantlets on basal medium (B1) for two to three weeks before taking nodal segments for further micropropagation.

Rooting and transplantation

Steps in the procedure

1. Transfer the somatic embryos to Medium B3L and maintain on a shaker with slow rotation (20 rpm) for two to three days.
2. Transfer to solidified medium B1 for further growth.
3. Culture the shoot tips obtained from micropropagation on Medium B1 for two to three weeks for rooting.

4. When the plantlets are 8 to 10 cm in height, remove them from the containers, wash the agar thoroughly from the roots and transfer to autoclaved vermiculite in jars.
5. Irrigate with sterilized water for 8 to 10 days.
6. Plant in pots containing soil, peatmoss and farm yard manure (2 : 1 : 1) and grow in growth chambers with 14 hour, 25 °C/10 h, 15 °C day/night cycles and 60% relative humidity. The plants should be initially covered with beakers to maintain high humidity around the plants.
7. Remove the beakers for increasingly long periods during the first week, after which the cover can be removed permanently. The plants can then be transferred to field conditions and grown to maturity in a net house. Irrigate the plants every alternate day with tap water.

Notes

1. The age of the plant used for carrying out pollinations appears to be important. Pollinations performed during peak of flowering, rather than at initiation or towards termination, yield better results.
2. The procedure described above is well suited to the wide hybrids of brassicas. However minor changes in the auxin and cytokinin concentrations may be required depending upon the response of individual embryos.
3. Addition of higher concentrations of kinetin, casein hydrolysate or coconut milk may also help in the growth of the embryos. However, after initial growth of embryos for 3—4 weeks, reduction in hormone concentrations is imperative for normal development.
4. Sequential culture (successive culture of ovary/ovule/embryo) is useful for the crosses where embryos abort early. However the age of ovary, ovule and embryo should be standardized for culture of each individual cross.
5. Hybrids are generally obtained in small numbers. It is therefore desirable to multiply the hybrids through clonal propagation and/or embryogenesis. Also, the hybrids are often obtained when the season is over and it helps to maintain them *in vitro* until the next growing season.
6. Transfer of somatic embryos to solid media sometimes results in precocious germination. A passage through liquid medium helps to maintain normal growth. This may be because liquid medium permits rapid leaching out of hormones accumulated in the embryo during their growth on high hormone medium.
7. Sterility of F_1 hybrids is a limiting step to the application of wide hybridization programs for crop improvement. Amphidiploidy is generally induced through application of colchicine to the shoot apex/axillary buds. Another way of inducing amphidiploidy is a passage of the hybrid embryo through a callusing stage to produce more embryos (amphidiploids).
8. Matromorphy is common in wide hybridization and therefore obtaining embryos after pollination is not necessarily an indication of successful hybridization. Thus it is essential to establish the genetic identity of each embryo. In the absence of genetic markers the seedlings have to be grown to maturity for conventional screening. DNA analysis at an early stage would enable the identification of hybrids at the seedling stage. It is rapid, unambiguous and saves considerable effort by selectively raising only the hybrids.

Solutions

MS (Murashige and Skoog 1962, e.g. from Sigma) basal medium with 3% w/v or 1% w/v sucrose. 0.7% extra pure bacteriological grade agar should be used for solidified media.

Medium	Growth supplements (mg/L)					Sucrose (%)	Agar (%)
	Kn	NAA	2.4-D	GA_3	CH		
B3	–	–	–	–	–	3	0.7
B1	–	–	–	–	–	1	0.7
B3L	–	–	–	–	–	3	–
E1	1.0	0.1	–	1.0	10	3	0.7
E2	0.1	0.01	–	0.1	10	3	0.7
E3	0.05	0.005	–	–	10	3	0.7
E4	0.05	0.005	–	–	50	3	0.7
E5	1.0	–	1.0	–	–	3	0.7

Kn = kinetin.
NAA = naphthaleneacetic acid.
2,4-D = 2,4-dichlorophenoxyacetic acid.
GA_3 = gibberellic acid.
CH = casein hydrolysate.

Acknowledgements

I thank Dr. V. Jagannathan, Tata Energy Research Institute, New Delhi and Dr. K.R. Shivanna, University of Delhi, Delhi for useful discussions.

References

1. Agnihotri A, Gupta V, Lakshmikumaran MS, Shivanna KR, Prakash S, Jagannathan V (1990a) Production of *Eruca-Brassica* hybrids by embryo rescue. Plant Breeding 104: 281–289.
2. Agnihotri A, Lakshmikumaran M, Shivanna KR, Jagannathan V (1990b) Embryo rescue of interspecific hybrids of *Brassica spinescens* × *Brassica campestris* and DNA analysis. In: Nijkamp HJJ, VanDerPlas LHW, VanAartrijk J (Eds.) Progress in Plant Cellular and Molecular Biology, pp 270–274, Kluwer Academic Publishers, Netherlands.
3. Agnihotri A, Shivanna KR, Raina SN, Lakshmikumaran M, Prakash S, Jagannathan V (1990c) Production of *Brassica napus* × *Raphanobrassica* Hybrids by Embryo Rescue: An attempt to introduce shattering resistance into *B. napus*. Plant Breeding 105: 292–299.
4. Ayotte R, Harney PM, Souza Machado V (1987) The transfer of triazine resistance from *Brassica napus* L. to *B. oleracea* L. I. Production of F1 hybrids through embryo rescue. Euphytica 36: 615–624.
5. Batra V, Prakash S, Shivanna KR (1990) Intergeneric hybridization between *Diplotaxis siifolia*, a wild species and crop brassicas. Theor Appl Genet 80: 537–541.
6. Batra V, Shivanna KR, Prakash S (1989) Hybrids of wild species *Erucastrum gallicum* and crop brassicas. Proc 6th Internatl Congr of SABRAO, pp 443–446.
7. Brar DS, Khush GS (1986) Wide hybridization and chromosome manipulation in cereals. In: Evans DA, Sharp WR, Ammirato PV (Eds.) Handbook of Plant Cell Culture: Techniques and Application, Vol 4, pp 221–263, Macmillan, New York.
8. Collins GB, Grosser JW (1984) Culture of embryos. In: Vasil IK (Ed.) Cell culture and somatic cell Genetics of Plants, Vol I, pp 241–257, Academic Press, New York.
9. Dunwell JM (1986) Pollen, ovule and embryo culture as tools in plant breeding. In: Withers LA, Alderson PG (Eds.) Plant Tissue Culture and its Agricultural Applications, pp 375–404, Butterworths, London.
10. Eenink AH (1975) Matromorphy in *Brassica oleracea* L. VI. Research on ovules, embryos and endosperms after prickle pollination. Euphytica 24: 33–43.
11. Evenari M (1984) Seed physiology: from ovule to maturing seed. Bot Rev 50: 143–170.
12. Goodman RM, Hauptli H, Crossway A, Knauf VC (1987) Gene transfer in crop improvement. Science 236: 48–54.
13. Harberd DJ (1969) A simple effective embryo culture technique for *Brassica*. Euphytica 18: 425–429.
14. Harberd DJ, McArthur ED (1980) Meiotic analysis of some species and genus hybrids in the Brassiceae. In: Tsunoda S, Hinata K, Gomez-Campo C (Eds.) *Brassica* Crops and Wild Allies: Biology and Breeding, pp 65–87, Japan Scientific Societies Press, Tokyo.
15. Hossain MM, Haruhisa Inden, Tadashi Asahira (1988) Intergeneric and interspecific hybrids through *in vitro* ovule culture in the *cruciferae*. Plant Science 58: 121–128.
16. Inomata N (1985) Interspecific hybrids between *Brassica campestris* and *B. cretica* by ovary culture *in vitro*. Cruciferae Newslett 10: 92–93.
17. Inomata N (1986) Interspecific hybrids between *Brassica campestris* and *B. bourgeaui* by ovary culture *in vitro*. Cruciferae Newslett 11: 14–15.
18. Laibach F (1925) Das Taubwerden von Bastardsamen und die kunstliche Aufzucht früh absterbender Bastard-embryonen. Zeist f Bot 17: 417–459.
19. Luo P, Li XF, Wang ZC, Lan ZQ (1989) A study on distant hybridization between rapeseed (*Brassica napus* L.) and oil radish (*Raphanus sativus* var. *oleifera Makino*). Proc 8th Internatl Congr of SABRAO, pp 467–470.
20. Maheshwari P, Rangaswamy NS (1965) Embryology in relation to physiology and genetics. In: Preston RD (Ed.) Advances in Botanical Research, Vol 2, pp 219–312, Academic Press, London.
21. Mohapatra D, Bajaj YPS (1987) Interspecific hybridization in *Brassica juncea* × *Brassica hirta* using embryo rescue. Euphytica 36: 321–326.

22. Monnier M (1984) Survival of young immature *Capsella* embryos cultured in vitro. J Plant Physiol 115: 105–113.
23. Monnier M (1990) Zygotic embryo culture. In: Bhojwani SS (Ed.) Plant Tissue Culture: Applications and Limitations, pp 366–393, Elsevier publishers, Netherlands.
24. Murashige T, Skoog F (1962) A revised medium for rapid growth and bioassays with tobacco tissue cultures. Physiol Plant 15: 473–497.
25. Nanda Kumar PBA, Prakash S, Shivanna KR (1989) Wide hybridization in *Brassica*: Studies on interspecific hybrids between cultivated species (*B. napus, B. juncea*) and a wild species (*B. gravinae*). Proc 8th Internatl Congr of SABRAO, pp 435–438.
26. Nanda Kumar PBA, Shivanna KR, Prakash S (1988) Wide hybridization in *Brassica*: crossability barriers and studies on the F1 hybrid and synthetic amphidiploid of *B. fruticulosa* × *B. campestris*. Sex Plant Reprod 1: 234–239.
27. Nishi S, Kawata J, Toda M (1959) On the breeding of interspecific hybrids between two genomes, c and a of *Brassica* through the application of embryo culture techniques. Jpn J Breeding 8: 215–222.
28. Prakash S, Tsunoda S (1983) Cytogenetics of Brassica. In: Swaminathan MS, Gupta PK, Sinha U (Eds.) Cytogenetics of crop plants, pp 481–513, Macmillan Publ, New Delhi.
29. Quazi MM (1988) Interspecific hybrids between *Brassica napus* L. and *B. oleracea* L. developed by embryo culture. Theor Appl Genet 75: 309–318.
30. Raghavan V (1977) Applied aspects of embryo culture. In: Reinert J, Bajaj YPS (Eds.) Applied and Fundamental Aspects of Plant Cell, Tissue and Organ Culture, pp 375–397, Springer-Verlag, Berlin and New York.
31. Raghavan V (1986) Variability through wide crosses and embryo rescue. In: Vasil IK (Ed.) Cell Culture and Somatic Cell Genetics of Plants, Vol 3, pp 613–633, Academic Press, Orlando.
32. Ripley VL, Arnison PG (1990) Hybridization of *Sinapis alba* L. and *B. napus* L. via embryo rescue. Plant Breeding 104: 26–33.
33. Sastri DC (1984) Incompatibility in Angiosperms: significance in crop improvement. Adv Appl Biol 10: 71–111.
34. Stalker HT (1980) Utilization of wild species for crop improvement. Adv Agron 33: 111–147.
35. Takahata Y, Takeda T (1990) Intergeneric (intersubtribe) hybridization between *Moricandia arvensis* and *Brassica* A and B genome species by ovary culture. Theor Appl Genet 80: 38–42.

Plant Tissue Culture Manual E5, 1–20, 1995.

In vitro culture of *Brassica juncea* zygotic proembryo

CHUN-MING LIU[a,1], ZHI-HONG XU[a,2] & NAM-HAI CHUA[b]

[a]*Institute of Molecular & Cell Biology, National University of Singapore, Singapore 0511, Singapore;* [b]*Laboratory of Plant Molecular Biology, The Rockefeller University, 1230 York Ave., New York, NY10021–6399, U.S.A.;* [1]*To whom correspondence should be addressed. Current address: Department of Applied Genetics, John Innes Centre, Norwich NR4 7UH, U.K.;* [2]*Current address: Shanghai Institute of Plant Physiology, Chinese Academy of Sciences, Shanghai, 200032, China*

Introduction

One of the most important approaches to studying embryo development is through embryo manipulation, which depends critically on the availability of an *in vitro* system. Three systems may be used for the manipulation – cultivation of zygotic embryos [10, 11], somatic embryos [3, 4, 19, 20, 25] and pollen embryos [8]. The advantages of using zygotic embryos rather than embryos induced from the somatic cells and pollens are their uniform pattern formation and the presence of a suspensor, which can be used as a visual indicator for the polarity of a globular proembryo. Somatic embryogenesis at the early stage is often characterised by cell proliferation that bears little resemblance to the polarized and highly regulated cell division in zygotic embryogenesis [2].

Proembryos are globular and heart-shape embryos preceding cotyledon initiation. Compared to the torpedo-shape and cotyledonary stage embryos, proembryos are heterotrophic in nature, dependent on the nutrients supplied by ovules and the surrounding endosperm for growth and development. By growing the embryos outside the environment of the ovules, it is possible to identify their nutritional requirements essential for the continued growth, cell differentiation and morphogenesis, which are difficult to determine while the embryos are enclosed in the ovules. Ideally, we would expect to follow the progressive embryogenesis *in vitro* starting from the fertilisation of an egg. Unfortunately, this goal is hard to achieve, especially in the stage from zygote to embryo when comprising only a few cells [5].

Cultivation of plant zygotic embryos was started by Hannig, who used a simple medium to culture 2 mm embryos at the turn of this century [7]. A practical usage of culturing embryos at this stage was to rescue embryos from interspecific hybridisation. A procedure for embryo rescue was contributed by Agnihotri [1]. This technique has been applied to a wide-range of species and many interspecific and intergeneric hybrids have been obtained through embryo culture. The first success in proembryo culture was not achieved until 1941 by Van Overbeek *et al.*, who observed that the growth of 150 μm-long *Datura* embryos was dramatically promoted by the addition of non-autoclaved coconut water to the culture medium [24]. Osmotic pressure of the medium is another critical consideration for culturing young embryos. Rietsema *et al.* (1952) found that the younger the embryo excised, the higher the medium osmolarity required for embryo culture [18]. Based on these observations, Ranghavan and Torrey (1963) and Nostog and Smith (1963) devised culture media that allowed the development of early heart-shape staged embryos into normal plants *in vitro* [15, 16, 17]. The importance of osmolarity in proembryo culture was confirmed by the measurement of the osmotic pressure of the milieu in the embryo sac of *Phaseolus vulgaris*. In this plant the osmolarity of the endosperm liquid was 0.7 mol/l when the embryos were at the heart-shaped stage, which decreased to 0.5 mol/l at the late cotyledonary stage [21]. Another major success in attempts to culture proembryos was achieved by Monnier [12, 13], who designed a system in which two media with different compositions were placed in juxtaposition in a Petri dish, ensuring a continual variation in the composition of the medium for the cultured embryos with time. Using this system, Monnier was able to culture *Capsella* embryos larger than 50 μm, but smaller embryos were still unable to develop under these conditions. Taking into considerations the requirements of both the osmolarity and nutrition, we successfully cultured early globular embryos (35 μm, containing 8–36 cells) of *Brassica juncea* (Indian Mustard) with high efficiency and the mature embryos germinated into fertile plants [10]. The culture system includes a double-layer culture system and a highly nutritious medium. This technique has also been used successfully in elucidating the role of auxin polar transport in embryonic pattern formation [11].

There are two major reasons for using *B. juncea* as a material for zygotic embryo culture. The first is the presence of a long suspensor connecting the embryo proper with the ovule tissue (Fig. 1), which makes dissection easier. In our previous work, we have observed that damage to the suspensor has no obvious effect on subsequent embryo development in culture [10]. Secondly, the development of *B. juncea* embryos follows that of a typical crucifer type in which the morphogenesis pattern has been well documented [22, 23].

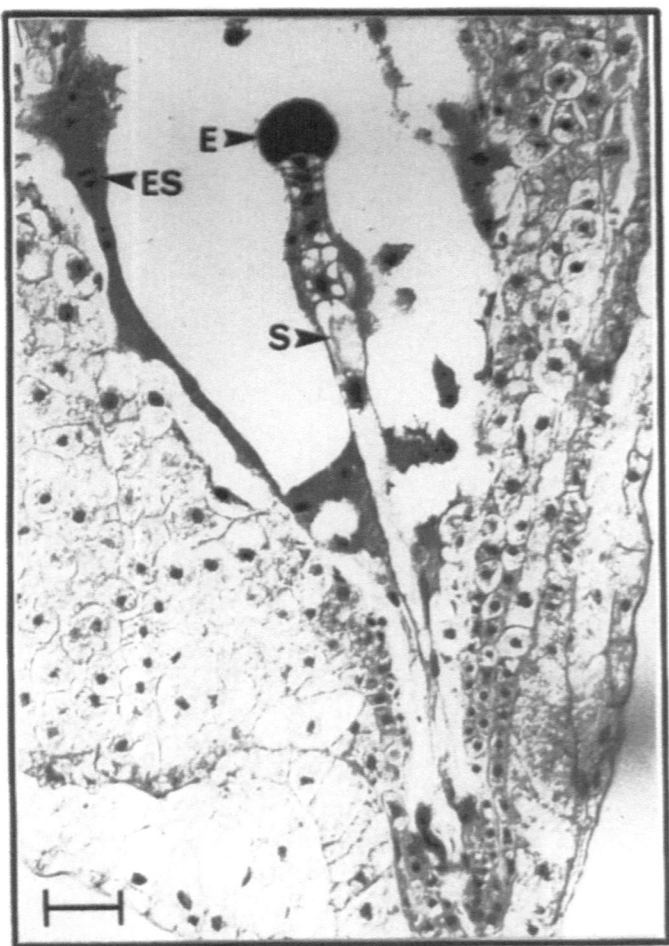

Fig. 1. Longitudinal section of a globular embryo of *B. juncea* in its ovule. The proembryo is connected to the maternal tissue through a long suspensor. Embryo dissection can be facilitated by holding the long suspensor, since damage to the suspensor has no evident effect on embryo development *in vitro*. E: embryo; S: suspensor; ES: endosperm. Bar = 50 μm.

Procedures

1. Major setups for embryo dissection

a) All dissection work is carried out in a sterile environment in an air-flow cabinet. A 30 W UV light mounted in the cabinet is used to sterilise the equipment such as the dissection microscope.
b) A dissection microscope. We used one from Nikon in which the light is supplied from the bottom of the dissection stage and which also has an adjustable reflection mirror. To observe the tiny embryo, we found it necessary to be able to adjust the angle of the mirror to achieve a dark-field effect.
c) A micropipette for embryo transfer, which is made by connecting a hand-drawn capillary pipette (ref. to this book, PTCM-A10/8) with a latex tube. The inner diameter of the open tip of the pipette should be 150–200 μm. The other end of the tube is blocked with a glass bead. The embryo can be sucked in and out by squeezing the tube (Fig. 2). The micropipette can be sterilised by immersing in

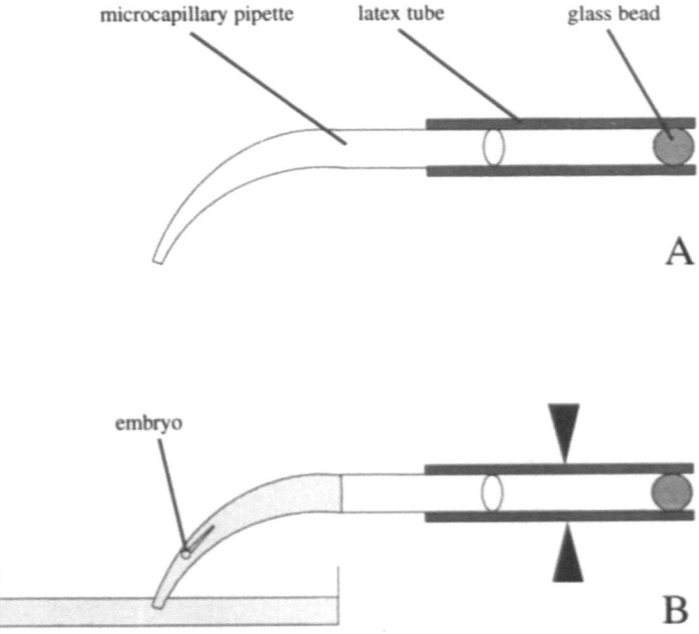

Fig. 2. Preparation of a micropipette for the embryo transfer. A. The micropipette is made by connecting a microcapillary pipette with a latex tube. The open end of the latex tube is blocked by a glass bead; B. Transfer the embryo by squeezing the tube. To prevent the embryo from sticking to the side wall of the micropipette, the micropipette should be filled up with the dissection solution (9% glucose) before sucking up the embryo.

70% ethanol for 10 min followed by air-drying in an air-flow cabinet.

2. Media preparations

The embryo culture medium (ECM, ref.10) is based on the KM8P's medium [9] with several modifications. Six stock solutions are made before preparing the medium.

Steps in the procedure
1. Prepare stock solutions

Stock 1   Macroelements (20X)

Components	Final concentrations	Weigh out for stock
NH_4NO_3	200 mg/l	2 g
KNO_3	1500 mg/l	15 g
$CaCl_2 \cdot 5H_2O$	850 mg/l	8.5 g
$MgSO_4 \cdot 7H_2O$	400 mg/l	4 g
$KH_2PO_4 \cdot 2H_2O$	100 mg/l	1 g
$Na_2EDTA \cdot 2H_2O$	37 mg/l	0.37 g
$FeSO_4 \cdot 7H_2O$	28 mg/l	0.28 g

Add distilled water to final volume 500 ml.

Stock 2   Microelements, as in B_5 medium [6], 1000X stock solution.

Stock 3   Sugar mixture (1000X)

Components	Final concentrations	Weigh out for stock
mannose	0.1 mg/l	10 mg
fructose	0.1 mg/l	10 mg
ribose	0.1 mg/l	10 mg
xylose	0.1 mg/l	10 mg
rhamnose	0.1 mg/l	10 mg
cellobiose	0.1 mg/l	10 mg
sorbitol	0.1 mg/l	10 mg

Add distilled water to final volume 100 ml.

Stock 4   Organic acids, as KM8P's [9], 100X stock solution (a powdered mixture is available from Sigma, K3004).

Stock 5   Vitamins and amino acids (200X)

Components	Final concentrations	Weigh out for stock
inositol	500 mg/l	25 g
glutamine	200 mg/l	10 g
thiamine · HCl	1 mg/l	50 mg
nicotinic acid	0.1 mg/l	5 mg
pyridoxine · HCl	0.1 mg/l	5 mg
d-biotin	0.01 mg/l	0.5 mg
casein hydrolysate	100 mg/l	5 g

Add distilled water to final volume 250 ml.

Stock 6   Coconut water
Final concentration is 300 ml/l.

2. Preparation of 2XECM
To prepare 200 ml medium, mix the appropriate amount of stock solutions as shown below, and adjust the volume first to 100 ml (2XECM), stir the mixture well.

stock 1	10 ml
stock 2	200 μl
stock 3	200 μl
stock 4	2 ml
stock 5	1 ml
stock 6	60 ml
sucrose	8 g
glucose	4 g

Add distilled water to final volume 100 ml, adjust the pH to 5.4.

3. Take 50 ml of the 2XECM solution, add 0.6 g agarose (low gelling temperature, SeaPlaque), dissolve by incubating in a water bath at 90 °C, and add distilled water to final volume 100 ml. This is the medium used for the bottom-layer in the two-layer culture system.

4. Sterilise the medium immediately by vacuum filtration using a 0.45 μm disposable filterware (Nalgene, U.S.A.).

5. Take the remaining 50 ml 2XECM solution, add 6 g sucrose, 0.6 g agarose (same as used in 2.3), dissolve by heating in a 90 °C water bath and then sterilise as in step 2.4. This medium will be used for the top-layer in the double-layer culture system.

6. Aliquot the media and store the aliquots at 4 °C. The media can be used for up to 2 months.

Notes

2.1. To make the stock solutions, all chemicals must be dissolved separately before mixing. In addition, $FeSO_4 \cdot 7H_2O$ and $Na_2EDTA \cdot 2H_2O$ have to be dissolved separately and mixed entirely before addition to the stock solution.

2.2. The coconut water is obtained from green coconuts bought from local markets in Singapore, which is filtered through a layer of Whatman No. 1 filter paper before sterilization by passing through a 0.45 μm filter (Nalgene, U.S.A.).

2.3. Stock solutions 3, 4, 5 and 6 should be stored in a −20 °C freezer to prevent contamination.

2.4. As measured by a vapour pressure osmometer, the osmotic pressure of the bottom-layer medium is 0.45 mol/l and the top-layer 0.63 mol/l.

3. Maintenance of plants

Plants of *B. juncea* are grown in a growth chamber under 14 hr light/10 hr dark at 24 °C day/20 °C night, respectively. The light intensity is about 4000 lux and relative humidity 70%. Under these conditions, the plants normally begin to flower in 2 months, and an additional 30 days are required to obtain mature seeds.

4. Selection of ovules and sterilisation

Steps in the procedure
1. Collect the siliques 6–7 days after anthesis. At this stage the length of the ovules varies from 1000 to 1200 μm.
2. Immerse the whole siliques in 70% (V/V) ethanol for 20 s and agitate by gentle swirling.
3. Wash twice with sterile water.
4. Surface-sterilise the siliques with 10% (V/V) commercial bleach (Clorox, sodium hypochlorite solution) for 10 min with shaking at 100 rpm.
5. Rinse four times with sterile water to remove all trace of bleach.

Notes
4.1. Based on our data comparing the size of the ovules with that of the embryos inside, we found that the size of the embryo at a particular developmental stage is quite constant and shows a good correlation to the ovule size. To obtain proembryos from 30 to 80 μm in length, ovules between 1000 and 1200 μm long are selected.

5. Embryo dissection and culture

To avoid osmotic shock to the proembryos, all the dissection work is carried out in a sterile 9% (W/V) glucose solution.

Steps in the procedure
1. Melt both the bottom- and top-layer ECM media gently in a micro-wave oven. Keep the melted top-layer medium in a 38 °C water bath for future use.
2. Transfer 300 μl bottom-layer of ECM medium to each well of a 24-well multiplate (Nuclon, Denmark) and allow agar to solidify before proceeding.
3. Split the silique into two with a pair of forceps and a dissection needle. Pick up ovules of the appropriate size and suspend in a 9% glucose solution in a 6 cm plastic Petri dish (Nuclon, Denmark).
4. Cut the ovule with a sharp needle following the steps shown in the diagram (Fig. 3A). Normally a zygotic embryo can be seen along the cutting edges at this stage.
5. Isolate the embryo from the mother tissue with dissection needles by holding the ovule with the forceps.

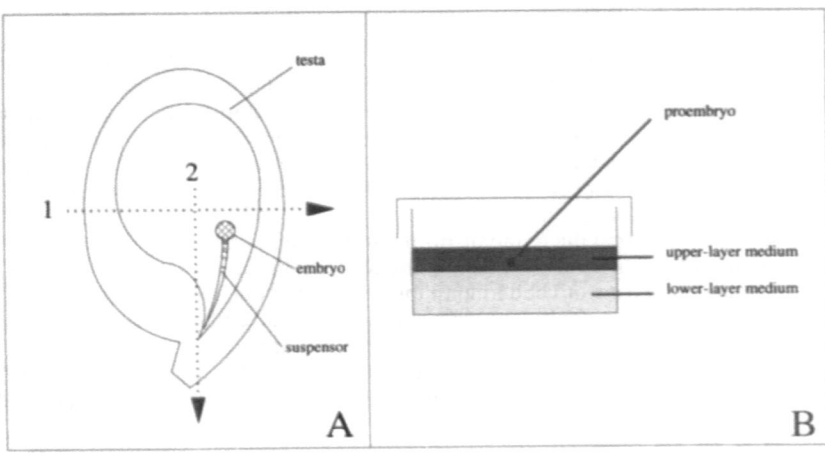

Fig. 3. A. A diagram shows the procedure for dissection of proembryo from *B. juncea.* Hold the ovule with a pair of sharp-point forceps, cut it with a dissection needle transversely, then cut the lower portion of the ovule longitudinally. Normally the embryo can be seen from the cut edge. Pick up the embryo by holding the suspensor with a pair of forceps; B. A diagram shows the double-layer culture system. 300 μl of ECM agarose medium is used for the lower-layer and a proembryo is embedded in 100 μl of upper-layer ECM medium which contains an additional 6% sucrose compared with the medium used in the lower layer.

6. Transfer one embryo onto the surface of the medium in each well of the multiplate using a micropipette (Fig. 2) and suck out the excess solution.
7. Continue dissection to obtain a suitable number of embryos for the experiments planned. In our work, 48 embryos were normally dissected for an experiment and 6 or 12 embryos were used for each treatment.
8. Transfer 100 μl of the top-layer medium (38 °C) to each well to embed the embryo (Fig. 3B).
9. Seal the plates with Nescofilm (Kobe, Japan).
10. Culture the plates in dark for the first 4 days in a 28 °C incubator, and then in a tissue culture room at the 1500 lux light for 18 hr each day, at the temperature of 28 °C/20 °C (day/night).
11. Observe the embryo development periodically with a Zeiss inverted phase-contrast microscope.

Notes
5.1. Tools for the dissection: a pair of sharp-point forceps (No. 4, Dumond, 72695-D, Switzerland); a home-made dissection needle assembled by connecting a sowing needle to a wooden handle.
5.2. As the proembryo is very fragile, it is important to avoid any direct mechanical damage to the embryo proper during the isolation. Great care should be taken not to directly touch the embryo proper with dissection needles or forceps. However, since damage of suspensor has no apparent effect on the subsequent embryo development in this species [10], the suspensor can be held with the forceps during embryo isolation.
5.3. Embryo culture plate: 24-well multi-plates (Nuclon) are used for the embryo culture. Plasticware from Nuclon was used because of its lower surface tension.
5.4. The culture system described above is suitable for culture proembryos larger than 35 μm in which 65% of 35–45 μm early globular- and 85% 45–60 μm globular stage embryos can develop into mature stages. Embryos larger than 80 μm at the early heart-shape stage grow very rapidly and develop into mature embryos at a frequency approaching 100%.
5.5. Before transferring the embryos from the dissection plate to the culture plate, we suggest firstly to fill up the micropipette with the dissection solution (Fig. 2) to prevent the embryos from adhering to the side wall of the micropipette.
5.6. Under the culture conditions described, a globular embryo will develop to the heart-shape stage after 2 days' culture, 6 days to the torpedo-shape stage (at this time the embryos can be seen by eye), and 8 days to the cotyledonary stage. Normally the embryos will reach the mature stage after 10 to 14 days in culture (Fig. 4A, B).

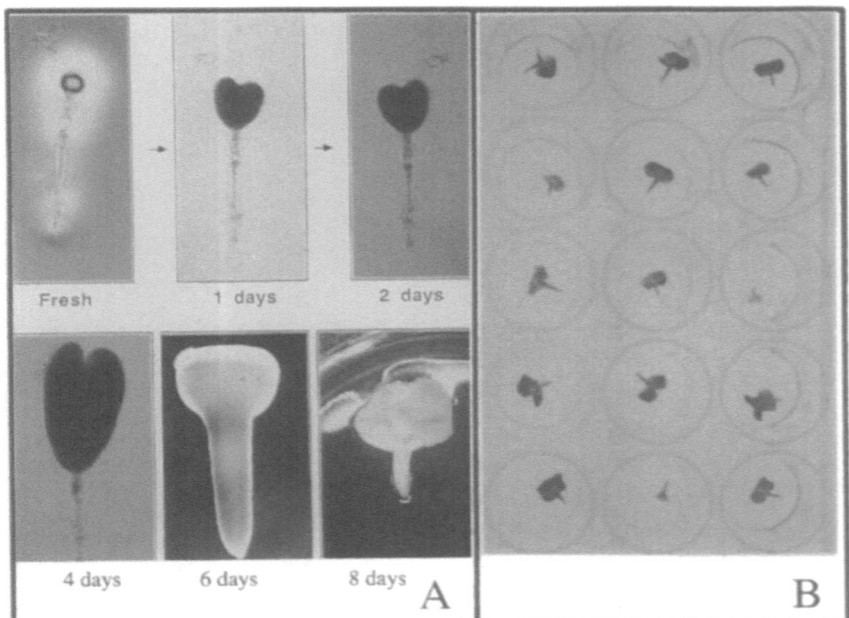

Fig. 4. Development of zygotic proembryos of *B. juncea in vitro.* A. Sequential development of a globular embryo (45 μm in length) in culture. In most cases, a globular embryo develop to the cotyledonary stage after 8 days' culture; B. Part of a 24-well multiplate shows the mature embryos developed *in vitro* from the early globular stage after 10 days culture.

6. Germination of mature embryos

Germination of the mature embryo occurs only after transfer to fresh medium; failure to do so results in the mature embryos turning yellow and eventually dying. We found that the best medium for the embryo germination is the B_5 basal medium containing 1% sucrose. A higher sucrose content promotes precocious germination which results in abnormal plants.

Steps in the procedure
1. After 14–16 days culture in the initial medium (ECM), transfer the embryos onto solid B_5 basal medium containing 1% sucrose and 0.7% agar (extra pure bacteriological grade) by inserting the radical part of the embryos into the agar.
2. Seal the plate with Nescofilm and culture them in the tissue culture room for one month.
3. Transfer the seedlings into pots to obtain fertile plants.

Notes
6.1. It is important to insert the radicle of the mature embryo into the agar medium and avoid contact of the cotyledon with the culture medium.
6.2. Germination of the seedlings will be very slow in the first 10 to 15 days, which is necessary for obtaining normal plants. Precocious germination will cause the production of abnormal plants.

Acknowledgments

We would like to thank Drs. D.G. Neuhaus and B. Kost of Institut für Pflanzenwissenschaften, ETH Zurich, Switzerland, G.Y. Wang of Shanghai Institute of Plant Physiology, China for valuable discussion, and Dr. M. Leech for critical reading of the manuscript.

References

1. Agnihotri A (1993) Hybrid embryo rescue. In: Lindsey K (Ed.) Plant Tiss Cult Man E4, pp 1–8, Kluwer Academic, The Netherlands.
2. Carman JG (1990) Embryogenic cells in plant tissue cultures: occurrence and behaviour. In Vitro Cell Dev Biol 26: 746–753.
3. De Jong AJ, Cordewener J, Lo Schiavo F, Terzi M, Vanderckkove J, Van Kammen A, De Vries SC (1992) A carrot somatic embryo mutant is rescued by chitinase. Plant Cell 4: 425–433.
4. De Jong AJ, Heldstra R, Spaink HP, Hartog MV, Meijer EA, Hendriks T, Lo Schiavo F, Terzi M, Bisseling T, Van Kammen A, De Vries SC (1993) *Rhizobium* lipooligo-saccharides rescue a carrot somatic embryo mutant. Plant Cell 5: 615–620.
5. Dumas C, Mogensen HL (1993) Gametes and fertilisation: maize as a model system for experimental embryogenesis in flowering plants. Plant Cell 5: 1337–1348.
6. Gamborg OL, Miller RA, Ojima K (1968) Nutrient requirements of suspension cultures of soybean root cells. Exp Cell Res 50: 151–158.
7. Hannig E (1904) Physiology of plant embryos. I. The culture of cruciferous embryos outside the embryo sac. Bot. Ztg. 62: 46–81.
8. Huang B, Keller WA (1990) Microspore culture technology. J Tiss Cult Methods 12: 171–178.
9. Kao KN, Michayluk MR (1975) Nutrient requirements for growth of *Vicia hajastana* cells and protoplasts at very low population density in liquid medium. Planta 126: 105–110.
10. Liu CM, Xu ZH, Chua N-H (1993) Proembryo culture: *in vitro* development of early globular-stage zygotic embryos from *Brassica juncea*. Plant J 3: 291–300.
11. Liu CM, Xu ZH, Chua N-H (1993) Auxin polar transport is essential for the establishment of bilateral symmetry during early plant embryogenesis. Plant Cell 5: 621–630.
12. Monnier M (1976) Culture *in vitro* de l'embryon immature de *Capsella bursa-pastoris* Moench (l). Rev Cytol Biol Veg 39: 1–120.
13. Monnier M (1978) Culture of zygotic embryos. In: Thorpe TA (Ed.) Frontiers of Plant Tissue Culture, pp 277–286, University of Calgary Press, Calgary.
14. Murashige T, Skoog F (1962) A revised medium for rapid growth and bioassays with tobacco tissue cultures. Physiol Plant 15: 473–497.
15. Norstog K (1961) The growth and differentiation of cultured barley embryos. Am J Bot 48: 867–884.
16. Norstog K, Smith J (1963) Culture of small barley embryos on defined media. Science 142: 1655–1656.
17. Raghavan V, Torrey JG (1963) Growth and morphogenesis of globular and older embryos of *Capsella* in culture. Am J Bot 50: 540–551.
18. Rietsema J, Satina S, Blakeslee AF (1952) The effect of sucrose on the growth of *Datura stramonium* embryos *in vitro*. Am J Bot 40: 538–545.
19. Schiavone FM, Racusen RH (1990) Microsurgery reveals regional capabilities for pattern re-establishment in somatic carrot embryos. Dev Biol 141: 211–219.
20. Schiavone FM, Racusen RH (1991) Regeneration of the root pole in surgically transected carrot embryos occurs by position-dependent, proximodistal replacement of missing tissues. Development 113: 1305–1313.
21. Smith JG (1973) Embryo development in *Phaseolus vulgaris* II. Analysis of selected inorganic ions, ammonia, organic acids, amino acids and sugars in the endosperm liquid. Plant Physiol 51: 454–458.
22. Tykarska T (1976) Rape embryogenesis I. The proembryo development. Acta Soc Bot Pol 45: 3–16.
23. Tykarska T (1979) Rape embryogenesis II. Development of embryo proper. Acta Soc Bot Pol 48: 391–422.

24. Van Overbeek J, Conklin ME, Blakeslee AF (1941) Factors in coconut water milk essential for growth and development of very young *Datura* embryos. Science USA 94: 350–351.
25. Zimmerman JL (1993) Somatic embryogenesis: A model for early development in higher plants. Plant Cell 5: 1411–1423.

Plant Tissue Culture Manual **E6**, 1–17, 1997.

Production of haploids in *Brassica* spp. via microspore culture

A. M. R. FERRIE & W. A. KELLER
Plant Biotechnology Institute, National Research Council of Canada, 110 Gymnasium Place, Saskatoon, Sk, S7N 5C3 Canada

Introduction

Haploids can occur naturally at low frequencies or be produced via culture of anthers, isolated microspores, unfertilized ovules, chromosome elimination with wide crosses, or pollination with irradiated pollen. The method chosen depends on the species to be studied. The development of embryos from microspores or anthers is termed androgenesis. This technique can be applied to many species and will be discussed further in this chapter.

Gynogenesis is the development of haploid embryos from unfertilized ovaries or ovules. This technique has been used in a number of species in both monocots and dicots. Given certain culture conditions, the cells of the embryo sac will divide and develop into an embryo. A number of factors influence the induction of embryogenesis of unfertilized ovules, including growing conditions of donor plants, pretreatments, developmental stage of the female gametophyte, genotype, culture medium and culture environment. This topic has been reviewed by several authors so will not be covered here [1, 2]. The number of haploid embryos that develop is low when compared to androgenesis, however these may be of better quality as no albino plants have been regenerated when using unfertilized ovaries or ovules. Gynogenesis may be beneficial in producing haploids when species are unresponsive to microspore or anther culture or where male sterility prevents pollen development.

Chromosome elimination protocols have been used extensively for haploid production in *Hordeum vulgare* L. Several barley cultivars have been developed, for example, Rodeo [3], Gwylan [4], and Mingo [5]. When *Hordeum vulgare* is crossed with *Hordeum bulbosum*, haploids of *H. vulgare* are recovered and the *H. bulbosum* chromosomes are eliminated [6]. Other wide crosses, e.g. wheat × maize [7] and durum wheat × maize [8] have resulted in the production of haploids.

Genes that confer the ability of cells to develop directly into haploid plants are present in some species and have resulted in the production of haploids. The haploid initiator gene (hap) in barley has been used to produce haploids [9]. This method does not use *in vitro* or aseptic tissue culture techniques. However the efficiency of haploid recovery is low [10].

This chapter deals with androgenesis which is the culture of male gametophytic cells. Both anthers and isolated microspores can be cultured. *In vitro* haploid production via anther culture was first demonstrated by Guha and Maheshwari [11] in 1964 with *Datura innoxia* and now haploids have been produced in over 200 species [12–16]. Isolated microspore culture is the preferred technique as it is less labour intensive and often more efficient than anther culture. Siebel and Pauls [17] reported microspore culture to be ten times more efficient than anther culture for *B. napus* embryo production. The lower efficiencies in anther culture may be due to competition among the microspores within the anther [18] or release of inhibitory compounds from the anther wall [19]. Each microspore has the capability of developing into an embryo and culture of a heterozygous population of microspores presents the opportunity to obtain a wide array of genetic recombinants.

Isolated microspore culture of *Brassica* species was first demonstrated by Lichter [20] in 1982. However, the frequency of embryogenesis was low. Since then, microspore culture techniques have been optimized for *B. napus* to the extent where the methodology can be used routinely in laboratories around the world to develop advanced breeding lines and cultivars (e.g. Cyclone and Quantum). The efficiency of embryogenesis has increased with up to 50% of the *B. napus* microspores undergoing embryogenesis [21]. For the other *Brassica* species including *B. rapa*, *B. oleracea*, *B. carinata*, *B. juncea*, *Sinapis alba* and *B. nigra*, the frequency of embryogenesis is lower and significant differences in embryogenesis exist between the genotypes [22–27]. Response to anther or microspore culture of the different species is shown in Table 1.

Table 1 Anther or microspore culture of *Brassica* species and relatives (modified from Palmer et al., 1995 [28]).

Species	Culture systems	References
A. thaliana	anther	29, 30
B. carinata	anther	31
	microspores	32
B. juncea	anther	25, 33, 34
B. napus		
Winter	anther	35, 36
	microspores	20
Spring	anther	37, 38
	microspores	39, 40
Rutabaga	anther	41
B. nigra	anther	7, 42
	microspores	43–45
B. oleracea		
Broccoli	anther	46–48
	microspores	23, 49
Brussels sprouts	anther	50, 51
Cabbage	anther	52–55
Cauliflower	anther	56, 57
Kohlrabi	anther	57
Kale	anther	58

Species	Culture systems	References
B. rapa		
Spring canola	anther	59
	microspores	22, 60
Chinese cabbage	anther	61
	microspores	62
Rapid cycling	anther	63, 64
Sinapis alba (*B. hirta*, *B. alba*)	anther	26, 65
Raphanus sativum	microspores	43

A number of factors influence microspore embryogenesis including genotype, donor plant growing conditions, developmental stage of the pollen, pretreatments, media constituents, and culture conditions. Conditions and treatments can vary depending on the species.

As with many tissue culture systems, genotypic differences are reflected in microspore response *in vitro*. Genotypic screening studies have been conducted in *B. napus* [17, 39, 63, 66], *B. rapa* [67], and *B. oleracea* [57]. Plant to plant variation within a genotype also exists [56, 68].

The donor plant growing conditions play a major role in embryogenesis. A lower growing temperature, 10/5 °C, was beneficial for *B. napus*, *B. rapa*, and *B. oleracea* microspore culture [22, 23, 69]. A higher temperature, including field and greenhouse conditions, may be used but frequencies of embryogenesis are reduced and response may not always be consistent. Photoperiod, light intensity, age of the plant, moisture, and nutrient supply may also influence embryogenesis.

The optimum developmental stage of the pollen can vary depending on the species, genotype, donor plant conditions, or technique used (anther or microspore culture). For *Brassica* species, the mid-late uninucleate stage has proven to be the most responsive for embryogenesis. However, one report stated that the most responsive stage for *B. oleracea* was the binucleate to trinucleate stage [70]. Cells outside of this window of embryogenesis may produce toxic substances which could inhibit embryogenesis or result in abnormal embryos [71, 72].

Pretreatment of the plants, buds, anthers, or microspores with a chemical or physical treatment may increase embryogenesis, but are not used as frequently in the *Brassica* species as they are in other species. Pretreatments evaluated in *Brassica* include decreased atmospheric pressure [26], gamma irradiation [73], colchicine treatment [74], and ethanol stress [73] which all enhanced embryogenesis.

The composition of the culture medium is an important factor influencing embryogenesis. The carbohydrate source and concentration, pH, nitrogen source and concentration, and growth regulators are among the most critical factors. The type of culture is also important, i.e. liquid or solid and the type of gelling agent used. Liquid cultures are the most commonly used in *Brassica*. Sucrose is the most commonly used carbohydrate in *Brassica* microspore culture. The concentration range is usually 10–14%. However, it has been observed in *B. napus* that 20% sucrose gave the highest embryogenesis induction frequencies [75]. Elevated sucrose levels were also beneficial in *B. rapa* for induction of embryogenesis, but lower levels are required for further embryo development [22].

The incubation environment is important, the major factor being temperature. An initial elevated temperature treatment of 32–35 °C for 24 to 72 hours is required by *Brassica* microspore cultures [22, 47, 54, 59]. Other factors influencing embryogenesis include light intensity and CO_2 levels. Most cultures are initiated in the dark. Little research has been conducted on the benefits of light regimes, aeration, or CO_2 levels. Microspore culture optimal density varies with species, although the general range is 10^3–10^5 microspores per ml [72, 76].

There are a number of important applications for anther/microspore culture technologies. A major application is the use of haploids to develop new cultivars. A conventional *B. napus* breeding program usually takes about ten years to develop a variety, from the initial cross to commercialization. With the use of haploids, this time can be reduced by three to four years [77]. The basis for this advantage is related to the fact that, through haploidy, homozygous lines can be produced within one generation and traits rapidly fixed rather than several generations of extensive selfing in a conventional plant breeding program. Recessive genes are easily identified as there is no masking by dominant genes and therefore selection can be more efficient. When screening for a desirable genotype, a smaller population is required when using doubled haploids compared to conventional diploid populations [78]. Genetic studies to estimate genetic variance, the number of genes involved in quantitative characteristics, linkage, and gene interaction have exploited haploid technology [79].

The haploid system can also be used in mutation breeding and selection. The advantage of this system in mutation is the equal expression of both dominant and recessive genes. Very little somaclonal variation is associated with microspore and anther culture as compared to other *in vitro* systems. Mutagenesis of microspore cultures has been used in *Brassica* for selection of mutant lines with herbicide resistance and altered fatty acid composition [80–82].

The haploid system would be beneficial in gene transfer studies as a large number of uniform cells or embryos capable of regeneration can be produced. With chromosome doubling, the introduced gene will be duplicated and therefore homozygous in the doubled haploid plants. Transgenic *B. napus* plants have been derived from microspore-derived embryos via *Agrobacterium* [83] and microinjection of DNA into the embryo [84]. Transformed plants with herbicide tolerance and antibiotic resistance have been reported [83, 85]. Transformation of freshly isolated microspores have also resulted in transformed *B. napus* and *B. rapa* plants [86].

Haploid embryos can also be used for biochemical studies. It has been shown that the developmental pathway and many biochemical pathways of a zygotic embryo and a microspore-derived embryo are similar [87]. Biochemical studies on lipid storage and biosynthesis [88, 89], oil quality [90], storage protein [88,

91], and glucosinolate metabolism [92] have utilized the haploid system. This has also been used in studies of chlorophyll metabolism [93], freezing tolerance [94, 95, 96], desiccation tolerance [97, 98] and physiological aspects of embryo maturation [99].

Procedures

The procedure outlined is for *B. napus* microspore culture. A similar procedure, with some modifications, could be used for *B. rapa*.

Steps in the procedure
1. Plant seeds in pots (15 cm) containing commercial greenhouse soil (e.g. Redi-Earth soil-mix) or a mix of soil, sand, and peat moss (4:1:1). Place pots in growth cabinets with a 16 hour photoperiod and a day/night temperature of 20/15 °C.
2. Prior to bolting, lower the temperature in the growth cabinet to 10/5 °C. Fertilize plants with 14–14–14 Nutricote 100 (slow release fertilizer) and water with 0.35 g l^{-1} 15–15–18 (15% N, 15% P, 18% K). Healthy, vigourous plants are essential, ie. free of disease and insects.
3. Determine the developmental stage of the microspore using 2μg ml^{-1} DAPI (4',6-diamidino-2-phenylindole) [71]. Apply a drop of DAPI to the microspore preparation on a slide. Leave the slide for 20–30 minutes then observe under a fluorescence microscope at 365 nm excitation. Developmental stage can be correlated with bud size for ease of bud selection. For *B. napus* and *B. rapa*, the mid-late uninucleate stage appears to be the most responsive.
4. Select buds, approximately 50–75, on the basis of size (usually 3–4 mm for *B. napus* and 2–3 mm for *B. rapa*).
5. Place buds in histological tissue sample baskets and surface sterilize in 6% sodium hypochlorite for 15 minutes on a shaker followed by three five-minute washes with sterile water.
6. Macerate buds in 5 ml of half strength B5-13 medium (B5 medium supplemented with 13% sucrose) with a glass rod.
7. Filter through 44 μm nylon screencloth into a 50 ml tube. Rinse the filter and beaker three times with 5 ml of B5-13 for a total of 20 ml.
8. Centrifuge the crude microspore suspension at 130–150 g for 3 minutes.
9. Decant the supernatant and add 5 ml B5-13 to the pellet. Repeat this procedure twice for a total of three washes.
10. Determine the number of microspores using a hemacytometer.
11. Add the required amount of modified Lichter medium (NLN) [20] to achieve a density of 10^5 microspores ml^{-1}. 12. Dispense 10 ml of microspore suspension into 100 × 15 mm sterile Petri plates.
13. Incubate plates in the dark at 32 °C for 72 hours (*B. napus*) or 48 hours (*B. rapa*) then incubate at 25 °C for the remainder of three weeks.
14. For *B. rapa*, the NLN culture medium (pH 5.8) has a higher concentration of sucrose (17%) for initial culture. After 48 hours,

change media to a lower concentration of sucrose (10%). Pipette the medium into a centrifuge tube, centrifuge for 3 minutes at 130–150 g, decant the supernatent, and add the same amount of medium containing the lower sucrose concentration.

15. After three weeks, count the embryos and place the Petri plates containing embryos on a gyratory shaker (70 rpm) under continuous light at 22 °C.
16. Once embryos turn green, transfer embryos to solid B5 medium free of growth regulators and maintain at 22 °C with a 14 hour photoperiod and a light intensity of 150 μmol m^{-2} s^{-1}.
17. Subculture embryos every 3–4 weeks if necessary until sufficient root and shoot growth is achieved.
18. Prior to transferring resultant plantlets to soil, submerge the roots of the plantlets in a 0.34% solution of colchicine for 1.5 hours. After colchicine treatment, rinse the roots in water and transfer the plantlet to a soil-less mix and grow in the greenhouse or growth cabinet.
19. Initially cover the plantlets with plastic to maintain a high humidity. Gradually remove this covering as the plants harden.
20. Shake or agitate the plants and flowers during flowering to ensure suitable levels of fertilization and seed set. Doubled-haploid and haploid plants are determined by the production or lack of production of pollen.
21. Because *B. rapa* is generally self-incompatible, seed production requires bud pollination, treatment of flowers with sodium chloride, or carbon dioxide.
22. Doubled-haploid plants are grown to maturity and seed is harvested.

Notes
1. Plants are usually grown in growth cabinets, however greenhouse conditions could be used.
2. Formation of embryos visible to the naked eye will start taking place about ten days after induction. Torpedo-shaped embryos will be formed in three weeks. Most genotypes of *B. napus* will produce embryos, however there are differences in embryo yield between genotypes. The line Topas 4079 is very responsive and will produce thousands of embryos/100 buds. *B. rapa* is more recalcitrant when compared to *B. napus*, but embryos can be produced from different genotypes. A *B. rapa* line that consistently produces thousands of embryos/100 buds has been identified [67].

Solutions (sterile)

B5 Medium [100]
Stock Solutions:
Vitamins: (1000X)

–Myo-inositol	100.0 g
–Nicotinic Acid	1.0 g
–Pyridoxine HCl	1.0 g
–Thiamine HCl	10.0 g

Dissolve each compound before adding the next vitamin and bring up to a final volume of 1 l with glass distilled water. Freeze in 10 ml aliquots.

Micronutrients: (1000X)

–$MnSO_4.H_2O$	10.000 g
–H_3BO_3	3.000 g
–$ZnSO_4.7H_2O$	2.000 g
–$Na_2MoO_4.2H_2O$	0.250 g
–$CuSO_4.5H_2O$	0.025 g
–$CoCl_2.6H_2O$	0.025 g

Dissolve each compound before adding the next micronutrient and make up to 1 l with glass distilled water. Freeze in 10 ml aliquots.
Potassium iodide:

–KI	0.750 g

Bring up to 1 l. Store refrigerated in the dark.

B5 Medium, (10 × Concentrate)

–KNO_3	30.00 g
–$MgSO_4.7H_2O$	5.00 g
–$NaH_2PO_4.H_2O$	1.50 g
–$CaCl_2.2H_2O$	1.50 g
–$(NH_4)_2SO_4$	1.50 g
–Sequestrene 330 Fe	0.28 g

–Add 10 ml of Vitamin Stock (1000X)
–Add 10 ml of Micronutrient Stock (1000X)
–Add 10 ml of KI Stock

Dissolve each compound before adding the next and bring up to a final volume of 1 l with glass distilled water. Freeze in 100 ml aliquots.

Notes
1. 100 ml of 10 × concentrate makes 1 l of media
2. The stock solutions are dissolved in glass distilled water. Sucrose (13%) is added and

the pH is adjusted to 6. The solution is made to volume and dispensed into 500 ml bottles. The medium is autoclaved at 121 °C, 15 psi for 20 minutes. For B5 solid medium, use the same procedure as for B5 medium but add 0.8% agar.

NLN Medium [20]
Stock Solutions:
Vitamins (1000X)

–Thiamine HCl	0.50 g
–Nicotinic Acid	5.00 g
–Pyridoxine HCl	0.50 g
–Glycine	2.00 g
–Biotin	0.05 g
–Folic Acid	0.50 g
–Myo-inositol	100.00 g

Dissolve each compound before adding the next vitamin and bring up to 1 l with glass distilled water and freeze in 10 ml aliquots.

Micronutrients [101]: (1000X)

–$MnSO_4.4H_2O$	22.300 g
–H_3BO_3	6.200 g
–$ZnSO_4.7H_2O$	8.600 g
–$Na_2MoO_4.2H_2O$	0.250 g
–$CuSO_4.5H_2O$	0.025 g
–$CoCl_2.6H_2O$	0.025 g

Dissolve each compound before adding the next micronutrient and bring up to 1 l with glass distilled water and freeze in 10 ml aliquots.

NLN Medium: (10X)

–KNO_3	1.25 g
–$MgSO_4.7H_2$	1.25 g
–KH_2PO_4	1.25 g
–$Ca(NO_3)_2.4H_2O$	5.00 g
–Sequestrene 330 Fe	0.40 g
–Glutathione	0.30 g
–L-serine	1.00 g
–L-glutamine	8.00 g

–Add 10 ml of Vitamin Stock
–Add 10 ml of Micronutrient Stock
–Add 10 ml of KI Stock

Dissolve each compound before adding the next and bring up to 1 l with glass distilled water. Freeze in 100 ml aliquots.

1. 100 ml of 10 × concentrate makes 1 l of media.
2. The stock solutions are dissolved in glass distilled water. Sucrose (13%) is added and the pH adjusted to the required level. The solution is made to volume and is filter sterilized using 0.2 μm filter units.

Colchicine solution

A 0.34% colchicine solution is made by dissolving 3.4 g of colchicine in 1 l of water. This solution can be stored in the fridge (4 °C) in the dark. *Note*: Colchicine must be used with caution. Avoid inhalation and skin contact with the dry powder. The liquid form is not volatile. However protective clothing, such as gloves and eye cover should be worn.

Trouble shooting

Problem	Possible cause or solution
Poor growth of donor plants	– temperature too high – insect infestations (e.g. aphids, thrips) – nutrient deficiency – adjust photoperiod – lighting (quality, intensity)
Inadequate bud sterilization	– use tightly closed buds – sterilize with ethanol for 30 sec – supplement with detergents or mercuric chloride
Media sterility	– check filters and stock solutions – use of antibiotics
Culture environment	– constant temperature, time
Embryo culture	– plates are sealed properly – timing of transfer onto solid media and subculturing
Colchicine	– time and concentration
Plant regeneration	– adaptation to external environment, humidity is essential – disease and insect control – maintain optimum growing conditions, temperature, fertility and water availability
Seed set	– agitation of plants – temperature too high – insect or disease problems

References

1. San LH, Gelebart P (1986) Production of gynogenetic haploids. In: Vasil IK (ed) Cell Culture and Somatic Cell Genetics of Plants, pp. 305–322. New York: Academic Press.
2. Yang HY, Zhou C (1990) In vitro gynogenesis. In: Bhojwani SS (ed) Plant Tissue Culture and Applications and Limitations, pp. 242–258. New York: Elsevier.
3. Campbell KW, Brown RI, Ho KM (1984) 'Rodeo' barley. Can J Plant Sci 64: 203–205.
4. Baenziger PS, Kudirka DT, Schaeffer GG, Lazar MD (1985) The significance of doubled haploid variation. In: Gustafson JP (ed) Gene Manipulation in Plant Improvement, pp. 385–414. New York: Plenum Press.
5. Ho KM, Jones GE (1980) Mingo Barley. Can J Plant Sci 60: 279–280.
6. Kasha KJ, Reinbergs E (1981) Recent developments in the production and utilization of haploids in barley. In: Asher MJC (ed) Barley Genetics IV, Proc 4th Int Barley Genet Symp, pp. 655–665. Edinburgh.
7. Laurie DA, Bennett MD (1988) The production of haploid wheat plants from wheat × maize crosses. Theor Appl Genet 70: 100–105.
8. O'Donoughue LS, Bennett MD (1994) Durum wheat haploid production using maize wide crossing. Theor Appl Genet 89: 559–566.
9. Hagberg A, Hagberg G (1980) High frequency of spontaneous haploids in the progeny of an induced mutation in barley. Hereditas 93: 341–343.
10. Powell W, Wood W (1984) An assessment of the hap initiator gene for haploid production in *Hordeum vulgare*. J Agric Sci Camb 103: 253–255.
11. Guha S, Maheshwari SC (1964) *In vitro* production of embryos from anthers in *Datura*. Nature 204: 497.
12. Ferrie AMR, Palmer CE, Keller WA (1994) Biotechnological applications for haploids. In: Shargool PD, Ngo TT (eds) Biotechnological Applications of Plant Cells, pp. 77–110. Boca Raton, FL: CRC Press.
13. Ferrie AMR, Palmer CE, Keller WA (1995) Haploid embryogenesis. In: Thorpe TA (ed) In Vitro Embryogenesis in Plants, pp. 309–344. Dordrecht: Kluwer Academic Publishers.
14. Keller WA, Arnison PG, Cardy BJ (1987) Haploids from gametophytic cells – recent developments and future prospects. In: Green CE, Somers DA, Hackett WP, Biesboer DD (eds) Plant Tissue and Cell Culture, Proc 6th Intl Tissue Cult Congr, pp. 223–241. New York: Alan R. Liss.
15. Maheshwari SC, Rashid A, Tyagi AK (1982) Haploids from pollen grains – retrospects and prospects. Am J Bot 69: 865–879.
16. Morrison RA, Evans DA (1988) Haploid plants from tissue cultures: new plant varieties in a shortened time frame. Biotechnology 6: 684–690.
17. Siebel J, Pauls KP (1989) A comparison of anther and microspore culture as a breeding tool in *Brassica napus*. Theor Appl Genet 78: 473–479.
18. Hoffmann F, Thomas E, Wenzel G (1982) Anther culture as a breeding tool in rape. 2. Progeny analyses of androgenetic lines and induced mutants from haploid cultures. Theor Appl Genet 61: 225–232.
19. Heberle-Bors E (1984) Genotypic control of pollen plant formation in *Nicotiana tabacum* L. Theor Appl Genet 68: 475–479.
20. Lichter R (1982) Induction of haploid plants from isolated pollen of *Brassica napus*. Z Pflanzenphysiol 105: 427–434.
21. Pechan PM, Keller WA (1988) Identification of potentially embryogenic microspores in *Brassica napus*. Physiol Plant 74: 377–384.
22. Baillie AMR, Epp DJ, Hutcheson D, Keller WA (1992) *In vitro* culture of isolated microspores and regeneration of plants in *Brassica campestris*. Plant Cell Rep 11: 234–237.
23. Takahata Y, Keller WA (1991) High frequency embryogenesis and plant regeneration in isolated microspore culture of *Brassica oleracea* L. Plant Sci 74: 235–242.

24. Arora R, Bhojwani SS (1988) Production of androgenic plants through pollen embryogenesis in anther culture of *Brassica carinata* A. Braun. Biologia Plantarum 30: 25–29.
25. Sharma KK, Bhojwani SS (1985) Microspore embryogenesis in anther cultures of two Indian cultivars of *Brassica juncea* (L.) Czern. Plant Cell Tiss Org Cult 4: 235–239.
26. Klimaszewska K, Keller WA (1983) The production of haploids from *Brassica hirta* Moench (*Sinapis alba* L.) anther cultures. Z Pflanzenphysiol 109: 235–241.
27. Govil S, Babber SB, Gupta SC (1986) Plant regeneration from *in vitro* cultured anthers of Black mustard (*Brassica nigra* Koch). Plant Breeding 97: 64–71.
28. Palmer CE, Keller, WA, Arnison PG (1996) Experimental haploidy in Brassica. In: Jaine SM, Sopary SK, Vielleux RE (eds) In vitro Haploidy of Higher Plants. Dordrecht: Kluwer Academic Publishers, Dordrecht (in press).
29. Scholl RL, Amos JA (1980) Isolation of doubled-haploid plants through anther culture in *Arabidopsis thaliana* (analyzed genetically and cytologically). Z Pflanzenphysiol 96: 407–414.
30. Gresshoff PM, Doy CH (1972) Haploid *Arabidopsis thaliana* callus and plants from anther culture. Australian J Biol Sci 25: 259–264.
31. Renu-Arora R, Bhojwani SS, Arora R (1988) Production of androgenic plants through pollen embryogenesis in anther cultures of *Brassica carinata* A. Braun. Biologia Plantarum 30: 25–29.
32. Chuong PV, Beversdorf WD (1985) High frequency embryogenesis through isolated microspore culture of *B. napus* and *B. carinata* Braun. Plant Sci 39: 219–226.
33. Yadav, RC, Yadav NR, Kumar PR, Sharma DR (1988) Differential androgenic response in *Brassica juncea* (L.) Czern and Coss. Cruciferae Newsletter 13: 76.
34. George L, Rao PS (1983) In vitro induction of pollen embryos and plantlets in *Brassica juncea* through anther culture. Plant Sci Lett 26: 111–116.
35. Dunwell JM, Cornish LM (1985) Influence of preculture variables on microspore embryo production in *Brassica napus* ssp. *oleifera* cv. Duplo. Ann Bot 56: 281–289.
36. Thurling N, Chay PM (1984) The influence of donor plant genotype and environment on production of multicellular microspores in cultured anthers of *Brassica napus* ssp. *oleifera*. Ann Bot 54: 681–695.
37. Lichter R (1981) Anther culture of *Brassica napus* in a liquid culture media. Z Pflanzenphysiol 103: 229–237.
38. Keller WA, Armstrong KC (1978) High frequency production of microspore-derived plants from *Brassica napus* anther cultures. Z Pflanzenzuchtg 80: 100–108.
39. Gland A, Lichter R, Schweiger HG (1988) Genetic and exogenous factors affecting embryogenesis in isolated microspore culture of *Brassica napus* L. J Plant Physiol 132: 613–617.
40. Swanson EB, Coumans MP, Wu SC, Barsby T, Beversdorf WD (1987) Efficient isolation of microspores and the production of microspore-derived embryos from *Brassica napus*. Plant Cell Rep 6: 94–97.
41. Souza Machado V, Shupe J, Keller WA (1985) Cytoplasmic inherited atrazine resistance transmitted through anther culture in rutabaga. Z Pflanzenzuchtg 95: 179–184.
42. Leelavathi S, Reddy VS, Sen SK (1987) Somatic cell genetic studies in *Brassica* species. II Production of androgenetic haploid plants in *Brassica nigra* (L.) Koch. Euphytica 36: 215–219.
43. Lichter R (1989) Efficient yield of embryoids by culture of isolated microspores of different *Brassicaceae* species. Plant Breed 103: 119–123.
44. Hetz E, Shieder O (1991) Direct embryogenesis and plant regeneration through microspore culture of *Brassica nigra*. Crucifer Newslett 14/15: 102–103.
45. Margale E, Chevre AM (1991) Factors affecting embryo production from microspore culture of *Brassica nigra* (Koch). Crucifer Newslett 14/15: 100–101.
46. Keller WA, Armstrong KC (1983) Production of haploids via anther culture in *Brassica oleracea* var. *italica*. Euphytica 32: 151–159.

47. Arnison PG, Donaldson P, Ho LCC, Keller WA (1990) The influence of various physical parameters on anther culture of broccoli (*Brassica oleracea* var. *italica*). Plant Cell Tiss Org Cult 20: 147–155.

48. Arnison PG, Donaldson P, Jackson A, Semple C, Keller W (1990) Genotype-specific response of cultured broccoli (*Brassica oleracea* var. *italica*) anthers to cytokinins. Plant Cell Tiss Org Cult 20: 217–222.

49. Duijs JG, Voorrips RE, Visser DL, Custers JBM (1992) Microspore culture is successful in most crop types of *B. oleracea* L. Euphytica 60: 45–55.

50. Ockendon DJ (1984) Anther culture in Brussels sprout (*Brassica oleracea* var. *gemmifera*). I. Embryo yield and plant regeneration. Ann Appl Biol 105: 285–291.

51. Lelu M-A, Ballon G (1985) Obtention d'haploides par culture d'anthers de *Brassica oleracea* L. var. *capitata* et var *gemmifera*. CR Acad Sci Paris 300 serie III 2: 71–76.

52. Lillo C, Hansen M (1987) Anther culture of cabbage. Influence of growth temperature of donor plants and media composition on embryo yield and plant regeneration. Norwegian J Agric Sci 1: 105–109.

53. Chiang MS, Frechette C, Kuo G, Chong C, Delafield SJ (1985) Embryogenesis and haploid plant production from anther culture of cabbage (*Brassica oleracea* var. *capitata*). Can J Plant Sci 65: 1033–1057.

54. Roulund N, Hansted L, Anderson SB, Farestveit B (1990) Effect of genotype, environment on carbohydrate on anther culture response in head cabbage (*Brassica oleracea* L. convar. *capitata* Alef). Euphytica 49: 237–242.

55. Bagga S, Bhalla-Sarin N, Sopory SK, Guha-Mukherjee S (1982) Comparison of in vitro plant formation from somatic tissues and pollen grains in *Brassica oleracea* var. *botrytis*. Phytomorphology 32: 152–156.

56. Phippen C, Ockendon DJ (1990) Genotype, plant, bud size and media factors affecting anther culture of cauliflower (*Brassica oleracea* var. *botrytis*). Theor Appl Genet 79: 33–38.

57. Arnison PG, Keller WA (1990) A survey of the anther culture response of *Brassica oleracea* L. cultivars grown under field conditions. Plant Breed 104: 125–133.

58. Keller WA, Armstrong KC (1981) Production of anther derived dihaploid plants in autotetraploid marrowstem kale (*Brassica oleracea* var. *acephala*). Can J Genet Cytol 23: 259–265.

59. Keller WA, Armstrong KC (1979) Stimulation and embryogenesis and haploid production in *Brassica campestris* anther cultures by elevated temperature treatments. Theor Appl Genet 55: 65–67.

60. Burnett L, Yarrow S, Huang B (1992) Embryogenesis and plant regeneration from isolated microspores of *Brassica rapa* L. ssp. *oleifera*. Plant Cell Rep 11: 215–218.

61. Sato T, Nishio T, Hirai M (1989) Varietal differences in embryogenic ability in anther culture of Chinese cabbage (*Brassica campestris* ssp. *pekinensis*). Jpn J Breed 39: 149–157.

62. Sato T, Nishio T, Hirai M (1989) Plant regeneration from isolated microspore cultures of Chinese cabbage (*Brassica campestris* ssp. *pekinensis*). Plant Cell Rep 8: 486–488.

63. Aslam FN, MacDonald MV, Louden PT, Ingram DS (1990) Rapid-cycling *Brassica* species: Inbreeding and selection of *Brassica napus* for anther culture ability and an assessment of its potential for microspore culture. Ann Bot 66: 331–339.

64. Aslam FN, MacDonald MV, Ingram DS (1990) Rapid-cycling *Brassica* species: anther culture ability of *Brassica campestris* L. and *Brassica napus* L. New Phytol 115: 1–9.

65. Leelavathi S, Reddy VS, Sen SK (1984) Somatic cell genetics in *Brassica* species. I. High frequency production of haploid plants in *Brassica alba* (L.). Plant Cell Rep 3: 102–105.

66. Chuong PV, Deslauriers C, Kott LS, Beversdorf WD (1988) Effects of donor genotype and bud sampling on microspore culture of *Brassica napus*. Can J Bot 66: 1653–1657.

67. Ferrie AMR, Epp DJ, Keller WA (1995) Evaluation of *Brassica rapa* L. genotypes for microspore culture response and identification of a highly embryogenic line. Plant Cell Rep 14: 580–584.

68. Ockendon DJ (1985) Anther culture in Brussels sprouts II. Effect of genotype on embryo yield. Ann Appl Biol 107: 101–104.
69. Keller WA, Fan Z, Pechan P, Long N, Grainger J (1987) An efficient method for culture of isolated microspores of *Brassica napus*. In: Proc 7th Int Rapeseed Congr, Poland, pp. 152–157.
70. Cao MQ, Charlot F, Dore C (1990) Embryogenese et regeneration de plantes de chou a choucroute (*Brassica oleracea* L. ssp. capitata) par culture in vitro de microspores isolees. CR Acad Sci Paris Serie III 310: 203–209.
71. Fan Z, Holbrook L, Keller WA (1988) Isolation and enrichment of embryogenic microspores in *Brassica napus* L. by fractionation using percoll density gradients. In: Proc 7th Int Rapeseed Congr, Poland, pp. 92–96.
72. Kott LS, Polonsi L, Ellis B, Beversdorf WD (1988) Autotoxicity in isolated microspore cultures of *Brassica napus*. Can J Bot 66: 1665–1670.
73. Pechan PM, Keller WA (1989) Induction of microspore embryogenesis in *Brassica napus* L. by gamma irradiation and ethanol stress. In Vitro Cell Dev Biol 25: 1073–1074.
74. Zaki MAM, Dickinson HG (1991) Microspore-derived embryos in *Brassica*: the significance of division symmetry in pollen mitosis I to embryogenic development. Sex Plant Reprod 4: 48–55.
75. Dunwell JM, Thurling N (1985) Role of sucrose in microspore embryo production in *Brassica napus* ssp. *oleifera*. J Exp Bot 36: 1478–1491.
76. Huang B, Bird S, Kemble R, Simmonds D, Keller WA, Miki B (1990) Effects of culture density, conditioned medium and feeder cultures on microspore embryogenesis in *Brassica napus* L. cv. Topas. Plant Cell Rep 8: 594–597.
77. Ulrich A, Furtan WH, Downey RK (1984) Biotechnology and rapeseed breeding: some economic considerations. Sci Counc Can Rep, 67 pp, Ottawa.
78. Rajhathy T (1976) Haploid flax revisited. Z Pflanzenzuchtg 76: 1–10.
79. Snape JW, Wright AJ, Simpson E (1984) Methods for estimating gene numbers for quantitative characters using doubled haploid lines. Theor Appl Genet 67: 143–148.
80. Swanson EB, Coumans MP, Brown GL, Patel JD, Beversdorf WD (1988) The characterization of herbicide tolerant plants in *Brassica napus* L. after in vitro selection of microspores and protoplasts. Plant Cell Rep 7: 83–87.
81. Swanson EB, Herrgesell MJ, Arnoldo M, Sippell DW, Wong RSC (1989) Microspore mutagenesis and selection: canola plants with field tolerance to the imidazolinones. Theor Appl Genet 78: 525–530.
82. Turner J, Facciotti D (1990) High oleic *Brassica napus* from mutagenized microspores. In: McFerson JR, Kresovich S, Dwyer SE (eds) Proc 6th Crucifer Genetics Workshop, p. 40. Geneva, NY: USDA-ARS.
83. Swanson EB, Erickson LR (1989) Haploid transformation in *Brassica napus* using an octopine-producing strain of *Agrobacterium tumefaciens*. Theor Appl Genet 78: 831–835.
84. Neuhaus G, Spangenberg G, Mittelsten Scheid O, Schweiger H-G (1987) Transgenic rapeseed plants obtained by the microinjection of DNA into microspore-derived embryoids. Theor Appl Genet 75: 30–36.
85. Huang B (1992) Genetic manipulation of microspores and microspore-derived embryos. In Vitro Cell Biol 28: 53–58.
86. Dormann M, Wang H-M, Datla N, Ferrie AMR, Keller WA, Oelck MM (1995) Transformation of freshly isolated *Brassica* microspores and regeneration to fertile homozygous plants. In: Proc 9th Int Rapeseed Congr, pp 816–818, July 4–7, 1995, Cambridge, England.
87. Crouch ML (1982) Nonzygotic embryos of *Brassica napus* L. contain embryo-specific storage proteins. Planta 156: 520–524.
88. Taylor DC, Weber N, Underhill E, Pomeroy MK, Keller WA, Scowcroft WR, Wilen RW, Moloney MM, Holbrook LA (1990) Storage protein regulation and lipid accumulation of microspore embryos of *Brassica napus* L. Planta 181: 18–26.

89. Taylor DC, Ferrie AMR, Keller WA, Giblin EM, Pass EW, MacKenzie SL (1993) Bioassembly of acyl lipids in microspore-derived embryos of *Brassica campestris* L. Plant Cell Rep 12: 375–384.

90. Wiberg, E, Rahlen L, Hellman M, Tillberg E, Glimelius K, Stymne S (1991) The microspore-derived embryo of *Brassica napus* L. as a tool for studying embryo-specific lipid biogenesis and regulation of oil quality. Theor Appl Genet 82: 515–520.

91. Wilen R, Mandel RM, Pharis RP, Holbrook LA, Moloney MM (1990) Effects of abscisic acid and high osmoticum on storage protein gene expression in microspore embryos of *Brassica napus*. Plant Physiol 94: 875–881.

92. McClellan D, Kott L, Beversdorf W, Ellis BE (1993) Glucosinolate metabolism in zygotic and microspore-derived embryos of *Brassica napus* L. J Plant Physiol 141: 153–159.

93. Johnson-Flanagan AM, Singh J (1991) Degreening and its inhibition by stress in haploid embryos of *Brassica napus* cv. Topas and Jet Neuf. In: McGregor DI (ed) Proc 8th Intl Rapeseed Congr, pp. 743–748, Saskatoon.

94. Cloutier S (1990) In vitro selection for freezing tolerance using *Brassica napus* microspore culture. MSc Thesis, Dept of Crop Sci, Univ of Guelph, Guelph, Ontario.

95. Orr W, Keller WA, Singh J (1986) Induction of freezing tolerance in an embryogenic cell suspension culture of *Brassica napus* by abscisic acid at room temperature. J Plant Physiol 126: 23–32.

96. Orr W, Johnson-Flanagan AM, Keller WA, Singh J (1990) Induction of freezing tolerance in microspore-derived embryos of winter *Brassica napus*. Plant Cell Rep 8: 579–581.

97. Anandarajah K, Kott L, Beversdorf WD, McKersie BD (1991) Induction of desiccation tolerance in microspore-derived embryos of *Brassica napus* L. by thermal stress. Plant Sci 77: 119–123.

98. Senaratna T, Kott L, Beversdorf WD, Mckersie BD (1991) Desiccation of microspore-derived embryos of oilseed rape (*Brassica napus* L.). Plant Cell Rep 10: 342–344.

99. Eikenberry EJ, Choung PV, Esser J, Romero J, Ram R (1991) Maturation, desiccation, germination and storage lipid accumulation in microspore embryos of *Brassica napus* L. In: McGregor DI (ed) Proc 8th Intl Rapeseed Congr, pp. 1809–1814, Saskatoon.

100. Gamborg OL, Miller RA, Ojima K (1968) Nutrient requirements of suspension cultures of soybean root cells. Exp Cell Res 50: 151–158.

101. Murashige T, Skoog F (1962) A revised medium for rapid growth and bioassays with tobacco tissue cultures. Physiol Plant 15: 473–497.

Section F:
Mutant Selection

Plant Molecular Biology
Manual

Section F Mutant Selection

Chapters planned for this section include:-
  Use of chemical and physical mutagens *in vitro*
  Screening for biochemical mutants *in vitro*
  Screening for disease resistance *in vitro*
  Screening for herbicide resistance *in vitro*
  Screening for drought resistance *in vitro*
  Screening for resistance to aluminium *in vitro*
  Screening for hormone auxotrophs *in vitro*

Plant Tissue Culture Manual **F1**: 1–17, 1993.
© 1993 *Kluwer Academic Publishers*.

Use of Chemical and Physical Mutagens *In vitro*

PHILIP J. DIX
Department of Biology, St. Patrick's College, Maynooth, Co. Kildare, Ireland

Introduction

In the twenty years since the first unequivocal reports (including inheritance data) on the selection of mutants *in vitro* [1, 2], a large number of mutants have been isolated, and reviews have frequently updated the progress, most extensively in a recent treatise [3]. The earliest reports made no use of deliberate mutagenesis, relying instead on the wide genetic variation exhibited by cultured cells, based on spontaneous mutation rates and possibly enhanced by genetic instabilities of rapidly dividing cells in culture [4]. The fact that the numbers worked in favour of selection of spontaneous mutants, when a dominant monogenic trait with a strong positive selection pressure was involved, probably delayed a critical evaluation of the role of chemical and physical mutagens in enhancing "mutation frequency". This was exacerbated by difficulties in reliable quantification of the latter term, given that single cell cultures were generally not in use, and colony forming potential (or plating efficiency) was usually profoundly influenced by viable cell density.

Whatever the precise reasons, many reports on *in vitro* selection of mutants have not used mutagens, and many others have had a "just for luck" attitude to their use. Mutagenic agents have been applied to cultures prior to selection, but little effort was made to optimise the treatment, or to evaluate its success in enhancing the yield of the desired mutant lines. Fortunately there have been sufficient exceptions to demonstrate how effective mutagenesis treatments can be. The most detailed and rigorous studies on both physical [5] and chemical [6] mutagenesis, benefited from the use of the efficient single cell plating procedures, and low density media, available for *Nicotiana* mesophyll protoplasts. While these investigations have clearly supported the value of mutagenic treatments for the isolation of biochemical mutants, other developments show a more critical dependence on their use.

The advent of efficient culture procedures for haploid protoplasts of several species led to the realistic methodical screening (total selection) of colonies for deficiency mutations, particularly auxotrophs [7, 8]. The laborious nature of this approach renders the use of mutagenesis procedures important to bring the number of colonies to be screened down to a manageable level. Additionally the increasing interest in minimising the callus step in *in vitro* selection, by selecting in organised cultures [9, 10], means that the number of selective units (cells with potential to develop into shoot primordia or embryos) is much lower than in callus, protoplast, or cell suspension culture. Again mutagenesis may be important.

The choice of mutagenic agent is informed by the ease with which the plant material can be handled, the size and nature of the plant inoculum (protoplast, cell suspension, callus, explant etc.), the nature of the lesion sought, and the mechanisms of action of the individual mutagens. Negrutiu [11], in a long overdue review on all aspects of *in vitro* mutagenesis, gives careful consideration to these topics, with a particular emphasis on the action of the mutagens. In the hope of simplifying (perhaps oversimplifying!) the field to a level appropriate to the ensuing protocols, and a practically-based review, the current author will restrict his observations to a few general statements. The most frequently applied chemical mutagens, *in vitro*, are alkylating agents, ethyl-methanesulfonate (EMS), 1-methyl-3-nitro-1-nitrosoguanidine (MNNG), and the nitrosoureas, N-ethyl-N-nitrosourea (NEU), and N-methyl-N-nitrosourea (NMU). The last of these is particularly effective at inducing plastome mutations, both *in vivo* [12] and *in vitro* [9], but induces nuclear mutations as well [13]. The most commonly used physical mutagens are ζ-rays, x-rays and UV-radiation. The last of these is most widely available, easy to use, and can be extremely efficient for the induction of single gene mutations [5]. In this respect the spectrum of lesions achieved probably comes closer to those obtained with chemical agents, than ζ- or x-rays, which may produce a higher incidence of large deletions and other chromosomal abnormalities. The disadvantage of UV-radiation is its low penetration which may restrict its use to single cell cultures, such as protoplasts or pollen grains.

The use of chemical mutagens generally involves a short exposure to relatively high concentrations, followed by thorough washing to remove the mutagen. The extra washing steps may be to the detriment of the more delicate protoplast cultures, again making UV-radiation a more attractive alternative. NEU and NMU have short half lives in aqueous solution leading to protocols [14] which avoid the washing step by using lower concentrations and relying on the breakdown of the mutagen during culture.

The following protocols place an emphasis on the use of freshly isolated mesophyll protoplasts, or large tissue explants, chemical mutagens EMS, NMU or NEU (with or without washing steps), and the use of UV-radiation. The procedures using protoplasts are described for *Nicotiana plumbaginifolia* which is an amenable, widely used, model species. Haploid or diploid shoot cultures can be used as sources of protoplasts, but the former should be employed for total selection for auxotrophs. The same procedures should be applicable to other species provided suitable conditions for the isolation and maintenance of protoplasts are substituted. The leaf strip mutagenesis procedure described has been successfully used to obtain chloroplast mutants of five Solanaceous species: *Nicotiana plumbaginifolia*, *N. tabacum*, *N. sylvestris*, *Lycopersicon peruvianum*, and *Solanum nigrum* [9]. It is described for *S. nigrum* because the greatest frequencies have been achieved with this species, but culture media differences (the only point on which the protocol differs) are indicated for the other species. Again the procedure should be amenable to other species for which efficient regeneration can be achieved from explants. The protocols are restricted to the preparation of the plant material, and the mutagenesis treatment. The downstream handling of cultures, to select mutants, is dealt with in succeeding chapters.

Finally, no consideration is given here to insertion mutagenesis, or "gene tagging". While beyond the scope of this chapter, this approach to mutagenesis is going to be extremely valuable for rapidly identifying genes associated with "loss of function" mutations, something which cannot be accomplished by conventional mutagenesis.

Procedures

Preparation and culture of protoplasts from shoot cultures of N. plumbaginifolia *Viviani*

The procedure is based on that of Maliga [15], and can also be used with shoot cultures of *N. tabacum*.

Steps in the procedure
1. Remove healthy, fully expanded, leaves from 4−6 week old axenic shoot cultures of *N. plumbaginifolia*. Finely slice the leaves with a scalpel and forceps and transfer to enzyme solution in sterile 100 ml Erlenmeyer flasks (15 ml per flask).
2. Incubate at 25 °C, in the dark, overnight (12−18 h). Some improvement in yield can be achieved by continuous slow rotary shaking (ca. 30 rpm), but this is not essential.
3. Swirl the flasks gently and filter through 60 μm nylon bolting cloth to remove partially digested material.
4. Transfer the suspension to sterile screw cap centrifuge tubes and spin at 500 rpm for 3 min.
5. Carefully remove the green surface layer of floating protoplasts, using a sterile Pasteur pipette, and transfer to fresh centrifuge tubes.
6. Fill the tubes with W5 solution, cap, mix thoroughly by gentle inversion, and spin at 500 rpm for 3 min.
7. Remove the supernatant and resuspend the pellet in a small volume (1−2 ml) of K_3 medium, containing 0.4 M glucose as osmoticum.
8. Pool the contents of the centrifuge tubes. Remove a sample and count intact protoplasts using a haemocytometer slide. Add sufficient K_3 medium (0.4 M glucose) to give a density of intact protoplasts of 10^5 ml^{-1}.
9. Mix gently by inversion and transfer to 5 cm Petri dishes (4−5 ml per dish). Seal with parafilm and incubate under low light intensity (ca. 100 lux).
10. After 8−10 days, when protoplasts have gone through 2−3 cell divisions, transfer the contents of each 5 cm dish to a 9 cm Petri dish and add 5 ml K_3 medium (0.4 M glucose). Culture as before.
11. After 7−10 days, remove half the contents of each dish to a fresh dish and add 5 ml K_3 medium (0.3 M glucose) to both.
12. After 7−10 days repeat step 11 using K_3 medium (0.2 M glucose).
13. After 7−10 days, if the colonies are at a high density but still quite small (< 1 mm diameter) repeat step 12.
14. Concentrate protoplast-derived colonies from step 12 or 13 by pooling the dishes and spinning in 50 ml sterile polypropylene centrifuge tubes at 500 × g for 3 min and discarding the supernatant.
15. Plate out colonies by adding to RMOP medium, containing 0.2 M glucose (instead of 2% sucrose) and 6.5 g l^{-1} Difco bactoagar, held molten at 40 °C. Add sufficient colonies to give a final density of 100−200 per dish when plated.

Swirl the flasks of medium plus colonies and pour into 9 cm Petri dishes (15–20 ml per dish).

16. Allow the agar to set, seal the dishes with parafilm and incubate in a culture room at 25 °C, 1000–1500 lux illumination, 16 h day.

Notes

1. Shoot cultures of *N. plumbaginifolia* are obtained by surface sterilising seed (30 s in 70% (v/v) ethanol, followed by 10 min in 20% (v/v) domestic bleach "Domestos", followed by two washes in sterile distilled water) and placing on RM medium in Petri dishes (20 ml per dish). When first true leaves are ca. 0.5 cm long, seedlings can be removed to RM medium in individual containers, to be used as a source of shoot cultures, maintained by transfer of single node cuttings to fresh RM medium every 4–6 weeks.

2. Protoplast yields vary greatly, even using fairly uniform shoot cultures. Typically one shoot culture shoot provides sufficient leaf material for one flask of enzyme solution and four such flasks provide $5-10 \times 10^6$ protoplasts, sufficient for small scale mutagenesis treatments.

10–13. The dilution steps may need to be modified depending on the efficiency of initiation of cell division (and hence the colony density) and the rate of colony growth. These factors vary between preparations and it is difficult to be precise about the interval between dilutions. Progress of cultures must be monitored daily. There should be substantial increase in colony size between dilutions, but any sign of browning of colonies should result in immediate dilution.

Solutions and media

Most solutions and media can be sterilised by autoclaving but those containing enzymes must be filter-sterilised through 45 or 22 μm filters. Final concentrations are given in Table 1. pH of all solutions should be adjusted to 5.6, by dropwise addition of 0.1 M KOH.

Enzyme solution must be freshly prepared. Other solutions and media can be stored for several weeks in the cold room, provided they remain clear, and are sealed to restrict evaporation.

— Enzyme solution: 0.5% Driselase, or alternatively 1% cellulase "Onozuka" R10 plus 0.5% Macerozyme R10 (all w/v), in K_3 medium (Table 1) containing 0.4 M sucrose (Table 2)

— W5 solution [15], containing (per l) 9.0 g NaCl, 1.0 g glucose, 18.4 g $CaCl_2 \cdot 2H_2O$, 0.4 g KCl, pH 5.6

— K_3 medium (Table 1) containing 0.4 M glucose (Table 2)
— K_3 medium (Table 1) containing 0.3 M glucose (Table 2)
— K_3 medium (Table 1) containing 0.2 M glucose (Table 2)

— RMOP medium: RM solution (Table 1) plus 100 mg/l m-inositol, 1 mg/l thiamine-HCl, 1 mg/l BAP, 0.1 mg/l NAA, solidified with 6.5 g/l Difco Bactoagar and containing 0.2 M glucose (Table 2)

— RM medium: RM solution (Table 1), solidified with 6.5 g/l Difco Bactoagar, for maintenance of shoot cultures

Table 1. Basal media for tissue and protoplast culture of *Nicotiana plumbaginifolia*. All components are listed in mg/l final concentration

Medium	K3[a] (15, modified from 16)	RM[b] (17)
NH_4NO_3	240	1650
KNO_3	2400	1900
$CaCl_2 \cdot 2H_2O$	900	440
$MgSO_4 \cdot 7H_2O$	250	370
KH_2PO_4		170
$NaH_2PO_4 \cdot 2H_2O$	120	
$(NH_4)_2SO_4$	130	
$FeSO_4 \cdot 7H_2O$	27.8	27.8
Na_2EDTA	37.3	37.3
H_3BO_3	3.0	6.2
KI	0.75	0.83
$MnSO_4 \cdot 4H_2O$	6.7	22.3
$NaMoO_4 \cdot 2H_2O$	0.24	0.25
$ZnSO_4 \cdot 7H_2O$	2.3	8.6
$CoCl_2 \cdot 6H_2O$	0.025	0.025
$CaSO_4 \cdot 5H_2O$	0.025	0.025
m-inositol	100	100
nicotinic acid	1.0	
pyridoxine · HCl	1.0	
thiamine · HCl	10	
xylose	250	
sucrose	see Table 2	30000
glucose	see Table 2	
BAP[c]	0.2	
NAA[c]	1.0	
2,4-D[c]	0.1	
pH	5.6	5.8

[a] K_3 medium, excluding sugars (sucrose or glucose) and phytohormones[c], can be prepared at $10 \times$ final concentration and stored in suitable aliquots (e.g. 50 or 100 ml) at $-20\,°C$. Medium can be prepared by thawing, diluting, and adding remaining components before adjusting pH.

[b] RM and RMOP media are normally prepared using MS salts (Flow Laboratories) at 4.6 g/l, which supply all the mineral salts listed in Table 1. When preparing from individual salts three stock solutions should be used: macroelements, first 5 in Table 1 (final concentration $\times 4$), $(Na_2EDTA + FeCl_3 \cdot 6H_2O)$ ($\times 200$) and remaining microelements ($\times 100$). Each litre of medium will contain 250 ml, 5 ml, and 10 ml, respectively, of these stocks which can be stored for up to 4 weeks in a cold room.

[c] Abbreviations of phytohormones: 2,4-D = 2,4-dichlorophenoxyacetic acid; NAA = 1-naphthalene acetic acid; BAP = 6-benzylaminopurine. Hormones should be prepared as 1 mg/ml stock solutions, in 0.1 M KOH (2,4-D, NAA) or 0.1 M HCl (BAP), and stored in refrigerator.

Table 2. Concentrations of glucose or sucrose (g/l) to provide different molarities of K3 or RMOP required in the protocol

	Concentration (M)		
	0.4	0.3	0.2
Sucrose	136.8		
Glucose	72.0	54.0	36.0

Mutagenesis with low concentrations of N-ethyl-N-nitrosourea (NEU)

This is an extremely simple technique, taken from Marton *et al.* [14]. They reported an increase in frequency of chlorate-resistant (nitrate reductase-deficient) mutants, from haploid *N. plumbaginifolia* protoplasts, from 5.8×10^{-5} to 1.1×10^{-3} (greater than two orders of magnitude), using 0.3 mM NEU, a concentration giving 52% protoplast survival (compared to control), and sufficiently low to allow the natural decay of the mutagen, without washing. A lower (0.15 mM) concentration of NEU also gives substantial improvements in mutation frequency over controls (about 10 fold), and may lead to the recovery of fertile plants from a greater proportion of the mutants.

Steps in the procedure

1. Isolate *N. plumbaginifolia* protoplasts as described in preceding section.
2. When transferring freshly isolated protoplasts into K_3 medium (0.4 M glucose) for culture, add mutagen stock to give a final concentration of 0.15 or 0.3 mM before plating.
3. Culture mutagenised protoplasts, alongside non-mutagenised controls, as described in preceding section.

Notes

1. Safe handling of mutagen is paramount in all these procedures. It is important to avoid skin contact, and protective apron and gloves should be used, in addition to a respirator. Wash down all work areas after use, in case of spillage. Spillage in the laminar flow can be retained by working on absorbent paper backed with aluminium foil. Exclude other workers from the area while manipulations using mutagens are in progress, and use a vertical flow, or containment unit, if available.
2. Wash hands after use, and immediately wash any area of skin where contact with mutagen is suspected. Use gentle washing motions (avoiding excess rubbing), and soapy water.
3. Contaminated paper should be incinerated. Glassware, and other materials as well as leftover mutagen solutions, should be decontaminated overnight with 5% NaOH in the fume hood, and given a second soaking (longer than 1 h) with 5% NaOH before washing in tap water, followed by distilled water.
4. Concentrations of NEU may need to be modified if a species other than *N. plumbaginifolia* is used. A concentration giving about a 50% reduction in plating efficiency should be effective.

Solutions

— Mutagen solution: 10 mM (1.17 mg/ml, adjusted to take account of acetic acid added by manufacturer as a stabiliser) NEU (Sigma), in K_3 medium (Table 1) with 0.4 M glucose.

Mutagenesis with high concentrations of EMS, NEU, or NMU

Alkylating agents are most effective when cells are undergoing DNA replication, and mutagenesis of cell suspension cultures is generally carried out with cultures in the exponential phase of growth. Efficient mutagenesis can also be achieved with freshly isolated mesophyll protoplasts (believed to be in G1) however, with the attraction that the risk of generating chimeric colonies is minimised. This is the protocol described here. More efficient mutagenesis may be achieved by culturing the proto-plasts for 36—48 h in K_3 (0.4 M glucose) before re-collecting and using the same protocol. This should enrich the population in S-phase cells, but to really optimise this will require additional series of experiments, particularly if a species other than *N. plumbaginifolia* is used.

The following protocol is a general one, applicable to several different alkylating agents, and includes a preliminary test to determine effective concentrations of the mutagen.

Steps in the procedure

1. Isolate mesophyll protoplasts from shoot cultures of *N. plumbaginifolia*, as described previously.
2. When transferring freshly isolated protoplasts to K_3 medium (0.4 M glucose) add mutagens at a range of concentrations from filter-sterilised stock solutions (except EMS — see Note 1) prepared in the same medium. As a rough guide suitable concentration ranges would be: 0.1—3% v/v EMS, or 0.3—10 mM (NEU or NMU). The effective concentration may be influenced by cell density which should be standardised at 10^5 protoplasts/ml.
3. After 60 min incubation under gently rotary shaking (ca. 30 rpm), sediment the protoplasts by centrifugation at 500 rpm for 3 min. Carefully remove the mutagen solution with a sterile Pasteur pipette, and resuspend the protoplasts in K_3 (0.4 M glucose) to a final density of 10^5/ml and plate in 5 cm Petri dishes. Seal and incubate under low light intensity (ca. 100 lux) for 8 days.
4. Score the cultures for percentage cell division and determine the effect of each mutagen concentration on division frequency. Select a mutagen concentration giving 10—50% division compared to the non-mutagenised control (taken as 100%).
5. Repeat steps 1—3 using the chosen mutagen treatment only, and continue culturing the protoplasts as described previously.

Notes

1. EMS is a volatile liquid, hence the recommended concentration range is given as % (v/v). The bottle should be opened in a fume hood and the required amount added to K_3 medium by automatic pipette, using autoclaved disposable tips.

   Solutions contaminated with EMS should be inactivated by gradual addition to a large excess of 3 M KOH in 95% ethanol, heated under reflux. The mixture should be refluxed and stirred for 2 hours before cooling, diluting with tap water, and disposal down the drain, chased by a large volume of tap water.

The safe disposal of NEU and NMU, and general comments on the safe handling of these mutagens, are provided in notes 1–3 of the preceding section.

2. The arbitrary duration of mutagen treatment (60 min) works well in most cases but may need to be modified downward (eg. to 30 min) if dividing cells are used. Longer (90–120 min) treatments can also be used for NMU and NEU.

3. EMS is poorly miscible with water. It is therefore a better mutagen for use with cell suspension cultures, than protoplasts, where more vigorous agitation (100 + rpm) can be used to ensure proper mixing.

Solutions

– All solutions required for protoplast isolation and culture are described in the first section and Table 1.

Leaf strip mutagenesis of Solanum nigrum *to obtain chloroplast-encoded antibiotic resistant mutants*

This simple procedure is based on the use of NMU as an efficient plastome-targeted mutagen. It is particularly effective with *S. nigrum*, but also gives good results with *Lycopersicon peruvianum, Nicotiana tabacum, N. sylvestris,* and *N. plumbaginifolia* [9, 18]. Attempts to adapt it to *Brassica* species have so far failed (O'Neill and Dix, unpublished).

Steps in the procedure

1. Remove leaves from axenic shoot cultures, and cut into strips (5–15 mm × 2–3 mm). Add 200 strips to 100 ml mutagen solution in a 250 ml Erlenmeyer flask.
2. Incubate on a rotary shaker (ca. 50 rpm).
3. Decant the mutagen solution, and wash the leaf strips four times with 100 ml RM solution, or sterile distilled water, pH adjusted to 5.6.
4. Transfer leaf strips, lower surface downwards to the surface of selective medium, 5 strips per 9 cm Petri dish.
5. Seal the dishes with parafilm and incubate in a culture room (1,500 lux, 16 h photoperiod) until green (resistant) adventitious shoots appear on bleached leaf strips (40–60 days).

Notes
1. For safe handling and disposal of NMU, and NMU–contaminated materials see notes for preceding sections.
2. The same procedure can be used for *Lycopersicon peruvianum* and *Nicotiana* species, but in the latter case a better selective medium is RMB:RM (Table 1) plus 100 mg/l m-inositol, 1 mg/l thiamine-HCl, 1 mg/l BAP, and selective levels of the antibiotics.

Solutions

— RM solution (Table 1)
— Mutagen solution: RM solution containing 5 mM NMU from a freshly prepared stock solution (80 mg NMU in 20 ml RM). For 100 ml mutagen solution 12.9 ml stock solution is added to 87.1 ml RM solution
— Selective medium: RM solution (Table 1) with sucrose reduced to 20 g/l, and the addition of (per l) 100 mg meso-inositol, 1 mg thiamine-HCl, 0.5 mg nicotinic acid, 0.5 mg pyridoxine-HCl, pH 5.6, solidified with 0.65% (w/v) Difco Bactoagar, plus selective levels of antibiotics added from filter-sterilised stock solutions (100× final concentration, prepared in distilled water), to autoclaved, molten medium, prior to pouring. Selective levels are as follows: Streptomycin sulphate: 500 mg/l, spectinomycin: 50 or 100 mg/l, or lincomycin hydrochloride: 100 mg/l

UV mutagenesis of N. plumbaginifolia *protoplasts*

This protocol closely follows the procedure developed by Grandbastien *et al.* [5] for tobacco protoplasts. The protocol describes the determination of a suitable UV dose, based on reduction in colony-forming ability, which can subsequently be used for mutagenesis and selection. This dosage is likely to be in the order of 1,000 ergs/mm^2.

Steps in the protocol

1. Isolate *N. plumbaginifolia* protoplasts and culture at 10^5/ml in K3 (0.4 M glucose) medium.
2. After 24 h, expose the protoplasts to UV doses in the range 200—2000 ergs/mm^2. Place dishes under the UV source and remove the lids before turning the source on.
3. Culture the treated protoplasts in total darkness for 48 h.
4. Carry out all remaining culture steps as before, taking care to use uniform dilution steps for all treatments.
5. 4 weeks after plating protoplast-derived colonies in solid medium, count the growing colonies and determine the effect of the mutagen treatment on the colony-forming ability of protoplasts.
6. Select a UV-dose giving 10—50% colony forming of non-mutagenised controls, for subsequent mutagenesis treatments carried out as above.

Notes

1. Safe handling of UV is straightforward. Safety glasses should be worn and direct exposure of the skin should be avoided.
2. If unable to calibrate the UV source, suitable treatments can still be determined by varying either the distance from the lamp or the period of exposure, and determining the effect on colony-forming ability.
3. Effect of UV dose on protoplasts can be determined at an earlier stage of culture, while protoplast-colonies are still in liquid K_3 medium, but care should be taken to exclude residual (1—2) cell divisions, of many UV-treated protoplasts, which are unable to go on to form colonies.
4. The same procedure can be used with both haploid and diploid protoplasts.

References

1. Maliga P, Breznovits A, Marton L (1973) Streptomycin resistant plants from callus cultures of tobacco. Nature New Biol 244: 29–30.
2. Maliga P, Marton L, Breznovits A (1973) 5-Bromodeoxyuridine-resistant cell lines from haploid tobacco. Plant Sci Lett 1: 375–383.
3. Dix PJ (Ed.) Plant Cell Line Selection. VCH, Weinheim, 1990.
4. Bayliss MW (1980) Chromosomal variation in plant tissues in culture. Int Rev Cytol Suppl 11A: 113–114.
5. Grandbastien MA, Bourgin JP, Caboche M (1985) Valine-resistance a potential marker in plant cell genetics. 11. Optimization of UV mutagenesis and selection of valine resistant colonies derived from tobacco mesophyll protoplasts. Genetics 109: 409–425.
6. Negrutiu I, Jacobs M, Caboche M (1984) Advances in somatic cell genetics of higher plants – the protoplast approach in basic studies on mutagenesis and isolation of biochemical mutants. Theor Appl Genet 67: 289–304.
7. Sidorov V, Menczel L, Maliga P (1981) Isoleucine-requiring *Nicotiana* plant deficient in threonine deaminase. Nature 294: 87–88.
8. Gebhardt C, Schnebli V, King PJ (1981) Isolation of biochemical mutants using haploid mesophyll protoplasts of *Hyoscyamus muticus*. 11 Auxotrophic and temperature sensitive clones. Planta 153: 81–89.
9. McCabe PF, Timmons AM, Dix PJ (1989) A simple procedure for the isolation of streptomycin resistant plants in *Solanaceae*. Mol Gen Genet 216: 132–137.
10. Freytag AH, Wrather JA, Erichsen AW (1990) Salt tolerant sugarbeet progeny from tissue cultures challenged with multiple salts. Plant Cell Reports 8: 647–650.
11. Negrutiu I (1990) *In vitro* mutagenesis. In: Dix PJ (Ed.) Plant Cell Line Selection, pp 19–38, VCH, Weinheim.
12. Fluhr R, Aviv D, Galun E, Edelman M (1985) Efficient induction and selection of chloroplast encoded antibiotic resistant mutants in *Nicotiana*. Proc Natl Acad Sci USA 82: 1485–1489.
13. Hosticka LP, Hanson MR (1984) Induction of plastid mutations in tomatoes by nitrosomethylurea. J Hered 75: 242–246.
14. Marton L, Dung TM, Mendel RR, Maliga P (1982) Nitrate reductase deficient cell lines from haploid protoplast cultures of *Nicotiana plumbaginifolia*. Mol Gen Genet 182: 301–304.
15. Maliga P (1984) Cell culture procedures for mutant selection and characterization in *Nicotiana plumbaginifolia*. In: Vasil IK (Ed.) Cell Culture and Somatic Cell Genetics of Plants, Vol 1, pp 552–562, Academic Press, Orlando.
16. Kao KN, Constabel NF, Michayluk MR, Gamborg OL (1974) Plant protoplast fusion and growth of intergeneric hybrids. Planta 120: 215–227.
17. Murashige T, Skoog F (1962) A revised medium for rapid growth and bioassay with tobacco tissue cultures. Physiol Plant 15: 473–497.
18. Timmons AM, Dix PJ (1991) Influence of ploidy on plastome mutagenesis in *Nicotiana*. Mol Gen Genet 227: 330–333.

Plant Tissue Culture Manual **F2**, 1–43, 1996.
© 1996 *Kluwer Academic Publishers.*

In vitro culture, mutant selection, genetic analysis and transformation of *Physcomitrella patens*

DAVID COVE
Department of Genetics, University of Leeds, Leeds LS26 8JF, UK

Introduction

The suitability of mosses for the study of plant development and genetics has been appreciated for a considerable time [19]. Early studies extended to a number of species, including *Funaria hygrometrica* and *Physcomitrella patens*. More recent studies, using both classical and molecular genetic analysis to study development, have concentrated principally on *Physcomitrella* and most of the techniques described in this chapter have been developed for this species. However, some of these methods are suitable, with little modification, to other moss species, in particular to *Ceratadon purpureus*.

The main phase of the life cycle of mosses is the haploid gametophyte. Spore germination or tissue regeneration gives rise to a system of cell filaments, the protonema, from which are produced gametophores, the leafy shoots which comprise the more familiar part of most mosses. Gametes are produced on the gametophores. *Physcomitrella* is monoecious and both male and female gametes are produced on the same gametophore. Gamete fusion leads to the production of the diploid sporophyte which in turn produces haploid spores following meiosis. Conventional genetic analysis is possible but many developmentally-abnormal mutant strains are infertile and so must be analysed by parasexual methods, using somatic hybrids obtained by protoplast fusion.

The protonemal phase of the *Physcomitrella* life cycle, comprises two types of cell, chloronema and caulonema. Filaments of both cell types extend by the serial division of the apical cell. Sub-apical cells of both chloronemal and caulonemal filaments can also divide, usually no more than twice, to produce further filaments. Chloronemal filaments develop following spore germination or as a result of regeneration from all types of tissue. The apical cells of chloronemal filaments, which have a cell cycle of about 24 h under standard conditions (see below), may after a few days of growth, give rise to the second cell type, caulonema. The apical cells of caulonemal filaments, under comparable conditions, have a cell cycle time of only about six hours, and so comprise the adventitious phase of protonemata. Branching of the sub-apical cells of caulonemal filaments can give rise to three types of side branches. Most commonly, chloronemal filaments are produced. Less commonly, side branches may develop into either further caulonemal filaments, or gametophores.

PIPS 99928

In vitro culture

The regenerative capacity of moss tissue, allows vegetative tissue to be propagated without difficulty in a number of ways.

Growth on agar medium

Agar medium in Petri dishes may be inoculated either with spores or with somatic tissue. For routine sub-culture using somatic tissue, it is best to use protonemal tissue from a vigorously growing culture. Chloronemal tissue, the growth of which is enhanced when ammonium is provided as nitrogen source, is easiest to sub-culture. A fragment of tissue 1 to 2 mm in diameter is sufficient to establish a new culture. Tissue other than chloronemata, *e.g.* leaf cells, may take a long time to regenerate. For growth in tubes or jars, it is important not to use lids which seal tightly.

If contamination is a problem, Petri dishes may be taped with "Micropore" medical plaster strip, without affecting the growth or development of cultures. This is a porous material that slows evaporation from the dish slightly and reduces contamination, principally because it limits air exchange during handling. Sealing cultures with paraffin film slows growth and prevents rapid regeneration. It is however satisfactory for long-term cultures which have been started with a protonemal inoculum (see below).

Growth on agar medium overlaid with cellophane

Protonemal tissue, free of agar, may be obtained by inoculating protonemal fragments onto cellophane overlaying solid medium in a Petri dish. Petri dishes containing solid minimal medium + 5mM di-ammonium tartrate are overlaid with sterilized cellophane discs (type 325P, Cannings, Avonmouth, Bristol, U.K.; the specification of the cellophane appears to be critical, but other sources may be suitable). Suspensions of protonemal fragments are prepared by blending protonemal tissue in sterile distilled water. The exact procedure will depend on the type of blender used. 500 mg fresh weight of tissue is suspended in 20 ml of water and blended until the tissue is cut into fragments containing approximately 20 to 100 cells. About 2 ml of suspension is required to inoculate a 90 mm Petri dish. After 7 days incubation in standard conditions, about 500 mg (fresh weight) of vigorously-growing protonemal tissue is obtained from a wild-type culture on a 90 mm Petri dish. Tissue may be harvested using a sterile spatula.

Growth in liquid medium

Physcomitrella can be grown in liquid culture either in shaken flasks or in a fermenter. For shaken flask culture, a tissue inoculum, prepared as for the inoculation of cellophane-overlay plates (see above), may be used to inoculate liquid minimal medium. Vigorous agitation is not necessary for growth, but growth rates in shaken liquid cultures are not as great as those obtained on Petri dishes or in a fermenter supplied with CO_2-enriched air. Full details of fermenter culture are given in Boyd *et al* [6]. The maintenance of fermenter cultures, free of contamination is difficult and the production of tissue in quantities up to about 100 g is usually easier using cellophane-overlay plates.

Temperature

Physcomitrella grows on solid media in temperatures up to about 28 °C, possibly somewhat higher in liquid medium. Little difference in growth rate is observed in the temperature range 20 °C to 26 °C. Temperatures between 24 °C and 26 °C are used for routine culture. Growth is slower but still satisfactory at 15 °C and this has been used as the permissive temperature when temperature-sensitive mutants have been sought. Because of the radiant heat from the light source it is usually necessary to keep the air temperature below the desired culture temperature. The temperatures given are those of the medium on/in which cultures are grown.

Light

For routine culture, the exact quality of light provided is not critical. Most studies to date have used continuous light from fluorescent tubes at an intensity of between 5 and 20 W m^{-2}. Some laboratories use intermittent light, the most commonly used cycle being 16 h light + 8 h dark. Intermittent light results in slower development. No developmental effects of intermittent light have yet been reported, although it may allow synchronization of the cell cycle.

Media

The medium which has been used most commonly is ABC medium. However, this medium, when prepared in the way described here, has a heavy precipitate and so, notwithstanding its high calcium content, is probably variable in the amount of calcium, and possibly of other nutrients, that are available. There has therefore recently been agreement among many *Physcomitrella* workers to try a modified ABC medium, BCE the recipe for which is also given below.

"Standard" conditions

Throughout this chapter, standard conditions refer to growth at 25 °C, in continuous white light at intensities greater than 5 W m^{-2}.

Procedures

Culture maintenance

Cultures can be kept in a healthy state for a considerable time (at least 6 months) after they have developed to a required stage by sealing the Petri dish or test tube with paraffin film. Storage is prolonged further by keeping the sealed cultures at 15 °C. For very long term storage, moss strains can be stored in liquid nitrogen. Recovery from cryo-preservation is not reliable and it is advisable to preserve a number of replica cultures. Some developmentally-abnormal mutants appear to be less tolerant of cryopreservation than developmentally-normal strains.

Cryopreservation [12]

1. Tissue of the strains to be preserved, is grown for at least 7 days under standard conditions on 90 mm cellophane-overlay plates of appropriately-supplemented minimal medium plus ammonium.
2. The tissue is transferred on the cellophane, to fresh appropriately-supplemented solid minimal medium plus ammonium plus 500 mM mannitol.
3. 1 ml of liquid medium plus ammonium plus 500 mM mannitol is added to the surface of the tissue.
4. The culture is incubated for a further seven days under standard conditions.
5. 2 ml of 5% v/v DMSO + 10% w/v glucose is added to a series of sterile Eppendorf tubes.
6. Tissue (about 1/10th of plate per tube) is added to each tube.
7. Incubate at 20 °C for 1 h.
8. Freeze the tubes at a rate of 1 °C per min to −35 °C, then place into liquid nitrogen for storage.

To retrieve cryopreserved tissue, tubes are thawed at room temperature, and the contents of each tube added to 10 ml of sterile distilled water and left for 30 min. The tissue is then inoculated onto appropriate solid medium.

Protoplast isolation and regeneration [9, 10, 11]

Protoplast isolation and regeneration are easy to achieve in *Physcomitrella*, and protoplasts therefore provide a way of establishing a clone from a single cell, a procedure that is often crucial for both transformation and somatic mutagenesis. Protoplasts can also be fused and so provide a method of somatic hybridization.

Protoplasts are isolated from young protonemal tissue obtained from cellophane-overlay plates.

1. Protonemal tissue is added to a sterile aqueous solution of Driselase in 8% D-mannitol (for preparation, see below), at about the rate of 100 mg of tissue (fresh weight) per ml of Driselase solution.
2. Incubate for 20 to 30 min at 25 °C with occasional gentle shaking.
3. Remove undigested tissue, by filtering sterilely through a stainless steel or nylon mesh (pore size: approx. 100 μm × 100 μm).
4. Sediment protoplasts by gentle centrifugation (100 to 200 × g for 3 min).
5. Remove supernatant carefully, and resuspend the protoplasts in about the same volume of sterile 8% (w/v) D-mannitol. Sediment as in step 4.
6. Repeat step 5 once more and finally resuspend the washed protoplast and adjust to a concentration appropriate to the procedure involved. About 10^6 viable protoplasts are obtained from 1 g fresh weight of tissue.

Protoplasts are usually regenerated embedded in soft agar medium overlaying cellophane, in turn overlaying osmotically-buffered medium. Once regenerated, the cellophane and protoplast can be transferred to medium without mannitol, where development is more rapid.

7. 1 volume of protoplast suspended in 8% D-mannitol solution is added gently to 10 volumes of molten PRT medium (see below) which has been maintained at 42 °C.
8. The mixture is pipetted or poured gently but quickly, onto cellophane overlaying PRB medium. 1 ml of top layer covers a 90 mm Petri dish, 400 μl covers a 50 mm dish.
9. After incubation under standard conditions for 3 to 4 days, the protoplasts will have regenerated cell walls and may be transferred to fresh medium without mannitol. Regeneration of *Physcomitrella* protoplasts requires a high light intensity (> 5 Wm^{-2}).

Mutagenesis

N-methyl-N'-nitro-N-nitrosoguanidine (NTG) mutagenesis

N-methyl-N'-nitro-N-nitrosoguanidine is a very potent mutagen and should only be used by those who have been trained in mutagen handling. Since it is a powder, extreme caution must be taken when weighing this compound and local safety rules for mutagen handling must be observed.

Using spores [1]

1. Prepare 10 ml of a spore suspension containing about 10^5 spores per ml in sterile Tris-maleate buffer (pH 6).
2. Dissolve 1 mg of NTG in a separate 10 ml of Tris-maleate buffer (pH 6) and incubated for 30 min at 25 °C.
3. Mix the spore suspension with the NTG solution.
4. Incubate for 30 min at 25 °C, shaking gently from time to time.
5. Centrifuge at 2500 × *g* for 5 min Discard the supernatant.
6. Resuspend the spores in 20 ml distilled water.
7. Repeat steps 5 and 6 twice more.

The spores are now ready for plating. About 10% of the originally viable spores survive this treatment. The mutagenic treatment results in an initial delay in development but once germinated, the sporelings grow as usual.

Using somatic tissue [5]

The method used for somatic mutagenesis is identical to that employed with spores except that the NTG solution is added to 1 g (fresh weight) of young protonemata obtained by growth on cellophane-overlay plates which has been pre-incubated for 30 min at 25 °C in 10 ml of Tris-maleate buffer. The mixture is then incubated for a further 60 min at 25 °C with gentle agitation. The treatment is terminated by filtering the mixture sterilely using a stainless steel or nylon mesh (pore size: approximately 100 μm $\times$ 100 μm) which retains the protonemal tissue and by washing thoroughly, with sterile, distilled water.

This procedure results in a cell survival rate of about 10%. If individual cells are required these may be obtained by protoplasting the tissue using the method described above.

UV mutagenesis

Although not as an effective mutagen as NTG, UV irradiation is less hazardous if used with the correct safety precautions. The mutagenic UV source will need to be standardized before use. Most studies aim for spore/cell survival around 5%.

Using spores

20 ml of a spore suspension in water is irradiated in an open 90 mm Petri dish. The spore suspension must be kept in darkness for 24 h following UV irradiation to limit repair.

Using somatic tissue

Seven-day old tissue growing on cellophane-overlay plates can be irradiated directly with UV. Tissue must be held in darkness for 24 h following UV irradiation to limit repair. If individual cells are required, the tissue can be protoplasted following the dark incubation.

Mutant isolation from either spores or protoplasts

Auxotrophic mutants [1, 5]

No satisfactory selective procedure for obtaining auxotrophic mutants is yet available. All such mutants have been obtained so far by non-selective isolation. This procedure involves culturing protonemata derived from mutagenized spores or protoplasts, initially on medium containing the substances which are required for the growth of the classes of nutritionally-deficient mutant which are being sought. After 1–2 weeks' growth on this medium, the protonemata can be tested for auxotrophies by transferring half of each culture on to supplemented medium and the other half on to minimal medium. 25 protonemata can be tested in this way in pairs of 9 cm petri dishes. In order to isolate a particular kind of auxotroph, it may be necessary to test several thousands of somatic clones, each of which has grown from a single mutagenized spore or protoplast. Analogue-resistant strains could also be obtained using total isolation but it is easier to isolate them selectively by including the analogue in the growth medium either from the outset, as for example for resistance to D-serine or p-fluorophenylalanine, or as soon as mutagen-treated spores/protoplasts have germinated/begun to regenerate, as for resistance to 8-azaguanine.

Isolation of developmentally-abnormal strains [2, 3, 8]

The most straightforward way to isolate a wide range of morphologically-altered mutants is to inoculate mutagenized spores or protoplasts at a density of about 100 survivors per 9 cm petri dish and incubate them for 3 to 4 weeks under standard conditions. At this time, abnormal strains blocked or altered in various stages of gametophytic development will be readily observable and can be saved for further study by subculturing on to fresh medium.

Procedures for genetic analysis

Sexual crossing [1, 2, 4]

Gamete production in *Physcomitrella* requires a temperature below 18 °C. Since *Physcomitrella* is monoecious, it is self-fertile but crossing can be ensured by the inclusion of auxotrophies, *eg* requirements for nicotinic acid or p-amino benzoic acid, in the strains to be crossed. Such auxotrophic strains are cross-fertile, but self-sterile on normally-supplemented medium [7]. The two strains to be crossed are inoculated near to each other, on minimal medium. After about 3 weeks growth in standard conditions, the culture is transferred to a temperature of 15 °C to 18 °C. After a further 3 weeks, sterile water is added to the culture to allow the male gametes to swim to the female gametes and effect fertilization. Sufficient water is added to make the culture thoroughly damp but not submerged. Sporophytes should be visible within a further two weeks, but if none are formed, further irrigation may be necessary. The timings given here are relaxed, and successful crosses have been made using shorter intervals. It is probable that the protocol for crossing could be improved.

Somatic hybridization [4, 8, 9]

Many developmentally-abnormal strains are sexually sterile, but can be analysed genetically by the use of somatic hybrids, which can be used to test for dominance of mutant phenotypes and for complementation between two phenotypically-similar developmental mutants [11]. Somatic hybrids involving recessive characters are fertile and can be used for further genetic analysis.

   The method using polyethyleneglycol (PEG) -induced protoplast fusion, to obtain somatic hybrids, first described by Grimsley et al [9, 10], is still used widely. Electrofusion of protoplast has also been used [18]. Either method relies on the use of complementing auxotrophies in the strains to be hybridized to allow the selection of the hybrid.

Transformation

Transformation of *Physcomitrella* can be carried out using plasmid DNA, containing a gene coding for resistance to either G418, or hygromycin or sulfadiazine. Selection for transformants is made using G418 at 50 μg/ml, hygromycin at 30 μg/ml and sulfadiazine at 150 μg/ml. Following transformation three classes of antibiotic-resistant regenerant are obtained:

transient: Do not retain resistance upon sub-culture.

unstable: Grow slowly on selective medium. Resistance lost when selection is relaxed. Probably not transmitted through meiosis.

stable: Grow on selective medium almost as fast as on non-selective medium. Resistance retained when selection is absent. Transmitted in a regular Mendelian manner through meiosis.

Transformation procedures are still being refined, but pBR-derived vectors appear to be most reliable, while the physical state of the DNA does not seem to be a significant factor influencing transformation rates. Two methods for transformation are used routinely. PEG-mediated DNA uptake by protoplasts is reliable and requires no special apparatus but micro-projectile bombardment gives higher transformation rates and this method can probably be improved further.

Transformation using PEG-mediated DNA uptake by protoplasts [17]

1. Protoplasts are prepared as described above.
2. After the second wash, estimate protoplast density using a haemocytometer. Centrifuge (100 to 200 × g for 3 min) and resuspend in sufficient D-mannitol/MgCl$_2$/MES solution to give a final protoplast density of 1.6×10^6/ml.
3. Meanwhile, prepare DNA to be used in transformation by dispensing 20 to 50 μg of DNA dissolved in no more than 30 μl TE, into sterile 10 ml tubes. Centrifuge gently to bring DNA solution to bottom of tube.
4. Add 300 μl protoplast suspension from step 2 to DNA from step 3.
5. Add 300 μl PEG/T solution.
6. Heat for 5 min at 45 °C. Return to room temperature (20 °C) for 10 min.
7. Add 1 ml 8% D-mannitol solution. Invert gently to mix. Wait 1 min.
8. Add 2 ml 8% D-mannitol solution. Invert gently to mix. Wait 1 min.
9. Add 7 ml 8% D-mannitol solution. Invert gently to mix. Wait 1 min.
10. Centrifuge (100–200 × g for 3 min). Remove most of supernatant, retaining about 500 μl, gently resuspend protoplast in the residual supernatant.
11. Add protoplasts to 10 ml molten PRT medium (at 42 °C). Pour immediately onto 3 Petri dishes containing PRB medium overlaid with cellophane.
12. Incubate in standard conditions for 3 days
13. Transfer regenerating protoplasts on top layer to appropriate selective medium.

Transformation using micro-projectile bombardment

This method was developed using a gun to the design of Lonsdale *et al.* [14], which uses polycarbonate macroprojectiles and 0.22 inch calibre blank launcher cartridges. This design gives best results with a distance of 150 mm between the stopper plate and the tissue, and with two discharges per petri dish of tissue (the position of the dish is moved between discharges).

1. Suspend 50 mg of dry tungsten powder (M17 grade, Sylvania, Towanda, PA 18848, USA) in 300 μl of absolute ethanol, mix vigorously, and centrifuge at 10,000 $\times$ g for 5 min. Withdraw the ethanol carefully.

2. Wash the tungsten by the addition of 1.5 ml of sterile distilled water. Resuspend and centrifuge. Repeat twice more and finally resuspend in 1 ml of 50% (w/v) sterile glycerol.

3. To 25 μl of the tungsten suspension add in turn:
   5 μl of a solution of 1 μg plasmid DNA/μl TE
   25 μl sterile 2.5 M $CaCl_2$ solution
   10 μl sterile 0.1 M spermidine (free base) solution
   Mix gently and allow to stand for 10 min. Carefully withdraw 35 μl of the supernatant and discard. Transfer to ice until used for bombardment.

4. The addition of $CaCl_2$ and spermidine causes DNA-tungsten complexes to from. These must be dispersed immediately before use. Flicking the tube with the finger is effective. Immediately after dispersal, place 3 μl of DNA-tungsten complex on the macroprojectile (or as appropriate for the method of micro-projectile bombardment used).

5. Bombard 6 to 7 day old tissue grown on cellophane-overlay plates containing minimal medium with ammonium as nitrogen source. Remove the lid from the Petri dish and cover with sterile stainless steel mesh (aperture size 1 mm $\times$ 1 mm). Evacuate the chamber to 28 mbar before discharge.

6. Incubate the tissue under standard conditions for 48 h.

7. After 48 h incubation, the tissue may be transferred on the cellophane directly to selective medium, or may instead be harvested and either blended and plated onto selective medium (as described above for the preparation of cultures on cellophane-overlay plates), or, if single-cell clones are required, protoplasted before plating onto osmotically-buffered selective medium.

Media, supplements and solutions [13]

Unless stated otherwise, sterilization is by autoclaving at 121 °C for 20 min.
Analytical grade inorganic chemicals should be used where possible.

Stock solutions for growth media

These are best dispensed into convenient aliquots and stored frozen. Under these
conditions there is no need to sterilize these solutions before use.

Solution A

$Ca(NO_3)_2.4H_2O$	118 g
$FeSO_4.7H_2O$	1.25 g
distilled H_2O	to 1 l

Solution B

$MgSO_4.7H_2O$	25 g
(**or** *anhydrous* $MgSO_4$	12 g)
distilled H_2O	to 1 l

Solution C

KH_2PO_4	25 g
distilled H_2O	500 ml

Adjust pH to 6.5 with minimal volume of 4 N KOH and make up to 1 l with
additional distilled H_2O.

Solution E

KNO_3	101 g
Ferric citrate.H_2O	263 mg
distilled H_2O	to 1 l

Hoagland's A-Z trace element solution (TES):

H_3BO_3	614 mg	$MnCl_2.4H_2O$	389 mg
$Al_2(SO_4)_3.K_2SO_4.24H_2O$	55 mg	$CoCl_2.6H_2O$	55 mg
$CuSO_4.5H_2O$	55 mg	$ZnSO_4.7H_2O$	55 mg
KBr	28 mg	KI	28 mg
LiCl	28 mg	$SnCl_2.2H_2O$	28 mg
		distilled H_2O	to 1 l

Hoagland's TES is widely used but the exact composition of the TES is probably not important.

Growth supplements

substance	concentration in the medium	weight per litre
adenine	500 μM	67.5 mg
p-aminobenzoic acid	1.8 μM	247 μg
di-ammonium (+) tartrate	5 mM	920 mg
nicotinic acid	8 μM	1 mg
D-sucrose	15 mM	5 g
thiamine HCl	1.5 μM	0.5 mg

All the above supplements, with the exception of adenine, may be kept as aqueous $100 \times$ concentrated stock solutions, sterilized by autoclaving, and added to growth media as required to give the concentrations listed above. Adenine is best added as a solid, and medium containing adenine may be autoclaved.

Growth media

Agar

It is probable that any high grade agar, such as Sigma High Gel Strength Agar (cat. # A9799), can be used to gel moss media. The quantities in the recipes below refer to Sigma Agar #A9799, and may need to be altered for other agars. Either heat media containing agar, to dissolve agar before aliquoting and autoclaving, or add the appropriate amount of agar to each individual aliquot, autoclave and disperse/dissolve the agar before medium solidifies.

ABC minimal medium

This is a modified Knop's medium that contains a high level of calcium and which precipitates on autoclaving. Although ABC medium has been used in many past studies, BCE medium may be give more reproducible results.

solution A	10 ml
solution B	10 ml
solution C	10 ml
TES (either)	1 ml
(agar	8g)
distilled H$_2$O	to 1 l

The pH of the autoclaved medium will be between 5.3 and 5.9, the actual value depending upon the type of agar used, and is not usually adjusted.

BCE minimal medium

This medium contains no calcium, which must be added (as $CaCl_2$) as required. 1mM calcium is now being assessed for routine use, but note that protoplast regeneration requires at least 5mM calcium.

solution B	10 ml
solution C	10 ml
solution E	10 ml
TES (either)	1 ml
(agar	8 g)
distilled H_2O	to 1 l

PRB (protoplast regeneration medium, bottom layer)

liquid BCE medium	1 l
$CaCl_2.6H_2O$	2.19 g (= 10mM)
D-mannitol	60 g
di-ammonium(+)tartrate	920 mg (=5mM)
agar	8 g

PRT (protoplast regeneration medium, top layer)

liquid BCE medium	1 l
$CaCl_2.6H_2O$	2.19 g (= 10mM)
D-mannitol	80 g
di-ammonium(+)tartrate	920 mg (=5mM)
agar	8 g

Other solutions.

1M $Ca(NO_3)_2$

$Ca(NO_3)_2.4\,H_2O$	236.1 g
distilled water	1 l

Sterilize by autoclaving. Store at 4 °C.

Driselase solution

Driselase	1 to 2 g (depending on batch)
D-mannitol	8 g
distilled H_2O	to 100 ml

1. Stir to mix but do not shake vigorously.
2. Leave to stand at room temperature for 15 min.

3. Centrifuge at 2500 × *g* for 5 min.
4. Remove the clear supernatant and filter sterilize.

D-mannitol/Ca(NO₃)₂ solution

8% (w/v) D-mannitol solution	9 ml
1M $Ca(NO_3)_2$ solution	1 ml
1M Tris buffer, pH 8.0	100 ml

Make up fresh, on day of use. Filter sterilize

D-mannitol/MgCl₂/MES solution

D-mannitol	9.1 g
distilled water	8.85 ml

Sterilize by autoclaving and store at room temperature. On day of use, add:

1M $MgCl_2$ (203.3 g $MgCl_2.6H_2O$/ l)	150 μl
1% MES pH5.6 solution	1ml

Filter sterilize

1% MES pH 5.6

Use 1% (w/v) 2-[N-morpholino]ethanesulphonic acid in distilled water. Adjust to pH 5.6 with 0.1M KOH. Sterilize by autoclaving. Store at 4 °C.

PEG/T (PEG solution for transformation)

1. Autoclave 2 g PEG 6000 in a glass container.
2. On day of transformation, melt PEG in microwave.
3. Add 5 ml D-mannitol/Ca(NO₃)₂ solution and mix well.
4. Leave at room temperature for 2 to 3 h before use

TE

Tris-(hydroxymethyl)-aminomethane	1.21 g
diaminoethanetetra-acetic acid	372 mg
distilled water	1 l

Adjust to pH 8.0 with 0.1M HCl. Sterilize and store at 4 °C.

1M Tris buffer (pH 8.0)

Tris-(hydroxymethyl)-aminomethane	121.1 g
distilled water	1 l

Adjust to pH 8.0 with 0.1M HCl. Sterilize and store at 4 °C.

Tris-maleate buffer (pH 6)

Tris-(hydroxymethyl)-aminomethane	6 g
maleic acid	6 g
distilled H$_2$O	to 1 l

Adjust pH to 6.0 with 10M NaOH or KOH. Store at 4 °C.

References

1. Ashton NW, Cove DJ (1977) The isolation and preliminary characterisation of auxotrophic and analogue resistant mutants of the moss, *Physcomitrella patens*. Molec Gen Genet 154: 87–95.
2. Ashton NW, Cove DJ, Featherstone DR (1979) The isolation and physiological analysis of mutants of the moss, *Physcomitrella patens*, which over-produce gametophores. Planta 144: 437–442.
3. Ashton NW, Grimsley NH, Cove DJ (1979) Analysis of gametophytic development in the moss, *Physcomitrella patens*, using auxin and cytokinin resistant mutants. Planta 144: 427–435.
4. Ashton NW, Boyd PJ, Cove DJ, Knight CD (1988) Genetic analysis in *Physcomitrella patens*. In: Glime JM (Ed.) Methods in Bryology, pp 59–72, Hattori Botanical Laboratory, Nichinan, Japan.
5. Boyd PJ, Grimsley NH, Cove DJ (1988) Somatic mutagenesis of the moss, *Physcomitrella patens*. Molec Gen Genet 211: 545–546.
6. Boyd PJ, Hall J, Cove DJ (1988). An airlift fermenter for the culture of the moss *Physcomitrella patens*. In: Glime JM (Ed.) Methods in Bryology, pp 41–45, Hattori Botanical Laboratory, Nichinan, Japan.
7. Courtice GRM, Ashton NW, Cove DJ (1978) Evidence for the restricted passage of metabolites into the sporophyte of the moss, *Physcomitrella patens*. J Bryol 10: 191–198.
8. Courtice GRM, Cove DJ (1983) Mutants of the moss, *Physcomitrella patens* which produce leaves of altered morphology. J Bryol 12: 595–609.
9. Grimsley NH, Ashton NW, Cove DJ (1977) The production of somatic hybrids by protoplast fusion in the moss, *Physcomitrella patens*. Molec Gen Genet 154: 97–100.
10. Grimsley, NH, Ashton NW, Cove DJ (1977) Complementation analysis of auxotrophic mutants of the moss, *Physcomitrella patens* using protoplast fusion. Molec Gen Genet 155: 103–107.
11. Grimsley, NH, Featherstone DR, Courtice GRM, Ashton NW, Cove DJ (1979) Somatic hybridization following protoplast fusion as a tool for the analysis of development in the moss, *Physcomitrella patens*. In: Advances in Protoplast Research, pp 363–376. Proceedings of the 5th International Protoplast Symposium, Szeged, Hungary. Academiai Kiado, Budapest, Hungary.
12. Grimsley NH, Withers LA (1983) Cryopreservation of cultures of the moss *Physcomitrella patens*. Cryoletters 4: 251–258.
13. Knight CD, Cove DJ, Boyd PJ, Ashton NW (1988) The isolation of biochemical and developmental mutants in *Physcomitrella patens*. In: Glime JM (Ed.) Methods in Bryology, pp 47–58, Hattori Botanical Laboratory, Nichinan, Japan.
14. Lonsdale D, Onde S, Cuming A (1990) Transient expression of exogenous DNA in intact, viable wheat embryos following particle bombardment. J Exp Bot 41: 1161–1165.
15. Sawahel W (1994) Genetic transformation of the moss, *Physcomitrella patens*. PhD thesis, University of Leeds, U.K.
16. Sawahel W, Onde S, Knight CD, Cove DJ (1992) Transfer of foreign DNA into Physcomitrella patens protonemal tissue by using the gene gun. Plant Mol Biol Rep 10: 315–316.
17. Schaefer D, Zryd J-P, Knight CD, Cove DJ (1991) Stable transformation of the moss *Physcomitrella patens*. Molec Gen Genet 226: 418–424.
18. Watts JW, Doonan JH, Cove DJ, King JM (1985) Production of somatic hybrids of moss by electrofusion. Molec Gen Genet 199: 349–351.
19. Wettstein, Fr von (1932) Genetik. In: Vedoorn F (Ed.) Manual of Bryology, pp 233–272, Martinus Nijhoff, The Hague, Netherlands.

Section G:
Secondary Metabolites

Plant Molecular Biology
Manual

Section G Secondary Metabolites

Chapters planned for this section include:-
  Cell line selection and screening
  Analytical procedures for specific groups of secondary metabolites
  Immobilisation techniques
  Hairy root cultures
  Fungal elicitor effects
  Assays for key enzymes in secondary metabolism
  Cell permeabilisation
  Biotransformations
  Bioreactor design
  Downstream processing

Plant Tissue Culture Manual **G1**, 1–17, 1995.

Tropane alkaloid biosynthesis in root cultures

RICHARD J. ROBINS[1] & BIRGIT DRÄGER[2]
[1] *Genetics and Microbiology Department, Institute of Food Research, Norwich Laboratory, Norwich Research Park, Colney, Norwich NR4 7UA, U.K.*
[2] *Institut für Pharmazeutische Biologie und Phytochemie, Westfälische Wilhelms-Universität, Hittorfstrasse 56, D-48149 Münster, Germany*

Introduction

The tropane alkaloids accumulate principally in the Solanaceae (e.g. *Atropa, Datura, Hyoscyamus*) and Erythroxylaceae (*Erythroxylon*) but are also found in a number of other families, notably the Convolvulaceae (*Calystegia*) and Proteaceae [8]. These compounds have been of interest to humans for many years as lethal poisons, for the induction of hallucinations in mystical rites, and as mild stimulants [4]. The use of three tropane alkaloids in particular in modern pharmacy, namely hyoscyamine, scopolamine and cocaine, has led to considerable activity both in (1) the elucidation of the pathways of their biosynthesis *in vivo* and, (2), *in vitro* cultures for their production. As no biosynthetic work on cocaine has been done in tissue culture, this compound will not be considered further.

Considerable progress in the description of the chemical intermediates in the tropane alkaloid pathway (Figure 1) has been made by feeding experiments *in planta* [7]. However, only since the establishment of alkaloid-producing *in vitro* cultures has it been possible to isolate and describe some of the biochemistry of the pathway [12]. In part, this success has been due to the ease with which it is possible to manipulate the culture conditions, to feed precursors or inhibitors and to extract the tissue for enzymes and products. This has enabled aspects of the sequence of reactions, the genetic and biochemical regulation of the pathway, and the morphological association of pathway activity with root-organ integrity to be examined in tissue cultures. These results have been summarized recently.

While it is possible to obtain alkaloid production in disorganized cell callus or suspension cultures, such systems have proved ineffective at accumulating tropane alkaloids [12]. In whole plants, these are biosynthesized in the roots and it was only following the establishment of root cultures that progress was made. In this chapter, therefore, we will:
a) describe the procedures required for the establishment and propagation of sterile *in vitro* root cultures of several tropane-alkaloid-forming cultures;
b) give procedures for the extraction and analysis of the alkaloids formed in such cultures;
c) give examples of the extraction, purification and characterization of enzymes of tropane-alkaloid biosynthesis from root cultures;

d) give examples of the value of substrate-feeding and inhibitor-treatment experiments in helping understand the regulation of the pathway.

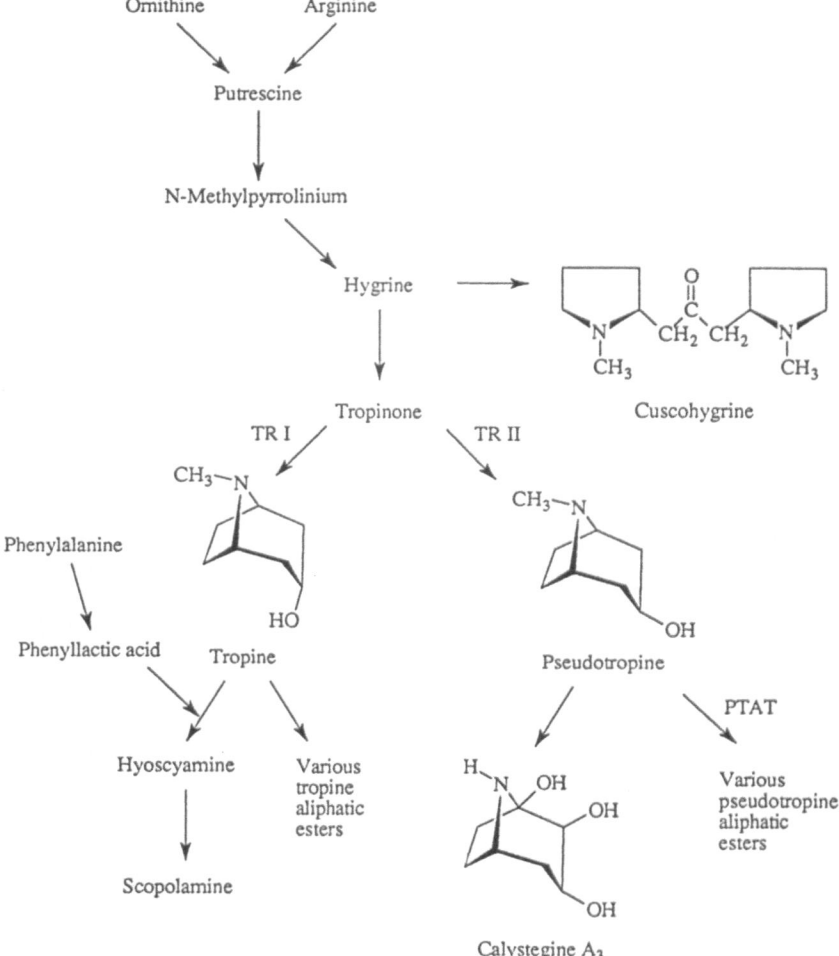

Fig. 1. The pathway by which tropine alkaloids are formed. Arrows may represent more than one enzymic step. PTAT = pseudotropine : acyl-CoA acyl transferase (see Recipe 5); TR I and TR II = tropinone reductases I (tropine-forming) and II (pseudotropine-forming) respectively (see Recipe 6).

Procedures

Generation of root cultures of Hyoscyamus niger *and* H. albus

When excised roots of *Hyoscyamus* species are established in culture, tropane alkaloid production at levels of about 1% dry mass (DM) can be achieved. The highest levels are typically obtained with *H. niger* and *H. albus* [6].The cultures produce alkaloids continuously during the culture period, with a maximum content of about 1% DM in *H. albus* and 0.1% DM in *H. niger*. The level of production is affected by the phytohormone regime applied. Supplementing the medium with indole butyric acid (IBA) at 10 µM increased the growth rate of both cultures by about 50%, apparently due to the stimulation of the initiation of lateral root primordia. However, a concomitant decrease in alkaloid accumulation occurred.

Steps in the procedure
1. Germinate sterile seeds to obtain sterile seedlings.
2. Excise a 5 mm-long section of the lower cotyledon with part of the root from each seedling and place in Linsmaier and Skoog medium [see Section **A1**] with 30 g/l sucrose at 25 °C in the dark on a reciprocal shaker at 60 strokes/min.
3. Leave to grow. Once established, sub-culture about 1 g fresh mass (FM) roots at 2-weekly intervals into fresh medium. The mass of the culture should increase about 5-fold between sub-cultures.

Note
Hyoscyamus root cultures are not typical, in that they are readily established without the presence of exogenous phytohormones. For many species, the presence of 10^{-6} M IBA is required for the initiation of root cultures, although this can often subsequently be withdrawn.

Establishing transformed root-organ cultures of Datura

Transformation with *Agrobacterium rhizogenes* provides a method by which root cultures can be generated and cultured in axenic conditions from a wide range of species. It has the advantages that
— cultures do not need added phytohormones
— cultures are frequently fast growing (though not always)
— genetic material can easily be introduced.
With many Solanaceae, these cultures are easy to establish and fast growing. They provide a good system for precursor and inhibitor feeding and are an excellent tissue from which to extract enzymes.

Steps in the procedure
1. Surface sterilize seeds with 10% (v/v) commercial bleach (e.g. Domestos) for 30 min. Wash 6 times with 100 ml sterile deionized water.
2. Germinate sterile seeds on 10 g/l agar B5 medium [see Section **A1**] containing no sucrose and no phytohormones.
3. Prepare a 48-h suspension of *A. rhizogenes* LBA9402 in YMB medium at 25 °C to 26 °C.
4. Wound the hypocotyl of 2- to 3-week-old seedlings with a narrow hypodermic needle containing the *A. rhizogenes* suspension.
5. When emergent roots are 5 to 10 mm long (7 to 14 days), excise the hypocotyl segment bearing roots and place in 8 ml B5 medium in a 50 ml sterile pot or Erlenmeyer flask containing 30 g/l sucrose and 0.5 mg/ml ampicillin sulphate. Place on a reciprocal shaker at 25 °C and about 70 strokes per min.
6. Passage rapidly growing roots into 50 ml of the same medium in 250 ml Erlenmeyer flasks. Place on a reciprocal shaker at 25 °C and about 90 strokes per min.
7. Continue sub-culturing every 2 weeks. Passage 2 to 4 lengths of root (3 to 4 cm long and bearing numerous side-branches) into fresh medium.
8. After about 8 sub-cultures it should be possible to omit ampicillin from the medium.

Notes
4. Seedlings of different species require different periods of growth to reach a suitable size.
   Leaf or stem tissue from greenhouse-grown plants can be used as explant material.
   Care must be taken not to apply more bacterial suspension than can be held in the surface scratch on the tissue, otherwise bacterial growth can over-run the emergent roots.
5. The time taken for roots to emerge from the wound site can vary considerably, from a few days to several months. Roots are unlikely to be produced once the parent tissue is fully necrotic but can emerge at a late stage of tissue decay.
   If roots are too late emerging, or do so too sparingly, or do not develop at all, acetosyringone or other phenolics that trigger the virulence response of the bacteria may be helpful [13]. Acetosyringone 10 µM should be added to the 48 h suspension culture of *Agrobacterium rhizogenes* immediately before it is used for transformation.

Solutions

YMB Medium:

KH$_2$PO$_4$	0.5 g/l
MgSO$_4$	2.0 g/l
NaCl	0.1 g/l
Mannitol	10.0 g/l
Yeast Extract (Difco)	0.4 g/l

pH 7.0

For solid medium, add:

Agar (Oxoid)	15.0 g/l

To maintain a pBIN 19-based plasmid, add:

Kanamycin sulphate	50 mg/l

Cultures should be maintained on this medium at 26–28 °C.

Extraction and analytical procedures for non-polar tropane alkaloids (e.g. tropine, hyoscyamine, scopolamine)

Tropane alkaloids can be extracted from aqueous solution readily either by liquid-liquid partitioning or by adsorption to a diatomaceous earth, such as Extrelute. The former method is more suitable for large amounts of tissue but, especially with root cultures, is prone to problems in separating the phases due to the carbohydrate present causing an emulsion to form. It also requires exhaustive sequential extractions to obtain reproducible quantitative recoveries of all the bases. The latter is cleaner and quicker and can readily be performed on a small scale on multiple samples (up to 20) simultaneously. It has the disadvantage of being unsuitable for handling large amounts of aqueous extract, making it a problem to apply for multiple extractions on a large scale.

Steps in the procedure
Method A. Liquid-liquid extraction
1. Homogenise (Ultraturrax) root material in 0.2 M H_2SO_4 (5 ml/g FM) and leave to stand 90 min.
2. Vacuum filter through a glass microfibre A/miracloth sandwich.
3. Extract the filtrate with an equal volume of $CHCl_3$. It may prove necessary to centrifuge the emulsion at $RCF_{max} = 700 \times g$ for 5 min to separate the phases. Retain the aqueous phase.
4. Basify the aqueous phase with 35% (v/v) NH_3 solution and extract with an equal volume of $CHCl_3$. It may prove necessary to centrifuge the emulsion at $RCF_{max} = 700 \times g$ for 5 min to separate the phases. The partitioning may need to be repeated 4 or 5 times for good quantitation.
5. Pool the organic phases, evaporate to dryness and dissolve the residue in 1 ml MeOH.
6. Dilute 50 μl MeOH into 1 ml ethyl acetate. Inject 1 to 5 μl onto a DB-17 (25 m × 0.32 mm, film thickness 0.23 mm: J & W Scientific) capillary GC column at 65 °C, fitted to a gas chromatograph with a cold on-column injector (e.g. Carlo Erba Mega MR5160), with a helium flow of 1.5 ml/min. Develop the chromatogram from 65 to 120 °C at 4 °C/min, from 120 to 280 °C at 8 °C/min and from 280 to 310 °C at 12 °C/min with 4 min at 310 °C. Record the separation by phosphorus-nitrogen detection (PND). Under these conditions the alkaloids elute in the order tropine (13.8 min), tropinone (14.0 min), pseudo-tropine (14.5 min), acetyltropine (17.0 min), cuscohygrine (23.7 min), tiglyl-tropine (25.0 min), littorine (32.0 min), hyoscyamine (33.5 min), scopolamine (35.4 min).

Note
6. A cold-on-column injector is preferable for analyzing tropane alkaloids as at the high temperature of a split-splitless injector (e.g. 250 °C) hyoscyamine and the other high-molecular weight esters (e.g. scopolamine), degrade to some extent. The degree of degradation is variable, leading to inaccurate quantitation of the sample being applied.
Quantitation can be improved by the use of an internal standard (e.g. spartein) added to the crude extract, provided this compound does not interfere with the detection of any of the products of interest.

Method B. Liquid-solid extraction

1. Harvest, blot and freeze-dry root tissue.
2. Grind to a fine powder with sand. Add 1 ml 0.05 M H_2SO_4 per 0.1 g root DM and leave for 15 min.
3. Filter and retain the filtrate. Wash the retentate twice with 0.5 ml 0.05 M H_2SO_4.
4. Make alkaline with approx 0.2 ml 35% (v/v) NH_3 solution and apply to an Extrelute® column (Merck) using 1 g solid phase per 1 ml aqueous phase.
5. After 15 min elute with $CHCl_3$:*iso*-propanol (95 : 5) using 5 ml per g solid phase.
   6.Evaporate to dryness and dissolve the residue in 1 ml MeOH. Analyze by GC as described above.

Extraction and analytical procedure for polar tropane alkaloids (e.g. calystegines)

Recently, a group of polar tropane alkaloids, the calystegines, derived from pseudotropine (Fig. 1), has been described [3]. These alkaloids are very water soluble and require different methods for their isolation and quantitation.

Steps in the procedure
1. Homogenise (Ultraturrax) fresh or deep-frozen root material in MeOH/H_2O 50 : 50 (5 ml/g FM).
2. Centrifuge (approx. 5000 × g) and remove debris. Evaporate the MeOH/H_2O phase *in vacuo* until the MeOH has disappeared. Centrifuge the remaining aqueous solution and discard the debris.
3. Load the neutral aqueous solution to a cation exchange column (Dowex 50 WX11) and wash with water to remove non-binding compounds.
4. Elute basic compounds by washing with 2 M NH_3 solution.
5. Neutralize and concentrate the eluate by evaporation *in vacuo* to approx. 0.1 ml/g FM.
6. This solution can be directly used for thin-layer-chromatographic analysis. TLC-conditions: Merck silica gel 60 plates; solvent system MeOH/$CHCl_3$/35% (v/v) NH_3 solution (66/11/23); detection by dipping the dried plate into i) silver nitrate solution (2 g/l acetone, 1 min) and, after drying, into ii) 2% (w/v) NaOH in EtOH until brown-grey spots become visible. R_f-values: calystegine A-group 0.48, calystegine B-group 0.6. Desoxynojirimycin (DNM, Boehringer) can be taken as reference compound, R_f 0.66.
7. For GC-analysis 50–100 µl of the extract (step 5) is lyophilised in sample vials. The dry residue is dissolved in dry pyridine and derivatised with a mixture of hexamethyldisilazane and trimethylchlorosilane (10 : 1) by heating to 50 °C for 15 min.
8. GC-conditions are: 250 °C split injector temperature; 1 : 20 split ratio; 1 µl injection volume; column DB5 (30 m × 0.25 mm ID, film thickness 0.2 µm, J&W Scientific) temperature programme 160 °C, 2 min isothermal, then 5 °C/min increase up to 240 °C, carrier gas helium 35 cm/sec at 100 °C, simultaneous flame-ionisation detection (FID) and PND detection. Retention times: calystegine A_3: 11.9 min, calystegine B_1: 15.4 min, calystegine B_2: 17.2 min; silylated DNM as internal standard: 14.1 min; octadecane as internal standard (for FID detection only): 13.1 min.

Extraction and assay of tiglyl-CoA : Pseudotropine acyl transferase activity

Root cultures provide good quality material for the preparation of enzymes [9, 10]. There are many different ways to extract enzymes from plant tissue. This and the following method give two examples of generally applicable sets of conditions that appear to give good reproducibility and recovery. The buffers and specialist additives required will vary with the enzyme being prepared. In general, it is advisable to:
— use a neutral or slightly alkaline buffer of an ionic strength of 100 to 200 mM
— include a chelating agent (usually ethylenediaminetetraacetic acid, EDTA) at 5 to 20 mM
— include a reducing agent (e.g. dithiothreitol, mercaptoethanol) at 3 to 20 mM
— add an adsorbent (e.g. Polyclar AT, XAD-7) to remove phenolics
— carry out all stages of the extraction in a 4 °C cold room.

Steps in the procedure
1. Harvest the roots and blot thoroughly on a paper towel.
2. Weigh the tissue and make into packets containing approx. 20 g (small scale preparation) or 60 g (large scale preparation).
3. Freeze rapidly in liquid nitrogen.
4. Store in ultra-cold conditions, preferably −70 or −40 °C. All the enzymes we have examined seem to be stable for at least one year in these conditions.
5. Remove packets from freezer into liquid nitrogen.
6. Prepare a beaker of buffer A, containing 3 ml buffer/g FM of tissue to be extracted, stirring gently on a magnetic stirrer.
7. Weigh out Polyclar AT at 1 g Polyclar AT/5 g FM roots.
8. Remove a packet of frozen roots from the liquid nitrogen and crush vigorously on a hard, cold surface. We use a marble pastry board and rolling pin.
9. Tip the crushed root material into the cold bowl of a coffee grinder, adding the Polyclar AT. Grind — preferably shaking the grinder — for 5 to 10 sec.
10. Tip the powder into the stirring buffer and mix in gently with a glass rod. Avoid lumps forming.
11. Repeat steps 8 to 11 until all the tissue is disintegrated.
12. Leave 15 to 30 min, stirring gently.
13. Press out the debris from the suspension. We use a tough linen sheet or miracloth for small volumes and a mechanical fruit press for larger amounts.
14. Centrifuge the liquid at 15 000 × g for 15 min at 4 °C.
15. Decant the supernatant.
16. Apply 2.5 ml aliquots to PD-10 columns (Pharmacia) pre-equilibrated in buffer B and elute in 3.5 ml buffer B.
17. Enzyme activity is determined by the extraction of the products and quantitation by GC. Incubate 200 µl de-salted extract (from step 16) with 60 µl glycine (1 M, pH 9.0), 20 µl pseudotropine (20 mM) and 20 µl tiglyl-CoA (20 mM) for 45 min. Stop the reaction with 100 µl 35% (v/v) NH_3 solution and apply to a 1 g Extrelute column. After 15 min elute with 12 ml $CHCl_3$: *iso*-propanol (95 : 5). Evaporate to dryness and dissolve the residue in 1 ml ethylacetate. Analyze by GC as described above.

Buffers

Buffer A: for every 100 ml mix:

    20 ml 1 M K-phosphate solution (pH 7.0) — final concentration = 200 mM
    10 ml 0. 2 M EDTA solution   — final concentration = 20 mM
    4.3 g sucrose   — final concentration = 125 mM
    45 mg DTT   — final concentration = 3 mM

Adjust to pH 7.0 at room temperature; add water to 100 ml.

Buffer B: for every 100 ml mix:

    20 ml Buffer A
    80 ml water
    2.4 g sucrose
    19 mg DTT

Adjust to pH 9.0 at room temperature; add water to 100 ml.

Extraction and assay of tropinone reductase activities

1-7. Harvesting and storage of the roots is done as in the previous method (steps 1-7).

8. Crush the roots and mix with cold buffer C (1 ml/g roots), Polyclar AT and quartz sand in a mortar, homogenise and leave to macerate for 20 min at 4 °C with occasional stirring.

9. Pour the suspension into cotton or linen tissue that is fixed onto a beaker in an ice bath and let the extract drip into the beaker. Press out the rest of the liquid from the extract and centrifuge (10 min, 10 000 × g, 4 °C).

10. Concentrate and fractionate the extracted protein by ammonium sulphate precipitation. The first fraction is precipitated between 0 and 40% ammonium sulphate saturation (centrifugation 30 min, 10 000 × g, 4 °C) and discarded. The second fraction, that precipitates between 40 and 75% saturation (centrifugation as above), contains tropinone reductase activity. The pellet is dissolved in buffer D (1 ml/10 g FM roots used initially).

11. Apply 2.5 ml aliquots to PD-10 columns (Pharmacia) pre-equilibrated in buffer D and elute in 3.5 ml buffer D.

12. Enzyme activity is determined spectrophotometrically by following NADPH consumption at 334 nm and 30 °C in a double-beam spectrophotometer. Incubation mixture (1 ml) contains 10–50 µl protein solution (from step 11), 200 µM NADPH, 5 mM tropinone, 0.1 M potassium phosphate buffer (pH 6.2). The reference cuvette contains the same mixture without tropinone, thus unspecific NADPH-consumption is automatically subtracted. The measured decrease of absorbance should be stable for at least 20 min. Enzyme activity is calculated based on the NADPH consumption observed per unit time (1 mM NADPH gives $E_{334} = 6.2 \times 10^3$).

Buffers

Buffer C: for every 100 ml mix:

10 ml 1 M K-phosphate solution (pH 7.6)	— final concentration = 100 mM
1 ml 0.2 M EDTA solution	— final concentration = 2 mM
8.6 g sucrose	— final concentration = 250 mM
45 mg DTT	— final concentration = 3 mM

Adjust to pH 7.6; add water to 100 ml.

Buffer D: for every 100 ml mix:

2 ml 1 M K-phosphate solution, pH 7.0	— final concentration = 20 mM
25 ml glycerol 85% (v/v)	— final concentration = 21.2% (v/v)
6.8 g sucrose	— final concentration = 200 mM
15 mg DTT	— final concentration = 1 mM

Adjust to pH 7.0; add water to 100 ml.

Use of natural substrates, inhibitors or substrate analogues to examine tropane alkaloid biochemistry (feeding compounds to cultured roots)

Tissue cultures are excellent systems in which to examine the influence of exogenous effectors or substrates on a pathway, or on specific enzymes in a pathway. Precursors, substrates, inhibitors and substrate analogues can easily be introduced and their metabolism, or effect on the metabolism of other compounds, examined. Such compounds appear to be taken up readily by the tissue; thus, transport barriers do not seem to be a problem. Very high levels of specific incorporation of tracers tagged with radio- (e.g. ^{3}H or ^{14}C) or heavy isotopes (e.g. ^{2}H, ^{13}C or ^{15}N) can be achieved. Provided the tracer is rapidly absorbed, the kinetics of its incorporation and turnover can be determined using *in vivo* nuclear magnetic resonance spectroscopy (NMR) [5]. As the tissue is sterile, interference by contaminating microorganisms is avoided.

   This procedure covers a common method and its use is exemplified in the study of three aspects of tropane alkaloid biosynthesis, viz :

a) *The incorporation of labelled intermediates.* The feeding of phenyl[1,3-^{13}C]lactic acid to root cultures of *D. stramonium* has been used to examine the pathway from phenylalanine to the tropate moiety of hyoscyamine [11]. Because tropane alkaloids can be analysed by GC/MS, it is possible to perform very sensitive incorporation experiments with heavy-labelled stable isotopes. The incorporation into all the major products can simultaneously be followed. From this work, it was possible to conclude that free tropic acid is not an intermediate in this pathway.

b) *Inhibitor studies.* Inhibitors highly specific to particular enzymes can prove to be particularly valuable tools to the biochemist who wishes to test the importance of that particular activity *in vivo*. When fed to *D. stramonium* roots, it was found that a tropinone analogue, thiobicyclonortropinone (TBON) inhibited the formation of hyoscyamine. From this data it was possible to infer that TR I was inhibited more than TR II. This was confirmed by testing the effect of TBON and related inhibitors against extracts of TR I and TR II [2].

c) *Analogue incorporation.* The feeding of chemically synthesized analogues of pathway intermediates can give valuable information about the plasticity of the biosynthetic apparatus present. This is useful both as a potential way of making novel metabolites and in giving some indications *in vivo* of the substrate specificity of reactions. When *N*-alkylnortropinone analogues of tropinone were fed to cultures of a *Brugmansia* hybrid, it was found that these were readily reduced to the respective alcohols which were then esterified to varying degrees [1]. From the data obtained, it could be inferred that the enzyme system forming littorine readily handles the analogue while that forming hyoscyamine does not.

Steps in the procedure

1. Sub-culture root cultures and leave 3 or 4 days to allow growth to be established and to confirm sterility.
2. Prepare a stock solution of feedant at 55 or 110 mM in distilled water. Neutralize and make to volume. Store frozen.

3. Through a sterilizing filter, add 1 ml stock solution to each flask.
4. If desired, repeat feeding twice more at 2 or 3 day intervals.
5. Harvest tissue at 10 to 12 days old and extract alkaloids.
6. To analyze by GC/MS, we use a DB1 or DB17 column (30 m × 0.32 mm, film thickness 0.25 mm) fitted to Hewlett Packard 5890 linked to a TRIO 1S mass spectrometer.

Notes

2. Potential precursors should be prepared (if possible) as a 20- to 50-fold concentrated stock solution in water, so that only 1 or 2 ml/flask need be added. This must be sterilized by filtration (preferably) or autoclaving prior to addition to the flasks.

   For less water-soluble compounds, a non-aqueous solvent such as methanol, ethanol or dimethylsulphoxide can be used. The final concentration of solvent should not exceed 5 to 10% (v/v) in the culture. A 'solvent-only' control should always be included.

   Stock solutions must be neutralized or buffered.

3, 4. Appropriate time periods for feeding during the growth course of the culture have to be established. Therefore, some prior knowledge of the time course of the accumulation of the metabolite(s) under investigation is advantageous to maximize incorporation or the effect of inhibitors.

References

1. Boswell HD, Eagles J, Robins DJ, Robins RJ, Walton NJ (1993) Biotransformation of tropinone analogues *in vivo* forms novel tropane alkaloids. Poster presented at: Phytochemistry of Plants used in Traditional Medicine, Phytochemical Society of Europe meeting, Lausanne, Switzerland, 28 September to 1 October 1993.
2. Dräger B, Portsteffen A, Schaal A, McCabe P, Peerless ACJ, Robins RJ (1992) Levels of tropinone-reductase activities influence the spectrum of tropane esters found in transformed root cultures of *Datura stramonium* L. Planta 188: 581–586.
3. Dräger B, Funck C, Höhler A, Mrachatz G, Portsteffen A, Schaal A, Schmidt R (1994) Calystegines as a new group of tropane alkaloids in Solanaceae. Plant Cell Tiss Org Cult (in press).
4. Evans W (1990) *Datura*, a commercial source of hyoscine. Pharm J 244: 651–653.
5. Ford Y-Y, Fox GG, Ratcliffe RG, Robins RJ (1994) *In vivo* ^{15}N NMR studies of secondary metabolism in transformed root cultures of *Datura stramonium* and *Nicotiana tabacum*. Phytochemistry 36: 333–339.
6. Hashimoto T, Yukimune Y, Yamada Y (1986) Tropane alkaloid production in *Hyoscyamus* root cultures. J Plant Physiol 124: 61–75.
7. Leete E (1990) Recent developments in the biosynthesis of the tropane alkaloids. Planta Medica 56: 339–352.
8. Lounasmaa M (1993) The tropane alkaloids. In: Cordell GA (Ed.) The Alkaloids, vol. 44, pp. 1–114, Academic Press, San Diego.
9. Portsteffen A, Dräger B, Nahrstedt A The reduction of tropinone in *Datura stramonium* root cultures by two specific reductases. Phytochemistry (submitted).
10. Robins RJ, Bachmann P, Peerless, ACJ, Rabot S (1994) Esterification reactions in the biosynthesis of tropane alkaloids in transformed root cultures. Plant Cell Tiss Org Cult (in press).
11. Robins, RJ, Woolley JG, Ansarin M, Eagles J, Goodfellow BJ (1994) Phenyllactic acid but not tropic acid is an intermediate in the biosynthesis of tropane alkaloids in *Datura* and *Brugmansia* transformed root cultures. Planta 194: 86–94.
12. Robins RJ, Walton, NJ (1993) The biosynthesis of tropane alkaloids. In: Cordell GA (Ed.) The Alkaloids, vol. 44, pp 115–187, Academic Press, San Diego.
13. Stachel SE, Messens E, Van Montagu M, Zambryski P (1985) Identification of the signal molecules produced by wounded plant cells that activate T-DNA transfer in *Agrobacterium tumefaciens*. Nature 318: 624–629.

Plant Tissue Culture Manual **G2**, 1–23, 1995.

Anthocyanin biosynthesis in vitro

KOICHI KAKEGAWA[1] & ATSUSHI KOMAMINE[2]
[1] *Forestry and Forest Products Research Institute, P.O. Box 16, Tsukuba Norin Kenkyu Danchi-nai, Ibaraki 305, Japan*
[2] *Department of Chemical and Biological Sciences, Faculty of Science, Japan Women's University, 2-8-1 Mejirodai, Tokyo 112, Japan*

Introduction

Anthocyanin, a red pigment, is widely distributed in various parts of higher plants. It provides the scarlet to blue colors of flower petals in association with other flavonoids and/or organic acids. Many plant biochemists have made efforts to elucidate its chemical structure and the details of its biosynthesis.

Cell cultures producing anthocyanin have been derived from many kind of plants, for example, *Haplopappus gracilis* [1], *Daucus carota* [2, 3, 4], *Catharanthus roseus* [5], *Vitis* sp. [6] and *Centaurea cyanus* [7]. Such cell cultures have been used successfully for studies of anthocyanin biosynthesis. In particular, successful studies related to the biosynthetic enzymes and regulation mechanisms of anthocyanin synthesis have taken advantage of these cell culture systems [8, 9].

In this chapter, we will first describe methods for the quantification and identification of anthocyanin and essays for enzymes involved in the biosynthesis of anthocyanin, namely, phenylalanine ammonia-lyase and chalcone synthase. Subsequently, some studies of anthocyanin biosynthesis and its regulation, mainly in cultured cells will be reviewed.

Chemistry of anthocyanin

Anthocyanin is a compound that is classified as a flavonoid. Most of the biosynthetic pathway to anthocyanin is common to pathways to other flavonoids and the basic structure of anthocyanidin, the aglycon of anthocyanin, is a C_6-C_3-C_6 unit (Fig. 1). This unit consists of a malonate-derived A-ring (C_6) plus a B-ring and a C-ring (C_6-C_3) derived from phenylpropane [8]. The major anthocyanins contain a hydroxy group at position 3 in the C-ring. The 3-deoxy form is rare. Positions 5 and 7 of the A-ring are mainly substituted by hydroxy groups, but methylation of such hydroxy groups occurs occasionally. The B-ring has been found to have several patterns of substitution at positions 3', 4' and 5' with hydroxy and/or methoxy groups. Anthocyanidins in flower pigments are usually pelargonidin (4'-OH), cyanidin (3', 4'-OH), delphinidin (3', 4', 5'-OH) and their methyl derivatives. About twenty species of anthocyanidins have been identified in nature.

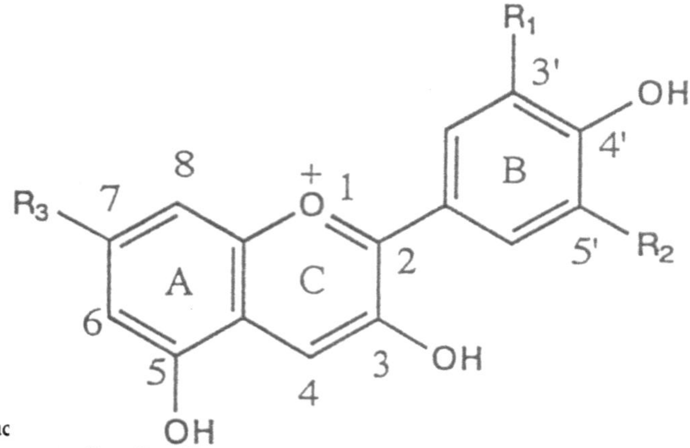

Fig. 1. Struc
pelargonidin:	$R_1 = R_2 = H$, $R_3 = OH$
cyanidin:	$R_1 = R_3 = OH$, $R_2 = H$
delphinidin:	$R_1 = R_2 = R_3 = OH$
peonidin:	$R_1 = OCH_3$, $R_2 = H$, $R_3 = OH$
rosinidin:	$R_1 = R_3 = OCH_3$, $R_2 = H$
petunidin:	$R_1 = OCH_3$, $R_2 = R_3 = OH$
malvidin:	$R_1 = R_2 = OCH_3$, $R_3 = OH$
hirsutidin:	$R_1 = R_2 = R_3 = OCH_3$

Anthocyanidins exist as O-glycosides, a more stable form than aglycons, *in vivo* and acylation *via* a sugar moiety is quite common. Frequently, the attached sugars are glucose, galactose and rhamnose but also xylose and arabinose have been reported [10]. The glycosyl group at position 3 is the most primitive form. Acylation by organic acids derived from phenylpropane appears to represent a major substitution pattern. However, a number of anthocyanins that are acylated by dicarboxylic acids, such as malonic, oxalic and succinic acid, have been found in nature [10]. Acyl groups render anthocyanin more stable in neutral solution [11].

In aqueous solutions, anthocyanin can be transformed to other compounds, depending on pH (Fig. 2). Under strongly acidic conditions (pH 1–2), anthocyanin exists as a stable flavilium cation or salt and the solution is a vivid red color. However, the flavilium cation turns into colorless anthocyanin, a pseudobase, in only slightly acidic solution (pH 3–5). At high pH (pH 6 and above), a quinoidal base is formed. This form is also colorless and unstable.

Anthocyanins exist as colored flavilium cations in plant vacuoles, even though vacuolar pH values range from about 4 to 6, at which the color of anthocyanin disappears. Acylation of anthocyanin affects the stability of the flavilium cation as, for example, in the case of platyconin which is highly acylated with caffeic acids [12]. Copigmentation with free organic acids or other flavonoids also stabilizes the colored flavilium structure [11].

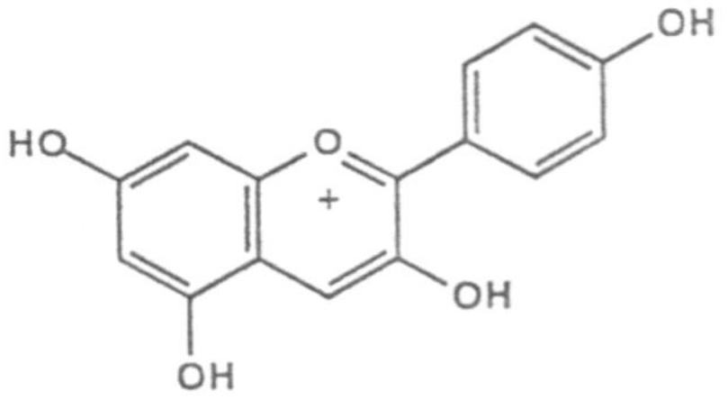

flavilium cation (pH 1-2) 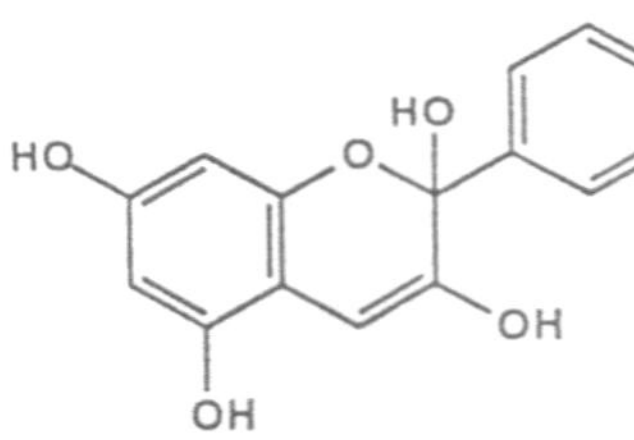   pseudobase (pH3-5)

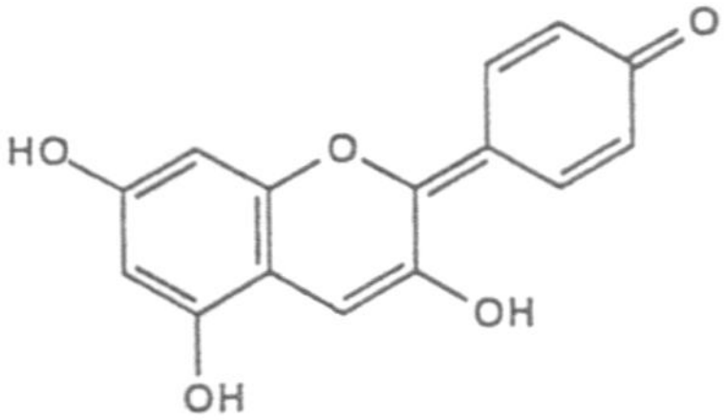

quinoidal base (pH 6 and above)

Fig. 2. Structures of anthocyanidin in water at various pH values.

Biosynthesis of anthocyanin

In the past twenty years, studies of flavonoid biosynthesis have made remark-able progress. The biosynthetic pathways to the main flavonoids have been clarified, with the exception of a few steps in the biosynthesis of anthocyanin which remain to be characterised. Tracer and supplementation experiments using mutant flowers of *Petunia hybrida* [13], *Matthiola incana* [14] and *Antirrhinum majus* [15] contributed to the elucidation of the biosynthetic pathway to anthocyanin. By contrast, it has been studies with cell cultures that have enhanced our knowledge of the enzymology and mechanisms of regulation of flavonoid biosynthesis [9]. In this section, a general overview of biosynthetic pathways to flavonoids, including anthocyanin, is presented.

A summary of the biosynthetic pathways to flavonoids and the enzymes involved is shown in Fig. 3. Direct precursors of the flavonoid skeleton are malonyl-CoA and 4-coumaroyl-CoA [8]. Malonyl-CoA is derived from acetyl-CoA which is an intermediate in glycolysis. 4-Coumaroyl-CoA is synthesized *via* the general phenylpropanoid pathway from phenylalanine. Phenylalanine ammonia-lyase (PAL; 1 in Fig. 3) is the first enzyme in this pathway and acts as the rate-limiting enzyme in the biosynthesis of phenolic compounds [16]. Phenylalanine is transformed to *trans*-cinnamic acid by the reaction catalyzed by PAL, *via* elimination of an amino group. The details of this enzyme have been reviewed by Jones [17]. Cinnamic acid 4-hydroxylase (2 in Fig. 3)

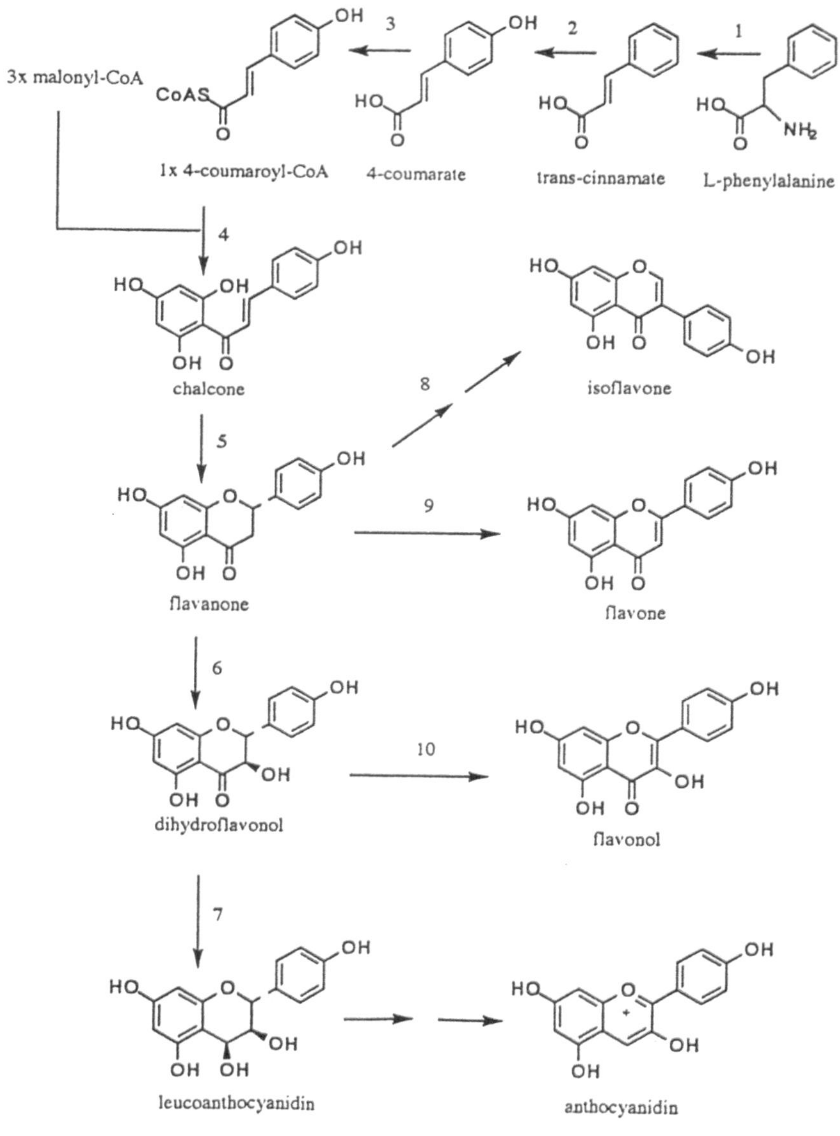

Fig. 3. Biosynthetic pathway for anthocyanidin. All of the flavonoids in this Figure are shown in the 4'-OH form. Numbers (1–10) refer to enzymes as follows: 1, phenylalanine ammonia-lyase; 2, cinnamate 4-hydroxylase; 3, 4-coumarate:CoA ligase; 4, chalcone synthase; 5, chalcone-flavanone isomerase; 6, (2S)-flavanone 3-hydroxylase; 7, dihydroflavonol 4-reductase; 8, iso-flavone synthase; 9, flavone synthase; 10, flavonol synthase.

catalyzes aromatic hydroxylation at position 4 and generates 4-coumaric acid from *trans*-cinnamic acid. This enzyme is reported to be cytochrome P-450 hydroxylase [18]. 4-Coumaric acid is further modified to 4-coumaroyl-CoA by a reaction catalyzed by 4-coumarate: CoA ligase (3 in Fig. 3) [19].

The most important step in the biosynthesis of flavonoids is the formation of the C_6-C_3-C_6 skeleton by condensation of three malonyl-CoA and one 4-coumaroyl-CoA molecules [8, 16]. Chalcone synthase (CHS; in Fig. 4) catalyzes this reaction, generating chalcone. Because of its relatively low specific activity and its pattern of induction, CHS is considered to act as a key enzyme in flavonoid biosynthesis [20, 21]. Other hydroxycinnamoyl-CoAs, such as caffeoyl- and feruloyl-CoA, are acceptable as substrates [22, 23] but 4-coumaroyl-CoA is the exclusive substrate *in vivo* [24, 25]. Therefore, naringenin chalcone (4,2′,4′,6′-tetrahydroxychalcone) is the central intermediate in the biosynthesis of flavonoids. This compound is transformed to the corresponding flavanone (naringenin) by a stereo-specific reaction catalyzed by chalcone-flavanone isomerase (5 in Fig. 3) [26]. This reaction also advances spontaneously *in vitro* [27]. There are branched pathways from flavanone to isoflavone and flavone, with reactions catalyzed by isoflavone synthase (8 in Fig. 3) and flavone synthase (9 in Fig. 3), respectively [28]. Modification of flavanone at 3 position also occurs *via* a reaction catalyzed by flavanone 3-hydroxylase (6 in Fig. 3) and dihydroflavonol is formed [29]. This latter enzyme is a cytosolic 2-oxoglutarate-dependent dioxygenase and requires Fe^{2+} ions and ascorbate. Flavonol and anthocyanidin are both synthesized from this compound. Flavonol synthase (10 in Fig. 3) catalyzes the reaction to flavonol [28]. However, the pathway from dihydroflavonol to anthocyanidin has been open to question for a long time. Recently, results of supplementation experiments with genetically defined white flowers of *Matthiola* provided evidence that leucoanthocyanidins are intermediates in the biosynthesis of anthocyanin [30]. The reaction to leucoanthocyanidin is catalyzed by dihydroflavonol 4-reductase (7 in Fig. 3) which requires NADPH [31]. However, the subsequent steps, from leucoanthocyanidin to anthocyanidin remain to be characterized. Pseudobases and their glycosides are postulated to be direct intermediates because of their occurrence in the flowers of several plants [32]. Furthermore, in the *am* mutant of *Pisum sativum*, the pseudobase malvidin 3-rhamnoside-5-glucoside accumulates in the colorless petals [33].

Modification by hydroxylation, methylation, glycosylation and acylation of the A- and B-rings occurs at the C_{15} level. Some of these modifications, for example hydroxylation of B-ring, occur on intermediates, such as flavone and dihydroflavonol, as well as on end products. Thus, the portions of pathways related to modification of flavonoid skeleton form a grid [32]. However, other modifications occur exclusively on the end products.

In the case of plant cell cultures, flavonoid biosynthesis can be induced by various external factors, such as light, phytohormones and elicitors. In parsley cells, which are a very useful plant material for such studies, flavone and flavonol glycosides are synthesized upon UV irradiation [9] and the activities of enzymes related to flavonoid biosynthesis are induced concomitantly. There are significant differences between patterns of induction of group I enzymes, involved in general phenylpropanoid metabolism, and group II enzymes, involved in the flavonoid pathway beyond CHS. The enzymes in group I are induced earlier than those in group II. Maximum activities are recorded at 20 h and 27–40 h, respectively, after the start of induction. Furthermore, activities of enzymes in group I are transiently induced as a result of dilution effects. This phenomenon is not observed with enzymes in group II. From these results, it appears that activities of enzymes in group I and group II are regulated separately [16].

Both PAL and CHS are important enzymes in the flavonoid-biosynthetic pathway. The changes in activities of these two enzymes correspond to those in the levels of several mRNAs [34]. It has been become clear that the accumulation of flavonoid occurs after transcription of the mRNAs that correspond to the biosynthetic enzymes and resultant induction of the activities of these enzymes.

Procedures

Quantification of anthocyanin

Anthocyanin is stable at acidic pH and solutions of anthocyanin are red, as described above. Therefore, extraction and quantification of anthocyanin are performed with 1% HCl-methanol. Harvested plant materials should be extracted with the solvent as soon as possible since anthocyanin in rapidly oxidized and degraded. For preservation, plant materials should be stored in a frozen or dry state. Anthocyanin is extracted repeatedly until samples become colorless. Extracts are filtered and brought to a predetermined volume. The amount of anthocyanin is determined from the absorbance at the wavelength of maximum absorbance by anthocyanin, usually 530 nm.

Identification of anthocyanins

1) Extraction and purification of anthocyanin

Steps in the procedure

1. Extract plant materials with 1% HCl-methanol for a few hours to overnight, and repeat two or three times.
2. Evaporate filtered extracts at 30 °C *in vacuo* and pass through a column of Sephadex LH-20 (Pharmacia) eluted with the same mixture.
3. Purify the anthocyanin fraction further by preparative paper chromatography in two or three solvent systems, such as n-butanol/acetic acid/H_2O (4 : 1 : 2, v/v), 15% acetic acid, and acetic acid/HCl/H_2O (15 : 3 : 82, v/v).
4. Cut out the red band and elute anthocyanin with the same mixture as used for column chromatography.
5. Separate the eluates to dryness. The anthocyanin obtained should be pure.
6. Determine the purity of anthocyanin by paper or thin-layer chromatography (TLC) on cellulose plates with various solvent systems. At this time, care should be taken to monitor the presence of invisible compounds, such as organic acids or other flavonoids. These compounds can be detected under UV light.

Note

1. If the anthocyanin is acylated, storage in 1% HCl-methanol for a long time leads to a loss of the acyl residue. In particular, in the case of zwitterionic anthocyanins, which are acylated by dicarboxylic acids, the acyl moiety is easily dissociated from anthocyanin. Since zwitterionic anthocyanins can be detected easily by simple paper electrophoresis [7, 35], their existence should be monitored. Successful extraction of these types of anthocyanin is dependent on replacement of HCl by acetic, formic or perchloric acid at low concentrations. The extraction of malonyl cyanidin from cell cultures of *C. cyanus* was successfully performed using a mixture of methanol, acetic acid and water (10 : 1 : 9, v/v) [7].

2) Analytical methods

Identification of anthocyanin is usually performed by chromatographic and spectrophotometric methods. In both cases, Rf values and absorbance maxima of aglycons should be compared with those of authentic samples. If no authentic samples are available, they can be obtained from various plants. Many anthocyanins and their sources are listed in 'The Flavonoids. Advances in Research Since 1980' [10]. They are available to serve as standards for identification. However, there are many anthocyanins with similar Rf values and patterns of absorbance. Therefore, various analytical methods must be used to obtain information about components of anthocyanins for their identification. In this section, some of the most readily available methods are described.

1. Acid hydrolysis

Upon acid hydrolysis, an anthocyanin gives rise to its aglycon, sugars and organic acids, if it is an acylated anthocyanin. Therefore, information about the components of an anthocyanin can be obtained by acid hydrolysis.

Steps in the procedure

1. Dissolve purified anthocyanin in 2 N HCl and hydrolyze in boiling water for an appropriate time (e.g. 30–40 min).
2. After cooling, add a volume of distilled water to the reaction mixture that is equal to the initial volume.
3. Remove organic acids to an ethyl ether layer from the inorganic phase and subsequently extract the aglycon with a small amount of isoamyl alcohol.
4. The remaining inorganic phase, containing sugars, is evaporated to dryness over solid NaOH *in vacuo*.
5. Extract sugars from the residue with distilled water.
6. Characterize the aglycon by TLC and spectrophotometry. Solvent systems for TLC are listed in Table 1 and measurement by spectrophotometry is usually carried out in 0.01% HCl-methanol.

Table 1. Solvent systems for thin layer chromatography of aglycons

Solvent system	Ratio (v/v)
acetic acid/HCL/H_2O	30:3:10
formic acid/HCl/H_2O	5:2:3
n-butanol/acetic acid/H_2O	4:1:5 (upper phase)
n-butanol/HCl/H_2O	7:2:5
acetic acid/HCl/H_2O	5:1:5

7. Identify sugars by TLC. The solvent systems listed in Table 2 are used for identification of sugars. Development of sugar spots on TLC plates is achieved by standard methods.

Table 2. Solvent systems for thin layer chromatography of sugars

Solvent system	Ratio (v/v)
n-butanol/acetic acid/H_2O	4:1:2
n-butanol/pyridine/H_2O	6:3:1
ethyl acetate/pyridine/H_2O	2:1:2
isopropanol/pyridine/acetic acid/H_2O	8:8:1:4

2. Alkaline saponification

Anthocyanins acylated with hydroxycinnamic acids and dicarboxylic acids are widespread in nature. Acyl residues can be dissociated from anthocyanins by alkaline saponification.

Steps in the procedure

1. Dissolve anthocyanin in a few drops of a 2 N solution of NaOH and incubate at room temperature for an appropriate time (e.g. 30–40 min) under a stream of nitrogen.

2. Acidify the reaction mixture by addition of 2 N HCl; the reaction mixture turns red.
3. Extract free organic acids with ethyl ether and identify by TLC with some of the solvent systems listed in Table 3.
4. Visualize spots of organic acids by standard methods and phenolic acids can be detected under UV light.

Table 3. Solvent systems for thin layer chromatography of organic acids

Solvent system	Ratio (v/v)
benzene/acetic acid/H_2O	6:7:3
isopropanol/28% aq. ammonia/H_2O	8:1:1
ethyl acetate/acetic acid/H_2O	3:1:1
ethanol/H_2O/aq. ammonia	16:3:1

Note
2. Deacylated anthocyanin remains in the aqueous layer of the reaction mixture. This inorganic phase is evaporated to dryness over solid NaOH *in vacuo* and deacylated anthocyanin is extracted with 1% HCl-methanol. The pigment obtained is used for partial acid hydrolysis (see the section of partial acid hydrolysis, below) and identified by TLC.

3. Degradation by H_2O_2

The chemical bond between carbon atoms at positions 2 and 3 of anthocyanin can be cleaved by H_2O_2. Upon further treatment with weakly acidic or alkaline solution, only the sugars attached at position 3 are dissociated from the main structure. Therefore, this method is suitable for the identification of sugars connected at position 3 of anthocyanidin. If organic acids are attached to the sugar at position 3, acyl sugars are obtained.

Steps in the procedure
1. Dissolve anthocyanin in a 5% solution of H_2O_2 and incubated at room temperature for a few minutes.
2. When the reaction mixture becomes colorless add an equal volume of 1 N NH_4OH to the reaction mixture.
3. Incubate the reaction for a few minutes at room temperature.
4. When the pigment is hard to dissolve in H_2O_2, add 1 N NH_4OH immediately to initiate the reaction.
5. Evaporate the reaction mixture to dryness and identify free sugars by TLC. In the case of acyl sugars, the same solvents can be used as for sugar analysis. In addition, n-butanol/toluene/pyridine/H_2O (5:1:3:3, v/v) and n-butanol/ethanol/H_2O (4:1:2.2, v/v) can also be used.

4. Partial acid hydrolysis
This analytical method provides information about the number and position of sugar moieties attached to anthocyanin and was established by Abe and Hayashi [36].

Steps in the procedure

1. Dissolve anthocyanin, usually deacylated anthocyanin, in 1% HCl-methanol and add an equal volume of 20% HCl.
2. Incubate the reaction mixture at 70 °C.
3. At fixed intervals, remove a portion of the reaction mixture and apply to a TLC plate. Suitable results can be obtained using acetic acid/HCl/H_2O (3 : 1 : 8, v/v) and n-butanol/HCl/H_2O (7 : 2 : 5, v/v; upper phase) as solvent systems. Chromatography using these solvent systems gives the number and the position of sugar moieties, respectively.

Note

3. Under these mild conditions, sugar moieties are gradually dissociated from anthocyanidin. For example, anthocyanidin diglycoside is changed to the monoglycoside and subsequently to the aglycon with time.

3) Identification

Most anthocyanins without complex structures, or unknown anthocyanins, can be identified by their absorbance spectra and Rf values during chromatography. Since the structure of an anthocyanin can be predicted from information obtained by the various analytical techniques described above, the anthocyanin to be identified can be compared with an authentic sample. Analysis by spectrophotometry should be performed in the standard solvent, for example, 0.01% HCl-methanol, because the absorption maximum is dependent on the composition of the solvent, and chromatography should be carried out in three different solvent systems, as shown in Table 4, in order to avoid ambiguities due to similar Rf values.

Table 4. Solvent systems for thin layer chromatography of anthocyanins

Solvent system	Ratio (v/v)
acetic acid/HCl/H_2O	15 : 3 : 82
n-butanol/acetic acid/H_2O	4 : 1 : 2
n-butanol/HCl/H_2O	7 : 2 : 5
1% HCl	
15% acetic acid	

In the cases of anthocyanins with complex structures or similar Rf values, the techniques described above are not sufficient for unequivocal identification. High-performance liquid chromatography (HPLC), nuclear magnetic resonance (NMR) and mass spectrometry (MS) are useful analytical methods for unambiguous identification of such anthocyanins. The outlines of these techniques are given below.

1. High-performance liquid chromatography

High-performance liquid chromatography (HPLC) is a sensitive, rapid technique for the identification, quantification and semi-preparative purification of anthocyanins. It is often the case that anthocyanins with very similar Rf values can be separated by HPLC systems. The usefulness of HPLC has been proven in the identification of anthocyanins from geranium and *Gerbera* by Asen [36]. For the analysis of anthocyanins, a reverse-phase column is particularly suitable, and various solvent systems can be used, for example, a $HClO_4$-methanol gradient and 1.5% H_3PO_4-20% acetic acid, 25% acetonitrile in 1.5% H_3PO_4 gradient [36, 37, 38]. From the retention times of various anthocyanins, it has been shown that hydroxylation and glycosylation increase mobility, while the retention time is increased by the presence of methyl and acyl groups [10].

2. Nuclear magnetic resonance

Nuclear magnetic resonance (NMR) spectroscopy was applied to the elucidation of anthocyanin structures by Goto *et al.* They determined the structures of awobanin, violanin and shisonin by ^{1}H-NMR [38]. Subsequently, the stereostructures of many complex anthocyanins were characterized [39, 40, 41].

In general, anthocyanins are analyzed as flavilium salts in acidic solvents, such as CD_3OD/DCl of D_2O/DCl. Since signals of aglycon protons appear at low field with characteristic features, it is easy to distinguish them from signals of sugar protons. Furthermore, the position of attached glycosyl group can be determined by application of the nuclear Overhouser effect (nOe) and configurations of sugar and the positions of acyl groups can be clarified. However, it is difficult to assign all the proton signals. One successful example can be found in the study of succinylcyanin, present in the blue flowers of *Centaurea cyanus* [42]. Signals from all of the protons, including those of two glucose and a succinic acid moiety, were assigned.

^{13}C-NMR has also be applied to anthocyanin since 1985 [41]. Signals of aglycon carbons appear at a distance from those of sugar carbons, as is the case of ^{1}H-NMR. The main advantage of ^{13}C-NMR over ^{1}H-NMR is the simple spectral pattern. Each carbon atom is recognized from a single resonance line as a result of the decoupling of ^{13}C atoms from proton atoms. The review by Markkam and Chari provides useful information on this topic [43]. However, little information is available about the application of ^{13}C-NMR to identification of anthocyanins. Therefore, the accumulation of spectral data is now required.

Recent advances in NMR techniques are reflected by the availability of two-dimensional NMR. This technique allows correlations between atoms, as well as coupling constants to be estimated more easily. For instance, coupling partners can be clarified immediately by COSY, and NOESY is valuable for elucidation of distances between nuclei that affect one another.

3. Mass spectrometry

In general, samples to be analyzed must be vaporated *in vacuo*. Anthocyanidin and its glucosides should be converted to suitable derivatives by methylation, acetylation or trimethylsilylation because of their low volatility. However, anthocyanins have been difficult to analyze since the molecular weights of samples after derivatization exceeds 1,000. Recently, new techniques for which derivatization of the sample is not required have been established. In particular, fast atom bombardment mass spectrometry (FAB-MS) is useful for studies of anthocyanin. FAB-MS is a straightforward and rapid analytical method which was first applied to anthocyanin by Saito *et al.* [44]. Since 1983, utilization of FAB-MS for the elucidation of structures of anthocyanins has increased. In particular, studies of malonated anthocyanins have progressed markedly using this technique [37, 38, 41, 42].

Samples are applied in a fluid matrix, such as glycerol, and usually a few μg of samples are dissolved in 1 μl of matrix. The solution is irradiated with a beam of fast atoms of Xe, Ar or Cs. At this time, positive or negative molecular and fragment ions are observed. From these data, information about the order and attachment style of sugars and of acyl residues can be obtained. For instance, pelargonidin 3 − (6″-malonylglucoside)-5-glucoside is expected to yield the following fragment ions; as

molecular ion	[M + 1]	681
loss of malonate	[M + 1 − 86]	595
loss of 5-glucose	[M + 1 − 162]	519
loss of malonate and glucose (position 3 or 5)	[M + 1 − 248]	433
aglycon + 1		271

Thus, FAB-MS is suitable for the characterization of anthocyanins. However, mass spectrometry alone is not sufficient for characterization of complex anthocyanins. Therefore, the combination of data from mass spectrometry and other analytical methods, such as NMR, is required.

Assays of the activities of key enzymes

In the biosynthesis of anthocyanin, the key enzymes are PAL and/or CHS. PAL is the enzyme that links primary and secondary metabolism and it regulates not only flavonoid metabolism but also many other aspects of secondary metabolism. CHS catalyzes the condensation reaction between malonyl-CoA and 4-coumaroyl-CoA to synthesize the first flavonoid compound, chalcone. In this section, assays of these two enzymes are described.

1) Preparation of enzymes
All steps for extraction of enzymes are carried out at 4 °C.

Steps in the procedure
1. Homogenize the sample in a chilled mortar with quartz sand and 0.1 M potassium-phosphate buffer (pH 7.5) that contains 1.4 mM 2-mercaptoethanol.
2. Centrifuge the homogenate at 18,000 × *g* for 15 min and stir the supernatant with polyvinylpolypyrrolidone (PVPP) in order to remove phenolic compounds that interfere with the enzymatic reactions.
3. Remove PVPP by centrifugation at 18,000 × *g*.
4. The supernatant is used for assays after exchange of buffer with reaction buffer on a gel-filtration column, for example, Sephadex G-10 (Pharmacia).

2) Assay for PAL
The assay for PAL is performed spectrophotometrically [45]. The reaction mixture consists of 0.1 M Tris-HCl buffer (pH 8.8) that contains 14 mM 2-mercaptoethanol (200 μl), 40 mM L-phenylalanine dissolved in the same buffer (200 μl) and enzyme solution (200 μl).

Steps in the procedure
1. Incubate the reaction mixture at 30 °C for 1 hr.
2. Terminate the reaction by the addition of 200 μl of 25% (w/v) trichloroacetic acid.
3. Remove the denatured protein by centrifuging the mixture at 18,000 × *g* for 15 min.
4. Measure the absorbance of cinnamic acid in the supernatant 280 nm in a spectrophotometer.

3) Assay for CHS
The activity of CHS is determined by using the method described by Kreuzaler and Hahlbrock [8]. The reaction mixture contains 0.2 M potassium phosphate buffer (pH 8.0) plus 2.8 mM 2-mercaptoethanol (50 μl), 1 nmol 4-coumaroyl-CoA dissolved in HCl (pH 4.0; 5 μl), 2.44 nmol [2-^{14}C]-malonyl-CoA in HCl (pH 4.0; 5 μl; 740 Bq) and 50 μl of enzyme solution.

Steps in the procedure

1. Incubate the reaction mixture at 30 °C.
2. After 10 min, add 200 µl of ethyl acetate and agitate the mixture vigorously for extraction of reaction products.
3. Mix 100 µl of the ethyl acetate layer with 4.5 ml of scintillation cocktail and measure radioactivity in a liquid scintillation counter.
4. Analyze an additional 100 µl of the organic phase by thin-layer chromatography on a cellulose plate with 5% acetic acid or a mixture of chloroform, acetic acid and water (10:9:1, v/v) as the solvent system.
5. If side products are observed on the TLC plate in addition to naringenin, which is the main product of the reaction, fractionate the reaction mixture by TLC. The spot of naringenin is then scraped off and naringenin is eluted with ethyl acetate. Then the radioactivity of naringenin is measured as described above.

Further notes

Studies of anthocyanin biosynthesis in cultured cells

Cell cultures have been used for studies of plant physiology and biochemistry because of their many advantages, for example, the ease with which environmental factors are controlled, exclusion of effects due to other tissues or microbial interactions and rapid growth. Now, many types of cell suspension culture that produce various secondary metabolites have been established and successfully exploited for studies of the biosynthesis of these compounds. In particular, studies of the biosynthesis of flavonoids, including anthocyanin, have made major progress as a result of the use of plant cell cultures. In many cases, flavonoid biosynthesis can be induced or enhanced by some external conditions. In this section, some studies of the biosynthesis of anthocyanin in cells in suspension culture are described.

1) Photoregulation
There are three photoreceptors related to flavonoid biosynthesis in parsley cells, namely, phytochrome, cryptochrome and the UV-B receptor. Induction and/or enhancement of the accumulation of anthocyanin by light irradiation has also been reported in various cell cultures.

In cell cultures of *Haplopappus gracilis* that produce cyanidin 3-glucoside, formation of anthocyanin was observed in darkness and was stimulated only by UV light below 345 nm [1]. Concomitant increases in the enzymatic activities of PAL and FS (flavanone synthase; chalcone synthase + chalcone-flavanone isomerase) were also observed under continuous illumination. FS was partially purified and characterized. It is of interest that FS has different pH optima for 4-coumaroyl-CoA and caffeoyl-CoA as substrates. At pH 8, naringenin is synthesized from 4-coumaroyl-CoA while the synthesis of eryodictiol occurs at pH 6.5–7 [40].

Cell suspension cultures of *Centaurea cyanus* were derived from stems of the intact plant which has blue flowers [7]. These cells accumulated no anthocyanin in darkness. However, under irradiation by UV and white light, about 70% of cells synthesized anthocyanin. This anthocyanin was identified as cyanidin 3–(6″-malonylglucoside) which is found in leaves and stems in autumn and winter, but the flower pigment, cyanidin 3–(6″-succinylglucoside)-5-glucoside, found in the parent plant, was not detected [7].

Enzymatic activities of PAL, CHS and CHFI are also induced and/or increased in *C. cyanus* cells, prior to the accumulation of anthocyanin, by illumination. The pattern of induction of these enzymes is very similar to that in parsley cells. Therefore, the possibility of separate regulatory mechanisms is indicated for the enzymes in group I and group II. In darkness, activities of both PAL and CHFI can be observed, but no CHS activity is detected prior to illumination. Furthermore, CHS activity is much lower than that of the other two enzymes. From these results, it is suggested that CHS plays an important role as the key enzyme in the biosynthesis of anthocyanin in *Centaurea* cell cultures [21].

Two cell lines of *Daucus carota* L. were established by Gleitz and Seitz [3]. An anthocyanin-accumulating cell line (DCb) produced anthocyanin even in darkness and increasing amounts of anthocyanin were obtained upon UV (315–420 nm) irradiation. The other cell line (DCs) was isolated from DCb by selection of anthocyanin-free cells. This cell line accumulated no anthocyanin under illumination. However, CHS activity was induced in both cell lines upon illumination. CHS from DCb had an isoelectric point of 6.5 and a molecular mass of 40 kDa per subunit. These properties of CHS are very similar to those of CHS from carrot cell cultures, in which anthocyanin synthesis is induced by transfer to a medium without 2.4-D [47]. By contrast, CHS with an isoelectric point of 5.5 and 43 kDa subunits was obtained from the DCs cell line. These two types of CHS are also found in carrot plants. In the purple flower petals of wild European carrot, only the 40 kDa subunit has been detected and the leaves contain only the 43 kDa subunit. These results indicate that tissue-specific regulation of CHS occurs in intact plants [3].

Another cell suspension culture of *Daucus carota* has been reported. It required transfer of cells to a medium without 2,4-D for induction of the synthesis of anthocyanin prior to light irradiation [4]. Without preculture in 2,4-D-free medium in darkness, no accumulation of anthocyanin was observed. In these cells in culture, exclusion of 2,4-D from the medium changed the state of cells from light-insensitive and the synthesis of anthocyanin occurred by upon subsequent irradiation with visible light. This result suggests that anthocyanin synthesis in this line of carrot cells is regulated by both 2,4-D and light irradiation.

2) Hormonal regulation

Phytohormones, such as auxins, cytokinins and gibberellins, affect not only the growth of plants and plant cells but also the biosynthesis of flavonoids. Both stimulatory and inhibitory effects have been reported in various intact plants and cell cultures. The cell suspension cultures of *Daucus carota*, established by Ozeki and Komamine, provide a unique model system for biosynthesis of anthocyanin that is regulated by 2,4-D [2]. It is well known that carrot cell cultures can be induced to undergo embryogenesis by transfer to auxin-free medium [48]. In this system, cells of low density are selected by density gradient centrifugation and formation of anthocyanin is induced by the same procedure, namely, transfer to auxin-free medium, as the induction of embryogenesis. In this case, the synthesis of anthocyanin can be inhibited by auxins, abscisic acid and gibberellin and promoted by cytokinins in a similar manner to embryogenesis [49]. When the synthesis of anthocyanin is induced, concomitant induction of enzymatic activities related to the biosynthesis of anthocyanin is observed [50]. However, there is no difference in terms of the timing of the initiation of group I and II enzymes, unlike the clear difference seen in parsley and *Centaurea* cells. This difference between plant cells may be due to a difference in the trigger for induction. The induced activities in carrot cells were suppressed by readdition of 2,4-D, and in particular the activities of PAL and CHS fell abruptly. These results suggest that PAL and CHS play important roles in the biosynthesis of anthocyanin that is controlled by 2,4-D [50]. Changes in activities of PAL and CHS are closely correlated with the changes in the levels of the respective enzyme proteins and, furthermore, of their mRNAs. Thus, anthocyanin biosynthesis in carrot cells is regulated at the transcriptional level by 2,4-D [51]. It has been shown that the mRNAs for PAL that are induced, respectively, during the synthesis of anthocyanin and by the dilution effect are transcribed from different genes [51].

3) Relationship between cell growth and the biosynthesis of anthocyanin
As shown for the accumulation of anthocyanin in suspension cultures of *Daucus carota* [52], *Catharanthus roseus* [5] and *Vitis* sp. [6], maximal accumulation of anthocyanin occurs during the stationary phase of growth as does that of most secondary metabolites. In particular, there is a clear negative correlation between cell division and the accumulation of anthocyanin in *Vitis* cells [53].

Nutrients in the medium are very important factors for both cell growth and anthocyanin synthesis. Enhancement of the accumulation of anthocyanin has been observed in suspension cultures of *Populus* [54] and *Vitis* [6]. Furthermore, in *Vitis* cells, Yamakawa *et al.* [6] showed that anthocyanin formation is affected by the ratio of levels of inorganic nitrogen and sucrose in the medium. Phosphate also has a critical effect on cell growth and anthocyanin synthesis. As shown in *Vitis* cells, a high concentration of phosphate in the medium promotes cell division and inhibits the accumulation of anthocyanin [6]. By contrast, when cells are cultured in medium that contains reduced levels of phosphate, cell division is limited and anthocyanin formation is stimulated [53]. Furthermore, Hirose *et al.* [53] demonstrated that the accumulation of anthocyanin was induced by the cessation of cell division upon treatment of cells with aphidicholin which is a specific inhibitor of DNA polymerase α. In *Vitis* cells, it appears that anthocyanin synthesis is regulated by aspects of the physiological state of cells that results from the cessation of cell division [53].

References

1. Wellman E, Hrazdina G, Grisebach H (1976) Induction of anthocyanin formation and of enzymes related to its biosynthesis by UV light in cell suspension cultures of *Haplopappus gracilis*. Phytochemistry 15: 913–915.
2. Ozeki Y, Komamine A (1981) Induction of anthocyanin synthesis in relation to embryogenesis in a carrot suspension culture: correlation of metabolic differentiation with morphological differentiation. Physiol Plant 53: 570–577.
3. Gleits J, Seitz HU (1989) Induction of chalcone synthase in cell suspension cultures of carrot (*Daucus carota* L. ssp. *sativus*) by ultraviolet light: evidence for two different forms of chalcone synthase. Planta 179: 323–330.
4. Takeda J (1990) Light-induced synthesis of anthocyanin in carrot cells in suspension. J Exp Bot 41: 749–755.
5. Hall RD, Yeoman MM (1986) Temporal and spatial heterogeneity in the accumulation of anthocyanin in cell suspension cultures of *Catharanthus roseus* (L.) G. Don. J Exp Bot 145: 1055–1065.
6. Yamakawa T, Kato S, Ishida K, Kodama T, Minoda Y (1983) Production of anthocyanin by *Vitis* cells in suspension culture. Agric Biol Chem 47: 2185–2191.
7. Kakegawa K, Kaneko Y, Hattori E, Koike K, Takeda K (1987) Cell cultures of *Centaurea cyanus* produce malonated anthocyanin in UV light. Photochemistry 26: 2261–2263.
8. Kreuzaler F, Hahlbrock K (1975) Enzymatic synthesis of an aromatic ring from acetate units. Partial purification and some properties of flavanone synthase from cell suspension cultures of *Petroselinum hortense*. Eur J Biochem 56: 205–213.
9. Hahlbrock K, Grisebach H (1979) Enzymic controls in the biosynthesis of lignin and flavonoids. Ann Rev Plant Physiol 30: 105–130.
10. Harborne JB, Grayer RJ (1988) The anthocyanins. In: Harborne JB (Ed.) The Flavonoids. Advances in Research Since 1980, pp. 1–20. Chapman and Hall, New York.
11. Brouillard R (1988) Flavonoids and flower colour. In: Harborne JB (Ed.) The Flavonoids, Advances in Research Since 1980, pp. 525–538. Chapman and Hall, New York.
12. Goto T, Kondo T, Tamura H, Kawahori K, Hattori H (1983) Structure of platyconin, a diacylated anthocyanin isolated from the Chinese bell-flower *Platycodon grandiflorum*. Tetrahedron Lett 24: 2181–2184.
13. Forkmann G, Vlaming P, Spribille R, Wiering H, Schran AW (1986) Genetic and biochemical studies on the conversion of dihydroflavonols to flavonols in flowers of *Petunia hybrida*. Z Naturforsch 41c: 179–186.
14. Teusch M, Forkmann G, Seyffert W (1986) UDP-glucose: anthocyanidin/flavonol 3-O-glucosyltransferase in enzyme preparation from flower extracts of genetically defined lines of *Matthiola incana* R. Br. Z Naturforsch 41c: 699–706.
15. Forkmann G, Stotz G (1981) Genetic control of flavanone 3-hydroxylase and flavonoid 3'-hydroxylase activity in *Antirrhinum majus* (snapdragon). Z Naturforsch 36c: 411–416.
16. Hahlbrock K, Knobloch KH, Kreuzaler F, Potts JRM, Wellmann E (1976) Coordinate induction and subsequent activity changes in two groups of metabolically interrelated enzymes. Light-induced synthesis of flavonoid glycosides in cell suspension cultures of *Petroselinum hortense*. Eur J Biochem 61: 199–206.
17. Jones H (1984) Phenylalanine ammonia-lyase: regulation of its induction and its role in plant develoment. Phytochemistry 23: 1349–1359.
18. Russell DW (1971) The metabolism of aromatic compounds in higher plants. X. Properties of the cinnamic acid 4-hydroxylase of pea seedlings and some aspects of its metabolic and developmental control. J Biol Chem 246: 3870–3878.
19. Knobloch KH, Hahlbrock K (1977) 4-Coumarate: CoA ligase from cell suspension cultures of *Petroselinum hortense* Hoffm. Partial purification, substrate specificity, and further properties. Arch Biochem Biophys 184: 237–248.
20. Ozeki Y, Komamine A, Noguchi H, Sankawa U (1987) Changes in activities of enzymes

involved in flavonoid metabolism during the initiation and suppression of anthocyanin synthesis in carrot suspension cultures regulated by 2,4-dichlorophenoxyacetic acid. Physiol Plant 69: 123–128.

21. Kakegawa K, Hattori E, Koike K, Takeda K (1991) Induction of anthocyanin synthesis and related enzyme activities in cell cultures of *Centaurea cyanus* by UV-light irradiation. Phytochemistry 30: 2271–2273.
22. Spribille R, Forkmann G (1982) Chalcone synthase and hydroxylation of flavonoids in 3'-position with enzyme preparations from flowers of *Dianthus caryophyllus* L. (carnation). Planta 155: 176–182.
23. Beerhues L, Wierman R (1984) Two different chalcone synthase activities from spinach. Z Naturforsch 40c: 160–165.
24. Hrazdina G, Kreuzaler F, Hahlbrock K, Grisebach H (1976) Substrate specificity of flavanone synthase from cell suspension cultures of parsley and structures of release products *in vitro*. Arch Biochem Biophys 175: 392–399.
25. Spribille R, Forkmann G (1982) Genetic control of chalcone synthase activity in flowers of *Antirrhinum majus*. Phytochemistry 21: 2231–2234.
26. Hahlbrock K, Wong E, Schill L, Grisebach H (1970) Comparison of chalcone-flavanone isomerase heteroenzymes and isoenzymes. Phytochemistry 9: 949–958.
27. Mol JNM, Robbins MP, Dixon RA, Veltkamp E (1985) Spontaneous and enzymic rearrangement of naringenin chalcone to flavanone. Phytochemistry 24: 2267–2269.
28. Heller W, Forkmann G (1988) Biosynthesis. In: Harborne JB (Ed.) The Flavanoids, Advance in Research Since 1980, pp. 399–425. Chapman and Hall, New York.
29. Britsch L, Grisebach H (1986) Purification and characterization of (2S)-flavanone 3-hydroxylase from *Petunia hybrida*. Eur J Biochem 156: 569–577.
30. Heller W, Britsch L, Forkmann G, Grisebach H (1985) Leucoanthocyanidins as intermediates in anthocyanidin biosynthesis in flowers of *Matthiola incana* R. Br. Planta 163: 191–196.
31. Heller W, Forkmann G, Britsche L, Grisebach H (1985) Enzymic reduction of (+)-dihydroflavonols to flavan-3,4-cis-diols with flower extracts from *Matthiola incana* and its role in anthocyanin biosynthesis. Planta 165: 284–287.
32. Stafford HA (Ed.) (1990) Flavonoid Metabolism. CRC Press Inc., New York.
33. Crowden RK (1982) Pseudobase of malvidin 3-rhamnoside-5-glucoside in *am* mutants of *Pisum sativum*. Phytochemistry 21: 2989–2990.
34. Chappell J, Hahlbrock K (1984) Transcription of plant defense genes in response to UV light or fungal elicitor. Nature 311: 76–78.
35. Abe Y, Hayashi K (1956) Further studies on paper chromatography of anthocyanins, involving an examination of glycoside types by partial hydrolysis. Studies on anthocyanins XXIX. Bot Mag Tokyo 69: 577–585.
36. Asen S (1984) High-pressure liquid chromatographic analysis of flavonoid chemical markers in petals from *Gerbera* flowers as an adjunct for cultivar and germplasm identification. Phytochemistry 23: 2523–2526.
37. Takeda K, Harborne JB, Self R (1986) Identification and distribution of malonated anthocyanins in plants of the Compositae. Phytochemistry 25: 1337–1342.
38. Goto T, Takase S, Kondo T (1978) PMR spectra of natural acylated anthocyanins: determination of stereostructure of awobanin, shisonin and violanin. Tetrahedron Lett 2413–2416.
39. Cornuz G, Wyler H, Lauterwein J (1981) Pelargonidin 3-malonylsophoroside from the Iceland poppy, *Papaver nudicaule*. Phytochemistry 20: 1461–1462.
40. Bridle P, Loeffler RST, Timberlake CF, Self R (1984) Cyanidin 3-malonylglucoside in *Cichorium intybus*. Phytochemistry 23: 2968–2969.
41. Saito N, Abe K, Honda T, Timberlake CF, Bridle P (1985) Acylated delphinidin glucosides and flavonols from *Clitoria ternatea*. Phytochemistry 24: 1583–1586.
42. Tamura H, Kondo T, Kato Y, Goto T (1983) Structures of a succinyl anthocyanin and a malonyl flavone, two constituents of the complex blue pigment of cornflower *Centaurea cyanus*. Tetrahedron Lett 24: 5749–5752.

43. Markham KR, Chari VM (1982) Carbon-13 NMR spectroscopy of flavonoids. In: Harborne JB, Mabry TJ (Eds.) The Flavonoids: Advances in Research, pp. 19–134. Chapman and Hall, New York.
44. Saito N, Timberlake CF, Tucknott O, Lewis IAS (1983) Fast atom bombardment mass spectrometry of the anthocyanins violanin and platyconin. Phytochemistry 22: 1007–1009.
45. Tanaka Y, Kojima M, Uritani I (1974) Properties, development and cellular-localization of cinnamic acid 4-hydroxylase in cut-injured sweet potato. Plant Cell Physiol 15: 843–854.
46. Saleh NAM, Fritsch H, Kreuzaler F, Grisebach H (1978) Flavanone synthase from cell suspension cultures of *Haplopappus gracilis* and comparison with the synthase from parsley. Phytochemistry 17: 183–186.
47. Ozeki Y, Sakano K, Komamine A, Tanaka Y, Noguchi H, Sankawa U, Suzuki T (1985) Purification and some properties of chalcone synthase from a carrot suspension culture induced for anthocyanin synthesis and preparation of its specific antiserum. J Biochem 98: 9–17.
48. Fujimura T, Komamine A (1979) Synchronization of somatic embryogenesis in a carrot cell suspension culture. Plant Physiol 64: 162–164.
49. Ozeki Y, Komamine A (1986) Effects of growth regulators on the induction of anthocyanin synthesis in carrot suspension cultures. Plant Cell Physiol 27: 1361–1368.
50. Ozeki Y, Komamine A, Noguchi H, Sankawa U (1987) Changes in activities of enzymes in flavonoid metabolism during the initiation and suppression of anthocyanin synthesis in carrot cell suspension cultures regulated by 2,4-dichlorophenoxyacetic acid. Physiol Plant 69: 123–128.
51. Ozeki Y, Komamine A, Tanaka Y (1990) Induction and repression of phenylalanine ammonia-lyase and chalcone synthase enzyme proteins and mRNAs in carrot cell suspension cultures regulated by 2,4-D. Physiol Plant 78: 400–408.
52. Noè W, Langebartels C, Seitz U (1980) Anthocyanin accumulation and PAL activity in a suspension cultures of *Daucus carota* L. Planta 149: 283–287.
53. Hirose M, Yamakawa T, Kodama T, Komamine A (1990) Accumulation of betacyanin in *Phytolacca americana* cells and of anthocyanin in *Vitis* sp. cells in relation to cell division in suspension cultures. Plant Cell Physiol 31: 267–271.
54. Matsumoto T, Nishida K, Noguchi M, Tamaki T (1973) Some factors affecting the anthocyanin formation by *Populus* cells in suspension culture. Agric Biol Chem 37: 561–567.

Plant Tissue Culture Manual **G3**, 1–23, 1995.
© 1995 *Kluwer Academic Publishers.*

Biosynthesis of Monoterpene Indole Alkaloids *in vitro*[†]

WOLFGANG G.W. KURZ, FRIEDRICH CONSTABEL & ROBERT
T. TYLER*
*Plant Biotechnology Institute, National Research Council of Canada, Saskatoon, Saskatchewan, S7N
0W9, Canada; *Department of Applied Microbiology and Food Science, University of Saskatchewan,
Saskatoon, Saskatchewan, S7N 0W0, Canada*

Introduction

The medicinal value of monoterpene indole alkaloids continues to attract interest in their phytochemistry, chemical structure, biosynthesis, and chemothera-
peutic activity. Structural complexity and non-profitability of total synthesis,
perceived or real shortages in the supply of botanicals, and recent concerns for
the environmental impact of wild cropping, made production of these alkaloids
in vitro, i.e. by plant cells cultured in bioreactors, a viable prospect. The synthesis and accumulation of monoterpene indole alkaloids in cell cultures received
great attention when commercial production of compounds like quinine (1), ajmalicine (2) and catharanthine (6), vinblastine (8) and vincristine (9) appeared
attainable. Technologies aimed at increasing alkaloid accumulation by employment of production media, by precursor feeding, elicitation, semi-continuous
culture, and by application of enhanced bioreactors were designed to achieve
this goal. At present, however, efforts to further promote cell cultures as medici-
nals have softened somewhat due to regulatory barriers, but also due to biological barriers. Vinblastine and vincristine, the most desirable *Catharanthus* alkaloids, did not occur in concentrations which would warrant commercial exploi-
tation of an *in vitro* process. Research will be required to overcome these barriers. Research into the biosynthesis of monoterpene indole alkaloids, when
supplemented with molecular biological approaches, appears to be the most sensitive approach.

Plant cell culture

The response of tissues excised from stems and leaves of *Catharanthus roseus*
(L.) G. Don to *in vitro* culture has been most favorable: crown gall tissue, habituated, and regular callus tissue have been grown *in vitro* since 1945. Culture
in bioreactors and occurrence of various alkaloids have been demonstrated as
early as 1969 [5]. DeLuca and Kurz [11] reviewed efforts to produce alkaloids
in *Catharanthus* cell cultures by one- and two-phase culture systems and produced a list of 30 plus alkaloids found. Single cell clones obtained with protoplasts derived from leaves showed extreme variation in alkaloid spectrum [7].
Cryopreservation allowed stability of such clones, potentially over several years

[†] NRC No. 38006.

[6]. Plant regeneration from cells [8], of importance for genetic engineering with this plant, may have gained in efficiency recently through embryogenesis in callus derived from anthers of specific germplasm, i.e. cv Little Delicata, using seeds from Takii & Comp., Tokyo [21].

Biosynthesis and production

Since the discovery of the hypoglycemic effects of *Catharanthus* alkaloids, but in particular since the demonstration of the antileukemic effect of vincaleucoblastine (VBL) in 1958, the structure of *Catharanthus* alkaloids has been under

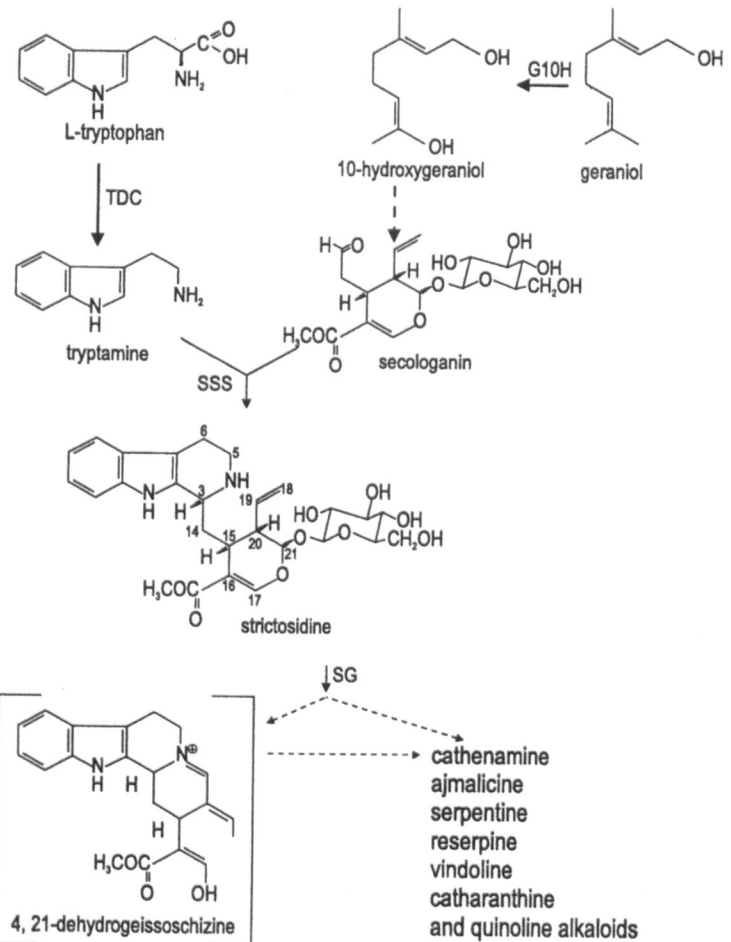

Fig; 1. Schematic representation of the biosynthesis of monoterpenoid indole and quinoline alkaloids (L.H. Stevens, 1994).

Fig; 1. Continued

investigation [44]. Today, the number of compounds isolated and elucidated has exceeded 100 by far.

The biosynthesis, as presented in Fig. 1, has largely been investigated by using radio-labeled precursors. The focus has been on the first steps leading from tryptamine and secologanine to cathenamine and ajmalicine, the transformation of tryptophan/tryptamine to ajmalicine (2), serpentine (3), reserpine (4), vindoline (5), and catharanthine (6), and the condensation of catharanthine and vindoline to 3'4'-anhydrovinblastine (7) and vinblastine (8).

Tryptamine

The rate of decarboxylation of tryptophan to tryptamine was thought to critically affect the production of monoterpene indole alkaloids in *C. roseus* cell cultures. Determination of the tryptamine pool size, but more so the kinetics of
tryptophan decarboxylase (TDC), were early targets. Interestingly, in *C. roseus*
seedlings, the strictosidine synthase (SSS) enzyme activity appeared several days
earlier in seedling development than TDC, and no traces of tryptamine could be
found in young seedlings until the appearance of TDC activity. In cell cultures,
TDC activity coincided with that of the SSS enzyme. Importantly, in cultured
cells, the induction of TDC did not always result in subsequent production of
indole alkaloids [12].

The TDC gene has been isolated, characterized, and expressed in transgenic
tobacco plants, now showing varying levels of tryptamine [40]. The question
remains whether similar transgenic *C. roseus* plants or cells might respond with
varying levels of alkaloid production due to TDC gene over-expression.

Strictosidine

The elucidation of the structure of the first key alkaloidal intermediate in monoterpene indole alkaloid biosynthesis, 3-alpha(S)-strictosidine, has been reviewed
recently [24]. The indole moiety of this alkaloid was demonstrated to be derived
from tryptophan and tryptamine. Its monoterpene origin was confirmed by specific incorporation of geraniol and, subsequently, of loganine and secologanine
in strictosidine and cathenamine/vindoline of *C. roseus* plants. Elucidation of
strictosidine synthesis was corroborated by employing cell-free extracts obtained
with *C. roseus* plants. The enzyme S-adenosyl-L-methionine:loganic acid methyl transferase was partially purified and characterized and, thus, marked the
opening of enzyme identification along the indole alkaloid biosynthestic pathway. The importance of this approach was recognized through characterization
of the tryptamine-secologanine coupling enzyme, dubbed strictosidine synthase
(SSS). The gene for this enzyme has been cloned and expressed heterologously.
Ajmalicine was the first alkaloid for which the biosynthesis was completely
clarified at the enzyme level. Strictosidine synthase activity has been demonstrated for *Catharanthus roseus* and for *Cinchona* spec. cell cultures as well [47].

Ajmalicine (2)

Schübel *et al.* [37] demonstrated a 10-step pathway from strictosidine to ajmalicine. They found raucaffricine, a glycoalkaloid, as intermediate in ajmalicine
biosynthesis. In *R. serpentina* cell cultures, this alkaloid occurred at concentrations of up to 1.6 g/l. *Rauvolfia* cell cultures have shown to be the best produc-
ers of ajmalicine, at levels up to 0.5% of dry weight.

Catharanthine (6)

Feeding radioactively labeled tryptophan to young plants of *Rauvolfia serpentina* Benth. and of *C. roseus* led to incorporation of label into ajmalicine (2), serpentine (3), reserpine (4), vindoline (5), and catharanthine (6). Increased accumulation of catharanthine was reported for cell cultures of *C. roseus* exposed to elicitors. Depending on the cell line, up to 5 and 30 mg/l of catharanthine was recorded with cells exposed to fungal elicitors and to vanadyl sulphate, respectively [14, 38]. Employment of cell lines selected for resistance to 5-methyltryptophan raised the catharanthine level in cells to over 200 mg/l [18].

Vindoline (5)

Vindoline has been shown to occur in *C. roseus* leaves, and appears not to accumulate in cell cultures. As long as it cannot be demonstrated for cells cultured *in vitro*, chances for synthesis of vinblastine through its coupling with catharanthine are dim. The elucidation of the pathway leading to vindoline was in part motivated by hopes to identify and rectify the enzymatic dysfunction in cultured cells. As a result, the synthesis of vindoline has become known to originate from tabersonine by a sequence: aromatic hydroxylation, O-methylation, hydration of 2,3-double bond, N(1)-methylation, hydroxylation at C-4, and 4-O-acetylation. All of these steps have been confirmed by isolation of the respective enzymes and determination of their activity [9, 10]. Studies with germinating seedlings have suggested that the three last steps are expressed in later developmental stages and the last two steps in light-grown seedlings only [1]. Strict developmental control may hinder vindoline synthesis under *in vitro* conditions.

Vinblastine (8)

Chemical coupling of catharanthine and vindoline has been accomplished by a modified Polonovski reaction in which 3'4'-anhydrovinblastine is formed. Biological coupling of the two monomers to vinblastine in *C. roseus* has been demonstrated for seedlings and with cell-free extracts of leaves. Finally, enzymatic conversion of anhydrovinblastine to vinblastine has been reported for cell-free extracts from cell suspension cultures. An economic process for dimer synthesis has yet to be developed. For example, enzymatic coupling of catharanthine and vindoline has been accomplished by catalysis using horseradish peroxidase; anhydrovinblastine yields of between 40 and 50% were obtained under optimized conditions [19]. Also, studies by Kutney [25] have shown that vindoline and catharanthine are being coupled to dimeric indole alkaloids by an enzyme complex which can be isolated from plant cell cultures and which can be immobilized.

The formation of vinblastine in leaves of multiple shoots of *C. roseus* cultured in vitro (15 µg/g dry weight) would confirm tissue and organ development as a prerequisite for vinblastine formation [29], and appears not to be a viable technological prospect.

Monoterpene indole alkaloids – analytical methods

Monoterpene indole alkaloids have been quantitatively determined in plant tissues or tissue/cell cultures from several genera and numerous species. *In vitro* studies have emphasized *Catharanthus roseus, Rauvolfia* spp. and *Tabernaemontana* spp. The various methodologies which have been employed seem generally applicable to any or all species/culture systems, although the presence of particular alkaloids or chlorophyll, the occurrence of alkaloids of interest at exceedingly low concentrations, or a requirement to screen large numbers of samples would influence the choice of methodology. Most assays of monoterpene indole alkaloids are best envisioned as consisting of four steps, namely extraction, purification, separation, and detection/identification. With some methods/species/cultures, however, one or more steps might be substantially reduced in significance, eliminated, or combined with other steps. Examples would include radioimmunoassay methods, where little or no purification of components is required (given suitably low cross-reactivities) and species/cultures accumulating only a single alkaloid in significant concentration, which would obviate the need for separation from similar compounds.

Extraction

The solvent of choice for extraction of monoterpene indole alkaloids from fresh or dried material has usually been methanol, either hot or cold, and with or without prior or simultaneous maceration or sonication of tissue [4, 13, 17, 23, 26, 27, 28, 31, 38, 46]. Other extraction solvents employed include ethanol [32], 70%-aqueous-ethanol [35], 80%-aqueous-ethanol [2], ethyl acetate [39], and methylene chloride [41].

Purification

Crude extracts of monoterpene indole alkaloids from plant tissues or tissue/cell cultures generally require partial purification or "clean-up" prior to separation and detection/identification of individual alkaloids. Typically, purification has taken advantage of the marked effect of pH on the solubility of alkaloids in polar organic solvents. Kurz and Constabel evaporated methanol extracts from *C. roseus* to dryness under vacuum [23]. Acid (1N HCl) extracts of the residue were washed with ethyl acetate, following which the pH of the acid fraction was adjusted to 10 prior to extraction with ethyl acetate. The ethyl acetate fraction was evaporated to dryness under vacuum and the residue taken up in a small volume of ethyl acetate in readiness for separation and identification of the alkaloids present. Smith *et al.* [38] and Loyola-Vargas *et al.* [27] employed more or less similar purification procedures with extracts from *C. roseus*, as did Asada and Shuler [4], Payne *et al.* [32] and Sasse *et al.* [35], who substituted methylene chloride for ethyl acetate, and Morris *et al.* [31], who used chloroform.

Purification of crude extracts may be simplified and made more rapid through the use of Sep-Pak C18 cartridges (Millipore Corporation, Bedford, MA) or similar solid phase extraction technology, as has been described by Morris *et al.* [31], van der Heijden [46] and Lee and Shuler [26]. An extract containing the alkaloid(s) of interest is applied to a miniature liquid chromatography column of appropriate chemistry. The alkaloids adsorb relatively strongly to the column, which allows impurities to be washed from the column prior to elution of the alkaloids with an appropriate solvent and subsequent column regeneration and reuse.

Separation and detection/identification

Although assays of total indole alkaloid content have been performed on occasion [32], plant tissues and tissue/cell cultures typically yield mixtures of monoterpene indole alkaloids which require chromatographic separation prior to detection/identification of individual components.

Thin-layer chromatography (TLC) has been widely employed as a separation technique, both as a preliminary screen prior to some other analytical method such as high pressure liquid chromatography (HPLC) and in TLC-based separation-quantitation methods in their own right. Typically, alkaloid mixtures in methanol, ethyl acetate or some other solvent are spotted on silica-gel-coated TLC plates containing a fluorescent indicator, which are then developed with an appropriate solvent mixture. The method of Kurz and Constabel [23] employs ethyl acetate:methanol (9 : 1, v/v), although many other combinations have been used successfully [4, 13, 26, 28, 30, 31, 36, 39]. Spots corresponding to individual alkaloids are then detected/identified on the basis of one or more of several characteristics, including their UV fluorescence, chromogenic reaction with ceric ammonium sulphate spray reagent [16] or Dragendorf's reagent and sodium nitrite [31], and R_f value. Rapid, direct quantitation on TLC is possible via fluorescence or absorbance densitometry scanning [13, 27, 36]. Alternatively, concentrations of individual alkaloids may be estimated visually by comparison to known concentrations of standards [26] or by spectrophotometric or other analysis of material scraped from TLC plates and eluted with an appropriate solvent.

HPLC appears to be the method most commonly employed for quantitation of monoterpene indole alkaloids. A number of systems have been described, most employing reverse-phase columns, a variety of mobile phases and gradients, and UV or fluorescence detection [4, 13, 22, 31, 38, 39, 46]. Gas-liquid chromatography (GLC) has not been widely applied to the analysis of monoterpene indole alkaloids, although its use for determination of ajmaline in *Rauvolfia vomitoria* has been reported [17].

Radioimmunoassay (RIA) methods have been applied to the quantitation of various constituents in plants and cell cultures, including determination of ajmalicine and serpentine in *C. roseus* [3, 48] and ajmaline in *Rauvolfia* spp. [2]. Stated advantages include precise measurements of very low concentrations of compounds, even in crude extracts, and an ability to be mechanized and applied to the analysis of hundreds of samples per day, making it suitable for large-scale screening of whole plants and cell cultures [3].

Enzymology of indole alkaloid synthesis *in vitro*

Enzymes involved in the biosynthetic pathways of indole alkaloids have been much studied in the past two decades. Due to the regulatory blocks associated with cell differentiation, however, only a limited number of enzymes involved in the synthesis of indole alkaloids could be studied in cell and tissue culture.

The synthesis of indole alkaloids involves the condensation of tryptamine and secologanine to strictosidine. Tryptamine, the indole part of these alkaloids, is formed by decarboxylation of L-tryptophan and catalyzed by tryptophan decarboxylase (TDC-E.C.4.1.1.28) while secologanine, the monoterpenoid part, is derived from geraniol. The subsequent condensation of tryptamine and secologanine is catalyzed by strictosidine synthase (SSS-E.C.4.3.3.2) [30, 45]. Roewer *et al.* [34] found that both TDC and SSS were regulated in a coordinated manner. The following step in the biosynthetic pathway of indole alkaloids involves the removal of the glucose moiety from strictosidine catalyzed by strictosidine-β-D-glucosidase (SG), which has been found to be specific to Apocynaceae producing indole alkaloids [20]. It is believed that the resulting dialdehyde undergoes several intramolecular rearrangements resulting in the formation of 4,21-dehydrogeissoschizine [43], although this compound has so far not been isolated from any *in vitro* bioreaction.

Procedures

Enzyme extraction for TDC and SSS assays [15]

Steps in the procedure
1. Thaw cells (0.3–1.2 g fresh weight) in 1.25 ml grinding buffer containing 0.1 M HEPES, 1 mM DTT and 5 mM EDTA and homogenize for 30–40 seconds with an Ultra-Turrax drive equipped with a 1–10 ml shaft.
2. Transfer the extract to 1.5 ml microfuge tubes and centrifuge for 3 min in a microcentrifuge.
3. Apply the clear supernatants (1 ml aliquots) to Pharmacia PD-10 columns (12.5 ml bed size) which have been pre-equilibrated with 20 mM HEPES, pH 7.6 and 1 mM DTT.
4. Collect protein samples for enzyme assays in 2 ml portions.

Assays for TDC and SSS [15]

The assays for both tryptophan decarboxylase (TDC) and strictosidine synthase (SSS) activities follow basically the ones described by Sasse *et al.* [36] and Mizukami *et al.* [30], respectively.

TDC

Steps in the procedure

1. The reaction mixture for the TDC assay includes 50 nM L-tryptophan containing 2.1×10^{-3} MBq L-[methylene -^{14}C] tryptophan, 4 nM pyridoxal phosphate, 5 μM HEPES, pH 7.5 and enzyme in a total volume of 100 μl.
2. Incubate the mixture for 30 min at 30 °C and stop the reaction by basifying to pH 10 with 10% K_2CO_3.
3. Extract the tryptamine with ethyl acetate and separate by TLC on silica gel using a mixture of $CHCl_3$: MeOH: 25% NH_3 (5 : 4:1) and measure its concentration either directly at 275 nm or, after spraying the plates with 0.1% ninhydrin solution, at 395 nm using a TLC-scanner.
4. Prepare a calibration curve for tryptamine.

SSS

Steps in the procedure
1. The reaction mixture for the SSS assay includes 85 nM tryptamine containing 4.6×10^{-3} MBq tryptamine [side chain-2-^{14}C], 2.8 μM D-gluconic acid lactone, 1.3 μM secologanine, 5 μM HEPES, pH 7.5 and enzyme in a total volume of 120 μl.
2. Incubate the mixture for 30 min at 30 °C and stop by basifying to pH 10 with 10% K_2CO_3.
3. Extract the strictosidine by using three 2 ml portions of ethylacetate.
4. Combine the portions and evaporate.
5. Analyze the extract by thin-layer chromatography on silica gel G using acetone: MeOH: diethylamine (7 : 2:1) for development.
6. Scrape off the band corresponding to strictosidine and count in a scintillation counter.

Enzyme extraction for SG [33]

1. Homogenize washed and frozen cells for 1 min in a Warring Blender at maximum rpm for 1 min.
2. Determine the weight of the frozen cell powder.
3. Add to the frozen cell powder, per gram cell fresh weights, 0.05 g polyvinylpyrrolidone and 1 ml of 0.1 M Tris buffer, pH 8.0 containing 3 mM DTT and 1 mM EDTA.
4. Thaw the material and remove cell debris by centrifugation for 30 min at 10,000×g.
5. Dissolve the pellet in 0.1 M NaH_2PO_4 buffer, pH 6.8, containing 1 mM DTT and 0.02% sodium azide.
6. Desalt this preparation with Sephadex G-25 equilibrated with the same buffer.

Assay for SG [42]

Steps in the procedure
1. Make up the reaction mixture to a total volume of 100 μl, by combining 25 μl of enzyme preparation with 0.625 mM of strictosidine dissolved in 0.1 M NaH_2PO_4 buffer, pH 6.3.
2. Incubate for 60 min at 30 °C.
3. Stop the reaction by adding 100 μl of 5% trichloroacetic acid.
4. Add 20 μl of internal standard (8 mM Codeine HCl) and clarify and analyze by HPLC using an 8 μl flow cell and a wavelength of 280 nm.
5. For the analysis, use a LiChrosorb RP-8 select B column, 7-μm particle size with an inner dimension of 4 mm and a length of 250 mm.
6. Analyse at room temperature at a flow rate of 1 ml/min.
7. Elute with a filtered and degassed mixture of 7 mM sodium dodecyl sulfate and 25 mM NaH_2PO_4 in methanol: water (68 : 32 v/v) pH 6.2.

References

1. Aerts RJ, DeLuca V (1992) Phytochrome is involved in the light regulation of vindoline biosynthesis in *Catharanthus*. Plant Physiol 100: 1029–1032.
2. Arens H, Deus-Neumann B, Zenk MH (1987) Radioimmunoassay for the quantitative determination of ajmaline. Planta medica 53: 179–183.
3. Arens H, Stöckigt J, Weiler EW, Zenk MH (1978) Radioimmunoassay for the determination of the indole alkaloids ajmalicine and serpentine in planta. Planta medica 34: 37–46
4. Asada M, Shuler ML (1989) Stimulation of ajmalicine production and secretion from *Catharanthus roseus*: effects of absorption in situ, elicitors and alginate immobilization. Appl Microbiol Biotechnol 30: 475–481.
5. Carew DP (1975) Tissue culture studies of *Catharanthus roseus*. In: Taylor WI, Farnsworth NR (Eds.) The *Catharanthus* Alkaloids, pp 193–208, Marcel Dekker Inc, New York.
6. Chen THH, Kartha KK, Kurz WGW, Chatson KB, Constabel F (1984) Cryopreservation of alkaloid-producing cell cultures of periwinkle (*Catharanthus roseus*). Plant Physiol 75: 726–731.
7. Constabel F, Rambold S, Chatson KB, Kurz WGW, Kutney JP (1981) Alkaloid production in *Catharanthus roseus* (L.) G. Don. VI. Variation in alkaloid spectra of cell lines derived from one single leaf. Plant Cell Rep 1: 3–5.
8. Constabel F, Gaudet-LaPrairie P, Kurz WGW, Kutney JP (1982). Alkaloid production in *Catharanthus roseus* cell cultures. XII. Biosynthetic capacity of callus from original explants and regenerated shoots. Plant Cell Rep 1: 139–142.
9. De Carolis E., DeLuca V (1993). Purification to homogeneity and characterization of a 2-oxoglutarate dependent dioxygenase involved in vindoline biosynthesis in *Catharanthus roseus*. J Niol Chem 268: 5504–5511.
10. DeLuca V, Balsevich J, Tyler RT, Eilert U, Panchuk BD, Kurz WGW (1986) Biosynthesis of indole alkaloids: Developmental regulation of the biosynthetic pathway from tabersonine to vindoline in *Catharanthus roseus*. J Plant Physiol 125: 147–156.
11. DeLuca V, Kurz WGW (1988) Monoterpene indole alkaloids (*Catharanthus* Alkaloids) In: Constabel F, Vasil IK (Eds.) Cell Culture and Somatic Cell Genetics of Plants, Vol 5, Phytochemicals in Plant Cell Cultures, pp 385–402, Academic Press Inc, New York.
12. DeLuca V, Fernandez JA, Campbell D, Kurz WGW (1988) Developmental regulation of enzymes of indole alkaloid biosynthesis in *Catharanthus roseus*. Plant Physiol 86: 447–450.
13. Duez P, Chamart S, Vanhaelen M, Vanhaelen-Fastré R, Hanocq M, Molle L (1986) Comparison between high-performance thin-layer chromatography-densitometry and high-performance liquid chromatography for the determination of ajmaline, reserpine and rescinnamine in *Rauvolfia vomitoria* root bark. J Chrom 356: 334–340.
14. Eilert U, Constabel F, Kurz WGW (1986) Elicitor-stimulation of monoterpene indole alkaloid formation in suspension cultures of *Catharanthus roseus*. J Plant Physiol 126: 11–22.
15. Eilert U, DeLuca V, Constabel F, Kurz WGW (1987) Elicitor-mediated induction of tryptophan decarboxylase and strictosidine synthase activities in cell suspension cultures of *Catharanthus roseus*. Arch Biochem Biophys 254: 491–497.
16. Farnsworth NR, Blomster RN, Damratoski D, Meer WA, Cammarato LV (1964) Studies on *Catharanthus* alkaloids. VI. Evaluation by means of thin-layer chromatography and ceric ammonium sulfate spray reagent. Lloydia 27: 302–314.
17. Forni GP (1979) Gas chromatographic determination of ajmaline in the bark of the root of *Rauvolfia vomitoria*. J Chrom 176: 129–133.
18. Fujita Y, Hara Y, Morimoto T, Misawa M (1990) Semisynthetic production of vinblastine involving cell cultures of *Catharanthus roseus* and chemical reaction. In: Nijkamp HJJ *et al.* (Eds.) Progress in Plant Cellular and Molecular Biology, pp 738–743, Kluwer Academic Publishers, Dordrecht.

19. Goodbody AE, Endo T, Vukovic J, Kutney JP, Choi LSL, Misawa M (1988) Enzymatic coupling of catharanthine and vindoline to form 3'4'-anhydrovinblastine by horseradish peroxidase. Planta Med 54: 136–140.
20. Hemscheidt T, Zenk MH (1980) Glucosidases involved in indole alkaloid biosynthesis of *Catharanthus roseus* cell cultures. FEBS Lett 110: 187–191.
21. Kim SW, Song NH, Jung KH, Kwak SS, Liu JR (1993) High frequency plant regeneration from anther-derived cell suspension cultures via somatic embryogenesis in *Catharanthus roseus*. Plant Cell Rep 13: 319–322
22. Kurz WGW (1984) Isolation and analysis of alkaloids. In: Vasil IK (Ed.) Cell Culture and Somatic Cell Genetics of Plants, Vol 1, pp 644–650, Academic Press, San Diego.
23. Kurz WGW, Constabel F (1982) Production and isolation of secondary metabolites. In: Wetter LR, Constabel F (Eds.) Plant Tissue Culture Methods, 2nd Revised Edition, pp 128–131, National Research Council of Canada, Ottawa.
24. Kutchan TM (1993) Strictosidine:from alkaloid to enzyme to gene. Phytochem 32: 493–506.
25. Kutney JP (1991) Plant Cell Cultures and synthetic chemistry: a potentially powerful route to complex natural products. Synlett 1: 11–19.
26. Lee CWT, Shuler ML (1991) Different shake flask closures alter gas phase composition and ajmalicine production in *Catharanthus roseus* suspensions. Biotechnology Tech 5: 173–178.
27. Loyola-Vargas VM, Méndez-Zeel M, Monforte-González M, de Lourdes Miranda-Ham M (1992) Serpentine accumulation during greening in normal and tumor tissues of *Catharanthus roseus*. J Plant Physiol 140: 213–217.
28. Merillon J-M, Chénieux JC, Rideau M (1983) Time course of growth, evolution of sugar-nitrogen metabolism and accumulation of alkaloids in a cell suspension of *Catharanthus roseus*. Planta medica 47: 169–176.
29. Miura Y, Hirata K, Kurano N, Miyamoto K, Uchida K (1988) Formation of vinblastine in multiple shoot culture of *Catharanthus roseus*. Planta Med: 54, 18–20.
30. Mizukami H, Nordlöv H, Lee SL, Scott AI (1979) Purification and properties of strictosidine synthetase (an enzyme condensing tryptamine and secologanin) from *Catharanthus roseus* culture cells. Biochem 18: 3760–3763.
31. Morris P, Scragg AH, Smart NJ, Stafford A (1985). Secondary product formation by cell suspension culture. In: Dixon RA (Ed.) Plant Cell Culture: A Practical Approach, pp 127–167, IRL Press, Oxford.
32. Payne GF, Payne NN, Shuler ML (1988) Bioreactor considerations for secondary metabolite production from plant cell tissue culture: indole alkaloids from *Catharanthus roseus*. Biotech Bioeng 31: 905–912.
33. Pennings EJM, van den Bosch RE, van der Heijden R, Stevens LH, Duine JA, Verpoorte R (1989). Assay of strictosidine synthase from plant cell cultures by high-performance liquid chromatography. Anal Biochem 176: 412–415.
34. Roewer IA, Cloutier N, Nessler CL, DeLuca V (1992) Transient induction of tryptophan decarboxylase (TDC) and strictosidine synthase (SS) genes in cell suspension cultures of *Catharanthus roseus*. Plant Cell Rep 11: 86–89.
35. Sasse F, Buchholz M, Berlin J (1983a) Site of action of growth inhibitory analogues in *Catharanthus roseus* cell suspension cultures. Z Naturforsch 36C: 910–915.
36. Sasse F, Buchholz M, Berlin J (1983b) Selection of cell lines of *Catharanthus roseus* with increased tryptophan decarboxylase activity. Z Naturforsch 38C: 916–922.
37. Schübel H, Ruyter CM, Stöckigt J (1989) Improved production of raucaffricine by cultivated *Rauvolfia* cells. Phytochem 28: 491–494.
38. Smith JI, Smart NJ, Misawa M, Kurz WGW, Tallevi SG, DiCosmo F (1987) Increased accumulation of indole alkaloids by some cell lines of *Catharanthus roseus* in response to addition of vanadyl sulphate. Plant Cell Rep 6: 142–145.
39. Smith JI, Amouzou E, Yamaguchi A, McLean S, DiCosmo F (1988) Peroxidase from bioreactor-cultivated *Catharanthus roseus* cell cultures mediates biosynthesis at α-3',4'-anhydrovinblastine. Biotech Appl Biochem 10 568–575.

40. Songstad DD, DeLuca V, Brisson N, Kurz WGW, Nessler CL (1990) High levels of tryptamine accumulation in transgenic tobacco expressing tryptophan decarboxylase. Plant Physiol 94: 1410–1413.
41. Stevens LH, Schripsema J, Pennings EJM, Verpoorte R (1992) Activities of enzymes involved in indole alkaloid biosynthesis in suspension cultures of *Catharanthus, Cinchona* and *Tabernaemontana* species. Plant Physiol Biochem 30: 675–681.
42. Stevens LH (1994) Formation and conversion of strictosidine in the biosynthesis of monoterpenoid indole and quinoline alkaloids. Thesis, University of Leiden, The Netherlands.
43. Stöckigt J (1980). The Biosynthesis of Heteroyohimbine-Type Alkaloids. In: Phillipson JD, Zenk MH (Eds.) Indole and biogenetically related alkaloids, pp 113–141, Academic Press, London.
44. Svoboda GH (1975) Introduction. In: Taylor WI, NR Farnsworth (Eds.) The *Catharanthus* Alkaloids, pp 1–7, Marcel Dekker Inc, New York.
45. Treimer JF, Zenk MH (1979) Purification and properties of strictosidine synthase, the key enzyme in indole alkaloid formation. Eur J Biochem 101: 225–233.
46. van der Heijden R, Lamping PJ, Out PP, Wijnsma R, Verpoorte R (1987) High performance liquid chromatographic determination of indole alkaloids in a suspension culture of *Tabernaemontana divaricata*. J Chrom 396: 287–295.
47. Verpoorte R, van der Heijden R, Schripsema J, Hoge JHC, ten Hoopen HJG (1993) Plant cell biotechnology for the production of alkaloids: present status and prospects. J Nat Prod 56: 186–207.
48. Zenk MH, El-Shagi H, Arens H, Stöckigt J, Weiler EW, Deus B (1977) Formation of the indole alkaloids serpentine and ajmalicine in cell suspension cultures of *Catharanthus roseus*. In: Barz W, Reinhard E, Zenk MH (Eds.) Plant Tissue Culture and its Biotechnological Applications, pp 27–43, Springer-Verlag, Berlin.

Abbreviations

TDC, tryptophan decarboxylase; SSS, strictosidine synthase; DTT, dithiothreitol; EDTA, ethylenediaminetetraacetic acid; HEPES, 4-(2-hydroxyethyl)-1-piperazineethanesulfonic acid; SG, strictosidine-β-D-glucosidase.

Section H:
Tissue Culture Techniques
for Fundamental Studies

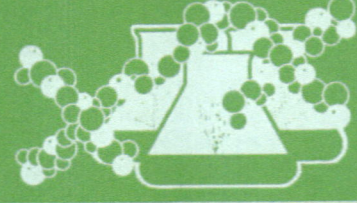

Plant Molecular Biology
Manual

Section H Tissue Culture Techniques for Fundamental Studies

Chapters planned for this section include:-
  Cell cycle studies
  Photoautotrophy
  Membrane permeability and transport
  Biochemical and metabolic studies
  NMR analysis
  Intercellular recognition
  Plant-pathogen interactions
  Tissue-specific gene expression

Plant Tissue Culture Manual **H1**, 1–30, 1995.

Establishment of Photoautrophic Cell Cultures

WOLFGANG HÜSEMANN
*Institut für Biochemie und Biotechnologie der Pflanzen, Universität Münster, Hindenburgplatz 55,
D-48143, Münster, Germany*

Introduction

Under the appropriate light and nutritional conditions, *in vitro* cultured plant cells form chlorophyll, develop functional chloroplasts and thus gain photosynthetic competence.

The mode of nutrition is termed photoheterotrophic or photomixotrophic, if carbon and energy provision of light-grown chlorophyllous cells is mainly from an exogenous sugar in the culture medium, whereas during photoautotrophic growth, carbon and energy provision of the cells is exclusively by photosynthesis.

Sustained photoautotrophic growth of *in vitro* cultured plant cells, in the absence of an exogenous sugar but in the presence of CO_2-enriched air (1%–2% CO_2; v/v) has been achieved now for about 30 different plant species.

The culture technique has advanced to a stage that it is possible now to propagate plant cell cultures under photoautotrophic conditions as a callus on the surface of a nutrient agar or as cell suspensions in small glass flasks, in two-tier culture vessels, as well as in various bioreactor and continuous culture systems (For review see: 8, 20).

The ultimate goal in the work with photoautotrophic plant cell cultures is to eliminate all organic compounds from the medium and to reduce the level of CO_2 to atmospheric concentrations so that the cells are growing in a simple mineral salt medium with ambient CO_2 as the only source of carbon and true photoautotrophism is reached. This has been achieved now for cell suspension cultures of *Chenopodium rubrum* [11], *Dianthus caryophyllous* [15], *Arachis hypogaea* [6], though cell culture growth under atmospheric air is very low. For example, biomass increase in photoautotrophic cell suspensions from *Chenopodium rubrum* is 500–600% within 2 weeks under high CO_2 (2% (v/v) CO_2), but only about 30% increase in biomass under ambient CO_2 concentration.

So far, most photoautotrophic cell cultures have been established from C_3 plants. The only photoautotrophic cell cultures established from C_4-plants are the *Amaranthus* species [21], but without exhibiting a typical C_4 photosynthesis. On the other hand, photosynthetic differentiation leading to the expression of the crassulacean acid metabolism (CAM) as has been achieved for photoheterotrophically growing callus cultures of *Kalanchoe blossfeldiana* [1, 12].

Despite any progress in establishing plant cell cultures, some unresolved problems still exist. This is the complexity of interactions of nutritional, physiological, biochemical and molecular-biological factors including the responsible photoreceptor(s) in controlling chloroplast differentiation in cultured plant cells and the difficulty in inducing chlorophyll formation and chloroplast differentiation in morphologically unorganized cell cultures from the *Gramineae*. Finally, the question of why cultured plant cells in general need high CO_2 concentrations far above ambient air level (1%–2% CO_2; v/v) for sustained photoautotrophic growth has not been answered up to now.

Culture conditions favouring chloroplast development and photosynthetic competence of the cells

As a general rule, lowering the sugar content in the nutrient medium and simultaneously increasing the CO_2 partial pressure far above ambient air level will stimulate photosynthetic development. Under this selection pressure only cells with high photosynthetic capacities will survive. Careful selection of viable green cells during subculture will finally result in photoautotrophically growing cell cultures. However, the culture conditions required for the formation of rapidly growing, friable, nonorganized callus cultures may not always be favourable for the induction of greening. The factors that favour chloroplast development and photosynthetic competence of the cells will be explained briefly.

A. *Culture medium*

Most often the Murashige/Skoog medium [14], and to a lesser extend the B5-medium according to Gamborg [5] are used. Modifications in the composition of the nutrient media are mostly restricted to the use of growth regulators, sugar and other organic components such as vitamins and amino acids.

B. *Phytohormones*

In general, auxin concentrations in the culture medium should kept below 10^{-7} M or auxin activity should be reduced for example by the change from 2,4-dichlorophenoxyacetic acid (2,4-D) to naphthaleneacetic acid (NAA), because auxin concentration optimal for cell growth may suppress chlorophyll formation. The use of indoleacetic acid (IAA) should be avoided, because this phytohormone is readily destroyed in the light.

The demand of cytokinins for chlorophyll formation in cultured cells is species-specific. Kinetin can exert its beneficial effect on chlorophyll formation and photosynthetic capacity of the cells, because it can stimulate chloroplast replication and maturation without inducing cell division.

The inhibitory effect of ethylene on chlorophyll formation in cultured cells can partially be abolished by elevated CO_2-concentrations in the culture atmosphere.

C. Carbohydrates

Based on numerous experimental studies, we know about the inhibitory effect of exogenous sugars of chlorophyll formation and photosynthesis in cultured plant cells. Chlorophyll formation in cultured cells is permitted as sugar concentration in the batch-culture medium is depleted and the specific growth rate declined [2, 3]. The mechanism of this effect is poorly understood. The concentrations of readily utilized sugars such as glucose or sucrose should be reduced to about 1.0%–0.5%.

D. Light

White light from fluorescent tubes (for example: Osram L 36W/11, day light) or from mercury vapor lamps (Osram HQLR de Luxe, 85 W) is used at about $100-120 \ \mu\text{mol m}^{-2} \ \text{s}^{-1}$.

Continuous illumination as well as light/dark cycles (for example: 16 h light/8 h darkness) can be used. As the process of greening and development of chloroplasts in cultured cells is specifically stimulated by the blue region of the visible spectrum [16, 17], in some cases it may be necessary to put the callus cultures exclusively under blue light (400–450 nm; Philips TLD 36W/18) to initiate chloroplast development.

Procedures

Selection and maintenance of photoheterotrophic cell cultures

Experimental trials have now to be undertaken to initiate or stimulate chlorophyll formation in cultured cells. In the author's experience, callus cells growing on the surface of a nutrient agar may be used in preference to suspension cultures. Plating of callus cells on agar medium in Petri dishes permits the separation of even small individual green parts of the callus during subculture. Once chlorophyllous callus cells have been sufficiently multiplied during several passages of subculture, a chlorophyllous suspension culture can be established.

Steps in the procedure

1. Place approximately 3–5 g cells into 50 ml liquid medium in a 200 ml Erlenmeyer flask, seal with aluminium foil and incubate on a gyrotory shaker (120 rpm; Certomat R, B. Braun Biotech International, Melsungen, Germany) in white light at 100–120 μmol m^{-2} s^{-1} and 26 ± 1 °C.
2. Separate cells from early stationary growth phase from the culture medium by filtering on a glass filter (Schott D 2) or through a 0.5 mm nylon gauze.
3. Select by visual monitoring only the greenest cell groups or microcalluses for subculturing.

This process of serial selection has to be continued until a rapidly growing (300%–400% increase in biomass within 14 days) and finely dispersed chlorophyllous cell suspension (50–60 μg chlorophyll/g fresh weight; approximately 5–10 μg chlorophyll/10^6 cells) has been obtained, that can be used as the starting material for the selection of photoautotrophic cell lines.

Selection and growth of photoautotrophic cell cultures

Selection procedures

Major prerequisites for establishing photoautotrophic cell cultures are the presence and maintainance of a high chlorophyll content and photosynthetic competence of the cells even in the phase of active cell division. Usually, the change from the photoheterotrophic to the photoautotrophic mode of nutrition and cell growth is accompanied by a transient drastic reduction in the growth rate, in the chlorophyll content and in the viability of the cells. Therefore, rapidly growing (300–400% increase in fresh weight within 2 weeks) and highly chlorophyllous cell cultures (approximately 5–10 μg chlorophyll/10^6 cells) should be available for this selection procedure.

As a general rule, lowering the sugar content in the nutrient medium and simultaneously increasing the CO_2 partial pressure far above ambient air level (1%–2% CO_2; v/v) will stimulate photosynthetic development. Under this selection pressure only cells with high photosynthetic capacities will survive and take up photoautotrophic growth.

Establishing a photoautotrophic cell culture is a long-term process, involving several months of serial selection under the appropriate selection pressure. For example, it took more than 6 months to obtain photoautotrophically growing cell cultures from *Chenopodium rubrum* [10], *Lycopersicon esculentum* [18], *Glycine max* [7], *Mesembryanthemum crystallinum* (Hüsemann, unpublished).

Principally, 2 different methods are used for selecting cells capable of sustained photoautotrophic growth.

A. Callus induction under photoautotrophic conditions

Callus induction under photoautotrophic conditions was first reported by Yasuda *et al.* [23]. This technique is based on callus induction from explanted leaves in the light at elevated CO_2 concentrations on a sugar-free nutrient agar in the presence of phytohormones.

B. Sequential change of cultured cells from photoheterotrophic to photoautotrophic growth

Most often, photoautotrophic cell lines are selected from highly chlorophyllous, photoheterotrophically growing callus cells. During the transition of the cells from the photoheterotrophic to the photoautotrophic mode of nutrition and growth, sugar may be omitted at once or in several steps from the culture medium. The slow adjustment of the cells to reduced growth rates by gradually lowering the sugar content in the nutrient medium (2%–1%–0.5%–0.25%–0% sugar) in the presence of elevated CO_2-concentrations (1%–2% CO_2; v/v) is obviously more beneficial and even perhaps necessary for plant cell cultures to survive.

For some plant species it may be necessary to keep the oxygen level low and to prevent accumulation of volatile compounds in the culture atmosphere. In this case, the cell cultures must continuously be aerated with a gas mixture of known composition (Fig. 3). Otherwise, the cells can be grown in a closed culture system with CO_2 coming from a carbonate/bicarbonate-buffer (Fig. 1, 2).

The establishment of photoautotrophic cell suspensions from *Chenopodium rubrum* will be described as an example.

Steps in the procedure

1. Highly chlorophyllous and rapidly growing photoheterotrophic cell suspensions (mean chlorophyll content: 50–70 µg chlorophyll/g fresh weight; 5–10 µg chlorophyll/10^6 cells; about 300% increase in fresh weight/10 days) are the starting material for the selection of cells capable of photoautotrophic growth.
2. The two-tier culture vessel system is used with 2% (v/v) CO_2 coming from a 2 M $KHCO_3$/2 M K_2CO_3 buffer solution (Table 1; Fig. 4). Usually 50 ml of the carbonate buffer solution is used to establish the desired CO_2 concentration in the gaseous atmosphere of the culture vessel.
3. Transfer 2 to 4 grams of cells into 30 ml Murashige/Skoog medium supplemented with 10^{-7} to 5×10^{-8} M 2,4-D.
4. Lower the sucrose content of the culture medium stepwise from initially 2% sucrose via 1%, 0.5%, 0.25% to complete omission of sucrose.
5. Grow the cells under continuous light (100–120 µmol m^{-2} s^{-1}) on a gyrotory shaker (Certomat R, B. Braun Biotech International, Melsungen, Germany) at 120 rpm and 26 ± 1 °C.
6. Carefully monitor the cell culture visually during subculture to select highly chlorophyllous cells.

7. Mechanically isolate or separate by filtering through nylon gauze of different pore size (200–600 μm) small green cell groups and subsequently collect on a 100 μm close-meshed nylon gauze.
8. Only transfer the greenest cells into fresh culture medium.

Notes
3. The culture medium for *Chenopodium rubrum* cells contains the mineral salts and vitamins according to Murashige and Skoog [14] supplemented with 10^{-7} M 2,4-dichlorophenoxyacetic acid (2,4-D) and varying amounts of sucrose.
The pH of the culture medium is adjusted to 5.8 prior to autoclaving.
Sterilize separately by autoclaving the empty two-tier culture vessel (openings sealed with aluminium foil), the culture medium (30 ml, Erlenmeyer flask sealed with aluminium foil) and the carbonate buffer solution (50 ml–100 ml, screw-cap Duran glass flask).
Nylon gauze is purchased from W. Babendererde, Hamburg, Germany.

CO_2-supply and composition of the culture atmosphere

As already pointed out, elevated CO_2 concentrations far above ambient air level (1%–2% CO_2, v/v) are a prerequisite for selecting and maintaining photoautotrophic plant cell cultures.

Two different methods can be used to increase CO_2 partial pressure in the culture atmosphere.

A. Supply of CO_2-enriched air in an open culture system

The cells are aerated with a gas mixture of known composition enriched in CO_2. By this procedure, the accumulation of volatile compounds like ethylene in the culture atmosphere can be prevented and the oxygen concentration can be kept at a desired low level to reduce photorespiration.

A gas mixture of known composition can be established by mixing together the appropriate volumes of CO_2, N_2 and O_2 using specifically designed gas flow-meters. Otherwise, the appropriate volumes of CO_2-free air (air can be kept CO_2-free by passing through a column of soda lime, carbosorb brand, 10–16 mesh) are mixed together with pure CO_2 to give the desired CO_2-concentration in the air.

For example, a gas flow rate of about 6 ml min^{-1} will sufficiently aerate 50 ml cell suspension growing in 200 ml Erlenmeyer flasks.

B. CO_2-supply from a bicarbonate/carbonate-buffer mixture

High CO_2-concentrations varying from 0.04% to 3% CO_2 (v/v) can be established by a 2 M $KHCO_3$/2M K_2CO_3 buffer solution [19] in the closed system of a two-tier culture vessel [9, 10].

Under these conditions the oxygen level in the closed culture unit will be adjusted by photosynthesis and respiration of the cells and volatile compounds may accumulate.

The CO_2 partial pressures above the buffer solutions can be calculated for 25 °C after the formula $(KHCO_3)_2/(K_2CO_3 \times CO_2) = K$. The value for $K_{25 °C}$ has been determined by interpolating between $K_{20 °C} = 3.35 \times 10^{-2}$ and $K_{30 °C} = 1.78 \times 10^{-2}$ (mol/liter/mm Brodie) according to Warburg [19].

Some $KHCO_3$/K_2CO_3 ratios for establishing different CO_2 concentrations in the culture atmosphere are given in Table 1.

Table 1. Establishment of high CO_2 concentrations by bicarbonate/carbonate buffer mixtures

Buffer mixtures: ml 2M $KHCO_3$ / ml 2M K_2CO_3	CO_2 concentration* % (v/v)
78/22	2.0
60/40	0.7
45/55	0.5
40/60	0.25
30/70	0.13
20/80	0.05

* CO_2 concentration in the gaseous atmosphere above the buffer solution in a closed glass flask has been measured using an infrared gas analyser (Finor, Maihak, Hamburg, Germany).

Light

White light from fluorescent tubes (Osram 36W/11, day light; Philips 36W/84) or from mercury vapor lamps (Osram HQLR de Luxe, 85W) is used. Photon flux density of the photosynthetic active radiation (PAR, 400–700 nm) ranging between 100–300 μmol m^{-2} s^{-1} are used. Both continuous illumination as well as light/dark cycles (for example: 16 h light; 8 h darkness) can be used. As continuous illumination supplies more energy to the cells, it will possibly reduce loss of carbon during dark respiration and therefore may be more beneficial to achieve higher growth rates of *in vitro* cultured cells. On the other hand, light-dark-changes will allow 'normal' metabolic processes as they occur in the intact plant.

Culture vessels used for photoautotrophic cell culture growth

A number of culture vessels different in size and shape, such as multiwell dishes, small conical flasks (50–200 ml), two-tier culture vessels, bubble tubes or airlift fermenter systems are used to grow photoautotrophic cell cultures.

A. Closed culture systems

Multi-well dishes for selecting and growing callus cells under photoautotrophic conditions

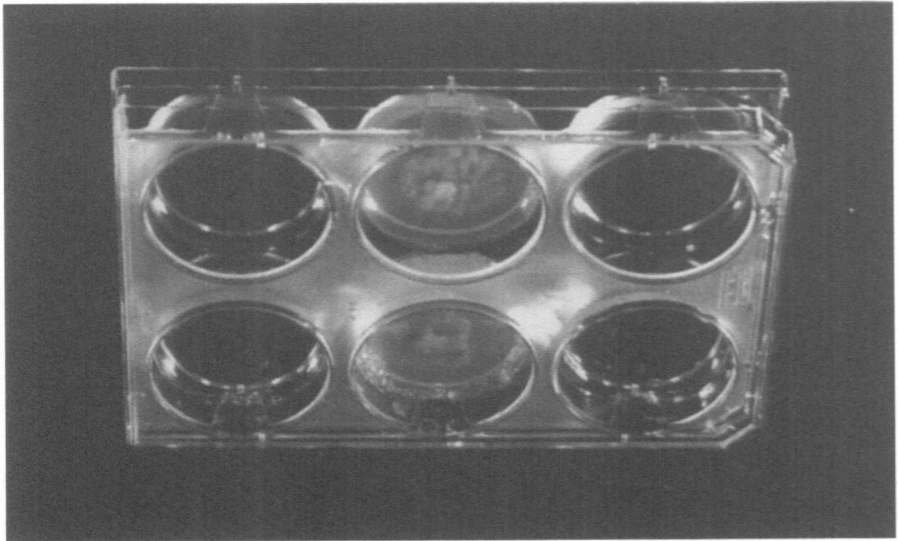

Fig. 1. Multiwell dish for growing photosynthetically active callus cells under elevated CO_2 concentrations

Steps in the procedure
1. Use presterilized transparent flat bottom plates with lid, containing 6 wells (well diameter 35 mm; depth 17 mm) to select and/or maintain callus cells capable of photoautotrophic growth (Fig. 1).
2. In three or four wells place 5 ml sterile 2 M $KHCO_3$/2 M K_2CO_3 buffer mixtures to establish elevated CO_2 concentrations (0.05%–2% CO_2, v/v) in the gas phase of the dishes.
3. In the remaining wells place 5 ml nutrient agar which contains no or reduced amounts of sucrose.
4. As an inoculum, plate out 0.5–1.0 g cells on nutrient agar. Selection

and/or maintenance of callus cells capable of photoautotrophic growth occurs under sugar famine at elevated CO_2 concentrations.

5. Seal the plates tightly with Parafilm to prevent loss of CO_2 and desiccation and put under continuous or intermittent white light $(100–120\ \mu\text{mol m}^{-2}\text{s}^{-1})$ at $26 \pm$ °C.

Notes

1. Presterilized multi-well plates can be purchased from Sigma Chemical Company, Deisenhofen, Germany as well as from C.A. Greiner, Labortechnik, Nürtingen, Germany.

3. Composition of the nutrient agar: The appropriate sugar-free culture medium solidified with 0.9% Difco-agar.

The two-tier culture vessel

The two-tier culture vessel has been developed for the photoautotrophic growth of cell suspensions of *Chenopodium rubrum* [9, 10]. In the author's laboratory, this culture method has been used successfully for the propagation of photoautotrophic cell cultures from 10 different plant species.

Fig. 2. The two-tier culture vessel.

The two-tier culture vessel (Fig. 2) is constructed of two 200 ml small-necked Erlenmeyer flasks, that are connected top-to-bottom, via taper joints 29/32. The upper flask contains a central glass tube (2.0 cm inner diameter; 3.5 cm length within the flask; 3.5 cm length outside the flask including taper joint 29/32. The lower flask contains 50 ml of a 2 M $KHCO_3$/2 M K_2CO_3 buffer solution for establishing the desired CO_2 concentrations in the culture atmosphere. The liberated CO_2 passes through the central glass tube into the upper compartment, that serves as the culture vessel, for propagating the cells in 30 ml sugar-free culture medium.

The two-tier culture flasks (the openings closed by aluminium foil), the culture medium (30 ml), and the bicarbonate/carbonate-buffer solution are separately sterilized by autoclaving for 20 min at 120 °C. The buffer mixture must be filled into a tightly screw-capped glass flask (to prevent loss of CO_2) made of Duran glass (Schott, Germany) that will withhold the pressure during autoclaving.

For routine subculturing 1–2 g cells are inoculated in 30 ml culture medium. For transferring the cells, a stainless steel spoon or pipette may be used.

Sufficient CO_2 (initially 2% CO_2, v/v) is produced by 50 ml of a 2 M $KHCO_3$/2 M K_2CO_3-buffer to allow sustained photoautotrophic cell culture growth (increase in fresh weight up to 500%) for 2–3 weeks. At the end of a 14 days subculture period the CO_2 content in the culture atmosphere has changed from initially 2% (v/v) CO_2 to approximately 0.5% (v/v) CO_2, still sufficiently high to support photoautotrophic cell culture growth.

The openings of the culture flask are tightly sealed with sterile aluminium foil and finally with Parafilm to reduce gas exchange with the atmosphere. Using the two-tier culture vessel system, photoautotrophic cell suspensions of *Chenopodium rubrum* are propagated on a gyrotory shaker (120 rpm) illuminated with continuous white light (100–120 μmol m^{-2} s^{-1}) at 26 $\pm$ 1 °C.

B. Open culture systems

The composition of the gas phase developed in the culture vessels may severely affect cell growth and chlorophyll formation. In cell suspension from *Spinacia oleracea* ethylene accumulation and oxygen concentrations above air saturation inhibited greening [2, 4, 13]. Reducing the oxygen content to about 20% of air saturation (photorespiration is reduced without inhibiting mitochondrial respiration) and increasing the CO_2 concentrations to abolish the inhibitory effect of ethylene, favoured sustained chlorophyll formation. In cases where the cultured cells are susceptible to enriched oxygen levels, ethylene or other volatile metabolic compounds, that may accumulate in the closed two-tier culture vessel, it will be necessary or at least beneficial to grow the cells in an open culture system.

Aeration of the cells with CO_2-enriched air in small culture volumes

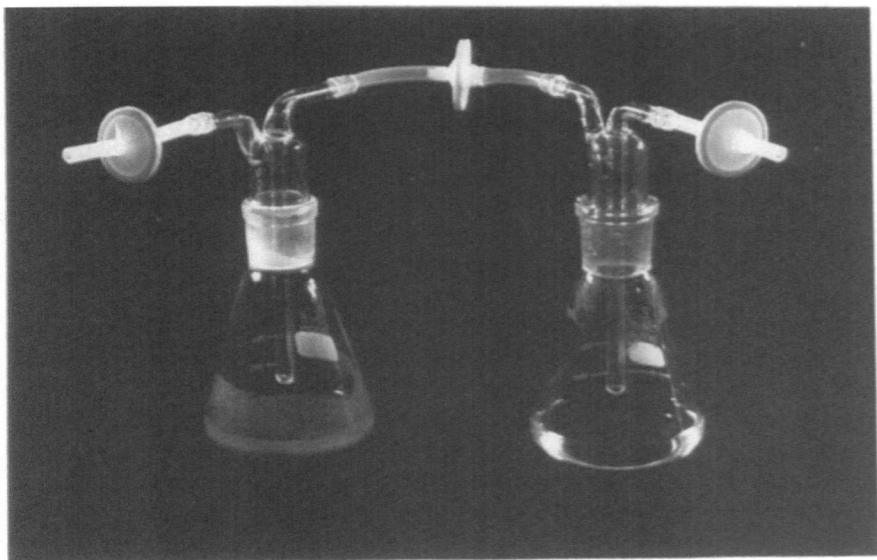

Fig. 3. An open culture system for aerating cells with CO_2-enriched air.

Steps in the procedure

1. Suspend the cells in 50 ml liquid culture medium or grow as callus masses on a nutrient agar in 200 ml Erlenmeyer flasks (Fig. 3).
2. Continuously flush with CO_2-enriched air (1%–2% CO_2, v/v) or with a gas mixture of known composition (1%–2% CO_2, v/v; reduced oxygen levels, 5%–10% O_2, v/v) at 26 ± 1 °C.
3. Keep the gas mixture aseptic and wetted by passing a sterile filter (Milipore filter unit, 0.45 μm pore size) and a water-filled glass flask (humidifier) before entering the culture vessel via small glass tubes.
4. Attach 2 flasks to each humidifier by branching the tubings.
5. Keep the cell cultures under white fluorescent light (continuous or intermittent illumination; 100–120 μmol m^{-2} s^{-1}) at 26 ± 1 °C.
6. Agitate cell suspensions at 120 rpm on a gyrotory shaker (Certomat R, B. Braun Biotech International, Melsungen, Germany).

This culture technique as has been applied successfully for photoautotrophic cell cultures of *Spinacea oleracea* [4], *Nicotiana tabacum* [22] and *Glycine max* [7].

Culture vessel for photoautotrophic cell culture growth under ambient air

The reasons that cultured plant cells in general require highly elevated CO_2 concentrations far above ambient air level for sustained photoautotrophic growth are still unknown. Meanwhile, cell cultures from *Arachis hypogaea, Euphorbia characias, Dianthus caryophyllous* and *Chenopodium rubrum* can be grown photoautotrophically under ambient CO_2 concentrations but at drastically reduced growth rates compared to cells growing under high CO_2.

For photoautotrophic cell culture growth under ambient air the culture flask (200 ml Erlenmeyer flask) is closed by silicon sponge closures (Fig. 4).

Fig. 4. Culture vessel for photoautotrophic cell culture growth under ambient air.

These closures (silicon rubber rings with sealed-in silicone sponge filters) fit well around the neck of 200 ml Erlenmeyer flasks. They permit efficient exchange of gases, but reduce evaporation and prevent passage of aerosols, are reusable and autoclavable and are available from Sigma-Techware, Sigma Chemical Company, St. Louis, U.S.A.

Using this culture technique as described, photoautotrophic cell suspensions from *Chenopodium rubrum* have been grown under atmospheric air for several years.

Determination of photoautotrophic growth parameters

Cell culture growth is monitored throughout the growth cycle by measuring increase in fresh biomass, cell number, packed cell volume, accumulation of chlorophyll, photosynthetic oxygen development and respiratory oxygen consumption.

Procedures

Fresh weight
Determine cell fresh weight after collecting the cells on a fiberglass filter under vacuum.

Packed cell volume
Sediment cells from 1 ml aliquots of the cell suspension by centrifugation (swinging buckets; $g = 200$; 10 min) using calibrated conical glass tubes (Schlee, Witten, Germany).

Cell number
Steps in the procedure
1. Suspend 40 mg cells (fresh weight) in 5 ml chromic acid (10%, w/v) and incubate at 70 °C for 5–10 min.
2. Disperse the cells by drawing the suspension through a needle (1 mm; 40 mm) attached to a syringe.
3. Count the cells in a Fuchs-Rosenthal counting chamber (depth: 0.2 mm) under an inverted microscope (Leitz-Diavert, Leitz, Wetzlar, Germany).

Chlorophyll content
Steps in the procedure
1. Weigh out 200 mg fresh cell biomass into 2 ml round-bottom plastic centrifugation tubes, rapidly freeze, resuspend and extract chlorophyll from the cells with 80% acetone (v/v) by stirring on a magnetic stirrer using magnetic stirring bars.
2. Centrifuge the tubes and pour the clear chlorophyllous supernatant into a graduated glass tube.
3. Repeat this procedure until the pellet is completely free of chlorophyll.
4. Measure absorbance of the chlorophyllous acetone extract in a spectrophotometer at 647nm and 664 nm and calculate concentration of chlorophyll a and b according to the formula of Ziegler and Egle [24]:
$C_{Chl.a} = 11.78 \times E_{647} - 2.29 \times E_{664}$ (μg chlorophyll a/ml acetone extract);

$C_{Chl.b} = 20.05 \times E_{664} - 4.77 \times E_{647}$ (μg chlorophyll b/ml acetone extract).

Measurements of photosynthetic and respiratory activities
Steps in the procedure
Photosynthetic oxygen production and respiratory oxygen consumption are measured using a Hansatech oxygen meter, equipped with a Clark-type electrode purchased from Bachofer, Reutlingen, Germany.

1. Resuspend 100 mg cells in 1 ml of the same medium in which they had been grown before, supplemented with 20 μM KHCO$_3$, and place into the electrode chamber.
2. After 1 min preincubation, determine dark respiratory oxygen consumption and finally photosynthetic oxygen production in the light (600 μmol m^{-2} s^{-1}).

References

1. Brulfert J, Mricha A, Sossoutov L, Queiroz O (1987) CAM induction by photoperiodism in green callus cultures from a CAM plant. Plant, Cell and Environment 10: 443–449.
2. Dalton C (1980) Photoautotrophy of spinach cells in continuous culture: Photosynthetic development and sustained photoautotrophic growth. J Exp Bot 31: 791–804.
3. Dalton CC (1984) The effect of sugar supply rate on photosynthetic development of *Ocimum basilicum* (sweet basil) cells in continuous culture. J Exp Bot 35: 505–516.
4. Dalton CC, Street HE (1976) The role of the gas phase in the greening and growth of illuminated cell suspension cultures of spinach (*Spinacia oleracea*). In Vitro 12: 485–493.
5. Gamborg O, Miller RA, Ojima K (1968) Nutrient requirements of suspension cultures of soybean root cells. Exp Cell Res 50: 151–158.
6. Gross U, Gilles F, Bender L, Berghöfer P, Neumann KH (1993) The influence of sucrose and elevated CO_2 concentration on photosynthesis of photoautotrophic peanut (*Arachis hypogaea* L.) cell cultures. Plant Cell Tiss Org Cult 33: 143–150.
7. Horn ME, Sherrard J, Widholm JM (1983) Photoautotrophic growth of soybean cells in suspension culture. Plant Physiol 72: 426–429.
8. Hüsemann W (1985) Photoautotrophic growth of cells in culture. In: Vasil IK (Ed.), Cell Culture and Somatic Cell Genetics of Plants, Vol II, Cell Growth, Nutrition, Cytodifferentiation, and Cryopreservation, pp. 213–252, Academic Press, New York.
9. Hüsemann W (1984) Photoautotrophic cell cultures. In: Vasil IK (Ed.) Cell Culture and Somatic Cell Genetics of Plants, Vol I, Laboratory Procedures and their Applications, pp. 182– 191, Academic Press, New York.
10. Hüsemann W, Barz W (1977) Photoautotrophic growth and photosynthesis in cell suspension cultures of *Chenopodium rubrum*. Physiol Plant 40: 77–81.
11. Hüsemann W, Fischer K, Mittelbach I, Hübner S, Richter G, Barz W (1989) Photoautotrophic plant cell cultures for studies on primary and secondary metabolism. In: Kurz WGW (Ed.) Primary and Secondary Metabolism of Plant Cell Cultures II, pp. 35–46, Springer-Verlag, Berlin, Heidelberg.
12. Kluge M, Hell R, Pfeffer A, Kramer D (1987) Structural and metabolic properties of green tissue cultures from a CAM plant, *Kalanchoe blossfeldiana*, hybr. *Montezuma*. Plant, Cell and Environment 10: 451–462.
13. Laulhere JP, Aguettaz P, Lescure A (1984) Regulation of the oxygen exchanges and the greening by controlled supplies of sugar in photomixotrophic spinach cell suspensions. Physiol Veg 22: 765–773.
14. Murashige T, Skoog F (1962) A revised medium for rapid growth and bioassays with tobacco tissue cultures. Physiol Plant 18: 473–497.
15. Rebeille F (1988) Photosynthesis and respiration in air-grown and CO_2-grown photoautotrophic cell suspension of carnation. Plant Sci 54: 11–21.
16. Richter G, Hundrieser J, Groß M, Schultz S, Bottländer K, Schneider Ch (1984) Blue light effects in cell cultures. In: Senger H (Ed.) Blue Light Effects in Biological Systems, pp. 387–396, Springer-Verlag, Berlin, Heidelberg.
17. Richter G, Dudel A, Einspanier R, Dannhauer I, Hüsemann W (1987) Blue-light control of mRNA level and transcription during chloroplast differentiation in photomixotrophic and photoautotrophic cell cultures (*Chenopodium rubrum* L.). Planta 172: 79–87.
18. Stöcker S, Guitton M-Ch, Barth A, Mühlbach HP (1993) Photosynthetically active suspension cultures of potato spindle tuber viroid infected tomato cells as tools for studying viroid-host cell interaction. Plant Cell Reports 12: 597–602.
19. Warburg O, Geissler AW, Lorenz S (1962) Neue Methode zur Bestimmung der Kohlensäuredrücke über Bicarbonat/Carbonatgemischen. In: Warburg O (Ed.) Weiterenwicklung der zellphysiologischen Methoden, pp. 578–581, Thieme-Verlag, Stuttgart.
20. Widholm JM (1992) Properties and uses of photoautotrophic plant cell cultures. In: International Review of Cytology, Vol 132, pp. 109–175, Academic Press, New York.

21. Xu Ch, Blair L, Rogers SMD, Godvindjee K, Widholm JM (1988) Characteristics of five new photoautotrophic suspension cultures including two *Amaranthus* species and a cotten strain growing on ambient CO_2 levels. Plant Physiol 88: 1297–1302.
22. Yamada Y, Imaizumi K, Sato F, Yasuda T (1981) Photoautotrophic and photomixotrophic culture of green tobacco cells in a jar-fermenter. Plant Cell Physiol 22: 917–922.
23. Yasuda T, Hashimoto T, Sato F, Yamada Y (1980) An efficient method of selecting photoautotrophic cells from cultured heterogenous cells. Plant Cell Physiol 21: 929–932.
24. Ziegler P, Egle K (1965) Zur quantitativen Analyse der Chloroplastenpigmente. Beitr Biol Pflanzen 41: 11–37.

Plant Tissue Culture Manual **H2**, 1–15, 1995.
© 1995 *Kluwer Academic Publishers.*

Zinnia mesophyll culture system to study xylogenesis

MUNETAKA SUGIYAMA & HIROO FUKUDA
Biological Institute, Faculty of Science, Tohoku University, Aoba-yama, Sendai 980–77, Japan

Introduction

Xylem cells are formed from the procambium of the root and shoot in the primary xylem and from the vascular cambium in the secondary xylem. The main components of xylem cells are tracheary elements, which are characterized by the formation of secondary cell walls that thicken with annular, spiral, reticulate or pitted patterns. At maturity, differentiating tracheary elements lose their nuclei and cell contents, leaving behind hollow tubes that form vessels and tracheids. *In vitro*, tracheary elements can be induced from the parenchymatous cells of various plant species by wounding and/or the application of phytohormones [1].

Fukuda and Komamine [2] established an *in vitro* experimental system in which single mesophyll cells of *Zinnia elegans* can redifferentiate directly into tracheary elements without cell division, based on the work of Kohlenbach and Schmidt [3]. The *Zinnia* system is considered to be very well suited for studies of redifferentiation into tracheary elements, because it possesses various characteristic features, which include a high frequency of redifferentiation and a high degree of synchrony (Table 1). A number of new physiological, biochemical and molecular markers, by which the physiological stages of the redifferentiation can be defined, have been found in the *Zinnia* system [1, 4, 5, 6]. Expression of some of these markers has been demonstrated to occur in close association with xylem development *in situ*, as well as *in vitro* [7]. Thus, the *Zinnia* cell system has contributed greatly to the study of xylogenesis. It is anticipated that this system will be employed more extensively in the future. Therefore, we describe here the basic method for culture of isolated *Zinnia* mesophyll cells, together with cytological methods for monitoring the process of differentiation of mesophyll cells into tracheary elements [2, 8].

Table 1. Merits of the *Zinnia* system

1. Single-cell system
   A) Little cell-cell interaction
   B) Homogeneous stimulation
   C) Ability to follow visually the sequence of cytodifferentiation
2. Homogeneous starting materials that are previously uncommitted to cytodifferentiation
3. High synchrony and frequency of cytodifferentiation
4. Strict hormonal regulation

Plant material

The first true leaves of young seedlings of *Zinnia elegans* L. are used as the source of mesophyll cells that are to be cultured *in vitro*. The use of healthy seedlings is essential for obtaining viable cells from leaves. Therefore, *Zinnia* seedlings should be grown with great care.

Steps in the procedure
1. Sow seeds of *Zinnia elegans* cv. Canary bird or Envy in moisten vermiculite.
2. Grow seedlings at 25 °C for 14 d under a cycle of 14 h of light (approx. 100 μmol/m²/s, white light from fluorescent lamps) and 10 h of darkness.
3. Water seedlings daily or every other day and feed with 150-fold-diluted HYPONeX (Murakamibussan, Tokyo, Japan) once in the first 5 d.

Notes
(Numbers refer to steps listed above)
1. Seeds of other varieties of *Zinnia elegans* can be used, although there is a slight variation among the rates of differentiation in mesophyll cells isolated from different varieties of seedlings.
2. White fluorescent light should be used. The use of other types of light, such as that from a xenon lamp, sometimes results in failure in the isolation of viable cells, probably because such light induces a high level of phenolics in leaves. Vermiculite should not be allowed to dry up. To avoid bacterial contamination of subsequent cell cultures, it is important to maintain low humidity (approx. 50%) during the growth of seedlings.
3. Water should not come in contact with leaves. Wet leaves are very often associated with bacterial contamination in the subsequent cell culture.

Isolation and culture of mesophyll cells

Since attachment between cells is weak in mesophyll tissues of *Zinnia elegans*, single mesophyll cells can easily be isolated by mechanical maceration of leaves. Steps in the procedure for the isolation and culture of mesophyll cells are described below and are summarized in Fig. 1.

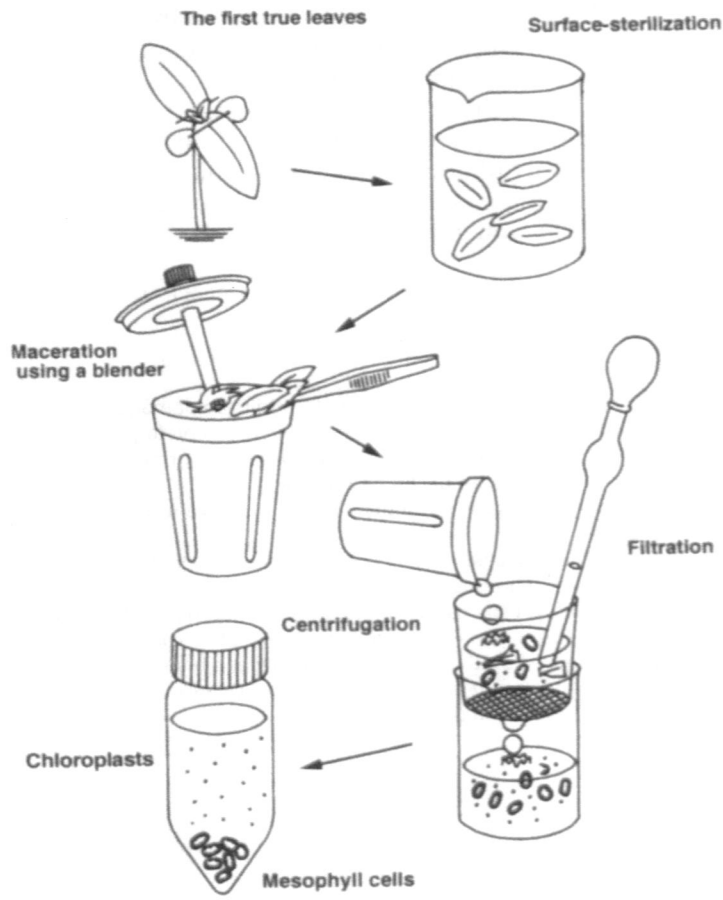

The first true leaves

Surface-sterilization

Maceration using a blender

Filtration

Centrifugation

Chloroplasts

Mesophyll cells

Fig. 1. Procedure for the isolation of mesophyll cells of *Zinnia elegans*.

Steps in the procedure
1. Harvest the first true leaves (60–80 leaves) that are 3 to 4 cm in length.
2. Surface-sterilize the leaves by soaking them for 10 min in 1 liter of

a solution of 0.15% sodium hypochlorite supplemented with 0.001% Triton X-100. Immerse floating leaves occasionally.

3. Rinse the leaves three times with autoclaved, distilled water.
4. Transfer the leaves to the 100 ml stainless-steel cup of a Waring-type homoblender containing 60–80 ml of culture medium.
5. Macerate the leaves at 8,000 rpm for 30 s.
6. Filter the homogenate through a nylon screen with a pore size of 50 to 80 μm. Agitate the homogenate on the screen by repeated pipetting during this filtration.
7. Centrifuge the filtrate at 200 × g for 1 min.
8. Remove the supernatant with a pipette and discard it. Suspend the pelleted cells in 80 ml of medium.
9. Centrifuge again at 200 × g for 1 min.
10. Resuspend the pelleted cells in medium (normally 300–500 ml) at a cell density of about 5×10^4 cells/ml.
11. Distribute the suspension of cells into culture tubes (20 ml for a tube of 30 mm i.d. × 200 mm and 3 ml for a tube of 18 mm i.d. × 180 mm).
12. Incubate culture tubes in darkness at 27 °C on a revolving drum (10 rpm).

Notes

2. Leaves should be handled gently to minimize damage to tissues.
3. Sodium hypochlorite should be removed completely.
5. It is also possible to release mesophyll cells by grinding the leaves gently in medium with a smooth pestle and mortar.
6. Pipetting during the filtration is a very effective method for increasing the yield of cells.
8. Organelles released from broken cells, such as chloroplasts, are removed together with this supernatant.
12. Alternatively, cells can be incubated in Erlenmeyer flasks (4 ml of cell suspension per 50-ml flask) on a gyrotary shaker at 75 rpm [9].
2–11. These steps should be carried out on a clean bench.
6–10. These steps should be carried out quickly. Delay during these steps results in low viability of isolated cells.

Culture medium

Table 2 shows the composition of the culture medium for induction of tracheary-element differentiation. This medium is the original medium reported by Fukuda and Komamine [2]. Various modified versions of this medium have been reported [e.g., 10,11]. Medium is prepared by standard method, and sterilized by autoclaving or filtration. Stock solutions for this medium are shown in Table 3. Media without 6-benzyladenine (BA) and/or 1-naphthaleneacetic acid (NAA) can be used for control cultures in which tracheary-element differentiation does not occur.

Table 2. Medium for the culture of *Zinnia* mesophyll cells

Constituents	Concentration (mg liter^{-1})
Macroelements	
$\quad$ KNO_3	2,020
$\quad$ NH_4Cl	54
$\quad$ $MgSO_4 \cdot 7H_2O$	247
$\quad$ $CaCl_2$	147
$\quad$ KH_2PO_4	68
Microelements	
$\quad$ $MnSO_4 \cdot 4H_2O$	25
$\quad$ h_3BO_3	10
$\quad$ $ZnSO_4 \cdot 7H_2O$	10
$\quad$ $Na_2MoO_4 \cdot 2H_2O$	0.25
$\quad$ $CuSO_4 \cdot 5H_2O$	0.025
$\quad$ Na_2EDTA	37
$\quad$ $FeSO_4 \cdot 7H_2O$	28
Organic growth factors	
$\quad$ Glycine	2
$\quad$ *myo*-Inositol	100
$\quad$ Nicotinic acid	5
$\quad$ Pyridoxine hydrochloride	0.5
$\quad$ Thiamine hydrochloride	0.5
$\quad$ Biotin	0.05
$\quad$ Folic acid	0.5
Phytohormones	
$\quad$ 1-Naphthaleneacetic acid (NAA)	0.1
$\quad$ 6-Benzyladenine (BA)	1
Sucrose	10,000
D-Mannitol	36,400
pH	5.5

Table 3. Stock solutions for preparation of the medium

Stock A (10×)	
KNO_3	20,200 mg
NH_4Cl	540
$MgSO_4 \cdot 7H_2O$	2,470
$CaCl_2 \cdot 2H_2O$	1,470
KH_2PO_4	680
	1,000 ml

Stock B (400×)	
$MnSO_4 \cdot 4H_2O$	2,500 mg
H_3BO_3	1,000
$ZnSO_4 \cdot 7H_2O$	1,000
$Na_2MoO_4 \cdot 2H_2O$	25
$CuSO_4 \cdot 5H_2O$	2.5
	250 ml

Stock C (400×)[a]	
Na_2EDTA	3,700 mg
$FeSO_4 \cdot 7H_2O$	2,800
	250 ml

Stock D (400×)	
Glycine	200 mg
myo-Inositol	10,000
Nicotinic acid	500
Pyridoxine hydrochloride	50
Thiamine hydrochloride	50
Biotin	5
	250 ml

Stock E (400×)[b]	
Folic acid	50 mg
	250 ml

[a] Stirred for at least several hours while being heated (100 °C).
[b] Dissolved by adding a small amount of KOH solution.

Determination of the frequency of tracheary-element differentiation

After 2 d in culture, 30–40% of cells differentiate synchronously into tracheary elements. Tracheary elements can easily be identified from the characteristic patterns of their secondary cell walls, which can be seen under a light microscope. The number of tracheary elements is determined using a haemocytometer without any treatment of cells prior to counting. The number of tracheary elements relative to the total number of cells represents the frequency of differentiation to tracheary elements. During culture of mesophyll cells, cell division occurs independently of, but concurrently with, tracheary-element differentiation. The frequency of cell division can be estimated from the number of septa, determined using a haemocytometer, since all the initial mesophyll cells are single cells.

Cytological techniques for monitoring the process of tracheary-element differentiation

Staining of lignin with phloroglucinol

During the differentiation of mesophyll cells to tracheary elements, lignin, a macromolecule composed of hydroxycinnamyl alcohols as monomeric units, accumulates in the secondary cell walls. Lignified tracheary elements can be detected by staining lignin with phloroglucinol-HCl [12]. Phloroglucinol is dissolved in 20% HCl at about 1% (w/v). A drop of this solution and a drop of cell suspension are mixed on a slide. After a 10-min incubation at room temperature, ligninified cell walls are stained reddish purple.

Staining of nuclei with DAPI

Upon the maturation of tracheary elements, intracellular components, including nuclei, chloroplasts, and mitochondria, are lysed autonomously. This stage of differentiation can be monitored by staining nuclei with various dyes. Among these dyes, DNA-specific fluorescent dyes, such as 4',6-diamidino-2-phenylindole (DAPI) are particularly useful. The procedure for DAPI staining, as described here is
based on the method of Kuroiwa *et al.* [13]. Dye solution contains 1
μg/ml DAPI, 0.25 M sucrose, 0.05% 2-mercaptoethanol, 0.6 mM spermidine, 1 mM EDTA, 0.4 mM phenylmethylsulfonyl fluoride (PMSF),
and 10 mM Tris-HCl at pH 7.6. Cell suspension is mixed with equal
volumes of a 5% solution of glutaraldehyde and the dye solution. The
preparation is ready to be examined after a brief incubation (more
than 3 min) at room temperature. Under ultraviolet light, nuclei emit
blue-white fluorescence. The number of anuclear tracheary elements
relative to the total number of tracheary elements is recorded as an
index of the extent of autolytic differentiation.

References

1. Fukuda, H (1992) Tracheary element differentiation as a model system of cell differentiation. Int Rev Cytol 136: 289–332.
2. Fukuda H, Komamine A (1980) Establishment of an experimental system for the tracheary element differentiation from single cells isolated from the mesophyll of *Zinnia elegans*. Plant Physiol 65: 57–60.
3. Kohlenbach HW, Schmidt B (1975) Cytodifferenzierung in Form einer direkten Umwandung isolierter Mesophyll-Zellen zu Tracheiden. Z Pflanzenphysiol 75: 369–374.
4. Fukuda, H (1994) Redifferentiation of single mesophyll cells into tracheary elements. Int J Plant Sci 155: 262–271.
5. Sugiyama M, Komamine A (1990) Transdifferentiation of quiescent parenchymatous cells into tracheary elements. Cell Differ Dev 31: 77–87.
6. Fukuda H, Yoshimura T, Sato Y, Demura T (1993) In: Komamine A *et al.* (Eds.) Molecular mechanism of xylem differentiation, pp 93–107, J Plant Res Special Issue 3, The Botanical Society of Japan, Tokyo.
7. Demura T, Fukuda H (1994) Novel vascular cell-specific genes whose expression is regulated temporally and spatially during vascular system development. Plant Cell 6: 967–981.
8. Fukuda H, Komamine A (1982) Lignin synthesis and its related enzymes as markers of tracheary-element differentiation in single cells isolated from the mesophyll of *Zinnia elegans*. Planta 155: 423–430.
9. Church DL, Galston AW (1988) Kinetics of determination in the differentiation of isolated mesophyll cells of *Zinnia elegans* to tracheary elements. Plant Physiol 88: 92–96.
10. Sugiyama M, Fukuda H, Komamine A (1986) Effects of nutrient limitation and gamma-irradiation on tracheary element differentiation and cell division in single mesophyll cells of *Zinnia elegans*. Plant Cell Physiol 27: 601–606.
11. Lin Q, Northcote DH (1990) Expression of phenylalanine ammonia-lyase gene during tracheary-element differentiation from cultured mesophyll cells of *Zinnia elegans* L. Planta 182: 591–598.
12. Siegel, SM (1953) On the biosynthesis of lignins. Physiol Plant 6: 134–139.
13. Kuroiwa T, Suzuki T, Ogawa K, Kawano S (1981) The chloroplast nucleus: distribution, number, size, and shape, and a model for the multiplication of the chloroplast genome during chloroplast development. Plant Cell Physiol 22: 381–396.

Plant Tissue Culture Manual **H3**, 1–31, 1995.
© 1995 *Kluwer Academic Publishers.*

Cell cycle studies: Induction of synchrony in suspension cultures of *Catharanthus roseus* cells

H. KODAMA[1] & A. KOMAMINE[2]

[1]*Department of Biology, Faculty of Science, Kyushu University 33, Fukuoka 812, Japan;* [2]*Department of Chemical and Biological Sciences, Japan Women's University, 2–8–1 Mejirodai, Bunkyo-ku, Tokyo 112, Japan*

Introduction

Synchronous cultures have the potential to be very useful tools for studies of the cell cycle because they can provide homogeneous populations of cells at specific phases of the cell cycle, allowing us to study specific biochemical and molecular biological events during the cell cycle. Synchronous cultures of *Catharanthus roseus* (L.) cells are particularly well suited for such studies since it is easy to achieve high reproducibility of the degree of synchrony with such cultures and procedures for synchronization are straightforward. Various aspects of the molecular events during the cell cycle have been investigated with synchronous cultures of *C. roseus*, and several cell-cycle-dependent genes, such as genes for proliferating-cell nuclear antigen, *cyc07* and cyclins, have been isolated [6].

Synchronization of suspension-cultured cells of higher plants is achieved by the arrest of almost all cells at a specific point in the cell cycle, with subsequent release of cells from growth arrest. The techniques that are used to arrest growth of *C. roseus* cells at the G_1 phase of the cell cycle require only the removal and refeeding of phosphate [1] or auxin [9]. As well as such manipulation of the availability of the phosphate and auxin, removal and resupply of other components of the growth medium, such as sucrose and nitrogen, have also been examined for their ability to induce synchronous cell division. In sucrose-limited cultures, cells with 2C DNA level, namely, cells in the G_1 phase, and cells with 4C DNA level, which correspond to cells in the G_2/M phase, accumulate (Fig. 1), as previously reported in suspension cultures of *Acer pseudoplatanus* [3]. However, the viability of cells in sucrose-starved cultures declines rapidly after cell proliferation has ceased [5]. Thus, manipulation of the availability of sucrose is unsuitable for the synchronization of *C. roseus* cells. The accumulation of cells with 2C DNA level can be observed in nitrogen-limited cultures of *C. roseus*, as well as in phosphate-starved cultures (Fig. 1). When nitrogen is added to nitrogen-limited cells, a step-wise increase in cell number is observed (Fig. 2). However, the duration of the first cell cycle (about 45 hours), namely, the period between the addition of nitrogen and the first cell division, is very much longer than that of the second cell cycle (about 35 hours), the period between the first cell division and the second cell division. The additional time required for the first cell cycle may correspond to the time required for the entry of cells from the so-called G_0 state to the G_1 phase. The proportion of cells that

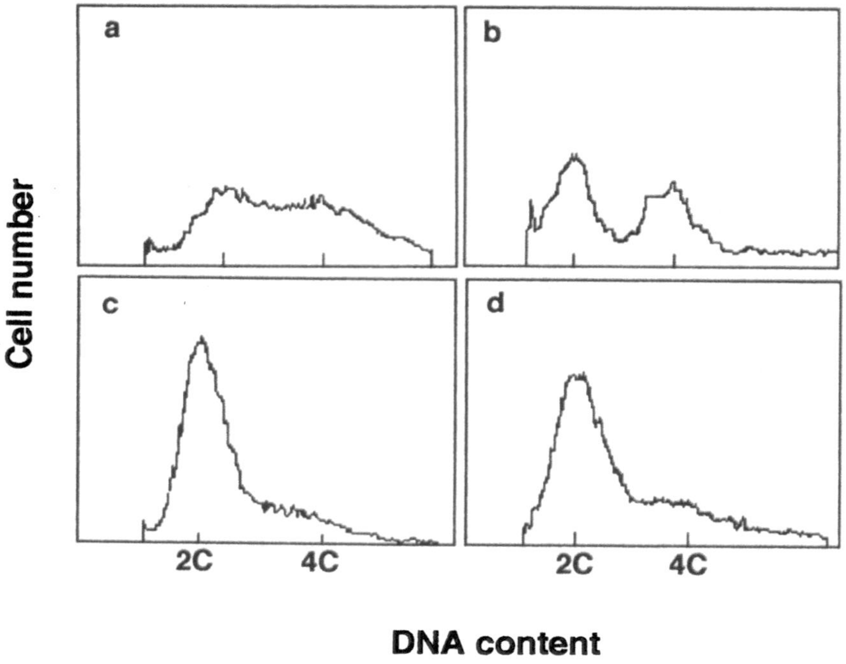
a
b
c
d
Cell number
2C
4C
2C
4C
DNA content

Fig. 1. Cell cycle analysis of nutrient-starved cultures by flow cytometry. The gross alterations in the distribution throughout the cell cycle of cells arrested under particular nutrient-starvation conditions were characterized in strain A by flow cytometry. Flow-cytometric analyses are shown of acridine orange-stained protoplasts from cells at the logarithmic phase (a), from a sucrose-limited culture (b), from a nitrogen-limited culture (c), and from a phosphate-starved culture (d). In the sucrose-limited culture, cells were subcultured in sucrose-limited medium that contained about 0.5% (w/v) sucrose. In the nitrogen-limited culture, cells were subcultured in nitrogen-limited medium that contained 1.2 mM nitrogen. Standard MS medium contains 3% (w/v) sucrose and 60 mM nitrogen. Growth of almost all cells ceased 4 days after subculture in the sucrose- or nitrogen-limited medium, and the distribution of cells throughout the cell cycle was determined by flow cytometry 4 days (sucrose-limited culture) and 5 days (nitrogen-limited culture) after subculture. In the phosphate-starved culture, cells were arrested at the G_1 phase by the phosphate starvation method, and flow-cytometric analysis was performed with protoplasts prepared from cells just before the second phosphate feeding of phosphate (see also text for details of steps in synchronization by the phosphate starvation method).

divide during one round of the cell cycle in such cultures is relatively low (about 30 to 50%) and it takes about 10 hours for cells to divide. These properties of cultures subjected to starvation and subsequent addition of nitrogen indicate that the synchrony obtained in such cultures is inadequate for investigations of particular events during the cell cycle.

In synchronous cultures induced by manipulating the availability of phosphate [1] or auxin [9], progression of the cell cycle is rapidly reinitiated, without any delay, after the addition of phosphate or auxin. Cell numbers usually increase by

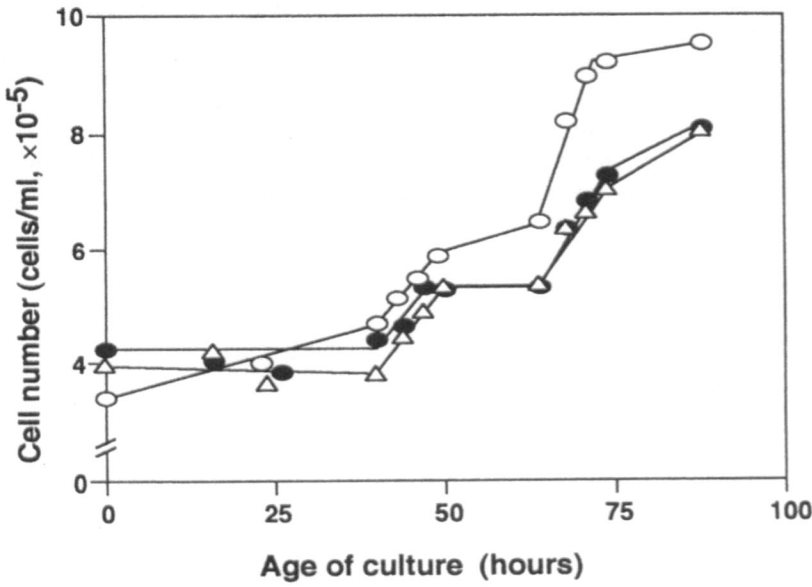

Fig. 2. Changes in cell number after the manipulation of levels of nitrogen in the culture medium. Cells were subcultured in nitrogen-limited medium (total nitrogen, 1.2 mM) and proliferation of cells ceased about 3 days after subculture. Then, nitrogen was added directly to the medium at a final concentration of 60 mM on the 3rd (○), 4th (●) and 5th days (△) after subculture. The age of culture refers to the time after the addition of nitrogen.

70 to 80% within 3 to 4 hours. These results indicate that starvation of phosphate or auxin does cause limited damage to cells. In this chapter we provide details of the procedure for the synchronization of suspension cultures of *C. roseus* cells by the phosphate starvation method, as well as the procedure for synchronization by the auxin starvation method.

Although synchronization by the phosphate starvation method had been considered to be a method whose applicability is limited to suspension-cultured cells of *C. roseus*, recent studies have shown that synchronous cell division can be induced by this method in suspension cultures of *Medicago varia* [4] and *Populus alba* [Dr. H. Maki, personal communication]. The composition of the culture medium may be an important factor for establishment of synchrony by this method. All suspension cultures amenable to synchronization by the phosphate starvation method are maintained in Murashige-Skoog (MS) medium [8]. Standard MS medium contains a relatively low concentration of phosphate, in contrast to elevated levels of nitrogen sources. It seems likely that, in this type of medium, the availability of phosphate limits the division of cells. Therefore, when suspension cultures of other higher plant species are to be synchronized, it is possible that, if cells can be maintained in MS medium and the cultures contain homogeneous and finely dispersed cells, a high degree synchrony may be attainable by this method.

Only a few strains appear to be suitable for synchronization by the auxin starvation method. Among three strains of *C. roseus* cells maintained in the authors' laboratory, cells of only one strain (TN21) can be well synchronized by the auxin starvation method. In the case of cells that have been in culture for long periods of time, this method of synchronization may be unsuitable because of habituation of cells to exogenous auxin.

Procedures for determination of percentages of living cells in cultures

A high percentage of actively dividing cells is required if a culture is to be used for successful synchronization because synchrony is strongly influenced by the percentage of non-dividing cells. The proportion of non-dividing cells can be roughly estimated by measuring the viability of cells in a rapidly growing culture (e.g., a culture in the logarithmic phase of growth). Cell viability can be determined by staining the living cells with a 0.12 mM solution of fluorescein diacetate by the method of Widholm [10] and we usually examine at least 500 cells (see also chapter A3, this volume, for details).

Procedures for determination of cell numbers

Cell clusters of all strains of *C. roseus* can easily be macerated by wall-degrading enzymes to yield protoplasts with an efficiency of almost 100%.

Steps in the procedure
1. Transfer three ml of a suspension of cells at a concentration of about 1×10^5 to 5×10^5 cells/ml to a 10-ml test tube, and then pellet cell clusters by centrifugation at $800 \times g$ for 2 min.
2. Remove the supernatant and replace by a mannitol solution.
3. After cell clusters have been suspended in a uniform manner, transfer 2 ml of the suspension to a 10-ml Erlenmeyer flask, and then add 0.5 ml of enzyme solution I (for strain A or B) or II (for strain TN21).
4. Incubate the mixture for 60 min at 27 °C on a shaker operated at 60 strokes/min.
5. Estimate cell numbers by counting protoplasts with a haemocytometer.

Procedures for determination of mitotic indices (Feulgen stain method)

The mitotic index is determined by counting the number of Feulgen-stained nuclei or nuclei stained with 4-6-diamidino-2-phenylindole (DAPI) that are observed to be in mitosis.

Steps in the procedure
1. Mix one volume of cell suspension and five volumes of fixative (10% formaldehyde in 30 mM sodium-phosphate buffer, pH 7.0).
2. After 24 h, wash cells with distilled water and hydrolyze in 1 M HCl for 10 min at 60 °C.
3. Stain in Schiff's reagent for 3 h and wash with 2% (w/v) sodium bisulfite and then with water.
4. Determine the mitotic index in the squash of Feulgen-stained nuclei (more than 1,000) that are clearly recognized to be at metaphase or anaphase in a given population of cells.

Procedures for determination of mitotic indices (DAPI stain method)

Steps in the procedure
1. Transfer 200 μl of a suspension of cells to a 1.5-ml microcentrifuge tube and pellet cell clusters by centrifugation at 3,000 × g for 1 min.
2. Resuspend in 2% (w/v) glutaraldehyde in 30 mM potassium phosphate buffer (pH 7.0). The fixed cells can be stored for several days at 4 °C.
3. Remove the supernatant by centrifugation at 3,000 × g for 1 min and hydrolyze in 1 M HCl for 15 min at 60 °C.
4. Collect cells by centrifugation at 3,000 × g for 1 min and resuspend in 2% (w/v) glutaraldehyde in 30 mM potassium phosphate buffer (pH 7.0). Place approximately 10 μl of a suspension of fixed cells on a slide, mix with an approximately equal volume of a DAPI solution and stain for a few minutes.
5. Determine the mitotic index in the squash of DAPI-stained nuclei (more than 1,000) that are clearly at metaphase or anaphase in a given population of cells. Examine stained cells under an fluorescence microscope fitted with a DApo 100 UV PL objective lens (Olympus Kogaku Kogyo) with fluorescence illumination at 360 nm for excitation of DAPI.

Procedures for determination of incorporation of [³H]thymidine

Steps in the procedure

1. The rate of DNA synthesis is measured by labeling cells with [³H]thymidine (111 kBq/ml suspension, 1.67 TBq/mmol, for example) for 60 min at 27 °C. After incubation, add 1.6 ml of ice-cold ethanol to 0.4 ml of the suspension of labeled cells.
2. Transfer the cells to a 1.5-ml microcentrifuge tube and centrifuge at 2,000 × *g* for 10 min at 4 °C.
3. Wash the pellet twice with ice-cold 80% (v/v) ethanol and then with ice-cold 0.2 M perchloric acid (PCA).
4. Solubilize nucleic acids by heating in 0.5 M PCA at 80 °C for 15 min.
5. Centrifuge the extract at 10,000 × *g* for 10 min, and measure an aliquot of radioactivity of the supernatant, after combination with a commercial scintillant, in a liquid scintillation counter.

Procedures for flow cytometry

Analyze the DNA content of protoplasts that have been stained with acridine orange by flow cytometry as described in Ando *et al.* [2].

Steps in the procedure
1. Isolate protoplasts of *C. roseus* as described for the procedures of determination of cell numbers.
2. Collect protoplasts by centrifugation at $500 \times g$ for 3 min, wash once with a mannitol solution and fix in 70% (v/v) ethanol.
3. Treat protoplasts (about 10^6/sample) with 10 ml of a solution containing 0.08 M hydrochloric acid, 0.1% Triton X-100, and 0.08 M NaCl in ice bath for 5 min.
4. Collect protoplasts by centrifugation at $500 \times g$ for 3 min and stain with 12 ml of a solution containing 4 mg/l acridine orange, 1 mM EDTA-2Na, 0.15 M NaCl, and 0.12 M phosphate-citrate buffer (pH 6.0), in an ice bath for about 2 h.
5. Histograms of green fluorescence are obtained from a total 20,000 protoplasts at a flow rate of 100 to 150 cells/second with an flow cytometer fitted with a 100 μm nozzle. The laser output is 480 nm at a 600 mW and barrier filters are LP515 and SP530.

Solutions

Enzyme solution I (for cells of strain A or B)
10% (w/v) Cellulase "Onozuka" R-10 (Yakult Honsha, Tokyo, Japan)
5% (w/v) Macerozyme R-10 (Yakult Honsha, Tokyo, Japan)
   Dissolve both enzymes in H_2O. The solution of enzymes can be stored for several months at $-20\ °C$.

Enzyme solution II (for cells of strain TN21)
10% (w/v) Cellulase "Onozuka" R-10
5% (w/v) Macerozyme R-10
5% (w/v) Pectolyase Y-23 (Seishin Seiyaku, Tokyo, Japan)
   Dissolve all three enzymes in H_2O. Insoluble materials should be removed from the solution by centrifugation at 10,000 × g for 10 min at 4 °C. The supernatant can be stored for several months at $-20\ °C$.

Mannitol solution
0.6 M mannitol containing 1% (w/v) $CaCl_2 \cdot 2H_2O$

DAPI solution
Dissolve DAPI at a final concentration of 1 μg/ml in S buffer [7] which contains 20 mM Tris-HCl (pH 7.6), 0.25 M sucrose, 1 mM EDTA, 0.6 M spermidine, 7 mM β-mercaptoethanol, 1 mM phenylmethylsulphonyl fluoride.

MS medium
Murashige and Skoog basal medium [8]
supplemented with:
2.2 μM 2,4-dichlorophenoxyacetic acid (2,4-D) for cells of strain A (or B), or 4.4 μM 2,4-D for cells of strain TN21, and 3% (w/v) sucrose.
   Adjust the pH to 5.7 with 1 M KOH. In our laboratory, some mineral components, namely, NH_4NO_3, KNO_3, H_3BO_3, $MnSO_4 \cdot 4H_2O$, $ZnSO_4 \cdot 7H_2O$, KI, $Na_2MoO_4 \cdot 2H_2O$, $CoCl_2 \cdot 6H_2O$ and $CuSO_4 \cdot 5H_2O$, are dissolved together as a concentrated stock solution (× 50). Solutions of KH_2PO_4, Fe-EDTA, $CaCl_2 \cdot 2H_2O$ and $MgSO_4 \cdot 7H_2O$ are prepared separately as individual stock solutions (× 200). By preparing the stock solution of KH_2PO_4 separately from those of other mineral components, we can easily prepare phosphate-free MS medium.

Procedures for synchronization of cell division by the phosphate starvation method

Establishment of suspension cultures of strain A (or B) of C. roseus cells

The available laboratory strains of *C. roseus* cells consist of homogeneous and finely dispersed cells. Suspension cultures of strain A (or B) of *C. roseus* (L.) G Don can be synchronized by the phosphate starvation method. These strains were initiated originally from a culture of stem tissue in 1969. Cells are maintained at 27 °C, in darkness, in MS medium that contains 3% (w/v) sucrose and 2.2 μM 2,4-D. Cells are subcultured at 7-day intervals by the transfer of 7 ml of the suspension of cells into 43 ml of fresh medium in a 300-ml Erlenmeyer flask, and flasks are shaken 80 to 90 strokes/min on a shaker.

Concentration of phosphate and proliferation of cells

The growth cycle of suspension-cultured *C. roseus* cells of strain A (or B) consists of a lag phase of about 1 day, a logarithmic phase of 4 days and a stationary phase that is reached about 6 days after subculture.

The growth of *C. roseus* cells in MS medium is limited by the concentration of phosphate in the medium. The relationship between the concentration of phosphate and the proliferation of cells should be determined as follows. Cells at the stationary phase (about 6 to 9 days after subculture) are cultured for 3 days in phosphate-free MS medium at an initial density of about 3×10^5 cells/ml. Then phosphate is added as KH_2PO_4 to cultures to final concentrations of 0 to 1.25 mM. Cells are cultured for a further 9 days (except in the case of the culture with 0 mM phosphate) and the number of cells is counted. The number of cells in the phosphate-free culture is determined 3 days after the addition of phosphate to the other cultures. A linear relationship should be obtained between the concentration of phosphate in the medium (0 to 1.25 mM; standard MS medium contains 1.25 mM phosphate) and the number of cells 9 days after the addition of phosphate. Figure 3 shows that the population of phosphate-starved cells at an initial density of 3×10^5 cells/ml can double in cell number after the addition of phosphate at a final concentration of 0.14 mM.

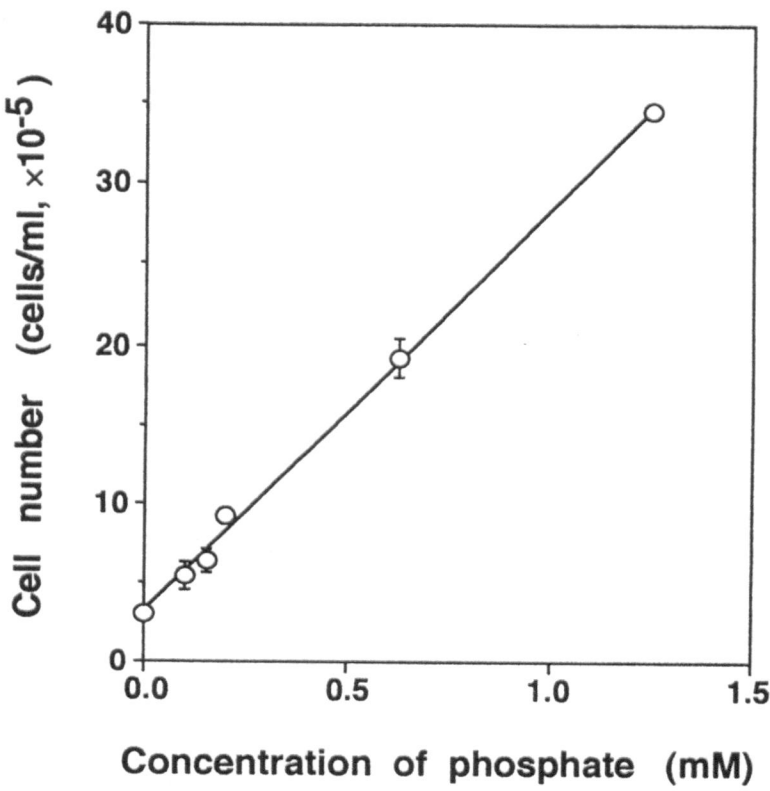

Fig. 3. Effects of the concentration of phosphate on the proliferation of *C. roseus* cells (strain A). Nine-day-cultured cells were transferred to phosphate-free medium. After culture for 3 days, phosphate was added to the medium at various concentrations, as indicated. After proliferation of cells had ceased, cell numbers were determined. Vertical lines indicate standard deviations. Their absence in this and other Figures indicates that standard deviations fell within symbols.

Steps in the procedure

A diagram of the procedure for implementation of the phosphate starvation method is shown in Fig. 4.

1. First phosphate starvation (S1)

During all steps described, cells are cultured at 27 °C in darkness, with shaking at 80 to 90 strokes/min on a shaker. Transfer cells at the stationary phase (about 6 to 9 days after subculture) directly to phosphate-free medium and culture for 3 days in 100 to 150 ml of phosphate-free medium at an initial density of about 3×10^5 cells/ml in a 500-ml Erlenmeyer flask.

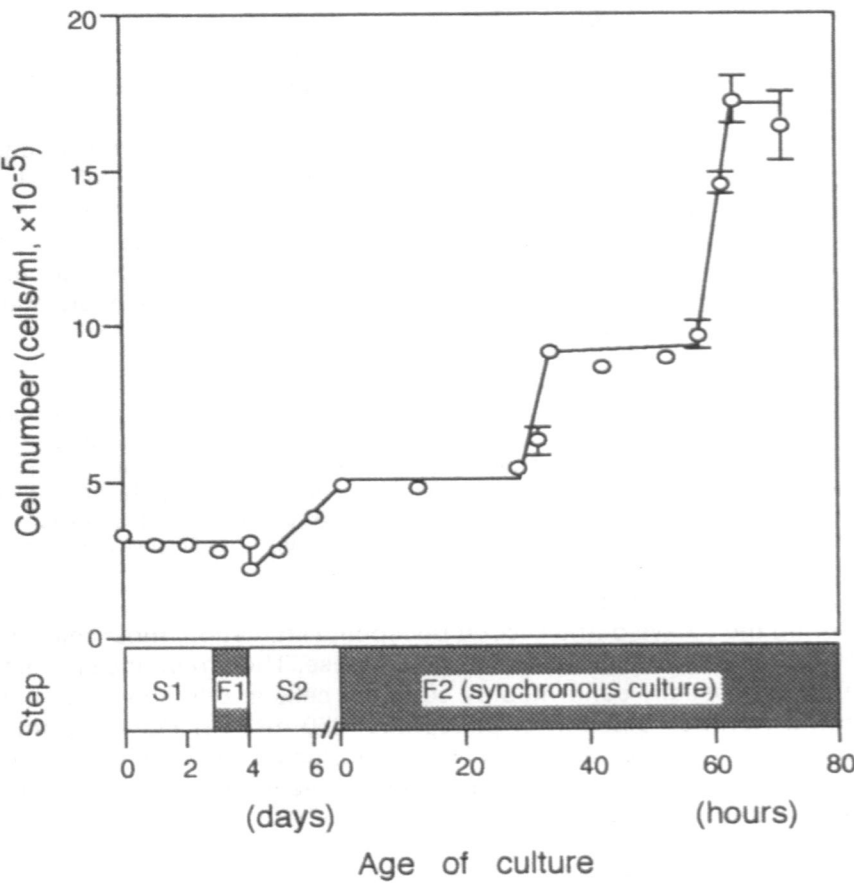

Fig. 4. Diagram of the steps required for synchronization by the phosphate starvation method. The abbreviations, namely, S1, F1, S2 and F2, are explained in the text. Vertical lines indicate standard deviations.

Note
This culture period is designated the first phosphate starvation (S1). It is important that no apparent proliferation of cells be observed during this culture period. If numbers of cells increase during this culture period, cells at later stationary phase, for example, 10 or 11 days after subculture, can be cultured in the phosphate-free medium.

2. First phosphate feeding (F1)
Add a sterilized solution of phosphate (250 mM KH_2PO_4) directly to the medium to a final concentration of 0.14 mM. Cells are cultured for 14 to 18 h. This culture period is designated the first phosphate feeding (F1).

3. Second phosphate starvation (S2)

1. Fourteen to eighteen hours after the start of F1, transfer the cells to a 50-ml plastic centrifuge tube and collect by centrifugation at $800 \times g$ for 2 min.
2. Wash the cells twice with phosphate-free medium and resuspend in phosphate-free medium at a density of about 2×10^5 cells/ml.
3. Mix this suspension of cells gently by stirring with a magnetic stirrer for homogeneous dispersion of cells and then divide into 10 ml aliquots in 50-ml Erlenmeyer flasks. The cell number should double during a 3-day culture under these second phosphate-starvation conditions (abbreviated as S2 in Fig. 4).

Note

It is essential, at this step, that excess phosphate be eliminated initially by washing cells with phosphate-free medium. It is important that the cells be washed before they begin to divide. Cells in the process of mitosis are quite liable to suffer from effects that are unfavorable for subsequent synchronization during the washing step. The number of cells increases gradually within 20 to 24 h after the first phosphate feeding. Therefore, we usually wash cells about 14 to 18 h after the start of the first phosphate feeding.

4. Second phosphate feeding (F2)

During the 3 days of the second phosphate starvation, most cells are arrested preferentially at the early G_1 phase. The arrest of cell proliferation is usually confirmed by counting cells at 8-h intervals. Then add a sterilized solution of phosphate (250 mM KH_2PO_4) to a final concentration of 0.625 mM and synchronized cell division can be observed. This culture period is referred as second phosphate feeding (F2), as shown in Fig. 4.

The degree of synchrony

Synchronous division of cells occurs 27 to 31 h after the second feeding of phosphate. The cell number increases by 70 to 80%. The S phase, which is determined by monitoring incorporation of [³H]thymidine into the DNA fraction, is 6 to 17 h in length after the second feeding of phosphate. A sharp increase in mitotic index is observed 26 to 30 h after the second feeding of phosphate (Fig. 5).

A DNA histogram of protoplasts prepared from the cells in the G_1 phase shows only one peak (peak A), which corresponds to a 2C level of DNA. Almost all protoplasts prepared from cells in the G_2 phase have a 4C level of DNA (peak B). In a histogram of protoplasts from cells in cytokinesis, two peaks are observed and they correspond to 2C (peak C) and 4C (peak D) levels of DNA, respectively. The former seems to correspond to cells after cytokinesis and the latter to cells before cytokinesis (Fig. 6).

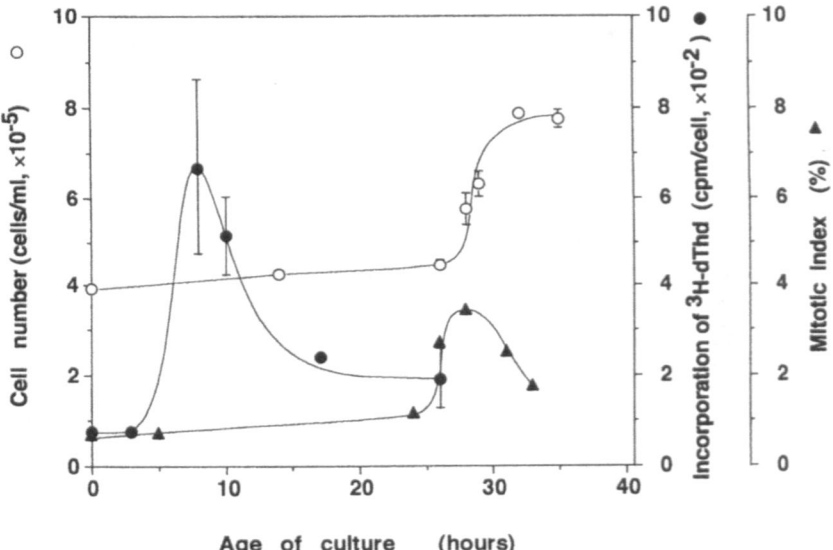

Fig. 5. Changes in cell number, incorporation of [³H]thymidine (³H-dThd) into the DNA fraction and mitotic indices in a synchronous culture of *C. roseus* (strain A), in which synchronization was achieved by the phosphate starvation method. The age of culture refers to the time after the second feeding of phosphate. Vertical lines indicate standard deviations.

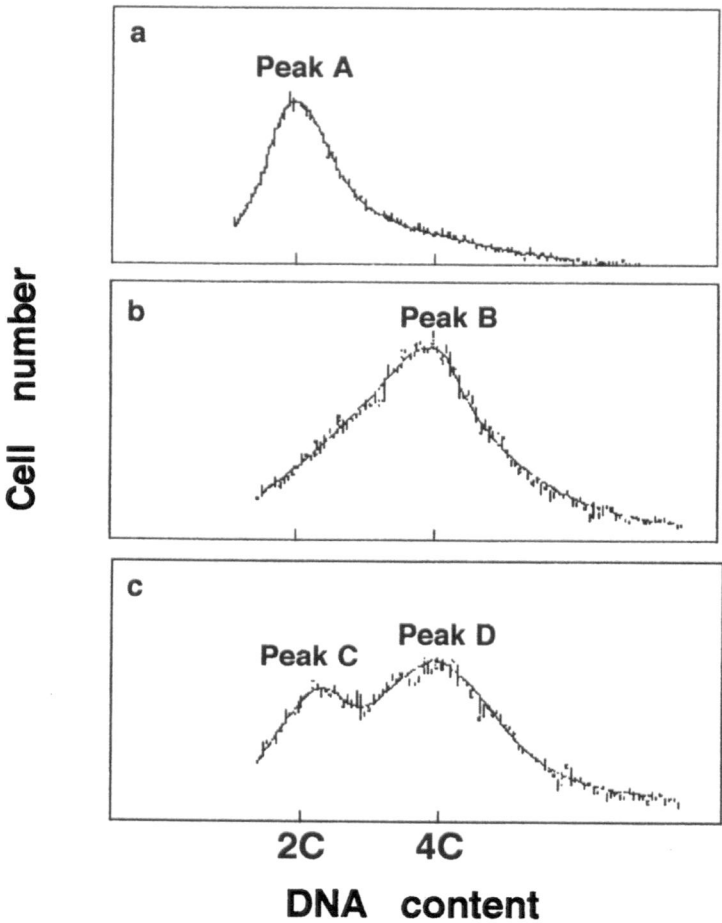

Fig. 6. Distribution of the DNA content in a synchronous culture of *C. roseus* cells in which synchronization was induced by the phosphate starvation method. Flow-cytometric analysis is shown of acridine orange-stained protoplasts from the cells in the G_1 phase (a), in the G_2 phase (b) and during cytokinesis (c). Peaks A through D are discussed in the text.

Procedures for synchronization of cell division by the auxin starvation method

Establishment of suspension cultures of strain TN21 of C. roseus *cells*

Suspension cultures of strain TN21 of *C. roseus* (L.) cv. Little Pinky can be synchronized by the auxin starvation method. Although this strain was originally initiated from a culture of anthers in 1988, almost all cells in cultures are diploid. Cells of strain TN21 are maintained under the same culture conditions as described for the phosphate starvation method (see *establishment of suspension cultures of strain A (or B) of* C. roseus *cells*), with the exception of a concentration of 2,4-D (4.4 μM).

Effects of auxin starvation on the proliferation of cells of the strain TN21

The growth cycle of cells of strain TN21 consists of a lag phase of about 1 day, a logarithmic phase of 5 to 6 days and a stationary phase that is reached after about 8 days of subculture. High cell viability (about 80%) is observed in the culture at the stationary phase.

A unique property of the cells of strain TN21 becomes apparent when cells at the stationary phase are transferred to fresh medium without 2,4-D. After several washes with auxin-free medium, cells are cultured under auxin-starved conditions. No increase in cell number is detectable in auxin-starved cultures of strain TN21 for at least 8 days (Fig. 7). If 2,4-D is added to the medium at a final concentration of 2.2 μM on days zero, 2, and 4 after 2,4-D is eliminated, rapid restoration of cell growth can be observed within 2 days after the addition of 2,4-D (Fig. 7). Thus, high cell viability is clearly maintained in cultures free of auxin for at least 4 days. Synchronization by the auxin starvation method is based on these results: cells of strain TN21 at the stationary phase can be arrested by transfer to fresh, auxin-free medium and refeeding of auxin rapidly induces the cell growth. By contrast, cells of strain A or B, used for synchronization by the phosphate starvation method, increase in number even under auxin-starved conditions. Thus, synchronization by the auxin starvation method can be applied only to cells of strain TN21 in our laboratories. However, the phenomenon of auxin-induced rapid restoration of cell division is very attractive for studies of the role of auxin in the control of cell proliferation. Therefore, it is worth examining the possibility of synchronization by this method in suspension cultures of other plant cells, if cells at the stationary phase show no increase in cell number after transfer to auxin-free fresh medium.

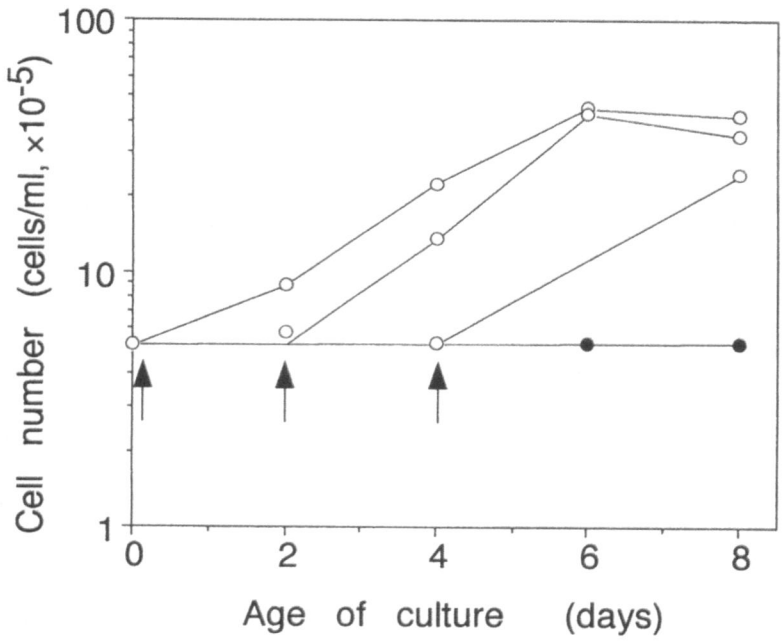

Fig. 7. Effects of 2,4-D on the proliferation of *C. roseus* cells (strain TN21) after auxin starvation. Changes are shown in the numbers of cells in auxin-free medium (•) and in medium supplemented with auxin (○). The age of the culture refers to the time after the transfer of cells to auxin-free medium. Arrows indicate the time at which 2,4-D was added.

Steps in the procedure
A diagram of the procedure for implementation of the auxin starvation method is shown in Fig. 8.

1. Auxin starvation
1. Transfer eight-day-cultured cells of strain TN21 to a 50-ml plastic centrifuge tube and wash 4 times by centrifugation at 800 × *g* for 2 min with auxin-free MS medium.
2. Resuspend in auxin-free MS medium at a density of about 5×10^5 cells/ml.
3. Divide the suspension of cells into 10 ml aliquots in 50-ml Erlenmeyer flasks, as described for the phosphate starvation method (step 3), and culture for 2 days.

Note
With strain TN21, the density of cells during auxin starvation is important for the maintenance of high cell viability. A density of cells below 5×10^5 cells/ml often results in a decline in cell viability.

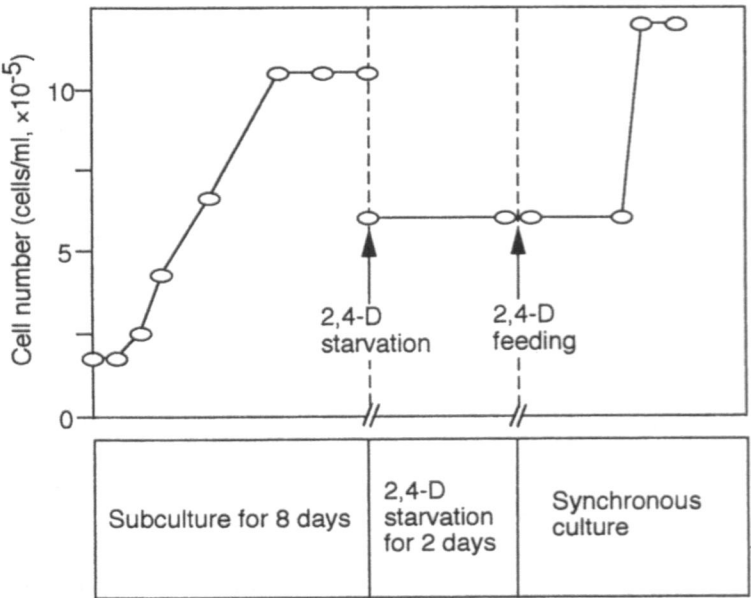

Fig. 8. Diagram of the steps required for synchronization by the auxin starvation method.

2. Auxin feeding

During 2 days of culture with auxin starvation, no apparent proliferation of cells is observed. Add 2,4-D directly to the medium at a final concentration of 4.4 μM to induce synchronized division of cells.

The degree of synchrony

Synchronous division of cells occurs 12 to 15 h after the refeeding of 2,4-D to cultures and cell numbers increase by 75% within 3 to 4 h. A single and clear peak of radioactivity due to incorporation of [³H]thymidine can be observed 9 h after the addition of auxin. The presence of a period of active synthesis of DNA (S phase) indicates that the auxin-starved cells are arrested preferentially at the G_1 phase. The mitotic index reaches a maximum value about 14 h after the addition of auxin, and it also shows a clear peak (Fig. 9).

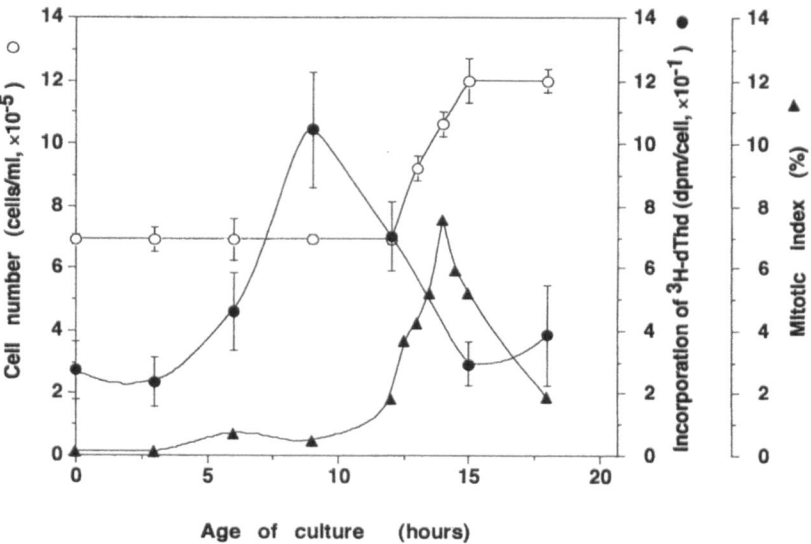

Fig. 9. Changes in cell number, incorporation of [³H] thymidine (³H-dThd) into the DNA fraction and mitotic indices in a synchronous culture of *C. roseus* cells (strain TN21) in which synchronization was induced by the auxin starvation method. The age of the culture refers to the time after the addition of auxin. Vertical lines indicate standard deviations.

Material and suppliers

1. Fluorescein diacetate: Aldrich Chemical Company, Milwaukee, WI, U.S.A.
2. Haemocytometer: Tatai type (or Fuchs-Rosenthal type); Kayagaki Irika Kogyo, Tokyo, Japan.
3. Schiff's reagent: Wako Pure Chemical Industries, Osaka, Japan
4. DAPI: Sigma, St Louis, MO, U.S.A.
5. Fluorescence microscope: Olympus BH-2 fluorescence microscope; Olympus Kogaku Kogyo, Tokyo, Japan.
6. [^{3}H]thymidine: Amersham, Buckinghamshire, England.
7. Liquid scintillation counter: 1216 RACKBETA II; LKB, Finland.
8. Flow cytometer: EPICS V flow cytometer; Coulter Electronics, Hialeah, FL, U.S.A.
9. 50-ml plastic centrifuge tubes: Sumilon centrifuge tubes; Sumitomo Bakelite, Tokyo, Japan.
   These tubes can be autoclaved several times at 121 °C for sterilization.

Acknowledgments

The authors thank Mr. K. Kusumi and Mr. H. Inada (Kyushu University) for their assistance in the preparation of Figures.

References

1. Amino S, Fujimura T, Komamine A (1983) Synchrony induced by double phosphate starvation in a suspension culture of *Catharanthus roseus*. Physiol Plant 59: 393–396
2. Ando S, Shimizu T, Kodama H, Amino S, Komamine A (1987) Flow cytometric analysis of the cell cycle in synchronous culture of *Catharanthus roseus*. Agric Biol Chem 51: 1443–1445
3. Gould AR, Everett NP, Wang TL, Street HE (1981) Studies on the control of the cell cycle in cultured plant cells. I. Effects of nutrient limitation and nutrient starvation. Protoplasma 106: 1–13
4. Kapros T, Bögre L, Németh K, Báko L, Gyögyey J, Wu SC, Dudits D (1992) Differential expression of histone H3 gene variants during cell cycle and somatic embryogenesis in alfalfa. Plant Physiol 98: 621–625
5. Kodama H, Ando S, Komamine A (1990) Detection of mRNAs correlated with proliferation of cells in suspension cultures of *Catharanthus roseus*. Physiol Plant 79: 319–326
6. Kodama H, Komamine A (1994) Studies of the plant cell cycle in synchronous cultures of *Catharanthus roseus* cells. Plant Cell Physiol 35: 529–537
7. Kuroiwa T, Suzuki T, Ogawa K, Kawano S (1981) The chloroplast nucleus: distribution, number, size, and shape, and a model for the multiplication of the chloroplast genome during chloroplast development. Plant Cell Physiol 22: 381–396
8. Murashige T, Skoog F (1962) A revised medium for rapid growth and bio assays with tobacco tissue cultures. Physiol Plant 15: 473–497
9. Nishida T, Ohnishi N, Kodama H, Komamine A (1992) Establishment of synchrony by starvation and readdition of auxin in suspension cultures of *Catharanthus roseus* cells. Plant Cell Tissue Organ Culture 28: 37–43
10. Widholm JM (1972) The use of fluorescein diacetate and phenosafranine for determining viability of cultured plant cells. Stain Technol 47: 189–194

Plant Tissue Culture Manual **H4**, 1–25, 1996.
© 1996 *Kluwer Academic Publishers.*

Thin cell layer (TCL) method to programme morphogenetic patterns

K. TRAN THANH VAN & C. GENDY
Institut de Biotechnologie des Plantes, CNRS Université Paris-Sud, 91405 Orsay, France

Introduction

The establishment of plant organ shape, structure and function proceeds from the embryo stage to the adult stage according to an ordered sequence of events. This implies that the genetic information is expressed or repressed according to a temporal and spatial order.

Due to the fact that growth and development events can be changed, to a certain extent, by a large array of factors which apparently lack specificity, the analysis at the biochemical molecular level cannot be easily conducted especially when developmental mutants are not available, as is the case for most plant species.

As an alternative to the study of developmental mutants, it would be valuable to be able to "programme" different morphogenetic patterns experimentally.

To date, the tobacco Thin Cell Layer (TCL) experimental system, consisting of a few cell layers (the epidermal layer and 3 to 6 cortical layers of differentiated cells) with reduced sizes (1mm × 5 or 10mm), is a unique system in which not only all patterns of morphogenesis known in plants can be programmed directly (without an intermediate callus phase), but also patterns which are new for the species studied. For example, in tobacco, somatic proembryos can be induced from subepidermal cells [14, 15, 16, 17, 19].

Markers of morphogenetic differentiation can be delineated in experimental systems in which different patterns can be : i) followed at the cell level from the early stage, ii) controlled by specific molecules, iii) inhibited by inhibitors of specific metabolic pathways and iv) recovered upon addition of such compounds. The identification of cytological and biochemical markers can lead to the identification of the gene(s) involved in specific steps of morphogenetic differentiation and to the study of their function(s) [3, 4, 7, 9, 12, 18, 20].

Application of the TCL method to recalcitrant species has allowed us and other researchers to overcome the difficulty in obtaining regeneration. For the study of gene function as well as for the production of transgenic plants, especially crop plants, for improvement purposes, the control of regeneration constitutes, at the present time, a serious limitation in plant biotechnology despite the availability of refined modern techniques of analysis at the molecular level.

The TCL method, originally developed on *Nicotiana tabacum* has been successfully applied and/or adapted to other species. Longitudinal TCLs are used in order to allow the analysis of mechanisms of cell differentiation and organogenesis from a defined cell layer whereas transverse TCL (tTCL) is used when the first objective is to overcome the difficulty in obtaining organ regeneration and/or somatic embryogenesis. As both longitudinal and transverse TCLs consist of one or a few cell layers, the content of their endogenous factors which are most of the time unknown, is assumed to be minimal, compared to the exogenous conditions applied in which the cells are cultured. Therefore, they are more medium-dependent than the classical type of explant which is larger and more complex.

In summary, the advantages of TCLs are as follows : i) when comprising only differentiated cells (epidermal and cortical cells), as in the case of longitudinal TCLs, the TCL method allows the analysis of the programmed patterns from the early stages to the full development of the organ, and therefore allows its study at the temporal level; ii) the minimal size of TCL and the well defined cell layer from which morphogenetic events start facilitate this study at a spatial level; iii) the rapidity (12, 14 days) in obtaining different patterns of morphogenesis including fully developed flowers has led to a reduced time in obtaining transgenic plants and transgenic seeds *in vitro* [2]; iv) the high number of organs obtained per TCL explant (up to 50 flowers or 800 vegetative buds) has also led to the high frequency in the production of transgenic plants or seeds.

To date, our group has succeeded, by using TCL methods, in reprogramming morphogenesis in several species including species of *Nicotiana, Torenia, Petunia, Brassica, Nautilocalyx, Saintpaulia, Begonia, Arabidopsis, Soja biloxi, Vicia faba, Psophocarpus*. For recalcitrant species such as *Phalaenopsis* and crop plants as well as for non-recalcitrant species with small sized organs (such as in *Arabidopsis*), one can use TCL to control: i) organogenesis in *Arabidopsis*, ii) both organogenesis and somatic embryogenesis in wheat, *Sorghum* (in collaboration with M. Sené, L. Bui, J. Vidal and P. Gadal), iris (in collaboration with H. Schricke and D. Joulain) and iii) somatic embryogenesis in other monocotyledons (bamboo [5]), and *Digitaria* (in collaboration with L. Bui, T. Do, J. Vidal and P. Gadal).

Other groups have successfully applied TCL methods to *Nicotiana tabacum* [1, 11], *Helianthus* [10] and *Pisum sativum* [6] in order to obtain programmed patterns including flower development, somatic embryogenesis and regeneration of vegetative buds respectively.

Procedures

I - Longitudinal TLC method

Model system: *Nicotiana tabacum* cv Samsun, Wisconsin and Xanthi.

1. *Culture conditions of the donor plants:*

Steps in the procedure
1. Grow plants from seeds in pots (10 cm diameter) in vermiculite.
2. Water with a mineral solution three times a week and on every other days, with deionized water (for both types of watering, 700 ml per day).
   Mineral solution:
   (in mM) 2.71 KNO_3, 1.00 KH_2PO_4, 1.11 $MgSO_4$, 1.04 $(NH_4)_2 SO_4$, 4.64 $Ca(NO_3)_2$,
   (in µM) 110.00 Na_2 Fe-EDTA, 36.75 KCl, 48.52 $H_3 BO_3$, 10.06 $MnSO_4$, 0.95 $ZnSO_4$, 0.55 $CuSO_4$, 0.22 $(NH_4)_6 Mo_7O_2$
   and (in nM) 2.55 H_2SO_4.
3. The environmental conditions are : 24 °C $\pm$ 2 °C for temperature, 65% relative humidity, 16h photoperiod of natural light complemented by artificial irradiation of 25 W. m^{-2} when the natural irradiance decreases below 50 W. m^{-2}
4. After three months, the plants reach the floral stage. When the terminal flower of the inflorescence becomes a green fruit and the donor plant bearing 5 – 80 green fruits, the physiological stage of the plant is appropriate for TCL.

2. Preparation of TCL

A. Sampling

Steps in the procedure
1. At this precise physiological stage, select 2–3 floral branches of 5 to 10 cm of length below the top of the inflorescence from each donor plant. Three to four donor plants i.e. 6 to a 12 floral branches in total are used.
2. Cut the floral branches from the inflorescence.
3. Wash in water, then in water with teepol and surface sterilize with 7% of calcium hypochlorite for 10 minutes.
4. Rinse thoroughly (five times) in sterile water and store in sterile water for the next phases of the procedure.

B. Excision of TCL

Steps in the procedure
1. Excise several long ribbons of TCL (1mm width and 4 to 8 cm length from each floral branch; Fig 1), using special microscalpels

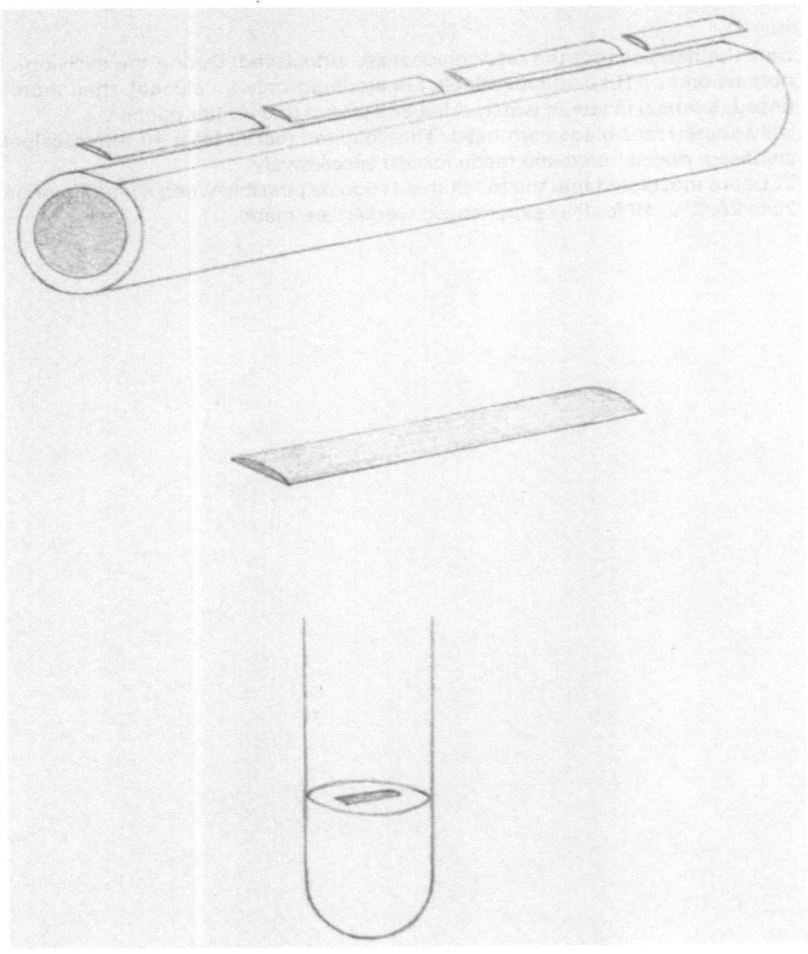

Fig. 1. Schematic of floral branches and excision of TCL.

   made in the laboratory : triangular pieces (7mm × 2mm) of razor blade mounted on special metal handles of the type used in micro-surgery (supplier: Moria).
2. Put these ribbons in a 10 cm diameter Petri dish and cover them with a film of sterile water.

3. Cut TCLs of 6 to 8 mm in length from each TCL ribbon in the Petri dish or directly on the floral branch (Fig 1). A pool of 20–25 TCLs from different floral branches is made and kept under a fine film of water in Petri-dishes.
4. Inoculate TCLs randomly onto a solidified medium and randomly distribute them to different types of culture medium.

Notes
- Only the handles (not the razor blades) are autoclaved. During the excision of TCL ribbons and of TCLs, microscalpels are sterilized only by alcohol, then thoroughly rinsed (5 times) in sterile water, dried and stored under filter paper.
- Only sharp razor blades are used. This requires that at least 10 microscalpels are sterilised, rinsed, dried and ready for use successively.
- TCLs are inoculated into the medium as soon as possible when a small number e.g. 20 to 25, (5 to 10 for less experienced worker) are made.

3. Inoculation of TCL

Steps in the procedure
1. Only a TCL method using solidified medium is described here. Use test tubes (15 cm length, 2.5 cm diameter) as containers for medium solidified with gelose Difco (10g/l) or gelrite (3g/l) except for flower programme.
2. Pour 25 ml of culture medium into each test tube.
3. Use forceps to pick up gently TCL (stored in Petri-dishes) and place them on the surface of the culture medium, **the epidermis side upwards**.

Notes
- Use only fine forceps with tips of 1 or 2 mm wide. Select forceps for which the distance between the tips does not exceed 1 cm, so as to reduce the pressure exerted on the TCL while manipulating it.
- Test tubes are covered by a metal lid (Bellco type) which allows gas exchange through a cotton wool ball placed at the bottom of the lid. Hermetic types of closure should be avoided. In order to ensure that the TCL is closely applied on the surface of the culture medium (but not immersed in the medium), a slight pressure is made on the TCL, the two tips touching the two poles of the TCL.
- Container shape and volume as well as the type of sealing, and so gas exchange, are important factors for "flower programme". Therefore, Petri dishes of 1 cm height, and sealing with parafilm, should be avoided.
- Multiwell (1 cm diameter) plates of 20 wells (supplier: Falcon) can be used, and so the volume of the medium can be reduced to 1.5 ml.

4. Culture conditions for TCLs

Once inoculated in test tubes, incubate in a growth chamber at 24 °C ± 1 °C under an irradiance of 20 W. m^{-2}. The TCL should receive the light vertically and not laterally. The relative humidity is 70%, the photoperiod can be 16h or 24 h. For root programme, the TCLs must be kept in the dark when solified medium is used.

5. Culture medium

A. Composition:
the culture medium contains macro and microelements of Murashige and Skoog [8], and myo-inositol (100 mg/l). IBA, kinetin and glucose are added at different concentrations (Table 1) to induce separately different morphogenetic programmes (Fig. 2). The medium pH is adjusted to 5.6. Gelose is added at 10 g/l, gelrite at 3g/l.

Table 1. Glucose, IBA and Kinetin concentrations for different morphogenetic programmes on tobacco thin cell layer (TCL).

Morphogenetic Programme	Glucose (g/L)	IBA (molar)	Kinetin (molar)
Flower	30	10^{-6}	10^{-6}
Vegetative Bud	30	10^{-6}	10^{-5}
Root	10	10^{-5}	10^{-7}
Callus	30	3.10^{-6}	10^{-7}

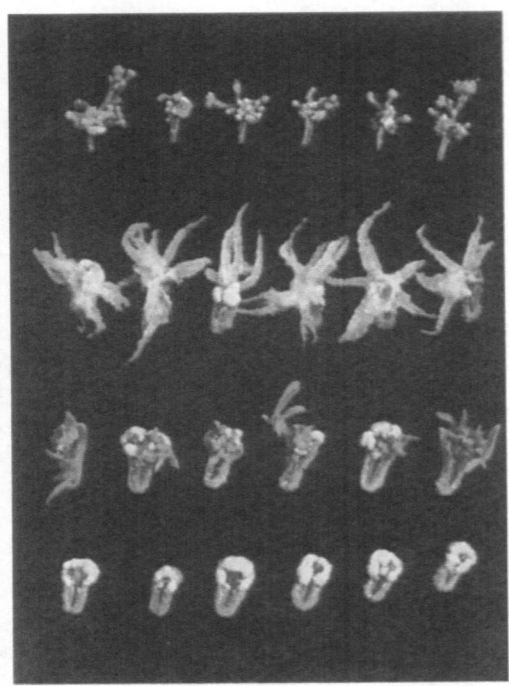

Fig. 2. Four morphogenetic programmes from tobacco TCL (stage: 14 days).
From the top : first line : Flower Programme (flowers are formed directly on the surface of the TCL)
Second line : Root Programme
Third line : Vegetative buds Programme
Fourth line : Callus Programme

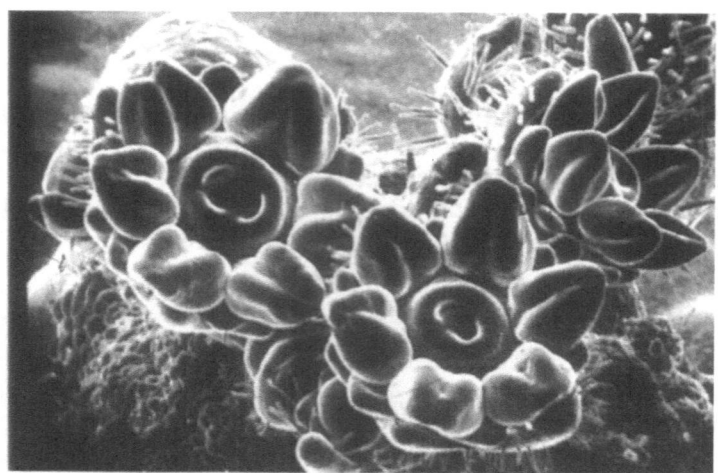

Fig. 3. Young flowers (approximately 50 in total) formed directly on one tobacco TCL (stage: 14 days).

Fig. 4. Fully developed flower from Petunia TCL (stage : 4 weeks).

Functional flowers are obtained within 12–14 days (Fig. 3). Fully developed flowers are obtained within 3 weeks for tobacco TCL, 4–5 weeks for Petunia TCL (Fig. 4).

B. Autoclaving:
The medium is autoclaved at 115 °C for 20 minutes under 1 bar pressure.

6. Observations

Morphogenetic changes are scored every two days (starting from day 4 to day 28 after the beginning of the culture) by observation under a stereo microscope and if necessary scanned with a scanning electron microscope.

The number of organs formed (flowers, vegetative buds, roots), per TCL and the number of the TCLs responding to the treatment are analysed statistically.

II – Transverse TCL (tTCL) method

Model system : *Iris pallida, Phalaenopsis, Oryza sativa, Sorghum, Digitaria, Arabidopsis, Panax ginseng*

This method is applied to recalcitrant species (monocotyledons, woody species, legumes) and non-recalcitrant species of small size.

Steps in the procedure
1. Grow donor plants in appropriate environmental conditions (see above, section I.1).
2. Select the optimal size for tTCL method, by making transverse sections of various sizes [e.g. from 1 mm, 1.5 to 2 mm, 500 μm–1 mm or 300–200 μm (using a vibratome)]. All plant organs can be used: root, mesocotyl, hypocotyl, epicotyl, cotyledon of immature/mature/germinating embryos, petiole, leaf blade, leaf vein, internode, node, stem, immature inflorescence, inflorescence, filament of anther, anther, carpel, ovary style, petal, sepal, etc.
3. Inoculate the tTCL into Petri dishes (10 cm diameter) or multiwell plates, on the surface of media solidified by gelose (10 g/l). Gelrite (2 g/l) or agarose (8 g/l) are used for monocotyledons (e.g. *Sorghum*, wheat, barley, rice, *Phalaenopsis*).
4. Inoculate the tTCL in rows of different sizes. For the first phase of the investigation, the proximal or distal poles can be put at random on the surface of the medium. However, the order of each tTCL on the original organ should be respected, in order to detect the influence of the gradient.
5. From this phase of the study, observations can be made on i) the influence of the polarity and ii) the optimal size of the tTCL to programme morphogenesis. If polarity between the basal and proximal poles is shown (by the difference in the responses obtained when inoculated at random), either reduce the size of the tTCL or consider the polarity for the next phase of the study. Tissues can be arranged in rows with the proximal pole placed on the medium, or rows with the distal pole, on the medium.
6. Other parameters such as cell proliferation, cell expansion, organogenesis and/or embryogenesis can be scored between 4 to 28 days.

Acknowledgments

We thank all scientists of our group, especially M. Drira, Dien N., H. Chlyah, A. Chlyah-Arnason, A. Cousson, F. Jullien, H. Trinh, M. Mulin, V. Marfa, P. Toubart, L. Richard, Van Lê Bui, In Ok Ahn and Thao Do and of other groups who have contributed to the development of the TCL method. Yan Norry and C. Jehanno are deeply acknowledged for the illustrations of this chapter.

References

1. Altamura MM, Capitani F, Serafini-Fracassini D, Torrigiani P, Falasca G (1991) Root histogenesis from tobacco thin cell layers. Protoplasma 161: 31–42.
2. Ammirato V (1987) Speeding transgenic plants. Bio/Technology 5: 1015.
3. Cousson A, Trân Thanh Vân K (1993) Influence of ionic composition of the culture medium on de novo flower formation in tobacco thin cell layers. Can. J. Bot. 71: 506–511.
4. Kay LE, Basile DV (1987) Specific peroxidase isoenzymes are correlated with organogenesis. Plant Physiol. 84: 99–105.
5. Jullien F, Trân Thanh Vân K (1994) Micropropagation and embryoid formation from young leaves of Bambusa glaucescens "Golden Goddess". Plant Science 98: 199–207.
6. Jullien F, Wyndaele R (1992) Precocious in vitro flowering of soybean cotyledonary nodes. J. Plant Physiol. 140: 251–253.
7. Meeks-Wagner D, Dennis E, Tran Thanh Van K, Peacock W (1989) Tobacco genes expressed during in vitro floral initiation and their expression during normal plant development. Plant Cell 1: 25–35.
8. Murashige T, Skoog F (1962) A revised medium for a rapid growth and bioassay with tobacco tissue culture. Physiol. Plant. 15: 473–479.
9. Neale A, Wahlleithner J, Lund M, Bonnet H, Kelly A, Meeks-Wagner D, Peacock W, Dennis E (1990) Chitinase, β-1, 3-Glucanase, osmotin, and extensin are expressed in tobacco explants during flower formation. Plant Cell 2: 673–684.
10. Pélissier B, Bouchefra O, Popin R, Freyssinet G (1990) Production of isolated somatic embryos from sunflower thin cell layers. Plant Cell Reports 9: 47–50.
11. Rajeevan MS, Lang A (1993) Flower-bud formation in explants of photoperiodic and day-neutral Nicotiana biotypes and its bearing on the regulation of flower formation. Proc. Natl. Acad. Sci. USA 90: 4636–4640.
12. Richard L, Arro M, Hoebecke J, Meeks-Wagner DR, Trân Thanh Vân K (1992) Immunological evidence of thaumatin-like proteins during tobacco floral differentiation. Plant Physiol. 98: 337–342.
13. Tiburcio AF, Gendy CA, Trân Thanh Vân K (1989) Morphogenesis in tobacco subepidermal cells : putrescine as marker of root differentiation. Plant Cell Tissue and Organ Culture. 19: 43–54.
14. Trân Thanh Vân K (1973) In vitro and de novo flower, bud, root and callus differentiation from excised epidermal tissue. Nature 246: 44–45.
15. Trân Thanh Vân K (1981) Control of morphogenesis in in vitro cultures. Ann. Rev. Plant Physiol. 33: 291 – 311.
16. Trân Thanh Vân K ,Toubart P, Cousson A, Darvill AG, Gollin DJ, Chelf P, Albersheim P (1985) Manipulation of the morphogenetic pathways of tobacco explants by oligosaccharins. Nature 314: 615–617.
17. Trân Thanh Vân K, Richard L, Gendy C (1990) An experimental model for the analysis of plant/cell differentiation: thin cell layer. Concept, strategy, methods, records and potential. In: Durzan, Rodriguez (Eds.) NATO Biotechnology Series, pp. 215–224. Plenum Academic Press, NY.
18. Trân Thanh Vân K (1991) Molecular aspects of flowering. In: Harding, Singh, Mol JNM (Eds.) Genetics and Breeding of Ornamental Plants. Plenum Academic Press.
19. Trân Thanh Vân K (1992) In vitro organogenesis and somatic embryogenesis. Acta Horticulturae Propagation of Ornamental Plants 314: 27–39.
20. Trân Thanh Vân K, Gendy CA (1993) Relation between some cytological, biochemical, molecular markers and plant morphogenesis. In: Roubelakis-Angelakis, Trân Thanh Vân (Eds.) Nato Asi Series, pp. 39–54. Plenum Academic Press, NY.

Plant Tissue Culture Manual **H5**, 1–17, 1996.
© 1996 *Kluwer Academic Publishers*.

In vitro infection of *Arabidopsis* with nematodes

JOKE C. KLAP & PETER C. SIJMONS[1]
MOGEN international NV, Einsteinweg 97, 2333 CB Leiden, The Netherlands.[1] present address:
Agrotechnological Research Centre, P.O. Box 17, 6700 AA Wageningen, The Netherlands.

Introduction

Sedentary plant-parasitic nematodes establish intricate relationships with their hosts. The interaction reveals several unique features and is a prime object to study host-pathogen recognition, cell-cell communication and induction of feeding structures.

Economically the most devastating species are root-knot nematodes (e.g. *Meloidogyne incognita*) and cyst nematodes (e.g. *Heterodera schachtii*). Host ranges can be very wide and in areas of intensive agriculture, population densities may increase in 2–3 years to such levels that crop yields are severely affected. Although plant-nematode research is difficult under laboratory conditions, the recent infection protocols for *Arabidopsis* have given new momentum to this area of phytopathology [1, 2, 4, 6, 8, 9].

Nematode juveniles remain dormant in the soil until proper host roots grow in their vicinity. Diffusable root factors induce hatching and the juveniles will migrate towards young roots. They invade roots preferably behind the zone of cell elongation and move up to specific sites of the developing vascular cylinder. Here they carefully select cells for the initiation of feeding structures. This process is induced by nematode secretions [3] and will eventually lead to large, multinucleate and hypertrophic structures that provide a constant source of nourishment for the then sedentary nematode. A more detailed description of the life cycle is given in [7, 8].

The protocols described in this chapter provide complete details for the in vitro infection of *Arabidopsis* with *Heterodera schachtii*. The first section describes the maintenance of nematode starter cultures and the preparation of a sterile inoculum, followed by the infection procedure for *Arabidopsis* with freshly hatched second-stage juveniles.

Procedures

In vivo *stock of* Heterodera schachtii

Grow *Brassica oleracea* in sand and inoculate the young plants with second stage juveniles of *Heterodera schachtii*. (Other suitable hosts can be used to maintain stock cultures). The plants can be grown at standard greenhouse or growth chamber conditions although the soil temperature should be kept below 20 °C. Harvest the *Brassica* plants after 4 weeks and store the sand containing roots with cysts at 4 °C in the dark. As long as the sand remains moist, the cysts can be stored this way for several years. (Note: in some countries, quarantine regulations are necessary to grow certain species of cyst nematodes). Have *in vitro* plantlets available when you start the hatching procedure.

Collecting cysts from the sand culture

Steps in the procedure

1. Put some sand containing the roots and cysts in a beaker and add tap water. The sand will settle while the cysts, along with root debris, will float. Pour the cysts off on a 0.5 mm mesh screen that is placed in a kitchen sieve or similar device. Repeat this until you have collected all the cysts from the sand. Next, rinse the cysts on the 0.5 mm mesh screen to remove foam.
2. For easier handling, rinse the cleaned cysts from the 0.5 mm onto a 0.25 mm mesh screen and place the whole screen flat in a large Petri dish. Using fine forceps and possibly a binocular, collect only the large and clearly egg containing cysts and put them in a small sieve of 20 μm gauze (see note 1). For sterilization and hatching you can keep the cysts in this sieve (about 150 cysts per sieve).

Sterilization of the cysts and hatching of the juveniles

Steps in the procedure
1. Work from now on in a laminar flow cabinet. Put the sieve with the cysts with the aid of a sterile pair of tweezers in a 50 ml beaker and add 0.05% $HgCl_2$. Leave this for 3 minutes. Rinse the cysts 3 times in 50 ml beakers containing distilled water.
2. For hatching place the sieve in a glass funnel to which a silicon tube is attached that is closed with a clamp and place the whole unit in a beaker (Fig. 1). Add sterile 3 mM $ZnCl_2$ until the cysts are just covered. Close the top of the beaker with sterile aluminium foil and store at 23 °C in the dark.

Fig. 1. Device that is used for hatching nematodes under axenic conditions. The top part is made from a syringe as described in note 1. The filter is placed in a glass funnel that is closed at the bottom with a silicon tube and a clamp. The entire unit is autoclaved before use. The hatching solution is added to just cover the cysts that are placed on the nylon filter.

3. After 3 to 4 days second stage juveniles will be visible at the bottom of the tube just above the clamp. All handling must be done in a laminar flow cabinet. Collect the juveniles by opening the clamp when the tube is above a 15 μm gauze sieve (see note 1) placed on a 25 ml beaker. Keep the juveniles on the sieve and rinse them 3 times by placing the sieve in sequential beakers filled with sterile distilled water. To surface sterilize the juveniles, place the sieve in a beaker with 0.05% $HgCl_2$ for 3 minutes and rinse again 3 times in clean beakers with sterile water. After the last rinsing the juveniles on the sieve in the beaker must be covered with water.
4. Transfer the juveniles with a sterile Pasteur's pipette from the sieve to a watch glass. Let the worms settle and decrease the volume by removing most of the water. A Pasteur's pipette that is heat-pulled into a thin capillary can be used to prevent worms being sucked up with the water. Small samples can be taken for counting the number of worms per μl. Suspend the juveniles in the watch glass in 0.5% Gelrite (see note 2) and dilute until you have the desired concentration.

In vitro stock of Heterodera schachtii on Sinapis alba plants

Steps in the procedure

1. Collect *Sinapis alba* seeds in a stainless steel teasieve or similar device. Soak for 2 minutes in 70% EtOH and for 10 minutes in 0.2x Teepol (0.8% active chlorite)+ 0.1% Tween 20. Rinse 3 × 10 minutes in sterile distilled water.
2. Pour 20 ml KNOP medium (0.6% agar) in 25 cm Petri dishes and let it cool to room temperature. Place two *Sinapis* seeds per dish and incubate at 23 °C (16h L, 8h D). After two weeks the plants can be inoculated with *Heterodera* second stage juveniles.
3. Inoculate the two weeks old axenic *Sinapis alba* plants with about 1000 juveniles divided in droplets of 1 μl containing about 10 juveniles per droplet. Store the inoculated *S. alba* plates at 23 °C in the dark.

Sterilisation and in vitro *culturing of* Arabidopsis *seeds*

Steps in the procedure
1. *Arabidopsis* seeds can be sterilized in several ways. We routinely use the protocol of Lluis Balcells; small seed batches (not more than a few hundred) are folded into filterpaper discs and closed with a plastified paperclip. The packages are immersed for 2 min in 70% EtOH and sterilized for 15 min in pure Teepol (4% active hypochlorite) + 0.1% Tween 20. Rinse 4 times for 10 min in sterile distilled water.
2. Pour 10 ml of solid KNOP medium (0.6% agar) in 9 cm Petri dishes (see note 3). Let the plates cool down to room temperature and place 10–20 sterilized seeds on a straight line at ca. 1/3 from the top. To synchronize germination, plates can be placed at 4 °C for 3–5 days before transfer to the growth room. During growth, the plates should be tilted at an angle of 10–30° for the roots to grow down. Two weeks after germination the roots can be inoculated.

Inoculation of *Arabidopsis* with sterile juveniles

Steps in the procedure
1. Collect cysts from the *Sinapis alba* roots by using a fine pair of sterilized tweezers (see note 4).
2. Hatch the juveniles without further sterilisation of the cysts. Follow for hatching the same procedure as described above.
3. As a precaution, a second sterilisation step is performed on the hatched juveniles by immersing them in 0.025% $HgCl_2$ for 3 minutes, followed by 3 rinses in sterile distilled water. Follow the same procedure as described in above.
4. Inoculate about 25 juveniles per plant suspended in droplets of 1 μl.
5. Transfer the inoculated plates back to the growth room. They can now be placed horizontal.

The infection process can be followed daily with an inverted or binocular microscope. The first signs of feeding structure development can be observed at the second or third day after inoculation. Normally, the *Heterodera* life cycle will be completed at 2–3 weeks after inoculation, when the cysts start to turn brown. Metabolic stains or detection of marker gene activities such as GUS can be done at any stage without disturbing the root system mechanically. Isolation of feeding structures from the agar-grown roots is also possible although the quantity by weight is limited. Very detailed *in situ* observations can be performed with the help of specially constructed observation chambers (Grundler, Univ. of Kiel, Germany), which allows for both root- and nematode development under high magnification.

Notes
1. Construction of the 20 μm and 15 μm gauze sieves (see also figure 1):
   - cut the top of a 25 ml syringe in a ring of about 1.5 cm in height
   - cut from 15 μm respectively 20 μm mesh gauze nylon cloth circles that are a bit larger in diameter than the syringe rings.
   - melt the gauze on the syringe by preheating the cut surface of the syringe on a hot plate of about 150 °C and quickly move the melting surface onto the nylon gauze. To improve the weld, the nylon gauze can be placed flat on a small glass plate which is heated on a second hot plate to ca. 80 °C.
   - check carefully whether the nylon gauze is completely welded onto the syringe. The juveniles can be lost during the sterilisation procedure if openings remain.
2. The gelrite is used to keep juveniles in suspension, permitting fairly even distribution during inoculation steps. The 0.5% gelrite is autoclaved for 15 minutes at 120 °C and cooled to room temperature before use.
3. It is very important not to pour more than 10 ml in standard size Petri dishes. The nematodes need root exudates to locate the roots. A larger volume will dilute this signal and leads to lower infection rates.
4. Use a binocular microscope in the laminar flow cabinet and avoid damage to the cysts as much as possible.

Solutions

Knop medium

This medium originated from W. Knop [5] and was modified for hydroponic culture. For growth of *Arabidopsis*, the medium is used at 0.2× strength. Stock solutions are stable when stored at 4 °C.

Stock solutions for Knop medium:

– KNO$_3$	120 gram	
MgSO$_4$.7H$_2$O	19.7 gram	together in 1 liter H$_2$O
– Ca(NO$_3$)$_2$.4H$_2$O	30 gram	100 ml H$_2$O
– KH$_2$PO$_4$	13.6 gram	100 ml H$_2$O
– Fe–Na EDTA	0.73 gram	100 ml H$_2$O
– Micronutrients/100 ml H$_2$O:		

– H$_3$BO$_3$	286 mg
– MnCl$_2$	181 mg
– CuCl$_2$.2H$_2$O	5 mg
– ZnSO$_4$	3 mg
– Na$_2$MoO$_4$.2H$_2$O	7 mg
– NaCl	585 mg

To obtain 10 liter of 0.2×Knop solution, add to ca. 5 liter of water:

– KNO$_3$ + MgSO$_4$.7H$_2$O	20 ml
– Ca(NO$_3$)$_2$.4H$_2$O	8 ml
– KH$_2$PO$_4$	4 ml
– Fe-Na EDTA	4 ml
– Micronutrients	2 ml

Mix and fill up to 10 liter. The pH must be 6.4 (adjust with 1 N KOH; there is hardly any buffering capacity in this medium; adjust the pH carefully). Store at 4 °C in the dark.

Solid Knop medium:

Add 0.6% Daichin (or similar tissue culture grade) agar and 1% sucrose to 0.2× Knop solution. Autoclave 20 minutes at 120 °C.

Acknowledgments

We would like to thank Maria Gagiy for practical advice and Ronny Krutwagen for the drawing. This research was partially funded by Avebe, The Netherlands.

References

1. Böckenhoff A, Grundler FMW (1994) Studies on the nutrient uptake by the beet cyst nematode *Heterodera schachtii* by in situ microinjection of fluorescent probes into the feeding structures in *Arabidopsis thaliana* . Parasitology 109: 249–254.
2. Goddijn OJM, Lindsey K, Vanderlee FM, Klap JC, Sijmons PC (1993) Differential gene expression in nematode-induced feeding structures of transgenic plants harbouring promoter gusA fusion constructs. Plant J 4: 863–873.
3. Hussey RS (1989) Disease-inducing secretions of plant-parasitic nematodes. Annu Rev Phytopathol 27: 123–141.
4. Niebel A (1994) Molecular and genetic approaches to plant-nematode interactions. Thesis, State Univ. Gent, Belgium.
5. Knop W (1860) Über die Ernährung der Pflanzen durch wässerige Lösungen unter Ausschluss des Bodens. Landwirtsch Versuchsstat 2: 65–99 and 270–293.
6. Sijmons PC (1993) Plant-nematode interactions. Plant Mol Biol 23: 917–931.
7. Sijmons PC, Atkinson HJ, Wyss U (1994) Parasitic strategies of root nematodes and associated host cell responses. Annu Rev Phytopathol 32: 235–259.
8. Sijmons PC, Grundler FMW, Vonmende N, Burrows PR, Wyss U (1991) *Arabidopsis thaliana* as a new model host for plant-parasitic nematodes. Plant J 1: 245–254.
9. Wyss U, Grundler FMW (1992) *Heterodera schachtii* and *Arabidopsis thaliana*, a model host-parasite interaction. Nematologica 38: 488–493.

Plant Tissue Culture Manual **H6**, 1–19, 1997.
© 1997 *Kluwer Academic Publishers.*

Asparagus cell cultures as a source of wound-inducible genes

R. M. DARBY & J. DRAPER
Department of Botany, University of Leicester, University Road, Leicester LE1 7RH, UK

Introduction

During its life cycle a plant is subjected to various external stimuli, some often beneficial, but many, such as wounding, often detrimental or life threatening. Unlike the majority of animals, where movement offers some degree of protection from wound damage, the options available to plants are somewhat more limited. Without the benefit of movement a plant must be able to tolerate any changes in the external environment – this requires plasticity of development and an ability to dedifferentiate.

The term "wound response" is very general and obscures the many types and levels of response which occur to protect the plant. Much of our understanding here is only very limited, and there are certain to be more responses which we have yet to discover. Although the purpose of this chapter is not to review our current understanding of wounding, it is worth summarising some of the areas where knowledge is weak, and how mechanically isolated asparagus cells can improve our understanding.

It must be remembered that all wounding stimuli are not the same. They can range from the invasion of just a few cells by a feeding aphid, through to complete decapitation by grazing or harvesting; and the severity of damage may well affect the types and level of response induced.

When a wound has been created there are certain primary responses which must occur to prevent dehydration and pathogen invasion. Between monocots and dicots fundamental differences exist; in dicots, cells at the wound-site will proliferate into a callus to prevent dehydration and pathogen entry, whereas in monocots, cell proliferation is much reduced and greater reliance is placed upon production of water insoluble compounds such as lignin and suberin [1]. These responses have been known for many years and are well understood at the physiological level, however the same is not true for the molecular and biochemical changes. For such changes to occur the normal route of cellular development must be altered i.e. dedifferentiation of affected cells must occur. There must be a reactivation of the cell cycle; cell wall modification; increase in ribosome synthesis, transcription and translation; and an increase in requirement for energy.

To activate the many different responses to a wound stimulus the expression of genes involved is complex, the pathway, coordination and transduction of signals being poorly understood. Matters are complicated further if wounding is associated with pathogen invasion, which may also involve the coordinated expression of a whole new range of resistance/antipathogen genes, such as induction of PR proteins following TMV infection [10].

One must also consider the hierarchy of perception and action of a wound response. Cells do not all behave in the same way following wounding. The cells at or local to a wound site offer the first line of defence to damage and their function may be to propagate a signal transduction pathway, both locally and systemically. Studies of systemic responses have generally produced more definitive results, whereby chemicals such as jasmonic acid and a small protein, systemin, have been implicated in the systemic induction of insect gut protease inhibitor (pin) genes [9]. Depending on the severity of the damage, signal transduction may only involve a limited number of cells local to the wound-site and not those in systemic tissue; for instance it is known that production of wound-healing and antimicrobial compounds in the damaged cell can diffuse to local cells only, producing a non-systemic response. The precise nature of local signalling is poorly understood, but signals can include such things as oligosaccharides, active oxygen species, ion equilibrium changes (often caused by decompartmentalization) and ethylene.

Much of the early work used whole plant systems, such as wounded artichokes and carrots [7], and although providing useful information on the physiological aspects of wounding, these systems have been difficult to adapt to a molecular biology study. Reasons for this include the small amount of tissue responding; difficulties in gaining access to tissue; production of compounds which interfere with molecular analysis; the timing of the response may be variable; and complex interactions with unwounded tissue must be considered. Established cell cultures have also been used to study changes in gene expression, in response to wounding, but these systems require the addition of endogenous substrates, such as cell wall fragments, to generate a response [3]. Generally, this tends to narrow the type and number of new genes detected. Also with these systems, gene expression can be altered by constituents of the medium, for instance salt stress, presence of sugars causing a change in general metabolism, presence of hormones, and osmotic stress. To maintain cell viability sub-culture is required, which when repeated continually can perpetuate genetic mutations.

The lack of information about local wound responses has generally been hampered by the lack of a good model to investigate. When studying complex interactions it is always favourable to reduce the study to basic building blocks, and in the case in question this is individual wounded cells. To model wound-site gene expression using single cells requires a system whereby cells can be isolated rapidly, in large numbers, and still retain high viability. We have found that several plant species are suitable for mechanical cell isolation, including peanut, *Calystegia sepium*, and *Asparagus officinalis*. *Asparagus* has been the system of choice, due to the high viability of isolated cells, and also being a monocot, is of direct relevance to important crop species.

Mechanically isolated asparagus cells have been used for studies of photosynthesis [5, 6], cellular dedifferentiation [8], and wounding [2, 4]. In our hands at Leicester, we have found asparagus cells to be an ideal system to study early events in local wound responses. The advantages offered are numerous and include:

(a) Large numbers of cells can be isolated, approximately 1×10^8 cells from 25 g of starting material, by homogenisation of whole plants in a pestle and mortar.

(b) Cytological, physiological and biochemical measurements have shown the cells behave in a highly synchronised manner, which eliminates many of the problems associated with previous studies undertaken to investigate differentially expressed genes within heterogeneous cell populations.

(c) Cells are maintained in an isotonic medium so expression of genes induced by osmotic shock is reduced to a minimum.

(d) The open structure of an asparagus cladophyll affords large air spaces between mesophyll cells, limiting cell to cell contact to small areas at the apex and base of each cell, in contrast to the more common structure in other species where cells are almost completely attached to one another by shared middle lamellae. For this reason mechanical shearing of cladophylls releases only intact mesophyll cells. Other cell types such as epidermal and vascular cells are more closely associated and so are ruptured more easily by the shearing forces created during isolation. Moreover not only is the cell population homogeneous, but due to the radial arrangement of the cells within each cladophyll the wounding stimulus received is similar across the whole population.

(e) As for reasons explained in (d), the cell viability is high, routinely greater than 65%.

(f) Large amounts of good quality RNA and protein are easily extracted from these cells. At Leicester, mechanically isolated cells have been studied for a number of years and we have been able to compile much information, characterising the cytological and physiological responses to wounding. Using this information and the technology of molecular biology we have been able to instigate an extensive study into the changes in gene expression associated with wounding. This has been achieved by differential screening of several different asparagus cDNA libraries, using poly A+ RNA extracted from unwounded cladophylls as a non-induced control probe, and poly A+ RNA extracted from mechanically isolated cells at different times after isolation as the induced probe. This yielded many cDNA clones, the temporal expression of which was then investigated using northern gel blot analysis of RNA extracted from mechanically isolated cells. The expression patterns of these cDNAs could be loosely arranged into three separate categories: i) expression level peaking at one day post isolation, ii) expression level peaking at three days post isolation, and iii) expression reaching a peak by the third day post isolation then remaining high. To examine whether the cDNAs were wound-inducible or merely induced by some response associated with cell isolation but not with wounding, northern gel blots were carried out against RNA extracted from wounded asparagus seedlings. It was found that the expression of the majority of cDNAs was also similar in wounded plants, although to a reduced level. The DNA sequence of these cDNAs is currently being investigated. The expression of the cDNA clones is also being investigated in established suspension cultures.

Procedures

Growth of asparagus *plants*

Asparagus plants are usually easy to grow, however depending on precise conditions, growth rates may be faster or slower than the those indicated below. Generally there is a broad time scale when the plants will yield satisfactory cells, but it is important to bear in mind that young plants ($<$ 3 weeks after sowing) will yield cells of low viability, and although old plants ($>$ 5 weeks after sowing) will produce cells of high viability it becomes increasingly difficult to obtain sterile isolations. Young plants can be identified by their pale colour and incompletely expanded cladophylls. Plants which are too old are more easily identified by the growth of secondary stems from the base of the seedling.

Steps in the procedure
1. Imbibe 35 g of seed of *Asparagus officinalis* cv. Connover's colossal (Johnsons Seeds, UK) in tap water for a period of 2–3 days.
2. Sow into a 20 cm × 35 cm seed tray containing a good quality potting compost.
3. Grow for five weeks in good light at 23 °C, with a 16-hr day length.

Notes
3. After 10–14 days seeds will begin to germinate. Over the preceding five days cladophylls will begin to form on the stem, and then within the following week will expand fully, at which point the plants are ready to use.

Mechanical isolation of cells

To avoid bacterial contamination, experiments lasting longer than 16 hrs must be carried out aseptically, the manipulations for which must be undertaken in a suitable sterile laminar air flow cabinet. It is assumed that operators will be fully conversant with aseptic techniques for use in plant tissue culture.

Steps in the procedure
1. Harvest a half tray of five week old asparagus fronds.
2. Strip cladophylls from fronds by running stem between index finger and thumb.
3. Split equally and place into two 500 ml powder rounds, or other similar vessel with water-tight lid, containing 10% commercial bleach.
4. Sterilise for 20 minutes with occasional agitation.
5. With sterile forceps move cladophylls to a sterile 64 μm, 11 cm diameter sieve.
6. In a sterile receptacle (a glass casserole dish is ideal) rinse cladophylls with 4×400 ml of sterile tap water, taking great care to make sure all of the bleach is rinsed away.
7. Place sieve, containing cladophylls on to suitable sterile surface, such as a casserole dish lid, and allow excess moisture to drain away.
8. Using sterile forceps take approximately one fifth of the material and place into a sterile mortar. Add approximately 20 ml of sterile tap water and begin to homogenise cells with pestle.
9. When homogenisation is complete, pour contents of mortar into a sterile 5 cm diameter 64 μm seive, contained within a sterile crystallizing dish, and allow to drain. Cells are contained within the filtrate and so the retentate can be discarded. Repeat until all material has been so treated.
10. Transfer cells from the crystallising dish to a 250 ml GSA centrifuge bottle (Nalgene), using a 20 ml glass pipette (Volvac), connected to an automated pipette.
11. Pellet cells at 800 rpm in a swing out rotor.
12. Remove supernatant, with automated pipette, and resuspend in 100 ml of sterile tap water.
13. Pellet cells as in 11. Remove supernatant as before and resuspend cells in 80 ml of sterile ASP medium. Repeat once more.
14. Remove 1 ml of cell suspension, and place onto a haemocytometer. Count number of cells as described in instruction supplied with haemocytometer.
15. Dilute cells to a density of 4×10^5 cells per ml with sterile ASP

medium, allowing for the volume of glutamine to be added as calculated.

16. Add sterile glutamine to a final concentration of 1 mg/ml.
17. Using an automated pipette, take 10 ml aliquots, and dispense cells into standard 9 cm Petri dishes (Sterilin). To maintain an even plating density keep cells agitated whilst dispensing.
18. Seal dishes with sealing film, such as Nescofilm (Nippon Shoji Kaisha Ltd., Japan).
19. Incubate cells in the dark at 25 °C on an orbital shaker (40 rpm).

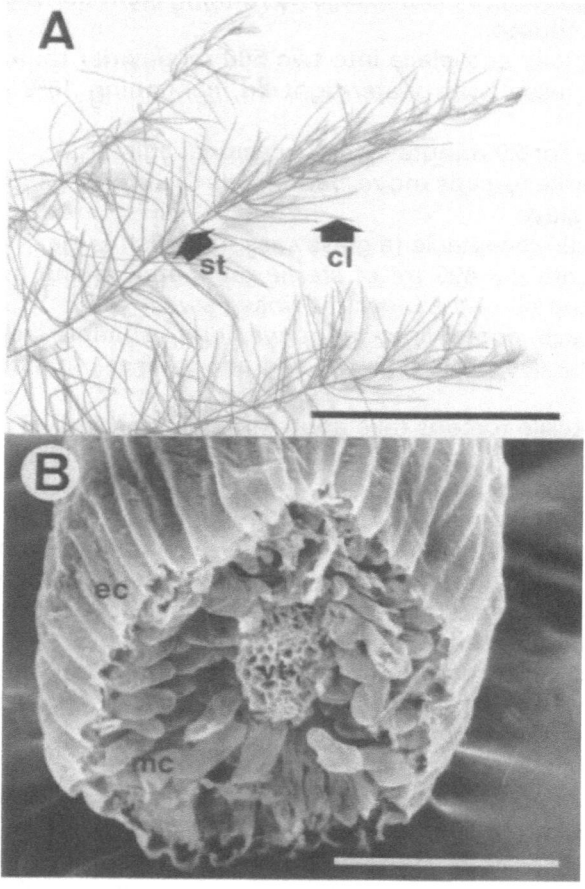

Fig. 1. Macro and micro structure of *Asparagus cladophylls*. A) Fronds from 4-week-old seedlings, with main stem (st) and cladophylls (cl) arrowed. Scale bar represents 10 cm. B) SEM of a transverse section through an *Asparagus* cladophyll, with mesophyll cells (mc), epidermal cells (ec), and vascular tube (vt) indicated. Scale bar represents 100 μm.

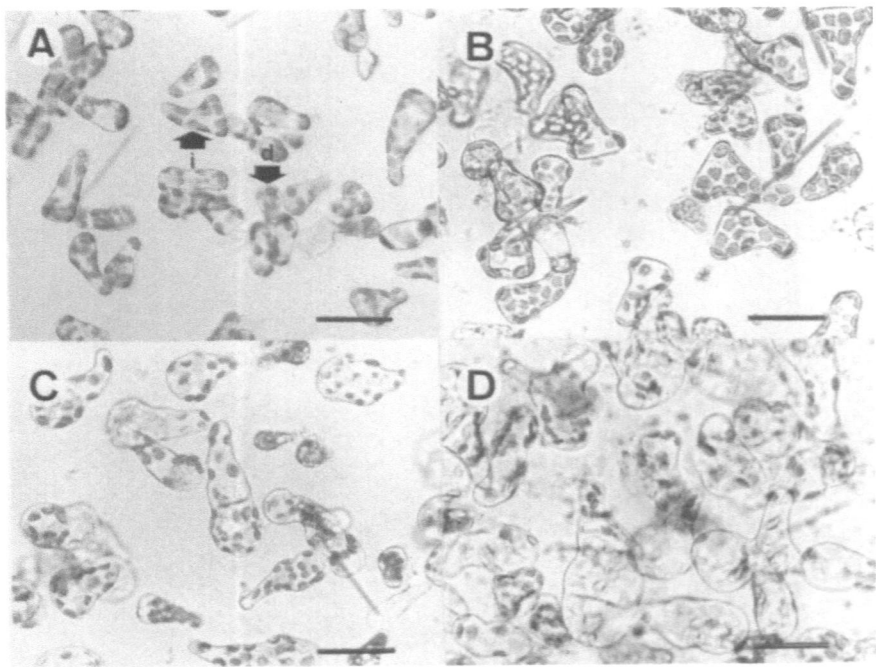

Fig. 2. Bright field microscopy of mechanically isolated *Asparagus* cells. A) Freshly isolated cells, with intact (i) and damaged (d) cells arrowed. Also, visible in this frame are raphids (long crystals of calcium oxalate). B) Mechanically isolated cells one day following cell isolation. C) Mechanically isolated cells three days following cell isolation. D) Mechanically isolated cells five days following cell isolation. All scale bars represent 30 μm.

Notes
2. Cladophylls are the asparagus equivalent of a leaf. They are the needle-like structures branching from the main stem, and together with the stem form the complete frond (Fig. 1A).
3. At this stage it has been found that if large volumes of air are trapped inside the powder round then sterility can be compromised. Thus the vessel must be filled until it overflows. Care must also be taken to ensure that cladophylls are not trapped between the neck and lid of vessel, as this again can be a source of contamination.
5. Sieves can be of any construction, the only criteria being that they are autoclavable and have a 64 μm mesh fitted. We have found those made from Nalgene tripour beakers, as illustrated in Fig. 3, to be of a durable and reliable nature.
8. The homogenisation stage of the isolation procedure is the most critical if good quality, high viability cells are to be obtained. During homogenisation it is advisable to apply the least amount of pressure possible to the pestle, which avoids unnecessary cell damage. Homogenisation is complete when the water has become dark green and the white vascular tissue becomes visible on the cladodes. This is approximately 3–5 minutes following commencement of homogenisation. Under these gentle shearing conditions there will be considerable material remaining intact.

Fig. 3. An example of a 64 μm sieve used in mechanical cell isolations.

9. The sieve may become partially blocked with large pieces of cell debris. If this occurs, hold the sieve body with forceps and gently apply pressure to the sieve membrane with the pestle to aid flow of filtrate.

14. When counting cells it is important to differentiate between those which are alive and those which are dead. Figure 2 shows cells immediately after cell division; with live and dead cells indicated by arrows. In living cells, the chloroplasts are clearly visible as distinct disc shapes, and the cytoplasm appears clear, but in dead cells, the chloroplasts have ruptured and the cytoplasm has a granular appearance. Further evidence for cell viability can be obtained by staining cells with fluorescein diacetate (see section Vital staining).

16. During mechanical cell isolation asparagus cells lose their ability to synthesise glutamine, through the disruption of glutamine synthase activity. To prevent cell death in longer term experiments, the medium must be supplemented with glutamine.

Generation of suspension cultures

Steps in the procedure

1. Take 20 ml of mechanically isolated cells approximately 4–6 days old and add to a 500 ml Erlenmeyer flask containing 80 ml of ASP medium supplemented with glutamine at a final concentration of 1 mg/ml.
2. Incubate in the dark at 25 °C on an orbital shaker set at 200 rpm. Allow 2–4 weeks for cells to become established.
3. Subculture every 2 weeks, by adding 20 ml suspension to 80 ml fresh medium supplemented with glutamine.

Vital staining

When first working with asparagus cells it is sometimes difficult to discern between those which are alive and those which are dead. To help with identification it is possible to use the vital stain fluorescein diacetate (FDA). When observed under UV light (B filter) viable cells will fluoresce, whereas dead ones will not.

Steps in the procedure
1. Add 20 μl of 5% FDA to 1 ml of cells.
2. Incubate cells for 5 minutes to allow uptake of dye.
3. Place cells onto glass slide, place on coverslip and view under UVB light.

Sampling of tissue

In general, a minimum of 10 plates are required to give sufficient material for each time point, this will give approximately 200 μg total RNA and 1 mg of soluble protein. To provide sufficient total RNA for poly A+ RNA extraction and subsequent library construction and screening 50 plates were harvested.

Steps in the procedure

1. Unwrap Petri dishes, and using a glass spreader gently scrape over the bottom plate. This will loosen any cells which have adhered to the surface of the dish.
2. Transfer suspension to a screw-capped centrifuge tube. Once all plates have been treated spin tubes at 2,000 rpm for 5 minutes, remove supernatant, resuspend cells in 1 ml of distilled water, transfer to foil packet, and flash freeze in liquid nitrogen. Store at −80 °C until required.

Solutions

- ASP Medium
- Major salts $mg\ l^{-1}$
  - NH_4NO_3 825
  - KNO_3 950
  - $CaCl_2.2H_2O$ 220
  - $MgSO_4.7H_2O$ 1233
  - KH_2PO_4 680
  - Na_2EDTA 37.3
  - $FeSO_4.7H_2O$ 27.8

- Minor salts
  - H_3BO_3 6.2
  - $MnSO_4.H_2O$ 22.3
  - $ZnSO_4.H_2O$ 10.58
  - KI   0.83
  - $NaMoO_4$ 0.25
  - $CuSO_4.5H_2O$ 0.025

  - Thiamine   1.0
  - NAA   3.0
  - 6-BAP   0.3

  - Glutamine   1000

  - Myo inositol   100
  - Sucrose   10000
  - Mannitol   30000

To avoid time consuming work weighing out many different medium constituents, it is simpler to make up concentrated stock solutions, which can be stored at $-20\ °C$ until required. Due to formation of precipitate when certain concentrated salts are mixed separate stock solutions must be made as follows:

- Stock solutions:		Stock concentrations (g/100 ml)	Volume of stock (ml/l media)
A	NH_4NO_3	8.25	10 ml
	KNO_3	9.50	
B	$CaCl_2.2H_2O$	2.20	10 ml
C	$MgSO_4.7H_2O$	12.33	10 ml
	KH_2PO_4	6.80 (mg/100 ml)	

D	Na$_2$EDTA	3730.0	10 ml
	FeSO$_4$.7H$_2$0	2780.0	
E	H$_3$BO$_3$	620.0	1 ml
	MnSO$_4$.H$_2$O	2230.0	
	ZnSO$_4$.H$_2$O	1058.0	
	KI	83.0	
	NaMoO$_4$	25.0	
	CuSO$_4$.H$_2$O	2.5	

– Vitamins and growth regulators

Thiamine	1 mg/ml	1 ml
NAA	1 mg/ml	1 ml
6-BAP	3 mg/ml	0.1 ml

– Sugars

Sucrose	10 g
Myo inositol	100 mg
Mannitol	30 g

– Heat labile supplements

Glutamine	2.35 g/100 ml	42.5 ml

Note that glutamine is heat labile and cannot be autoclaved. Filter sterilize and store in 20 ml aliquots at −20 °C.

Adjust pH to 5.8 with KOH, aliqot into convenient volumes such as 100 ml, and autoclave for 20 minutes at 121 °C, 15 psi. Following autoclaving it is normal for there to be precipitate present. ASP medium can only be autoclaved once.

– FDA Stain: Dissolve 5 g of FDA in 100 ml of acetone. Store at −20 °C.
– Useful equipment:
  – Casserole dish (Pyrex)
  – 6 cm and 11 cm diameter 64 μm sieves (Wilson Sieves, 2 Long Acre Lane, Hucknall, Nottingham NG15 6QD, UK)
  – Crystallizing dish
  – Pestle and mortar
  – 20 × 20 ml glass pipettes (Volvac)
  – Automatic pipette
  – 2 × 250 ml GSA centrifuge tubes (Nalgene)
  – Haemocytometer
  – Disposable 9 cm Petri dishes, standard laboratory grade (Sterilin)

References

1. Bostock RM, Stermer BA (1989) Perspectives on wound healing in response to pathogens. Annu Rev Phytopathool 27: 343–371.
2. Darby RM, Warner SAJ, Harikrishna K, Draper J (1996) Wound response and dedifferentiation in mechanically separated asparagus mesophyll cells. Submitted.
3. Felix G, Boller T (1995) Systemin induces rapid ion fluxes and ethylene biosynthesis in *Lycopersicum peruvianum* cells. Plant J 7: 381–389.
4. Harikrishna K, Paul E, Darby R, Draper J (1991) Wound response in mechanically isolated *Asparagus* mesophyll cells: a model monocotyledon system. J Exp Bot 42: 791–799.
5. Harikrishna K, Darby R, Draper J (1992) Chloroplast dedifferentiation in mechanically isolated *Asparagus* mesophyll cells during culture initiation. Plant Physiol 100: 1177–1183.
6. Hills MJ (1986) Photosynthetic characteristics of mesophyll cells isolated from cladophylls of *Asparagus officinalis* L. Planta 169: 38–45.
7. Lewis BG, Davies WP, Garrod B (1981) Wound healing in carrot roots in relation to infection by *Mycocentrospora acernia*. Ann Appl Biol 99: 35–42.
8. Paul E, Harikrishna K, Fioroni O, Draper J (1989) Dedifferentiation of *Asparagus officinalis* L. mesophyll cells during initiation of cell cultures. Plant Sci 65: 111–117.
9. Ryan CA (1992) The search for the proteinase inhibitor-inducing factor, PIIF. Plant Mol Biol 19: 123–133.
10. Van Loon LC, Gerritsen YAM, Ritter CE (1987) Identification, purification, and characterisation of pathogenesis-related proteins from virus-infected Samsun NN tobacco leaves. Plant Mol Biol 9: 593–609.

Plant Tissue Culture Manual **H7**, 1–45, 1997.
© 1997 *Kluwer Academic Publishers.*

Use of video cell tracking to identify embryogenic cultured cells

MARCEL A.J. TOONEN & SACCO C. DE VRIES
Department of Molecular Biology, Agricultural University Wageningen, Dreijenlaan 3, 6703 HA Wageningen, The Netherlands

Abbreviations: 2,4-D = 2,4-dichlophenoxyacetic acid; B5 = Gamborgs B5 medium; B5-0.2 = B5 medium supplemented with 0.2 μM 2,4-D; CCD = Charge Coupled Device; CSLM = Confocal Scanning Laser Microscopy; FITC = Fluorescein Isothiocyanate; IMA = Argus specific image-file format; PBS = Phosphate buffered saline; PC = Personal Computer; PIC = BioRad specific image-file format; PICT = Apple specific image-file format; SIT = silicon intensified tube; TIFF = Tagged Interchange File Format

Introduction

Plant somatic embryos develop from embryogenic cells, that in turn originate from somatic cells. Embryogenic cell formation usually commences when explants or suspension culture cells are exposed to exogenous plant growth regulators. A large number of cells in the explant or culture generally responds to the inducing treatment, while only a limited number of cells actually becomes embryogenic. There is a considerable gap in our knowledge of the events that take place between the moment the inducing treatment is applied and the first morphologically visible signs of somatic embryo development occur. This is due in part to the fact that changes observed after particular treatments with, for example growth regulators at the culture level are often interpreted based on effects in very few cells. It is therefore important to be able to identify the few responding cells in culture and follow their development into somatic embryos. For this purpose a cell tracking system was established based upon the system previously developed by Dr. H.A. Verhoeven (CPRO-DLO, Wageningen, The Netherlands). The cell tracking system allows recordings of protoplasts, cells or cell clusters over periods of several weeks. In this way development of individual cells into somatic embryos can be monitored. This cell tracking system has been used to identify single carrot suspension cells that are in the process of acquiring embryogenic capacity [11]. It has also been used to identify guard cells as the cell type that develops microcalli in the recalcitrant crop species sugar beet [4].

A second useful procedure combines cell tracking with Confocal Scanning Laser Microscopy (CSLM) as a method to test potential markers for their role in embryogenic cell formation. This was used to follow development of cells labelled with the monoclonal antibody JIM8, reported to be a marker for embryogenic cells [10]. However, cell tracking experiments showed no correlation between expression of the JIM8 cell wall epitope and the ability of that cell to develop into a somatic embryo [12]. A third application uses cell tracking in combination with a 2D luminometer system to detect *in vivo* gene expression during embryo development. For this purpose carrot suspension cultures were obtained from cells transformed with the embryo-expressed LTP1 promoter fused to the firefly luciferase coding sequence [13]. The 2D luminometer system is able to detect photons released by the luciferase and in this way enables the detection of gene expression in live, developing somatic embryos. Here we will describe the equipment needed and procedures used to allow cell tracking in combination with the different detection systems.

1. Instrumental set-up

Figure 1 shows a schematic overview of the cell tracking equipment and the various signal detection systems. The system is based upon a step motor driven cross table attached to an inverted microscope equipped with a step motor-controlled focus knob. MicroScan 2.0 cell tracking software has been developed to allow control of the plane of focus and movement of the cross table. The combination of the cross table with the MicroScan software allows recording of the same positions in culture dishes over periods of several weeks. The area of the culture dish at such a position that is visible on the video screen is defined as 'frame'. For detection of visible light images a video camera can be connected to the microscope. For detection of fluorescent signals a CSLM system can be used while the 2D luminometer system is used for bioluminescent signals. In all cases data can be stored either as video images or as digital images. For large numbers of images video recording is the preferred option, since a 3 hour video tape can store over 20 000 images in colour, which is beyond the capability of digital storage media commonly used at this moment. In the case of bioluminescent signals less images are obtained and they can be stored digitally on optical disks or CD-ROMs.

Features of each individual part of the equipment and software are described below.

1.1. Microscope and cross table

The cell tracking equipment is based upon the Märzhäuser EM-32 IM cross table (Fig. 1d; Märzhäuser, Wetzlar, Germany). The cross table itself is universal but it is built in such a way that it can only be attached to one specific type of microscope. Different holders can be placed into the cross table to fit culture dishes of various sizes. To perform cell tracking a microscope containing a video connector is required. We used a Nikon Diaphot inverted microscope (Fig. 1a; Nikon, Tokyo, Japan) or a Zeiss Axiovert 135-TV inverted microscope (Fig. 1b; Zeiss, Oberkochen, Germany) both equipped with their specific EM-32 IM cross table. Each type of microscope is equipped with a specific light path. In the Zeiss microscope we used, all light derived from the specimen is either directed to the eyepiece for visual observation or alternatively, through the video connection

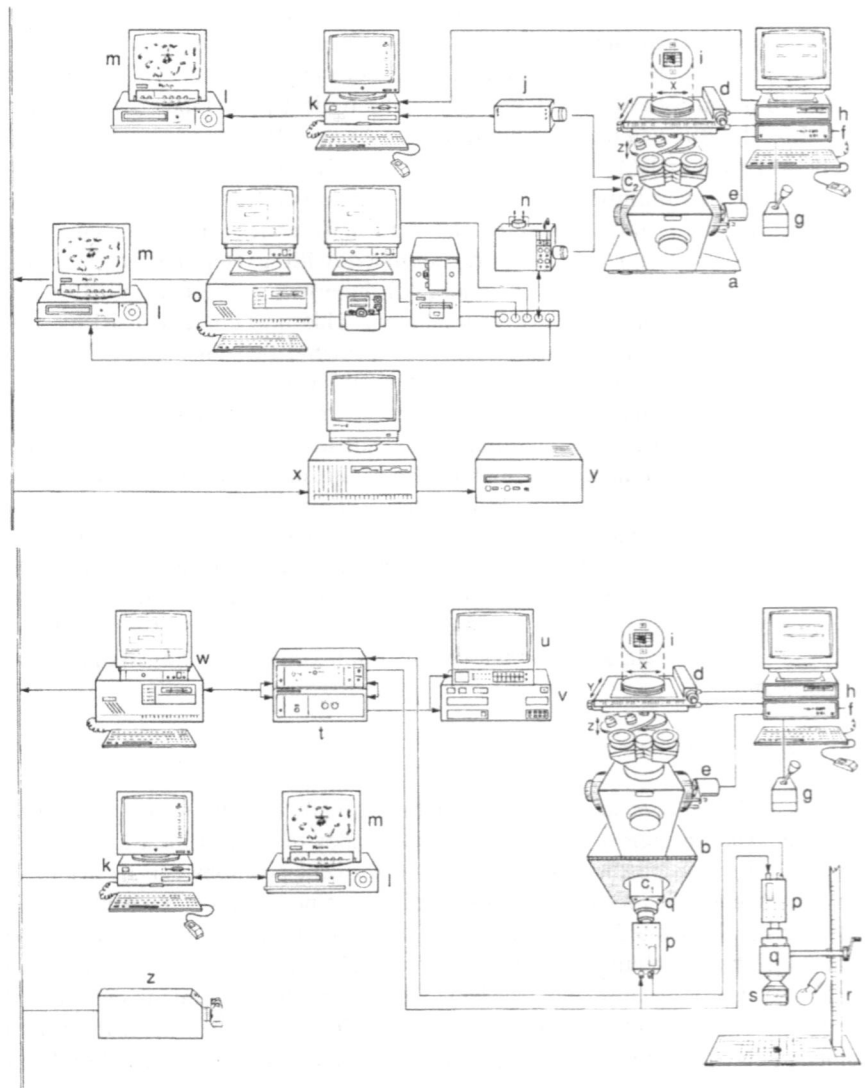

Fig. 1. Equipment employed for cell tracking and the detection of light images and images of fluorescent and bioluminescent signals. Each part is displayed in detail in the text.

(100%) at the bottom of the microscope (Fig. 1c1). The advantage of the video connection at the bottom of the microscope is that the distance between specimen and camera is as short as possible and little signal is lost in the optical path. In the Nikon Diaphot we also used, 20% of the light derived from the specimen is directed to the eyepiece and the remaining 80% is directed through the video connection (Fig. 1c2). For detection of low intensity signals such as produced in bioluminescence experiments, a 100% light video connection, as in the Zeiss microscope, is preferred. In addition, the microscope optics must be of first quality to avoid loss of signal strength in the optical path. Objectives ranging from 2.5× to 20× can be used, but routinely 4× or 10× objectives are applied.

The Märzhäuser EM-32 IM cross table (Fig. 1d) is equipped with two step motors with a 0.4 millimetre pitch enabling movements of the cross table in the X- and Y-direction with steps of 10 nanometres. To control the Z-movement of the objective both microscopes are equipped with a step motor (Fig. 1e; Märzhäuser) attached to the focus knob. All three motors are controlled by the MultiControl 2000 unit (Fig. 1f; Märzhäuser). This unit contains a RISC-processor controlled by the operating software VENUS-1 (Märzhäuser). Connected to the MultiControl unit is a Tulip DC486 sx PC (Fig. 1h; Tulip, 's-Hertogenbosch, The Netherlands). The computer acts as terminal to control the RISC-processor by the VENUS-1 communication software under MS Windows. By using VENUS-1 commands, settings of the MultiControl 2000 unit can be adjusted to individual preferences. Movement of the cross table and focus motor can be controlled using a joystick (Fig. 1g) or entering coordinates. The step motors are driven by sequential pulses, each pulse allowing the cross table to make one step of 10 nanometres. After calibration of the cross table every position of the table is exactly defined and can be retrieved using the VENUS-1 software or MicroScan 2.0 software (Agricultural University, Wageningen, The Netherlands).

1.2. Microscan 2.0

MicroScan 2.0 allows scanning of immobilised cells or cell clusters at defined positions. The program uses VENUS-1 commands to control the MultiControl unit. Either single positions or a matrix of adjacent positions can be scanned. For convenience the area of the culture dish at one position that is visible on the video screen is defined as 'frame' (Fig. 2a). Each frame position can also be scanned in the Z-direction in a number of steps. The position of the culture dish relative to the cross table does not have to be fixed, because this is determined by the two mark-points present in each dish (Fig. 2b). Based upon these two points a new set of mathematically defined axes is calculated by the MicroScan program. All positions in the culture dish are determined relative to these axes. This feature allows to retrieve exactly the same position in the culture dish at any desired time point. In addition, it makes it possible to exchange dishes between the different microscopes equipped with a Märzhäuser EM-32 IM cross table and to retrieve particular single frame positions in a given dish. Individual frame positions can be programmed manually or a matrix of adjacent frames can be calculated by the MicroScan program. The programmed positions or the set of frame information calculated, are stored to the computer hard disk by the program. To follow development into somatic embryos of large numbers of cells distributed in a petri dish the matrix of adjacent frames is used. Frame images

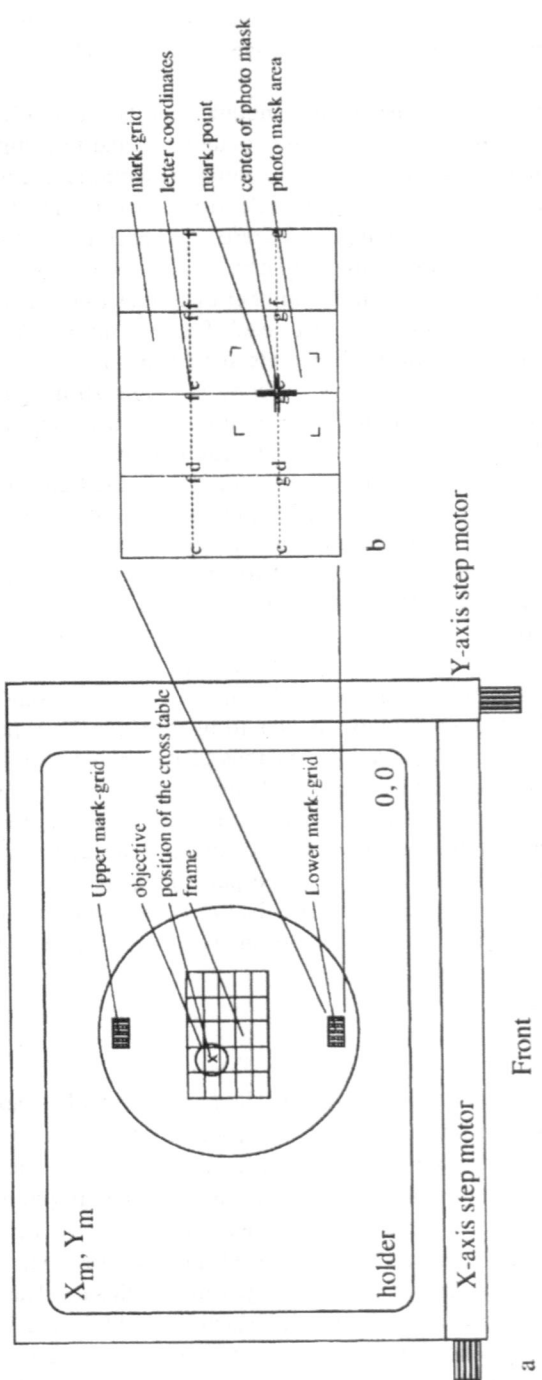

Fig. 2. Schematic drawing of a top view on the cross table. The table consists of two step motors to control movement in X and Y direction. The axes defined by the VENUS software. The basis of the set of axes is the (0, 0) position at the right bottom of the cross table. The maximal XY position (X_m, Y_m) is located at the outer right frame top. The culture dish fitted with two mark-grids is placed in the holder is. The calculated matrix of adjacent frames is in the centre of the culture dish. One frame is the area that is visible on the monitor with a given objective. The position of the cross table with respect to the objective is defined as the position of the cross table (x). *b* The mark-points are located on the mark-grid. The grid consists of a number of lines in horizontal and vertical direction. The distance between the two lines is 400 nm. On crossings of the lines letter coordinates are given. One letter coordinate is selected as mark-point and placed in the centre of the photo mask (represented as a set of black lines in the form of a cross on the Nikon and Zeiss microscopes used).

are recorded on video tapes during embryo development. Placing the images of the same frame into sequence then gives an overview of the development of all individual cells visible in that frame.

The MicroScan 2.0 software consists of five modules and two data files: profiles.dat and msconfig.dat. The msconfig.dat file contains standard settings such as the cross table movement speed and settings to define the communication port (COM port) used for connection with the MultiControl unit (Fig. 1f), the Macintosh computer (Fig. 1k) and the computer mouse. The profiles.dat file contains pre-defined settings used for scanning a matrix. These settings are the size of a frame visible on the monitor screen, the number of frames to be scanned in X-, Y- and Z-direction and the time to stop at each frame position. For video microscopy the cross table stops 350 milliseconds at each position to allow recording of one image. For CSLM it takes 1.1 second to obtain an image of the fluorescent signal and this is the minimal time to stop at each frame position. Both data files can be changed using the MS-DOS edit command.

Individual frames to be scanned can be programmed using the PoinT-init and PT-edit modules. A matrix of frames for cell tracking can be calculated by the MatriX-init module. For both type of scans a frame is selected by the user where it is possible to define a low and a high plane of focus, between which the program will scan in the Z-direction. The modules PT-scan and MX-scan are used to scan the programmed frames. For each new culture dish the cell tracking system has to be calibrated with a full calibration. Hereby the effective area of the cross table that can be scanned is determined. First the cross table is sent to the 0, 0 position (Fig. 2a). Thereafter the table moves to its maximal X, Y position (Fig. 2a; X_m, Y_m). In this way the exact dimensions to be scanned are known and the basic set of axes are defined by the MicroScan program. Half calibration will send the cross table to the 0,0 position. After calibration of a culture dish a new set of axes is defined by the MicroScan program. As long as the specific dish is on the cross table no calibration has to be performed. If a new culture dish is placed on the cross table the set of axes has to be recalibrated. This is done by a calibration of the cross table to restore the basic set of axes followed by the calibration of the new dish.

1.3. Titlegen 1.1

The TitleGen 1.1 program (ID Systems, Vaassen, The Netherlands) has been developed to record the position number on the corresponding frame image. This feature considerably speeds up the retrieval of recorded images of the same frame after the cell tracking experiment has been performed. The program uses standard Macintosh software modules and requires a Macintosh 7100 AV Power-PC (Fig. 1k; Apple, Cupertino, CA, USA). Using additional video settings such as brightness and contrast the video image can be adjusted. The video signal is obtained from the video camera by the Macintosh video-card-in port and the position numbers are obtained from the MicroScan 2.0 program via the serial port of the Macintosh connected to the RS232 port of the Tulip cell tracking computer (Fig. 1h). Position numbers are visualised by a text generator, merged into the video signal and send to the video recorder (Fig. 1l; Hitachi F780E, Tokyo, Japan) via the Macintosh video-card-out port.

1.4. Video recording

To record frame images a video camera (Fig. 1j; KP-C503, Hitachi) is attached to the video connection (Fig. 1c2) of the Nikon Diaphot microscope (Fig. 1a). This microscope is equipped with a 50 W halogen lamp. Eighty percent of the light from the frame image is available for the camera. Due to the low sensitivity of the Hitachi KP-C503 camera visible light images can be obtained with up to a 20× objective. With the TitleGen video settings, image quality can be improved and a 40× objective can be used. Alternatively, more sensitive CCD cameras can be used for low light conditions.

Video images are stored on video tape by continuous recording of the images on a VHS video recorder (Fig. 1l) and displayed on a monitor (Fig. 1m; CM11342, Philips, Eindhoven, The Netherlands). Better image quality can be obtained using a Super-VHS recorder. This type of machine has an improved recording and replay system. Normally they have also additional options which facilitate for example mounting the tapes. This type of video recorder has a good performance/price ratio especially when compared to professional video recorders.

An average cell tracking experiment consists of a set of 2 to 6 culture dishes. In each dish 1 000 frame positions are scanned in XY direction and each frame position is scanned in 5 Z-steps. Such a scan of 5 000 images requires about 45 minutes scanning time. Frame images of four culture dishes can be recorded on one tape. Scanning at two-daily intervals for two weeks thus will generate seven video tapes for every four dishes. When the cell tracking experiment is completed, the video tape of the last recorded day is examined for the presence of frame images containing somatic embryos. Images of the corresponding frame positions of previous days are collected and then sequentially copied onto a tape in a second video recorder. Alternatively, all frame images from one tape where somatic embryos have developed are transferred to the Macintosh Power PC and digitised by the VideoMonitor program (Apple). Corresponding frame images recorded on previous days are also transferred and digitised, omitting the need of a second video recorder. Placing the corresponding images in their correct temporal sequence will give the desired developmental history of a particular embryo down to the original cell or cluster of cells that produced it.

1.5. Confocal scanning laser microscopy

To detect fluorescent signals from single cells labelled with specific antibody-FITC conjugates the Nikon Diaphot inverted microscope with the Märzhäuser cross table was connected to the BioRad MRC-600 Confocal Scanning Laser Microscopy system (Fig. 1n; BioRad, Cambridge, MA, USA). The advantage of

CSLM over a UV source is that it generates single wavelength light and it is possible to adjust the light intensity. In this way damage to the cells due to strong UV-light can be prevented. The argon laser light passes an excitation filter and the 488 nm light excites FITC. The emitted 515 nm light passes the 500–650 nm transmission filter and is detected by a photo multiplier. Visible light is detected by a second multiplier. The entire equipment is controlled by the MRC600 control software running on a 486 PC (Fig. 1o). The image of the fluorescent signal as well as the visible light image are displayed on a monitor. Brightness and contrast of the image can be adjusted using the software. The latest version of the control software allows images to be stored in program specific PIC-format as well as the MS-DOS Tagged Interchange File Format (TIFF). For cell tracking a video recorder (Fig. 1l) can be attached to the video-out connection of the BioRad MRC 600. The time (1.1 sec) the cross table stops at each frame position is set to the time the BioRad system requires to build one image in the Fast 1 (F1) scanning mode. Images from the fluorescent signal and visible light of each frame are continuously recorded onto the video recorder.

As an alternative for the CSLM system, Silicon Intensified Tube (SIT) camera systems can be used. For this type of camera a UV source attached to the microscope is required. SIT camera systems have about the same sensitivity as the human eye, a high resolution, are easy to use and are less expensive when compared to CSLM systems.

1.6. 2D luminometer

Transgenic plant cells containing promotor::luciferase constructs have been used to detect *in vivo* gene expression [6–8]. After application of luciferin (Promega, Madison, WI, USA) to the cells, the luciferase enzyme converts this substrate into oxyluciferin. Photons released in this reaction are visualised with the Hamamatsu ARGUS 50 system (Hamamatsu, Hamamatsu City, Japan) based on an intensified CCD camera. Photons detected by the CCD camera are displayed as spots on a monitor. By digitally accumulating the images of these spots over time an image of the bioluminescent signal can be obtained which is displayed on the monitor using false colours. Due to the intensifier the resolution of this system is fairly low.

The ARGUS luminometer system consists of a CCD camera (Fig. 1p; C2400-77, Hamamatsu) coupled to an intensifier (Fig. 1q; Hamamatsu). To obtain images from calluses ($\geq$ 1 millimetre in diameter) or whole plants ($\leq$ 1 metre) the camera is attached to a support (Fig. 1r) in a dark box and connected to a 55 mm f1.1 Nikon lens (Fig. 1s), optionally fitted with extension tubes for the smaller objects. To obtain microscopic images, the intensifier and CCD camera can be connected to the 100% light video connection at the bottom of the Zeiss Axiovert 135 TV inverted microscope (Fig. 1c1) placed in a dark box (Hansa, Tokyo, Japan). Both CCD camera and intensifier are controlled by the ARGUS 50 control unit (Fig. 1t; Hamamatsu). All functions of the control unit can be controlled by the ARGUS V3.3 software, except for the sensitivity of the intensifier which is set manually. In addition intensity levels of obtained images can be converted in various ways to allow correct interpretation. Images are displayed on a Trinitron monitor (Fig. 1u; PVM1444QM, Sony, Tokyo, Japan) and can be printed with a photo printer (Fig. 1v; CP100E, Mitsubishi, Tokyo, Japan). With use of the PC (Fig. 1w; DeskPro 66M, Compaq, Houston, TX, USA) the ARGUS software images can also be stored digitally either as ARGUS software specific IMA files or as interchangeable TIFF files. Use of IMA files is restricted to the ARGUS software but these files contain additional information when compared to the TIFF files. TIFF files however can be used by most image processing software. Both type of files can contain up to 0.5 Mbyte of data per image and are stored on a 1 Gbyte network server disk (Fig. 1x). A CD-ROM writer (Fig. 1y) is connected to this server for long time storage of the image files. A 2D-luminometer based upon an Astromed cooled CCD camera has recently been described by Kost et al. [5].

1.7. Macintosh 7100 AV Power PC

This computer is equipped with a Power PC 601-chip, 16 Mbyte ROM memory, 2 Mbyte video memory, ethernet network connection and an Audio/Video card. With the VideoMonitor program (Apple) images from the video recorder or camera can be digitised and stored as Macintosh interchangeable PICT file. Using the local ethernet, TIFF files stored by the ARGUS 50 luminometer system can be accessed. Quality of these images can be improved in Photoshop 3.0 (Adobe Systems Inc. Mountain View, CA, USA) by adjusting brightness and contrast or by expanding grey levels and colour levels. Subsequently, images can be combined in Pagemaker 6.0 (Adobe Systems Inc.) to make compositions for publication. These can be transferred to either photo or slide film using a film recorder (Fig. 1z).

Using the TitleGen program the computer is used as title generator, merging position numbers of the cell tracking into the video signal.

2. Embedding of cells

One of the main requirements for cell tracking is the ability to immobilise cells in the culture dish. In our experiments we use the phytagel system. Phytagel (Sigma, St. Louis, MO, USA) gives a strong and completely transparent matrix. The pore size allows diffusion of molecules as large as an IgM molecule. This facilitates the addition or removal of plant growth regulators at certain timepoints during the culture period. The temperature at which phytagel solidifies depends on the concentration of free calcium ions. When cooled down phytagel hardly solidifies at room temperature in the absence of Ca^{2+}. With 1 mM Ca^{2+} phytagel solidifies at approximately 30 °C and medium with 5 mM Ca^{2+} solidifies at approximately 55 °C. In the case of Gamborgs B5 medium the Ca^{2+} concentration is 1 mM [3]. Immobilisation of cells depends on two layers of phytagel.

The bottom layer consists of 0.2% phytagel in B5 medium with Ca^{2+} added to a final concentration of 5 mM. On top of this layer a layer containing 0.1% phytagel in B5 medium without Ca^{2+} is poured at room temperature. Ca^{2+} diffuses from the bottom layer to the cell layer allowing the phytagel to solidify. After two hours the phytagel matrix containing the cells has been solidified. On top of this layer of phytagel a layer of liquid B5 medium is placed to prevent dehydration of the phytagel (Fig. 3). The final Ca^{2+} concentration is below 2.4 mM and does not affect somatic embryo development [11].

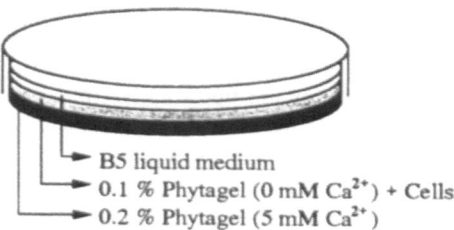

B5 liquid medium
0.1 % Phytagel (0 mM Ca^{2+}) + Cells
0.2 % Phytagel (5 mM Ca^{2+})

Fig. 3. Schematic drawing of a culture dish containing a bottom layer of 0.2% phytagel with 5 mM Ca^{2+} and a cell containing layer without Ca^{2+}. Ca^{2+} diffuses from the bottom layer to the cell layer and solidifies the phytagel. The liquid top layer prevents dehydration of the phytagel layers.

Procedures

Isolation of cell populations

Steps in the procedure
1. Sieve one to two litres of a 7 day old carrot suspension culture [1] first through a 200 and then through an 125 μm polyester sieve (Monodur-PES, Verseidag Industrietextilien, Kempen, Germany).
2. Transfer half of the filtrate to 50 ml tubes (Sarstedt, Nümbrecht, Germany) and centrifuge for 5 minutes at 200 × g.
3. Discard the supernatant and sieve the pelleted cells together with the remaining filtrate through a 50 μm polyester sieve (Monodur-PES).
4. Collect the cell population remaining on the sieve and transfer them to a tube. This is the 50–125 μm cell population.
5. Store one hundred ml of the filtrate obtained by the 50 μm sieve step and centrifuge the remaining filtrate at 200 × g for 5 minutes.
6. Discard the supernatant and sieve the pelleted cells together with the remaining 100 ml of the filtrate gently through the 22 μm sieve (Monodur-PES).
7. Collect the cell population remaining on the sieve and transfer them to a tube. This is the 22–50 μm cell population. Cells passing the 22 μm sieve are the < 22 μm cell population.
8. Wash all cell populations twice with B5 medium without Ca^{2+} in 50 ml tubes. Pellet them for 5 minutes at 200 × g.
9. Determine the cell density in each population using a haemacytometer and dilute the cell populations to the desired cell density.

Preparation of phytagel embedded cultures

Steps in the procedure

1. Pour one ml of B5 medium containing 5 mM Ca^{2+} and 0.2% phytagel (P 8196; Sigma, St. Louis, MO, USA) in a petriperm culture dish (Hereaus, Haneu, Germany) at 60 °C and allow the phytagel to solidify for 5 minutes at room temperature.
2. Prepare a cell population in B5 medium without Ca^{2+}. The cell density should be 10 times as high as the desired cell density in the embedded culture. This is 1 000 000 cells $\cdot$ ml^{-1} for the < 22 μm and 22–50 μm carrot cell populations and 200 000 cells $\cdot$ ml^{-1} for the 50–125 μm carrot cell populations.
3. Mix the diluted cell population with an equal amount of 0.2% phytagel in B5 medium without Ca^{2+}, and pour at room temperature 0.5 ml onto the phytagel bottom layer using a 5 ml wide-mouthed disposable pipette (Sarstedt).
4. After two to three hours the phytagel is polymerised. Add 1 ml liquid B5 medium on top of the cell containing phytagel layer to prevent dehydration of the phytagel.
5. The final cell density of the embedded cells is 100 000 cells $\cdot$ ml^{-1} for the < 22 μm and 22–50 μm cell populations and 20 000 cells $\cdot$ ml^{-1} for 50–125 μm cell populations.

Notes

1. For other types of media the concentrations of Ca^{2+} might be adapted to the Ca^{2+} concentration used in the medium. If media contain Ca^{2+} concentrations of 2.5 mM or higher, cells can be embedded as described above. The desired Ca^{2+} concentration of the medium can be reached by adding additional Ca^{2+} to the liquid medium added to the phytagel layers.
2. In most cases plant growth regulators are added to the media. For 2,4-D a 0.2 μM concentration is used for single cell embryogenesis. This amount is sufficient to induce embryo development and is easily removed from the cultures after one week. It is added before autoclaving to all media. Heat labile compounds can either be added to all media after cooling or in a 2.5 times excess to the liquid B5 medium that is added to the culture dish after solidification. The compounds then diffuse into the phytagel layer.
3. Using diffusion also other components can be transferred into the phytagel medium. In our experiments we added 2.5 times excess of the IgM class JIM8 monoclonal antibody conjugated to FITC or luciferin to the B5 medium. After 30 minutes the presence of all of these compounds in the cell containing layer was shown by labelling of cell wall epitopes or release of photons due to luciferase activity respectively (see below).
4. The cell containing phytagel layer is 0.25 mm thick. Large cell clusters or embryos stick out of this layer during their development. If this area becomes to large they might be released from the phytagel and move through the culture dish. To prevent this an additional 0.1% phytagel layer in B5 medium can be poured on top of the cell containing layer at 30 to 35 °C.
5. This protocol has also been applied for video cell tracking of protoplasts. To adjust the osmolarity of the culture medium 0.3 M mannitol has to be added to all media.

Washing the phytagel cultures

Steps in the procedure
1. Remove the liquid medium on top of the phytagel with a 1 ml disposable pipette (Sarstedt).
2. Add carefully 2 ml fresh medium with a 5 ml wide-mouthed pipette and incubate for 10 minutes to allow diffusion of molecules from the phytagel layer into the liquid medium.
3. Remove the liquid medium on top of the phytagel with a 1 ml disposable pipette (Sarstedt).
4. Repeat steps 2 and 3 for 9 times.
5. Add 1 ml fresh liquid medium on top of the cell containing phytagel layer.

Note
Using this procedure 98% of small molecules like Ca^{2+} or 2,4-D are removed from the phytagel medium. For 2,4-D this is sufficient to allow carrot embryo development beyond the globular stage [11].

Video cell tracking

1. Preparation of the culture dishes

Steps in the procedure
1. Stick a piece of double sided Sellotape (Sellotape GB LTD, Dun-stable, UK) on a Leica S54 cell finder grid (Leica, Wetzlar, Germany) and cut the grid in square pieces of about 5 to 5 mm.
2. Mount two pieces on the bottom side of a Petriperm culture dish (Hereaus) as indicated in Fig. 2a. These pieces of finder grid con-sist of a number of lines with coordinates consisting of two letters (Fig. 2b). From both pieces one letter combination is selected as mark-point for the dish. These letter combinations are referred to as Lower and Upper mark-point.
3. Place the culture dish in the holder of the cross table.

MatriX init Program

Roland van Zoest/I&D/MolBio/WAU/1994
----
enter the new profile name (or STOP)
file name, without extension 1–8 char.:

chose a general profile (from file profiles.dat)
by using $< \uparrow >$ and $< \downarrow >$ and $< enter >$ keys
press $< esc >$ to abort

2
    description: Nikon MicroScan 10×
Number X, Y, Z steps: 30/33/5
Size X and Y steps: 0.620/0.405
Z-step delay (ms): 350

> > > calibrate No Half Full [n/h/f]:

Fig. 4. Startup screen of the MicroScan 2.0 MatriX init module.

2. Initiation of a matrix video cell tracking experiment

Steps in the procedure
1. Start the MatriX-init module by typing the *mx-init* command, enter the profile name for the current experiment, select a general profile with the arrow keys and press < *enter*> to select the profile. Perform a full calibration by typing *f.*
2. Use the joystick to move the cross table so that the Lower mark-grid is visible through the eye piece. Pull the microscope photo mask into the view field and move the cross table so that one of the letter coordinates is in the centre of the photo mask. (Fig. 2b). This letter combination is defined as the Lower mark-point.
3. Write down the letter combination of the Lower mark-point. Type < *enter*> to automatically register the XY coordinates of the cross table for the Lower mark-point by the MX-init module.
4. Use the joystick to move the cross table so that the Upper mark-grid is visible in the photo mask. Move the cross table so that one of the letter coordinates is in the centre of the photo mask. (Fig. 2b). This letter combination is defined as the Upper mark-point.
5. Write down the letter combination of the Upper mark-point. Type < *enter*> to automatically register the XY coordinates of the cross table for the Upper mark-point by the MX-init module. The module then measures the distance between the two mark-points and displays it on the monitor.
6. Move the cross table with the joystick to a position in the area of the dish that will be scanned. This position is called focusing point and cells visible in this frame have to be present both in a low and in a high plane of focus. XY coordinates of the cross table are registered by the MX-init module by typing < *enter*>.
7. The MicroScan program stores the data registered on the computer's hard disk.
8. Start the MX-scan module by typing *mx-scan* and enter the desired profile name.
9. Select the 'no calibration' option by typing *n.*
10. The cross table moves to the Lower mark-point. Use the joystick to adjust the cross table so the mark-point is in the centre of the photo mask (Fig. 2b). Register the cross tables XY coordinates by the MX-scan module by typing < *enter*>.
11. The cross table moves to near the Upper mark-point. Use the joystick to adjust the cross table so the mark-point is in the centre of the photo mask (Fig. 2b). Register the cross tables XY coordinates by the MX-scan module by typing < *enter*>. The module measures the distance between the two mark-points and will compare it with the distance at the time of initiation. The difference is expressed as the measurement error and expressed in a percent-

age. If the error is above 1% the calibration procedure should be repeated.

12. The cross table moves to the focusing point.
13. With the Z-direction of the joystick the lowest plane of focus is selected and the Z-position is registered by the MX-scan module by typing < enter >.
14. With the Z-direction of the joystick the highest plane of focus is selected and the Z-position is registered by the MX-scan module by typing < enter >.
15. Activate the TitleGen program by double clicking the program icon on the Macintosh computer and start recording of the video recorder.
16. Start the cell tracking with the mx-scan module by typing < enter >.
17. The video monitor shows the frame images with the corresponding position numbers. Progress of the scan is also indicated by two bars on the monitor of the cell tracking computer.
18. If all programmed frames are scanned by the MX-scan module, 'Ready!' is shown on the monitor of the cell tracking computer. Stop the recording of the video and remove the culture dish.

Notes

1. Full calibration is required to reset the set of XY axes of the cross table and to determine the exact dimensions and position of the cross table. This is required for each new culture dish placed upon the cross table. Starting only a new module without replacing the culture dish does not require any calibration. After an error or power failure a full calibration has to be performed again.
2. Recording has to take place using normal playing speed of the video recorder. Some recorders are equipped with long play options that allow recording twice the data when compared to normal speed. However, when paused long play recordings give low image quality and cause problems with mounting the tapes (see above).

3. Matrix video cell tracking on subsequent days

Steps in the procedure

1. The MX-scan module is started by typing *mx-scan* and the desired profile name is entered.
2. Select the full calibration option by typing *f* and the cross table will move to the Lower mark-point position.
3. Turn the culture dish by hand until the letter coordinates of the mark-point are near the centre of the photo mask.
4. Use the joystick to exactly position the Lower mark-point in the centre of the photo mask and register the cross tables XY coordinates by the MX-scan module by typing < *enter*>.
5. The cross table moves to near the Upper mark-point. Use the joystick to exactly position the Upper mark-point in the centre of the photo mask and register the cross tables XY coordinates by the MX-scan module by typing < *enter*>. The module measures the distance between the two mark-points and will compare it with the distance at the time of initiation. The difference is expressed as the measurement error and expressed in a percentage. If the error is above 1% the calibration procedure should be repeated.
6. The cross table moves to the focusing point.
7. With the Z-direction of the joystick the lowest plane of focus is selected and the Z-position is registered by the MX-scan module by typing < *enter*>.
8. With the Z-direction of the joystick the highest plane of focus is selected and the Z-position is registered by the MX-scan module by typing < *enter*>.
9. Activate the TitleGen program by double clicking the program icon on the Macintosh computer and start recording of the video recorder.
10. Start the cell tracking with the MX-scan module by typing < *enter*>.
11. The video monitor shows the frame images with the corresponding position numbers. Progress of the scan is also indicated by two bars on the monitor of the cell tracking computer.
12. If all programmed frames are scanned by the MX-scan module, 'Ready!' is shown on the monitor of the cell tracking computer. Stop the recording of the video and remove the culture dish.

4. Initiation of a point scanning profile

Steps in the procedure
1. Start the PoinT init module by typing *pt-init.*
2. Enter the profile name for the current experiment.
3. Type your name and the plate name/number. The current date will be displayed. This date will be used to calculate the number of days the experiment is in progress. At the first day it will display 'day 1'. Via the TitleGen program these data will be shown to the video monitor before scanning of the frame images by PT-scan, facilitating identification of the tapes.
4. Type the number of Z-steps to scan (between 1 and 10).
5. Give the number of milliseconds to hold at each Z-step (between 100 and 5000 ms). For normal video cell tracking at least 350 ms are required.
6. Perform a full calibration of the cross table by typing *f.*
7. Use the joystick to move the cross table so that the Lower mark-grid is visible through the eye piece. Pull the microscope photo mask into the view field and move the cross table so that one of the letter coordinates is in the centre of the photo mask. (Fig. 2b). This letter combination is defined as the Lower mark-point.
8. Write down the letter combination of the Lower mark-point. Type < enter> to automatically register the XY coordinates of the cross table for the Lower mark-point by the PT-init module.
9. Use the joystick to move the cross table so that the Upper mark-grid is visible in the photo mask. Move the cross table so that one of the letter coordinates is in the centre of the photo mask. (Fig. 2b). This letter combination is defined as the Upper mark-point.
10. Write down the letter combination of the Upper mark-point. Type < enter> to automatically register the XY coordinates of the cross table for the Upper mark-point by the PT-init module. The module then measures the distance between the two mark-points and displays it on the monitor.
11. Move the cross table with the joystick to a focusing point. Cells visible in this frame have to be present both in a low and in a high planes of focus. XY coordinates of the cross table are registered by the PT-init module by typing < enter>.
12. The MicroScan program stores registered data to the computer hard disk.

Editing the point scanning profile

10. Type *pt-edit* command to select the PT-edit module.
11. Enter the profile name.
12. Continuing with the same culture dish does not require calibration. Select the no calibration option by typing *n*. If at later days points are added or removed and no calibration of the cross table was performed for this dish select a full calibration by typing *f*.
13. The cross table moves to the Lower mark-point. Use the joystick to adjust the cross table so the mark-point is in the centre of the photo mask (Fig. 2b). Register the cross tables XY coordinates by the PT-edit module by typing <*enter*>.
14. The cross table moves to the Upper mark-point. Use the joystick to adjust the cross table so the mark-point is in the centre of the photo mask (Fig. 2b). Register the cross tables XY coordinates by the PT-edit module by typing <*enter*>.
15. The module measures the distance between the two mark-points and will compare it with the distance at the time of initiation. The difference is expressed as the measurement error and expressed as a percentage. If the error is above 1% the calibration procedure should be repeated.
16. The culture dish is graphically represented on the monitor. The blinking cursor represents the present position of the cross table with respect to the objective (Fig. 2a). The cross table can be moved using the joystick. Points representing frame positions can be stored into the profile by typing the *a*, <*space*> or <*enter*> when the cross table is at the desired position. Points can be deleted by typing *d*. With the *r*, *n* or *p* command the table will respectively move to the nearest, next or previous point of the present position. Maximally 1 000 point cans be added per profile.
17. Typing *q* or <*esc*> will end the edit program. No additional save commands are required.

5. Scanning the point scanning profile

Steps in the procedure
1. Start the PT-scan module by typing *pt-scan*.
2. Enter the profile name of the profile to scan.
3. Perform a calibration. If a calibration was done for this dish with a previous module chose the no calibration option by typing *n*.
4. If the culture dish is just placed upon the cross table and no previous calibration was performed for this dish, perform a full calibration by typing *f*. Then, turn the culture dish by hand until the letter coordinates of the marker-point are near the centre of the photo mask.
5. Use the joystick to exactly position the Lower mark-point in the centre of the photo mask and register the cross tables XY coordinates by the PT-scan module by typing < *enter*>.
6. The table moves to the Upper mark-point. Use the joystick to exactly position the Upper mark-point in the centre of the photo mask and register the cross tables XY coordinates by the PT-scan module by typing < *enter*>. The module measures the distance between the two mark-points and will compare it with the distance at the time of initiation. The difference is expressed as the measurement error and expressed in a percentage. If the error is above 1% the calibration procedure should be repeated.
7. The table moves to the focusing point.
8. With the Z-direction of the joystick the lowest plane of focus is selected and the Z-position is registered by the PT-scan module by typing < *enter*>.
9. With the Z-direction of the joystick the highest plane of focus is selected and the Z-position is registered by the PT-scan module by typing < *enter*>.
10. Select the point you want to start.
11. Select automatic (*a*) or manual (*m*) movement. Manual movement will go to the next point by typing < *enter*>. Automatic will use the time defined in the PT-init module during initialisation.
12. Activate the TitleGen program by double clicking the program icon on the Macintosh computer and start recording of the video recorder.
13. Start the point tracking with the PT-scan module by typing < *enter*>.
14. First the name, plate name and date information are send to the video recorder. Hereafter the scanning of the programmed frames will start.
15. The video monitor shows the frame images with the corresponding position numbers. Progress of the scan is also indicated by a bar on the monitor of the cell tracking computer.

16. If all programmed frames are scanned by the PT-scan module, 'Ready!' is shown on the monitor of the cell tracking computer. Stop the recording of the video recorder and remove the culture dish.

6. Analysing recorded data by video recorder

Steps in the procedure

1. The tape recorded at the last day of cell tracking is played and stopped using pause at the frame in which one or more somatic embryos develop.
2. The corresponding frame images at tapes of previous days are retrieved aided by the position numbers included on the frame images.
3. The video recorder playing the primary tapes, is connected to a second video recorder.
4. The tape containing the selected frame image of day 1 is placed in the playing recorder and the image is recorded on a new tape in the second video recorder for about 6 seconds.
5. These steps are repeated for the corresponding frame images from subsequent days.

7. Analysing recorded data by computer

Steps in the procedure
1. The tape recorded at the last cell tracking day is played and stopped using pause at frames images in which somatic embryos develop.
2. Each frame image is stored digitally as PICT file using the 'copy' function of the VideoMonitor program.
3. Frames images of previous days are retrieved from tapes of previous days and also stored digitally.
4. PICT files are transferred to Photoshop 3.0 PICT files and sequentially mounted in Pagemaker 6.0.

8. Changing microscan 2.0 profile settings

Steps in the procedure
1. Select the objective you want to use.
2. Determine the size of the frame that is visible on the video monitor. (You can use the Cell Finder mark-grids for this purpose. The distance between two lines on these grids is 400 μm).
3. Type edit profiles.dat to start the MS-DOS edit program.
4. Use the mouse or arrow keys to move the cursor to the end of a line with settings.
5. Start a new row by typing < enter >.
6. Type the number of frames to be scanned in the X-direction (NX) and Y-direction (NY), the size of the frame in X (dX-mm) and Y (dY-mm) in millimetres, the number of planes of focus to be scanned in Z-direction (NZ) and the time to stop at each plane of focus at each frame (t_Z) in milliseconds. Finally you can a enter a description to describe the profile.
7. Leave the edit-mode by *File Exit*.

NX	NY	dX-mm	dY-mm	NZ	t_Z	Description
30	33	1.60	1.12	5	350	Nikon MicroScan 4×
30	33	0.62	0.405	5	350	Nikon MicroScan 10×
30	33	0.31	0.23	5	350	Nikon MicroScan 20×
30	33	0.62	0.405	10	120	Nikon SuperScan 10×
30	33	0.26	0.38	4	1100	Nikon Confocal 10

Fig. 5. Screen of the MicroScan profile.dat screen with profile settings.

```
10 = speed [mm/sec]
10 = max joystick speed [mm/sec]
50 = acceleration [mm/sec2]
2  = Com-port for Controller
1  = Com-port for TitleGen (0 = no computer connected)
0  = operation: 0 = normal 1 = test
-------------------------------------------------------
-Don't change the sequence of the lines above!
```

Fig. 6. Screen of the MicroScan msconfig.dat screen with standard settings for the MultiControl unit and MicroScan 2.0 program.

9. Changing microscan 2.0 settings in msconfig.dat

Steps in the procedure

1. Type *edit msconfig.dat* to start the MS-DOS edit program.
2. Use the mouse or arrow keys to move the cursor to the settings.
3. The first number represents the default setting. Replace the settings with the new settings. See the VENUS-1 manual (Märzhäuser) for more information about the settings.
4. Leave the edit-mode by *File Exit*.

Cell tracking combined with immersion immunofluorescence

1. Preparation of cultures

Steps in the procedure
1. Prepare a FITC conjugate of the monoclonal antibody of choice.
2. Prepare phytagel embedded cultures as described above without liquid medium.
3. Prepare a filter sterile solutions of 2.5% calf serum in B5 medium and an 1.25% calf serum, 10% FITC conjugated antibody solution in B5 medium.
4. To block non-specific binding sites add 1 ml B5 medium with 2.5% calf serum with a 5 ml wide-mouthed pipette to the culture dish.
5. Incubate 30 minutes.
6. Remove the B5 medium with a 1 ml pipette.
7. Add 1 ml B5 medium with 1.25% calf serum and 10% FITC conjugated antibody with a 5 ml wide-mouthed pipette.
8. Incubate one hour.
9. Remove excess unbound antibody by washing 10 times with basal B5 medium as described above.

2. Cell tracking

Steps in the procedure

1. The cell tracking system is coupled to the CSLM as described above. The image of the fluorescent and visible signal is optimised by adjusting the intensity of the laser and the gain control of the photo multipliers detecting the fluorescent and visual signals. The scan speed setting Fast 1 (F1), will require 1.1 second to build up one image. This has to be defined in the MicroScan profiles.dat file.
2. Perform a MatriX cell tracking experiment as described above. Select a general profile designed for confocal cell tracking (see above).
3. Cell tracking at subsequent days can be performed using the CSLM or a video recorder.

Notes

1. Antibodies can be isolated using Bakerbond 40 μm prescale ABx (7269-02, JT Baker Inc., Phillipsburg, NJ, USA) as described by the manufacturer. FITC conjugated antibodies can be prepared using the QuickTag FITC Conjugation Kit (1248 618, Boehringer Mannheim, Germany) according to the manufacturer's protocol.
2. For JIM8 cell tracking 20 mM Ca^{2+} is required for obtaining a JIM8 signal comparable to control labelling in PBS. For this reason the blocking solution used in step 2 contained an additional 40 mM Ca^{2+}. In the other steps 20 mM Ca^{2+} was added to the B5-0.2 medium. After the cell tracking excess Ca^{2+} was removed by 10 washing steps with basal B5-0.2 medium as described above (for details see [12]).
3. A number of methods is available to obtain the image of fluorescent signals. For specific applications we refer to the CSLM manual.

Cell tracking combined with bioluminescence

1. Preparation of cultures

Steps in the procedure
1. Prepare phytagel embedded cultures of cells containing promotor-::luciferase constructs as described above without liquid medium.
2. Prepare a filter sterile solution of 50 μM luciferin (Beetle luciferin E1603, Promega, Madison, WI, USA) in B5 medium.
3. Add 1 ml B5 medium with 50 μM luciferin with a 5 ml wide-mouthed pipette to the culture dish.
4. Incubate 30 minutes to allow diffusion of the luciferin. The final luciferin concentration in the culture will be approximately 20 μM.

2. Cell tracking

Steps in the procedure
1. The cell tracking system is coupled to the 2D-luminometer as described above.
2. Initiate a PoinT scan with the PT-init module as described above. Select cells or cell cluster and program their position in the PT-edit module as described above.
3. Perform a PT-scan with visible light as described above. Select the manual (*m*) option. To obtain a visible light image the ARGUS 50 system takes the average of several visible light images. Select the 'rolling average' option of the ARGUS software and save the obtained image to disk as a TIFF file.
4. Go to the next frame position by typing < *enter*> on the cell tracking computer.
5. Repeat step 7 and 8 until all programmed frame positions are recorded.
6. Perform a PT-scan with the ARGUS 'photon counting' option. The images obtained from the photon signals are digitally accumulated to generate an images of the bioluminescent signal. The time required to obtain one image depends on the promoter used, the number of positive cells, the microscope optics, ect. Therefore this time has to be determined empirically in a pilot experiment. Save the image to disk as a TIFF file.
7. Go to the next frame position by typing < *enter*> on the cell tracking computer.
8. Repeat step 10 and 11 until all programmed frame positions are recorded.
9. Cell tracking at subsequent days can be performed using the 2D-luminometer or normal video recording.

Notes
1. Two types of luciferase proteins are available. The bacterial protein uses an aldehyde as substrate. In our opinion this protein is not suited for cell tracking due to the toxic characteristics of its substrate. In our experiments we have used the firefly luciferase coding sequence [2]. The substrate of the encoded protein is luciferin. At concentrations of 50 μM or higher luciferin inhibits embryo development in carrot [13]. Luciferin concentrations above 400 μM are toxic to tobacco cells [9].
2. The time required to build up one image depends on a large number of factors. In our system it takes about half an hour to obtain a significant signal of CaMV 35S driven luciferase expression of a single cell. Significant signals of large cell clusters and somatic embryos can be obtained in 5 minutes.
3. A number of methods is available to obtain images of bioluminescent signals. For specific application we refer to the luminometer manual.

Analysing recorded data

Steps in the procedure

1. The server containing the stored files can be accessed via the local ethernet computer network.
2. Images are opened as TIFF files in Photoshop 3.0 on the Macintosh PowerPC.
3. Image contrast, brightness, and colour use can be adjusted.
4. Files can be stored as PICT files.
5. Sequential compositions of PICT files can be made in Pagemaker 6.0.

Acknowledgments

We gratefully acknowledge Ir. Roland van Zoest (Dept. of Information and Datacommunication, Wageningen Agricultural University (WAU)) for development of the MicroScan 2.0 program, Jack Burger and colleagues (ID Systems, Vaassen, The Netherlands) for development of the TitleGen 1.1 program, Dr. Harry Verhoeven (CPRO-DLO) for initial development of the cell tracking system that was based on equipment established by Dr. Hans-Ulrich Koop (Botany Department, University of Munich, Germany), Dr. Theo Hendriks (Dept. of Molecular Biology, WAU) for helpful advice and assistance, Dr. Sander van der Krol (Dept. of Plant Physiology, WAU) for introducing the 2D-luminometer and Piet Kostense (Duotone, WAU) for drawing Fig. 1. We also would like to thank the Research School Experimental Plant Sciences (WAU), the Centre for Plant Breeding and Reproduction Research (CPRO-DLO, Wageningen, The Netherlands) and the Department of Plant Cytology and Morphology (WAU) for the ability to use their equipment. M.A.J.T. is supported by the Technology Foundation (STW, Utrecht, The Netherlands).

References

1. de Vries SC, Booij H, Meyerink P, Huisman G, Wilde HD, Thomas TL, van Kammen A (1988) Acquisition of embryogenic potential in carrot cell-suspension cultures. Planta 176: 196–204.
2. de Wet JR, Wood KV, DeLuca M, Helinski DR, Subramani S (1987) Firefly luciferase gene: Structure and expression in mamalian cells. Mol Cell Biol 7: 725–737.
3. Gamborg OL (1992) Media preparation. In: Lindsey K (ed) Plant Tissue Culture Manual, pp. A1/1–A1/24. Dordrecht: Kluwer Academic Publishers.
4. Hall RD, Verhoeven HA, Krens FA (1995) Computer-assisted identification of protoplasts responsible for rare division events reveals guard-cell totipotency. Plant Physiol 107: 1379–1386.
5. Kost B, Schnorf M, Potrykus I, Neuhaus G (1995) Non-destructive detection of firefly luciferase (LUC) activity in single plant cells using a cooled, slow-scan CCD camera and an optimized assay. Plant J 8: 155–166.
6. Millar AJ, Short SR, Hiratsuka K, Chua N-H, Kay SA (1992) Firefly luciferase as a reporter of regulated gene expression in higher plants. Plant Mol Biol Rep 10: 324–337.
7. Millar AJ, Carre IA, Strayer CA, Chua N-H, Kay SA (1995) Circadian clock mutants in *Arabidopsis* identified by luciferase imaging. Science 267: 1161–1163.
8. Millar AJ, Straume M, Chory J, Chua N-H, Kay SA (1995) The regulation of circadian period by phototransduction pathways in *Arabidopsis*. Science 267: 1163–1166.
9. Ow DW, Wod KV, DeLuca M, de Wet JR, Helinski DR, Howell SH (1986) Transient and stable expression of the firefly luciferase gene in plant cells and transgenic plants. Science 234: 856–859.
10. Pennell RI, Janniche L, Scofield GN, Booij H, de Vries SC, Roberts K (1992) Identification of a transitional cell state in the developmental pathway to carrot somatic embryogenesis. J Cell Biol 119: 1371–1380.
11. Toonen MAJ, Hendriks T, Schmidt EDL, Verhoeven HA, van Kammen A, de Vries SC (1994) Description of somatic-embryo-forming single cells in carrot suspension cultures employing video cell tracking. Planta 194: 565–572.
12. Toonen MAJ, Schmidt EDL, Hendriks T, Verhoeven HA, van Kammen A, de Vries SC (1996) Expression of the JIM8 cell wall epitope in carrot somatic embryogenesis. Planta 200: 167–173.
13. Toonen MAJ, Verhees JA, Schmidt EDL, van Kammen A, de Vries SC (1997) *AtLTP1* luciferase expression during carrot somatic embryogenesis (submitted).

Index

AA medium (amino acid medium) A1/1, 2, 14, 17

ABC medium (for moss), and modifications F2/3, 38, 39

Abies spp. (firs) C3/1
- transformation B13/1

Abscisic acid
- component of media A1/3–5; B13/41; C3/11, 12; C7/3; G2/18

Acalypha
- endosperm E3/1

Acer pseudoplatanus
- cell suspension cultures H3/1

Aceto-carmine
- in cytology C4/3, 5, 7

Acetolactate synthase (ALS; enzyme, gene) B9/2

Acetosyringone
- component of media B5/6; B6/5; B8/6, 11, 15; G1/5

Acetylsalicylic acid
- in protoplast culture B9/1

Achras sp.
- endosperm culture E3/2

Acridine orange
- DNA stain H3/2, 15

Actin1 gene, promoter
- to drive transgene expression B12/3, 10

Actinidia spp.
- endosperm culture E3/2, 8

Activated charcoal
- component of media A6/13; B1/9; C2/4; C3/3, 7, 11; C8/6

Adenine
- component of media C7/3, 8; E3/13

Agar
- component of media A1/6, 7; A2/1, 6, 12; A4/7, 9, 11; A8/7, 9, 13–15; A9/19, 23, 29; B1/5, 12; B4/17, 21; B11/3, 4, 11; C1/4–6; C3/3, 7, 8, 11, 12; C7/9; C8/5, 11, 15, 17; D5/3; D7/17; E4/5, 6; E5/17; F1/5, 6, 13; G1/5, 6; H1/18

Agarose
- protoplast embedding A1/14; A10/21, 23–25; B1/13; A11/1, 3, 4, 7; B2/4, 13; B3/7, 9, 11, 14; B4/3, 4, 13, 15, 17; B10/4–6, 10
- for callus culture A1/7; A2/1; A5/3
- for embryo culture E5/8, 13, 14

Agrobacterium rhizogenes B4/1; C8/6, 8; G1/5

- see also 'hairy roots'

Agrobacterium tumefaciens
- in plant transformation A4/2, 9; A5/1; A6/1–17; A7/1; A8/1–17; B1/1; B2/1; B4/1, 2; B5/1–9; B6/1–9; B7/1–14; B8/1–18; B10/1; B11/1–18; B12/1; B13/1; C7/5, 10; D2/1, 2; D7/1; E6/4

Alkaloids
- tropane alkaloid biosynthesis in root cultures G1/1–17
- monoterpene indole alkaloids, biosynthesis G3/1–23

Amaranthus sp.
- photoautotrophic cultures H1/1

Amino acids
- components of media B10/9; E3/2, 3, 6, 13
- see also 'AA medium', 'casein hydrolysate'

Aminoglycoside phosphotransferase
- see 'neomycin phosphotransferase'

Ampicillin
- for bacterial selection C5/9
- to control culture contamination A2/16; D5/15; G1/5

Aniline blue
- fluorescent stain (pollen) D5/7, 9

Ancymidol
- gibberellin inhibitor C8/5

Annona sp.
- endosperm culture E3/2, 5

Anther culture: see 'Microspore culture'

Anthocyanin
- accumulation and biosynthesis in cell cultures A3/6; A10/17; G2/1–23
- extraction and identification G2/7, 14
- enzyme assays G2/15, 16
- hormonal regulation *in vitro* G2/18, 19
- photoregulation *in vitro* G2/17, 18
- screenable gene markers D1/9; E3/8, 9

Antirrhinum majus
- anthocyanins G2/3

Apiaceae A9/1

Aphidicolin (inhibitor of DNA polymerase)
- effect on anthocyanins *in vitro* G2/19

Apple: see *Malus pumila*

Arabidopsis thaliana
- microspore/anther culture E6/2
- nematode infection H5/1–17
- pollen gene expression in transgenic D2/2
- protoplasts: isolation, transformation, regeneration A6/3; A7/1–20; D5/3

– regeneration and transformation
  A6/1–17; A8/1–17
– thin cell layers H4/2
Arachis hypogaea
– photoautotrophic cultures H1/1, 25
Artificial seeds C3/6; C8/8
Artichoke, Jerusalem: see '*Helianthus tuberosus*'
Ascocentrum sp.
– *in vitro* culture C1/3
Asimina sp.
– endosperm culture E3/2, 5
ASP medium (for *Asparagus*)
– H6/17
Asparagus sp.
Asparagus officinalis
– endosperm culture E3/2, 4, 5
– mechanical isolation and culture of cells
  H6/1–19
Atropa sp.
– alkaloid biosynthesis G1/1
Automated culture systems C2/2, 5, 7–9;
  C8/9
Auxotrophic mutants
– selection *in vitro* F1/1–17; F2/23
Avena sativa
– transformation B12/2
7-Aza-indole (AZI)
– inhibitor of auxin synthesis C7/3
8-azaguanine F2/23

B5 medium (Gamborg's B5) A1/1, 2, 7, 9–11,
  14, 15, 17; A6/6; A7/4, 5, 13, 14; A8/15;
  B6/6, 7; C6/7; C8/17; D2/11; E5/17; E6/7–9;
  G1/5; H1/2; H7/9, 10, 11, 13, 35, 39
Bamboo: see *Bambusa* sp.
Bambusa sp.
– thin cell layers H4/2
Barley: see *Hordeum vulgare*
BASTA: see 'phosphinothricin'
Bar gene: see 'phosphinothricin
  acetyltransferase'
Begonia sp.
– thin cell layers H4/2
Benlate
– fungicide C8/19
Benomyl
– to prevent culture contamination C8/4
Benzalkonium chloride (Zephiran) C8/11
6-Benzyladenine; 6-benzylaminopurine (BA,
  BAP)
– component of media A1/4; A6/1, 8, 12;
  A7/5; A11/4, 5, 7; B4/17; B5/2, 7, 8;

B7/8, 9; B10/10; B11/4; B13/41; C3/3, 5,
  7, 8, 11; C7/3, 8; C8/4, 6, 11, 13, 17;
  C9/21, 22; D1/7; D4/11, 13; D5/8; D7/17;
  D9/12; E3/2, 3, 5, 7, 8, 13; F1/7, 13;
  H2/7
Beta vulgaris
– cytological analysis C4/3
– protoplasts D3/3
– shoot cultures B7/3, 5, 8
– sterilization of seed A2/16; B7/3
– transformation B7/1–14
Betaine phosphate
– component of media B8/11, 15
Betula sp. C7/1
Biotin
– non-radioactive DNA labelling D8/6
Brassica sp.
– somatic hybrids D3/3; D6/2, 5
– thin cell layers H4/2
– microspore culture E6/1–17
Brassica campestris
– in interspecific cross E4/1, 2
Brassica carinata
– microspore/anther culture E6/2
Brassica fruticulosa
– in interspecific cross E4/2
Brassica gravinae
– in interspecific cross E4/2
Brassica juncea
– microspore/anther culture E6/2
– in interspecific cross E4/2
– *in vitro* culture of zygotic embryos
  E5/1–19
Brassica napus
– microspore/anther culture E6/1–17
– cytological analysis C4/3
– in interspecific cross E4/2
– protoplast transformation and regeneration
  A10/2, 3, 17–19; B4/1–24
– shoot cultures B4/9, 21
– somatic hybrids D3/3, 5
– transformation of somatic embryos E6/4
Brassica nigra
– microspore/anther culture E6/2
– isolation of nuclei D7/9, 13
Brassica oleracea
– microspore/anther culture E6/2, 3
– host for *Heterodera schachtii* H5/3
– in interspecific cross E4/1
– microprojectile bombardment D1/5, 6
Brassica rapa
– microspore/anther culture E6/2–4, 7, 8
– transformation of somatic embryos E6/4

2

Brassica spinescens
- in interspecific cross E4/2
Brassica spinescens
- in interspecific cross E4/2
α-bromonaphthalene
- to arrest mitosis C4/3, 9; D8/2, 5
Brugmansia spp.
- tropane alkaloids G1/15
Buffers A1/24; A8/15; A9/24; A11/6; B9/9;
    B11/14; B12/17; C4/9; C5/4, 6, 8, 11; C6/8;
    C7/10; D6/3, 4, 6; D7/11, 12; D8/2, 3, 4, 6;
    E2/7, 8; F2/40, 41; G1/12, 13; H1/9, 13, 17,
    19, 20; H3/11, 17
Butomopsis sp.
- endosperm culture E3/1

3C5ZR medium B11/3–5
Caffeine
- production *in vitro* E3/9
Calcium hypochlorite
- as sterilant C7/7; H4/7
Callus cultures
- embryogenic A5/5, 9, 12; A9/3, 4, 9–11,
    23, 24, 27–30; B1/1–16; B9/1; C7/2–4,
    7–9; E2/3
- initiation and maintenance A2/1–19;
    A9/3, 4, 9–11, 23, 24
- eucalyptus C8/7, 8, 17
- potato B5/2, 5
Calystegia sepium
- mechanical cell isolation H6/2
Calystegia spp.
- alkaloid biosynthesis G1/1
Capsella spp.
- embryos E5/2
Carbenicillin
- to control *Agrobacterium* A4/9; B5/5–8;
    B6/3–5; B11/4, 5, 10
- to control culture contamination A2/16
Carbol fuchsin
- in cytology C4/11
Carbon dioxide
- regulation/determination in cultures: see
    'photoautotrophic cultures'
Carnation: see *'Dianthus caryophyllus'*
Carnoy's fixative
- in cytology C4/7
Carrot: see *'Daucus carota'*
Casein hydrolysate
- component of media A1/6, 17; A5/9, 12;
    A9/9, 11, 24; B1/5, 11; B5/8; B9/10, 11;
    B13/41; C3/11; C7/2, 7, 8; C8/17; C9/21;

    D4/11; D9/12; E3/2–8, 11–13; E4/5, 6;
    E5/7
Catharanthus roseus
- alkaloid biosynthesis G3/1–23
- anthocyanin production A3/6, 12; G2/1,
    19
- callus culture E3/9, 11
- growth in suspension culture A3/6, 9, 19,
    12
- induction of synchrony in suspension
    culture H3/1–31
Cattleya
- *in vitro* culture C1/3; C6/7
Cauliflower: see *Brassica oleracea*
Cauliflower mosaic virus (CaMV) 35S RNA
    gene promoter B2/1; B3/14; B4/2, 3;
    B10/2; B12/3; D1/9; D2/3
Cefotaxime (Claforan)
- to control *Agrobacterium* A4/9;
    A6/11–13; B5/5–8; B7/2, 7; B8/3, 12–14
- to control culture contamination A2/1
Cell cycle A3/2; C4/3; E1/1; H3/1–31
Cell number, determination of A3/15, 17;
    H2/9; H1/27; H3/7
- see also packed cell volume
Cell suspension cultures
- biosynthesis of indole alkaloids G3/1–23
- cytology C4/9, 10
- embryogenic A9/11–24, 29; B1/1–16;
    B2/3, 5, 7, 9, 11; B3/1–3, 8, 14; B9/1, 3;
    B10/3, 4; B12/1–20; B13/1–46; C3/11,
    12; D1/9
- in eucalypts C8/7, 8, 17
- inducton of synchrony H3/1–31
- growth curve A3/2, 5, 6, 13–17
- initiation and maintenance A2/2–21;
    A3/1–21; B6/5; D7/2, 9; E1/3, 4, 9
- microprojectile bombardment B12/1–20;
    D1/9; D2/2
- mutagenesis C7/4
- transformation B12/1–20; C7/10
Cell viability, determination of
- fluorescein diacetate staining A3/15, 17;
    B9/3; C9/19; D5/3; H3/5; H6/13, 18
Centaurea cyanus
- anthocyanins G2/1, 9, 13, 17, 18
Ceratadon purpureus F2/1
Cereals
- cytology C4/3, 7
- embryogenic cultures B1/1–16; B9/1–13;
    B12/1–20
- endosperm culture E3/1–5, 9, 10, 16, 17

3

– transgenic B1/1–3; B9/1–13; B10/1–11;
B12/1–20
– see also *Hordeum vulgare*, *Oryza sativa*,
Sorghum, *Triticum aestivum*, *Zea mays*
Chalcone synthase
– assay G2/15, 16
– see also 'anthocyanins'
Chenopodium quinoa
– in virus testing C6/3
Chenopodium rubrum
– photoautotropic cultures H1/1, 7, 9, 10,
19, 20, 25
Chlamydomonas
– microprojectile bombardment D1/1
Chloramphenicol
– inducer of embryogenesis C7/3
Chloramphenicol acetyltransferase (CAT)
gene/enzyme B13/2; C7/5
Chloramphenicol acetyltransferase (CAT) assay
B4/2; B7/7, 8; B10/2
2-Chloroethyltrimethyl ammonium chloride
(CCC)
– inhibitor of gibberellin synthesis C7/3
α-Chloronaphthalene
– to arrest mitosis D8/5
p-Chlorophenoxyacetic acid
– component of media B6/5
Chlorophyll content
– determination H1/27, 28
– in cultures H1/5, 7
Chloroplasts
– DNA (plastome) D4/1–3, 13, 15, 16;
D6/1–4
– microprojectile bombardment D1/1; D2/5
– mutants F1/2, 13
– transfer E1/1; see also 'Cybrid'
Chlorsulphuron
– selective agent A8/9; D7/5
Chromium trioxide solution
– for cell counts A3/15, 17
Chromosomes
– cytological techniques C4/1–13
– inactivation, elimination for cybrid
production D4/2, 7
– *in situ* hybridization A11/1; D8/1–15
– see also 'nuclei, isolation and uptake'
Chrysanthemum sp.
– virus elimination C6/7
Cinchona sp.
– alkaloids G3/4
Citropsis gilletiana
– in somatic hybridization C7/4, 13
Citrus spp.

– clonal propagation C7/1–18
– endosperm culture C7/3; E3/2, 5, 16
– somatic embryogenesis C7/1–4, 7–10, 13
Claforan: see 'cefotaxime'
Clover: see *Trifolium* sp.
Coconut water/milk
– component of media A1/6; A9/1, 9–11;
B1/11; C1/2, 5, 6; C7/8; C8/17; D4/11;
E3/2–5; E4/5; E5/7
Cocos nucifera (coconut palm)
– propagation C2/1
– endosperm culture E3/2, 5
Codiaeum sp.
– endosperm culture E3/2, 9
Coffea arabica
– endosperm culture E3/2, 9
Colchicine
– to arrest mitosis C4/3, 9, 11; D8/2
– in microspore culture E6/3, 11
– in ovule culture C7/3
Conditioned medium
– for protoplasts A9/20; A10/1; B3/7, 11
Confocal Scanning Laser Microscopy (CSLM)
– for cell tracking H7/1, 2, 6–8, 37
Conifers
– cell suspension cultures C3/11, 12
– clonal propagation C3/1–16
– cryopreservation C9/1–27
– transformation B13/1–46
Contamination of cultures by microorganisms
– methods of control A2/7, 13, 15, 16;
A3/20
– see also 'sterilization'
Cotton: see '*Gossypium hirsutum*'
4 CCPU
– component of media C8/17
CPW salts solution D4/5–7; D9/3, 5, 9–11
Croton sp.
– endosperm culture E3/2, 6
Cryopreservation (freeze preservation)
– suitable tissues A5/11; C/4, 12
– of *Catharanthus roseus* G3/1
– of conifers B13/37; C9/1–27
– of moss F2/7
Cucumis sativus
– endosperm culture E3/2, 5
Cybrid (cytoplasmic hybrid)
– production and selection A10/3;
D4/1–17; E1/1
Cymbidium
– *in vitro* culture C1/3
Cytoplast
– see 'Cybrid'

Dactylis glomerata
- protoplast transformation, regeneration A5/1; B3/1–15
- somatic embryogenesis A5/1–15; E2/2
DAPI stain (4',6-diamidino-2-phenylindole)
- DNA stain H2/13; H3/9, 11, 17; E6/7
Datura spp.
- alkaloid biosynthesis G1/1, 5, 15
- embryo culture E5/1
Datura innoxia
- anther culture E6/1
- protoplast fusion D3/6
- auxotrophic mutant D7/5, 9, 13, 17
Daucus carota
- anthocyanins G2/1, 18, 19
- callus culture A2/5–9
- cell suspension culture A3/7, 8
- somatic embryogenesis A9/1–32; E2/1, 2
- video cell tracking H7/1–45
Dendrobium
- *in vitro* culture C1/3
Dendrophthoe sp.
- endosperm culture E3/2, 6, 7, 12
Denhardt's solution
- in nucleic acid hybridization reactions D8/3, 8
Deoxyribonucleic acid (DNA)
- carrier DNA, in direct gene transfer A7/2, 8, 9; B3/11, 13; B4/13, 16; B10/5, 6
- DNA-DNA hybridization and autoradiography C5/13, 14, 16; D7/5; D8/1–15
- isolation B11/13, 14; C5/2–4, 15; see also organellar DNA
- organellar DNA D4/15, 16; D6/1–8
- probe isolation, oligolabelling, nick translation C5/9–11, 16; D8/2, 3, 6, 7
- plasmid DNA
- *Agrobacterium* A6/2; A8/1–17; B8/1, 2, 5–7; B7/7; B11/1, 3, 9, 13–15
- direct gene transfer A7/1, 7, 8; B2/13; 14; B3/11, 13, 14; B4/2, 5, 13, 16; B9/5; B10/2, 5, 6, 10; B12/3, 7, 9, 10, 18; B13/7, 25; C7/4, 5, 10; D1/5, 9; D2/3, 7, 11
- restriction endonuclease digestion C5/2, 5, 6, 15; D6/3
- Southern blotting A6/2; C5/7, 8, 15; C6/4; D4/16
- see also 'transient gene expression'
Dextran sulphate
- as protoplast fusogen D9/1

4',6-Diamidino-2-phenylindole (DAPI)
- DNA stain H2/13; H3/9, 11, 17
Dianthus caryophyllus
- photoautotrophic cultures H1/1, 25
- virus elimination C6/7
3,6-Dichloro-*o*-anisic acid (DICAMBA)
- component of media A5/3, 9, 11, 12; B1/5
2,4-Dichlorophenoxyacetic acid (2,4-D)
- component of media A1/4; A2/6; A3/8–10, 20; A5/11; A6/1, 5, 7, 8; A7/5; A8/1, 9, 15; A9/1, 9, 11, 13, 29, 30; B1/1, 2, 5, 11–13; B2/4; B3/13; B4/17; B5/2, 8; B9/3, 9–11; B10/5, 10; B11/4; B12/17; B13/41; C2/4; C3/4, 11; C7/8, 9; C8/17; C9/21, 22; D5/5; D7/17; D9/12; E1/3, 9; E2/3; E3/2, 3, 5–8; E4/6; F1/7; G2/19; H1/2, 9, 10; H3/19, 25–27; H7/15
- resistance to C7/4, 13
Diethyldithiocarbamate
- inhibitor of phenol oxidases D7/3
Digitaria sp.
- thin cell layers H4/2, 21
Digoxigenin
- non-radioactive DNA labelling D8/36, 7, 11
Dihydrofolate reductase
- selectable marker A6/2
6-(γ,γ-dimethylallylamino)purine
- component of media E3/6, 7, 13
Dimethyldichlorosilane (repel silane)
- siliconization of glassware E1/7, 9
Dimethylsulphoxide
- cryoprotectant C9/6, 11, 17, 22, 23
- in protoplast fusion D5/13; D7/13
- solvent for plant growth substances A6/12, 13; A8/16; B11/4
2,4-Dioxohexahydro-1,3,5-triazine (DHT)
- antiviral chemical C6/2
Diphenylurea
- component of media E3/5
Diphenylurea thidiazuron
- component of media B8/8
Direct gene transfer A5/11; A6/3; A7/1–20; B1/1, 3; B2/1–17; B3/8, 11–14; B4/1–24; B7/1; B8/1; B9/1–13; B10/1–11; B12/1–20; B13/1–46; D1/1–12; D2/1–14; E1/1
- see also 'electroporation', 'microinjection', 'microprojectile bombardment', 'nuclei, isolation and uptake', 'polyethylene glycol', 'polyvinyl alcohol'

5

Doritis
  - *in vitro* culture C1/3
Dragendorf's reagent
  - for alkaloid detection G3/7
Dry weight, determination of A3/16, 17

Elaeis guineensis (oil palm)
  - *in vitro* culture C2/1, 2
Electrofusion
  - of gametes E1/1–12
  - see also 'Protoplasts'
Electroporation A7/1; B2/3, 5, 9, 11, 13, 15;
  B3/11, 13, 14; B4/2; B7/1; B9/1–13; B12/1;
  B13/2; C8/8; D3/1–3; E1/1
Emblica sp.
  - endosperm culture E3/2, 5
Embryo culture
  - zygotic embryos E5/1–19; see also
  'somatic embryos'
Embryo rescue
  - hybrids E4/1–8
Endosperm culture E3/1–21
  - in *Citrus* C7/3
Enucleation of protoplasts D4/2
Enzyme assays
  - in anthocyanin biosynthesis G2/15, 16
  - in tropane alkaloid biosynthesis
  G1/11–16
  - in indole alkaloid biosynthesis G3/8–19
Enzyme-linked immunosorbent assay (ELISA)
  - hybridoma screening E2/5–7
  - in virus testing C6/2–5, 8, 9
  - NPTII assay B13/7
Epidendrum
  - *In vitro* culture C1/3
Equipment
  - general A1/5, 6, 14; A2/2; A3/2, 3; A4/1;
  B8/4; D2/7, 8
  - automated micropropagation C2/7–9
  - electrofusion D3/3, 7, 8; E1/2–9
  - electroporation B4/15
  - embryo microdissection E5/5, 6, 14
  - fluorescence-activated cell sorter
  D5/1–19; H3/29
  - micromanipulation A10/1–28
  - microprojectile bombardment B12/1–20;
  B13/9, 10; D1/1–12; D2/1–11
  - photoautotrophic cultures H1/17–25, 28
  - protoplast isolation A11/2; D9/13
Erigeron sp.
  - endosperm culture E3/1
Erucastrum gallicum
  - in interspecific cross E4/2

Erucastrum sativa
  - in interspecific cross E4/2
Erythroxylon spp.
  - alkaloid biosynthesis G1/1
Ethanol
  - as sterilant A2/5, 6, 11, 12, 15; A3/10;
  A4/5, 7; A5/3; A6/5; A7/8; A8/5; B1/5, 7;
  B2/7; B4/9; B7/3; B11/3; B12/5; C1/2;
  C3/11; C6/7; C8/3, 11; D2/7; D5/11; E1/2,
  7; E3/15, 16; H5/9, 11
Ethephon
  - component of media C7/3
Ethylene
  - effect on chlorophyll synthesis in cultures
  H1/2, 21
Ethyl-methanesulphonate (EMS)
  - use as mutagen F1/2, 11, 12
N-Ethyl-N-nitrosourea (NEU)
  - use as mutagen F1/2, 9–12
Eucalyptus spp.
  - clonal propagation C8/1–24
Euphorbia sp.
  - endosperm culture E3/2, 4, 5, 15
Euphorbia characias
  - photoautotrophic cultures H1/25
Evan's Blue stain D7/4, 8, 15
Exocarpus sp.
  - endosperm E3/1, 2, 6, 11

Feeder layers: see 'nurse cultures'
Fertilization, *in vitro* A10/3
Feulgen stain
  - in cytology C4/3–5, 11; H3/9
Fixatives
  - in cytology C4/1–13; D8/2, 5
Flounder antifreeze protein
  - in transgenic plants C7/13
Flow cytometry: see 'fluorescence-activated cel
  sorting'
Fluorescein diacetate (FDA)
  - fluorochrome marker D5/9
  - see also 'cell viability, determination of
Fluorescein isothiocyanate (FITC)
  - fluorochrome marker D5/9, 13, 14; H7/7,
  8, 13, 35, 37
Fluorescence-activated cell sorting (FACS),
  flow cytometry A11/1; D5/1–19; D9/2;
  H3/2, 15, 24, 29
p-fluorophenylalanine F2/23
Fragaria sp.
  - virus elimination C6/5, 7
Freund's adjuvant E2/4

Fresh weight, determination of A3/16, 17; H1/27
Fritillaria sp.
– endosperm E3/1
Funaria hygrometrica
– protoplasts A10/2, 3, 17–19; F2/1

G418 (geneticin)
– selective agent A6/2, 11, 12; A7/1, 7, 13, 17; B3/14; B7/2, 8; B10/2, 6, 7, 10; B13/3; F2/31
Gamborg's medium: see 'B5 medium'
Gametes
– isolation E1/1–12
– see also 'Electrofusion'
Gamma radiation
– use as mutagen F1/2
Gas chromatography/mass spectrometry
– in alkaloid analysis G1/15
– see also 'mass spectrometry'
Gelose
– component of media H4/11, 15
Gelrite
– component of media A1/7, 14; A2/1; A5/3; A9/29; B11/4; C3/7, 11; C8/5, 11, 15; H4/11, 15; H5/7, 13
Geneticin: see 'G418'
Gentamycin
– selective agent B6/5
Gentamycin acetyltransferase
– selectable marker A6/2
Geranium spp.
– anthocyanins G2/12
Gerbera spp.
– anthocyanins G2/12
Gibberellins
– components of media A1/4–6; A6/8, 12; B4/17; B5/2, 7, 8; B11/4; C7/7–9; C8/5, 13; D4/3; E3/2–5, 7, 8, 12, 15, 16; E4/6; G2/18
Giemsa stain
– in cytology D8/2, 6, 9
Glucose
– component of media B10/6, 10; H4/15
β-glucuronidase (GUS) assay, reporter gene A6/11; A7/8; B2/11, 13, 14; B4/2–5, 7, 19, 21; B5/5; B7/7, 8; B8/9, 12; B10/2; B12/3, 10, 11, 15, 18; B13/2, 3, 11–15, 42; D1/5, 6, 9; D2/3, 8, 9, 11; H5/13
– as enzyme in gamete isolation E1/2
Glycine: see 'vitamins'
Glycine max
– microprojectile bombardment D1/1

– photoautotrophic cultures H1/7
Gomphrena globosa
– in virus testing C6/2
Gossypium hirsutum
– microprojectile bombardment D1/1
Grasses (Gramineae)
– in vitro culture A5/1–15; B1/1, 3, 4
– transformation B2/1; B3/1–15; B10/1
Green banana, homogenized
– component of media C1/2, 4, 6
Growth measurements: see 'cell number', 'dry weight', 'fresh weight', 'packed cell volume'
Gymnosperms
– conifers, clonal propagation C3/1–16

[³H]thymidine
– determination of incorporation into DNA H3/13, 23, 28
Hairy roots A4/1; C8/6, 8; G1/5, 6
Haploid production
– microspore culture E6/1–17
Haplopappus gracilis
– anthocyanins G2/1, 17
Helianthus sp.
– thin cell layers H4/2
Helianthus tuberosus
– wounding E6/2
Herbicide resistance
– transfer by protoplast fusion D6/1
– see also 'phosphinothricin acetyltransferase'
Heterodera schachtii
– infection of plants H5/1–17
Hibiscus cannabinus
– protoplasts C8/8
High performance liquid chromatography (HPLC)
– of anthocyanins G2/12, 13
– of indole alkaloids G3/7, 19
Hoechst dye D7/4, 8, 15
Hoogland nutrient solution A1/23; F2/37
Hordeum bulbosum
– haploid production E6/1
Hordeum marinum
– protoplasts B10/2
Hordeum murinum
– protoplasts B10/2
Hordeum vulgare
– endosperm culture E3/2, 4, 5, 17
– haploid production E6/1
– protoplasts B10/1–11
– thin cell layers H4/21

– transformation B2/1; B10/1–11; B12/2; D2/1

Hyacinth
– cytological analysis C4/3
Hybridomas E2/2–7
Hydrogen peroxide
– as sterilant C3/7
5-Hydroxynitro-benzyl-bromide (HNB)
– inhibitor of auxin synthesis C7/3
Hygromycin B
– selective agent A6/2, 11, 12, 15; A7/1, 2, 13, 17; A8/9; B2/3, 13; B3/14; B4/2, 5, 6; B5/5–8; B6/5; B10/7; B12/2; F2/31
8-Hydroxyquinoline
– to arrest mitosis C4/3
Hygromycin phosphotransferase (enzyme, gene) A6/2, 11, 15; A7/2, 7, 8; B2/3; B3/14; B9/2; C7/5
Hyoscyamus spp.
– alkaloid biosynthesis G1/1
Hyoscyamus albus
– root cultures G1/3
Hyoscyamus niger
– root cultures G1/3

Immunization of mice E2/4
Indole-3-acetic acid (IAA)
– component of media A1/1, 6; A3/20; A7/5; A8/15; B5/6; C7/3, 8; D1/7; D4/13; E3/2, 3, 6–9, 11–13; H1/2
IAA-aspartic acid
– component of media B11/4
Indole alkaloids
– biosynthesis in *Catharanthus roseus* G3/1–23
– extraction and purification G3/6–8
– enzyme assays G3/8–19
Indole-3-butyric acid (IBA)
– component of media A1/4; A6/8, 13; B5/6; B8/3, 13, 14; B12/13; C3/3, 8, 9; C8/6, 13; E3/2, 3, 7, 8, 11, 13; G1/3; H4/15
In situ hybridization: see 'Chromosomes'
In vitro fertilization
– in maize E1/1–12
Iodoacetate
– antimetabolite D4/7, 9
Iris pallida
– thin cell layers H4/2, 21
Irradiation of protoplasts (cybrid production) D4/2, 7
2-isopentenyladenine,
2-isopentenylaminopurine (2-iP, IPA)

– component of media A1/4; A6/2, 8, 12; A7/5; A8/1, 2, 15; B4/17; C2/4; C3/7, 8; C7/3; E3/13

Jatropha sp.
– endosperm culture E3/2, 6
Juglans sp.
– endosperm E3/1, 2, 5, 8, 16
Juniperus spp. (Juniper) C3/1
K3 medium
– protoplast culture A11/3, 6, 7; I/5–11, 15
Kalanchoe blossfeldiana
– callus culture H1/1
Kanamycin
– selective agent A6/2, 11, 12, 15; A7/1, 7, 13; A8/1, 3, 9, 10, 13–15; A10/25; B3/14; B4/1, 2, 5; B5/5–8; B6/3–6; B7/1, 2, 7–9; B8/1, 5, 7–9, 11, 12, 14, 15; B9/2, 9; B11/3–5, 7, 10, 11; B13/3, 31–39; C7/4, 10; D1/7; D2/5; D7/5; G1/6
– to control culture contamination A2/16
Kao and Michayluk's medium B1/13; B3/3, 7, 9, 13
– modified for embryo culture E5/7, 8
Kinetin
– component of media A1/4; A2/12; A3/8, 9; A6/7, 8; A7/5; A8/15; C7/3, 8, 9; C8/4, 6, 13, 17; D4/13; D9/12; E1/3, 9; E2/3; E3/2, 3, 5–8, 11–13; E4/5, 6; H1/2; H4/15
Knop medium H5/9, 11, 15
Knudson C medium C1/1

L3 medium B12/17
– and further modifications B12/5, 13, 17
Larix spp. (larch, tamarack) C3/1
– cryopreservation C9/3, 5, 7, 11, 17
– transformation B13/1, 2, 37
Leaf strip mutagenesis F1/2, 13
Leptomeria sp.
– endosperm culture E3/2, 6, 7, 11
Light regime for cultures, plants A2/5; A3/7, 9, 12; A4/1, 7, 9, 10; A5/3, 5, 7–9; A6/5; A7/3, 13–15, 17; A8/5; B2/7, 9, 13; B3/7, 11; B5/5; B7/3, 7; B9/7, 8; B10/3–5, 7; B11/5, 9, 10; B12/13; C1/2, 5; C3/7, 11; C6/7; C7/7, 9, 10; C8/8, 9, 13; D4/3, 5, 9, 15; D5/3; D9/8; E1/8; E3/15, 16; E5/14; F2/3; H1/3, 9, 15, 18, 20, 23, 28; H2/3; H4/5, 13
Libocedrus sp.
– transformation B13/1
Lilium sp.
– pollen transfection B13/3

8

– virus elimination C6/2

Lincomycin
 – chloroplast-encoded resistance D4/13; F1/13

Linsmaier and Skoog medium A1/1; G1/3

Linum austriacum
 – in interspecific cross E4/1

Linum perenne
 – in interspecific cross E4/1

Litvay's (LM) medium B13/5, 41; C9/11, 17, 21

Lolium spp.
 – endosperm culture E3/2, 5, 9
 – seed sterilization A2/16

Loranthaceae
 – endosperm culture E3/4

Luciferase
 – reporter gene, enzyme assay B13/2, 3, 11, 19, 42; D2/11; H7/8, 39, 41

Luria -Bertani (LB) medium (Luria broth) A8/9, 10; B5/5, 7; B6/7; B7/7, 8; B8/11, 15; B11/3, 11

Lycium sp.
 – endosperm culture E3/2, 5

Lycopersicon esculentum
 – endosperm culture E3/3, 15
 – photoautotrophic cultures H1/7
 – pollen genes D2/2–5
 – pollen transfection D2/3
 – roots A4/2
 – transformation B6/1–9

Lycopersicon pennellii
 – hybrid with *L. esculentum* B6/1, 4

Lycopersicon peruvianum
 – chloroplast mutants F1/2, 13
 – hybrid with *L. esculentum* B6/1

Lygus sp.
 – pest of carrot A9/4

Maize: see *Zea mays*

Male sterility
 – cytoplasmic B1/3; D6/1

Malt extract
 – component of media C7/3, 7, 8; E3/5

Maltose
 – component of media B10/6, 9

Malus pumila
 – endosperm culture E3/3, 5, 15
 – regeneration and transformation B8/1–18
 – virus elimination C6/2

Malus prunifolia
 – shoot regeneration B8/5

Mangifera sp.

– somatic embryogenesis C7/1

Mannitol
 – as osmoticum: see 'protoplasts', '*Zinnia elegans*', '*Asparagus officinalis*' (*mechanical isolation and culture of cells*)

Mass spectrometry
 – in anthocyanin analysis G2/12–14

Matthiola incana
 – anthocyanins G2/3, 5

McCown woody plant medium C8/4

Medicago varia
 – cell suspension cultures H3/3

Melandrium album
 – chromosome spread D8/11

Meloidogyne incognita
 – infection of plants H5/1–17

Mercaptobenzthiazol
 – inhibitor of phenol oxidases D7/3

Methotrexate
 – inducer of embryogenesis C7/3
 – selective agent A8/9; B4/2; B13/3; D7/5

1-Methyl-3 nitro-1-nitrosoguanidine (MNNG, NTG)
 – use as mutagen F1/2; F2/11–17

N-Methyl-N-nitrosourea (NMU)
 – use as mutagen F1/2, 10–13

Mercuric chloride
 – use as sterilant C8/15; E4/3; H5/7, 8, 13

Meristem culture A4/2; C6/1, 7

Mesembryanthemum crystallinum
 – photoautotrophic cultures H1/7

Metroxylon sp.
 – propagation C2/1

Microinjection A7/1; B4/1; D7/6; E1/1
 – see also 'micromanipulation'

Micromanipulation A10/1–28
 – see also 'microinjection'

Micronuclei D7/6

Microprojectile bombardment
 – somatic tissues A5/11; B1/3; B3/2; B8/1; B10/1; B12/1–20; B13/1–46; D1/1–12; F2/35
 – pollen B13/3, 4, 23, 25; D2/1–14

Micropropagation C6/1, 5, 7
 – apple B8/1, 3–6
 – conifers C3/1–16
 – eucalypts C8/1–24
 – orchids C1/1–7
 – palms C2/1–14
 – use of embryogenic cultures A5/11; C3/2, 4, 6, 11–13

Microspore culture
 – *Brassica* spp. E6/1–17

Miltonia
- *in vitro* culture C1/3
Minimal medium (for *Agrobacterium*) B11/10
Mitochondria
- DNA D6/1, 2, 5, 6
- transfer E1/1; see also 'Cybrid'
Mitotic index
- determination of H3/9, 11
Monoclonal antibodies
- to markers for embryogenesis E2/1-9;
  H7/13, 35, 37
Morel vitamins A11/6, 7
Moss F2/1-43
MSG medium C9/11, 22
Murashige and Skoog medium A1/1, 2, 7, 9,
  10, 13, 16; A2/6, 11, 12; A3/8-10; A4/7, 9,
  11; A6/5, 6, 11; A8/15; A9/9-11; A11/2, 6,
  7; B1/5, 11, 12; B2/7; B5/2, 5, 7, 8; B6/3-7;
  B7/3, 8, 9; B8/3, 5, 11, 13-15; B9/1, 9;
  B11/3-6, 9, 10; B12/13, 17; C6/7; C7/7, 9;
  C8/4, 6, 11-14; D1/7; D2/8, 11; D5/3, 5, 13;
  D7/17; D9/12; E1/3, 9; E2/3; E3/2-5, 7, 8,
  12; E4/4, 5; F1/7; H1/2, 9, 10; H3/2, 3, 17,
  19, 26; H4/15
Murashige and Tucker medium C7/7, 9
Mutagenesis *in vitro* F1/1-17; F2/1-43
Myeloma cells: see 'Hybridomas'

N6 medium A1/1, 2; B9/1, 3, 9-11
α-naphthalene acetic acid (NAA)
- component of media A1/4, 5; A2/12;
  A3/20; A6/8, 12; A9/1, 9, 11, 13; A11/5,
  7; B3/13; B4/17; B5/2, 8; B7/9; B11/4;
  C2/4; C3/4; C7/3, 8; C8/6, 11, 13, 17;
  D4/11, 13; D5/8; D9/12; E3/2, 3, 5, 7, 13;
  E4/6; F1/7; H1/2; H2/7
Nautilocalyx sp.
- thin cell layers H4/2
Neomycin phosphostransferase (enzyme, gene)
  A6/2, 15; A7/1, 2, 7, 13; B2/3; B5/5; B7/1,
  7, 8; B8/1, 9; B9/2; B10/2; B11/1, 5, 13, 14;
  B13/1, 3, 11, 17; C7/4, 5, 10
Nematodes
- infection of *Arabidopsis* H5/1-17
Nicotiana bigelovii
- protoplasts A10/19
Nicotiana debneyi
- protoplasts A10/19
Nicotiana plumbaginifolia
- protoplasts A7/1; F1/1, 2, 5-7, 9, 10, 13,
  15
- mutagenesis F1/1-17
- shoot cultures F1/5, 6, 11, 13

Nicotiana suaveolens
- protoplasts A10/19
Nicotiana sylvestris
- chloroplast mutants F1/2, 13
Nicotiana tabacum
- callus culture A2/11-13
- cell suspension culture A6/5, 7; B6/5;
  B11/3, 5, 10; D5/5
- chloroplast mutants F1/2, 13
- cytological analysis C4/3
- microprojectile bombardment D1/7
- photoautotrophic cultures H1/23
- pollen transfection B13/3; D2/1-9
- protoplasts A4/2, 9; A7/1, 7; A10/2, 3,
  15, 17-19, 25; A11/1-11; D4/3-17; D5/3,
  5, 6, 13, 17
- shoot and root cultures A4/1-14
- thin cell layers H4/1-25
- transformation A4/2, 9; A6/1; B4/1;
  B6/1; D1/7
Nicotiana undulata
- organellar genomes D4/3
Nicotinic acid: see 'vitamins'
Nigella sp.
- endosperm culture E3/3, 5
Nitsch medium A1/1, 2; D4/3, 13, 15
NLN medium (*Brassica*) E6/7, 8, 10
Nopaline
- produced in transformed cells B8/1
Nopaline synthase (*nos*) gene
- screenable marker B8/1, 9, 12
- termination sequence B11/13, 14; D1/9
Norflurazon
- component of media D5/13
Nuclei, isolation and uptake D7/1-20
Nuclear magnetic resonance spectroscopy
- in alkaloid analysis G1/15
- in anthocyanin analysis G2/12, 13
Nurse cultures/feeder layers A6/5, 7, 11;
  A10/1-28; B2/3, 5, 11; B5/6; B6/3, 5; B7/8;
  B8/7; B9/1, 7, 9; B10/4; B11/3, 5, 10; D4/7,
  11; D5/11; E1/3, 5, 9
Nuytsia sp.
- endosperm culture E3/3, 7
Nystatin
- to control culture contamination A2/16

Oats: see 'Avena sativa'
Oenothera sp.
- genes isolated D2/2
Oilseed Rape: see 'Brassica napus'
Oligomycin
- chloroplast-encoded resistance D4/13

Oncidium
- *in vitro* culture C1/3
Orange: see '*Citrus* spp.'
Orange G stain
- in cytology D8/6
Orchard grass: see *Dactylis glomerata*
Orchidaceae (Orchids)
- clonal propagation C1/1–7
Organ culture A4/1–14; G1/1–17
Organogenesis
- apple B8/5, 6
- *Citrus* spp. C7/1
- conifers C3/2, 3, 7–9
- from endosperm culture E3/1–21
- from thin cell layers H4/1–25
Oryzalin
- to arrest mitosis D8/2, 5
Oryza sativa
- embryogenesis A5/11; E2/2
- endosperm E3/1, 3, 5, 17
- protoplasts B1/3; B2/1–17; B3/1, 8
- thin cell layers H4/21
- transgenic B1/3; B2/1–17; B3/1; B12/1,
  2
Ovule culture
- in *Citrus* C7/1–3, 7–9, 13
- hybrid embryo rescue E4/2, 4

Packed cell volume A3/16, 17; H1/27
Palms (Arecaceae)
- clonal propagation C2/1–14
Panax ginseng
- thin cell layers H4/21
Pantothenate: see 'vitamins'
Papaver sp.
- chromosome isolation D7/6
Paper chromatography
- of anthocyanins G2/9
Paphiopedilum
- *in vitro* culture C1/2
Paraffin wax
- use during tissue sterilization A2/11–13
Paromomycin
- in selection of transformed cells B4/4–6;
  B8/8; C7/5
Parsley: see '*Petroselinum*'
Patatin gene promoter
- B11/13, 14
Pea: see '*Pisum sativum*'
Pelargonium sp.
- virus detection C6/3
Peperomia sp.
- endosperm E3/1

Petroselinum sp.
- flavonoid metabolism G2/5, 6, 17, 18
- endosperm E3/1, 3, 5, 8, 16
Petunia sp.
- chromosome isolation D7/6
- genes isolated D2/2
- pollen nuclease D2/1
- thin cell layers H4/2, 16, 17
- transformation by Agrobacteria B4/1;
  B6/1; B8/5
Petunia hybrida
- anthocyanins G2/3
- protoplast isolation and fusion D9/2,
  5–13
Petunia parodii
- protoplast isolation and fusion D9/2–4,
  7–13
Phalaenopsis sp.
- *in vitro* culture C1/2, 3
- thin cell layers H4/2, 21
Phaseolus vulgaris
- embryo sac E5/2
Phenylalanine ammonia-lyase
- assay G2/15
- see also 'anthocyanins'
Phleomycin
- selective agent B8/8
Phloroglucinol
- component of media B8/3
- stain for lignin H2/11
Phoenix dactylifera (date palm)
- clonal propagation: see 'palms'
Phosphinothricin (PPT; BASTA)
- selective agent B4/2, 5, 6; B12/3, 13, 15,
  18
Phosphinothricin acetyltransferase (PAT; *bar*
  gene) (enzyme, gene) A7/2; B9/2; B12/3,
  15
Photoautotrophic growth
- in eucalypts C8/9
- establishment of cultures H1/1–30
Photosynthetic activity
- determination H1/28
Physcomitrella patens
- protoplasts A10/2, 17
- tissue culture, transformation, mutagenesis
  F2/1–43
Phytagel
- for immobilizing cells H7/9, 10, 13, 15,
  39
Phytophthora cinnamomi C8/9
Picea spp. (spruces)
- *in vitro* culture C3/1, 11, 12, 13

– cryopreservation C9/3, 5, 7, 11, 17
– transformation B13/1–3, 9, 37
Picloram
– component of media A1/4; B1/5; C3/11
Pinus spp. (pines)
– clonal propagation C3/1, 3, 5–9
– cryopreservation C9/3
– transformation B13/1, 2
Pisum sativum
– anthocyanin mutant G2/5
– somatic embryogenesis E2/2, 3
– thin cell layers H4/2
Plating efficiency
– protoplasts B10/5, 7
Pollen
– embryogenesis E5/1
– source of male gametes E1/1–12
– standard for cell sorting D5/7, 9–11
– transformation B13/3, 4, 23, 25; D2/1–14
Polyethylene glycol (PEG)
– for direct gene transfer A1/13; A7/1, 2,
  7–9; B2/3; B3/11, 13, 14; B10/1, 2, 5, 6,
  10; B12/1; B13/2; C7/4, 5, 10, 11;
  F2/31–33, 40
– for hybridomas E2/4, 5
– for protoplast fusion A1/13; B9/1, 2;
  D3/2, 9; D4/1, 2, 9, 11; D5/13; D9/1–15;
  F2/29
– for uptake of nuclei D7/4, 13
– osmoticum in cryopreservation C9/6
– to improve protoplast viability A11/5, 9,
  10
Polymerase chain reaction (PCR)
– for detecting transgene integration B11/1,
  6, 13–15; B13/37, 39
– for detecting mRNAs E1/1
Polyvinylalcohol (PVA)
– in direct gene transfer A7/1
– in protoplast fusion D9/1
Polyvinylpyrrolidone
– inhibitor of phenol oxidases C8/3; D7/3
Poncirus trifoliata
– genetic marker C7/7
– organogenesis, protoplasts C7/1, 4
Populus sp. C7/1
– anthocyanins G2/19
– cell culture synchrony H3/3
Potato: see *'Solanum tuberosum'*
Potato extract
– component of media E3/3
Potting out plants
– E4/3, 5

Protein A-linked immuno-electron microscopy
  (PALIEM)
– in virus testing C6/4, 9
Protein determination B4/19; B13/21; E2/3
– antibody screening E2/5–8
Protoplasts
– chromosome preparations C4/1, 11;
  D8/5, 6, 8
– endosperm cells E3/9, 10
– in eucalypts C8/8
– fusion (chemical) A11/1–11; D3/1, 9;
  D4/1–17; D5/1, 13, 14; D7/2, 4–8, 13;
  D9/1–15; F2/29
– fusion (electrical) A10/2, 3, 6, 8, 17;
  D3/1–11; D4/1; D9/1
– KCl as osmoticum B10/3
– mannitol as osmoticum A1/15; A7/4,
  7–11; A11/7; B1/13; B2/14; B3/11, 13;
  B4/16, 17; B9/3, 9, 11; C7/9–11; D5/5, 7,
  8, 13; D7/3, 10, 13; D9/11, 12; H3/17
– micromanipulation A10/1–28
– moss F2/9, 23, 25, 29–33
– mutagenesis F1/1–17
– ovule cells C7/4, 5, 8–11
– regeneration A2/2; A7/1–20; A9/2;
  B1/1–16; B2/1–17; B3/1–15; B9/1–13;
  B10/1–11; D9/10
– source material A1/14–17; A4/2; A5/11;
  B4/1–24; B7/1
– stable transformation A7/1–20; B1/3;
  B2/1–17; B3/1–15; B4/3; B7/1; B9/1–13;
  B10/1–11; C7/4, 5, 10, 11; F2/33–35
– wall-digesting enzymes A1/15; A3/20;
  A7/3, 7; B1/13; A11/5; B2/14; B3/3, 4;
  B4/16; B9/9; B10/3, 10; C4/9; C7/9;
  D3/1; D4/5, 6; D5/3, 5, 6; D7/9, 10;
  D8/2; D9/11, 13; E1/2, 7, 8, 9; F1/6;
  F2/9, 39, 40; H3/17
– see also: 'cybrids', nuclei, isolation and
  uptake', 'K3 medium', 'plating
  efficiency', 'polyethylene glycol',
  'transient gene expression', 'somatic
  hybrids'
Prunus sp.
– endosperm culture E3/3, 5
– regeneration C7/1
Pseudotsuga (Douglas fir) C3/1, 5
– transformation B13/1, 2
Psophocarpus
– thin cell layers H4/2
Putranjiva sp.
– endosperm culture E3/3, 6, 13, 16
Pyridoxine-HCl: see 'vitamins'

Quoiri and Lepoivre medium   C8/4

Radioimmunoassay
– for indole alkaloids   G3/8
Raffinose as osmoticum   B13/5
Raphanobrassica
– in interspecific cross   E4/2
Raphanus sativus
– protoplasts   A10/19
Rauvolfia serpentina
– alkaloids   G3/4–6
Rauvolfia vomitoria
– alkaloids   G3/7
Respiratory activity
– determination   H1/27
Restriction fragment length polymorphism(s)
(RFLP)
– of organellar genomes   D4/15, 16;
D6/1–8
– of somaclones   C5/1–18
– of somatic hybrids   D9/2
RFLP: see 'restriction fragment length
polymorphism'
Rhodamine-6G
– antimetabolite   D4/7
Ribavirin: see 'virazole'
Rice: see *Oryza sativa*
Ricinus communis
– endosperm culture   E3/3, 4, 6, 16
Ridomil
– fungicide   C8/19
Rifampicin
– to control culture contamination   C8/4
RMOP medium   F1/5–8
RM medium   F1/6
RM solution   F1/6, 7, 13
Root culture   A4/1–14; G1/1–17
Rye: see *'Secale cereale'*

Saccharomyces cerevisiae
– microprojectile bombardment   D1/1, 3, 4
Saccharum officinarum
– in sugar production   B7/1
Saintpaulia sp.
– thin cell layers   H4/2
Salt tolerance
– *Citrus*   C7/4, 13
Santalum sp.
– endosperm   E3/1, 3, 6–9, 12, 15, 16
Schenk and Hildebrandt medium   A5/3, 5,
7–9, 11, 13; B3/7, 14; C3/7, 11
Scilla spp.
– cytological analysis   C4/3

Scurrula sp.
– endosperm culture   E3/3, 6, 11
Secale cereale
– meiosis   C4/7
Secondary metabolites
– cell line selection   A10/17
– phenylpropanoids/flavonoids: see
'anthocyanins'
– see also 'tropane alkaloids', 'indole
alkaloids'
Sequoia spp. (redwoods)   C3/1, 5
D-Serine   F2/23
Shoot culture
– sugarbeet   B7/3
– potato   B11/1, 9, 10
– tobacco   A4/1–14; A11/2
Shoot-tip grafting
– in *Citrus*   C7/1, 2, 13
Silicon carbide
– for direct gene transfer   B13/2
Silver nitrate
– component of media   B5/6; B8/8
Sinapis alba
– microspore/anther culture   E6/2
– host for *Heterodera schachtii*   H5/9, 13
Slow growth
– for germplasm storage   C9/1
Sodium alginate
– embedding protoplasts   A7/11, 13, 15
Sodium hypochlorite (bleach)
– as sterilant   A2/5, 6,11–13, 15; A4/5, 7,
9; A6/5; A7/3; A8/5; A9/5, 6; B1/5, 7;
B2/7; B4/9; B5/7; B6/3; B11/3; B12/5;
C1/2–4; C2/3; C3/7, 11; C4/4; C6/7;
C8/3, 11, 15; D4/3; D5/11; E3/15, 16;
E6/7; G1/5; H2/6; H5/9, 11
Sodium nitrate
– as protoplast fusogen   D9/1
Soja biloxi
– thin cell layers   H4/2
Solanum brevidens
– somatic hybrids   D3/3, 5, 6, 9; D6/2
Solanum nigrum
– chloroplast mutants   F1/2, 13
Solanum tuberosum   C5/2
– cytology   C4/3
– somatic hybrids   D3/3, 9; D6/2
– transformation by Agrobacteria   B5/1–9;
B11/1–18
– virus elimination   C6/2, 3, 7
Somaclonal variation
– in eucalypts   C8/7, 8
– in palms   C2/2

13

– in potato B5/1
– in rice B2/3
– in tomato B6/6
– RFLP analysis C5/1–18
Somatic embryos
– from microspore/anther culture E6/1–17
– carrot A9/1–32; G2/18
– *Catharanthus roseus* G3/2
– cereals/grasses A5/1–15; B1/1–16; B3/7; B10/5
– *Citrus* spp. C7/1–4, 7–10, 13
– eucalypts C8/8, 17
– conifers B13/2, 29, 33–39; C3/2, 4, 6, 11–13
– from endosperm culture E3/1–21
– from artificial hybrids E4/1, 4, 5
– induced by exogenous antibiotics, growth substance inhibitors C7/3
– palms C2/1–4, 7
– protein markers E2/1–9
– from thin cell layers H4/1, 2, 21
Somatic hybrids
– of *Citrus* C7/4, 13
– of moss F2/29
– RFLP analysis of organellar genomes D4/7; D6/1–8
– use of triploid protoplasts; E3/9
– see also 'Protoplasts'
Sorbitol
– as osmoticum A9/20; B13/5; C9/6, 11, 22, 23
Sørensen's buffer
– in cytology D8/2, 6
Sorghum
– suspension culture B1/11
– thin cell layers H4/2, 21
Southern blotting: see 'DNA'
Soybean: see '*Glycine max*'
Spectinomycin
– chloroplast encoded resistance B11/1, 3, 11; D4/13; F1/13
Spinacia oleracea
– photoautotrophic cultures H1/21, 23
Strawberry: see '*Fragaria* sp.'
Sterilization
– general procedures A1/6, 7, 13–15, 17; A2/2, 5, 6, 11–13; A8/16
– chemicals A2/2, 5, 6, 11–13, 15, 16
– equipment A2/2; B1/13; D5/11
– filter sterilization A6/7, 12, 15; A7/4; A8/16; B1/13; B3/3, 11; B10/9; B12/17; D4/5; D5/11, 13; F1/6
– fruits (*Citrus*) C7/7

– nematodes H5/7, 8, 13
– seeds A4/5, 7; A6/5; A7/3; A8/5; A9/4–6, 23; B1/5; B2/7; B4/9; B6/3; B7/3; C3/7, 8, 11; C4/4; C8/15; E3/15; F1/6; G1/5; H5/9, 11
– tissue explants A4/9; A5/3; B1/7; B4/21; B5/5; B11/3; C1/2–5; C2/3; C6/7; C8/3, 4, 11; E3/16; E4/3; H2/5, 6; H4/7; E6/7
Storage proteins
– markers of embryogenesis E2/2
– zeins E3/9
Streptomycin
– chloroplast-encoded resistance D4/3, 13; F1/13
– for bacterial selection B7/7
– to control culture contamination A2/16
Strictosidine-β-D-glucosidase
– enzyme assay G3/8, 17–19
– see also 'indole alkaloids'
Strictosidine synthase
– enzyme assay G3/8–11, 15
– see also 'indole alkaloids'
Succinic acid 2,2-methyl-hydrazide (ALAR)
– inhibitor of gibberellin synthesis C7/3
Sugarbeet: see '*Beta vulgaris*'
Sugarcane: see '*Saccharum officinarum*'
Sulfadiazine
– selective agent F2/31
Sycamore: see '*Acer pseudoplatanus*'
Synchrony
– in cell suspension cultures H3/1–31

Tabernaemontana spp.
– alkaloids G3/6
Taraxacum sp.
– endosperm E3/1
Taxillus sp.
– endosperm culture E3/3, 6, 7, 11
Taxodium spp. (bald cypress) C3/1
Taxus sp.
– transformation B13/1
Tentoxin
– chloroplast-encoded resistance D4/13
Thiamine-HCl: see 'vitamins'
Thidiazuron
– component of media C8/17
Thin cell layers (TCL)
– to study morphogenesis *in vitro* H4/1–25
Thin-layer chromatography (TLC)
– of anthocyanins G2/9–12
– of tropane alkaloids G1/9
– of indole alkaloids G3/7, 13, 15
Thula spp. (arbor vitae) C3/1

Tiglyl-CoA:pseudotropine acyl transferase
 - assay G1/11, 12
 - also see 'tropane alkaloids'
Tobacco: see *Nicotiana tabacum*
Tomato: see *Lycopersicon esculentum*
Tomato juice
 - component of media E3/3, 5
Torenia sp.
 - thin cell layers H4/2
Transient gene expression
 - in protoplasts A11/1–11; B2/3, 11; B4/4,
   13, 19; B10/1, 5, 6; B12/3, 11; C8/8;
   F2/31
 - following microprojectile bombardment
   B13/1–46; D1/1, 5, 6, 9; D2/1–14
 - see also 'chloramphenicol
   acetyltransferase (CAT) assay',
   'β-glucuronidase (GUS) assay'
2,4,5-Trichlorophenoxyacetic acid
 - component of media A1/4
Triacanthine
 - component of media E3/13
Trifolium sp.
 - somatic embryogenesis E2/2
Triticale
 - transformation B12/2
Triticum aestivum
 - chromosome isolation D7/6
 - embryogenic cultures B1/3
 - endosperm culture E3/3, 5, 17
 - meiosis C4/7
 - protoplasts B1/3
 - thin cell layers H4/2, 21
 - transformation B2/1; B12/1–20; D2/1
Tritordeum
 - transformation B12/2
Tropane alkaloids
 - biosynthesis in root cultures G1/1–17
 - enzyme assays G1/11–16
 - extraction and analysis G1/7–9
Tropinone reductase
 - assay G1/13
 - also see 'tropane alkaloids'
Tryptophan decarboxylase
 - enzyme assay G3/8–13
 - see also 'indole alkaloids'
Tsuga spp. (hemlock) C3/1
 - transformation B13/1
Tween
 - as surfactant A2/15; A4/5, 7, 9; A8/5;
   A9/5, 6; B1/5; B6/3; B7/3; B11/3;
   B12/15; C1/2–5; C2/3; C3/7, 11; C6/8;
   C8/3, 11; E3/16

UDP-glucose pyrophosphorylase cDNA
 - B11/13, 14
UM medium
 - for cell suspension cultures D9/12
Uracil
 - yeast selective agent D1/3, 4
UV radiation
 - use as mutagen F1/2, 15; F2/17–21

Vacin and Went medium C1/1, 2, 5, 6
Vancomycin
 - antibiotic A8/3, 10, 16
Vanda sp.
 - *in vitro* culture C1/3
Verticordia spp. C8/5
Vibarabine
 - antiviral chemical C6/2
Vicia faba
 - cytological analysis C4/3
 - thin cell layers H4/2
Vicia hajastana
 - protoplasts D7/3, 5, 6, 9, 13
Video cell tracking
 - of embryogenic cells H7/1–45
Virazole (ribavirin)
 - antiviral chemical C6/1, 2, 7, 8
Viruses
 - elimination A4/2; C6/1–12
 - testing C6/1–12
Viscaceae
 - endosperm culture E3/4
Vitamins
 - components of media A1/1–3, 6, 7, 9–11,
   15–17, 22; A4/2; A5/3; A6/6; A7/5;
   A8/15, 16; A9/9, 11, 15, 19, 24; B2/4;
   B3/13; B4/17; B5/2, 7, 8; B8/13–15;
   B9/10, 11; B10/9; B11/4; B12/17; B13/41,
   42; C2/4; C3/2, 12; C7/9; C9/21, 22;
   D1/7; D2/11; D5/8; D7/17; D9/12; E5/7;
   E6/9, 10; F1/6, 7, 13; F2/38; H2/7–8;
   H6/17, 18
 - see also 'Morel vitamins'
Vitis spp.
 - anthocyanins G2/1, 19
Vitis vinifera
 - endosperm culture E3/3
Vitrification A8/10; B8/3; C3/5, 7; C8/5
VKM medium D4/9–11

W5 salts solution A7/3, 4, 7–9; A11/3, 5, 10;
   D5/3, 6, 13; D8/2, 5; F1/5, 6
Wheat: see *Triticum aestivum*
White's medium A9/9–11; E3/4–8, 11–13, 16

15

X-rays
- use as mutagen F1/2
Xylogenesis
- in *Zinnia elegans* cell cultures: H2/1–15

Yeast: see *Saccharomyces cerevisiae*
Yeast extract
- component of media E3/2–6; G1/6
- see also 'Luria-Bertani medium', 'YMB medium'
YEB medium A6/9
YEP medium B8/11, 15
YEPD medium D1/3, 4
YMB medium G1/5, 6

Zea mays
- embryogenic cultures B1/2, 3; B9/1
- endosperm culture E3/3, 17
- feeder cells E1/9
- genes isolated D2/2, 4

- *in vitro* fertilization E1/1–12
- microfusion of gametes A10/3
- pollen transfection B13/3
- protoplasts A3/1; B1/3; B3/1; B9/1–13
- transient gene expression D1/9
- transgenic B1/3; B2/1; B3/1; B9/1–13; B12/1, 2; D1/1; D2/1
Zeatin
- component of media A1/4; B1/13; B6/5–7; B12/17; C8/4, 6, 13, 17; D9/12; E3/2, 6, 12, 13
Zeatin riboside
- component of media B11/4
Zeins: see 'Storage proteins'
Zephiran: see 'benzalkonium chloride'
Zinnia elegans
- mesophyll culture system H2/1–15
Zygotic embryos
- *in vitro* culture E5/1–19

please print or type

Please correct my address from (as now appears on shipping label and/or invoice

....

....

....

....

to the preferred mailing address

....

....

....

....

- -

PLANT TISSUE CULTURE MANUAL
Change of Address Card

please print or type

Please correct my address from (as now appears on shipping label and/or invoice

....

....

....

....

to the preferred mailing address

....

....

....

....

- -

Is your mailing address different from the address on your shipping label or invoice?

To guarantee swift delivery, change your address by simply completing this card. Be sure to includ the reference number found in the upper corner of the shipping label and/or on your invoice. Thi important number is uniquely yours. Refer to it in all correspondence.

Return the card promptly, either to USA or The Netherlands. Do not risk missing the latest an finest information available in plant tissue culture.

KLUWER ACADEMIC PUBLISHERS
P.O. BOX 17
3300 AA   DORDRECHT
THE NETHERLANDS

KLUWER ACADEMIC PUBLISHERS
101 PHILIP DRIVE
ASSINIPPI PARK
NORWELL, MA 02061
USA